Anonymous

Rocky Mountain Medical Review

Vol. 1

Anonymous

Rocky Mountain Medical Review
Vol. 1

ISBN/EAN: 9783337317331

Printed in Europe, USA, Canada, Australia, Japan

Cover: Foto ©berggeist007 / pixelio.de

More available books at **www.hansebooks.com**

Vol. 1.   COLORADO SPRINGS, SEPTEMBER, 1880.   No. 1.

ORIGINAL ARTICLES.

ASTHMA AS A REFLEX PHENOMENON.

By J. W. COLLINS, M. D.

The condition of system or set of symptoms present in a given case, denominated Asthma, are so varied and peculiar in different subjects, that it occupies in my mind the same relative position as the term Hysteria. The characteristics of the two so-called diseases are well known, and therefore do not demand any analytical description. That asthma, no more than hysteria, is a condition of pathological change, I think no one now admits, for the manifestations are so varied and so uncertain, that the pathologist cannot define the lesion in either condition. For the purpose of showing the reflex nature of asthma and its very probable relation to the term hysteria, I shall first proceed to relate a case, and afterwards endeavor to show its absolutely nervous origin. I shall not, however, contend that every case of asthma is traceable to pathological or functional changes, but that every case must stand upon its own merits. We must view it *de novo*, and treat it accordingly.

Mrs. —— came under my charge in June, 1879. I found her suffering the most intense asthma, long continued and rebellious to all known remedies. The history of her case was, that eleven years previously she was married; that at the age of 23 years, when six months pregnant with her first child, after a fatiguing walk of some considerable distance, she developed some shortness and difficulty of breathing, with some pelvic pain and pressure from pregnant uterus. This was her first symptom of asthma. Although several members of her family had been the subjects of hay asthma, yet no one of her immediate family or herself had suffered from the disease.

From that time on she was subject, after unusual exercise, to attacks of dyspnœa, of short duration and mild in character, until her confinement at full term, when she was delivered, by the aid of the forceps, of a male child weighing ten pounds. As a result of her accouchement, she suffered an extensive laceration of the cervix uteri and perineum, the latter extending through the sphincter vaginæ and a portion of the sphincter ani. The laceration of the cervix was bilateral and extended to the vaginal insertion. She had a long and serious "getting up," and from the thirteenth day developed the most intense asthma—so much so that her life was despaired of. She, however, slowly recovered; but from that time on suffered with asthma, especially at the menstrual epochs, and after any and the slightest irritation of the pelvic organs, such as the sexual act, horseback or carriage riding, walking, etc. Within the next ten years she bore three other healthy, living children, at each confinement suffering a fresh laceration of the cervix, accompanied by serious hemorrhage and an increase of asthma, until at the time she came under my observation the attacks were almost continuous and uncontrollable. She had had local treatment for "ulceration of the cervix uteri," at home, in New York city, and in Colorado, in which latter State she spent the winter of '78–9. After arresting the attack of asthma, by large hypodermic doses of morphia, I made a digital examination, and found, first, the laceration of perineum, and after introducing my finger within the vulva, I found the uterine neck and body quite low, very much engorged, and tender to touch. The uterine neck nearly filled the circumference of the vagina, was dense and cicatricial to the touch and seamed in several directions; one of the cicatricial seams involving the vaginal junction upon the left side ; pressure upon the uterine neck would cause symptoms of asthma. Upon using the speculum (Sims), the old laceration of the cervix was plainly visible, and the surface presented to the eye an outrolled, a partly eroded and partly cicatricial appearance, and the cervix was fully an inch and a half in diameter, the os. occupying an angle to the left. The uterus measured one and three-quarter inches in depth. By the aid of two tenaculi, I was enabled to inroll this tissue, which was flattened out upon the vaginal wall almost concentrically, and increase the depth of the uterus to two and one-quarter inches, and cover in most of the erosion. She had also a thick, tenacious, uterine discharge. My diagnosis was that the asthma was intensified and rendered uncontrollable by reflex irritation of the great sympathetic, through its cervical uterine branches. This was certainly sustained by the fact that any irritation applied to that part would induce an attack. So plain was this to the patient, that the marital rights were almost suspended ; she dared not ride in carriage or on horseback, or walk outside the house, for fear of an attack. My prognosis was that she could be much benefited, and her asthma probably cured, or at least greatly modified, by the performance of Emmet's well-known operation, for laceration of the cervix uteri and restoration of the perineum, as during her attacks of asthma the coughing was so intense and distressing that she could not lie down, and the uterine neck and posterior wall of the vagina would protrude from the vulva, showing that the uterus had lost its natural support. As I was at the time located at Jackson, Tenn., and the patient was never free from asthma in that climate, I endeavored to so

modify the attacks by means of Faradization and Galvanization of the great sympathetic, as to put her in condition for the operation ; but after six weeks daily application, with but little benefit, I desisted. Medicines were of no use. She had exhausted the whole range of remedies, under the hands of the most skillful medical men. She informed me, however, that while in Colorado the preceding winter, and while under local treatment for the uterine ulceration, she was almost free from asthma, except at the menstrual epochs, and then not so severely as in other localities. I wrote to Dr. Emmet, describing the condition of the patient and asking his opinion as to probable cause. He informed me that while he had never met a case that he thought due to that peculiar pathological condition, yet the reflex manifestations caused by it were so varied, that he at once sustained me in diagnosis, and strongly urged the operation. As the success of the operation depended upon its being performed in some climate where the patient would be comparatively free from asthma, it was decided that she should go to Manitou, Colorado, using some preparatory treatment, and that I should follow and perform the operation there.

She came to Colorado in July, but had asthma more or less severely, daily, up to the 12th of September, on which date I performed the operation for the restoration of the lacerated cervix, assisted by Dr. Anderson and several other physicians of this place. The operation consisted in taking out two V shaped pieces, one upon each side, down to the vaginal junction, and including the greater portion of the dense cicatricial tissue, bringing the sides of the incisions together by eight interrupted silver sutures, four upon each side, as practiced by Dr. Emmet. There was but little loss of blood, as the depth of the original and recurring lacerations had destroyed the circular artery. The sutures were removed on the eighth day—union was complete throughout, and the depth of the uterus was found to be 2¼ inches. What was quite remarkable just here was the fact that, although the patient had, up to the very day of the operation, more or less asthma, sometimes quite severe, yet for two weeks after the operation, and during the time of healing, she had none. But soon after this, and subsequent to taking severe cold, consequent upon change of room, she did have quite a severe attack, and, at intervals, some attacks during the winter following. I again visited her in November, to perform the operation for restoration of perineum, but found that preparatory treatment would be necessary, as she was having quite scanty menstruation, and there was considerable erosion of the cervix, attended by excessive engorgement of the uterus and pelvic viscera, accompanied by severe asthma at or before the menstrual molimen. This decided me to give her the Faradic current directly to the uterus, by placing an insulated copper electrode, attached to the negative pole, within the uterus, and the positive over the stomach, uterus, and ovaries. I also gave her suitable treatment for the remaining erosion. At the second menstrual epoch after using the electricity, which was during the first week in February, she menstruated freely, and for seven days, also, in the month of March; after which date (February) she had no asthma until the latter part of May, 1880, and that attack was induced by a slight prolapse of the uterus, brought on by lifting a heavy weight. On the 5th of April, 1880, I performed the operation for lacerated perineum, after Emmet, using six stout wire sutures, assisted by Drs. Reed, Strickler and Anderson, of this city. In four weeks the patient was out walking around, visiting, etc. She has had no asthma since the attack noticed above, and has regained her general health almost entirely, which had been lost for ten years. She is now living with her husband, rides, walks, and lives as other people, and has no symptom of the old enemy's presence. The prospects now are that she may return to her home in the following fall, relieved, if not entirely cured of the asthma.

The results in this case show, to my mind, that asthma frequently depends upon reflex conditions—most generally, perhaps, upon derangements of the digestive apparatus, frequently upon those very common diseases, nasal and pharyngeal catarrh, tumors, etc. Therefore, we should always look for the *cause*, near or remote. It was my good fortune, in the early part of the last decade, to cure entirely, and permanently, so far, a very distressing case of asthma of sixteen years standing, by the use of the continuous current, and replacing and curing a retroverted uterus.

Colorado Springs, August, 1880.

A Consideration of a Few of the Details in the Treatment of Diseases of the Ear, Nose and Throat.

Read before the Colorado State Medical Society at its Tenth Annual Convention, held in Denver, June, 1880.

BY A. WELLINGTON ADAMS, M. D.

It will be generally conceded that for the successful treatment of such affections, the *first* essential condition is a correct diagnosis, and *second*, such an adaptation and dexterous manipulation of mechanical appliances, as will render the application of medicaments as thorough and, at the same time, unobjectionable to the subject as possible.

Such being the case, let me first, then, briefly call your attention to the various methods of objective examination, for in this branch of practice, the latter virtually constitutes the diagnosis.

Of these there are two principal ones— examination by sunlight and examination by artificial light. Each of these again have numerous sub-divisions and modifications, varying in form according to the character of the particular apparatus employed.

Sunlight, as a means of illumination in such examinations, may be almost immediately withdrawn from consideration, as being obsolete, and as having *too* little in its favor; while its disadvantages are almost too numerous and palpable for mention. In fact, the respective merits of these two methods may be summed up by saying that the availability of sunlight is dependent upon its own unstableness, the time of day, location of operating room, etc., etc.; while with artificial light no such contingencies and dependencies can be argued.

Again, it is of the utmost importance that an operator should accustom himself to uniformly make all examinations by either one or the other of these methods; inasmuch as all tissues, whether in a normal or pathological condition, present widely dissimilar appearances under artificial and natural illumination respectively. And as long as one or the other of these two methods must be chosen for constant adoption, it is far preferable to select that which is *always* available.

Furthermore, if natural light be employed as the source of illumination, it is necessary to pose and contort the subject into all manner of shapes, in order to collect and properly focus the rays of light; whereas, if artificial light be used, the *subject* may remain stationary, and the *light* be adjusted to suit the purposes in hand.

These are the conclusions of all the world-renowned specialists — Tobold, Czermak, Cohen, Elsberg, Browne, and many others.

And now, with regard to the different appliances which serve as collectors and reflectors of the artificial light, whether resulting from the combustion of oil or gas.

First, and most simple of these, are the head and hand reflectors of Czermak and Von Troltsch. These, again, I hardly consider worthy of notice in the present state of the science, for they are founded upon absurd principles, and, in my opinion, there is no earthly reason for their longer use, especially when there are so many appliances so far superior to them in every particular—such, for instance, as the Tobold lamp, Tobold pocket illuminator, Bracket apparatus, and lastly, and best of all, the Tobold tripod arrangement for employing oil or gas.

The latter apparatus, laryngoscopists throughout the world, with one accord, acknowledge to be the best form at present in the market. Concerning it, Dr. Cohen, in his valuable work on "Diseases of the Throat," remarks : "This adaptation of the Tobold lamp leaves but little to be desired." Very true, there *seemingly* remained but little to be desired, so *great* did the present accomplishments appear from a retrospective point of view. How *natural* such an assertion is, under the circumstances, and how frequently we find examples of it ! It's the same old story, told over and over again with the invention of every new and startling mechanical device or process. How universal the remark, on the invention of the first Howe sewing machine, "There now remains little or nothing to be accomplished, for this is perfection personified ; one could hardly wish for anything better." On the other hand, how opposite would be the expressions from an experienced operator, since the invention of the "Singer improved." To carry the simile further: how often we hear the physician, like the old home-spun woman, with her first and only machine, remark, "I'm perfectly satisfied with this, I want nothing better—in fact, I think this the best." Why does he desire nothing better—why so well contented with what he has ? Simply because his experience has extended no further than the one instrument or appliance ; he is, therefore, deprived of the advantages of comparison, and is happy in his ignorance.

Gentlemen, the Tobold apparatus is clumsy, ungainly in appearance, needlessly complicated, and inconvenient. Why do I make this assertion ? Because, in order to adjust it, we are compelled to drop everything and use both hands in unscrewing and adjusting; it revolves hard ; its range of adjustment is too small; and lastly, its lower portion, or standard, is constantly in the way of everything, but particularly the operator's left arm.

And now, I will call your attention to an arrangement which obviates all of these disadvantages, and which is, therefore, far superior to the Tobold illuminator. This broad claim is not made owing to the fact of its being my own contrivance, but because experience has substantiated it, and, moreover, because I feel assured every gentleman present will recognize and acknowledge the claim to be a just one.

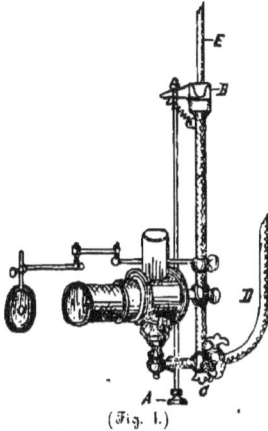

(*Fig.* I.)

A—Rod for operating eccentric lever.
B—Eccentric Lever.
C—Hand nut.
D—Rubber Tube to be attached to gas burner.
E—First Section of Tubing. Primary section or rod not shown.

This apparatus, as will be seen in fig. I, consists of a circular rod-like pendant, three feet long. suspended from the ceiling, and having two tubes of corresponding length, one inside the other, encircling it.

The lamp (if oil is used, or if gas, the Argand burner) is fastened, by means of a hand nut, to the lower extremity of the most external of these tubes. The concentrator and reflector also find support from this external tube. Of course the first section or rod may be made any desired length, to accommodate itself to the height of any room, but with the above described arrangement we have nine (9) feet of telescopic support pendent from the ceiling. The first section of tubing is fastened to the primary rod at the required point (this point depends upon the height of the room and, desired range of adjustment) by means of a screw. The second or most external section of tubing is held at any desired height, by

means of an eccentric lever, seen at B, and which is operated through the medium of the rod A. By slightly pressing this rod upward with the tip of the fore finger, the light may be placed at any height ranging from three to six feet, and turned in any direction with almost the rapidity of thought, and this with the use of but one hand.

The space beneath the lamp being entirely free, the arm of the operator may be brought up under the same with perfect liberty, or it may be extended toward the table placed at the operator's left hand. It may be used for the eye, ear, nose or throat, standing or sitting. Another desirable feature is the fact that when the lamp is not in use, it may be thrown up and out of the way.

But, gentlemen, perfect as would seem to be this apparatus, it nevertheless remains a fact that it, too, has its manifold imperfections. *In fact, the whole system of Laryngoscopy is an absurdity*—the *principles* are correct enough, but the way we've beaten the devil 'round the bush in applying them, is simply ridiculous. We've gone about things in the most roundabout manner imaginable.

Why is it general practitioners pay so little attention to Laryngoscopy and Rhinoscopy, but rather prefer relegating such subjects to the specialists ? Simply because they find an army of details in the examination alone, requiring close attention and delicate manipulation.

And all this inconvenience is brought about by the absence of a proper means of establishing a *permanent* relationship between the source of light for illumination and the laryngoscopic or throat mirror. This absence entails upon the operator the almost impossible task of regulating the illuminating apparatus so as to focus the light upon the laryngoscopic mirror, at the same time that the latter and the patient's head are being held in the proper position.

I will now begin attacking this method of examining the larynx and nares, by setting forth its disadvantages.

First, then, with its use we are obliged to have an intense heat at the side of, and in close contact with, our patient's head. which is both annoying and injurious.

Second. We are compelled to adjust the position of the light to suit the height of every individual to be examined.

Third. It is necessary to have the chair in which the patient sits, in close proximity to the table upon which is fastened the

Seventh. The intense light coming from the concentrator strikes the eye of the operator, which is placed at the opening in the center of the reflector, and causes a contraction of the pupil, thus rendering indistinct the less brilliant image in the throat mirror.

Eighth. There is too great a diffusion of the light, when, on the contrary, there

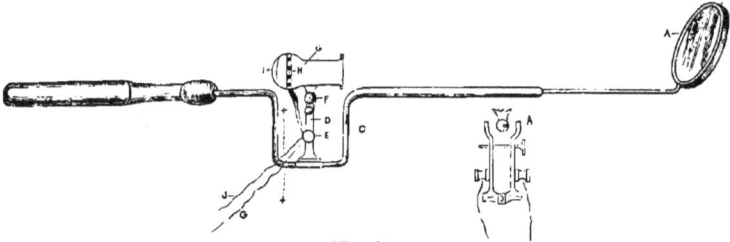

(Fig. 2.)

illuminator; and the first impulse of the patient, when about to sit down, is to pull the chair away, and arrange it to suit himself.

Fourth. The arm holding the reflecting mirror is constantly in the way of the operator, preventing the free use of his head and arms.

Fifth. Inasmuch as the operator is obliged to view everything through the small opening in the reflecting mirror, monocular vision is necessary, and this, for obvious reasons, can never be said to equal binocular vision.

Sixth. The slightest change in the position of the patient's head (which is occurring constantly, and perhaps at the most

should be great concentration, so that everything may be brought out in bold relief.

Ninth. The light, no matter whether it be from oil or gas, is of insufficient intensity.

Tenth. Owing to the double reflection, there is a great loss of illuminating power.

Never yet have I, in making a laryngoscopic or rhinoscopic examination, failed to experience one or all of the above enumerated disadvantages and deficiencies.

So it appears that for lo! these many years, we have been plodding on in darkness; we've had the cart before the horse, or, to speak more strictly, we have been trifling

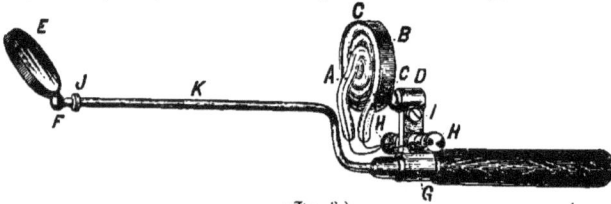

(Fig. 3.)

important moment,) changes the relationship between the source of light and the laryngoscopic mirror, which, as before intimated, must be just so. The same result follows the frequent accidental hitting of the reflector by the head or arm of the operator, who may be thus completely baffled in his attempt at securing a view of the passages requiring an examination.

with the delicate laws of reflection, which require *fine* and *permanent* adjustment.

But we will do so no longer, for I now have to present to, and bring before, the medical profession—for the first time—a *veritable* laryngoscope, for examining the larynx and nares—not a *method* for examining the same, but in the true sense of the word—a LARYNGOSCOPE.

This instrument as first designed, and described in the New York Medical Gazette of May 22d, is illustrated in fig. 2; while fig. 3 shows it as now perfected. There are two principal features governing its construction and contributing to its excellency, which require immediate attention. These are: *First*, the application of what is the nearest approach to sun light—the electric light—in such a way as to bring it under perfect subjection, that it may be easily maneuvred with. And, *second*, the establishment of a permanent, as well as adjust-

The aforesaid cap or case is for the reception of a concave mirror, and what is a slight departure from M. Gassiott's modification of Geissler's tubes. This consists of a glass carbonic acid vacuum tube, twisted in the form of a spiral, and having a fine platinum wire hermetically sealed within it.

Near the proximal end of this brass arm or upright are two "binding-posts," one on either side. These posts are in electrical communication with the platinum wires projecting from the extremities of the spiral vacuum tube. Attached to one end of the

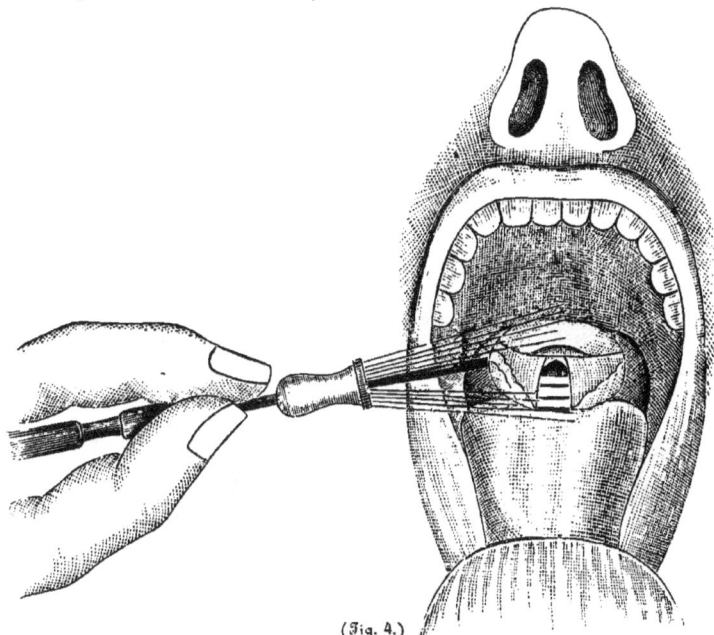

(Fig. 4.)

able, relationship between the source of light for illumination and the throat mirror. This is accomplished after the manner about to be described: Mounted upon a hard rubber handle, in any way most convenient to the maker, is a brass arm, slit down the center, and having a spherical socket drilled in its free extremity, the latter for the reception of a brass ball which projects from the periphery of a circular brass cap or case; thus creating a ball-and-socket joint between the arm and circular case above referred to.

aforesaid handle is a shank, made of some light metal, and bent in such a way as to bring the circular mirror, which we shall see is attached to its free extremity, upon a level with the spiral vacuum tube.

The circular mirror just spoken of is the "throat mirror," and it is attached to the distal end of the shank by means of a ball-and-socket joint; thus enabling an adjustment at any desired angle or position, the extremity of the shank forming the axis around which the mirror revolves.

A more lucid idea of the instrument as a whole, and of its various parts, may be had by reference to the perspective view in figure 3, where G represents the brass arm projecting upward from the small end of the handle; A—the spiral vacuum tube from whence the light is emitted; B—the brass cap serving as a receptacle or case for the vacuum tube and concave mirror, the latter being placed behind the spiral tube; C C—the little catches for retaining in place the vacuum tube and mirror, and which are formed by simply bending in

opposite side is only partially in sight. K—represents the shank which supports the throat mirror (E), by means of a delicate ball-and-socket joint at F; J—a sliding or screwing ring for regulating the mobility of this joint.

By this combination and arrangement, most any desired relationship between the source of light and the throat mirror may be attained in the "twinkling of an eye," so to speak, and there retained until the examination in any individual case shall have been completed. It must be borne in mind that

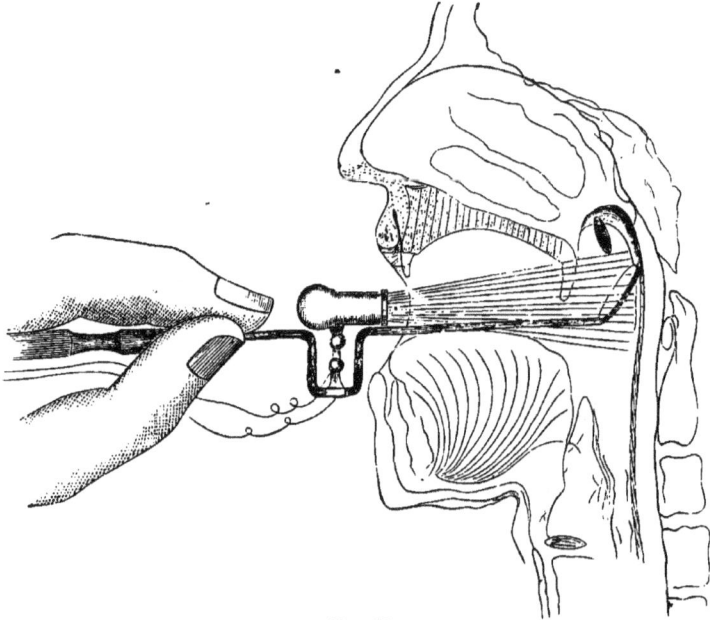

(*Fig. 5.*)

small segments of the case here and there; D—the ball-and-socket joint for furnishing universal motion to the light; I—the screw for regulating the mobility of said joint; H—one of the "binding posts" for receiving the flexible wires leading from the induction coil. It will be remembered I said there were two of these posts, one on either side, and that they were in electrical communication with the wires projecting from the two extremities of the spiral vacuum tube; in this figure the binding post of the

the entire arm supporting the light may be revolved upon the handle at G; so that when it is necessary in any given case to place the shank in the angle of the mouth, in order to secure a proper view, the beam of light will not be intercepted by the root of the tongue. When the flexible wires (wound with silk, similar to those used in connection with the telephone) leading from a Ruhmkorff coil, giving an inch spark, are connected with the binding posts H H, a BRILLIANT WHITE LIGHT is emitted from the glass spiral. THIS LIGHT HAS NEITHER HEAT NOR GAS, AND IS OF SUCH CONCENTRATION AND INTENSITY AS TO BRILLIANTLY ILLUMINATE EVERY PART OF THE RESPIRATORY TRACT LYING WITHIN THE REACH OF ITS PENETRATING RAYS; IN FACT, SO INTENSE IS THIS LIGHT, THAT EVERY DETAIL IN THE LARYNX AND TRACHEA LYING WITHIN ITS COURSE, MAY BE SHARPLY DEFINED AND BROUGHT OUT IN BOLD RELIEF IN THE THROAT MIRROR; AND IF THE LATTER BE LARGE AND SLIGHTLY CONCAVED, ANY PARTICULAR DETAIL REQUIRING SPECIAL STRUCTURAL EXAMINATION MAY THUS BE GREATLY MAGNIFIED.

The induction coil for this light may be worked with only three cells of the Grenet battery, and both of these may be placed any required distance from the Laryngoscope. I, however, prefer an equivalent

which the method is founded are erroneous, or the task very much more difficult of accomplishment than Laryngoscopy, but simply because the light was too crudely adjusted and of insufficient intensity to properly illuminate the much smaller and more tortuous passages of the nares.

In any case where the anatomy or condition of the parts will admit of an examination, the electric laryngoscope will furnish a brilliant and sharply defined rhinoscopic image that would startle and brighten the face of any skilled operator. All who have examined it declare it to be the greatest medical instrument ever invented—this, because of its accomplishments combined with its great *simplicity* and *novelty*. It is really the only *veritable* laryngoscope ever yet produced. Up to this time we have had but a

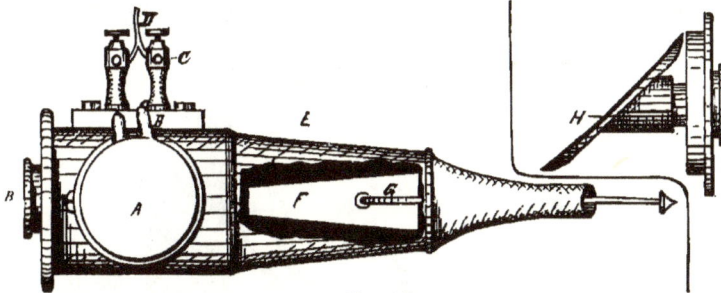

(Fig. 6.)

number of the Leclanche cells, as this form of battery requires no special attention or renewal from one month to several years, according to use. Again, with the Leclanche battery, in order to stop action it is only necessary to break the continuity of the circuit, whereas with the other form, the zincs must be removed.

Fig. 4 illustrates the instrument as applied to Laryngoscopy. Fig. 5, its application to Rhinoscopy.*

What this little instrument can accomplish in Rhinoscopy is truly marvelous.

Heretofore, the idea of procuring an image of the nares has been more theoretical than practical. Not that the principles upon

method for examining the larynx and nares, not an *instrument* we could pick up and say "this is a laryngoscope," or "this is a rhinoscope." It is light, easily managed, is not liable to get out of order, and, taking it all in all, costs much less in the end than the Tobold paraphernalia. Concerning it, Prof. J. Solis Cohen, the great American Laryngoscopist, in a letter to me says: "Your instrument is indeed a laryngoscope, and I congratulate you on making a great step in advance. I shall be only too glad to discard the paraphernalia connected with the Tobold or any other reflector."

All that I have said in regard to Laryngoscopy and Rhinoscopy holds most beauti-

---

* The instrument seen in these two illustrations is as first designed, it has since been slightly modified. Fig. 3 shows the improved and perfected ELECTRIC LARYNGOSCOPE. as now manufactured by Geo. Tieman & Co.

fully true in Otoscopy. Hence, without any further preliminary remarks, I shall proceed to introduce to you the ELECTRIC OPERATING OTOSCOPE, in the construction of which the same primary considerations are taken into account, and the same principle utilized.

It consists essentially, as seen in Fig. 6, of an electric light precisely similar to the one used in the laryngoscope, so placed and arranged with a reflector at its back, that its luminous rays are reflected directly downward upon the surface of a concavo-oblique mirror,

been removed. At E a portion of the upper surface of the otoscope has been removed, for the purpose of bringing into view a fenestra (F) in the under surface, thro' which instruments are introduced. At G a paracentesis knife is seen passed up through the fanestra and out at the small end of the otoscope. Different sized specula after Gruber's pattern may be introduced at will. Fig. 7 pictures a person in the act of operating through the medium of the electric otoscope. This instrument is undeniably

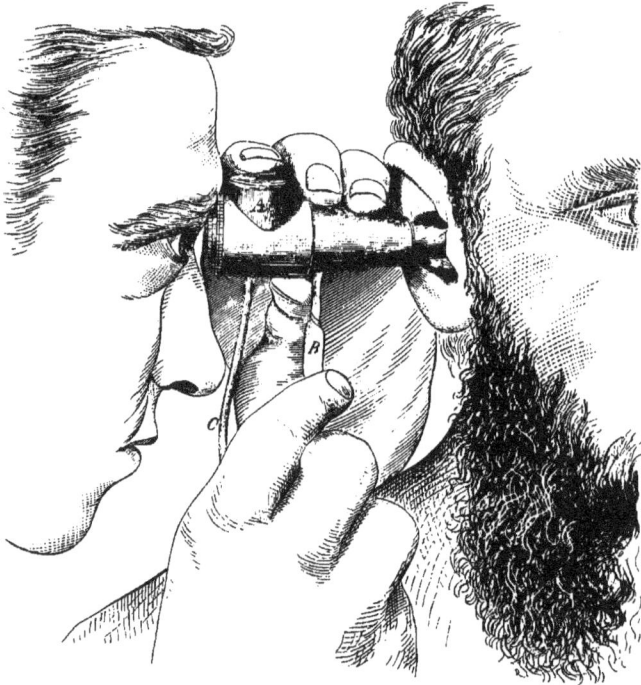

(*Fig. 7.*)

which thence reflects them in a horizontal direction. Fig. 6 is a top view of the entire Otoscope, in which A indicates the position of the electric light ; B—an eye piece having enclosed within it a magnifying glass of proper focal distance ; C C—two binding posts, which perform the same functions as those described in connection with the laryngoscope. H—the concavo-oblique mirror attached to the eye piece, which has

the *best* of its kind, and supplies a long felt want in otoscopy, for by means of it we are enabled to operate upon the drum membrane and middle ear under intense illumination and magnification. In treating diseases of the naso-pharyngeal space and nares, it is almost invariably necessary to resort to the use of a douche or spray of one form or another, most frequently, the post-nasal syringe. This instrument, as found in the shops, is very crude, and open to two objections: first, in order to throw the required quantity through the nares, we are compelled to remove and reinsert the instrument several times, which is not only annoying to the patient, but irritating to the pharynx; second, the inevitable jerk,

The medicine bulb (B) is elevated to compensate for the height of the curve (A); they should be blown out of glass by the physician himself, and modified in form according to the conformation of the pharyngo-nasal space in any individual case— it is impossible to adapt one form to all persons.

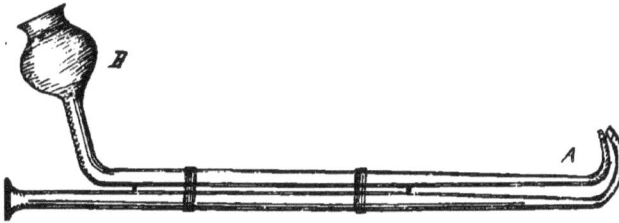

(Fig. 8.)

accompanying the expulsion of the fluid, is also disagreeable and irritating. To obviate these, I employ a post-nasal nozzle in conjunction with a fountain syringe. With this arrangement any quantity of fluid may be passed through the nares in one continuous and steady stream. Upon first thought this may appear to be almost too trivial to demand attention, but upon reflection, I am confident you will appreciate its importance.

The usual methods of spraying the nares, viz., from behind, with a straight tube throwing a spray upwards at an angle of 45 degrees, introduced through the mouth; and from in front, with Leffert's nasal spray, are inadequate to meet the requirements. The middle and superior meati can only be reached from behind, and in most cases this is utterly impossible with the ordinary straight tube, from the fact that upon the introduction of the instrument the palate either keeps bobbing up and down or clings to the pharynx. This difficulty may be overcome by the use of a spray producer constructed after the manner depicted in fig. 8. This instrument may be introduced up behind the palate when an opportune moment presents itself, thus placing the spray above and out of the way of the palate, while, at the same time, it holds the latter forward.

sons. If the pharynx should prove too sensitive for this procedure, as is frequently the case, we may resort to the use of a long, slender glass spray producer, introduced through the inferior meatus of the nares to the vault of the pharynx, and from that point direct a spray upon every portion of the nares; if a straight spray be used, we may in this way also thoroughly wash or medicate the vault of the pharynx. I have now resorted to these two procedures for over a year, and found them in every instance exceedingly efficacious.

(Fig. 9.)

I will next call your attention to a novel dry preparation of the human larynx, with the hyoid bone, thyro-hyoid membrane, a portion of the œsophagus, and a part of the trachea *in situ*. It is prepared in a novel way, and is intended as an agent upon which to perform mock operations preliminary to operating upon the living subject. In its preparation

no ante-putrefactive injections were used,
hence its strict fidelity to natural color.
It is simply hermetically sealed with twenty
or more coats of artists' varnish; while
these coatings were being made, shrinkage
and distortion were prevented by means of
weights and pulleys. When enclosed within
a box and mounted upon a small stage

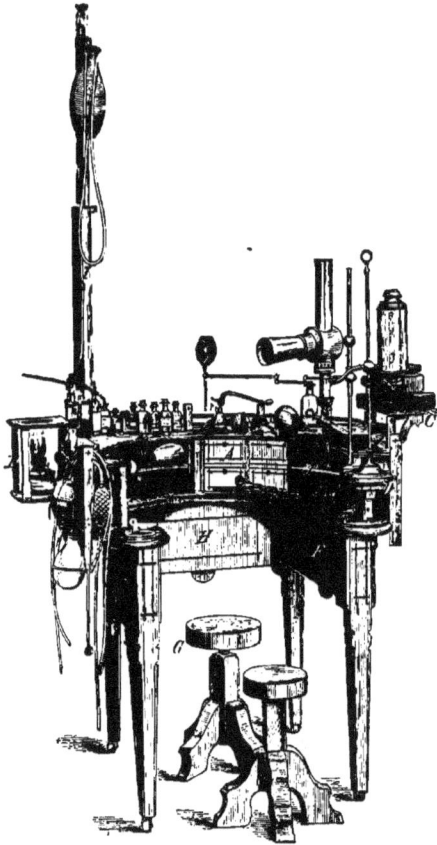

(Fig. 10.)

having universal motion, its relative parts
may be studied at various angles of reflection, and mock operations performed, with
surprising advantage, in the way of practice,
to the operator. Fig. 9 gives a perspective
view of this larynx, looking from behind
forward and from above downward.

Dentists have devoted much time and

study to the details of their science, such as
the construction of office furniture, etc.,
with a view to convenience. But the doctor,
greatly to his disadvantage, has almost
entirely neglected and ignored these important bearings. It may be laid down as a
general proposition, that *the degree of success
attending the treatment of diseases of the ear,
nose and throat, will depend essentially
upon the degree of perfection attained in
these details,* so that we may examine,
attend to, and dismiss a patient in the
shortest space of time, and with as little
bungling as possible. To accomplish
this, we must have a convenient place
for everything, and everything kept in
its place. To meet these requirements, I
have designed, had constructed, and now
exhibit to you, an ear, nose and throat
specialists' operating table (see fig. 10).
This table is constructed with a view to
being placed at the patient's right and
the operator's left side. The operator
has everything arranged in a circle at his
left hand. The open space in the center
admits of the operator's left leg being
placed therein, out of the way of everything, while, at the same time, it affords
a place for a slop jar to stand, within
easy reach of the operator, inconspicuous
and out of the way. It also admits of
the free and uninterrupted use of his left
arm, or should he wish to lean over to
reach an instrument that had accidentally fallen upon the floor, or for any
other purpose, he may do so with perfect
freedom. When about to make an examination or perform an operation, the
instruments likely to be used may be
removed from the two enclosed circular
tiers where they are kept, and placed at
suitable distances from each other upon the
ledge E, from which point they may be
conveniently and quickly reached when
required. The upper ledge with railing
around, is intended for the reception of spray
producers, and small bottles and glass jars
containing those medicaments which are
constantly resorted to. At H there is a

place for a test-tube stand ; at F, a place for a bracket supporting a spittoon, and revolving outward in the arc of a circle. A series of small drawers are seen at A, for holding small instruments, cotton, napkins, and other traps. I, represents an extension upright for supporting the fountain syringe at any desired height. On one side of the table there is a projecting bracket (C) upon which to keep a tank of cold water, with faucet for withdrawing the same; upon the opposite side (B) is a similar arrangement for hot water, with a lower shelf for supporting a lamp or Bunson burner beneath the tank. Just over the right leg there is an upright with a revolving top (D), the whole capable of being' raised or lowered. This supports the lamp, concentrator and reflector. (I employed this arrangement before I hit upon the pendent illuminator ; I now, however, find it very convenient as an adjustable standard and stage upon which to place medicaments for immediate use.) At J is seen the shelf dividing the two enclosed tiers. The advantages appertaining to this table must be realized to be fully appreciated. Inasmuch as it constitutes a beautiful piece of office furniture, it combines elegance with utility, which is a rare quality for anything.

Gentlemen, I heartily thank you for your close and courteous attention, and the only excuse I can offer for occupying so much of your valuable time, is the fact that I have presented nothing old.

Dr. STONE, of St. Paul, Minn., communicates to the Chicago Medical Review as follows : A lady weighing 197 pounds came to me complaining of superabundant weight. Remembering a "squib" seen in some medical journal, I advised her to restrict her diet to milk, of which she might drink ad lib. She followed the advice closely, not even taking a drink of water, and at the end of five weeks, weighed, in the same clothing, 176 pounds, a loss of 21 pounds. The patient is feeling better by far than when commencing the diet.

SELECTIONS FROM JOURNALS.

On Atmospheric and Climatic Influence in the Causation and Cure of Pulmonary Diseases.

BY DR. JOHN C. THOROWGOOD, F. R. C. P.,

Physician to the City of London Hospital for Diseases of the Chest, Victoria Park, etc.

Those of us who watch the weekly returns of the Registrar-General will not fail to have noticed how soon the effect of a week or two of cold weather is seen in the increased death-rate from such complaints as laryngitis, bronchitis, pneumonia, and pleurisy. That severe cold can greatly intensify the severity of pulmonary affections is shown by Dr. Hjaltelin's account of the pneumonia which was so prevalent and fatal in Iceland during the year 1863. During sixteen years spent in various foreign countries, Hjaltelin never saw any pulmonary inflammation so acute and so rapidly fatal as that which he was called upon to treat in his own native home of Iceland.

In the present day we hear a good deal said of inflammation as being the great destroying agent in pulmonary phthisis or consumption of the lungs. This statement, first put forth by that careful observer, the late Dr. Addison, of Guy's Hospital, finds exception in the fact that it is not in the cold countries, where inflammatory affections of the lungs are common, that we meet with most consumption. Dr. Schleisner says (Lancet, 1857), "there are countries where phthisis is unknown, as for instance, Iceland," the very place where pneumonia, as we have just seen, is often rampant. In Finmark no phthisis is seen ; and the Swedish physicians maintain that consumption becomes less common as we proceed northwards. Dr. Lombard, of Geneva, says he has never known a case of consumption among the monks of St. Bernard. In the civil population of Great Britain it is found that, in London, the deaths from phthisis are 18 per cent. ; in Edinburgh, 11·9 ; in Leith, 10·3 ; and in Aberdeen, 6·3. With regard to the prevalence of phthisis in hot countries, I get, from information published by Dr. Pollock, the facts that in the West Indies consumption is acute in form and rapid in progress. In Italy it is a malady of universal prevalence and great intensity, and the Italians regard consumption as a contagious and malignant disease. Without adducing further evidence of the prevalence of consumption in warm climates, and its comparative rarity in those that are cold, I will proceed to consider certain forms of phthisis.

First, we have knowledge enough already acquired to enable us to differentiate phthisis, or consumption of the lungs, into two large groups, and hospital experience appears to me clearly to bear out this distinction. We may recognize as group 1, cases of consumption originating in catarrh and cold, or in some inflammatory attack in the chest produced by cold. This is the kind of consumption prevalent in raw cold climates, common enough in England, and furnishing a large contingent of hospital cases.

Inflammatory action is a very active agent in this form of phthisis, but the disease may run a chronic course ; its tendency, under judicious treatment, being oftentimes to more or less fibrosis of lung and arrest of active symptoms.

Patients affected with this catarrhal phthisis derive benefit from going to a mild climate, such as Ventor or South Devon ; and, after what I have just stated, I see nothing paradoxical in Dr. Walshe's words in the last edition of his work on the lungs, page 687 : "While Southern Devonia is a favorable resort in certain cases for phthisical Londoners, it seems a very hot-bed of the disease for its own native population." I also understand how it comes about that Mr. A. extols the climate of Davos Platz as having saved his son's life, while Lady B., who went to Davos with a consumptive daughter, cannot bear to hear the name of the place mentioned.

Group No. 2 comprises that form of disease known as true tubercular consumption.

To use the words of Dr. Clifford Allbut, "the despair of the physician and the terror of the public." This form of consumption comes on insiduously, often from no cold caught, from no privation of food, but simply from some inherent, often hereditary, vice in the system. It is a febrile disease, having much the character of rapid blood-poisoning, it attacks other organs besides the lungs, and lasts only a few months, sometimes only a few weeks. Boudin has described it in its intensity as it prevails in the littoral of Peru and in the West India islands. According to M. Comeiras, it carries off one-third of the population of Tahiti, prevailing more among women than among men ; just the reverse is seen in this country, among our cases of catarrhal phthisis. Dr. C. T. Williams was informed by Dr. Busk of the Dreadnought that he made a number of post-mortem examinations on South Sea Islanders who had been attacked with acute tuberculosis on their voyage to London, and in whom the disease proved rapidly fatal.

As a practical inference from the foregoing statements, are we to send all persons who are in no matter what stage of consumption to the coldest place we can find, and expect a wonderful cure to result ? By no means. Where the chief agent for mischief that is to be combated is of an inflammatory nature, a mild soothing air is best. Where cough is severe and bronchial spasm frequent, a mild air also is suitable. For a patient recovering from a recent pleurisy or pneumonia, also, I hold that a mild air is eminently beneficial. That cases of catarrhal phthisis arising from cold caught improve wonderfully at Hastings, Ventnor and Bournemouth, is a matter with most of us of no rare experience. The advantage of such a climate is this: The patient can pass much time out of doors with safety, and can sit with open windows without risk of incurring another attack of bronchitis or pneumonia from the exposure.

Suppose, for the sake of experiment, we wished to make a bronchitis run into a true phthisis, then one way might be to expose the patient freely to cold air while in the weakness of convalescence ; but a tolerably prolonged observation among dispensary and hospital patients would make me name, as the most certain way of setting consumption of lungs going, to confine the patient closely to one room, and let him breathe over and over again the same atmosphere, while silence is temporarily imposed on the incessant cough by frequent doses of opium. In this way I am in the habit of saying that consumption may be cultivated and developed out of group 1 into the more serious form described in group 2, so that it becomes a fearfully destructive malady.

I expect one reason why the Italians speak of phthisis as a contagious disease of great malignity, is that they shut up their consumptive patients within doors slowly to smother in their own exhalations. The tendency of confinement in a close atmosphere to cause blood-spitting and consumption has been convincingly demonstrated by the statistics obtained by Dr. Gray when engaged in investigating the effects of certain trades on the health of those employed.

In the cases of young children who are kept very close in heated rooms, and who are said to be always taking cold, we often see most obstinate cough and catarrh due to the throwing off from the air-passages of a weak poorly-nourished epithelium which in time may choke the air-cells, and so lead to pulmonary consumption. The cure consists in laying aside paregoric and squills while we feed the epithelium with a pure healthy air. Appetite soon returns, and the cough speedily takes its flight.

In removing lingering inflammation after an acute attack on the chest, I have seen excellent results come from a sojourn at Torquay, Ventnor, and similar mild warm health-resorts ; but when the disorder has passed from the inflammatory stage to one that involves the general nutrition, and that is marked clinically by softening and breaking down of lung-tissue. with night sweats and copious purulent expectoration, I never saw any good of a residence in a mild sedative climate.

I cannot from my experience give you one example of a patient well on in true consumption who has found benefit from a relaxing air like that of Torquay. Dr. Symes Thompson once told me of a patient of his who had got well apparently of consumption during a winter abroad, and in whom cough, expectoration, and other morbid symptoms all returned during a few months' sojourn at Torquay.

To cure a softening lung, dryness of the air is essential. Whether a dry, cold, pure antiseptic air does really exercise its curative action by checking some septic process that is going on in the lung, or by annihilating certain morbific germs which engender tubercle is not yet proved ; though for myself, I confess, the idea is attractive, for I cannot help thinking that germs may be given off in the expirations of people in advanced tubercular consumption that may set up the disease in others who are vulnerable in their pulmonary epithelium, or, in other words, who have a predisposition to consumption.

A French observer, M. Bergeret, believes the expired air of tuberculous patients transports the tuberculous elements from lung to lung. If this be true, all the more need for the tuberculous patient to live in an atmosphere that is hostile to the propagation of morbific germs. A mild warm air hardly fulfills this indication, and practical experience proves it does not ; while practical experience certainly shows that tubercular consumption may be arrested by a clear cold atmosphere at a high altitude.

At Davos Platz and in the upper Engadine, 5,000 to 6,000 ft. above the sea-level, we are assured by the local physician that among the natives consumption is quite unknown.—*Practitioner, March, 1880.*

WE'RE a "blarsted" young country, to be sure ; but we're conceited enough to think we can turn out an "a number one" medical journal.

THE MEDICINAL LEECH.

By Dr. A. Berghaus.

Many swamps and ponds, which are now considered utterly worthless, might be made sources of great profit by devoting them to the production of a worm which is exceedingly valuable, and the cultivation of which requires no expensive outlay. This worm is the medicinal leech ; formerly esteemed of no value, and hated and hunted on account of its bloodthirstiness, it has commanded extremely high prices since its useful qualities have been recognized. Its general appearance is familiar, its internal structure is very wonderful. Its body forms a cylindrical sac, composed of a course of about one hundred rings. The terminal ring of the hinder part is broader and stouter than the others, and serves as a foot. At the front extremity, which is more pointed than the hinder part, are two fine, separated lips, which, when brought together, form a closed ring. Several straight lines run along the back for the whole length of the body, while the belly is of a clearer color and is motled with irregular dark spots. The body of the leech is so elastic that it can stretch itself out to a length of nearly ten inches, and draw itself up again to within the dimensions of an olive. Within and back of the lips are three thick membranous pads covered with a thin, horny mass bearing several rows of microscopic teeth ; they may be described as the jaws. Between the jaws passes the very narrow throat, which can be opened and closed at will by means of a transverse muscle. The animal derives its importance to man from the close aggregation of the movable lips, the narrow throat, and the toothed jaws, for it is enabled by this peculiarity to break through the skin and suck the blood from it. The mechanical operation is as follows: When the lips close in a circle upon the air-tight skin, the jaws are also brought down to it and their saw-like teeth are pressed tight upon the cuticle. The throat having now become fast closed,

the worm is drawn back a little, and the lips are thereby given the form of an exhausted cupping-glass, which is divided internally, by the jaws still fastened to the skin, into three distinctly separated parts. The skin is powerfully sucked up into these three divisions of the cupping apparatus till it is torn, and rents are formed corresponding to the three spaces between the jaws, the inner ends of which run into each other and form a larger, still three-parted wound. It follows that the sucking of the leech must be without effect on the hairy parts of the body, where a cupping-glass could not be made air-tight, and this is the case. When the space between the skin and lips which answers to the interior of the cupping-glass, is filled with blood, the throat is opened, the blood is drawn by sucking movements of the body into the maw, and the mouth of the worm is filled anew with blood. The long, narrow maw is competent, by means of twenty-six peculiarly formed sacs or valves, which are arranged in two rows, to retain an immense quantity of blood without any being driven back by the muscular activity of the body ; and, if a hole is pricked in the body of the leech at the rear end of the maw, all the blood that has been sucked up may be made to flow out. On account of the narrowness of its throat, the leech can not take solid food. Its usual nourishment consists of animal and vegetable infusoria, which it swallows in masses as the whale does herrings ; and while the whale spouts out through its nose the water it has swallowed and only retains the herring, the leech exudes the excess of water by means of a peculiar glandular apparatus in its skin, and keeps the infusoria in its maw. It also readily drinks the blood of cold-blooded and warm-blooded animals, and fills itself so greedily with the latter that it can not endure the surfeit, and dies soon afterward. Leeches were formerly abundant in the bogs and ponds of Germany, where, by reason of their great fruitfulness, they increased to millions, and were considered so worthless, even noxious, that the owners of the lands permitted the traveling dealers to fish them out at first for nothing, afterward for a small price. Finally the ponds were cleared of them; the dealers had sold the leeches for an immense profit, and millions on millions of them had been exported from Hamburg to America, and wherever else this costly and irreplaceable medical apparatus was needed, while the land of its production had none. The useful leech is not found in all countries, but its abode is limited to central Europe, Asia Minor, and a small part of the northern coast of Africa. In some of these relatively confined regions it has been exterminated. The demand for it has become very great; France and Germany, for example, use about thirty million, and the exportation from Hamburg alone has been thirty million in a year.* It is not surprising that so important a demand has raised the price of leeches till they have become a very profitable article of trade.

The successful stocking of a pond with leeches is a work requiring considerable care; the animal has many enemies, against which it must be protected, and will not thrive except under specially favorable conditions. The mother-leeches, when planted in the pond, lay their cocoons (which contain the eggs) as in nature, and the young brood is hatched out at the proper time; but this brood, besides protection, requires its natural food, sickens if it does not find it, and can not be fed artificially. The young worms will not thrive in artificial ponds; neither can they be transplanted from other countries and left to themselves without having first undergone a process of acclimatization. The most suitable ponds for acclimatizing leeches should be dug in bog-lands to a depth of about six feet, and should have from about six to ten inches of bog-soil on the bottom. The pond should

contain about three feet of water, and should be provided with an inflow of fresh water and be surrounded with a wall two or three feet high. If the leeches are put in the pond in May or June, they will deposit their cocoons toward September in funnel-shaped holes in the peaty bottom; in the course of a few days some ten or fifteen young leeches will come out from the cocoon, and will attach themselves to the old one to suck from it till they are large enough to seek food for themselves. For food, the pond should be furnished with calamus and other reed-like plants; and duck-weed, little fishes, snails and frogs should be put into it. Toward the latter part of the fall the animals should be taken out of the propagating pond and put into a smaller pond with a solid bottom of clear loam or sand, from which they may be taken in the spring to the marshes and bogs, in which, from that time, they will increase quite rapidly.

The question whether leeches can be cultivated on a large scale with profit, may be answered decisively in the affirmative by pointing to a few examples in which the business was carried on successfully. The brothers Bechade hired a swamp from Baron Pichon, near Bordeaux, as grass-land, for a rent of three hundred francs; after stocking it with leeches they were able to have the rent gradually raised to 25,000 francs without feeling overcharged. Since they began their enterprise, in 1835, leech-culture has risen at Bordeaux to be a source of great profit, involving the application of 5,000 hectares (12,500 acres) of land to the purpose, employing a great many workmen, and representing a capital of several million francs. A land-owner in Mecklenburg is said to receive an income of not less than 18,000 marks ($4,284) from his share of the rent of a leech-farm. A physician at Liegingen, in Wurtemburg, stocked a marsh of

---

* Although the application of leeches has been diminished in consequence of the adoption of new practices in medicine, which permit bloodletting only in a limited degree, the use of the animal is still considerable, and always will be so. A few years ago, when bloodletting played an important part in sickness, leeches enough could not be had, and it was hard to satisfy the demand in the ordinary way. Five to six million leeches, costing a million and a half of francs, were used in the hospitals of Paris yearly from 1829 to 1836, and 187,000 pounds of blood were drawn annually, or 1,496,000 pounds in the eight years!

two and a half hectares (six and a quarter acres) with leeches in 1827, and succeeded so well with it that he was able to sell his worms by the hundred-weight. — *Translated from Die Natur for the Popular Science Monthly.*

Recent Progress in the Treatment of Diseases of the Throat.

By F. I. KNIGHT, M. D.

EXTERNAL INCISIONS FOR THE REMOVAL OF BENIGN INTRA-LARYNGEAL NEOPLASMS.

Dr. J. Solis Cohen (Archives of Laryngology, vol. I. No. 2) gives illustrations of several of the methods of operating and comments thereon. The importance of restriction of the operation, which Dr. Cohen speaks of, cannot be mentioned too often till surgeons stop making those sweeping incisions from the hyoid bone to the trachea, which unfortunately are still too common. Dr. Cohen says that section of the cricoid cartilage is to be avoided, if possible, as its division is apt to impair the solidity of the laryngeal skeleton, and is liable, though but rarely, to be followed by necrosis. Section of the thyroid cartilage is likewise to be avoided if other means suffice for ample access to the neoplasm, as the consequent agglutination of the anterior portions of the vocal bands, in cicatrization, cannot be effectually prevented, and the resulting diminution in the length of the vibrating portions of the vocal bands necessarily impairs the quality of the voice.

Small growths, immediately beneath the vocal bands or upon their edges, and inaccessible to intra-laryngeal maneuvres, are not infrequently accessible to direct attack through an incision in the middle crico-thyroid ligament. This operation, the easiest of the series, involves the minimum risk incurred in artificial openings into the air-passage, leaves but an insignificant external cicatrix, and does not engender any impairment of vocal function. Dr. Cohen gives a case in which this restricted operation was performed on a gentleman for the destruc-

tion by galvano-cautery of nodules of papilloma beneath the anterior commissure of the glottis and upon the lower surface of the vocal bands.

In regard to section of the thyroid cartilage, Dr. Cohen says that its extreme liability to impair the voice irretrievably should always be borne in mind, and for this reason it may be questionable, in certain cases, especially in children, when the growth is in the upper portion of the larynx, whether the more conservative operation of sub-hyoid pharyngotomy should not be performed in preference. Dr. Cohen, however, farther on gives a case which shows that sub-hyoid pharyngotomy, that is, simple section of the thyro-hyoid membrane, does not always give sufficient room in children for a successful operation. A child, five years old, aphonic from birth, was seen in consultation with Dr. J. H. Packard. A neoplasm was discovered at the anterior commissure of the vocal bands. As it was impossible to have the repeated interviews necessary for a laryngoscopic operation, an external operation was decided upon. In order to avoid injury to the vocal bands, sub-hyoid pharyngotomy was chosen instead of thyotomy. Dr. Packard performed the operation, but when the epiglottis was drawn out through the wound it was found impossible, either by direct sunlight or reflected light, to illuminate the interior of the larynx beyond a meagre area just below the upper border of the aryteno-epiglottic folds, so that the growth could not be exposed to vision. Attempts were made to remove the growth by forceps, but the opening was too small to admit the passage of both finger and forceps, and the growth could not be distinguished from the supra-arytenoid cartilages without either digital aid or illumination; so the operation was abandoned. [It should be borne in mind that papilloma not unfrequently disappear spontaneously in children, so that it may be better oftentimes not to open the larynx at all, but insert a tracheotomy tube, if necessary to relieve the dyspnœa, and wait, as was strongly urged by Dr. H. K. Oliver some years ago.—REP.]

EXTIRPATION OF THE LARYNX.

Dr. Max Schuller (Billroth and Lucke's Deutsche Chirurgie, No. 37, Stuttgart, 1880) gives a very interesting account of this operation. The number of operations known to the author is nineteen, three for sarcoma, fifteen for carcinoma, and one for perichondritis from unknown cause. The three patients operated on for sarcoma not only survived the operation, but have remained free from disease. Of the fifteen from whom the larynx was removed on account of carcinoma, five died of pneumonia and two of exhaustion within two weeks after the operation. Five others died from a return of the disease long after the operation. Two patients (one operated on in September, 1877, by Wegner, and one in July, 1878, by Billroth) were alive, but there had been a recurrence of the disease in the latter ; of one case there had been no late intelligence. The recurrence usually took place, not at the seat of the operation, but in the neighboring lymphatic glands. In our present knowledge the prognosis is most favorable in the case of sarcoma. It cannot be denied that the operations so far do not offer much encouragement as to the final result in cases of carcinoma, but we must not forget that we are yet in the infancy of the operation, and that the disease was far advanced in most of the cases subjected to it. If extirpation is practiced earlier, as soon as the diagnosis is assured, the prognosis will become more favorable. Schuller considers that complete removal of the larynx is indicated only in the case of malignant growth. It has been proposed also for organic stricture from inflammatory processes, in order to substitute a larynx which would perform its functions for one which would not, but laryngotomy, with partial resection, suffices for these cases, and is far less dangerous ; moreover, the preservation of a part of the laryngeal walls renders the fitting of an artificial larynx easier. Schuller gives an excellent description of the proper method of operating.

In Vol. I., No. 2, of the Archives of Laryngology, is a very full account of the first extirpation of the larynx done in this country, by the operator, Dr. Friedrich Lange, of New York.

ADHESIONS OF THE VOCAL CORDS AFTER DIPHTHERIA.

Dr. Jules Boeckel (London Lancet, Dec. 27, 1879) showed, at a recent meeting of the Medical Society of Strasburg, the larynx of a child on whom tracheotomy had been performed for diphtheria. The patient recovered, but it was impossible to withdraw the canula, suffocation being threatened on every attempt to do so. Some months later the child died of scarlatina. The pathological specimen showed that there was an adhesion of the vocal cords in their entire extent, except at one point, where an opening remained scarcely admitting the point of a probe. Dr. Boeckel thought that the union of the vocal cords was due to the agglutination of the abraded surfaces after the elimination of the pseudo-membranes.

BURSAL CYSTS OF THE THYROID CARTILAGE.

Dr. J. W. Robertson (Toledo Medical and Surgical Journal, October, 1879) reports a case of cystic degeneration of the thyroid bursæ. The tumor was flat, and covered the whole surface of the subcutaneous portion of the thyroid cartilage, extending backwards under the muscles on the right side, and pressing the larynx backward and toward the left. It was about the size of a large walnut, oblong and flattened in shape, and its fluid contents could be easily diagnosticated. The larynx, except in position, was apparently healthy. The lungs were emphysematous, with a small portion of consolidated tissue in upper part of right lung. It being found impossible to extirpate the cyst entire, it being closely adherent to the anterior surface of the cartilage, the sac was simply cut away with scissors, and the internal wall of the cyst destroyed. The hyoid bursa, which was similarly affected, was eradicated in the same manner. The cysts were filled with a thick, glairy fluid, similar to that of a ranula. The wound healed slowly by granulation. A few applications of tincture of iodine to the granulations beneath the loose skin soon dried up a peculiar serous discharge, which lasted some time after the operation. In a week's time the larynx had resumed its proper position, deglutition and respiration were normal, and the voice improved, but still hoarse from catarrh.

LARYNGOTOMIA SUBHYOIDEA VERA, SEU SUB-
EPIGLOTTICA.

Dr. Carl Langenbuch (Berl. klin. Woch., No. 5, 1880) calls attention to the fact, stated by Langenbeck, that the operation commonly called subhyoid laryngotomy is really a pharyngotomy, the opening being made above the epiglottis. Dr. Langenbuch then gives the details of an operation by himself, in which the opening was made under the epiglottis. The patient had a small polyp at the anterior commissure of the vocal cords. A transverse incision of the skin was made, the muscles were separated from the hyoid bone, the hyothyroid membrane was divided transversely close to the upper edge of the thyroid cartilage, and a median incision was made through the ligamentous tissue of the superior thyroid notch and the upper third of the thyroid cartilage (the latter was perhaps unnecessary). A transverse section was then made of the root of the epiglottis. The slight hemorrhage was easily checked. The larynx was now drawn downward and outward by means of two strong hooks (one on each side), and a good inspection of the interior of the larynx afforded, and the growth easily removed by the scissors. Communication with the external air continued only two days, and the voice, which had been hoarse previously for five years, became normal.

Roser subsequently called Langenbuch's attention to the fact that he (Roser) had studied this operation experimentally on animals in 1851, an account of which was published in Shuppert's thesis in that year.

A NEW SUBCUTANEOUS METHOD FOR THE RE-
MOVAL OF NEOPLASMS IN THE LARYNX.

Under this title Professor Rossbach (Berl. klin. Woch., No. 5, 1880) describes a case of growth on the anterior third of one of the vocal cords, which he removed by means of a narrow knife introduced through the lamina mediana of the thyroid cartilage, the subsequent movements of the knife being guided by using the laryngoscopic mirror. Rossbach thinks the method may become generally useful in laryngeal operations. This method of operating has been used before by Eysell, of Halle.

ELECTRIC LIGHT FOR LARYNGOSCOPY AND
RHINOSCOPY.

Dr. A. Wellington Adams, (Medical Gazette, May 22, 1880) of Colorado Springs, claims priority in the application of electricity to laryngoscopy and rhinoscopy, and describes an apparatus which seems very convenient for the purpose. The light is arranged on the stalk of the laryngeal mirror, and several advantages are claimed for it over the "electrical polyscope" of M. Trouve.

APHONIA SPASTICA.

Dr. M. A. Fritsche (Berl. klin. Woch., Nos. 15 and 16, 1880) communicates six cases of this affection, first described and named by Schnitzler in 1875. It consists in cramp of the muscular apparatus of the larynx on attempted phonation, an affection analogous to writer's cramp. There is total loss of voice. The great effort made by the patient to produce sound is noticeable, but it "sticks in his throat." On laryngoscopic examination, the vocal cords are seen to be pressed tightly together, not presenting the usual elliptical opening, but on forcible effort sufficient air escapes through the glottis posterior to the vocal processes to produce an interrupted, convulsive whisper. In the lighter form of the affection there is only a momentary spasmodic closure, limited to the anterior part of the glottis. For treatment, electricity, especially the constant current, is of the most value.

FUNCTIONAL SPASM OF THE LARYNX.

Professor C. Gerhardt (Archives of Laryngology) under this title relates a case of the same affection, which bears a closer resemblance to writer's cramp from the fact that it came on only when the patient pronounced certain words or syllables. Professor Gerhardt previously reported the case of a flute player who suffered from a spasmodic affection of the larynx whenever he produced certain notes.

EXTREME STENOSIS OF THE PHARYNX.

Dr. T. Gilbert Smith and Mr. W. J. Walsham (British Medical Journal, April 17, 1880) report a case of extreme stenosis of the lower pharynx resulting from syphilis. The tongue could not be protruded beyond the teeth. The right posterior pillar of the fauces was drawn backwards, and was adherent to the posterior wall of the pharynx. The uvula and a considerable portion of the soft palate had disappeared, leaving a clean semicircular border to the portion that remained. On laryngoscopic examination the epiglottis and vocal cords could not be seen, but an aperture one-eighth of an inch in diameter was visible at the bottom of a funnel-shaped depression to the left of the middle line, on a level with the epiglottis. This was separated by a thick cicatricial band from another and deeper depression to the right, which terminated in a cul-de-sac containing pus. This small aperture was the only entrance to the larynx and œsophagus. Dilatation was not effectual until several cicatricial bands had been divided, when dilatation was resumed by means of the finger and œsophageal bougie. When the patient left the hospital the aperture measured three quarters of an inch in diameter. She was seen nine months after, and the aperture had not contracted in the least. In regard to the treatment the following suggestions are made : (1.) Tracheotomy. (2.) Division of cicatrices with a guarded knife. (3.) Several small notches are preferable to a deep incision, and when possible the parts should be divided from above downwards, so as to allow a full view of the tissues divided. (4.) The aperture should be enlarged in a direction so as to allow the passage of liquid food clear of the entrance to the larynx.

ANGIOMA OF THE LARYNX.

Dr. O. Heinze (Archives of Laryngology, vol. i. No. 2) reports a case of this rare affection. The tumor projected from the right ventricle, and was removed by means of the simple wire loop followed by galvano-cautery. Profuse hemorrhage followed the use of the simple loop, and Heinze warns us that galvano-cautery should always be used in such cases to avoid possible danger from hemorrhage. The piece removed lost its black color and became bluish-gray, and showed in section a fine spongy structure, and under the microscope a net-work of connective tissue, with blood in the interstices and numerous blood-vessels. Heinze was able to find only five cases previously reported.

TREATMENT OF NASO-PHARYNGEAL POLYPI.

At a meeting of the Societe de Chirurgie, (Bull. gen. de Therapie, December, 1879, London Medical Record, April, 1880,) Verneuil showed a patient on whom he had divided the soft palate, and had then cut away a portion of the tumor with an ecraseur. He then applied cauterization of chromic acid every two or three days. The polypus gradually softened and atrophied. Duplay showed a patient who had been treated for the same affection with injections of chloride of zinc. Rochard mentioned a case in which the soft palate had been divided, the projecting portions of the tumor cut away, and the remainder removed by injections of chloride of zinc. Auger deprecated the removal of polypi by radical operations, and recommended injections of per-chloride of iron.

To insure early attention, manuscript must be sent in *immediately*. We earnestly solicit original articles, reviews, translations, correspondence, etc., etc.

EDITORIAL.

To the Medical Profession of Colorado.

The present and prospective importance of Colorado, as the sanitarium of America, requires a State Medical Journal, which shall collect and diffuse an adequate knowledge of our healing resources. Her broad borders comprise many localities equal or superior to the most celebrated resorts of Europe, but their specific qualities and differences are not yet generally appreciated. The eight hundred physicians of our State include many gentlemen of wide culture and experience, whose observations would be of the utmost value. Add to this the importance of elevating the standard of our profession, and affording a means of fraternal communication between its members, and ample warrant is found for the action of our recent State Medical Convention, authorizing and sanctioning the present enterprise.

The editors of the ROCKY MOUNTAIN MEDICAL REVIEW ask your cordial co-operation in their work. The collection and comparison of observations as to the modifying influences of this climate upon various diseases, (as well those to which it may be adverse as those for which it furnishes a palliative or a panacea ;) the influences of different localities ; the therapeutic effect of our many mineral and thermal springs ; these and like enquiries afford a broad and inviting field to the thinking physician; and his researches will yield an abundant harvest.

We trust that an honorable personal ambition, and a just pride in this State journal, will induce many physicians of Colorado to become contributors. Their articles will, in every instance, receive a prompt and candid examination. We ask them also to at once subscribe for the REVIEW, and to secure subscriptions from others. Private capital has placed the enterprise upon an assured basis, but every addition to our subscription list will enable us the better to serve our readers.

To the Medical Profession Abroad.

We ask the material support, and the considerate judgment of our brethren in other States ; as the inception of the REVIEW is largely the result of a demand, by many of their number, for definite and accurate information concerning the influence of the high altitude and dry climate of Western America. At this day no educated physician can afford to contract his view within the limits of any single locality. He must study the effect of various climates. Like the mariner he must know something of the landmarks, the currents and counter-currents of many seas, and allow the pilots of distant coasts to supplement his own partial knowledge. The extant medical charts of this region are but hasty compilations, from imperfect data, and in some instances made by those not fully qualified for their important task. Those who address you have been chosen for the work by the voice of a convention, composed of physicians who have come hither from all parts of the country. We have no ends to serve except to forward the interests of our profession, and to prove ourselves worthy of our trust.

The REVIEW will be the organ of no clique or sect. Able writers from abroad will contribute to its columns, and its managers intend to make it the equal of any medical journal in the country.

In every instance the typography will be of a superior character, executed upon the finest quality of tinted book paper.

To Editors and Publishers.

We request the usual courtesy of exchange with all reputable Medical and Scientific Journals, commencing with the date of our first number—September, 1880. Works sent for review will receive a candid and impartial examination, and full justice will be rendered to every valuable contribution to scientific medicine, or to any branch of general science in the least way related to medicine. Books and periodicals should be addressed to the editor of the REVIEW.

To Advertisers.

The ROCKY MOUNTAIN MEDICAL REVIEW affords you access to a large number of our profession, not easily reached by any other publication. It will address a cosmopolitan audience, composed of members from many lands.

The wealth of Colorado, and her giant strides in material prosperity command the services of eminent talent in all professions, and demand in every direction the best appliances of modern art. Advertisers who send us their favors promptly, will, in addition to our permanent subscription list, have the benefit of a large circulation of specimen copies, which will be distributed widely and judiciously.

Business communications should be addressed to the publishers—Messrs. Tribe & Jefferay, Colorado Springs, Colorado.

Plan of the Rocky Mountain Medical Review.

It is the intention of the editors to insert in each issue of the journal, under the head of "Original Articles," at least one original communication from some prominent physician or scientist of America or Great Britain.

Under the head of "Selections from Journals," will be given, in full, articles selected from other medical or scientific journals, for their interest, intrinsic value, or originality.

Essays, selections, condensations, notes, etc., etc., pertaining to hygiene and sanitation will be relegated to the department entitled "Sanitary and Hygienic News."

The services of several able translators having been secured, that portion devoted to "Translations" will be constantly made replete with interesting articles, emanating from the pen of the most highly cultured scientific and medical observers of France and Germany.

The "Department of Criticism," is intended as an original hit and distinguishing feature of the journal. Under its head, will be printed prominent articles, selected from the best journals, and accompanied by just criticisms. It is the intention also

under this head, boldly and fearlessly to discuss the fallacies of medicine, their origin, growth and influence ; extant theories, the claims and erroneous principles of the various so-called "Schools of Medicine," etc., etc. The headings of the other departments, as tabulated in the title page, sufficiently indicate their purport.

To sum up—the editors intend making the ROCKY MOUNTAIN MEDICAL REVIEW of unparalled value to the scientific physician at large, as being replete with medical and general scientific news, thus making it totally different and distinct in outline, object and aim, from any medical journal now in existance.

We hope to receive a hearty support from all lovers and promoters of Scientific Medicine.

For want of space, several of the departments herein named have been omitted.

MEDICAL LEGISLATION.

IT is obviously desirable that in our young State, where matters and customs are to a great extent still inchoate, that whatever pressure may be wisely used shoud be employed to mould matters in a proper form.

This is pre eminently necessary in regard to measures pertaining to the conservation of the public health, prevention here being worth not ounces but tons of cure. In this relation the proper regulation of the practice of medicine is most important, peculiarly so in Colorado, as, owing to efficient laws in the central States, our community is fairly flooded with quacks and charlatans against whose abuses there is no legal remedy.

The matter has received some attention from the medical fraternity, but unfortunately in a way not useful, not being calculated to increase the respect of the laity for the profession, local jealousies being allowed to interfere with the best interests of the community and of the faculty.

At the last meeting of the Legislature, a committee appointed by the State Society to present specified matters for legislation, submitted a bill for the regulation of the practice of medicine.

Upon the whole the bill was a fair one, but unfortunately it contained a vital defect, ensuring its certain defeat.

The powers were confided to the State Board of Health, a Board which by its constitution does not command the respect or confidence of the profession. With this, and in opposition to it, was presented Senate Bill 64. The author to us is unknown. If not printed in official style and absolutely presented to the Legislature, the bill would seem to be a squib, a parody, or, as Orpheus C. would say, a "sarkasm."

A more complicated piece of machinery could hardly be conceived, or a more singularly inefficient series of regulations constructed.

Five physicians are to be appointed by the Governor as an Examining Board ; they are to prepare one thousand questions and publish the same.

Each and every physician is to present himself for examination, and must answer six questions upon each branch of medical knowledge. These questions being chosen by lottery from among the thousand.

There is to be a paraphernalia of tin boxes, locks, keys, envelopes and seals, and provision is made that no candidate during examination shall be allowed to defecate without being attended by two members of the Examining Board, and even then the act must not occupy more than fifteen minutes. Finally, after the board has examined all the candidates and superintended their evacuations, they in turn are to be examined by a committee selected from the late examiners ! ! !

For this work the gentlemen appointed, presumably practitioners of standing, are to receive the handsome remuneration of three dollars per day, they being employed, it is calculated, about "thirty days in each year."

If the bill was meant as a parody upon the frequent and often ridiculous bills for

"the regulation of the practice of medicine," it is certainly a great success. Otherwise, a more absurd measure could not be conceived.

We request the attention of the profession to this subject, and invite correspondence. Bad legislation is infinitely worse than none at all, and we hope that by discussion, a mutual connection of ideas may be attained, and that at the next meeting of the Legislature the committee of the Society having this matter in charge may be enabled to propose a bill commanding our confidence and approval.

BOOK NOTICES AND REVIEWS.

The Treatment of Puerperal Septicemia by Intra-Uterine Injections.

BY EDWARD W. JENKS, M. D., LL.D.

The reasons for the long delay in publishing the proceedings of the Gynecological Society are doubtless known to the committee on publication, and are satisfactory to them. To the outside public they are a mystery. The consequences, however, are apparent. Members who have put much research and labor into the preparation of their papers will not have them buried for a year, as has been the custom. In the present rapid progress of medical science and art, the lapse of a year will frequently cause the best production to become stale and unprofitable, and render a volume, which, if published with due promptness, would be invaluable, almost valueless.

This condition of affairs will account for the rather anomalous imprint upon the pamphlet before us: 'Reprint from vol. 4, Gynecological Transactions,' said volume not having as yet been issued.

The author is, under the circumstances, perfectly justified in publishing, and if by these anticipatory publications of valuable papers the sale of the volume of transactions is injured, the injury is solely due to the inexcusable procrastination of the society's committee.

Dr. Jenks' paper is a consideration of the uses of antiseptic injections in cases of threatened puerperal poisoning.

As the author justly remarks, "the subject is by no means a new one." But a more important subject could hardly be found. The condition to be controlled, the disease to be combatted, is one that addresses itself peculiarly to the feelings of the physician. The feeling of responsibility is greater, the anxiety more acute in a case of puerperal poisoning than in almost any other position the conscientious doctor is called upon to assume. Added to this, until lately, these feelings were aggravated intensely by an almost utter sense of helplessness.

Happily, with the means now in our possession, we may have much hope to be able to cure, and an approximation to certainty of our power to prevent these fearful conditions.

As with many other measures in the treatment of disease, the use of intra-uterine injections has had its tides, now being used largely, and again falling into almost utter discredit and disuse.

The author gives a full and scholarly history of the measure, through which it is unnecessary to follow him. He has not, however, in this part of his brochure discriminated as carefully as could be desired between the use of the injections in the puerperal and non-puerperal uterus. This, however, is carefully done later on.

In the latter they are undoubtedly extremely dangerous, and should seldom, if ever, be used. In the former, the author asserts, and we think justifiably, when properly used, they are positively harmless, the harmlessness, however, not extending to the use of strong astringents, as Barns' Iron Solution and Churchill's Iodine. These are admissable only as a *dernier ressort*, as in cases of post-partum flooding, where death has to be fought at any risk. This assertion is based not only upon an apparently large personal experience, but also upon the authority of such men as Schuler,

with twelve hundred cases, and of Richter, "*who administered intra-uterine injections to three thousand lying-in women without a single accident.*"

The conclusions reached are, that in addition to the ordinary vaginal douche following delivery : "Whenever there is premature cessation of the lochia, with any constitutional disturbance ; if there exists a purulent or fetid discharge ; whenever there is abnormality of the lochia or offensive uterine discharge, attended by elevation of temperature or increased frequency of pulse ; when there are good reasons to believe that the uterus contains fragments of placenta, or is imperfectly contracted and contains clots or any animal matter," intra-uterine injections are imperative.

With *proper precautions*, they are absolutely safe. "They should always be given by the accoucheur. The mouth and neck of the womb should be well dilated. Air must not be admitted. The fluid injected slowly and without force. The fluid should not be of a lower temperature than the normal temperature of the body. (*Sic.*) Powerful astringents should under no circumstances be injected within the uterus." In short, common sense should be used generally.

Intra-uterine injections should be more generally used in the prophylaxis and treatment of puerperal diseases, because "they are devoid of danger and capable of accomplishing results for good which cannot be attained by any other means."

The fluid recommended is a weak solution of carbolic acid, or permanganate of potassium in tepid water. To all of which we heartily agree. We are rather surprised, though, at the absence of mention of the use of *hot* water in these cases. This may simply be an omission ; if so, a serious one. If not an omission, a mistake hardly creditable. Hot water, of say 110° to 120° fah., itself, and alone, will fill the requirements of most cases, and is not open to the objections raised by the anti-injectors ; for, should a little obtain entrance into the dilated sinuses, it can do no possible harm, thus doing away with the principal danger, mostly imaginary, of this operation. Certainly to use the injections recommended by the author at a high degree of temperature, would not only add to their efficacy, but much to the comfort of the patient.

We would cordially recommend this valuable paper to the careful study of the practitioner, as we are convinced that by following its teachings many valuable lives may be saved, and much mental distress avoided.

As illustrating the efficacy of antiseptic injections, in lowering the temperature in a case of autogenetic septicemia, we will give the following case, which, although not requiring intra-uterine injections, yet has a direct bearing upon local antiseptic treatment:

Mrs. B., after a trying labor, during which she and her husband refused to allow the child to be destroyed, was finally delivered of a living infant, but at the expense of a perineum ruptured to the sphincter, and a torn and sloughing vagina. Owing to the condition of the vagina it was thought useless to stitch the perineum.

On the second day, with chill, fever, suppression of all secretions and profound prostration, septicemia was pronounced. The treatment was simply quinine and antiseptic injections, with the following result :

June 9, 10:00 A. M..........Temperature	93.5°	10:30 Douche 11:30	... .Temperature	98.5°
" 9, 2:30 P. M. ... . "	102.	3: Douche 4:00	.... "	99.5
" 9, 7:00 P. M. .... "	105	7:30 Douche 8:00	.... "	101.5
June 10, 12:00 noon.......... ...... "	101.5	12:30 Douche 1:00 P M.................	"	100.6
" 10, 7:00 P. M. .... "	100.	7:30 Douche 8:00 P.M................	"	99.
June 11, 7:00 A. M .... "	100.	7:30 Douche 8:00	.... "	99.
" 11, 12:00 noon...... .... "	99.5	12:30 Douche 1:00 P.M	"	98.5
" 11, 4:00 P. M..................... "	101.	4:30 Douche 5:00 P.M............	"	100.
" 11, 8.00 P. M..................... "	101.	8:30 Douche 9:00	.... "	103.
June 12, 12:00 M. .... "	99.5	12:30 Douche 1:00 P.M.................	"	98.5
" 12, 4:00 P. M. .... "	100.5	4:30 Douche 5:00 P.M	"	100.5
" 12, 8:00 P. M. .... "	101.5	8:45 Douche 9:30	.... "	101.5

After this, although the illness lasted for some days, the temperature never exceeded 100.5°, and the patient did well.

JACOB REED.

ROCKY MOUNTAIN HEALTH RESORTS.

BY CHARLES DENISON.

Within the last decade there has probably more been written—mostly trash—concerning Colorado than of any other portion of the universe. Its mines, scenery and climate have each received their tribute of adulation from a tribe of careless observers and enthusiastic but generally interested admirers. Grace Greenwood has gushed, exhausting the vocabulary of adjectives in her efforts to compensate the Colony for a handsome gift. H. H., actuated simply by enthusiasm, has found, and with her wonderful power of word painting pictured, beauties that are not discernable to the common mind. The mighty statesman who bears upon his shoulders the weight of the iron interests of Pennsylvania—for a consideration—has puffed the country to the extent of his very small but noisy abilities also for a consideration. And the numerous army of tourists, "special" and "occasional" correspondents, each have contributed their quantum of nonsense to their village newspapers.

It would be difficult to say what charm of scenery or excellence of climate has not been attributed to these foot-hills. Its atmosphere "has the effect of champagne." It is necessary occasionally to shoot a man in order to encourage that useful class of public servants—the undertakers. Overshoes and umbrellas are useless articles in this region of eternal sunshine. Enthusiastic and imaginative minds find an exquisite "symphony" in the red of the rocks and the barren brown of the plains; and the writer has lately met with a most intelligent gentleman, who, seeking a change for the benefit of his health, from his rather extensive reading labored under the impression that the "procession of flowers" was perennial in this paradise.

But little of a scientific character has been written concerning this climate, outside of the masterly reports of the U. S. survey, the few articles contributed by the members of the local medical societies being generally characterized by the same loose observations and exaggerated claims as those of the general writers.

The first effort to consider scientifically the bearings of this altitude upon disease is contained in the work before us. We write "effort" advisedly, for necessarily the number of years for observation having been but few, and the possibilities for study but limited, all deductions must be but tentative, and there can be but an *effort* at conclusions. The present effort is however sufficiently comprehensive, comprising everything from the profoundest pathological disquisitions to—the price of *cheese*. A work so comprehensive is difficult to review. We will confine ourselves to a few salient and important points; for this purpose considering what in the work is practical, theoretical, and—what is left.

The work is based upon the observation by the author of three hundred and fifty years, spent by two hundred and two consumptives in this climate, being an average of about twenty months for each patient. Of these, seventy-five were in the first stage (deposit), and averaged one year and eight months stay each. Sixty-four improved, ten improved slightly, and one retrograded. Of forty-two arriving in the second stage (softening), sixteen improved greatly, twelve slightly, six showed "favorable resistance," and eight advanced; being under observation on an average of one year and five months each. Eighty-five cases sought the change after the formation of cavities, averaging one year and eight months residence.

Of this class fifteen improved greatly, twenty-two slightly, and thirty-one sank. Giving a total result of forty-seven per cent. "much improved;" twenty-two per cent. "slightly improved;" eleven per cent. "favorable resistance;" twenty per cent. "extension and advance." Certainly a most favorable result if we consider that in forty-two per cent. of these cases cavities existed upon arrival at this altitude.

Aside from this simple clinical classification the author endeavors to tabulate his cases pathologically, subdividing them into "inflammatory," "catarrhal" and "chronic tuberculous" affections. Classifications of this sort are necessarily unsatisfactory and imperfect; the method adopted by the author being perhaps as satisfactory as attainable.

The conclusions arrived at are, that "the inflammatory and hemorrhagic group are most favorably influenced by Colorado climate" while the catarrhal and tuberculous cases, especially when laryngeal complications exist," are much less favorably affected. The author next comments upon the absolute necessity for a prolonged residence in the climate, where improvement is found, giving some painfully interesting illustrations of the unfortunate result of a too early return to a lower altitude of patients who had improved here. "More than half of the patients going east, having done well here, became worse there, and seventeen per cent. died quickly" upon returning east. The aptitude for a quick recurrence of former hemorrhages, in those who had improved at this height in that respect, is easily and forcibly noticed. Finally an attempt is made to compare these results with those reported by other authors under different circumstances.

This at present is impossible, the problem being entirely too complicated to be solved authoritatively from a small number of cases. If space permitted it would be pleasant to follow the author carefully through his different tabulations of cases; they show a patient, careful industry which is highly creditable, and the result can but be useful to the profession, to whom we recommend their careful study. The general conclusions reached from the by no means small number of cases, are such as will be endorsed by those who have had experience in the behavior of invalids in this climate. And it is only by experience that the effect of climate can be decided.

This is painfully shown in the book before us, wherein, while statements of experience and deductions therefrom are most profitable, the efforts at explanation, and the pathological and physical speculations, are singularly crude and marvelously incorrect.

The work is full of these curious phenomena. At the very introduction, Phthisis is defined as "slow death commencing in the lungs." As consumption is not death, (if so why this work?); as consumption does not begin in the lungs; this definition, although striking to the ear, is probably as perfectly inaccurate as could be constructed. But this is a fair specimen of the author's pathology. The physics are worse, if worse be possible. P. 92 : "With the air one-fifth rarified, the respirations are deeper and more frequent. Then the density of the air in the lungs during inspiration would seem to be lessened in proportion to the greater quantity of air which has to be breathed. This increased approach to a state of vacuum in the lungs, tends to draw the blood quickly into the pulmonary vessels, which movement of the circulatory fluid is aided by the accompanying increased action of the heart." Here we have the delicious statement that the extremely tenuous membrane forming the air cells can act as an air pump with sufficient force to cause "an approach to a vacuum in the lungs." And yet the author denies that there is any predisposition to hemorrhage!

The lungs now having acted as an exaggerated cupping-glass, and having produced this "approach to a vacuum" and its consequent engorgement of the pulmonary capillaries, expiration follows. "It is during expiration that the respiratory muscles have the greater power. * * Then

the density of the air within the lungs is suddenly made greater than that outside the body. * * The blood vessels are pressed upon," etc. We feel sorry for these poor blood vessels, first cupped and then squeezed twenty times in a minute, and can but admire the author's wisdom in the statement that "possibly at very great elevations the pulmonary vessels may not bear" this treatment.

But, nevertheless, this wonderful process must go on, for only by it can the "stasis of blood, which is an early stage of inflammation," be remedied, and the "chronic hypertrophied endothelium, and the foci and products of inflammation in intercellular tissue, be crowded outwards, and their absorption accelerated." "The lungs thus purified and cleansed of morbid principles of lingering tendency, have no place for adventitious products, or the deposits of tubercle."

Truly we are fearfully and wonderfully made, and the atmosphere at this altitude moves in a most mysterious way its wonders to perform! It compresses and it expands; it causes absorption and expulsion; it removes "morbid principles of lingering tendency"—delicious phrase—and leaves no room for any peccant product to work its wicked will! Finally, it removes "foci." Wonderful atmosphere! reminding the irreverent of the high wind that blew away the Irishman's well hole.

On page 110 we have presented phenomena more remarkable still. The subject is asthma, which "it is pertinent to define as a paroxysmal contraction of the circular muscles of the smaller bronchial tubes." Owing to this contraction, "inspiration is especially difficult." "Now, lessen the density of the air breathed, and the lungs must labor with greater activity in order to get the same amount of oxygen as before. The increased respiratory labor compressing the air, gives a proportionately greater pressure on the inside of the lungs than before, compared with that of the surrounding atmosphere. This augmented force within, pressing outwardly, distends the constricted circular muscles of the bronchial tubes to a point where they lose their abnormal contraction." This is exquisite, and will doubtless commend itself to—the laity, being after the manner of the patent galvanic shield advertisement : "You see, gentlemen, the battery is formed of various blocks of metal, so placed that when the electricity is formed, it will be in *gimlet shape*, enter the system in that form, and pass on twisting until it spends its force."

It is useless to go further ; the fault of what should have been a useful work, is painfully apparent, its *raison d'etre*, evident. The work is addressed, not to the profession, but to the laity ; is in fact simply an advertisement after the order of the "Popular Guides" so common and so mischievous.

We regret to use such language, but the work is such a flagrant instance of a common abuse, that there seems no alternative. A gentleman of the author's education, who produces a work filled with false physics, physiology and pathology, resounding with resonant phrases calculated to catch the popular ear, and that alone, and who in addition, encloses a blank to be filled by the patient "and then referred to some physician specially informed on this subject," certainly exposes himself to criticism and to the charge of employing methods not calculated to command the respect of the profession.

We write strongly because we feel keenly the mischief being done by many in "writing up" this climate and—themselves.

This we believe is the climate offering the best chances for improvement to many cases of lung disease.

Its remarkable freedom from moisture ; wonderful number of hours of sunshine ; the absence, at this altitude, of organic germs common to other atmospheres ; the attractions of scenery inviting an out-door life, are sufficient to account for its beneficent influence.

If there are other causes, they are to be reached by careful observation and study, not by crude assumptions, hasty generalizations and speculations addressed to the vulgar, but disgusting to the physician.

We need a careful, serious study of our cases, and their honest, frank and conscientious record and report, and this not to the people, but to the faculty.

J. REED.

Johnson's New Universal Cyclopedia.

Editors-in-Chief, F. A. P. Barnard and Arnold Guyot, with a corps of some thirty or more eminent authors as associate and assistant Editors. Published by A. J. Johnson & Co., New York.

We are very much pleased with the plan of this work, inasmuch as it selects as writers upon special subjects such men as are eminently fit, by reason of their peculiar experience and distinguished positions, to write upon such special subjects as may be allotted them. Other good features are its compactness (it being in four volumes), its reliability and the recent date of its articles. It is printed with very clear, good type, and is profusely illustrated. Amongst the articles we noticed as being specially attractive and comprehensive was one on Light House Construction and Light House Illumination, by Barnard and Peter C. Hains ; also one on Lightning and Lightning Rods, by Joseph Henry, than whom there is no better authority. He says : "We have thought it necessary to dwell upon the subject of lightning rods because innumerable patents have been granted in this country for improved rods, most of which have been devised by persons ignorant of the principles of electricity." Porter C. Bliss figures conspicuously in the biographical department ; Henry Wurtz and Professor Chandler in chemistry. There are geographical and biographical sketches by A. H. Stevens. In the field of mechanics and mechanical engineering, we find R. H. Thurston. The department of physics we find filled by the noted physicist, Prof. A. M. Mayer. Paleontology is in charge of Prof. Theodore Gill.

Although the work before us is undoubtedly the cheapest, most compact, reliable, and in every way the best cyclopedia in existence, we must avail ourselves of this opportunity to enter a protest against this "one-sided" way of dealing with the theory of evolution. We refer to Professor Anderson's article on Man. True, he does Darwin the justice to say that "All that Darwin really claims for his hypothesis in its application to man, is that it *may* be true. It is noteworthy that Darwin is much less positive in his conclusions than his pupils and followers. He concedes that his hypothesis is an inference from premises which are not seldom speculative or doubtful. In all sciences in which we reason from phenomena to their causes we can only infer a possible past from the actual and ascertained present." Further on, however, the writer says: "When we take into account the nature of the inquiry, and the serious difficulties which have led so many naturalists of the highest reputation to reject it entirely, we can safely say that, whatever may become true in the future, at present the hypothesis is unverified, and has no legitimate standing among the settled theories of natural science." Most assuredly the writer is perfectly justified in making such a statement, and it's all correct enough in so far as it goes, but he neglects to go further and say that the theory of evolution, although not yet completely verified, is regarded by the great mass of scientists, with but few exceptions, as undoubtedly true, but hard to prove; yet every year that rolls by brings in an aggregation of additional proof. Surely the following shows a marked leaning toward one side of the question : "When we take into account the laws of man's mind, his use of articulate language, his social, moral and religious constitution, we are met by problems which are extremely difficult of explanation on any hypothesis of evolution. Speaking generally, the opinions of those who deny to man the possession of a mind, with its constitution and laws of action, as an inherent part of his original being, fall naturally into two classes. The first, which includes the immense majority of materialists, hold that man comes into the world endowed with sensibility alone, and that contact with the external world through the nervous system comes, in the course of his life, to generate the capacity for acquiring knowledge, the power of thought and all those fixed modes of mental action which psychologists of a different school call the primordial laws of action that are native to mind, whether necessary or contingent in their nature. With these the development of intelligence originates in the matter of the nerves, and is completed in the lifetime of each human being. They deny that the laws of mental and moral activity which limit and condition all possible thought are laws of the mind as a distinct entity. They find by analysis no such elements. They resolve these laws or limitations of all thinking into habits or associational residue which have developed by the activity of the nervous system. Whatever this school may attribute to the development of capacity by hereditary influences does not essentially modify these statements. This school is best represented by the elder Mill in his Analysis of the Human Mind. The second school of sensational (sic) psychologists is best represented by Spencer." Then after going on at some length to discuss the differences between the teachings of these two schools, he says : "However much these two schools differ regarding the facts of the human mind and their analysis, the systems are both liable to similar criticism and present similar *defects of method,*" etc., etc.

Now this we claim to be a palpable discussion of the respective merits of the two sides, on the part of the writer, a thing that is wholly out of place in a work of this kind. An evident leaning in either direction in such a case is both improper and uncalled for.

———

Carbonic acid water readily dissolves salicylic acid. The acid should first be placed in a glass and mixed thoroughly with but a small quantity of the water, then the glass should be filled and the solution drunk at once.—*Louisville Med. News.*

The Problem of Human Life,

Embracing "the Evolution of Sound" and "Evolution Evolved." With a review of the six Great Modern Scientists, Darwin, Huxley, Tyndall, Haeckel, Helmholtz and Mayer. By one A. Wilford Hall.

The author of this work wishes to prove fallacious the theory of evolution, and begins by undertaking to disprove the vibratory theory of sound. We must certainly conclude that either the writer of this book is seeking an easy avenue to nonsensical notoriety, or that he is a religious fanatic, who, resting under the impression that the theory of evolution is not only destroying the Mosaic fabrication, but taking the props from under the whole Christian system, boldly undertakes to prove fallacious the very fundamental principles of science—such, for instance, as the correlation and conservation of physical forces, which Farraday contends is "the highest law in physical science which our faculties permit us to perceive." Thus innocently acknowledging that in order to disprove the theory of evolution, it is necessary that the primary principles of science taught in all our institutions, and accepted as axioms by all classes, be shown to be erroneous. Now, sound is simply an *effect*, and the unconscious logic of common sense compels us to account for these *effects* by the *causes* which we *know* to be competent to produce them, for example—the action of the siren and that recent wonder, the phonograph, in mechanically moulding the air into sonorous vibrations; but of course there are yet to be found circle-squarers, perpetual-motioners, flat-earth men, table-turners, spirit-rappers, and the like, who will argue that such explanations are not sufficiently mythological in character. "To such men nothing can be more hopelessly vulgar, more unlike the majestic development of a system of grandly unintelligible conclusions from sublimely inconceivable premises, such as delights the magian heart." How any one possessed of a sane mind could, since the invention of that wonderful exemplification of the correlation and conservation of energy, the magneto-electric telephone, have the impudence to undertake such an absurdity, we can hardly conceive.

His idea seems to be a revival of the philosophy of Spinoza, for he contends that everything is primarily due to the influence of a subtle substance, and would give us to understand that this substance is God. This ideal philosopher of his, however, thought that substance, in order to be substance, must both be *conceived per se* and *exist in se ;* that is, must exist without a cause. According to Hall's conclusions, therefore, we ought to be able to receive and communicate ideas without any physical effort upon our part. Nothing would please us better than to give this book a more lengthy review, but it does not deserve it, even though the author does take up some five hundred double-column octavo pages in his discussion.

MEDICAL NEWS.

FROG POISON.—A poison producing effects identical with those of curare is said to be extracted by the natives of Columbia from a peculiar frog found in the *tierra templada.*

CLINICAL THERMOMETERS.—Leonard Waldo, S. D., (Harv.) the author of a paper published in the Medical Record, entitled "Note on the Errors of Clinical Thermometers," claims to have written the same in answer to inquiries from members of the medical profession. He thinks there should be some warning given that the thermometers in general use are not to be relied upon within one-half degree Fahrenheit. "Of course I except from this statement those instruments which have a Kew certificate not more than six months old. To show the errors commonly existing in clinical thermometers, I have selected the readings of sixty-eight thermometers verified in June of this year. Since each one had its errors determined at five points of its scale, $90°$, $95°$, $100°$, $105°$, $110°$ (except a few which were graduated from $95°$ to $110°$), we have about three hundred and twenty separate points upon whose errors we can base an estimate of the correctness of those in general use.

Seven different and leading makers, domestic and foreign, are represented in the thermometers used. An analysis of all the readings showed, on an average, a discrepancy of from ½ to 1°, Fah.

It is an accepted fact that mercurial thermometers, at the temperature used in medicine, always increase their readings with age, and much more rapidly in the first few months after the tubes have been made than at any subsequent time.

It is very much to be desired, therefore, that the tubes to be used for thermometers should first be filled with mercury, and then laid aside for two years before they are finally graduated.

It has been suggested that the observatory not only verify thermometers, but that it receive the ungraduated tubes, and after registering the numbers in them, seal them up in packages for the space of two years. That for such seasoned thermometers the observatory issue a certificate stating not only the usual scale of errors, but also that they were properly aged before graduation. Such thermometers would change their errors but slowly, and ought to command enough higher price to justify the dealers in the trouble taken.

It is much better to send clinical thermometers to some recognized observatory and obtain an investigation of its errors. So firmly is this view rooted in the medical mind in England, that in 1879 some 3,400 clinical thermometers were verified at Kew alone.

It is now pretty generally known, at least among the New England members of the medical profession, that the new observatory of Yale College has undertaken to afford to physicians an accurate statement of the errors of clinical thermometers sent to the observatory for such purpose."

———•———

CONTRIBUTORS may rely upon having their articles published in an artistic manner, free from typographical blemish. And this in a journal which will not only aim at superiority intellectually but æsthetically.

Artificial Inflation as a Remedial Agent in Diseases of the Lungs.

———

Under this caption, Dr. W. G. Gadbury, in a paper read before the American Medical Association, at its recent convention, by Dr. J. Solis Cohen, and published in the August number of the "St. Louis Courier of Medicine," proceeds in a very unassuming manner to describe a self-devised, cheap and convenient method of effecting pulmonary distention by artificial inflation. The apparatus improvised for this purpose is a double bulb Richardson atomizer, and the directions are : "Insert the tube into the mouth with the left hand, take a deep inspiration, and with the fingers of the same hand close the lips and nostrils, and work the hand-ball rapidly with the right hand, so long as the patient can bear it."

After detailing several interesting cases of lung difficulty, in which this procedure was resorted to as a remedial agent, with favorable results ; one being a case of phthisis, so diagnosed by two other physicians, and which he claims to have cured by this agent alone ; he sums up by saying that—

"Inflation forces fresh air into the lungs, expanding unused capillary tubes and air cells ; displaces the residual air and noxious gas ; excites cough and expectoration, which removes morbid secretions at once, thereby lessening the danger of infection from unhealthy accumulation, and obviates the necessity of expectorant medicines, which often disturb the digestive organs ; oxygenates the blood ; promotes absorption ; relieves dyspnœa ; gives impetus to pulmonary circulation ; reduces temperature in fever, and dessicates the fluids in the air passages.

Beneficial effects may be derived from it in croup, diphtheria, bronchitis, asthma, tuberculosis, whooping cough, asphyxia, chloroform poisoning, shock, foreign bodies in the air passages, and many other obstructive lesions of the pulmonary organs.

By inflation, vapors and gases may be introduced into the air passages, and it is believed that experiments in this line will open a new field of usefulness, and brighten hopes for the hapless sufferer from croup and diphtheria.

A recent distressing case of membranous croup suggested to my mind the propriety of inflating the lungs through a large aspirating needle or curved trocar passed into the trachea as a substitute for tracheotomy, or to prolong life until that operation could be performed; and it occurred to me also that foreign bodies in the larynx and trachea could be expelled by the same operation.

An inflator for such purposes should have two hand balls, so as to force into the lungs a sufficient quantity of air to substitute what would be taken in by the inspiratory act. Expiration would necessarily follow inflation from the elasticity of the thoracic walls, through an orifice too small to permit inspiration. The internal pressure of a large volume of atmosphere upon the strictured larynx would render it more patulous, and probably detach and expel any false membrane present. The main difficulty would be in preserving the natural rythm of the respiratory movement. This simple and harmless operation would be justifiable in chloroform poisoning, asphyxia from drowning and other causes. The instrument may be made small enough to be carried in the pocket.

The instrument I have used is imperfect. It should have a shield to close the mouth and nostrils, and a much larger and more elastic air-bulb.

The economic advantages of inflation especially commends it to the poorer classes, and its harmless nature and prompt action suggest the propriety of using it as a substitute for medicinal expectorants in cases of enfeebled digestion and irritable stomach. It is hoped that its use in persons with weak lungs may assist in developing these organs, and perhaps operate as a preventive of tubercular deposit. It is not urged as a remedy for tuberculosis, for all such are justly chargeable with suspicion, and the burden of proof in their favor must be submitted to many generations; but the faithful physician should never cease to contend with it so long as there is hope of success.''

Concerning this plan of treatment, Dr. J. Solis Cohen, in a paper published in the Medical and Surgical Reporter, says:

"I have given this plan a trial during the past year, in my private practice, and in the clinical practice of the hospitals with which I am connected; and am in a position to estimate its value and capabilities. It cannot be safely employed in all the cases in which insufflations of compressed air, as supplied from the apparatus of Waldenburg and others, are applicable; but it has a sufficiently wide range of utility to commend it to professional attention; and in a certain class of cases, to which allusion will be made presently, it is of greater service than the bulky machines adverted to.

In patients liable to hemoptysis, or other hemorrhages, and in certain cardiac and visceral dis-

orders, the intra-thoracic compression, if left to the patient, is apt to be too powerfully exercised, and thus to be absolutely detrimental; and it is to the hands of patients that the instrument is to be confided. It is seldom safe to employ compressed air with a pressure exceeding from one-sixtieth to one-thirtieth of an atmosphere, and quite delicate handling of the ball-compressor is requisite to keep within this limit, while the size of the compressor prevents access of air in large volume, or at constant pressure. Thus, for general purposes, this plan, unmodified, cannot supersede the use of more complicated appliances.

There is one use of the Gadbury method, however, to which I desire to call the attention of the profession prominently; and that is its employment as a mechanical expectorant. Time and again I have placed the little compressor in the hands of a patient with bronchioles and air cells clogged with mucous and pus, to see its use immediately followed by copious expectoration, to the great comfort of the patient. The process is repeated until it ceases to be followed by expectoration, and there is absolute or relative relief from the desire to cough, until re-accumulation indicates a renewal of the procedure at intervals of a couple of hours, or longer, according to circumstances. I have frequently availed myself of this method of clearing the air passages previous to careful physical examinations, when abundance of moist rales were present, and have been better able to estimate the actual conditions of the respiratory organs on auscultation afterward. Hence, in chronic bronchitis, of whatever origin, compressed air can be employed with advantage in this way, to discharge the mucous accumulations from the air passages, and spare them much of the topical irritation to which they are otherwise subjected. In a few instances I have seen chronic bronchitis relieved by the use of this method, without any medication whatever, and far more rapidly and effectually than follows the administration of medicinal expectorants, which are too often coupled with the disadvantage of interference with the processes of nutrition, by their nauseant influence upon the alimentary tract.

The physical action of this mechanical expectorant is simple. The hyper-distention of the air-cells permits the access of air under pressure to points beside and beyond the masses of mucous clinging to the walls of the bronchioles and alveoli, and excites effective cough, which removes the partially detached masses. Several of my consumptive patients clear their passages out at bedtime in the manner indicated, and secure a good night's rest, free from disturbance by cough, without the administration of opiates. When they rise to dress, they clear the parts of the accumulation over night in like manner, and attack their breakfast with relish. Some individuals have little or no occasion to expectorate during the intervals, and can pursue their vocations, relieved of the frequent and recurring plague of an annoying and harrassing cough. The therapeutic advantage of an agent capable of doing this much, is incontestable ; and it is for the purpose of drawing attention to this simple and inexpensive contrivance, and of having its merits tested on an extended scale, that this article has been written."

A New and Startling Pain Obtunder.—Is it Hypnotism Pure and Simple, or is it Analgesia ?

The medical profession generally, concede ether to be a safe but inconvenient anæsthetic, and chloroform to be an unsafe but convenient one. For this reason there have been many attempts at finding a substance uniting to the safety of ether the good qualities of. chloroform. Such an anæsthetic was recently reported to have been found in a compound possessing very feeble bonds of union, and known as the bromide of ethyl. It rapidly rose in favor, and many of our clinicians and experimental therapeutists took hold of it, but as soon as their reports began to come in, the agent as rapidly fell into disrepute. We now meet with the report of a new and safe pain obtunder in the form of rapid breathing. Dr. Benjamin Lee brought before the Philadelphia County Medical Society the results of some observations and experiments of his own upon the subject of analgesia, induced by forcible and rapid breathing. "From his paper published in the Philadelphia Medical Times, we (the St. Louis Courier of Medicine) gather the following facts concerning the subject:—

His attention was first called to the subject by the report of a servant who had been sent to Dr. Bonwill, a well-known dentist of Philadelphia.

She said that " Bonwill had pulled her tooth and did not hurt a bit," that " he had made her breathe as fast as ever she could, and before she knew it, the tooth was out." There was no pain, although she perceived the jerk, when the tooth was extracted.

Not long after this, he had occasion to open an abscess in the perineum of a young man about twenty-five years old, rather delicate and decidedly

nervous. After the young man had breathed rapidly for about three-quarters of a minute, the doctor made an incision about an inch long, and evacuated several ounces of pus. He continued the rapid breathing for at least a half minute longer, and was surprised to find that the operation was completed. He had felt nothing except a sensation of pressure upon the tumor. A fistulous communication with the urethra appeared in a couple of days, and it became necessary, ultimately, to lay open this fistula. Two bridles, each an inch broad, were divided with scissors on a grooved director ; and by the same method perfect freedom from pain was secured, although the operation was, of course, much longer ' than the former one.

In another case where he lanced a felon, there was not the same success. The patient's nerves were completely unstrung from the intense and protracted pain which she had undergone ; and she could not be made to breathe with sufficient force and rapidity to secure the desired effect.

In another case, also of a hyperæsthetic, hysterical lady, in whom he attempted to inject a pile with carbolic acid, the rapid respiration failed to produce analgesia.

Two members of Dr. Lee's family have had teeth drawn by Dr. Bonwill while they were under the effect of the rapid breathing. One of them spoke of a sensation of giddiness produced by the rapid respiration.

Dr. Lee does not undertake to explain how this effect is produced, whether it is a form of hypnotism or the result of a modification of the cerebral circulation, brought about by the respiratory act. He merely brings forward the result of his observations thus far, believing that they show that by a contin- ' uance of rapid and forcible respirations for a certain length of time, it is possible " to induce such a condition of the nervous system that pain shall not be appreciated by the sensorium."

Dr. Bonwill has made use of this mode of securing freedom from pain in dental surgery for several years past, and especially during the last five years. He informs his patients that they will be fully conscious of all that occurs, and perceive every touch, but will feel no pain if they keep up the inhalations energetically and steadily during the whole operation.

The inhalations must be at the rate of one hundred a minute. It is very difficult for a person to breathe more than one hundred times a minute, and "*for the minute following the completion of the operation the subject will not breathe more than once or twice.* Very few have force enough left to raise hand or foot." Dr. Bonwill claims that the results of his experience are such that there is no longer any necessity for chloroform, ether, or nitrous oxide in the *dental* office for the purpose of extracting teeth or deadening sensitive dentine.

Drs. Garretson and Hewson have made use of this system of rapid respiration in connection with the usual anæsthetics in major operations where time is needed, and find a much smaller quantity of the drug to suffice than when it is given in the usual way.

Dr. Hewson makes use of the rapid breathing to the exclusion of drug anæsthetics in midwifery practice.

Dr. Bonwill's theory of the effect of the rapid respiration is : First. That there is diversion of the will-force in the act of forced respiration at the rate of one hundred per minute, which involves such concentrated effort that ordinary pain would make no impression while this abstraction is kept up.

Second. That there is a speedy effect due to the excess of carbonic acid set free from the tissues by the rapid respiration.

Third. That hyperemia is caused by the rapid respiration retarding the flow of blood from the brain.

In the discussion of the papers of Drs. Lee and Bonwill, Dr. Hewson stated that he constantly employs this method of securing analgesia in minor surgical operations in his office. He does not accept the explanation of Dr. Bonwill, at least so far as the second point is concerned. He claims that the blood is *less* thoroughly oxygenated with one hundred respirations per minute than with the usual number.

Several other members of the society stated the results of their observations. Dr. Kite stated that " at first there is a decided sense of exhilaration during rapid breathing, just as from ether, then the senses become confused, there is blurring of sight, peculiar buzzing in the ears, and more or less vertigo."

Further observation and investigation are necessary to determine the scope as well as the *modus agendi* of rapid respiration in causing analgesia, but if it shall prove as efficient in the practice of the many as it has done with Drs. Lee and Bonwill, it will be a very valuable discovery."

PRIZE-ESSAY FUND.—At the recent meeting of the Kentucky State Medical Society, a resolution was offered providing for the creation of a prize-essay fund. It is to be hoped a similar resolution will not only be offered, but passed, at the next meeting of the Colorado Medical Society.

PARTIES receiving this number of the RE-VIEW, and desirous of having it continued from this on, should immediately remit the price of subscription.

TRANSLATIONS.

BY G. J. BULL, M. D.

TRAUMATIC TETANUS.

Dr. G. Trevisanello reports a case (Lo Sperimentale, Feb., 1880) of tetanus in a girl 13 years of age, caused by a punctured wound of the foot. On the sixteenth day of the disease the entire body became covered with a papular eruption. During twenty-two days of severe illness the patient took one hundred and sixty-four grams of hydrate of chloral, and made a good recovery. Though not a specific, chloral combats the hyperæsthesia of the nervous centres and consequent reflex action, and thus obviates the most common cause of death in tetanus, rigidity of the respiratory muscles.

The author maintains that tetanus is a disease of infection, the infectious principle originating within the body as in septicemia, or perhaps being absorbed from without. This conclusion is based on the following arguments :

1st. No constant or characteristic lesion has been found in the spinal cord on post-mortem examination.

2d. The fever of tetanus runs a course altogether different from inflammatory fever, and resembles rather the infectious fevers ; milder in its onset, after irregular remissions and intermissions, it reaches the highest degree, and the temperature rises, even after death, as in cholera.

3d. Tetanus presents its peculiar clinical features, however diverse may be the apparent causes, and is always preceded by trismus.

If the spinal hyperæsthesia were the effect of a nervous alteration propagated from a peripheral nervous lesion, the excitement of the 5th pair should be preceded by excitement of other centres nearer the origin of the wounded nerve.

4th. Cutaneous eruptions have been noted in tetanus (by Jaccoud and the author), as in the greater number of infectious diseases.

5th. Tetanus has many points of clinical resemblance to another infectious disease, hydrophobia.

6th. Tetanus is artificially produced by poisoning the organism with certain substances (strychnine, brucine, picrotoxine).

7th. Among the post mortem appearances, observed by Griesinger in a case of tetanus, were ischæmia of the liver and spleen; engorgement of Peyer's patches, and obstruction of the renal pyramids by recent fibrinous cylinders.

Unfortunately this is an isolated observation, yet the same appearances might have been observed if others had examined the internal organs more carefully.

8th. Billroth, Erichsen, and other surgeons, have observed that, while they sometimes pass years without seeing a case of tetanus, at other times they see it so commonly as to suspect an epidemic cause—especially during the oppressive heat of summer.

9th. Tetanus resembles infectious diseases in being endemic in certain tropical regions, and in predominating during certain months of the year—in Italy, during August and September.

CHLORAL AS AN ANÆSTHETIC.

Dr. Adolphus Paggi writes from Paris (Lo Sperimentale, March, 1880, p. 321) that Prof. Trelat has been using chloral as an anæsthetic, administering it by the stomach, in various operations. The amount required varied from four to six grams, or more, given in two or three doses, at intervals of twenty or thirty minutes, beginning one hour before the operation. In a few of the patients, after they had swallowed two-thirds of the chloral solution; in others after they had taken it all, an incoherence of ideas was observed, with some loquacity and a species of gay intoxication; others, on the contrary, were quiet, and did not speak unless interrogated, when they began to stammer. All were soon overcome with sleep, light at first, but soon profound, and accompanied with complete anæsthesia.

The sleep produced by chloral may last from one to five hours, according to the subject, and to the dose, and passes off without leaving pain or sense of heaviness in the head, or intellectual disturbance. Respiration is regular; the pulse becomes small, hard and somewhat accelerated. The digestive functions are not disturbed by the action of chloral hydrate; only the urinary secretion of the first twenty-four hours is notably diminished. In no case was there retching or vomiting.

Dr. Paggi holds that chloral cannot be used as a substitute for chloroform in general practice, neither can it be employed as an anæsthetic in ordinary cases, without error; but under certain circumstances it will prove of the greatest utility, especially in operations on the buccal and nasal cavities.

By reason of the long duration of its anæsthetic effect, it is to be recommended for the pains of parturition, and in obstetric operations.

Finally, it is said that Prof. Trelat has found it an advantage in operations on the genital organs, to precede the inhalation of chloroform, by giving a few grains of chloral to procure a more complete anaesthesia.

MISCELLANEOUS.

THE Colorado State Medical Society held its tenth annual session in the Supreme Court room, Denver, Colorado, from June 29th to July 1st, 1880. The proceedings virtually opened with the reading of a carefully prepared and eloquent address by the retiring president, Dr. B. P. Anderson. Papers were read by Drs. Bancroft, Adams, Hawes, Solly, Denison, J. W. Collins, Steele and W. W. Anderson. Some of these papers will be published in subsequent issues of the REVIEW, as opportunity offers. The following officers were elected for the ensuing year: President, F. J. Bancroft; First Vice President, S. E. Solly; Second Vice President, D. H. Dougan; Third Vice President, Jesse Hawes; Permanent Secretary, W. H. Warn; Recording Secretary, A. W. Adams; Assistant Recording Secretary, B. Johnson.

Among others, the following resolution was passed :

Resolved, That the Colorado State Medical Society hereby authorize and sanction the editing and publishing of a State journal, to be known as the Rocky Mountain Medical Review ; and said journal is hereby made the official medical organ of Colorado. Next year the society meets in Leadville.

PROFESSOR ALFRED M. MAYER, of the University of Technology, Hoboken, has recently invented a very wonderful and practical instrument for determining not only the direction, but the actual distance of the source of a given sound from the position of the observer.

The vibratory theory of sound and all of its involved minutiæ, has already been proved beyond peradventure ; but this instrument, in addition to presenting another example of the correctness of the theory, furnishes a beautiful exemplification of one of its details. The mental picture which has been formed of a sonorous wave under aerial propagation may now be demonstrated with the topophone, for such is the name of the new instrument. Its practicability rests upon its application by the mariner to navigation, in conjunction with the compass and sextant. "While the compass points out to the sailing-master at sea the position of a known point on the earth, and the sextant points out his position on the earth's surface, the topophone will prove of equal value in determining the position and the distance from an invisible source of sound, either on land or on another vessel. On approaching a coast in the night and observing a light, the compass indicates, by the aid of the chart and sailing directions, the course to be pursued in entering the port. In like manner, when in a fog, the sound of a fog-horn is heard, either on land or afloat, the topophone indicates to the navigator the precise direction from which the sound proceeds, and by simple experiment will give its exact distance."

THE British Medical Association met at Cambridge, August 13, 1880.

THE HELIOGRAPH IN PRACTICAL USE.— The heliograph has recently been put to the practical test by the transmission of a dispatch from General Stewart, in Afghanistan, announcing the result of an attack on the British troops, which was sent from Camp Ghuzni, April 22d, and was received at the India Office, London, on the following day. The news could hardly have been brought more speedily by electric telegraph. The heliograph, signaling right over the heads of the enemy, if necessary, to stations which may be few and far between, does not require any route to be kept open, and can not be interrupted. A ten-inch mirror, that being the size of the ordinary field-heliograph, is capable of reflecting the sun's rays in the form of a bright spot to a distance of fifty miles, where the signal can be seen without the aid of a glass. The adjustment of the instrument is very simple. If an army corps, having left its base where a heliograph station is established, desires to communicate with the other division from a distance of several miles, a hill is chosen and a sapper goes upon it with his heliograph-stand containing a mirror swung so as to move horizontally and vertically. A little of the quicksilver having been removed from behind the center of the mirror, a clear spot is made through which the sapper can look from behind his instrument toward the station he desires to signal. Having sighted the station by adjusting the mirror, he next proceeds to set up in front of the heliograph a rod on which is a movable stud, manipulated like the foresight of a rifle. The sapper, standing behind his instrument, directs the adjustment of this stud until the clear spot in the mirror, the stud, and the distant station are in a line. The heliograph is then ready to work, and the sapper has only to take care that his mirror reflects the sunshine on the stud just in front of him to be able to flash signals so that they may be seen at a distance.—[P. 716, P. S. M., September No.

THE American Medical Association library contains 3,258 volumes.

HYPODERMIC INJECTIONS OF CHLORAL HYDRATE.—At the Cook County Hospital, Chicago, remarkable success has attended the subcutaneous administration of chloral hydrate. Three cases are reported by Dr. Bridge. (M. R. Aug. 14th,) where it was administered for uremic convulsions and delirium tremens. In all these cases it is said there were no phlegmons or abscesses formed at any of the points of introduction of the medicine. In all, one hundred grains of chloral were injected under the skin. [In this connection we would say that we have frequently used chloral in this way, followed in every instance by excessive inflammation, but never by the formation of an actual abscess.]

THE RECENT PHARMACOPŒIAL CONVENTION.—The sixth Decennial Pharmacopœial Convention of the United States, by a close vote, adopted "parts by weight," as the mode of expressing the proportions between the several ingredients in the pharmacopœial working formulae, and thus virtually abolished fluid measures. This, from the fact that greater accuracy is claimed for weighing as against measuring, because temperature affects volume, and because undoubtedly we can approach theoretical precision more closely with a sensitive balance than with graduated volume measures.

M. TRONSSOINT has been investigating the question of the transmission of tubercle, by means of experiments on the hog. He caused animals to eat the lungs of tuberculous sheep, and tried inoculation by the blood and by milk, and found that the animals became diseased in every case. Similar effects were produced upon healthy animals living with tuberculous ones.—[P. S. Monthly.

PROF. G. J. MULDER, the eminent Holland chemist, who proved carbonic acid to be a normal constituent of the blood, died at Utrecht, in May last.

TRAINING SCHOOLS FOR NURSES.—Both Washington and Baltimore now have training schools for nurses. Let other large cities follow their example.

THE Society of Arts has recently conferred the Albert medal upon Dr. James Prescott Joule, for having "established, after most laborious research, the true relation between heat, electricity and mechanical work."

Books and Pamphlets Received.

"A Contribution to the Pathology of the Cicatrices of Pregnancy." By Samuel C. Busey, M. D., Professor of the theory and practice of medicine, Medical Department University of Georgetown; physician to the Children's Hospital, District of Columbia. Reprint from Vol. IV Gynecological Transactions. (Through the author).

"The Therapeutic Value of the Iodide of Ethyl." By Robert M. Lawrence, M. D., Boston. Reprint from New York M. Record. (Through the author).

"Missmanaged Labor the Source of much of the Gynecological Practice of the present day." By Joseph Taber Johnson, M. D., Washington, D. C. Reprint from Vol. IV. Gynecological Transactions. (Through the author).

"Second Annual Announcement of the College of Physicians and Surgeons, of St. Joseph, Missouri." (Through the secretary).

"Perineorrhaphy, with Special Reference to its Benefits in Slight Lacerations, and a Description of a New Mode of Operating." By Edward W. Jenks, M. D.; with eight woodcuts. (Through the author).

"Some Clinical Considerations on Access to Benign Intra-Laryngeal Neoplasms through External Incisions; as Illustrated in a Small Group of Personal Observations, Heretofore Unreported." By J. Solis Cohen, M. D., Philadelphia. Reprint from Archives of Laryngology. (Through the author).

"On Coccygodynia; a Lecture Delivered in Chicago Medical College, March 20th, 1880." By Edward W. Jenks, M. D. LL. D. (Through the author).

"The Cause of Sudden Death of Puperal Women." By Edward W. Jenks. M. D.

"The Southern Clinic;" C. A. Bryce M. D., Editor and Proprietor.

"The Independent Practitioner;" Harvey L. Byrd, A. M., M. D., and B. M. Wilkerson, D. D. S., M. D., Editors.

"The Medico-Literary Journal," Mrs. M. P. Sawtelle, M. D., editor and publisher.

"Rocky Mountain Health Resorts." By Charles Denison, A. M., M. D.

"Chicago Medical Review." "Johnson's Cyclopedia." "Archives of Laryngology." "Index Medicus." "Walsh's Retrospect."

FRONT VIEW.

COLORADO COLLEGE, - COLORADO SPRINGS.

ROCKY MOUNTAIN MEDICAL REVIEW.

Vol. I.   COLORADO SPRINGS, OCTOBER, 1880.   No. 2.

ORIGINAL ARTICLES.

ADDRESS OF PRESIDENT ANDERSON BEFORE THE COLORADO STATE MEDICAL SOCIETY.

GENTLEMEN :—Another year has been added to the past, and we are again brought together in annual convention. In appearing before you at this our tenth annual meeting, allow me to cherish the hope that the year passed has been one of continued prosperity, of genuine pleasure and happiness to you all, and that each and every one present can recall only pleasant remembrances of the time which has elapsed since our last meeting. We are particularly fortunate in having secured such a season and such a locality for our meeting, and I congratulate you, gentlemen, that we are privileged to come together under such favorable circumstances, that we are met in a city whose very name is the pride of Colorado, and whose beauty and attractiveness have won for it the admiration of all comers. Though we are not met for the enjoyment of æsthetic pleasure, still the location of our meeting in the city of Denver will doubtless constitute a most pleasurable feature. Its beautiful surroundings, its capacious and well appointed hotels, its hospitality and magnanimity of the people, the mildness of the climate, together with the efforts constantly being exerted for the comfort and protection of the invalid, will beyond any question of doubt make Denver as valuable a winter health resort as it is agreeable to the seeker of pleasure and recreation. While it is to be regretted that the sanitary condition of the city has gained a reputation as none of the best, yet it must be admitted that its

growth has only but recently warranted the expensive outlay necessary to make the sanitary system effective and perfect. To the members of the medical profession who are its residents are due largely the means set about for the accomplishment of this purpose and as the guardians and ever vigilant watchers over the health of the community, their well directed efforts, backed by the sanitary authorities, will eventually make their fair city the healthiest upon the globe and a monument to their zeal and labor.

In again coming together as a State Medical Society, let us inquire briefly, gentlemen, our object, our purposes. We meet as a professional brotherhood, not only to renew pleasant acquaintances and obtain that pleasure which is derived from social commingling, but for the far more important and laudable purpose of an interchange of professional opinions which are of paramount and common interest upon subjects and questions of vital importance to all. We meet for mutual enlightenment and instruction, "seeking more light," seeking to obtain that which will more surely aid and assist us in relieving the pangs of disease and the pain of suffering humanity. We meet to contribute individual experience and observation, which have been brought forth at the bedside of the sick and the suffering, to the common stock of scientific research and investigation. We meet as a society in the interest and for the common good of the profession, to renew our pledges and our efforts in behalf of the art of medicine and the health and life of society at large. With the strength and organization which ten years have given our society, with the earnest labor and intellectual attainments of its members, we should not

meet to-day in vain, and let us hope that
the endeavors of our body at this meeting
may shed that light and lustre upon its
proceedings which can only reflect credit
and honor upon the whole profession.

As we look back upon the inception and
progressive growth of our society, upon the
few earnest, zealous workers who first
brought it into existence and established its
claim to permanency and respectability, and
again note its increased development, review its transactions, which compare favorably with those of older State societies,
members have just cause for feelings of
pride and abundant encouragement for renewed effort. That its existence has been
the cause of much good, we have only to
note the large number of annual accessions.
It has been the means of bringing together
scattered members from every portion of
the State. Its discussions and papers have
awakened a lively interest among its members and prompted greater zeal and desire
for the attainment of professional knowledge and the acquirement of professional
information. While it is a deep source of
gratified pleasure when we consider the
progress made and the beneficial influence
of our organization, our work has only
fairly begun, and there remains much to be
accomplished. The world moves on, and
each annual revolution effaces, or at least
obscures, the old and well-trodden by-ways ;
old guiding posts are thrust aside and replaced by the new ; old remedies thrown as
rubbish from their exalted pedestal, and
their places supplied by the fruits of the
modern therapy ; the library of yesterday
becomes musty and untouched volumes.
Our pet and most cherished hobbies are
cast at our feet·a broken pillar, and we stand
amazed, even half inclined to turn back,
ere we enter that new field of development
and investigation which lies before us. The
science of medicine of to-day has caused
quite a disturbance of our early teachings,
and the many new and useful accessions to
our art bring us nearer and nearer the thresh-
old of an exact science and infuse that en-

thusiastic zeal and determination which only
comes of constant labor and unremitting
toil. But while the explorer and laborious
investigator have done much towards bringing "order out of chaos," and are accepted
as the instruments pointing the "better
way," questions vitally affecting the principles of our art are constantly presenting,
and remind us that the field is not yet traversed, the harvest not yet o'er. One of
the most important questions at present agitating the mind of the medical profession,
in truth the most important, is the one bearing upon the means of preventing disease,
and the subject of preventive medicine is to-
day attracting the deepest interest and attention. That great and enthusiastic laborer in
this portion of the field, Professor Bowditch,
of Boston, in writing upon this subject, says :
"The profession, joining heartily with the
laity, and aided by the material and intellectual resources of great States will study to
unravel the primal cause of all disease, with
the object of preventing it." Though insti-
gated by laudable and noble desires, though
we work without ceasing, it yet remains a
lamentable fact that those diseases which we
name preventable still stalk boldly through
every portion of the land, and the deaths
appearing regularly upon every mortuary
list speak in undisguised tones as to their
cause, and the hundreds of victims of the
dread visitor—the epidemic—tell us, in agonizing language, that the period of prevention has not arrived; that all our energies,
our labors, yea, even our lives, must be
devoted to the ceaseless fight, the ever-con-
stant pursuit of knowledge, means and
measures, which will surely bring their reward, and as surely accomplish the end in
view. It is a source of profound gratification to note the advancement made in this
important portion of the field, and to the
medical profession especially belongs the
honor and emolument of having awakened
so much interest, and of being the direct
cause of the existence to-day of more than
twenty State boards of health. Through
their untiring efforts a National Board of

Health is also a fixed and permanent organization, which, although possessing as yet inadequate means for the prosecution of its work and aided too meagerly by our National Congress to possess that encouragement it should have, yet, at its last session, held in Nashville, Tennessee, accomplished much good, and papers prepared by its members upon the subject of preventive medicine did not fail to elicit that interest and attention which their importance demanded. Their discussions and suggestions were telegraphed abroad and their published proceedings scattered broadcast, and it is devoutly to be hoped has aroused such an interest among the non-professional which will be productive of the great benefit that they were intended to obtain. It has been stated, and cannot be denied, as we are brought face to face with the evidence, every day, that those whose lives we labor to prolong and those whose health we direct our efforts to protect, are after all the greatest obstacles which we have to surmount in the successful accomplishment of our purposes. We not only labor to acquire the most available and successful preventive means, but our efforts must per force be divided in the often discouraging task of "exhortation and prayer" with a stubborn municipal father and the laborious campaign in the too-often vain effort of convincing an enlightened constituency.

Notwithstanding efficient State Boards of Health, notwithstanding the measures urged and placed in view, it must be confessed that there still remains an indifference, an utter disregard upon the part of the people at large, for those hygienic laws and those sanitary requirements which nature alone has imposed, and we are so often confronted with such entire depravity upon the subject, that it should not occasion surprise if the medical man turned away discouraged, or even if he sought solace in relinquishing his efforts and giving up entirely the work he has prosecuted so assiduously. "The public press some time since credited a city father of a city

of this country with having said in a debate upon quarantine that 'Quarantine was cruel, unjust and oppressive,' and his utterances met with approval from more than one correspondent throughout the country." Another of these city fathers, when appealed to for a sanitary appropriation, thought "that all such matters should be left to the providence of God." The president of an Eastern State Medical Society, in an annual address a year or two since, says : "Ask for municipal appropriation for building a public hall, a new market house, an engine house, or a public school, for a fire company, or for clothing the police force with great coats—all very commendable objects in their way—and the public ear is open and the public hand ready to dispense the public money. But ask for an appropriation or help to cleanse the Augean stables, whose stench, worse than the fumes from Tartarus, are welling up from street, from court and from alley, and carrying death and desolation to every home and house, chasing out the pure air which God gave man to breathe and choking with mephytic horror the innocent and the helpless, and the municipal father's heart grows very hard, and his ear very dull, and his understanding very obtuse, and his faith in the Lord very strong."

When we are enabled to so educate the masses that they will by pecuniary aid lend encouragement and join heartily in co-operation, we can then work with a degree of satisfaction, promising a glorious return and a speedy reward. But, gentlemen, it is a fact deeply humiliating and lamentable to confess, as we are compelled to do, that all the discouragement to well directed efforts, and all the obstacles in the way of honest intent and the desire single to the purpose of arousing an interest and obeyance of laws of hygiene, does not exist alone among the laity. Our own profession is not without fault. There are members, men it may be of character and local fame in their profession, who sustain municipal authorities in their refusal of aid, and who cry down the laudable efforts of their brethren as all "parade and fuss." When we arrogate to ourselves the prime right and boast of the proper knowledge of effective methods used in protecting the health of the people, the foregoing revelation seems the more astonishing, and we pause for an explanation sufficiently clear and commendable which can sweep away the charge, and can only bow our heads in shame when called upon for an apology for this the most culpable misdeed of an erring professional member. Can such men call themselves the guardians of the public health, supposed to be in possession of the principles of science and the proper knowledge of all agents for the protection of life? Are they actuated by that earnest desire for doing good, and prompted by those grand motives of zeal and self-sacrifice which characterize, "par excellence," the true physician? We must confess that the answer to both questions must be in the negative, and that the motive which inspires such dishonesty and contemptible hypocritical inconsistency, is their inordinate and base pursuit of gain and the desire of pleasing and conforming for transient popularity to the wishes and whims of the majority. As a profession, it becomes our duty to expel all such from the ranks and treat them with that contempt and loathing their actions merit. Gentlemen, those of you who embrace the majority of earnest workers and zealous laborers will not relax your efforts. Ever be on the alert to guard against the approach of danger, and be ever ready to co-operate with any society, with any body having for its aim and object the good of the public health. It is our duty. It should be our constant care, always ready to traverse the by-ways and the highways, to plunge into alley and into court, and see to it that the causes producing the epidemic, the cause of the avoidable scourge, is swept away even at any cost. Constantly aiding and assisting in this grand work, the time will come when the period of prevention can be proclaimed as a fixed and permanent epoch.

I would call your attention, gentlemen, to another topic which more closely concerns the good name of our profession, and the subject demands from each and all reputable physicians suggestions which will lead to a process of elimination. At no former period in the history of the Colorado profession has the necessity of regulating the practice of medicine more forcibly urged itself than at the present time, and the subject has assumed that importance which calls for the most prompt and speedy action. Owing to the laudable efforts of the profession in many of the Eastern States in arousing an interest in this subject and obtaining an efficient co-operation from their legislators, laws have been created which delegate the practice of medicine to the qualified physician, and as a consequence the unqualified have been compelled to "move on," and those States which have been so unfortunate as to delay in framing such laws prove a safe haven, an El Dorado, for the charlatan, the mountebank and the unprincipled pretender, and in such States these flagrant personifications of ignorance and cheek find and work a bonanza which rarely falls to the lot of true merit. Our own State, having made rapid strides in development and unprecedented progress in everything else, is sadly behind many of her sister States in this respect. As a profession having at heart the best interests of society, having in our hands the lives and health of the people among whom we reside, it should be our chief aim and endeavor to rid the State of this ingesta which has been so continuously vomited into her lap. The bill presented to our last Legislature and failing to pass from causes which I do not pretend to understand, should again, and if necessary, again, be brought forward, and the joint effort and influence of the profession should be used in urging its passage, and our efforts should not relax until this bill has secured a place upon our statute books, and we are convinced that the many base, merciless excrescences who disgrace the name of doctor, are driven from the community and forced to relinquish their nefarious calling. It has been frequently said that of all nations, the Americans were by far the most susceptible of humbuggery—and the greater the humbug the greater the success. However true this may be, it is doubtless often the case that our efforts and interferences may be thankless undertakings, and viewed as malicious, envious and unfair, and the public may consider it none of our business if the *vox populi* choose to cry out in favor of the "gilded pill," or the wonderful health-giving properties of the "decillionth trituration." This line of reasoning may appear forcible to those of the retaliating spirit, and we are often sorely tempted to relinquish our efforts when we devote a laborious life and meagre earnings endeavoring to acquire means whereby life may be protected and prolonged, and in return see palaces built by vile nostrums and patent medicines; and luxury, wealth, comfort and ease obtained by the shallow pretender and the arrogant charlatan. It must be confessed the temptation to stay our efforts grows strong and the inclination often possesses us to turn away forever. I very well remember the story told by a friend who visited the session of the American Medical association held in San Francisco some years since. This friend of mine enjoyed the acquaintance of a classmate, a graduate of one of our best medical schools, whom he had heard of as having settled for the purpose of practicing his profession in the above-named city. Of course much pleasure was anticipated from the meeting and no doubt felt but that the former classmate was a "bright particular star" of the San Francisco medical world. But alas, no one knew or had ever heard of such a man. All the "oldest inhabitants" were quite sure no one of the name had ever practiced in the city, and my friend had almost lost hope, when a few days after his arrival in the city, the former classmate presented himself at my friend's hotel, sent up his card, but instead of his own name, bearing that of one of the wealthiest and most notorious quacks and venders of nostrums in the city. My good old friend

remonstrated against the course he had adopted and censured him for his departure from former sound ethical teaching. The explanation came, and if not satisfactory, certainly convincing. Taking my friend to a street, this "sarsaparilla king" pointed out such and such a block, such and such a building, with the remark, "those are all mine; now what have *you* obtained as fruit of your laborious life? what have you to show me as a reward for your strict adherance and conformity to your code of ethics?" The answer may have been, it is true, nothing of worldly compensation, but honesty and a self satisfied conviction of having sacrificed pelf and gain to a life of noble purposes and conscientious duty.

However discouraging it oftentimes seems, however much our efforts are misconstrued and unappreciated, we should never lose sight of the fact that we live for the good of the health of the community, and as philanthropists, as the true humanitarians we claim to be, it becomes our sacred duty to interpose every obstacle and continue our remonstrances until we are assured that the laws are sufficient to protect the welfare and the health of the people from this army of base and ignorant pretenders who pursue their calling and gain their livelihood at the sacrifice of human life. But, gentlemen, while we are striving and laboring without ceasing to rid our profession of imposters, let us not forget our own shortcomings. Let us act in harmony, work shoulder to shoulder for the elevation of the noblest of all professions, having an "eye single" for its advancement, and a common desire to promote the interests and great good which our calling can accomplish. Let us constantly strive to be true to each other, cultivating instead of jealous bickerings, that honorable feeling of respect which the honest physician should command, no matter how humble his pretentions. Let us be ever ready to extend the "right hand of fellowship" to the young and it may be inexperienced brother just entering our field, welcoming him with encouragement instead of frowns; and, finally, let us so live and our demeanor toward each other always be such as will command the respect and admiration of all, and justly entitle us to membership in a profession of which it has been said "God never created a nobler."

A NEW METHOD FOR WITHDRAWING FLUIDS FROM THE MIDDLE EAR.

BY CHAS. DENISON, M. D.

I believe that otitis media may often be checked by the procedure I will now describe.

We all appreciate the liability to serious inflammation of the middle ear resulting from the use of the nasal douche : from forcing secretions through the Eustachian tube in the act of blowing the nose ; from an extension of catarrhal inflammation into the middle ear, etc.

How to relieve the pressure caused by this introduction of fluids into the middle ear and the congestion resulting therefrom, is a difficult problem, which if solved would often prevent the serious result of suppurative inflammation.

The books treat mainly of paliative means, and so far as any teaching I can gain from them, there is a strong probability that the membrana tympani will soon have to be perforated by the physician, or perforation may occur spontaneously. We have been sadly in need of means to forestall this result, by withdrawing the foreign material through the natural passages.

Last winter I believe I hit upon a procedure that is capable to a certain extent of accomplishing this result.

In the act of blowing my nose, having a cold at the time, catarrhal or other secretion was forced through the Eustachian tube into the middle ear, causing pain which was not relieved by the usual paliative measures, which were assiduously employed. The experiment which gradually overcame the difficulty was this: Attaching the flat nozzle used with Pollitzer's air bag, to the exhaust bottle of my aspirating apparatus, I first

exhausted the air from the bottle, and then compressing the alæ of the nose, with the nozzle introduced, so as to wholly exclude air from without, I closed my throat as in the act of swallowing, and kept it so, and at the same time directed the stop-cock to the vacuum bottle to be turned on. The effect of thus transferring the vacuum to the middle of the head was decided, seeming to draw every thing inwards toward the pharynx and nasal passages, creating a temporary congestion of those parts. But the pain in the ear was perceptibly lessened by the experiment, and on its repetition finally wholly disappeared.

I think the effect was mechanical, and that a small portion of the fluid was probably drawn into the Eustachian tube at each operation.

Since that time I have tried the experiment in practice as opportunity has offered, and mainly with results similar to those above mentioned.

In one instance I believe the stage of suppuration had set in, for I expected to puncture the membrana tympani the next day after this operation had been tried. But my patient did not return, and I found some time afterward that improvement had from that time continued till she wholly recovered.

A NEW SPLINT.

BY JESSE HAWES, M. D., GREELEY, COLO.

I would call the attention of the society to an appliance for contused and compound fractures of the leg. The principles involved in it have not the merit of novelty, for, to retain fractured bones in a desired position we have numerous splints, while to soothe contusions our professional brethren have given us a multiplicity of lotions, not the least valuable of which is water, but so far as I am aware heretofore no appliance has rendered easy the use of the latter with the former.

Dr. Hamilton of New York, a few years ago in several papers did as much perhaps as any American surgeon to call *especial atten-*

tion to the use of *warm water* in surgery, giving credit to German surgeons for its reintroduction. *Its value has been most fully established.*

It is not appropriate in this connection to discuss the value of warm water as contrasted with cold in their applications in surgery, nor the relative merits of submersion, fomentation or irrigation, since this appliance practically allows the use of either.

The splint when used with irrigation consists of a tin trough of a length sufficient to extend from a point several inches above the

is added to prevent the rotation of the splint, and is not always needed.

The accessories are a bucket—having a small perforation in its bottom for the delivery of water upon the limb—and a pail for the reception of water that will flow from the distal end of the splint, after performing its work of irrigation.

The limb with the fracture reduced should be placed in the splint—the distal end of which extends beyond the footboard—little pads covered with oilcloth between splint and limb retain it perfectly in position.

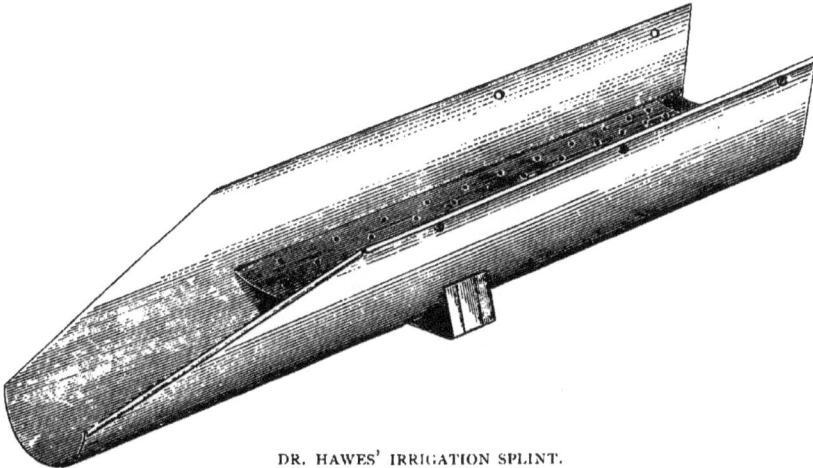

DR. HAWES' IRRIGATION SPLINT.

place of fracture to a foot beyond the end of the limb—the sides of the trough rising higher than the upper surface of the enclosed limb. It should have a diameter at least an inch greater than the limb. A perforated false bottom extends across and an inch above the true bottom; the perforations at the proximal end being so complete as to prohibit the possibility of water flowing backward over its surface and into the bed. At the proximal end the space between the true and false bottoms should be closed, that water may not flow backward into the bed. At the distal end this space is open for the escape of the water used in irrigation. On the under side of the true bottom, two or more stiff tin pedastals should be soldered, furnishing the splint with a base—this base

A pad between the tendo achillis and the false bottom raises the heel and avoids its pressure.

If desired, Buck's extension can be applied.

The quantity and temperature of the water is easily controlled, and must vary with the necessities of each case. Several thicknesses of porous cloth, like crinoline, kept thoroughly wet by streams delivered upon the cloth at many points, has proved an excellent substitute for the "bath" so highly and justly esteemed by many surgeons.

In cases of compound fracture and in gunshot wounds producing fractures of the leg or thigh, by using carbolized water, I have been relieved of much anxiety, while the duration of convalescence, and the amount of suffering endured by the patient have been reduced to a minimum.

A modification that will suggest itself, can be used in cases of fracture of the arm and forearm, if the patient is confined to the bed or chair.

The advantages of this appliance may be summarized as follows:

AS A SPLINT,

The limb is easily exposed to view.

As a retentive apparatus, its efficiency is second to none.

It admits of Buck's extension when that is desirable.

A single splint may be applied on limbs varying much in size.

AS A MEANS OF APPLYING WATER,

It is neat—the bed need never be soaked; the water is applied and disposed of most easily.

It has proved itself of much value in allaying inflammation; and, finally,

It can be made in a few minutes by any tinsmith, and very cheaply.

In the phrase of applicants for patents, "what I claim as new for this appliance" is an increased facility and neatness in the use of water where a splint is applied.

Vesical Calculus in a Female Child: Urethrotomy—Supra-pubic Cystotomy—Recovery.

BY DR. VITTORIO DE SEMO, OF CORFU.

Translated for the Review by Dr. G. J. Bull, Colorado Springs, Colorado.

About the close of the year 1872, I was called to give my professional care to a girl eight years of age, of lymphatic constitution, who had commenced to suffer two years before with symptoms referable to the urinary passages. She was often tormented with pains in the perineum, with a sense of weight, with scalding sensations during the passage of urine, and sometimes the flow of urine would suddenly stop, to continue only after many efforts, and upon change of position. The urine was abundant and clear, presenting but little mucus and nothing else of interest. This disturbance gradually increased in intensity, the pain became more severe and almost continuous, the dysuria more painful and the quantity of mucus more abundant in the urine, which was now ammoniacal, and a general fever set in, at first intermittent, then remittent and subcontinued. The fever was accompanied by repeated chills and often by profuse sweats. At no time was any blood found in the urine. The general condition of the patient was such as would be produced by the local suffering; she was pale, extremely thin, and unable to leave her bed.

This assemblage of symptoms pointed plainly to the existence of grave disease of the urinary tract, and caused a strong suspicion of the presence of stone in the bladder. The extreme obstinacy of the patient up to this time had not permitted me to resort to the use of the catheter, so necessary to a positive diagnosis.

One night, being called in haste, I found the patient absolutely unable to pass water; the bladder was full, and its pear-like shape could be made out very well by palpation above the pubes. Making use of some pretence, I succeeded in introducing a catheter, and hardly had it entered the bladder when it came in contact with a hard, resonant body, evidently a stone. The suppression of urine now recurred from time to time, so that I was constrained to pass the catheter, and on each occasion observed the same condition. I now informed the family that the only therapeutic resource to liberate the little patient from her suffering, and to save her life, was extraction of the stone. I called the eminent Prof. Lavrano in consultation, and we agreed to attempt extraction through the natural passage, after dilatation of the urethra, before proceeding to a more serious operation.

I dilated the urethra with prepared sponge, and being then able to pass the extremity of my little finger into the bladder, could feel a smooth, rounded surface, of not more than three centimeters in width. One of the diameters of the stone being but three centimeters, it therefore seemed possible to extract it through the urethra without rupture. This I proceeded to do, introducing forceps of various dimensions and grasping the stone, but it was impossible to extract it. I adopted this maneuvre on two subsequent days, but without success. In the meanwhile the patient's condition was hourly becoming worse ; a continued fever consumed her ; she was becoming very weak, and it was evident that I must operate without delay. I decided, therefore, to practice urethrotomy, and selected the incision of the upper wall of the urethra as that which most rarely leads to hemorrhage, and is least apt to induce incontinence of urine. I called Prof. Lavrano again to my assistance, and he approved of the proposed method of treatment The discovery that one of the diameters of the calculus was no greater than three centimeters, authorized me to hold that the opening to be obtained by superior urethrotomy might be sufficient for our purpose. I should have been able, it is true, to measure the calculus, but the state of the bladder and the general condition of the patient permitted of no further delay, and the maneuvres required for mensuration might have induced the gravest consequences.

On February 1st, 1872, I proceeded to operate, after having thoroughly chloroformed the patient. I passed the little finger of my left hand into the bladder as a guide, and incised the superior urethral wall with a probe-pointed bistoury. I then introduced a pair of strong forceps, seized the stone and brought it by gentle traction to the arch of the pubes ; but at that point the forceps slipped from the stone. This was repeated several times till it became evident that the stone was too large to come out from the sub-pubic region. To have made out that one of the diameters of the stone was equal to no more than three centimeters, signified but little, for the stone might have other diameters much greater. It occurred to me then to break the stone with the lithotrite, but it appeared absolutely impossible to

manipulate this instrument in the bladder, seeing that the bladder was spasmodically contracted and prevented its introduction ; moreover, the incision of the urethra made it impossible to retain an injection in the bladder for purposes of dilatation. Under these unfortunate circumstances the only expedient remaining was supra-pubic cystotomy, which Prof. Lavrano proposed, and which I at once accepted, though fully aware that I was about to undertake an operation of extraordinary and exceptional gravity. For reasons already mentioned, it was impossible to distend the bladder by injection. Dr. Lavrano, therefore, passed his index finger into the bladder through the dilated urethra, and pushed the stone against the abdominal wall in the supra-pubic region. I then made a transverse incision of the skin three and a half centimeters in length, parallel with the arch of the pubes, using a curved bistoury, and cut all the underlying tissues successively till I reached the bladder, which I recognized by its smooth surface. Cutting through this, I came upon the calculus and easily drew it forth with the forceps. The calculus was not unlike a large egg in shape, somewhat flattened. Its broadest surface looked upward, and its diameters were us follows :

The greatest diameter,	. .	centim. 6
Width above,	. . . . .	" 4½
Width below,	. . . . .	" 3

Its weight was 80 grams. The size of the stone greatly exceeded my expectations, although the information obtained through the digital exploration was correct. The width of the surface examined was exactly three centimeters, but that happened to be the smallest diameter of the calculus—the smaller end of the oval.

I closed the wound, using a simple dressing of dry lint, and carefully watched the progress of the patient. At first there was profound collapse, the skin was cold, the pulse small and fugitive, and the face pale. I prescribed good broth with brandy, and sinapisms to the extremities. At 6 o'clock that evening she had a chill, followed by high fever and abundant sweating. I gave twelve grains of sulphate of quinine, and the night was passed comfortably. On the following days there were recurrent febrile attacks, always combated with quinine. Nothing else noteworthy occurred, and one month after the operation the abdominal wound was perfectly cicatrized.

Now, in consequence of the urethral incision, there remained, as we had feared, a troublesome incontinence of urine. I employed a general tonic treatment, in conjunction with cold bathing, but without effect, till one morning (about the middle of April, 1872,) the child informed me she could retain the urine in the bladder at will. This sudden cessation of so grave a symptom surprised me not a little, and I confess I could not account for it. Soon, however, new symptoms appeared to demand my attention. The patient began again to suffer pain in the region of the bladder, difficulty in passing water, and sudden arrest of the flow ; symptoms, in short, analogous to those observed before the operation. I decided to make another examination of the bladder, and at the moment of passing the sound through the neck of the bladder, found an unusual resistance. Then passing the tip of my little finger into the urethra, I discovered the opposing body to be a tumor, of the size of a large lentil, attached by a small pedicle to the deeper orifice of the urethra. I easily removed it by a gentle rotatory movement of my finger, and found it to be a mucous polypus. This happened at Corfu, on July 10th, 1872, after which the incontinence of urine returned, and continued obstinately in spite of the many measures I employed to prevent it. An almost constant dropping of urine caused very troublesome erythema of the inner surface of the thighs and of the greater labiæ. The child could not hold her water more than four or five minutes. Thus she went on for three long years, when circumstances occurred which obliged me to go for a time to Naples, and I induced the little patient's parents to allow me to carry her

thither in the hope of finding a remedy for her condition in the natural resources of that country. Before adopting any treatment, I consulted the illustrious Prof. Gallozzi, telling him the history of the case, of the serious operation, of the temporary cessation of the incontinence, of the subsequent extraction of the polypus, and of· the immediate suppression of the loss of power. I said that the temporary relief of the incontinence had certainly depended on the polypus, which closed the urethra sufficiently to prevent the escape of urine. I maintained that the actual incontinence was the result of three principal causes : 1st. The incision and laceration of the fibres of the urethra and of the neck, rendered necessary by the operation. 2d. The erythema of the inner surface of the thighs propagated to the vulva, must have provoked a chronic irritative condition of the deep parts of the urethra, and hence the frequent contractions of the detrusor urinæ. 3d. By reason of the long duration of the incontinence it was probable the capacity of the bladder was greatly diminished, so that it could hold only a small quantity of urine. This, however, was only hypothetical, for I had been unable to induce the patient to submit to the catheterization necessary to a demonstration.

While we consulted together, Prof. Gallozzi passed the catheter and showed, contrary to my hypothesis, that the bladder was not diminished in its capacity. The incontinence, therefore, was due to the first two causes only. The curative indications were to give tonicity to the urethro-vesical apparatus, and to combat the uric acid diathesis in order that the formation of other calculi might be prevented. To meet the first indication, Prof. Gallozzi suggested a treatment by the Bagnoli waters ; and in case the symptoms showed no improvement, thought it would be desirable to use the electric current, and if that failed, to produce a slight constriction of the urethra by cauterization.

The patient was then put under regular treatment at the Patamia Establishment at Bagnoli under the direction of Dr. Villari ;

a treatment which consisted of douching the perineo-vulvar region with cold water, for from one to three minutes, and of bathing, continued daily for from fifteen to thirty-five minutes. After ten baths there was marked improvement ; the patient being able to hold her water an hour or more. After thirty baths the water was passed not oftener than once every two or three hours, the erythema of the thighs had disappeared, and the general condition was flourishing. At night the precaution was taken to wake the child every two or three hours, to prevent the involuntary passage of urine. The cure being now almost complete the treatment was suspended. Another course of baths next summer and the development of the child will render her condition perfect.

At the present time (Sept., 1879,) seven years after the operation, the patient enjoys excellent health, her only remaining symptoms being an occasional nocturnal incontinence.

SELECTIONS FROM JOURNALS.

On the Destruction of Infectious Germs.

BY DR. A. WERNICH.

The theory that contagious diseases as well as putrefaction and fermentation are developed and propagated by the agency of organisms allied to the bacteria, has been widely accepted, and is supported by the results of recent investigations. It becomes, then, of paramount importance to ascertain the most efficacious means of destroying these organisms.

With this object in view, these researches into the conditions under which bacteria may be destroyed have taken three directions: 1. To test an observation made by Ernst Baumann, that the putrefactive organisms in the course of their action develop carbolic acid, a deadly poison to them, and to inquire whether there are not other poisons to bacteria developed in a similar manner. 2. Investigations prosecuted during the prevalence of the plague to ascertain whether a dry disinfection of clothing and goods could be made effective wholly to destroy the infectious organisms. 3. Having transplanted active infectious organisms from one substance to another to which they are suited, to arrest them in the most rapid stage of their development, destroy them, or cause them to perish.

In order to make the experiments of real value, a sure means must be found of knowing whether the organisms are alive or dead ; they may seem dead when they are only passive. The only unfailing test is afforded by the reproductive faculty : when reproduction ceases, and can not be excited, organisms may be considered dead.

Particular investigators have doubted whether it is possible wholly to destroy these lower organisms. Naegeli* says it can not be fully done without the aid of heat, and even heat is not always equally effective. They are generally more easily destroyed by heat when moist than when dry, but even a boiling heat will not destroy some of them when they are in fluids of a neutral reaction. The more acid the reaction, the less is the degree of heat that is required. The degree of heat required to destroy the germs of infectious diseases is believed to be greater than it is practicable to apply by the dry process to clothing and similar materials. The capacity of many of the organisms to reproduce may, indeed, be destroyed by a more moderate temperature, but a question remains concerning the germs or spores which had been taken up into the materials and were carried away with them. These are believed to have some kind of a coating which enables them to resist what destroys the parent organisms.

In order to test the value of the dry process as applied to infected clothing, pieces of different clothing materials were impregnated with strong putrefying and bacteria-bearing fluids, then dried slowly, and kept for a long time without protection against external influences. Whenever the smallest

------

* Die niederen Pilze in ihren Beziehungen zu den Infektionskrankheiten und der Gesundheitspflege, s. 201.

piece of one of these materials was put into a suitable fluid, the perturbation invariably took place which is the sure sign of the active multiplication of bacteria. Clothing which had not been impregnated did not excite this perturbation, or only in an insignificant degree. Specimens of the defiled clothing which were placed in a similar solution after they had been exposed for five minutes to a temperature of from 125° to 150° C., or for one or two minutes to a higher temperature, produced no change. The capacity of the bacteria to resist heat varies widely among the different species, and appears to depend largely on the faculty of developing spores. The individuals are killed, but the spores remain vital. The increase of any one kind is limited by the presence of other kinds, with which a struggle for existence has to be maintained.

No increase of bacteria takes place without the presence of a suitable substance to support them. The most favorable of non-nitrogenous substances is sugar; among nitrogenous substances the most favorable are the albuminoids; among mineral matters, potash, phosphorus, magnesia and sulphur. If the supporting substance, even though it is needed in only a minute quantity, is consumed, or if it is present in great excess, a pause in the development, but not the death of the bacteria, takes place. A similar effect is produced by taking away the water, but when the water is restored, an increase of life again takes place.

The practical object of disinfection should be to go beyond securing a suspension of animation of the bacteria, and to seek to destroy the vitality of the spores. Neither years of dryness, nor months of exposure in foul water, nor repeated drying and moistening, will injure the fertility of these germs.

An excess of water produces a similar effect with desiccation upon the vital conditions of the bacteria. A great dilution of the supporting fluid by the infusion of pure water will in a short time produce a suspension of the process of decomposition. Pri-

vation of light has no effect. The operation of electricity has not been enough observed to justify the drawing of any conclusion. The effect of the privation of air has not been fully determined. It was once thought that the development of bacteria could be hindered by the removal of oxygen, but this is doubtful. Oxygen greatly speeds the development, but it can take place without it. Bacteria are not developed in nitrogen, hydrogen, carbonic oxide, carbonic acid, nitrous oxide, and illuminating gas.

The substances which are fatal to the life of the bacteria next demand attention. Among these, the concentrated mineral acids, iodine, bromine, chlorine, the sulphates of copper and zinc, corrosive sublimate, benzoic acid and its salts, salicylic and metasalicylic acids, quinia, many aromatic substances, and alcohol, have long been known as such. Carbolic acid is the highest in repute among these poisons; and it is an interesting fact that Mr. E. Baumann has discovered this very substance among the products of the bacterian fermentation to which it is so fatal. Alcohol is another substance similarly associated. The discovery of the curious relations of these two substances gives a new light upon the cause of the spontaneous destruction of bacteria in strongly fermenting fluids, and encourages us to look for other substances having a similar origin and a like action. As evidence of the possession of such properties by any substance, we should require—

1. That substances favorable to the development of bacteria should remain free from them when the substance to be tested is added to them.

2. That active bacteria, when transplanted into a supporting mixture to which a substance supposed to be poisonous to them has been added, should cease to propagate themselves and die out. This may be called the aseptic test.

3. That, when the supposed poison is introduced into a solution swarming with bacteria, all living examples should be killed. This may be called the antiseptic test.

Various aromatic substances, the products of fermentation, were added to a mixture of water and chopped meat at a temperature of 35° C. They proved efficacious in preventing, suspending, or wholly stopping decomposition in the following order, according to the strength of their working :

1. As PREVENTIVES OF DECOMPOSITION :

Indol.....................in a proportion of 1 :1,000 of the mixture.
Kresol............ ......   "   "   2 :1,000   "   "
Phenylacetic acid   "   "   2.5:1,000   "   "
Carbolic acid...........   "   "   5 :1,000   "   "

(The working of scatol and hydrocinnamic acid could not be satisfactorily ascertained, on account of the difficulty of dissolving them in water.)

2. As ASEPTICS—killing transplanted organisms by poisoning the supporting fluid :

Scatol.........................in a proportion of 0.4:1,600 of the mixture.
Hydrocinnamic acid   "   "   0.6:1,000   "   "
Indol..........   "   "   0.6:1,000   "   "
Kresol ....   "   "   0.8:1,000   "   "
Phenylacetic acid..   "   "   1.2:1,000   "   "
Carbolic acid.........   "   "   5 :1,000   "   "

3. As ANTISEPTICS—wholly destroying all living bacteria :

Scatol, in the proportion of 0.5:1,000 of the mixture, in twenty-four hours.

Hydrocinnamic acid, in a proportion of 0.8:1,000 of the mixture in twenty-four hours.

Phenylacetic acid, in a saturated solution (1:400) immediately.

Indol, in a saturated solution (1:900), in twenty-four hours.

Kresol, in the proportion of 5:1,000 of the mixture in twenty-four hours.

Carbolic acid, in the proportion of 20:1,000 of the mixture, immediately.

Two points strike us in this review : first, the difference in the amount of poison required to produce the aseptic and the antiseptic effect; again, it is curious that carbolic acid, the favorite antiseptic, appears to be the weakest on the list. It is at the same time one of the most soluble, while scatol, the most difficult of solution, is the strongest.

If we add the substances we have been examining to a saccharine solution exposed to fermentation, a slackening of the fermenting process will take place, and the different substances will, as before, exhibit their power to delay the process in the following order : scatol, hydrocinnamic acid, indol, phenylacetic acid, kresol, carbolic acid.

These facts seem to justify us in looking for specific disinfectants and prophylactics among the aromatic products of chemical decomposition. They also give a strong air of plausibility to the theory that the bacteria produce, through the chemical changes of which they are the direct cause, the most effective substances that can be used to destroy them. The idea is logically deducible from this theory that the germs of disease finally produce their own destruction by the operation of their growth and development, and helps us to comprehend the cynical course which is characteristic of most infectious diseases.—*P. S. Monthly.*

On Some Points in the Pathology and Treatment of Typhoid Fever.

BY DR. WILLIAM CAYLEY,

Physician to the Middlesex Hospital and to the London Fever Hospital.

What are the nature and properties of the poison which is supposed to give rise to the disease ?

Now, the typhoid poison has up to the present time eluded all attempts to isolate it or to demonstrate its nature either by microscopical examination or chemical analysis ; we are only conscious of its existence by the effects which it produces on the human organism. Nevertheless, some of its properties are known with tolerable certainty : 1. When introduced into the system it multiplies. 2. It is contained in the alvine discharges of persons suffering from the disease. 3. It retains its activity for an indefinite time after it has passed out of the body, when placed under favorable conditions, these conditions being the presence of decaying animal matter and moisture. Hence its usual habitats out of the body are drains, sewers, cesspools, dungheaps, wet manured soils. And there are some grounds for supposing that in these situations also it may possess the power of multiplying. Lastly, in all probability it is particulate, and not either liquid or gaseous.

The actual nature of the poison—whether it be, according to the hypothesis most generally accepted, some kind of fungus, or microzyme, or protoplasm, in a word, a *con-*

tagium vivum, or whether, as maintained by others, it is some derivative of albumen, capable of exercising a catalytic action on other albumen,—I do not propose to discuss, as it is a question at present rather of theoretical than practical interest, and it is one, moreover, for whose final determination the data are hardly yet sufficient.

A subsidiary question to this, but one of considerable practical importance, is whether the poison can be generated *de novo* from decaying organic matter, whether it be pythogenic, as was so ably maintained by Dr. Murchison, or whether it can only arise by continuous propagation, as was maintained no less ably by Dr. William Budd.

The argument on both sides may be very briefly stated.

In favor of its origin *de novo*, it is asserted that typhoid fever has often broken out in isolated situations—as solitary farm-houses, far removed from, and holding no communication with, places where the disease exists ; and many such instances are given by Dr. Murchison. On the other hand, it has been proved incontestably by many instances, both in this country and on the Continent, that all the conditions supposed to be required for its generation may be present for an indefinite time—as percolation of sewage into wells supplying drinking-water, —and yet the disease does not show itself till the poison is introduced by the arrival of an infected person, when an outbreak at once takes place.

Now, I would submit that this latter argument far outweighs the former ; for otherwise, if it be proved that all the conditions necessary for the origination of the poison are present, as shown by its subsequent development when the germs are introduced, and yet it does not develope, we should have to admit that the same causes are not always followed by the same effects.

The instances in which persons in the latent stage of typhoid fever, or actually ill with it, have carried the disease to distant places, and caused it to spread, are so numerous that I believe this mode of propaga-

tion is now universally admitted ; the communication of the disease taking place not by direct contagion from the sick to those brought into immediate contact with them, but by the ordinary mode of sewage contamination.

The apparently spontaneous origin of the disease in isolated places may, doubtless, be in part explained by the very long time the poison may lie dormant, still retaining its essential properties, and capable under favorable conditions of developing its activity.

The poison, too, may be introduced in quite unsuspected modes. It is well known that the typhoid poison by no means always produces what is usually recognized as typhoid fever. In very many cases its only effect is to cause some malaise, together with slight intestinal catarrh, which is not necessarily attended by diarrhœa. These cases are only recognized as typhoid when they occur during an epidemic, and among persons who have been exposed to infection ; but if we may judge by the analogy of scarlatina, diphtheria, and other contagious fevers, where similar slight unrecognizable forms are not uncommon, these apparently trivial cases should be as potent in communicating the disease as the severe well-marked forms. Hence it is quite possible that the typhoid poison may be introduced into a locality by a person whom no one would suspect of being infected with the disease.

It has been ingeniously suggested by my colleague, Dr. Robert King, that the typhoid poison is derived solely from decomposing albumen, which is not present in healthy stools, and that it can therefore only arise from morbid stools which contain albumen, as is the case in some forms of catarrhal diarrhœa, and he has published a case where the poison seemed to be generated in this manner. This theory might certainly account for long-continued sewage contamination of water without the production of typhoid fever, as the special material from which the typhoid poison is generated might be absent. But it is hardly likely that any such derivative of albumen would remain undecomposed in a drain or cesspool for so long a period as we have reason to believe is the case with typhoid poison.

We have next to consider how long the typhoid poison can thus retain its activity out of the body in a suitable locality. One of the clearest pieces of evidence on this point is the well-known instance related by Dr. von Gietl. To a village free from typhoid an inhabitant returned suffering from the disease, which he had acquired at a distant place. His evacuations were buried in a dunghill. Some weeks later five persons who were employed in removing dung from this heap were attacked by typhoid fever: their alvine discharges were again buried deeply in the same heap, and nine months later one of two men who were employed in the complete removal of the dung was attacked and died. Here we have distinct evidence that the poison retained its powers for nine months.

Dr. Murchison, in his work on Fever, gives an instance in which six cases were spread over a period of eight years. I have recently seen an instance in which an interval of two years occurred without apparently any fresh importation of the poison. Supposing the germ theory to be correct, there is, of course, no reason why the poison should not preserve its vitality, by continuous propagation, for an indefinite time.

On the whole, though the point cannot be regarded as finally determined, I think the weight of evidence is against the *de novo* origin of the disease.

We have now to consider a question about which much difference of opinion still prevails, viz., the contagiousness of typhoid fever. That the disease is communicable from the sick to the healthy is, I believe, universally acknowledged, the point in dispute being the mode in which the transmission takes place, whether directly by emanations from the patient, or from his fresh evacuations, or indirectly from eating or drinking, or inhaling the emanations from the stools, modified by their having undergone some kind of decomposition or fer-

mentation outside the organism. Although the point in dispute is a narrow one, it is of considerable importance, and erroneous views may lead, on the one hand, to the adoption of unnecessary restrictions, which may seriously incommode the patient and his attendants, while at the same time it is only too likely to cause a neglect of the really essential precautions.

I have myself witnessed an outbreak of typhoid where the belief in its contagiousness—shared, I may say, by the medical practitioners—was so strong as to excite quite a panic, so that difficulty was found in procuring attendants for the sick. In consequence, I had on one occasion myself to carry a patient from one room to another, and a man who volunteered to help me was regarded as having done something rather heroic. I need hardly say that, at the same time, the utmost recklessness is commonly shown with regard to the real causes of its dissemination.

On the other hand, a disbelief in the direct contagiousness of the disease might, if not well founded, lead to unnecessary exposure of the patient's friends.

The arguments against the direct contagiousness of the disease are in the main these : 1. In hospitals we rarely find that typhoid fever spreads either to the attendants on the sick or to the other patients. When it does thus spread, persons in other parts of the building, who have never been brought in contact with the typhoid cases, are attacked about as frequently as the immediate attendants themselves.

Dr. Murchison has published the evidence on this point afforded by the London Fever Hospital for a period of twenty-three years, up to 1870. During this period, 5,988 cases of typhoid were admitted, and seventeen of the resident staff took the disease, but of these seventeen only five were in communication with the typhoid cases, and twelve occurred at a time when there were serious defects in the drains ; twelve patients also admitted for other diseases became infected. But since 1861, when the patients were so classified that the typhoid cases and the patients suffering from other acute diseases, not fever, are placed in the same wards, not a single instance has occurred in which the infection has spread to the non-typhoid cases, though the ·same night-stools and water-closets have been used by both classes of patients, and the use of disinfectants has been exceptional. This shows pretty conclusively that the emanations from the recent stools are not capable of communicating the disease.—*Medical Times & Gazette.*

THE OPIUM HABIT.

A Statistical and Clinical Lecture.

BY CHARLES WARRINGTON EARLE, M. D., CHICAGO, ILL.

It is nine years since I was appointed physician to the Washingtonian Home. During the first four years of that time I have no remembrance of treating systematically any cases of the opium habit; but during the last five years my attention has been directed to this subject by more than a score of cases, and during the past year hardly a month has passed without a patient being placed under my care. I found that their symptoms, in nearly every case, were similar, and several months ago I commenced to study them more closely, with the view of making some definite deductions at the appropriate time.

It is proper for me to state that the Washingtonian Home was not intended by its founders as a place for the treatment of opium inebriety, yet during these latter years certain cases have presented themselves at our doors which seemed to demand our attention, and our institution being the only one in the Northwest, we could hardly refuse them admission. We have found it very difficult to care for these individuals ; they require more watching and more nurses than the alcoholic class ; notwithstanding this, we have taken them from time to time, and to the majority good results have been attained.

THE OPIUM TRAFFIC.

In an article entitled "The Impending Danger," (Medical Record, 1876,) Dr.

Mattison, of Brooklyn, states that he is assured by both dealers in the crude drug and manufacturers of the alkaloids, that the importation of opium is increasing rapidly every year, and that the supply may become insufficient for the demand. The question naturally suggesting itself is : What becomes of the vast amount yearly brought to our shores ? With the intention of finding to some extent the amount used non-medically, I have interviewed fifty druggists in various sections of the city. The questions have been in regard to : First, the number of regular opium customers ; second, sex ; third, nativity ; fourth, age ; fifth, social relations ; sixth, married or single ; seventh, occupation ; eighth, cause assigned ; ninth, kind of narcotic ; tenth, quantity. The pharmaceutists, without exception, have seemed willing to aid me in my investigation, and the great majority deprecate in unmeasured terms the opium trade in all its aspects. Four of those interviewed do not sell the drug unless prescribed, and a large number express themselves as dissatisfied with this part of their business, as liable to cause trouble ; and, from the fact that almost without exception, this class of customers sooner or later become irresponsible and unreliable. In order to discourage the trade, some druggists are driving from their doors those who would purchase the different opiates by asking a high price for the drug. For instance, one gentleman has ceased to sell to any by asking $1.25 per drachm for morphia, when it could be purchased for from 65 to 75 cents elsewhere.

I am aware that errors, especially in two particulars, may have been made in my investigations. The druggists may not have communicated to me all the facts in their possession in regard to the number of regular customers, and I may have counted the same individual more than once. In regard to the first possible error, I should say that the pharmaceutists have all seemed desirous to assist me, and quite a number have taken great pains to look over their records in search of data. To all these gentlemen I desire, in as public a manner as possible, to express my gratitude for their courtesy. In regard to the second possible error, i. e., counting the same individual more than once, I will say that great care has been exercised to include in only one enumeration any who were known to patronize more than one store, and whose description and habits have been given to me by those previously interviewed.

I certainly have no desire to exaggerate in regard to the number using this drug, and think it quite probable that my estimate may fall below that made by others.

RESULTS OF INVESTIGATION.

Inquiries were made at fifty different drug stores. The three divisions of the city were visited, and localities inhabited by the different classes and nationalities were thoroughly canvassed. I was greatly surprised to find that druggists on the West Side were patronized to a greater extent (excepting a few on Clark street) than in any other part of the city. Foreign druggists (German and Scandinavian), seem to exhibit more conscientious scruples in regard to the trade than our own nationality. I learned from some of these gentlemen that in Denmark, and, if I mistake not, in Norway and Sweden, the trade is absolutely forbidden. Fifty druggists have 235 customers, or an average of nearly five to each store.

SEX.

Among the 235 habitual opium-eaters, 169 were found to be females, a proportion of about 3 to 1. Of the 169 females, about one-third belong to that class known as prostitutes. Deducting these, we still have among those taking the different kinds of opiates, 2 females to 1 male. In one family I found the mother, at the age of 65, taking one drachm of gum opium each day, and her daughter, at the age of 30, consuming two drachms of the tincture. One lady, aged 50, has taken it since she was 13 years of age. Suffering from some painful sickness during her youth, she was given, by a physician, a box of powders on which was written " Morphia." She had the prescrip-

tion repeated, and gradually found herself in the power of the seductive drug, from which, in all probability, she will never be freed.

I am acquainted with an aged couple living on Harrison street, aged respectively 70 and 75, who take a drachm of morphia each every week, when by any means whatever they can procure it. The husband has suffered from fracture, and the wife with neuralgia and rheumatism.

NATIONALITY.

Entire number of cases......	235
American...	160
German..	7
Irish.............	17
Scotch..	10
Colored...	12
English...	6
Scandinavian.........................	5
Unknown..	18

It will be noticed that it is among our own people that we find the largest number yielding easily, and in considerable numbers, to the influence of this drug. The Germans and the Irish find relief from their troubles in the anæsthetic effect of beer and whisky, while the American takes a not less effective agent, but one whose effects for the time do not incapacitate the victim for business. And, in addition, I suppose it is true that it is more particularly among Americans, and the foreign class who come to be Americanized, that we find those neurasthenic people who bear pain badly, and demand relief from some source. I have always found it difficult to dissuade a certain class of ladies from taking more anodynes than I thought proper. They constantly demand to be relieved from pain.

AGE (APPROXIMATE).

Males.

From 20 to 30 years....................:...................................	5
From 30 to 40 years..	19
From 40 to 50 years..	11
From 50 to 60 years..	7
From 60 to 70 years..	1
From 70 to 80 years...	1
Unknown age	22
Total................... ....:.............	66

Females.

From 10 to 20 years...	2
From 20 to 30 years...	18
From 30 to 40 years...	39
From 40 to 50 years...	22
From 50 to 60 years...	14
From 60 to 70 years...	4
One-third entire number prostitutes, probably from 15 to 50,	56
Unknown age..	14
Total..	169

It is, as will be seen, a vice of middle life, the larger number, by far, being from 30 to 40 years of age. One woman, an octoroon, commenced using it when 13 years of age. Away from her friends, she became down-hearted and homesick, when an elderly lady, herself a morphia-eater, offered the young girl a powder, with the remark that it would cheer her up and cause her to forget her sorrows. This was repeated for several days, the morphia habit was established, which has clung to the woman to this day. She now takes thirteen grains of morphia in the morning and five grains in the evening, and has taken as high as sixty grains per day.

It is reported to me, although I cannot be held responsible for the statement, that a young woman, now 25 years of age, and following the occupation of prostitute, commenced to take morphia when only five years of age. While I cannot vouch for the truth of the above remark, I know, from my own observation, that many young children, even infants, become accustomed to and feel the stimulating effects of opiates. Not only this, but they experience the terrible depression, and have the symptoms which I always notice in an adult after the withdrawal. An infant at two weeks of age was given its first dose of soothing syrup. It took two bottles during the first month, six bottles during the second and third months, and four bottles each month during the remaining four months of its life. It died during its seventh month. During the last three months it was constantly nervous, it gradually became pale and slightly yellow, yet increased in flesh. Upon the rapid withdrawal of its morphia, which I assume to be the anodyne ingredient in its soothing syrup, it was taken with terrible diarrhœa, incessant vomiting, apparent unbearable muscular pains, prostration, and death. While a bronchitis with which the little sufferer was attacked, may have had something to do with its death, it has always seemed to me that many of its symptoms were due to the withdrawal of its customary soothing syrup. Of course

the entire history of this case was not given to me until after the death of the child. With all the facts in my possession I should have stimulated the child in every way possible, and gradually reduced the soothing syrup.

SOCIETY RELATIONS.

It is very rare to find a poor Bohemian or Swede habitually taking any kind of an opiate. It is equally rare to find a wealthy person, in the full enjoyment of his or her property, taking it. A large number of those who have formerly occupied high social standing and enjoyed wealth, but from different causes have become reduced in circumstances and position, are taking the drug. It is, however, among the middle class that we find the very great majority who are to-day our opium-eaters. There are a few exceptions, but in general I believe my statement is correct.

MARRIED OR SINGLE.

The great majority of morphia-takers either are or have been married. Many of both sexes, who have occupied this relation, are now separated, the unreliability and loss of respect, and untruthfulness, which the use of the drug usually produces, being the cause of the unhappy condition between husband and wife.

OCCUPATION.

From my notes, made while making these investigations, I take the occupation of 100 occurring at the head of the list :

Males.

Iron merchant	1	Capitalists	2
Newsdealer	1	Clerk	1
Business men	5	Contractor	1
Physicians	2	Insurance agent	1
Laborers	3	Book agent	1
Turner	1	Railroad men	1
Druggist	1	Attorney	1
Bookkeepers	2	Unknown	9

Females.

Housewives	45	Servant	1
Society ladies	3	Washwomen	2
Widow	1	Prostitutes	5
Sewing woman	1	Unknown	9

It will be observed that it is in those occupying middle stations, that we find the largest number taking the drug, although among thirty-three men, two are classed as capitalists. I find, also, with much regret, two of our profession, which leads me to remark that a few of the heaviest consumers of morphia are physicians and druggists. Their devotion to the seductive influences of the drug has, however, destroyed their business, and they wander around, in some instances at least, leading an aimless and abandoned life. A large proportion of the females are classed as "housewives," but they must necessarily include many widows or those abandoned by their former husbands. The small number of prostitutes numbered in the list arises from the fact that the locations where these observations were made, were among the best in the city, only a few of this class residing there.

CAUSE ASSIGNED.

In many cases it is difficult to ascertain the reason for taking the drug. The customer buys the preparation he desires and immediately takes his departure; and it is usually a delicate question to ask, "What do you take it for?" Dr. Joseph Parish says "that men take it not for social enjoyment, but for a physical necessity." With such an opinion I have no sympathy. When it is a physical necessity for men to steal and lie, and murder, and partake of alcohol to such an extent as to incapacitate themselves for work, and bring ruin on their families, then I will admit the same in regard to opium-eating. Quite a number, however, freely confess the reason for taking the drug. Eight say that it is for its stimulating and happy effect; four formerly were addicted to drink and seek quietude by taking opium; five are unhappily married; thirty-eight have had rheumatism and are now suffering from it to some extent, and an equal number assign neuralgia as the cause of their pains, for which some form of opiate is taken. Those diseases known by the incomprehensible name of "female complaints," are frequently given as the cause for taking an opiate. Previous sickness, and wounds received during the war, painful stumps after amputation, injuries to nerves, etc., loss of property and position in society, are given by a few. But the great majority confess that it was first given during some disease in the course of which pain was a prominent symptom. An opiate was prescribed by the physician, and ever afterward, when suffering pain, the little powder, or laudanum, or gum opium, has become their solace. However much we may desire to avoid this grave responsibility, the truth must be confessed, that the greater number of men and women who are now completely enslaved to the different preparations of opium, received their first dose from members of our profession.

KIND OF NARCOTIC.

Morphia was used in	120 cases.
Tincture opium "	30 "
Paregoric "	5 "
McMann's elixir "	2 "
Gum opium "	50 "
Dover powders "	1 "
Unknown "	27 "
	235

Ladies use morphia in the majority of cases, men, of the lower classes, gum opium, and a few, of both sexes, who have a desire for alcoholic stimulants, in addition to the opiate, use the tincture, and occasionally one is found taking large doses of paregoric. An American lady, aged 50, a widow, buys of one druggist half a gallon every week. One lady takes morphia and chloral; another morphia and Dover powders; a patient, under my care at this time, has nearly destroyed herself and ruined her friends by taking morphia and chloroform.

QUANTITY.

Morphia.

21 persons use	from 1 to 3	grains each day	
17 "	" 3 to 6	"	"
12 "	" 6 to 10	"	"
10 "	" 10 to 15	"	"
12 "	" 15 to 20	"	"
7 "	½ a drachm		"
6 "	1 drachm		"
20 "	1 bottle	per week	
5 "	2	"	"
1 "	.11	" each month	

Tr. Opium.

15 "	1 drachm each day		
4 "	3	"	"
7 "	4	"	"
12 "	1 ounce		"
4 "	2	"	"
1 "	3	"	"
1 "	4	"	"

Gum Opium.

3 "	10 grains each day		
5 "	20	"	"
9 "	½ drachm		"
12 "	1	"	"
4 "	2	"	"
2 "	3	"	"
1 "	4	"	"

In addition to those known and recognized as opium-eaters, a large number of ladies are in the habit of using from one-third to one grain of morphia daily. They have done this for years without imparting their secret to their nearest friends. It was commenced to allay some pain, and then continued for its stimulating effect. The lady I referred to as being under treatment for morphia and chloroform, took the first-named drug four years before her husband was aware of it.

MANNER AND TIME OF TAKING.

My observations do not agree with certain writers ; for instance, Dr. Kane, who has recently written a book on the hypodermic method, in regarding this way of taking morphia as particularly liable to cultivate the habit :

" A physician of the present day, without a hypodermic syringe in his pocket, or close at hand, would be looked upon as would have been a physician fifty years ago, did he not own and use a lancet. There is no proceeding in medicine that has become so rapidly popular; no method of allaying pain so prompt in its action and permanent in its effect. No plan of medication that has been so carelessly used and so thoroughly abused; and no therapeutic discovery that has been so great a blessing and so great a curse to mankind, as the hypodermic injection of morphia."—Morphia Hypodermically. By Dr. Kane; preface.

In some respects I agree with this author, but in my experience the hypodermic syringe is not resorted to with the frequency that Dr. Kane has stated. But very few of the 235 cases, not more than four or five, upon which I base my article, take the drug in this manner, and only two or three of those who have been placed under my care used the instrument.

The great bulk consumed is by the mouth, and, as regards time and frequency of dose, it varies from two to three times a day to a single large dose once in two or three days. It is a curious fact that some of the oldest opium-eaters take a large dose at intervals of from one to three days, and that an even and happy effect is experienced during all that time.

PHYSICAL EFFECTS.

In many cases the deleterious results are very long in making their appearance. Indeed, some take this drug for years and years, and seem to enjoy excellent health, and in no way can they be distinguished from those not addicted to the drug. Only a few days ago I met a gentleman on Carroll avenue who has taken morphia thirteen years, in doses of from two to fifty grains daily. He was well dressed, had a healthy color, and was in every respect as respectable looking as the majority I saw on the street. Some become fleshy, especially when commencing its use, while others emaciate. It makes one logy and sleepy, another vivacious and happy. Sooner or later, however, we find symptoms denoting disordered nutrition and enervation. The opium-eater's countenance, in the greater number of cases, betrays him. It becomes sullen, haggard and apathetic, and the eye loses its brilliancy. All these objective symptoms are most noticeable when the habituate is deprived for even a short time of his usual opiate. We soon find the appetite impaired, and digestion poor. The bowels are habitually constipated. There is also vesical and sexual torpor. Indeed, every function of the body is performed in a sluggish manner. If, from any cause, the victim is deprived of his accustomed drug, he soon begins to have intolerable cramps in the muscular system, with involuntary twitchings. The patient by this time is usually weak and feeble. A gentleman from the central part of our State, when presented for treatment, was a perfect picture of a marasmic patient.

MENTAL AND MORAL EFFECTS.

The man avoids society, and before he is fairly confirmed in his habit is tortured with the thought that he is becoming a victim to a habit that he can now only rid himself of by great will power and bodily suffering. He cannot make the resolution to stop to-day, but defers it until to-morrow. The memory of the poor man is already becoming poor. He neglects his business; he falls asleep for a few moments as he rides in the car, or as conversation lags for a moment with his associates.

Equally marked, and more so, if possible, is the change in the moral sentiment of the individual. He neglects his family. He will obtain his usual proportion of opium by making promises that he knows he is unable to fulfill. He becomes cross and irritable if attempts are made to deprive him of his opiate ; and, sooner or later, moral rectitude, every noble impulse, every generous thought, is swallowed up in this terrible fight to possess more and more of the narcotic, to obtain which, in almost every instance, the victim has become an inveterate prevaricator.

It is claimed by those having greater experience than myself, that the hypodermic use of morphia does not so rapidly, or in such a complete degree, gain the ascendancy over the physical condition and moral sentiments of the individual.—*Chicago Medical Review.*

SOCIETY TRANSACTIONS.

ANNUAL MEETING OF THE COLORADO STATE MEDICAL SOCIETY.

The society having under consideration Dr. Denison's paper on " A new method of withdrawing fluids from the middle ear " (see article on page 44) ;

Dr. Adams thought that, while the procedure described by Dr. Denison was novel and ingenious, it would be confined to just such cases as the one related in his paper, where it would be of incalculable value ; but that its usefulness would not extend to acute otitis media arising from any other cause, such as cold, of either fauceal or tympanic origin, or inflammation of the Eustachian tube and tympanic cavity as the result of chemical irritants, accidentally introduced into these passages by means of the nasal douche, although Dr. Denison's paper would seem to imply its applicability to just such cases. In a case of threatened inflammation of the middle ear, where the history would seem to indicate the presence of a mechanical irritant as the primary cause, as a prophylactic measure such treatment would undoubtedly be of great value, and

reflect much credit upon its originator ; but when the inflammation had fairly set in, or the affection was due to the existence or extension of an inflammatory or catarrhal process, it would exaggerate the very condition we should endeavor to relieve, namely, exhaustion or rarity of the air in the tympanic cavity ; for in such cases the pain, hardness of hearing, congestion and other concomitant symptoms, are not due so much to the presence of mucus in this cavity, but rather to the existence of a vacuum or tendency thereto, and consequent retraction of the drum membrane ; this condition having been produced by swelling and thickening of. the mucus lining of the Eustachian tube and tympanum, occlusion of the latter, absorption of the remaining air, and engorgement. Now, under these circumstances, a passive action such as this procedure represents, would, in the first place, be insufficient to meet the requirements ; and in the second place, contrary to the indications, which would naturally call for the. forcible use of Politzer's method, or, as a dernier resort, puncture of the drum membrane.

Dr. Denison replied that his views accorded perfectly with Dr. Adams', and that his intention was to claim for the method only this and nothing more.

Remarks after the reading of Dr. Hawes' paper on " A new splint " (see article on page 44) :

Dr. Denison spoke favorably of the facility for drainage and ventilation the apparatus affords, when these were required, as in severe compound fractures of the leg.

He wished, however, to suggest some additions which he thought would be necessary to make the splint complete for the purposes indicated.

First. The apparatus should not be allowed to rest on the bed or mattress, but should be slung so that the whole limb would move together when there was any motion of the body. This could be easily done by suspending this splint from the ceiling, or within a coop made of two light skeleton wooden frames, hinged together at one side.

Second. It is very desirable to combine *extension* with the use of such an apparatus when the fractures are oblique or the bones much comminuted. The pulley and weight will not answer the purpose, because their use necessitates the portion of the limb below the point of injury being stationary, while that above is not so, and the movements of the body are necessarily transmitted to the ends of the broken bones. The extending and counter-extending forces should be parts of the apparatus, which may be slung as already described. To be sure, the foot may be fastened to a footpiece fitted into this splint, and the body be made to act as a counter-extending force, but this force is very indefinite and unreliable.

For the purpose of furnishing extension, the speaker suggested fitting into this splint a footpiece, with holes in it for making the foot stationary thereon by adhesive strips, while he would make counter-extension with his own "Extension Windlass"* in the following manner : Slats about the size of laths and fastened one on each side of this tin splint, as the speaker had done to the sides of a fracture-box ; these slats reaching well up on the thigh, and having the windlasses fastened to their upper ends. Fanshaped adhesive strips are bandaged to the sides of the thigh and have their smaller ends converged toward the windlasses, where they are threaded through the slots in the winding rods. Thus we have stationary points connected with the splint, from which counter-extension can be made. By turning the winding-rods with the key, the plasters attached to the thigh are pulled upon, and the splint with its stationary footpiece is pushed *from* the body. In this way the desired extension is secured and maintained.

The muscles once used to this normal amount of extension cease abnormal contractions, and there is the nearest approach possible to desired perfect rest. The pulley and weight, being a constantly increasing force as the muscular resistance to it lessens, is a torture entirely unnecessary in most cases. Especially is this true, as Dr. D. had witnessed in his hospital experience where they used it, when patients have delirium tremens, or are exhausted by diarrhœa or great debility, as they frequently are. In the light of such experience, Buck's and other pulley and weight apparatuses can never successfully compete with the ratchet and pinion method, which asks no more when once the desired amount of extension is obtained.

The speaker illustrated the proposed attachment of the windlasses to this tin splint by his own similar treatment of a case with the fracture-box. This was a compound comminuted fracture of the leg, the soft parts being considerably lacerated and half an ounce of fragments from the tibia taken away. Yet, through this method of extension, which was used for a long time, there was complete union with less than a fourth of an inch shortening.

LIST OF COMMITTEES

and special appointments, introduced for the information of the members of our State Society :

STANDING COMMITTEES.

Executive—F. F. D'Avignon, D. H. Dougan, T. H. Hawkins, F. D. Sanford, J. W. Graham.

Finance—J. C. Davis, W. F. McClelland, A. Steadman, T. H. Hawkins, L. E. Lemen, R. G. Buckingham.

Publication—W. H. Warn, A. W. Adams, James A. Hart, S. E. Solly, W. E. Wilson, J. S. Smith.

Ethics—H. A. Lemen, W. Edmundson, G. S. McMurtrie, Alexander Shaw, D. Coates.

Medical Societies—W. H. Williams, G. S. McMurtrie, T. H. Hawkins, D. H. Dougan, F. F. D'Avignon.

SPECIAL COMMITTEES.

Surgery—H. K. Steel, J. C. Blickensderfer, A. W. Eyer, F. D. Sanford, J. W. Col-

---

* This Extension Windlass is a slotted winding-rod, governed by a ratchet and pinion, and supported by a brass bracket, by which it may be made stationary on most any kind of splint to furnish extensions as desired. It is fully described in the New York Medical Journal of May, 1875.

lins, G. Wohlgesinger, G. P. Allen, J. A. Dubois, C. B. Richmond.

Medicine—C. M. Parker, E. T. Schumaker, E. C. Kimball, C. S. Laughlin, W. E. Wilson, A. T. Poichet, W. H. Newman,

Obstetrics—F. F. D'Avignon, J. W. Collins, E. N. Cushing, J. C. McBeth, J. H. Bean, Drydon Johnson, S. D. Bowker, C. E. Edwards.

Medical Literature—B. P. Anderson, T. H. Craven, M. A. Wilson, A. W. Graham, Jacob Reed, John Russell.

Climatology—A. Steadman, W. M. Strickler, J. L. Prentis, G. S. McMurtrie, James Innis, B. St. George Tucker.

Materia Medica and New Remedies—Jno. Elsner, Joseph Anderson, F. B. Blake, F. C. Blachley, S. T. Floyd, D. M. Parker, W. W. Anderson.

Chemical and Pharmaceutical Preparations—B. St. George Tucker, T. E. Owens, S. T. Floyd, C. C. Lathrop, P. L. Rice, W. G. Scott, P. R. Thombs.

Public Health—Chas. Denison, J. W. Graham, J. H. Kimball, F. M. Ramey, John Russell, D. Heimberger, Samuel Rapp.

Legislation and Dissection—H. K. Steel, M. Beshoar, F. M. Ramey, H. O. Dodge, H. A. Lemen, A. Steadman, J. A. Gale, R. G. Buckingham, J. C. Blickensderfer.

Necrology—Joseph Anderson, D. H. Dougan, P. R. Thombs, L. E. Lemen, T. E. Owens, W. M. Strickler, W. F. McClelland.

SPECIAL SUBJECTS.

Pulmonary Consumption—J. J. McDonald.

Orthopædic Surgery—T. H. Hawkins.

Ophthalmology—Sam'l Cole, A. B. Kibbe.

Climatic Influence on Nervous System—F. D. Sanford.

Pneumonia in Relation to High Altitude—L. E. Lemen.

Brights Disease—J. C. Davis.

Gynæcology—R. G. Buckingham.

Otology—A. Wellington Adams.

Specific Medication—Jacob Reed.

Diphtheria—W. E. Wilson.

Uterine Forceps—G. W. Cox.

Diseases of Children—M. A. Wilson.

Venerial Diseases—W. H. Davis.

Fractures and Dislocations—J. P. Allen.

Railroad Injuries—T. H. Craven.

Mental Diseases—P. R. Thombs.

Miasma—Jesse Hawes.

Epidemic Diseases—H. A. Lemen.

Forensic Medicine—M. Beshoar.

Asthma—D. H. Dougan.

EDITORIAL.

Propagation of Sound Through the Medium of Light—The Photophone.

During the years 1875 and '76 the entire civilized world was startled and amazed by the announcement that articulate speech, with all its undulations, modulations and diversified timbres, could be transmitted by means of a simple wire, to almost unlimited distances. Although but comparatively few years have elapsed since the discovery of this fact, and the introduction of the system (telephony) by means of which it is accomplished, it now holds a prominent and permanent place in the commercial and manufacturing interests of our country, and bids fair to outstrip the Morse system in practicability and utility.

Through ignorance of scientific principles and superficial reading, the newspaper reporters of our country, ever ready to create a furor, regardless of *truth* and *facts*, have led the masses to believe that to Edison belongs the praise and glory for this advance, when, on the contrary, he is deserving of no praise whatever in this connection, he having accomplished nothing of practical value, and what he did do having been done after the system became an established fact. For the development of the telephone we are indebted indirectly to Profs. Henry, Page and Reiss, and directly to Profs. Dolbear, Gray and Bell. The latter—Prof. Alexander Graham Bell—in conjunction with Mr. Sumner Tainter, has now, after most laborious research, succeeded in developing a system by means of which vocal utterances may be transmitted to points far beyond the range of hearing power without any visible or tangible connection—an accomplishment far more wonderful than its precursor.

Some seventy years ago, Berzelius, the great Swedish chemist of Stockholm, while instituting research directed toward the discovery of a more direct method of preparing sulphuric acid, observed a peculiar sediment, which, upon separation and analysis, proved to be a new elementary substance, and which, owing to its resemblance to tellurium, he named Selenium. Hardly had Berzelius' ardor grown cold over his new discovery, than scientists generally began series upon series of experiments to determine what peculiar properties this new metal possessed. It was shown by different investigators successively to be a non-conductor, a conductor when fused, a conductor at ordinary temperatures when assuming one particular form—the " metalic ;" that it was allotropic, that exposure to sunlight hastened its changes from one allotropic form to another ; that its resistance to an electric current was extremely variable, and that the degree of resistance was proportionate to the amount of light to which it might be exposed. From these isolated and disconnected scraps of knowledge, resulting from the investigations of some twelve or thirteen scientists scattered over the four quarters of the globe, Prof. Bell conceived the idea of producing and reproducing sound through the agency of light. The intricate windings leading to a realization of this conception can be better imagined than described. Certain it is, however, that it, like all other grand practical accomplishments, was slowly evolved from a gradually increasing aggregation of disconnected discoveries, of, apparently, but little moment and practical value, yet all forming necessary and integral parts of its development.

In the first place it was necessary to reduce the natural resistance of crystalline selenium from 250,000 to 300 ohms in the dark. Af-

ter many experiments, this was found to be best accomplished by combining the selenium with brass, because with these two substances there was formed a more intimate bond of union than in any other combination.

According to Prof. Bell—" The fundamental idea, on which rests the possibility of producing speech by the action of light, is the conception of what may be termed an undulatory beam of light in contradistinction to a merely intermittent one. By an undulatory beam of light, I mean a beam that shines continuously upon the selenium receiver, but the intensity of which upon that receiver is subject to rapid changes, corresponding to the changes in the vibratory movement of a particle of air during the transmission of a sound of definite quality through the atmosphere." The most suitable apparatus devised by Prof. Bell for accomplishing this, consists—in his own language—

"Of a plane mirror of flexible material—such as silvered mica or microscope glass. Against the back of this mirror the speaker's voice is directed. The light reflected from this mirror is thus thrown into vibrations corresponding to those of the diaphragm itself.

" In arranging the apparatus for the purpose of reproducing sound at a distance, any powerful source of light may be used, but we have experimented chiefly with sunlight. For this purpose a large beam is concentrated by means of a lens upon the diaphragm-mirror, and, after reflection, is again rendered parallel by means of another lens. The beam is received at a distant station upon a parabolic reflector, in the focus of which is placed a sensitive selenium-cell, connected in a local circuit with a battery and telephone. A large number of trials of this apparatus have been made with the transmitting and receiving instruments so far apart that sounds could not be heard directly through the air. In illustration, I shall describe one of the most recent of these experiments. Mr. Tainter operated the transmitting instrument, which was placed on the top of the Franklin schoolhouse in Washington, and the sensitive receiver was arranged in one of the windows of my laboratory, 1325 L Street, at a distance of two hundred and thirteen meters. Upon placing the telephone to my ear I heard distinctly from the illuminated receiver the words, ' Mr. Bell. if you hear what I say, come to the window and wave your hat.' In laboratory experiments the transmitting and receiving instruments are necessarily within ear-shot of one another, and we have, therefore, been accustomed to prolong the electric circuit connected with the selenium receiver, so as to place the telephone in another room. By such experiments we have found that articulate speech can be reproduced by the oxyhydrogen light, and even by the light of a kerosene-lamp. The loudest effects obtained from light are produced by rapidly interrupting the beam by the perforated disk. The great advantage of this form of apparatus for experimental work is the noiselessness of its rotation, admitting the close approach of the receiver without interfering with the audibility of the effect heard from the latter; for it will be understood that musical tones are emitted from the receiver when no sound is made at the transmitter. A silent motion thus produces a sound. In this way musical tones have been heard even from the light of a candle. When distant effects are sought another apparatus is used. By placing an opaque screen near the rotating disk the beam can be entirely cut off by a slight motion of the hand, and musical signals, like the dots and dashes of the Morse telegraph code, can thus be produced at the distant receiving station."

In the light of Prof. Bell's experiments, the effect would appear to be due to some invisible form of radient energy, and not to either thermal or luminous rays; for when the sound-bearing beam of light is intercepted by an opaque substance, or by a solution of alum, the results are very nearly the same.

As yet, we can but conjecture as to the practical advantages likely to accrue from this wonderful discovery; we may be certain, however, of its opening up a broad and inviting field to the scientist, and, no doubt, one that will yield a rich harvest.

In this connection it would be well to refer to a plea set up against vivisection, by a few ridiculously tender-hearted and short-sighted individuals calling themselves anti-vivisectors. They ask us, "Of what practical value, pray tell, are all these isolated and disconnected facts you declare you have determined through the agency of vivisection; do they not represent so much time, labor and suffering sacrificed?" To such we would say: Were you acquainted with Berzelius, Knox, Hittorff, Smith, Sale, Draper—Galvani, Volta, Arago, Oersted, Page, Henry—Helmholtz, Konig, Lissajous, Scott, Blake, and

many others; had you been associated with them, seen them spend many a toilsome hour and hard earned dollar, year in and year out, in establishing some of the little— yes, little—abstruse points, which have since, however, led to those great inventions, the Photophone, Electric Telegraph, Telephone and Phonograph, you would, doubtless, have propounded the same silly question to them, and they would, probably, have simply replied, *"I am sorry for you."*

AN EXPLANATION.

The misconception of a sentence in the editorial upon "Medical Legislation," in our last number, has caused annoyance to some members of the "State Board of Health" whom we would regret to offend.

The statement that the Board "was so *constituted* as not to command the respect or confidence of the profession," was carefully considered, and, we believe, true. This statement, however, was by no means intended to reflect upon the character or capacity of the majority of the members constituting the Board.

That there are members who are unfit for the position occupied by them is evident to all, and most painfully so to their associates; an inconvenience common to all Boards appointed politically.

The majority of the Board are eminent members of the profession, commanding respect, and in every way worthy of the position. But according all due esteem to them, we cannot consider a Board containing at least one member who is unable to qualify for membership in the State Society, as fitted for the power confided to it in the bill referred to, or as commanding the confidence of the profession. This, and only this, was our meaning, and we sincerely regret this was not apparent to the worthy members of the Board.

PLEASE remember that the Rocky Mountain Medical Review solicits contributions from *all* quarters, of both a scientific and medical character.

THE BILL! WHAT DO YOU THINK OF IT?

This is the question now being agitated in medical circles throughout the State of Colorado. During the last session of our State Medical Society there was a resolution passed providing for the creation of a committee whose duty it should be to draft a bill providing for the protection of the public health and regulation of the practice of medicine, and present the same to the Legislature for enactment ; after having first, however, sent a copy of said bill to every physician within the State, with the request that he suggest such alteration as he may deem needful, and finally amended the bill in accordance, as far as possible, with the received suggestions. The president of the Society, in making up this committee, endeavored to unite all factions, that we might have no conflicting bill presented, as was the case last year. This committee is composed of members of the highest standing in our profession ; men who have given the subject of protective medical legislation considerable thought and study, who understand perfectly all questions pro and con, and conflicting contingencies ; men who are as desirous as any of us for the passage of the most stringent laws, yet who realize the impossibility of securing the passage of such laws under the existing state of affairs, and who accordingly believe in the truth of the trite maxim, " The half of a loaf is better than none at all."

This committee, after much hard work and many a fruitless effort in its endeavors to please everybody, sends us, as the result of its work, the following bill :

A BILL
FOR AN ACT TO PROTECT THE PUBLIC HEALTH
AND REGULATE THE PRACTICE OF MEDICINE
IN THE STATE OF COLORADO.

Be it enacted by the General Assembly of the State of Colorado :

SECTION 1. That a Board is hereby established which shall be known under the name and style of the State Board of Medical Examiners, to be composed of nine practicing physicians, of known ability and integrity, who are graduates of medical schools of undoubted respectability, giving each of the three schools of medicine (known as the Regular, Homœopathic, and Eclectic schools) a representation proportionate to the number of graduates of said schools within the State.

SEC. 2. The Chief Justice of the Supreme Court of this State, shall, as soon as practicable after this Act shall have become a law, appoint a State Board of Medical Examiners as provided in section 1 of this act, and the members first appointed shall be so designated by the Chief Justice that the term of office of three shall expire in two years from the date of appointment ; the term of office of three shall expire in four years from the date of appointment ; and the term of office of three shall expire in six years from the date of appointment. Thereafter the Chief Justice shall biennially appoint three members, possessing qualifications as specified in section 1, to serve for the term of six years, and he shall also fill all vacancies that may occur, as soon as practicable ; provided, that in making biennial appointments or filling vacancies, the ratio of representation of the medical schools in the Board shall not be changed from the original basis as in section 1.

SEC. 3. The Board of Medical Examiners shall as soon after their appointment as practicable, organize by the election of one of their members as president, one as secretary and one as treasurer, and adopt such rules as are necessary for their guidance in the performance of the duties assigned them.

SEC. 4. That every person practicing medicine in any of its departments shall possess the qualifications required by this act. If a graduate in medicine, he shall present his diploma to the State Board of Medical Examiners for verification, or furnish other evidence conclusive of his being a graduate of a legally chartered medical school in good standing. The State Board of Medical Examiners shall issue its certificate to that effect, signed by a majority of the members thereof, and such diploma or evidence and certificate shall be conclusive as to the right of the lawful holder of the same to practice medicine in this State. If not a graduate of a legally chartered medical institution in good standing, the person practicing, or wishing to practice, medicine in this State, shall present himself before said Board of Medical Examiners and submit himself to such examination as defined in section 7 of this act, and if the examination be satisfactory to the examiners, the said Board of Medical Examiners shall issue its certificate in accordance with the facts, and the lawful holder of such certificate shall be entitled to all the rights and privileges herein mentioned. All persons who have made the practice of medicine and surgery their profession or business continuously for the period of ten (10) years within this State, and can furnish satisfactory evidence thereof to the State Board of Medical Examiners, shall receive from said Board a license to continue practice in the State of Colorado.

SEC. 5. The State Board of Medical Examiners, within ninety (90) days after the passage of this act, shall receive through its president applications for certificates and examinations. The president of said Board of Medical Examiners shall have authority to administer oaths, and the said Board of Medical Examiners to take testimony in all matters relating to its duties. It shall issue certificates to all who furnish satisfactory proofs of having received diplomas from some legally chartered medical institution in good standing. It shall prepare two (2) forms of certificates—one for persons in possession of diplomas, the other for candidates examined by its members. It shall furnish to the county clerks of the several counties a list of all persons receiving certificates. Certificates shall be signed by a majority of the members of the said Board of Medical Examiners granting them.

SEC. 6. There shall be paid to the treasurer of the State Board of Medical Examiners a fee of one dollar ($1.00) for each certificate issued to graduates or practitioners of ten (10) years' standing, and no further charge shall be made to the applicant; candidates for examination shall pay a fee of ten dollars ($10) in advance.

SEC 7. All examinations of persons not graduates shall be made directly by the State Board of Medical Examiners; examinations may be in whole or in part in writing, and the subjects of examinations shall be as follows: Anatomy, Physiology, Chemistry, Pathology, Surgery, Obstetrics and Medicine (exclusive of Materia Medica and Therapeutics.)

SEC. 8. Every person holding a certificate from the State Board of Medical Examiners shall have it recorded in the office of the Clerk of the County in which he resides, and the record shall be endorsed thereon. Any person removing to another county to practice shall procure an endorsement to that effect on the certificate from the County Clerk, and shall record the certificate in like manner in the county to which he removes, and the holder of the certificate shall pay to the

County Clerk a fee of one dollar ($1.00) for making the record.

SEC. 9. The County Clerk shall keep in a book provided for the purpose a complete list of the certificates recorded by him. If the certificate be based on a diploma, he shall record the name of the Medical Institution conferring it and the date when conferred. This register shall be open to public inspection during business hours.

SEC. 10. The State Board of Medical Examiners may refuse certificates to individuals guilty of unprofessional conduct of a criminal nature, and they may revoke certificates for like causes.

SEC. 11. Any person shall be regarded as practicing medicine within the meaning of this act who shall profess publicly to be a physician and prescribe for the sick, or shall attach to his name the title "M. D.," or "Surgeon," or "Doctor," in a medical sense. But nothing in this act shall be construed to prohibit gratuitous services in cases of emergency.

SEC. 12. Any itinerant vender of any drug, nostrum, ointment, or appliance of any kind intended for the treatment of disease or injury, or who shall by writing or printing, or any other method, publicly profess to cure or treat disease, injury or deformity, by any drug, nostrum, manipulation or other expedient, shall pay to the treasurer of the State Board of Medical Examiners the sum of one hundred dollars ($100) quarterly for a license, to be collected by the treasurer of the State Board of Medical Examiners.

SEC. 13. Any person practicing medicine or surgery, in any of their departments, in this State, without complying with the provisions of this act, shall be punished by a fine of not less than fifty dollars ($50) nor more than three hundred dollars ($300), or by imprisonment in the county jail for not less than ten (10) days nor more than thirty (30) days, or by fine and imprisonment, for each and every offense; and any person filing, or attempting to file, as his own the diploma or certificate of another, or who shall give false or forged evidence of any kind, shall be guilty of a felony, and upon conviction shall be subject to such fine and imprisonment as are made and provided by the statutes of this State for the crime of forgery.

SEC. 14. All fees received by the treasurer of said Board of Examiners, and all fines collected by any officer of the law under this act, shall be paid into the State treasury; and all necessary expenses of the Board shall be paid for out of the fund of the State treasury not otherwise appropriated ; but no fee shall be required or accepted by any member of the Board for services.

SEC. 15. The State Board of Medical Examiners shall meet as a Board of Medical Examiners, in the city of Denver, on the first Tuesday of January, April, July and October of each year, and at such other times and places as may be found necessary for the performance of their duties.

SEC. 16. Justices of the peace and all courts of record in the State of Colorado shall have full jurisdiction over and power to enforce the provisions of this act.

While we very much deprecate the necessity for some of the provisions contained in the above, we think we but utter the opinion of every thoughtful member of the profession in declaring the bill, on the whole, an exceedingly good one, calling for the approval and support of every fair minded man.

After carefully studying over each section separately, we have come to the conclusion that, while there are many points which, to the casual observer, would appear to be objectionable, yet upon thoughtful consideration and judicious weighing, there is, we think, but one point seriously demanding correction.

After stipulating certain provisions to be complied with by every *person practicing medicine*, in default of which such persons are made amenable to punishment, they define in sec. 11 such persons to be those who "profess publicly to be physicians and *prescribe for the sick.*" Then in the very next section they say: "Any itinerant vender of any drug, nostrum, ointment, or appliance of any kind, intended for the *treatment of disease* or injury, or who shall by writing or printing, or any other method, *publicly profess to cure or treat disease*, injury, or deformity, by any drug, nostrum, manipulation or other expedient, shall pay to the Treasurer of the State Board of *Medical Examiners* the sum of one hundred dollars ($100) quarterly, for a license, to be collected by the Treasurer of the State Board of *Medical Examiners.*"

In other words, altho' we have at interest the *public health*, and wish to *protect* the peo-ple against imposters and their seductions, we—the State Board of Medical Examiners, composed, presumably, of members of the medical profession in high standing—will wink at you and your vile impositions; you, the lowest of all charlatans and vagabonds, venders of nostrums and "cure-alls" that we and all educated people know to be directly and indirectly more productive of sickness and premature death than any other known agent. We'll sanction your, what we know to be, thousands of murders, will give you a certificate implying as much, to flaunt before the public and make as much out of as possible, for the sum of four hundred dollars, to be paid annually into the hands of the treasurer of said Board—also presumably a good representative of our dignified profession. This treasurer is to have imposed upon him the very agreeable duty of receiving these characters, who in ordinary life he spurns because of the crime and filth they carry upon their hands, into his office, proffer them a chair, collect their money, and give them his sanction in return—this, *simply* for the love of his profession. O ! would we were the treasurer.

Surely the medical profession can never look upon such things with the least degree of complacency. It no doubt was an oversight on the part of the committee, and will promptly be corrected. We would recommend expunging the entire *section*, as it is not supposed to be a thing medical gentlemen could sanction or in any way control. If the State wishes to put down this damnable traffic, or impose a fine upon such venders, in the way of a heavy license fee, let it do so independently of the medical profession, such license to be made out in the name of the State, and paid for directly to it. If this license fee of one hundred dollars is intended as a protection, it falls far short of the mark. Men who carry away thousands of dollars in a quarter would be very willing to pay such a sum for the additional advantages thus conferred over such of their unfortunate confreres as would be unable to raise the stipulated sum of money.

The appointed time for the convening of the Legislature is near at hand, and now that the committee has so promptly and admirably performed its part, we, in turn, should follow suit—return the bill to them *immediately,* with as few suggestions as consistency will permit, and then when we receive the bill as finally amended, exert our utmost influence to secure its passage.

TRANSLATIONS.

BY G. J. BULL, M.D.

Experimental Researches Upon the Resistance and Manner of Laceration of the Membranes of the Human Ovum.

BY ALBAN RIBEMONT.

(*Archives de Tocologie, Nov., 1879.*)

The author refers to the works of Poppel and Duncan on this subject, and reports his own experiments.

The apparatus he employed consists of a reservoir of water connected by a long rubber pipe with one of the arms of a Y shaped tube, to the second arm of which is attached a mercurial manometer that indicates the pressure employed, and to the third arm the membranes of the ovum are closely applied and made fast.

The author's conclusions are as follows :

1st. Rupture of the membranes which constitute the walls of the bag of waters, occurs in two distinct ways. Sometimes the membranes break simultaneously, sometimes successively.

2d. In the greater number of cases, one of the membranes ruptures before the others, the first to give way being the amnion.

3d. The seat of the laceration is variable ; it may be central or peripheral.

4th. The form is also variable—it may be star-like, angular, minute, etc.

5th. The laceration may occupy the same seat and present the same form, in every one of the three membranes of the ovum ; this is observed in two-thirds of the cases in which the membranes are ruptured simultaneously.

6th. The seat and form of the laceration are different in each membrane, in two-thirds of the cases, when the rupture of one membrane precedes that of the others.

7th. The fœtal envelopes possess an extensibility which varies not only in different ova, but in every one of the membranes. The degree of extensibility is almost equal for each membrane, when laceration occurs in all simultaneously.

8th. The three envelopes of the human ovum present a variable resistance in different ova, and in different zones of the same ovum.

9th. The three membranes together have a greater resistance than that of each membrane taken alone.

10th. The resistance in cases of simultaneous rupture of the three membranes may be estimated at 10,302 kilograms.

11th. Each membrane has a special resistance ; the strongest is the amnion. The part of the amnion which covers the placenta is stronger than the other parts.

12th. The force of the uterine contractions, which have determined the expulsion of the fœtus in cases in which birth occurs within from 5 to 15 minutes after rupture of the membranes, may be estimated approximately at 11,178 kilograms. The minimum pressure employed was 7,125 kilograms, and the maximum was 17,301 kilograms.

TRANSLATOR'S NOTE.

It is no part of the translator's duty to point out the practical application of these conclusions ; but it appears not undesirable that he should say a word or two to indicate their importance.

The apparatus used by Dr. Ribemont was designed to make hydrostatic pressure on an exposed surface of the membranes of mature human ova, with the object of measuring the pressure required to rupture them. The author apparently accepts the hypothesis of Dr. J. Matthews Duncan, to the effect that the force required to rupture the membranes is equivalent to the power employed in the simplest cases of normal labor, in which birth occurs a few minutes after the escape of the waters. In these cases the greatest obstacle to be overcome by the maternal powers is found in the toughness of the membranes, hence the importance of measuring the pressure required to accomplish their rupture. In this connection, it is interesting to compare Dr. Ribemont's results with those obtained in the experiments of Duncan and Poppel, who made observations independently of each other. The translator has made many tests of the same nature during the last few years, and hopes to present his conclusions within a short time in the pages of this journal.

One word with regard to the first of the author's conclusions, viz: the membranes sometimes rupture simultaneously, sometimes separately. The truth of this is very apparent in the laboratory; but it remains yet to be demonstrated whether rupture of the several membranes occurs successively in the natural progress of labor. The amnion being the innermost of the membranes, as well as the strongest, it does not seem probable that its rupture can precede the giving way of the chorion and decidua by any considerable time.

Further observations on the subject are very desirable, and may help us also to understand those curious cases in which a quantity of liquor amnii escapes several weeks before the completion of labor.

It frequently becomes the duty of the obstetrician to rupture the membranes artificially. Here, also, he should recognize the fact that the membranes may be ruptured separately; for unless he is well informed on this subject, concerning which the text-books are silent, he may think he has obtained all the assistance to be had from letting off the waters, when he has succeeded only in breaking through the chorion and decidua.

The practitioner should be aware of this source of error in his observations, especially as he is the more likely to fall into it from the fact that a small gush of waters may occur upon rupture of the outer membranes. Whether the waters obtain a position between the amnion and chorion by means of

rupture of the inner membrane, or by a process of transudation, is a question for further consideration.

BOOK NOTICES AND REVIEWS.

"Mismanaged Labor the Source of Much of the Gynecological Practice of the Present Day."

By Joseph Taber Johnson, M. D., Washington, D. C. Reprint from Vol. IV, Gynecological Transactions. (Received through the Author.)

The writer's object is to draw attention to the fact that gynecological practice derives much of its prominence and importance from the mismanagement of obstetrical cases, and their faulty treatment during the puerperal month. He argues that the tendency of physicians of the present day is to study the treatment of diseases of women, rather than the means of prevention, with which every obstetrician should be familiar.

Speaking of mismanaged abortion as a fruitful source of death and uterine disease, he says that the lives of many women have been saved, when endangered by chills, high pulse and temperature, associated with a putrid discharge, by removing the retained placenta and washing out the uterus with antiseptic fluids. Many of the women treated by the old plan of leaving the placenta undelivered in cases requiring a redilatation of the cervix and manipulation for its removal, sooner or later require the services of the gynecologist for relief from fibroid tumors, subinvolution, uterine displacement, or other disease. Then, in labor at term, too early rupture of the amniotic sac, and too frequent manipulation will not unfrequently cause a prolonged dry labor, with laceration of the soft parts and subsequent disease. Prolonged pressure during the second stage of labor, which should have been prevented by the timely use of forceps or other appropriate means, furnishes many a case of cellulitis, laceration and fistula. The excessive use of ergot exhausts the uterus and leads to its subsequent disease. Portions of the secundines and blood clots left in utero when the afterbirth has been removed by pulling upon the cord, have been the cause of septicæmia, metrorrhagia, and fibroid tumors, which have called for treatment months after the attendant physician had ceased his visits.

Failure to operate at once for the relief of laceration of the perineum, has entailed endless trouble. Putrid discharges, permitted to run over abraded and lacerated surfaces should be purified with warm water injections. Symptoms which have greatly alarmed the family physician and friends, quickly and completely disappear under the use of antiseptic vaginal and uterine injections and antipyretic doses of quinine. Early getting up after delivery is a common cause of subinvolution, uterine displacement and hemorrhage even among the colored women of the South, who have been said to do housework soon after confinement without disadvantage.

While in favor of the early use of the forceps in skilled hands, the author emphasizes the statement of Dr. Roper that much of the gynecological work of the present day results from the frequent interference with the natural functions of the uterus in childbirth. He points out that podalic version is frequently attempted too soon in cases of placenta previa at the risk of causing rupture of the uterine walls. If the cervix is not sufficiently dilated it is better to rupture the membranes and plug the cervix with a dilator, and then, if necessary, resort to Braxton Hicks' method of combined external and internal manipulation.

The author closes his paper by urging the clinical teaching of obstetrics in our colleges, and greater caution among obstetricians lest their patients pass from their care in a condition to develop uterine or periuterine disease.

G. J. B.

A Contribution to the Pathology of the Cicatrices of Pregnancy.

By SAMUEL C. BUSEY, M.D., Washington, D. C. Reprint from Vol. IV., Gynecological Transactions. 1880.

The striæ or scars upon the abdomen of women who have borne children, are commonly said " to be due to the rupture of some one of the layers of the integument, caused by distention ; the red striæ being recent solutions of continuity, and the white those which have undergone the process of separation by cicatrization." It is the author's object in the present paper to disprove this theory of rupture and cicatrization, and to place the subject upon the basis of a true pathology.

From the observations of Kustner and Hecker, it appears that these scar-like striæ may mark the localities of accumulations of serum or lymph in the expanded lymph spaces of the cutis, and of cavities filled or collapsed beneath the corium. A similar condition has been observed in a few cases of disease of the lymphatic apparatus (Congenital Occlusion and Dilatation of Lymph Channels—Busey), in which pressure upon the lymph channels produced striæ like those on the abdomen of child-bearing women. It is evident that these striæ are not cicatrices in the pathological acceptation of the word ; neither do they represent any form or stage of vesicle formation. As Professor Langer states it, there is " no solution of continuity," " but only a permanent disarrangement of the tissue, produced by stretching."

MEDICAL NEWS.

SPONTANEOUS GANGRENE. — Dr. J. H. Booley, of Columbus, Ohio, gives an account in the August number of the Virginia Medical Monthly, of a case of spontaneous gangrene occurring in a child seven years old. The child was suddenly seized with a chill, followed by fever and headache. During the succeeding three or four days, she had chills or chilly sensations at irregular intervals, and more or less continuous fever. She complained of headache, swelling and stiffness of the knees, inability to go up and down stairs, followed by a sense of tingling, and later, decided pain in the legs. Just one week from her first chill, a discoloration made its appearance on the dorsum of each foot, and extended up and around the legs, as far as the knee on the right side ; on the left, to some distance above the joint. The color, at first a purple mottling, soon became uniform, and exactly resembled the hue of a negro's skin. Her condition gradually grew worse in every respect, and she died six weeks from the date of the initial chill.

Autopsy.—Sloughing at the line of demarcation had become quite extensive, the arteries alone resisted the general destruction. An occlusion, by a firm thrombus, was found in both arteries, at almost the same point, viz.: in the popliteal, including all that portion between the superior and inferior external and internal articular arteries. Spontaneous gangrene is sufficiently rare, especially in a young child, to deserve record. The extensive character of the mortification, and its somewhat protracted course, also add to the interest of the present case.

MORPHINISM.—Dr. Levinstein, of Berlin, has had eighty-two men and twenty-eight women under treatment for morphinism; of these one hundred and ten patients, thirty-two were physicians; eight, physicians' wives; seven were more or less intimately connected with the profession; eight, apothecaries, and one apothecary's wife. Of the thirty-two physicians, twenty-eight relapsed, as did all the apothecaries. — *Gaillard's Med. Jour.*

HYOSCYAMIA, the new alkaloid of hyoscyamus, is very highly recommended by Dr. Brown of Chicago (Chicago Medical Journal and Examiner), as a valuable remedial agent in the treatment of mania (particularly puerperal), delirium tremens, and paralysis agitans. He states that it may now be obtained in a state of comparative purity, and is a very valuable recent addition to the pharmacopœia, possessing the hypnotic, anodyne, and anti-spasmodic properties of the bulky and disagreeable preparations of hyoscyamus; being insoluble in cold water, and easily soluble in hot water, alcohol, ether and chloroform. Its taste can be concealed with coffee, tea or milk. He thus summarizes: The great value of the drug consists in its furnishing, in a small compass,

a powerful agent for overcoming the cerebral irritability and hyperæmia of mania, and in its being an agent that can, usually, be administered without the knowledge of the patient. Moreover, my experience with the drug, that is with Merck's chrystalized alkaloid, would lead me to apprehend fatal results from the administration of one-half or one-quarter grain doses.''

A POSITIVE SIGN OF EARLY PREGNANCY. —Dr. J. H. Caritens (Detroit Lancet) calls attention to the color of the mucous membrane of the vagina and cervix uteri as a positive sign of pregnancy during the first three months. He says: '' This I have always found of a purplish blue, or rather deep violet hue, in pregnant women, and I have depended on this peculiar color in making a diagnosis of pregnancy in the first, second and third months. I can say it has never failed, and it is not produced by any pathological condition; the different colors produced by uterine disease cannot be mistaken for this pathognomonic violet hue.''—*Maryland Medical Journal.*

A SUBSTITUTE FOR TRACHEOTOMY.—Dr. William Mac Ewen has succeeded in catheterizing the larynx. The procedure was first suggested (accidentally) by Desault. Schrotter has practiced dilatation of the larynx by means of triangular tubes in chronic cases of stenosis, for the past four years. But it remained for Hack to successfully catheterize the larynx in acute cases, such as œdema of the glottis, etc. Almost simultaneous with the developments of Hack came the published account of Dr. Mac Ewen's systematic observations. His first experiments were upon the cadaver, subsequently it was put to the practical test in cases of œdema of the glottis and bloody operations in the pharynx. It is accomplished by introducing the finger into the mouth and drawing the epiglottis close up against the root of the tongue, then the tube to be introduced (preferably a large size catheter) is guided over the back of the finger into the larynx, during a deep inspiration. This tube, he says, may be kept in place for twelve hours, and removed only to be cleansed. During a bloody operation in the pharynx, a sponge was placed in the glottis, surrounding the catheter, for the purpose of preventing the entrance of blood into the trachea ; the chloroform being inhaled through the tube. Dr. Mac Ewen's paper, which is published in the British Medical Journal for July, sums up with the following deductions *en seriatum :* 1. Tubes may be passed through the mouth into the trachea, not only in *chronic*, but in *acute* affections, such as œdema glottidis. 2. They can be introduced without placing the patient under an anæsthetic. 3. The respirations can be perfectly carried on through them. 4. The expectorations can be expelled through them. 5. Deglutition can be carried on during the time the tube is in the trachea. 6. Though the patient at first suffers from a painful sensation, yet this passes off, and the parts soon become tolerant of the pressure of the tube. 7. The patient can sleep with the tube *in situ*. 8. The tubes, in these cases at least, were harmless. 9. The ultimate results were rapid, complete and satisfactory. 10. Such tubes may be introduced in operations on the face and mouth, in order to prevent blood from gaining entrance to the trachea, and for the purpose of administering the anæsthetic, and they answer this purpose admirably. This procedure, if thoroughly practicable, would seem to be a valuable substitute for tracheotomy.

NARROW PELVES. — Dr. Aug. F. Erich, Professor of Diseases of Women, College of Physicians and Surgeons, etc., in a masterly bibliographical and clinical paper entitled " A contribution to the relative value of the different operations for delivery in narrow pelves ; with the history of eighteen cases" —published in the Maryland Medical Journal, thus concludes :

1. The propriety of inducing premature labor is still questionable.

2. That version, while it should never be the alternative of the forceps, should be tried in contracted flat pelves before resort-

ing to craniotomy, but that it is worse than useless in a uniformly contracted pelvis after the forceps have failed.

3. The forceps, when properly applied and used, are the safest means of delivery for both mother and child. After failure with them, craniotomy is indicated, except in cases of narrow flat pelves, where version should first be attempted.

4. When there is not room enough for the application of the forceps, and when the smallest diameter of the pelvis is less than two inches, laparo elytrotomy is indicated. Our methods of measuring the diameters of the pelvis and of estimating the size of the child's head in utero, are, however, so very inexact that it is amusing to see cases reported with diameters given down to one-twelfth of an inch. Considering that these estimates are at best rough guesses, it will generally be well to give the child the benefit of the doubt and attempt to apply the forceps whenever the smallest diameter of the pelvis seems to be somewhere above two inches.

5. In cases of rupture of the uterus where the child has escaped into the abdominal cavity, and in cases. of extensive carcinoma of the cervix, Porro's operation (gastro-hysterectomy) should be performed in the interest of the child.

6. The unmodified Cæsarean section (gastro-hysterotomy) has been superseded by Porro's operation, which meets all the indications, with less danger to the mother.

SOME POINTS IN THE TREATMENT OF BRIGHT'S DISEASE, ETC.—We extract the following from Lagorio's admirable letter to the Chicago Medical Journal and Examiner, October, 1880 :

Some cases of Bright's disease were brought to the clinic this year. In six, the course of the disease has been followed with exactness, noting the daily quantity of urine and albumen, the other principal phenomena and the influence of the various methods of treatment. Special methods were tried, and some of the remedies being quite new, deserve attention, more especially as regards their effects, since the treatment of Bright's disease often improved the condition of the patient and arrested the morbid process. Chronic Bright's disease, untreated, generally does not improve; therefore it must be excluded from the category of those diseases that sometimes present spontaneous relief. It was noted by Prof. DeRenzi that the patients in the first days after their entrance to the clinic, or when the treatment is interrupted, readily secrete a greater quantity of urine. But there are some exceptions. Fuchsine, recently introduced in the treatment of Bright's disease, produced a sensible diminution of albumen. In clinic it was used under two forms, diluted with water, and added to an extract in pillular mass, each in two and one-half centigr. As the intense coloring of the water with fuchsine is objectionable, the pill form is to be preferred. The daily dose of fuchsine may be far greater than that thus far advocated in the treatment of the disease. Ordinarily a small dose of five centigr. was gradually increased to twenty-five centigr. in twenty-four hours. No remarkable physiological action of the fuchsine on the principal functions of the organism has been discovered. According to the dose, the urine commences to present a more or less reddish tint, persisting during all the time of treatment. Generally the urine acquires such a tint about five days after administering the remedy, and loses it about three to five days after suspending the fuchsine. Very often in Bright's disease the urine shows some mucus, and then the fuchsine is of great benefit, the mucus disappearing from the urine. The mucous membranes of the digestive tract become intensely colored by the drug, and the plasma of the blood shows a well-marked coloring. The quantity of hæmoglobin in the blood, and the chromometric grade were examined with the apparatus of Bizzozero, and it became evident that the chromometric grade corresponded to a quantity of coloring matter in the sanguineous plasma, which far surpassed the proportion of hæmoglobin. Evidently, therefore, the more intense coloring is not due to an augment of hæmoglobin, but rather to the dissolution of the fuchsine in the blood. Many consecutive observations have been confirmed, according to Prof. DeRenzi, the efficacy of fuchsine as a remedy for Bright's disease. In exceptional cases, where fuchsine does not avail, its presence in the urine is wanting, and the red coloring notably absent. Therefore, when, after some days of treatment with the fuchsine, the urine does not become red, the existence of some obscure alterations of the kidneys must be admitted. Besides fuchsine, other methods have been used, mainly repose in bed, as an efficacious means of diminishing the increase of albumen, and to this is added the milk diet. Apomorphine has been generally well tolerated and prescribed in

higher doses than those ordinarily given, five to six centig. being given daily without the least disturbance. In one case under this remedy the patient's state ameliorated considerably. As a therapeutical remedy in the treatment of globular anæmia and absence of hæmoglobin in the blood, the professor had used the hæmoglobin prepared by Grinon, according to the process of Dr. Lebou. So far the observations have been insufficient; but in one case the red blood corpuscles had considerably increased after the use of the hæmoglobin.

Piperine and oil of *eucalyptus globulus* have been largely tried for intermittent fever and tumor of the spleen. The former was largely advocated by Celsus, Dioscorides, Muller, Franck, Altdœfer, Riedmiller, Pettagua, etc. The action of the latter has also been studied by Weber, Gubler, Ullersperger, Tristany, Lambert, Lovinser, Kesser, Mossler, and Gimbert. But the observations in clinic did not furnish such good results as those lauded by these authors, having been found far inferior to those of quinine. As a great number of practitioners attributed either to the piperine, or to the eucalyptus taken separately, a strong anti-febrile action, the two remedies were used together. The formula of Mossler was preferred, that recommended for the treatment of leukæmia, that is both remedies combined in pill form, each pill consisting of one drop of oil of eucalyptus and four centigr. of piperine. The efficacy of the eucalyptus associated with the piperine was soon confirmed, but yet did not equal that of quinine. The results were good, however, and the professor did not hesitate to admit that the two combined remedies, eucalyptus and piperine, were the most potent articles next to quinine.

Just at this moment of closing, my eye has fallen on the following in the "Caffaro," of Genoa: "Naples, 11th. Yesterday in Fontana dei Serpi, on Pendino street, a woman, called *Concetta la Pavattiera*, gave birth to six little creatures, four dead and two living!"   A. LAGORIO, M.D.

CHIAVARI, Italy, Aug. 14, 1880.

NOVEL MODE OF ADMINISTERING NAUSEOUS OILS.—Dip a tea-cup in ice water, so that its entire surface will be well wetted. A small quantity of water, simply enough to float the oil, should be left in the bottom of the cup, the oil carefully poured in the center, and a little nutmeg grated on it. The water prevents the sticking of the oil to the cup, lips or mouth; the cold blunts the sense of taste, and the nutmeg masks the odor, which, to some delicate persons, is so offensive as to be one of the chief obstacles to its administration.—(Dr. Wolfe, in V. Medical Monthly.)

LOCAL ANÆSTHESIA BY BROMIDE OF ETHYL. —M. Perier, of Paris, states that he has employed the bromide of ethyl several times as a local anæsthetic with considerable success. It has the advantage over ether of not being inflammable, and hence can be employed when the actual cautery is to be used. *La France Medicale—Med. Record.*

PURULENT OTITIS MEDIA, has recently been successfully treated with insufflations of powdered borax, by Drs. Todd and Spencer of St. Louis. These gentlemen contend that, as the parts are naturally bathed in mucus, and the involved tissues very much macerated, a dry is preferable to a moist treatment. The borax, they say, acts as an absorbent, readily taking up the discharges and helping to dry up the tissues.

HOW TO MASK THE ODOR OF IODOFORM.— According to Dr. Linderman, the balsam of Peru completely masks the odor of iodoform; two parts of this balsam neutralizes perfectly one part of iodoform. The best vehicles are lard, glycerine, and above all, vaseline. The Doctor recommends the following formulæ:

Formula No. 1.   R. Iodoform.................. 1 part.
                 Balsam Peru............. 3 "
                 Vaseline................... 8 "
Formula No. 2.   R. Iodoform .... 1 "
                 Balsam Peru............. 3 "
                 Alcohol, glycerine, or
                 collodion... ....12 "

First, mix thoroughly the iodoform and balsam, then add the other ingredients.— *Journal de Medicine et de Chirurgie.*

EFFECTS OF MEDICAL LEGISLATION IN ILLINOIS.—The Philadelphia *Medical and Surgical Reporter* presents the following facts regarding the effect of medical legislation in Illinois. In July, 1877, when the law first went into effect, there were 7,400 physicians in the State, of whom only 3,600 were licensed or qualified practitioners. In 1880

the number of qualified practitioners had increased to 4,825, while the unqualified practitioners had decreased to 1,500. The total number has thus decreased to 6,325. The number of itinerants has decreased from 73 to 9, and the number of cancer doctors from 23 to 4. [Of these 1,500 parasites, about one-fifth have migrated to Colorado, and are now ingrafting themselves upon its citizens.]

TREATMENT OF BURNS.—Dr. Shrady, of New York, recommends that burns be treated by applying a paste composed of three ounces of gum arabic, one ounce of gum tragacanth, one pint of carbolized water (one part to sixty), and two ounces of molasses. The paste is to be applied with a brush, the applications to be renewed at intervals; it is stated to be a successful method. Four applications are usually sufficient, the granulating surface being treated with simple cerate or the oxide of zinc ointment, as indicated.

TRAUMATIC TETANUS: ATROPIA HYPODERMICALLY.—A number of successfully treated cases of traumatic tetanus by the hypodermic use of atropia have been reported within the last year. In one case the treatment was commenced by injecting the sixtieth of a grain in the back three times a day, increasing it to the fortieth, which was the maximum quantity employed. In another case one-tenth of a grain was used at each injection. Whether this plan will prove superior to the chloroform and chloral treatment, lately so popular, remains to be tested.— *Nashville Journal of Medicine and Surgery.*

THE FORMATION OF BILIARY CALCULI is reported by Dr. B. N. Steenberg (Medical Annals) to have been prevented by the use of nitromuriatic acid. He bases his conclusions upon six severe cases. .All were cases of long standing, in which the diagnosis was verified by an examination of the stools. One case, an old gentleman subject to attacks at intervals varying from three to eight weeks, in which the seizures were so severe as to demand the exhibition of three teaspoonfuls of the solution of morphia (U. S. P.) and eight minims of Magendies solution hypodermically before relief could be secured, was entirely cured by the following prescription :

> R. Acidi nitromuriatici dil. 2 drachms.
> Ext. Taraxici fl. · } $a. a.$ 2 ounces.
> Tr. Gentianæ Co. }
> Misce et signa.
> One teaspoonful before each meal.

Taken continuously for some months. In this particular case there has been no recurrence of an attack since Nov., 1875. In all the other cases it is claimed the benefits were as marked. His conclusions are as follows : " I do not claim nitromuriatic acid to be a panacea for gall-stone, notwithstanding every case in which I have administered it has been benefited by it ; but it would appear to be a singular circumstance that all six cases mentioned had run their course, and were cured by limitation. In every case the general health improved immediately. It is not a new remedy, but I have never seen it recommended for the difficulty in question. Nor am I prepared to say how it acts ; but I might suggest that gall-stones do not form from healthy normal bile ; that the acid so acts upon the liver as to induce it to the secretion of healthy bile, thus preventing further formations. I would further suggest the possibility of healthy bile possessing the ability of dissolving or breaking down concretions already formed."

MISCELLANEOUS.

STILL ANOTHER ADVANCE FOR HARVARD.— Four years ago the Medical School of Harvard University added to its curriculum, extended its course of study to three years, and required a preliminary examination upon entering. This has resulted in the general elevation of the standard of medical education throughout the country, and Harvard College has reaped material advantage therefrom. Her classes, we are informed, have constantly grown larger, and the character of her students has improved in nearly the same ratio. And now her faculty very wisely concludes that the allotted three years is insufficient time in which to complete a thorough course of medical training, and have, therefore, deemed it advisable to add another year to the course. Furthermore, we learn from an article in the August number of the Popular Science Monthly, entitled "Recent Original Work at Harvard," by J. R. W. Hitchcock, A. B., that the Harvard Professors are instituting extensive experimental research of an exceedingly creditable character. Professors Hill and Jackson have published twenty-five papers since establishing the "organic laboratory," in 1875, giving the results of their work, and have discovered one hundred new compounds. The *composition* of uric acid has long been known to be C_5 H_4 N_4 O_3, but its *constitution*— the exact arrangement of the atoms—has been uncertain. Chemists all over the world had endeavored to settle the question, but their failures resulted in eleven different formulæ for this one substance. Professor Hill, taking this uric acid, C_5 H_4 N_4 O_3, marked out one part by replacing H by C H_3 (methyl) ; then treating the acid so as to split it up, he determined to which part the methyl was attached, and, by continuing his treatment, was enabled to reduce the possible formulæ from eleven to three, with strong probabilities in favor of one. This possesses a practical value, inasmuch as it will lead to a knowledge of the method of formation of uric acid in the animal body. Fully as interesting also are the investigations of Professor Jackson into the nature of anthracene, which is obtained from coal-tar, and yields alizarine (madder-dye), used in dyeing pink and purple calicoes, Turkey reds, etc. These, however, are but stray examples of their diversified researches. Dr. Asa Gray is doing a great work in classifying the flora of California, and in completing his work on "The Flora of North America"—a labor of great magnitude.

In the medical department the largest amount of original investigation is carried on in the physiological and chemical laboratories. In the former a number of new forms of apparatus are in use, which have been designed by Professor Bowditch and his assistants. Among these are an apparatus for keeping animals alive by artificial respiration, a dog-holder, canulæ for observations on the vocal cords of animals, without interfering with their natural respirations; unpolarizable electrodes used in studying certain problems in the physiology of the nervous system, a new form of apparatus for barometric measurements, and a novel plan for measuring the volume of air inspired and expelled in respiration. A new form of plethysmograph has been devised by Dr. Bowditch. This is an instrument for measuring the changes in the size of organs, either hollow or solid, which are produced by variations in the conditions to which they are subjected. The essential part of Dr. Bowditch's invention is a contrivance by which fluid is allowed to flow freely to and from the organ to be measured without changing its absolute level in the receptacle into which it flows, while at the same time a record is made of the volume of the fluid thus displaced. Other important experiments are being carried on in regard to respiration, with special reference to the functions of the glottis and epiglottis, and trials of disinfectants with a view to ascertaining the temperature necessary to kill germs. Accounts of the most important investigations carried on during the last year are contained in the following papers: "Growth as a Function of Cells;" "Preliminary Notice of Certain Laws of Histological Differentiations," by C. G. Minot; "Effects of the Respiratory Movements on the Pulmonary Circulation," by H. P. Bowditch, M.D., and G. M. Garland, M.D.; "Pharyngeal Respiration," by G. M. Garland, M.D.; "Functions of the Epiglottis in Deglutition and Phonation," by G. L. Walton. This latter paper shows that the removal of the epiglottis does not seriously affect deglutition, and therefore it is not necessary for that process. The epiglottis, however, plays an important part in forming and modifying the voice, taking different positions during vocalization, changes of pitch, quality, and intensity.

Professor Wood is investigating the extent to which arsenic is being used in the manufacture or ornamentation of articles in general use, such as wall-paper, confectionery, playthings, etc. The results of this work will be published in the next report of the State Board of Health. Professor Wood is also writing the addition to Ziemssen's Cyclopœdia on the subject of toxicology. Dr. William B. Hills is engaged upon a special investigation in regard to the localization of arsenic in the animal economy. The most important feature of original work at the school of late years has been Dr. Bigelow's introduction of the new operation of litholapaxy.

A number of interesting papers have been recently written by members of the faculty, some of which contain new discoveries of considerable scientific importance. I cite two: "Effects of Certain Drugs in Increasing or Diminishing Red Blood-Corpuscles," by Dr. Cutter; and "Alterations in the Spinal Cord in Hydrophobia," by Dr. Fitz.

It is very creditable to Harvard that so much outside of routine work is being accomplished within her walls. It is the kind of work we are desirous of heralding.

THE CLIMATE AND METEOROLOGY OF ZANZIBAR.—Considerable interest is attached to the climate and meteorology of Zanzibar, since that island is the starting-point of most of the expeditions which proceed into the interior of East Africa. Observations taken by Dr. John Robb, of the Indian army, during the five years from 1874 to 1878, show that the average rainfall, which they give at not more than sixty-one inches, or double that of England, has very materially decreased since the time when Dr. Christie and Captain Burton made their observations; and it is suggested that the decrease may be due to the destruction of the trees over the whole island by a cyclone which swept it in 1872. The average number of rainy days is one hundred and twenty in the year. The double seasons, which are of unequal duration, are marked out by the prevailing winds, and are less exactly determined by the so-called greater and lesser rains. The rainy seasons begin when the sun crosses the zenith of Zanzibar in passing to its northern and southern declinations, March 4th and October 9th. The greater rains fall in March, April, and May, the lesser rains from the middle of October to the end of the year. The dryest month is September. The mean temperature of the five years was 80.6°, the hottest months being February and March, with a mean temperature of 83.1° and 80.4° respectively, the cooler are July and August, with mean temperatures of 77.5° and 77.7°. These figures give a variation of less than 6° in a year, and to this limited range is ascribed the debilitating nature of the climate. The mean pressure of the barometer for four years differed but a thousandth of an inch from that indicated at the equator. The coast of the mainland of Africa, Dr. Robb says, is undoubtedly prejudicial to health, and both Europeans and natives of India who pass any considerable time there suffer severely from fever of a bad remittent type, and from dysentery. All seasons are bad, but some are better than others, and travelers going into the interior are usually advised to leave the coast-region before the heavy rains begin to fall. The seeds of disease are often sown by even a short residence on the coast, and the traveler dies before he has advanced many marches into the interior. Travelers, therefore, should always make a careful and quick march across the unhealthy belt of country along the coast, and pitch their camps in the higher and drier districts beyond; and, if they have to linger on the coast, they should take care to pass their nights in the safest places they can find.

AT a show of birds lately held in Berlin, several canaries were exhibited that attracted much attention on account of the peculiar colors of their plumage. Some were green, others red and light brown, and others of a soft gray tint, while all differed more or less from the light yellow of the common bird. These variations of color were produced by the daily use of Cayenne pepper in their food. The pepper is given in small quantities at first, and the birds appear to like it, but the immediate effects are anything but pleasing to the beholder. The feathers soon begin to fall, giving the bird very much the appearance of molting; in a short time, however, new feathers make their appearance, and it is then, as they attain full growth, that they exhibit the curious tints observed. —*P. S. Monthly.*

DR. CARPENTER has measured the variation in the intensity of light to which the eye is sensitive, and has found it to be equivalent to about seven or eight hundredths; that is, a given light, whether strong or weak, must be diminished or increased in that degree to give a new sensation distinct from the former one. The difference is essentially the same in direct and indirect vision and with light of every color.

PHOSPHORESCENT LIGHTING.—Dr. Phipson takes sulphide of barium, or some other substance which is rendered phosphorescent by the solar rays, and incloses it in a Geissler tube, through which he passes a constant electric current of a feeble but regular intensity. He claims to obtain in this manner a uniform and agreeable light, at a cost lower than that of gas.—*Les Mondes.*

WE regret to have to announce the death of M. Lissajous, professor of physics at Toulouse, and author of several valuable scientific memoirs. His name is indissolubly connected with acoustics.

THE ROCKY MOUNTAIN MEDICAL REVIEW is the only journal which unites *General Science* with medicine, and which, therefore, peculiarly addresses itself to the scientific physician as well as the ordinary practitioner. Articles and advertisements sent to it now will receive unsurpassed prominence, as the Review in being introduced will reach every physician in the United States.

BRIGADIER-GENERAL ALBERT J. MEYER, of world-wide reputation for having established and developed the present extensive and complete U. S. Weather Service, died at Buffalo, Aug. 24th.

A few Press Notices and Letters Commendatory of the "Review."

Dr. J. Solis Cohen, Philadelphia, says : "I congratulate you on the appearance of the first number of the Review."

"It is prepared with conspicuous care and judgment, and compares most favorably with some of the oldest medical journals on the exchange list."—*Gaillard's Medical Journal.*

ALBANY MEDICAL COLLEGE.

Dear Doctor :—Please to accept my sincerest thanks for the copy of the Review, which I have read with pleasure. I wish it all possible success. Yours,

W. G. TUCKER, M.D.

GAILLARD'S MEDICAL JOURNAL, NEW YORK.

Gentlemen :—I received with pleasure No. 1 of Vol. 1 of the Review, and congratulate you on its appearance and arrangement. I have placed it on the exchange list, and will notice it in my Nov. number. Truly yours,

E. V. GAILLARD, M.D.

LOUISVILLE, KY.

Dear Doctor :—You sent sample copy of your journal to my friend Dr. Cooms of this city. He speaks very highly of your first number. I should like to see it, he having misplaced his copy. Yours very truly,

J. O. DYER, M.D.

ALBANY, N.Y.

Dear Doctor :—I congratulate you on the first issue of your journal and its handsome appearance. I shall be happy to place it on the exchange list of "The Medical Annals," and will send next number, to be out this week. Yours truly, F. C. CURTIS, M.D.

ZANESVILLE, OHIO.

Dear Sir :—Please accept my best thanks for copy of R. M. Medical Review. I am much pleased with the faultless mechanical execution of the work, and the contents are very creditable for an initial number. I am particularly pleased with the tone of several of the articles. It almost seems that it is the reflex of a new and improved civilization on the virgin soil of the Rockies. I send you copy of one of my latest contributions to the medical press; please review it, and if possible, in the same spirit of independ-ence as characterizes that on page 31 of the Review. If you can kick it out as unceremoniously and with as much justice as Hall's work, I will not complain. Yours truly,

Z. C. MCELROY, M.D.

JOLIET, ILLINOIS.

Dear Doctor :—I think the first number of the Rocky Mountain Medical Review has more sense and science in it than any twenty journals I have received within a year. You start out well, and I am greatly pleased with the exhibition of enterprise, backed up by verified facts. * * If I should find anything worthy of publication, I know of no journal I would be prouder to see it in than yours. * * * Yours,

R. J. CURTIS, M.D.

DENVER, COLORADO.

My Dear Doctor :—I have read the Rocky Mountain Medical Review, and am delighted with it ; I feel a State pride in it already. * * * Very sincerely yours,

JOHN W. BRANNAN, M.D. (Harvd.)

Books and Pamphlets Received.

"A New and Practical Treatise on the Principles and Practice of Medicine." By Roberts Bartholow, M.A., M.D., LL.D., Professor of Materia Medica and General Therapeutics in the Jefferson Medical College of Philadelphia ; formerly Professor of the Practice of Medicine and Clinical Medicine in the Medical College of Ohio at Cincinnati, etc., etc. (Through the publishers, D. Appleton & Co.)

"A Practical Treatise on Tumors of the Mammary Gland ; Embracing their Histology, Pathology, Diagnosis, and Treatment." By Samuel W. Gross, M.D., Surgeon to, and Lecturer on Clinical Surgery in, the Jefferson Medical Hospital, and the Philadelphia Hospital. (Through the publishers, D. Appleton & Co.)

"The Sanitarian," a monthly magazine devoted to the preservation of health, mental and physical culture. A. N. Bell, A.M., M.D., editor and proprietor. T. P. Corbally, A.M., M.D., associate editor. New York. [Placed on our exchange list.]

"The Specialist and Intelligencer," a monthly journal of medical science. Devoted specially to the eye, ear, throat and skin, venereal diseases, etc. Edited by Charles W. Dulles, M.D. Presley Blakiston, publisher. Philadelphia. [Placed on our exchange list.]

"The Maryland Medical Journal," a semi-monthly journal of medicine and surgery; T. A. Ashby, M.D., editor. J. W. Borst & Co., publishers. Baltimore, Md. [Placed on our exchange list.]

"Science," a weekly record of scientific progress. Illustrated. John Michels, editor and proprietor. [Placed on our exchange list.]

"The Nashville Journal of Medicine and Surgery." C. S. Briggs, M.D., editor and proprietor. [Placed on our exchange list.]

"The North Carolina Medical Journal." M. S. DeRosset, M.D., and Thos. F. Wood, M.D., editors. Jackson & Bell, publishers. Wilmington, N.C. [Placed on exchange list.]

"The Ohio Medical and Surgical Review." Henry G. Cornwell, M.D., editor and proprietor ; M. S. Clark, M.D., associate editor.

"Prevention and Cure of Chronic Consumption." By David Wark, M.D., Professor of Obstetrics and Diseases of Women and Children in the United States Medical College, New York. Authors Publishing Company. (Through the publishers.)

"On Some Points in the Pathological-Histology of So-Called Phthisis Pulmonalis or Consumption." By Z. C. McElroy, M.D. Reprint from St. Louis Medical and Surgical Journal. (Through the author.)

"The Monthly Review of Medicine and Pharmacy." Editor, Richard V. Mattison, Ph.G., M.D. Publishers, Keasbey & Mattison, Philadelphia. [Placed on our exchange list.]

"Proceedings of the Alumni Association of the Albany Medical College." (Through the secretary.)

"The Forty-ninth Annual Announcement of the Albany Medical College." (Through the secretary.)

"Consumption and Tuberculosis." Notes on their Treatment by the Hypophosphites. Collated from Books, Periodicals, etc., and addressed to the Medical Profession. Second edition, completely rearranged. Revised and re-written by J. A. McArthur, M.D. (Harv.), Fellow of the Massachusetts Medical Society. A pamphlet after the manner of quack advertisements ; written in the interest of, and purely an advertisement for, Churchill's Hypophosphites ; a production which should reflect an everlasting stigma upon the name of the author.

"Louisville Medical News:" A weekly journal of medicine and surgery. Editors, Richard O. Cowling, A.M., M.D., and Lunsford P. Yandell, M.D. John P. Morton & Co., publishers; Louisville, Ky. [Placed on our exchange list.]

"Colorado ; and Homes in the New West." By E. P. Tenney, president of Colorado College. With numerous handsome illustrations. Price : Paper, 75 cents ; cloth, $1.00. Publishers : Lee & Shepard, Boston.

"The Saint Joseph Medical and Surgical Reporter." Editor, J. P. Chesney, M.D. C. P. Kingsbury, publisher. [Placed on our exchange list.]

"Gaillard's Medical Journal." E. S. Gaillard, A.M., M.D., LL.D., editor and publisher. New York. [Placed on our exchange list.]

"The Medical Annals:" A journal of the Medical Society of the County of Albany. (Through the secretary.) [Placed on our exchange list.]

"Announcement of the Twenty-second Annual Session of the Long Island College Hospital," Brooklyn, N. Y. Collegiate year, 1880-81. (Through the secretary.)

"The American Monthly Microscopical Journal." Romyn Hitchcock, F. R. M. S., editor and publisher. [Placed on our exchange list.]

"Pacific Medical and Surgical Journal." Henry Gibbons, M.D., and Henry Gibbons, Jr., M.D., editors and proprietors. [Placed on our exchange list.]

ROCKY MOUNTAIN MEDICAL REVIEW.

Vol. 1.　　　　COLORADO SPRINGS, NOVEMBER, 1880.　　　　No. 3.

ORIGINAL ARTICLES.

THE THERMOGRAPH—ITS EVOLUTION AND DESTINY.

By A. Wellington Adams, M.D.

The introduction or application to medical science of no *one* instrument has afforded a more forcible impetus toward the goal of perfection, and the establishment of an exact science, than has the ushering in of the thermometer by Sanctorius, and subsequently the foundation, by de Haen, of medical thermometry.

And as the mind reviews and takes cognizance of the successively progressive steps constituting its history, from its inception to its present stage of development, one is compelled to look forward to disclosures far transcending present realizations.

Hence I was led, in the early part of 1879, to conclude that, notwithstanding the present degree of perfection in the science of medical thermometry, and the important function already performed by the thermometer in clinical, diagnostic, prognostic and experimental medicine, there was yet still further room for improvement. And, in fact, a palpable *need* for a more delicate and accurate thermometric system as an auxiliary to physiological, therapeutical and pathological investigations.

At present the office and usefulness of the system is sadly limited by its imperfections and narrow range of application. There are, no doubt, a few who will, upon first impulse, stamp this as a bold assertion, tending to depreciate the value of this most important aid to modern therapeutics, and, indeed, to *all* branches of scientific medicine. Such a *petitio principii* cannot, however, for one moment be allowed, since it must be acknowledged by all, that with existing appliances we are but able to ascertain the morning and evening minimum and maximum of temperature, regardless of the momentary or even hourly remissions and exacerbations or minor waves which fall within these, as it were, medium tidal points. By this latter expression I would have you understand me as referring to the points representing the morning and evening thermometric observations, A A A, fig. 1.

Now, a visual revelation of these constant thermal changes would, even with our present degree of knowledge concerning the pathognomonic character of certain thermal phenomena, be of incalculable value, and, therefore, a consummation to be devoutly sought for. But, outside of the advantages accruing from the application of existing knowledge to these observations, the anticipation of new developments and additional advantages resulting therefrom, becomes extremely plausible and justifiable ; for it's as probable as it is possible that these momentary perturbations or minor waves, of which we as yet know almost nothing, are also conformable to law ; forming, in different diseases and under the administration of various drugs, characteristics *quite* as significant as do these medium waves, if I may be allowed the expression, when looked at collectively as preserved for a consecutive number of days.

From the results of these observations, taken, according to the present method, at comparatively irregular intervals, we can crudely construct a major wave, B B, fig. 1, analogous to the now so-called "fever curve," which may after a week or so prove characteristic of some peculiar form of the exanthemata, which we have become accustomed to recognize by comparison with a typical curve or major wave, whose conformation depends upon the position of these medium tidal points—that is to say, the points recording the morning and evening thermometric observations taken each successive day from the stage of pyrogenisis to that of defervescence.

I, therefore, concluded that in order to render the system more perfect and extend its range of application, it would be necessary to devise some means for procuring a thermometric record which should represent the *constant* condition of the subject under experimentation or treatment, for any desired length of time ; this record to assume the form of a continuous curve automatically inscribed *pari passu* with the development of the disease.

This, of course, I found impracticable with the instruments already at the command of the profession. A solution of the problem, then, evidently depended upon the invention of an instrument capable of fulfilling these requirements. No sooner did this proposition present itself than I began to turn my attention in that direction. And is it to be wondered at, that in looking for the *principia* upon which would depend the construction of such an instrument, I should follow the footsteps of Breschet and Becquerel, who laid the corner-stone of the science by the aid of delicate thermo-electric apparatus? No, but on the contrary, quite natural, that in looking for further boon, I should turn toward that force which has already been the means of marvelously facilitating commercial intercourse—of conferring untold benefits upon man of *general* application and lasting endurance—and of heaping upon science and its unfolders immensurable *eclat*.

Having already given much attention to experimental electricity and acoustics, as bearing upon otology and laryngology, I was peculiarly and specially prepared, both as regards practical or detail knowledge, and the possession of suitable apparatus, to un-

dertake the solution of so trying and complex a problem.

So, after adding to my laboratory much improvised paraphernalia for experimental research, I began operations, which, as before intimated, at first lay in the direction of thermo-electricity ; to which principle I ascribed the ability to meet the requirements indicated in the first steps, i. e., the contriving that portion of the embryo instrument capable of perceiving and responding physically to all variations in temperature with which it might be brought in contact, be they ever so slight. And, as regards this specific property, I was not in the least disappointed as to the inherent power of this principle ; for, after procuring several cylindrical tubes of antimony and bismuth 3-16 of an inch in external diameter, and cutting them up into segments one-quarter of an inch long, then soldering these pieces together in alternate order, and now bending the whole into a spiral one inch in diameter, to each end of which was attached a copper wire—one being fastened to a segment of bismuth at one end, and the other to one of antimony at the opposite end—I had an instrument (see fig. 2) which would, when placed in the axilla, and the terminal wires connected with the binding posts of a differential galvanometer, give evidence, through the varying deflections of the galvanometer needle, of the most minute and transient thermal fluctuations.

This, however, was not the only indication to be met. There was the recording part yet to be constructed, and that necessarily must bear a certain relationship to the thermometer proper.

In the device described above and figured in illustration No. 2, there were generated thermo-electric impulses sufficiently strong to induce an extensive range of deflections in a sensitive galvanometer, but entirely too weak to manifest themselves in any more crude manner ; while, of course, the delicacy of the galvanometer rendered impossible the application of mechanical appliances for graphically recording deviations ; for it must be borne in mind that with the graphic method, the first essential is a surface traveling in a straight line at a uniform rate of speed ; this may be accomplished after any of the known methods. The second essential consists in the provision of an automatic writer for inscribing the record upon said surface. This, as is generally the case, may be effected by means of a delicate lever, having attached to its longer extremity a marker so arranged as to move backward and forward over the traveling surface. When the lever is quiescent and the movable surface in motion, a horizontal line is made, known as the *abscissa.* This signifies time, and all deviations from this line are called *ordinates.* These may vary in direction, from the vertical, to any degree of obliquity, according as the impingements upon the shorter end of the lever, which give rise to them, are abrupt or sluggish.

It will readily be seen, then, that with the addition of these pieces of mechanism, we have a commensurate amount of friction to contend with, and in order to overcome this, our electric impulses must be stronger than could be generated by the small antimony and bismuth spiral seen in fig. 2.

Not in the least daunted, however, by the dissipation of all hopes in this direction, I still continued my researches, fully believing that even many more failures were in store for me, as the *res necessariae* to the proper evolution of a successful issue.

Experience seemed now to clearly indicate a necessity for a definite source from which to derive a constant and unvarying supply of electricity, of sufficient strength to undergo a mechanical metamorphosis equal to our requirements. This current, however generated, to be moulded into electrical waves corresponding to the thermotic changes, through the introduction, at any given point in the circuit, of some physical or chemico-physical device capable of accomplishing the same. To effect this, the instrument illustrated in fig. 3 was constructed. This consists virtually of an ordinary thermometer made of vulcanite and bent into

a spiral. For convenience of illustration this shape is not shown in the cut. The bulb was lined with a silver film in electrical communication with the lower binding-post, A. Up both sides of the shaft B, and at right angles to its axis, minute channels were drilled at uniform distances from each other, and leading into the shaft bore. These channels, instead of being placed opposite each other, were arranged one above the other, e. g., the first one on the right hand side (C) being placed 1-32 of an inch from the first one on the left hand side (D), while the *second* one on the right hand side (E) was placed 1-16 of an inch from C, and 1-32 of an inch from D ; and so on throughout the entire series. The shaft-bore was now temporarily filled with a wire of corresponding calibre, and the aforesaid channels filled with melted platinum. A No. 36 silk-covered copper wire was next soldered to the first platinum *fibrillus* (C), and then coiled several times around the shaft (B), finally to be soldered at its terminal extremity to the platinum *fibrillus* (D). Another corresponding piece of wire of like length, calibre, etc., was soldered to the platinum *fibrillus* (D), and now an equal number of lamilla convolutions made around the shaft (B), and its terminus soldered to the platinum *fibrillus* (E). When all these *fibrilli* were treated in like manner, and the last wire in the series connected with binding-post (F), a vacuum formed, and the mercury inserted in the bulb; the instrument was complete, and only awaited a trial to prove its efficacy or inefficiency, as the case might be. To an electrician the *modus operandi* of this apparatus becomes at once apparent. The binding-posts A and F being placed in an electric circuit and the thermometer bulb subjected to thermal variations—the instrument is in operation. Each of the above described coils of wire go to form a series of " resistance coils," and any number of these are introduced into the circuit or " short-circuited"—that is, cut out, and the strength of the current thus diminished or increased, by the rising and falling of the column of mercury. For, when the column of mercury rises to the first platinum *fibrillus* (C), the electric current will pass from the binding-post (A) through the column of mercury to the first platinum *fibrillus* (C), and thence through all of the "resistance coils" to emerge at the binding-post (F), on its way to complete the circuit. If, however, the column of mercury rises to the *fibrillus* (D), the first "resistance coil" is short-circuited, and the current is obliged to pass through the remaining coils *only*. Thus the resistance is diminished a certain number of ohms and the current accordingly strengthened a relative amount of volts. This instrument was experimentally a complete success, inasmuch as with it there could be produced electrical impulses varying in intensity in accordance with the character of the temperature changes brought in contact with it. Moreover, the current designed to be thus moulded into waves could have any desired electro-motive force, and this, as before explained, was a *sine qua non*.

When it came to the question of a practical application, though, this instrument also seemed to be deficient in many respects. The principal defect lay in the fact that the electric waves were not *always* in exact relationship with the thermometric indications. For, supposing the column of mercury to have reached a point half-way between C and D, the electricity would then meet with a resistance equal to that offered when the column stood no higher than C, hence we should have no increased strength in the current to mark this difference in temperature. Another thing rendering it somewhat impracticable, was its complicated nature and expensiveness.

My researches now remained in *statu quo* for some time, when one day, in the course of conversation with Dr. J. Harry Thompson, of Washington, D. C., he suggested the utilization of the newly discovered property of carbon to vary its conductivity under different degrees of pressure.

I immediately availed myself of this suggestion, developing the instrument as now

perfected and illustrated in figs. 4 and 5. This is the thermometer proper or responding portion of the instrument, and consists of a spiral spring made of two *lamellae* of brass and steel respectively, soldered together, the brass occupying the outer side. Of course this spring expands uniformly with equal increments of heat, and the brass, the most expansible of the two metals, will, upon a rise of temperature, give the platinum knob (a), attached to the free end of the spring, a concentric twist. In this way we produce a varying pressure upon the contents of the vulcanite tube (T), against which the platinum knob (a) impinges.

The other end of the substance contained in the hard-rubber tube (T) has for its abutment the platinum knob (b) attached to the hard-rubber bracket (C). The whole, as seen in fig. 4, is inclosed in a perforated German-silver case, with rounded edges, and having an external diameter of but $1\frac{1}{4}$ inches.

The binding-post (A), fig. 4, is in electrical communication with the platinum knob (a), and the binding-post (B) is in electrical communication with the platinum knob (b). When the apparatus is introduced into an electric circuit, by attaching the two poles to the two binding-posts, the current enters through one and emerges at the other, passing in its course, through the substance in the vulcanite tube (T). The two little handles (H) (H) are intended as a means of securing the instrument in its proper position in the axilla. The composition used in the vulcanite tube (T) may be either a solid stick of baked lamp-black, a series of thin carbon discs with intervening ones of silver, or a powder made of plumbago, gas-carbon and silver, finely divided. After receiving a communication from Thos. A. Edison in regard to this matter, I commenced a series of experiments to determine the most suitable composition for this purpose, and the best results were obtained from the powder already referred to. The salient feature of this instrument is the changing of its electrical resistance with

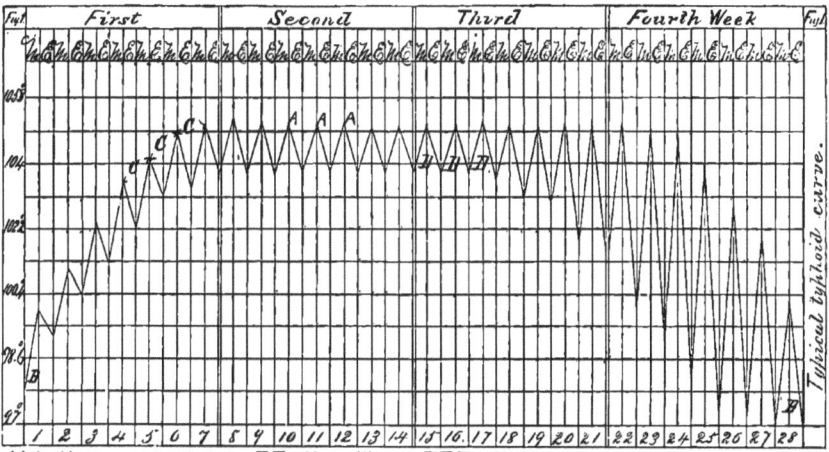

A.A.A.- Medium tidal points. B.B.- Major Wave. D.D.D.- Medium Waves.
C.C.C.- Intervals in which the minor waves occur.

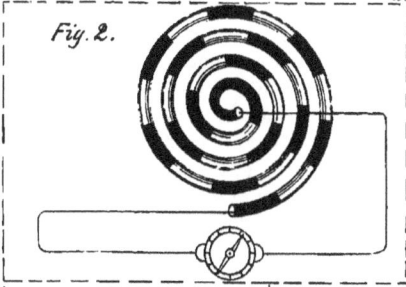

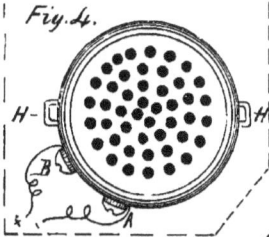

Fig. 2.

Fig. 3.

G

Range - 20°

Fig. 4.

Fig. 5.

S

Fig. 5.- Inside plan Fig. 4.
H.H.- Handles.
A & B- Binding posts.
T- Vulcanite tube.
a & b- Platinum knobs.

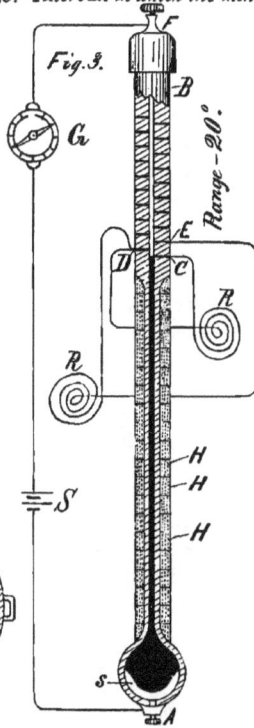

G- Galvanometer. S- Battery. H.H.H.- Coils in position.
R.R.- Diagram shewing arrangement of coils.
S- Silver-film lining bulb.

pressure, and the ratio of these changes, moreover, corresponding exactly with the pressure, the latter, in turn, being dependent upon and in unison with the rise and fall of temperature.

Here, then, was the true solution, for, by subjecting this instrument to varying degrees of temperature, the resistance of the powder would vary in precise accordance with the pressure exerted by the uniform expansion of the spiral spring under equal increments of heat, and consequently a proportionate variation would be produced in the strength of the current. The latter would thus possess all the characteristics of the heat waves, and by its reaction through the medium of some electro-magnetic piece of mechanism yet to be devised, these might be transferred to our movable surface, in the form of a sinuous line, whose rising and falling inflections would give a graphic representation of them.

Now, that I had satisfactorily reduced this portion of the problem, the next in order was the devising that part of the instrument intended for recording such variations as the other branch might be subjected to. This, I assure you, was no easy task, but one requiring a mint of patience and tedious application. For, first—it must be simple; second —there must be established a permanent relationship between the first and second branches of the instrument, in other words, there must exist throughout a strict interdependence; third—in order that the electromotive force required might be reduced as much as possible, it must be delicate; fourth —to render the latter possible, friction must be practically reduced to a minimum. To carry you through the almost endless and varied experiments necessary in developing means for meeting these indications would be as tiresome as it would be unnecessary. Hence, I shall confine myself to the result only.

If a number of coils of insulated wire be wound around a hollow reel, there is formed what is known to electricians as a *helix*. If this is now placed in an electric circuit and a current passed through its convolutions, it

is temporarily constituted a magnet, the two ends forming the poles; so that it may be said to possess all the properties of a permanent magnet during the passage of the current. Moreover, if such a helix, mounted in a vertical position in such a way that an iron rod can be introduced into it from below, be connected with a battery, the iron rod will be at once drawn up into it and be sustained oscillating in its axis, even though the rod may weigh considerable.

The depth the iron rod enters will also depend entirely upon the strength of the current and the amount of resistance offered by the iron rod. This principle is well known in physics as the "axial electro-magnetic force," and in it I found what I sought, namely—a combination of delicacy and strength in the proper proportions. A diagramatical illustration of its application may be seen in fig. 6, where H represents a helix of peculiar construction applied to the purpose in hand; D is a soft iron tube in connection with the short end of the lever (L); P, represents the movable surface or strip of paper; F, the fulcrum of the lever, and E its marker or stylus; T, the thermometer proper introduced into the circuit; X, the battery; C, the curve, and S, the brass drum over which the strip of paper (P) moves. . Having comprehended the principles, the action of this combination is obvious. If an electric current passes through the helix (H), the core (D) will be drawn into said helix, carrying with it the short end of the lever (L), to which it is attached. This movement naturally causes the marking end of the lever to make a still longer excursion in the opposite direction. Upon breaking the circuit, the attractive power of the helix is abolished, and the counter-action of the spring (s) returns the lever to its normal position.

The depth to which the core (D) is drawn into the helix (H) being dependent upon the strength of the current passing through the coils of wire, the excursions of the tracer or lever will also be great or small, according as the current is weak or strong.

The lever (L) is delicately made, and its fulcrum provided with jewel mountings. Its short end is connected with the core (D) by means of a universal joint, while its longer end has inserted in it a silver stylus reaching to the surface of the traveling paper. The latter moves over a brass drum forming a portion of the circuit. The strip of paper passing over the brass cylinder, having been saturated with a solution of chloride of sodium, pyrogallic acid, and ferrocyanide of potassium, the instrument is complete.

When this combination is in operation, a current of electricity will pass from one pole of the battery to the binding-post (A) of the thermometer proper, through the substance in the vulcanite tube to emerge at binding-post (B); thence through the helix to the lever, along this to the silver stylus; thence through the moistened paper and brass cylinder to the other pole of the battery—thus completing the circuit. Upon the application of varying degrees of heat to the thermometer proper, the resistance the current meets with during its course will be varied in precise accordance with the various changes of temperature. This waxing and waning current will now pass through the helix, and by the latter's peculiar action produce to and fro motions in the lever, passing, at the same time, through the lever, and chemically prepared paper, and producing as it passes a double chemical decomposition upon the paper; one of which decompositions renders the development of friction, during the movements of the lever, so slight as to be imperceptible; the second decomposition producing a change in color upon the paper, corresponding to the movements of the stylus, and affecting no larger surface than it covers, thus obviating the additional friction accompanying the use of an ordinary marker.

From this description you will understand that the lever is moved backward and forward by a difference in the attractive power exerted by the helix, this in turn being dependent upon the strength of the current, which has already passed through the ther-

mometer proper, and there been moulded into electric waves corresponding to the heat waves; the motion of the lever being facilitated by the lubricating action of the current, as the result of one of the chemical decompositions during its passage through the chemically moistened paper; while the other decomposition causes a discoloration, and thus produces a mark corresponding in outline to the movements of the lever. This mark will, therefore, form an irregular line, whose sinuosities will give a graphic representation of the heat variations. This apparatus is extremely sensitive and can be made to record 1-100 of a degree.

Now, after marking upon our strip of paper the minimum and maximum points representing respectively 90° and 110°, it becomes a very easy matter to determine the degree of heat represented by any point lying within this range. This is accomplished by dividing the intervening space into any number of equal parts, when any one of these divisions will represent a degree or any part of a degree, according to the number of divisions. These horizontal lines may be placed at such distances from each other as to represent 1-10 of a degree. Having provided the traveling paper with a uniform speed, it also becomes an easy matter to determine the time represented by any given distance upon its surface; for, supposing a certain amount of paper passes a given point in the instrument in one hour, to determine the amount passing the same point in five min., it is only necessary to draw vertical lines dividing this distance into twelve equal parts, each one of these will then represent five minutes.

After determining upon the principles it became very easy to work up the details that would place the instrument in a convenient form for manufacture and use. These may be seen, as applied, in figs. 7 and 8. Fig. 7 is a front elevation of the complete thermograph.

It consists of a cast-iron case having two departments, one for the recording mechanism and actuating clock movement, the other for the battery. In the upper part of the front

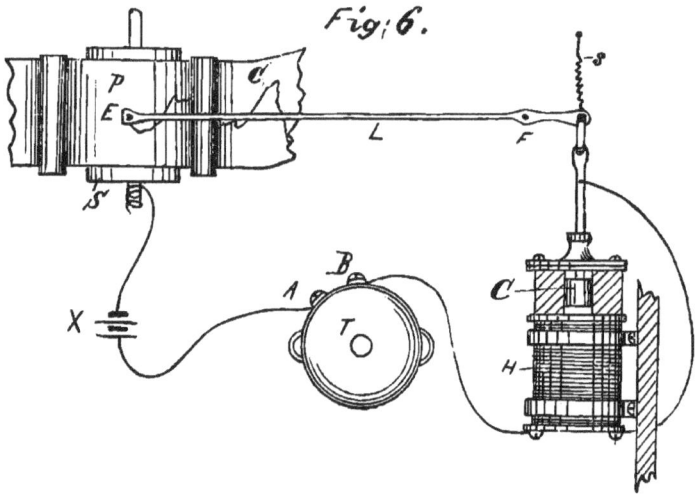

Fig: 6.

Fig: 7.

THERMOGRAPH.

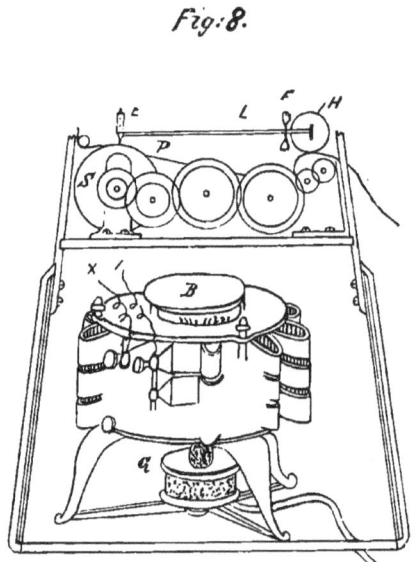

Fig: 8.

there is a circular depression for the reception of the thermometer proper or perceiving portion of the instrument when not in use.

This is held in place by means of two little catches, one on either side, as seen in the figure. On both sides of this are the binding-posts for the reception of the wires leading from the thermometer proper, when the latter is in position in the axilla. The open work in the lower portion is intended for the ingress of air and egress of gases. Fig. 8 is an interior elevation with the front removed ; above is seen the recording mechanism, and below the thermo-electric battery. This form of battery gives a continuous and unvarying current, requires no cleaning or recharging, and costs but little to run, hence it is the most available source from which to derive the current ; the heat for operating it may be supplied by either an alcohol lamp or gas-burner. Not only is it possible with this instrument to procure a continuous curve denoting the constant febrile condition of a subject, but, with the addition of certain accessories now in process of construction, and as suggested by Prof. Mayer, of the University of Technology, Hoboken, and Dr. Toner, of Washington, we may be able to procure, on the same strip of paper, at the same time and under similar conditions, a sphygmographic and a respiratory curve ; thus enabling pathologists, therapeutists, physiologists, and, in fact, general practitioners, to study the inter-relationship of these three cardinal symptoms under various modifying circumstances. These are the *possibilities*, but when we drift into the *probabilities*, we see in prospective the addition of that which will also furnish a moisture curve. Of the advantages of the graphic method as applied to medicine, I need hardly speak. It already promises for medicine what it has accomplished in physics.

Every physicist adores such familiar names as Leon Scott and Dr. Clarence Blake, to whom we are indebted for the application of the graphic method to the science of acoustics, through the medium of the phonautographs invented by themselves.

To the experimental therapeutist this instrument is of incalculable value, as affording a means of determining the precise character of the temperature changes under the administration of various therapeutic agents in different sized doses and modes of exhibition.

The experimental physiologist will find in it that which will materially facilitate accurate observation in his field. And the advantages accruing from its application in pathological investigations, and the possibility of thus elucidating hitherto obscure phenomena, must be patent to every one. An instrument of so much value as an aid to observations in these three important branches of scientific medicine, needs no further lauding ; but I cannot draw my paper to a close without setting forth the mode of application and the advantages attending its use in every day practice.

Take, for example, a suspected case of typhoid fever, experience and experimentation with the thermograph having already revealed a characteristic *minor* wave curve for typhoid fever. The physician is summoned. Upon arriving he applies his thermograph in the following manner : First, the perceiving portion, as seen in fig. 4, is fastened in the axilla by means of two elastic bands attached to the handles H H, one passing around the trunk, the other over the shoulder. Next, two fine and flexible silk covered wires are led from the binding-posts A B, fig. 4, to the binding-posts B B, of fig. 7, the latter having been previously placed upon a stand at the head of the patient's bed.

These wires, of course, are of sufficient length to admit of any degree of motion on the part of the patient without interfering with the position of the recorder. The instrument will now be ready for use, and, upon starting the battery, it will continue in operation for any desired number of days, with little or no attention outside of winding and replenishing with new rolls of paper.

The first benefit to be derived from its use in such a case, consists in the ability to determine upon a diagnosis much earlier than would ordinarily be possible ; second—the physician is furnished with a permanent record of the condition of his patient from hour to hour and day to day ; third—the slightest modification or variation by reason of an exposure, the exhibition of prescribed remedies at given hours, or the ingestion of prescribed food during the day, will be revealed to the physician when he makes his evening visit, thus affording him from time to time, a more definite idea of the immediate effect, good or bad, of his treatment ; fourth—it will give warning of danger from collapse during the crisis before it could be detected in any other way ; fifth—the physician is provided with a means of leaving more definite directions with the attendant or nurse, e. g. he will be able to say that "should the curve assume such or such a character, or the line rise to this or that point, you may discontinue this, that, or the other remedy, and proceed to exhibit *this*, according to the directions ; or, should such and such a thing take place it will indicate an emergency calling for this, that, or the other measure."

The science of meteorology, also, will find in the thermograph an instrument it has long felt the need of. Never before has there been invented an instrument capable of furnishing a curve representing the constant temperature of the atmosphere. To be sure, there are two or three instruments in the possession of the United States Weather Department at Washington, constructed upon an entirely different principle, which automatically produce a continuous curve, but the latter is only by reason of the velocity with which the cylinder revolves, besides, they are of an exceedingly complicated nature, cumbersome, and *very* expensive.

The simplicity and inexpensiveness (it will cost about $50) of the thermograph, places it within the reach of almost every physician, and will enable the United States Weather Department to furnish one or more of them to every one of its sub-stations.

This, gentlemen, is the instrument I have chosen to dignify with the title of *Thermo-*

graph, and which I have lately placed in the hands of Aloe & Hernstein to manufacture for the use of the medical profession.

With its introduction, I predict the dawn of a new era in medicine, marked by progress equal to that accompanying the introduction of the sphygmograph, myograph, cardiograph, and other important instruments of a similar character.

[Condensed from a paper read before the El Paso County Medical Society, Colorado Springs, Colorado.]

THE INFLUENCE OF ALTITUDE UPON RESPIRATION.

By S.Edwin Solly, M. R. C. S., Eng., Vice-President Colorado State Med. Society.

In this article it will be my endeavor, not to present new facts, but to try and arrange old ones ; the subject is one on which our actual knowledge when we take careful stock of it, at the present time, is very limited, but it is also one on which speculation has been peculiarly wild and unlimited, and the entanglement of fact and theory specially confusing.

Let us take first what I hold to be well established facts concerning persons or animals subjected to an ascending change of residence.

It has been proved beyond doubt that the inhabitants of elevated lands are remarkable for their great chest capacity, and further, it has been shown by actual measurement that the chest capacity is increased in individuals changing their residence to greater elevations.

The frequency of the respiratory act is increased by altitude quite irrespective of exertion, as shown by observations taken during balloon ascensions.

In men and animals coming from a low to a high altitude, and using great exertions as in a race, they cannot maintain the same speed, as proved by the time records of race-horses brought there to run.

It has been clearly demonstrated that, although it is doubtful whether a zone of absolute immunity from phthisis absolutely exists, yet phthisis is a very rare disease among the inhabitants of high lands, and that the disease is generally mitigated in those possessing it who change to the higher grounds ; on the other hand, acute croupous pneumonia is, other things being equal, more common and severe on high than on low land. We know that phthisis is commonly fostered by imperfect performance of the respiratory act and sluggishness of the pulmonary circulation leading to deficient chest expansion, and this may be due either to insufficient exercise on the part of the individual or animal, or on account of there being an insufficient supply of oxygen in the air breathed. With regard to acute pneumonia, we know that on the contrary it is accompanied in its initial stages by an increased excitement of the pulmonary circulation.

Asthma, taken apart from its cause, whatever that may happen to be, is relieved as far as its symptoms of irregular and incomplete respiration, in most cases, by altitude, and is rare as arising *de novo.*

These are briefly the facts concerning the individual that appear pertinent to our inquiry.

Let us now take some established facts in physics.

To maintain respiration at a normal point and to insure a continuance of healthy life, a certain amount of oxygen must be absorbed into the blood through the lungs by the act of inspiration, and to effect this there must be a certain percentage of oxygen present. Parkes' experiments have fixed the minimum at 14 per cent., and this percentage has been shown to be present at much greater elevations than we are concerned about for health purposes.

The superincumbent pressure of air being necessarily less the higher we ascend from the sea, there is less oxygen and nitrogen compressed into each cubic inch of the atmosphere, and, therefore, for the lungs to absorb the same amount of oxygen on the mountains as at the sea level, a greater number of cubic inches of air must be taken into the chest.

The experiments of Paul Bert have proved that even when an animal dies from oxygen starvation, there is always three to four per cent. of oxygen left unused in the air in which the animal is confined.

His experiments also point very clearly to the conclusion that it is the diminution of the oxygen and not of the atmospheric pressure that gives rise to the symptoms of mountain sickness, and that the equality between the pressure within and without the walls of the blood vessels is very quickly established, and therefore has little if any influence.

In effecting the main object of respiration, viz., the replacing of the carbonic acid accumulated in the blood, after its circulation through the body and after it has been pumped into the pulmonary vessels by the heart, by pure oxygen, the first cause acting to accomplish this is the law of diffusion of gases whereby the carbonic acid slowly leaves the lungs and diffuses itself into the surrounding atmosphere, and the oxygen tends to leave the air which is in contact with the pulmonary vessels, and pass through their walls and replace the carbonic acid which is passing out. Now, although this law of diffusion of gases will act independently of life, yet it can only come into play very imperfectly and feebly unless the pneumogastric nerve is acting in the capacity which gives it the first half of its name, and that is as a constant and regular excitant of the muscular contractions of the diaphragm. It is by the rising and sinking of the floor of the vital bellows that air is expelled and drawn into the chest, in ordinary respiration. But when from whatever cause the breathing is quickened, then the diaphragmatic contractions are aided by the muscles attached to the chest walls, in a small degree, by the intercostals drawing the ribs together and apart, and in a great degree by the pectorals raising the whole upper part of the frame work, to do which most effectually the hands and arms are fixed on some stationary object, so as to allow the muscles a fixed point from which to act.

The last fact in physics that it is well to note, is that the air in the chest after inspiration is of three kinds, viz.: breathing, supplemental and residual; the breathing being the air which is changed at each ordinary respiration, the supplemental, that which is changed only on extraordinary respiration, when the pectoral muscles come into play, and the residual, which is never changed, but under all circumstances remains in the lungs. Of course, according to the law of diffusion of gases, both the supplemental and residual air are indirectly changed, more or less, even in ordinary respiration.

Reviewing these facts we find that altitude increases the rapidity of respiration in the first place, and as a secondary and permanent effect, causes the lungs themselves to expand, which expansion can only be due to portions of the lungs in which the amount of air entering the cells was very small and seldom changed—receiving more air and changing it more frequently, or at least more directly—it is, I think, fair to assume that the amount of breathing air is permanently increased and the amount of complemental air diminished. Then by the law of diffusion of gases, more atmospheric air being drawn into the chest, will cause the carbonic acid in the residual air to be given off more rapidly and the oxygen from the complemental and breathing air to be absorbed more freely.

It is, therefore, right to assume that the osmosis of the pulmonary vessels is increased by altitude; by which means the venous blood is more rapidly and completely converted into arterial—and as it is shown that a larger area of lung surface is brought into contact with the atmosphere, therefore a greater amount of blood is oxygenated, and that more completely, at each respiratory act the higher we rise above sea level, until an altitude is reached where the per centage of oxygen in the air falls below 14.

On reaching such an elevation, although the rapidity of respiration continues to increase, the change of venous into arterial blood diminishes, on account of there being an insufficiency of oxygen to replace the carbonic acid. The exosmosis, however, continues to increase so that at last not merely the gases and watery vapor of the blood are given off, but the blood itself oozes through the walls of the capillaries.

We now see that increased chest expansion and increased rapidity of respiration are the inevitable results of altitude. Further, that up to the height at which the per centage of oxygen is sufficient to replace gases given off, both endosmosis and exosmosis are increased; but where there is not sufficient oxygen in the air to replace the gases exhaled, the endosmosis is lessened and the exosmosis increased to the point of hemorrhage.

We have so far, I believe, the effects of altitude pure and simple, as far as facts at present warrant us to go. These effects, though of course modified greatly, first by personal peculiarities, and secondly by the conditions of rest and exercise in the individual, and thirdly and most markedly by barometric and thermometric changes—sunlight and humidity—yet are exhibited more or less strikingly on the animal organism by ascending from sea level. That the respiratory movements are quickened and the chest expansion increased on elevated ground, both in a humid and a dry atmosphere, is shown by observations made in such a damp atmosphere as the highlands of Scotland, and such a dry atmosphere as the foothills of Colorado.

Now let us consider the secondary phenomena induced by the effects upon the nervous and circulatory systems. The most noticeable and constant accompaniment of increased respiration, is accelerated cardiac action. This occurs independently of exercise, as proved by the observations of aeronauts at various elevations. As a consequence of this increased vigor of the heart, more blood is thrown into the lungs in a given space of time, but if the respiration is too rapid for the heart's action, the venous blood will accumulate in the auricles and cyanotic symptoms will appear, or on the other hand, if the blood is pumped through the lungs by an over-excited heart faster than it can be oxygenated by respiration, similar symptoms will also occur. It is this want of balance between the supply of blood to the lungs and its aeration that causes the distress exhibited under violent exercise by untrained persons or animals, and the same distress is experienced at slight elevations on moderate exercise by those of feeble physique, and at great elevations by the strong. As training will overcome this distress under violent exercise in the healthy, so continued residence will modify or remove it when it occurs at elevations, after moderate exertions, by those of normal physique. In looking for the peculiar influence of altitude we must specially note the first effects, because the range of physical adaptability is so great that we find in healthy persons organs discharging their functions normally under the most varied conditions of climate and atmosphere. So far, then, we find the effects of altitude to be that the heart pumps the blood more rapidly through the lungs, and that the lungs receive more atmospheric air at each inspiration and change it more frequently. Therefore the results are that the air cells are more fully and equally expanded, that the pulmonary circulation is more active, and the supply of oxygen more frequent, and consequently the osmosis more perfect.

This activity of the normal functions of the lungs, necessarily antagonizes the conditions which give rise to or accompany phthisis, viz : the stasis of venous blood in the lungs and the imperfect performance of their excretory function, causing the accumulation of unoxygenated deposits and the consequent local congestions around them. This arrest of oxygenation causing the retention of irritating material in the lungs, is analagous to what happens in the elimination of urea when oxygenation is arrested at the uric acid stage. Low forms of inflammation, also, such as catarrhal pneumonia, are opposed by this more healthy pulmonary life. While, on the other hand, it is easy to understand that if any thing occurred to suddenly interfere with

the temperature of the blood in a certain portion of the chest, as is specially likely to happen at elevations where the atmospheric changes are sudden and extreme, the consequent inflammation of the lung tissue would be liable to run to a great height at the onset, by reason of its taking place in an organ so richly supplied with blood ; whereas in the subsidence of the disease, this same activity of circulation would come into play to diminish the effects of the disease by increasing and hastening the absorption of inflammatory products, and all this exactly coincides with clinical facts. Before we leave the consideration of the effect of altitude upon the pulmonary circulation, it may be well to notice that in this case also clinical observation confirms what has gone before. For instance, all cardiac symptoms arising from valvular disease are greatly aggravated by altitude. Where the muscle of the heart is increased in strength by disease, the heart's action will be inordinately hurried, and where it is weakened as by fatty degeneration, the action will become slower and feebler, the muscle being unable to force onward the larger bulk of blood it has to contend with.

The direct effect of altitude upon the nervous system, unconnected on the one hand with barometric and thermometric variations, and on the other with the influence of respiratory and circulatory changes, is very difficult if not impossible to ascertain. When we consider that so far as our present knowledge extends, the effects of altitude pure and simple arise entirely from two causes only, viz., lessened atmospheric pressure and lessened oxygen, it does not seem probable that there is any direct effect of altitude upon the nervous system, and that all the effects we witness are caused by vascular and respiratory changes.

The nerve supply to the respiratory apparatus, it need be scarcely stated, is of three kinds, sympathetic, sensory and motor. The sympathetic keeps up the contractions of the diaphragm so that there is a constant entrance and exit of air during life, independent of the will. The sensory nerves are affected by any deficiency in quality or quantity of the air taken in, and by reflex action stimulate the respiratory muscles to increased exertions and to take in more oxygen ; or on the other hand, they protect the lungs from receiving obnoxious gases, etc., by stimulating motor nerves to lessen the force of inspiration and contract the air passages. The motor supply is for increasing the force of respiration through the costal and pectoral muscles, and dilating or contracting the air passages as required by the sensory stimulus. The demand for oxygen necessitated by altitude does undoubtedly stimulate the sensory nerves to cause the respiratory muscles to act more intensely, and so expand the chest. The effect of altitude being to improve the pulmonary circulation, it therefore follows that the blood supply to the nervous system generally would be much better in quantity and quality, and therefore the respiratory nerves would act with more promptitude and vigor.

This would appear to embrace what we know of the influence of altitude upon respiration through the agency of the nervous system.

The more vigorous and perfect working of the respiratory nerves caused by altitude may help to account for the relief that mere elevation usually affords asthmatics ; the unequal and violent action of the sensory nerves of the respiratory tract being apparently the immediate cause of the prominent symptoms of their disease.

Thus far we have endeavored to deal with this problem as if rise in elevation from sea level was not invariably accompanied by more or less change in other atmospheric conditions than those of tenuity of the air and diminution of oxygen ; whereas we know that, practically, changes in the conditions of the sun's heat and light, the amount of water in the atmosphere in the form of clouds, mists, suspended humidity, rain and snow, the quantity of ozone, the amount of movement in the air, the variations and degrees of temperature, the quality of the soil and vegetation, all largely modify the phenomena we have been considering.

The adequate discussion of these various elements that qualify the influence of altitude cannot be undertaken within the limits of a single magazine article, even if the outlines of the matter are only roughly sketched, as in the portion of the subject already treated. I therefore propose to continue this enquiry in the next number of the Review.

SELECTIONS FROM JOURNALS.

HYPNOTISM.

By G. J. Romanes.

Considering the length of time that so-called "animal magnetism," "mesmerism," or "electro-biology," has been before the world, it is a matter of surprise that so inviting a field of physiological inquiry should have been so long allowed to lie fallow. A few scientific men in France and Germany have indeed, from time to time, made a few observations on what Preyer has called the "Kataleptic state" as artificially induced in human beings and sundry species of animals ; but anything resembling a systematic investigation of the remarkable facts of mesmerism has not hitherto been attempted by any physiologist in our generation. The scientific world will therefore give a more than usually hearty welcome to a treatise which has just been published upon the subject by a man so eminent as Heidenhain. The research of which this treatise is the outcome is in every way worthy of its distinguished author ; for it serves not only to present a considerable and systematic body of carefully observed facts, but also to lead the way for an indefinite amount of further inquiry along the lines that it has opened up.

Heidenhain conducted his investigations on medical men and students as his subjects, one of them being his brother. He found that, in the first or least profound stage of hypnotism, the patient, on being awakened, can remember all that happened during the state of mesmeric sleep ; on awakening from the second or more profound stage, the patient can only partially recollect what has happened ; while in the third, or most profound stage, all power of subsequent recollection is lost. But, during even the most profound stage, the power of sensory perception remains. The condition of the patient is then the same, so far as the reception of sensory impressions is concerned, as that of a man whose attention is absorbed or distracted ; he sees sights, hears sounds, etc., without *knowing* that he sees or hears them, and he cannot afterward recollect the impressions that were made. But the less profound stages of hypnotism are paralleled by those less profound conditions of reverie in which a passing sight or sound, although not noticed at the time, may be subsequently recalled by an effort of the will. Further on in his treatise, Heidenhain tells us that, even when all memory of what has passed during the hypnotic state is absent on awakening, it may be aroused by giving the patient a clew, just as in the case of a forgotten dream. This clew may consist only of a single word in a sentence. Thus, for instance, if a line of poetry is read to a patient during his sleep, the whole line may sometimes be recalled to his memory, when awake, by repeating a single word of the line. Again, we know from daily experience, that the most complicated neuro-muscular actions—such as those required for piano-playing—become, by frequent repetition, " mechanical," or performed without consciousness of the processes by which the result is achieved. So it is in the case of hypnotism. Actions which have been previously rendered mechanical by long habit, are, in the state of hypnotism, performed automatically in response to their appropriate stimuli. There being a strong tendency to imitate movements, these appropriate stimuli may consist in the operator himself performing the movements. Thus when Heidenhain held his fist before his hypnotized subject's face, his subject immediately imitated the movement ; when he opened his hand his subject did the same, provided that his

hand was visible to his subject at the time. Also, when he clattered his teeth, the hypnotized patient repeated the movement, even though the patient could only hear, and not see, the movement ; similarly, the patient would follow him about the room, providing that in walking he made sufficient noise to constitute a stimulus to automatic walking on the part of his patient. In order to constitute stimuli to such automatic movements, the sounds or gestures must stand in some such customary relation to the movements that the occurrence of the former naturally suggests the latter.

Another characteristic of the hypnotic state is that of an extraordinary exaltation of sensibility, so that stimuli of various kinds, although much too feeble to evoke any response in the ordinary condition of the nervous system, are effective as stimuli in the hypnotic condition. It is remarkable that this state of exalted sensibility should be accompanied by what appears to be a lowered, or even a dormant, state of consciousness. It is also remarkable that this exaltation of sensibility does not appear to take place with what may be called a proportional reference to all kinds of stimuli. Indeed, far from there being any such proportional reference, the greatly exalted state of sensibility toward slight stimuli is accompanied by a greatly diminished state of excitability toward strong stimuli. Thus, deeply hypnotized persons will allow themselves to be cut, or burned, or to have pins stuck into their flesh, without showing the smallest signs of discomfort. Heidenhain is careful to point out the interesting similarity, if not identity, between this condition and that which sometimes occurs in certain pathological derangements of the central nervous system, as well as in a certain stage of anæsthesia, wherein the patient is able to feel the contact of the surgical instruments, while quite insensible to any pain produced by the cutting of his flesh. Reflex sensibility, or sensibility conducing to reflex movements, also undergoes a change, and it does so in the direction of increase, as might be expected from the consideration that with the temporary abolition of consciousness the inhibitory influence, which we know the higher nerve centers to be capable of exerting upon the lower, is presumably suspended. But quite unanticipated is the remarkable fact that the state of exalted reflex excitability may persist for several days —perhaps for a week—after a man has been aroused from a state of profound hypnotism. Thus, Dr. Krener, after having been hypnotized by Professor Heidenhain, and while asleep made to bend his arm twice, for several days afterward was unable again to straighten it, on account of the flexor muscles continuing in a state of tonic contraction, or cramp. In these experiments Heidenhain found that a very gentle stimulation of the skin caused only the muscles lying immediately below the seat of stimulation to contract, and that on progressively increasing the strength of the stimulus, its effect progressively spread to muscles and to muscle-groups farther and farther removed from the seat of stimulation. It is interesting that this progressive spread of stimulation follows almost exactly Professor Pfluger's law of irradiation. But the rate at which a reflex excitation is propagated through the central-nerve organs is very slow, as compared with the rapidity with which such propagation takes place in ordinary circumstances. Moreover, the muscles are prone to go into tonic contraction, rather than to respond to a stimulus in the ordinary way. The whole hypnotic condition thus so strongly resembles that of catalepsy that Heidenhain regards the former as nothing other than the latter artificially induced. In the case of strong persons this tonic contraction of the muscles may make the body as stiff as a board, so that, if a man is supported in an horizontal position by his head and his feet only, one may stand upon his stomach without causing the body to yield. The rate of breathing has been seen by Heidenhain to be increased four-fold, and the pulse also to be accelerated, though not in so considerable a degree.

In a chapter on the conditions which induce the state of hypnotism, Heidenhain begins by dismissing all ideas of any special "force" as required to produce or to explain any of the phenomena which he has witnessed. He does not doubt that some persons are more susceptible than others to the influences which induce the hypnotic state, and he thinks that this susceptibility is greatest in persons of high nervous sensibility. These "influences" may be of various kinds ; such as looking continuously at a small bright object, listening continuously to a monotonous sound, submitting to be gently and continuously stroked upon the skin, etc.—the common peculiarity of all the influences which may induce the hypnotic state being that they are sensory stimuli of a gentle, continuous, and monotonous kind. Awakening may be produced by suddenly blowing upon the face, slapping the hand, screaming in the ear, etc., and even by the change of stimulus proceeding from the retina, which is caused by a person other than the operator suddenly taking his place before the patient. On the whole, the hypnotic condition may be induced in susceptible persons by a feeble, continued, and regular stimulation of the nerves of touch, sight, or hearing ; and may be terminated by a strong or sudden change in the stimulation of these same nerves.

The physiological explanation of the hypnotic state which Heidenhain ventures to suggest is, that a stimulus of the kind just mentioned has the effect of inhibiting the functions of the cerebral hemispheres, in a manner analogous to that which is known to occur in several other cases which he quotes of ganglionic action being inhibited by certain kinds of stimuli operating upon their sensory nerves.

In a more recent paper, embodying the results of a further investigation in which he was joined by P. Grutzner, Heidenhain gives us the following supplementary information :

The muscles which are earliest affected are those of the eyelids ; the patient is unable to open his closed eyes by any effort of his will. Next, the affection extends in a similar manner to the muscles of the jaw, then to the arms, trunk and legs. But even when so many of the muscles of the body have passed beyond the control of the will, consciousness may remain intact. In other cases, however, the hypnotic sleep comes on earlier.

Imitative movements become more and more certain the more they are practiced, so that at last they may be invariable and wonderfully precise, extending to the least striking or conspicuous of the changes of attitude and general movements of the operator. Professor Berger observed that, when pressure is exerted with the hand at the nape of the neck upon the spinous process of the seventh cervical vertebra, the patient will begin to imitate spoken words. It is immaterial whether or not the words make sense, or whether they belong to a known or to an unknown language. The tone in which the imitation is made varies greatly in different individuals, but for the same individual is always constant. In one case it was a hollow tone, "like a voice from the grave;" in another almost a whisper, and so on. In all cases, however, the tone is continued in one kind, i. e., it is monotonous. Further experiments showed that pressure on the nape of the neck was not the only means whereby imitative speaking could be induced, but that the latter would follow with equal certainty and precision if the experimenter spoke against the nape of the neck—especially if he directed his words upon it by means of a sound-funnel. A similar result followed if the words were directed against the pit of the stomach. It followed with less certainty when the words were directed against the larynx or into the open mouth, and the patient remained quite dumb when the words were directed into his ear, or upon any other part of his head. If a tuning-fork were substituted for the voice, the note of the fork would be imitated by the patient when the end of the fork was placed on any of the situations just mentioned as sensitive.

By exploring the pit of the stomach with a tuning-fork, the sensitive area was found to begin about an inch below the breastbone, and from thence to extend for about two inches downward and about the same distance right and left from the middle line, while the navel, breastbone, ribs, etc., were quite insensitive. Heidenhain seeks—though not, we think, very successfully—to explain this curious distribution of areas sensitive to sound, by considerations as to the distribution of the vagus nerve.

Next we have a chapter on the subjection of the intellectual faculties to the will of the operator which is manifested by persons when in a state of hypnotism. For the manifestation of these phenomena the sleep must be less profound than that which is required for producing imitative movements; in this stage of hypnotism the experimenter has not only the motor mechanism on which to operate, but likewise the imagination. "Artificial hallucinations" may be produced to any extent by rehearsing to the patient the scenes or events which it may be desired to make him imagine. A number of interesting details of particular cases are given, but we have only space to repeat one of the most curious. A medical student, when hypnotized in the morning, had a long and consecutive dream, in which he imagined that he had gone to the Zoological Gardens, that a lion had broken loose, that he was greatly terrified, etc. On the evening of the same day he was again hypnotized, and again had exactly the same dream. Lastly, at night, while sleeping normally, the dream was a third time repeated.

A number of experiments proved that stimulation of certain parts of the skin by hypnotized persons is followed by certain reflex movements. For instance, when the skin of the neck between the fourth and seventh cervical vertebræ is gently stroked with the finger, the patient emits a peculiar sighing sound. The similarity of these reflex movements to those which occur in the well-known "croak-experiment" of Goltz, is pointed out.

A number of other experiments proved that unilateral hypnotism might be induced by gently and repeatedly stroking one side or other of the head and forehead. The resulting hypnotism manifested itself on the side opposite to that which was stroked, and affected both the face and limbs. When the left side of the head was stroked, there further resulted all the phenomena of aphasia, which was not the case when the right side of the head was stroked. When both sides of the head were stroked, all the limbs were rendered cataleptic, but aphasia did not result. On placing the arms in Mosso's apparatus for measuring the volume of blood, it was found that, when one arm was hypnotized by the unilateral method, its volume of blood was much diminished, while that of the other arm was increased, and that the balance was restored as soon as the cataleptic condition passed off. In these experiments consciousness remained unaffected, and there were no disagreeable sensations experienced by the patient. In some instances, however, the above results were equivocal, catalepsy occurring on the same side as the stroking, or sometimes on one side and sometimes on the other. In all cases of unilateral hypnotism, the side affected as to motion is also affected as to sensation. Sense of temperature under these circumstances remains intact long after sense of touch has been abolished. As regards special sensation, the eye on the hypnotized side is affected both as to its mechanism of accommodation and its sense of color. While color-blind to "objective colors," the hypnotized eye will see "subjective colors" when it is gently pressed and the pressure suddenly removed. Moreover, if a dose of atropine be administered to it, and if it be then from time to time hypnotized while the drug is gradually developing its influence, the color-sense will be found to be undergoing a gradual change. In the first stage yellow appears gray with a bluish tinge, in the second stage pure blue, in the third blue with a yellowish tinge, and in the fourth yellow with a light-bluish tinge. The research

concludes with some experiments which show that in partly hypnotized persons imitative movements take place involuntarily, and persist until interrupted by a direct effort of the will. From this fact Heidenhain infers that the imitative movements which occur in the more profound stages of hypnotism are purely automatic, or involuntary.

In concluding this brief sketch of Heidenhain's interesting results, it is desirable to add that in most of them he has been anticipated by the experiments of Braid. Braid's book is now out of print, and, as it is not once alluded to by Heidenhain, we must fairly suppose that he has not read it. But we should be doing scant justice to this book if we said merely that it anticipated nearly all the observations above mentioned. It has done much more than this. In the vast number of careful experiments which it records—all undertaken and prosecuted in a manner strictly scientific—it carried the inquiry into various provinces which have not been entered by Heidenhain. Many of the facts which that inquiry yielded, appear, *a priori*, to be almost incredible ; but, as their painstaking investigator has had every one of his results confirmed by Heidenhain so far as the latter physiologist has prosecuted his researches, it is but fair to conclude that the hitherto unconfirmed observations deserve to be repeated. No one can read Braid's work without being impressed by the care and candor with which, amid violent opposition from all quarters, his investigations were pursued ; and now, when, after a lapse of nearly forty years, his results are beginning to receive the confirmation which they deserve, the physiologists who yield it ought not to forget the credit that is due to the earliest, the most laborious, and the hitherto most extensive investigator of the phenomena of what he called hypnotism.— *Nineteenth Century.*

WE should be pleased to open negotiations with medical and scientific writers desirous of being enrolled upon our list of Home correspondents.

ON THE NUTRITIVE VALUE OF FISH.

By PROF. W. O. ATWATER.

This paper (read before the A. A. A. S. Boston, 1880,) gives the results of an investigation made under the auspices of the Smithsonian Institution and the United States fish commission. They included analyses of a large number of specimens of more common food fishes, whose details, though quite extended, were mainly of theoretical value. Some of the applications, however, were of much practical interest. In 100 pounds of the flesh of fresh cod we have 83 pounds of water and only 17 pounds of solids, while the flesh of the salmon contains only 66½ per cent. of water and 33½ per cent. of solids ; that is to say, about one-sixth of flesh of cod and one-third of that of salmon consists of solid, that is, nutritive substances, the rest being water. Lean beef, free from bone, contains about seventy-five per cent. water and twenty-five per cent. solids. The figures for some of the more common sorts of fish were :

	Solids. per cent.		Solids. per cent.
Flounder	17.2	Halibut, fat	30.7
Cod	16.9	Mackerel	22.2
Striped bass	20.4	Shad	30.7
Bluefish	21.8	Whitefish	30.4
Halibut, lean	20.6	Salmon	33.6

If we take into account not the flesh only but the whole fish as sold in the market, including bones, skin and other waste, the actual percentage of nutritive material, is, of course, smaller. Thus the following percentages of edible solids were found in samples analyzed :

Flounders	7.1	Shad	14.8
Cod	10.5	Shad	18.7
Mackerel	11.4	Lake trout	13.6
Halibut, lean	15.6	Salmon	25.6
Halibut, fatter	27.2		

The subject has of late attracted unusual attention. The chemico-physiological investigation of the past two decades has brought us where we can judge with a considerable degree of accuracy from the chemical composition of a food-material what is its value for nourishment as compared with other foods. The bulk of the best late investigation of this subject has been made in Germany, where a large number of chemists

and physiologists are busying themselves in the experimental study of the laws of animal nutrition. They have already got so far as to feel themselves warranted in computing the relative values of our common foods, and arranging them in tables, which are coming into popular use. The valuations are based upon the amounts of albuminoids, carbohydrates and fats, each being rated at a certain standard, just as a grocer makes out his bill for a lot of sugar, tea and coffee, by rating each at a certain price per pound, and adding the sums thus computed to make the whole bill. A table was given showing the composition of a list of animal foods. Thus it appeared that, while medium beef has about three-fourths water and one-fourth solids, milk is seven-eighths water and one-eighth solids. Assuming a pint of milk to weigh a pound, and speaking roughly, a quart of milk and a pound of beefsteak would both contain the same amount—about four ounces—of solids. But the quart of milk would not be worth as much for food as the pound of steak. The reason is that the nutrients of the steak are almost entirely albuminoid, while the milk contains a good deal of carbo-hydrates and fats, which have a lower nutritive value. According to the valuations given, taking medium beef at 100, we should have for like weights of flesh free from bone :

Medium beef	100.0	Bluefish	85.0
Fresh milk	23.8	Mackerel	86.0
Skimmed milk	18.5	Halibut	88.0
Butter	124.0	Lake trout	94.0
Cheese	155.0	Eels	95.0
Hens eggs	72.0	Shad	99.0
Cod (fresh fish)	68.0	Whitefish	103.0
Flounders	65.0	Salmon	104.0
Halibut	88.0	Salt mackerel	111.0
Striped bass	79.0	Dried codfish	346.0

These figures differ widely from the market values. But we pay for our food according, not to their value for nourishing our bodies, but to their agreeableness. Taking the samples of fish at their retail prices in the Middletown, Conn., markets, the total edible solids in striped bass came to about $2.30 a pound, while the Connecticut river shad's nutritive material was bought at 44 cents per pound. The cost of the nutritive material in one sample of halibut was 57 cents, and in the other $1.45 per pound, though both were purchased in the same place at the same price—15 cents per pound, gross weight. In closing, Professor Atwater referred to the widespread but unfounded notion that fish is particularly valuable for brain food on account of its large per centage of phosphorus. Suffice it to say that there is no evidence as yet to prove that the flesh of fish is specially richer in phosphorus than other meats are, and that, even if it were so, there is no proof that it would be on that account more valuable for brain food. The question of the nourishment of the brain and the sources of intellectual energy are too abstruse tor speedy solution in the present condition of our knowledge.—*Science.*

A Third Case of Extra-Uterine Gestation Treated by Abdominal Section.

By Lawson Tait, F.R.C.S.

The case I have now to submit to the Society (read before the Royal Medical and Chirurgical Society, May 11, 1880,) differs from the two which I have previously narrated, in that the operation was performed before the death of the child, whose life was saved, and that the mother succumbed to the operation.

On January 23 I was asked by Mr. Walter Lattey, of Southam, to see with him Mrs. S——, aged thirty-three, whom he suspected to be suffering from extra-uterine pregnancy.

She had been married for fourteen years, and had had six children. The last menstruation occurred early in May, 1879, and since that time she had been in almost constant pain. Previous to May she was a strong, healthy woman, of good temper, but since then she had worn away, and her husband told us that her temper and manner had entirely changed.

In July a sudden attack of severe abdominal pain occurred, followed by a sharp illness. After her recovery from this, she applied to a physician, who treated her for uterine retroflexion, and fitted her with a pessary, which gave her intense pain. In

October she became satisfied she was pregnant, and applied early in January to Mr. Lattey, that he might attend her in her confinement. About the middle of January her pain became much more severe, and when Mr. Lattey examined her, he found such an unusual state of matters in the pelvis that he suspected the true nature of the case, and asked me to see her.

I found a large abdominal swelling, which was evidently a pregnancy. To the left of the umbilicus was a prominent mass rather larger than the closed fist, the centre of which coincided with the umbilical level. This was clearly the uterus, for the rhythmic contractions could be both seen and felt in it ; and it was evidently attached to the larger mass of the swelling which lay behind it, extending to the right crest of the ilium. All over its front the placental souffle could be heard and almost felt ; and above a ridge, which I regarded as the edge of the placenta, the limbs of the child could be felt, and the fœtal heart heard.

The vagina was elongated, and the cervix could not be reached, but the head of the child was readily detected lying between the rectum and the vagina.

A few days after I saw her, the pain became agonizing, and she consented to come to Birmingham to have something done. I admitted her into hospital on the evening of the 29th, and ordered her stimulants and opiates. On the 30th, the sister informed me that opium seemed to have no influence in relieving her pain, and therefore I determined to operate next morning. For this I obtained the concurrence and assistance of my colleague, Dr. Savage.

On the morning of the 31st she was so much exhausted as to lead me to regret that the operation had not been performed a week earlier, but as the child was still alive, I had no hesitation in proceeding.

Ether was administered by the sister in charge, and I made the usual median incision. I came at once upon the empty uterus, and pushing that over to the left, I opened the cyst, which was in the right broad ligament, just above the uterus, and clear of the large placental sinuses. I came immediately upon the legs of the child, and removed it without difficulty. I then secured the edges of the opening in the cyst to the edges of the parietal wound, leaving a free opening for the future discharge of the placenta, and then carefully closed the peritoneal cavity.

The child was small, but evidently mature. It was profoundly narcotised by the ether, and for hours its breath smelt strongly of that substance. It is now (three weeks after its Cæsarean birth) thriving well.

The shock of the operation to the mother was so great that I feared she would not rally. It may be doubted if she ever did really rally, for though death did not take place till the fourth day after the operation, she was so entirely restless and uncontrollable, and there was so little other cause discernible for her death, that it might be regarded as death from protracted shock. Dr. Saundby made a post-mortem examination, and found some morbid appearances in the liver, and ante-mortem clots in the heart, but no peritonitis. The pregnancy cyst was removed *en masse*, and on careful examination proved to be exactly what I had diagnosed it, the right broad ligament expanded into a cyst, with the placenta spread out on the anterior wall. In the empty uterus there was a well-marked deciduous lining in process of separation.

On the top of the cyst the Fallopian tube could be seen spread out, and when fresh it was evident that this tube had been ruptured at its lower aspect, the ovum thus escaping into the tissue of the broad ligament, separating its layers, and thus forming the cyst. This variety of extra-uterine gestation is what has been called by Deseimeris the " *sous peritoneo pelvienne,*" and my views of its origin I have already expressed at length in my book on " Diseases of Women." The present case is one which fully confirms these, and also the opinion I have in the same place advanced, that all extra-uterine pregnancies are originally tubal, that the

tube inevitably bursts between the eighth and twelfth weeks, the rupture in this case having probably occurred in July, when she had the illness ; and that in those comparatively few cases where the rupture is not fatal, an extra-uterine pregnancy is completed, the variety being due solely to the direction in which the tube bursts.

In the present case the unfavorable result to the mother was, I believe, entirely due to the fact that we could not persuade her to have anything done till her chances of recovery were almost gone. The case is still of interest, on account of the success in saving the child.

It would have been quite as easy to remove the child by vaginal section, but I do not regret having adhered to my usual practice of operating through the linea alba. Large venous sinuses spread over the cyst in every direction, injury to one of which would have been immediately fatal, and in vaginal section it is, of course, equally impossible to see what to avoid as it is to repair what has been damaged.

P. S.—August 5, 1880.—The child is still alive, and has grown into a fine, healthy and rather pretty infant.—*Obstetrical Jour.*

Chloral as an Anæsthetic in the Minor Operations on Children.

M. Bouchut (Gazette des Hopitaux—Medical Press and Circular) says that he uses chloral as an anæsthetic in the minor operations on children. He administers one, two, three and four grammes, according to the age ; two grammes might be given without danger between three and five years. The dose is given in four ounces of sweetened water, taken at once. In half an hour the child sleeps, and in an hour is perfectly insensible. This sleep lasts from three to six hours, and the child awakens as fresh as after natural sleep. Once insensibility arrives, a great number of operations can be performed, such as extraction of teeth, opening of abscesses, re-dressing of malformed limbs, etc., without any other inconvenience than that of leaving the children to sleep off the effects of the chloral. He says that he has administered this anæsthetic over ten thousand times without an accident.

TRANSLATIONS.

By G. J. Bull, M.D.

TREATMENT OF ABSCESS OF THE LIVER

At a meeting of the Academy of Medicine at Paris, Oct. 26, 1880, M. J. Rochard made an oral communication on the treatment of abscess of the liver by free incision under Lister's antiseptic method. The subject matter of this communication had been furnished him by Dr. Louis Stromeyer-Little, physician to the hospital of Shanghai, and by one of his patients, Dr. A——, and appeared to him to indicate considerable progress in the treatment of this grave disease.

The method consists in determining the seat of the purulent collection, verifying the diagnosis by means of the aspirator ; and then, with the aspirator needle as a guide, making a free incision into the abscess with a bistoury, evacuating its contents, and preventing subsequent accidents by the use of antiseptic injections, good drainage, and Lister's dressing.

The puncture is made with an aspirator needle of large caliber below the border of the ribs at the point of greatest tenderness on pressure. The part selected is first washed with a five per cent. solution of carbolic acid ; the needle, which has been immersed in antiseptic oil, is then plunged into the part to a depth sufficient to make it certain that the abscess has been reached. For this purpose it must penetrate to the depth of seven or eight centimeters, and it is often necessary to make several punctures before the pus is reached.

As soon as evidence of pus is obtained, it is the practice to open the abscess by means of a long bistoury without attempting to remove the pus with the aspirator. The incision should be made parallel to the ribs, and pierce at one stroke the entire thickness of the parieties. Then, to facilitate the escape of pus, it is well to introduce a strong forceps and separate its blades, at the same time making pressure on the lower surface of the liver, through the abdominal wall. After-wards the cavity is cleansed by irrigation with a one per cent. solution of carbolic acid, and a drainage tube is passed to its deepest part, and left in position. Lister's dressing is then applied with the utmost care.

Closure of the abscess cavity takes place with surprising rapidity ; the patient recovers his appetite, sleep and strength, and his functions become re-established.

The most remarkable fact is the disappearance of fever immediately after the operation, and the complete absence of febrile reaction during the following days.

In the three cases reported by the physicians of Shanghai, a cure was obtained in less than a month.—*La France Medicale.*

LEAD-POISONING.

Dr. Giulio Lepidi-Chioti has an article on lead-poisoning, in the " Morgagni" of July last, describing his clinical and anatomical researches on dogs and-other animals which he poisoned slowly with subacetate of lead.

The following are his conclusions :

1st. There are several sources of lead-poisoning that are not generally recognized.

2d. The symptoms of lead-poisoning are often difficult of interpretation.

3d. The diagnosis of saturnine poisoning may be made by the discovery of lead in the urine and fæces, as shown in these experiments.

4th. The experiments have also shown that the poison is eliminated in large part by the intestines and the kidneys, in the proportion of 3 to 2, whatever may be the date of the poisoning, or the quantity of poison eliminated from the organism.

5th. Poisoning by lead may take place through the respiratory tract, as was shown in the experiments on rabbits.

6th. In searching for lead in the tissues of animals that have been slowly poisoned, it is important to stop the administration of the poison some days before destroying life, and then wash the vascular system with water, after carefully removing the blood, in order to be able to recognize the tissues with which the metal has entered into combination. If this precaution is taken, lead will be found even in the marrow of the bones.

7th. The rarity of anatomical lesions, notwithstanding the gravity of the functional disturbances, establishes the fact that the alterations produced by lead are primarily chemical, and secondarily anatomical. Therefore we must study these changes chiefly from a chemical point of view, and afterwards make use of the microscope as much as possible.

8th. Clinical observation and these experiments have shown the great value of milk in the treatment of lead-poisoning ; also the value of iodide of potassium, and of baths containing sulphur and hypochlorite of soda. Nitrite of amyl is to be recommended in cases of lead-colic ; and nux vomica is especially useful in lead-palsy.—*Journal d' Hygiene.*

PURULENT INFECTION CAUSED BY CATGUT.

Professor Zweifel, of Erlangen, reports (Centralblatt fur Chirurgie, No. 12, 1879,) a case of death from purulent infection twelve days after an operation for a small vesico-vaginal fistula in which a catgut suture was employed. The autopsy showed that the purulent infection started in the pelvis. The instrument used in the operation had been carefully disinfected and all antiseptic precautions had been taken. The bad result could be attributed only to the catgut. Zweifel was confirmed in this opinion on reading an article in a foreign journal relating a case of ovariotomy in which the patient died of pyæmia, notwithstanding all the precautions of the antiseptic method. Microscopical examination showed the presence of bacteria in the catgut. Zweifel then proceeded to examine catgut that he kept soaked in carbolized oil for use in an ovariotomy, and found it full of bacteria. This would seem to show that bacteria enjoy an immunity against carbolic acid. However difficult it may be to say how these microscopic organisms penetrate the catgut liga-

ture, the author believes that they are developed even in well-stoppered bottles, by reason of the facility with which carbolic acid evaporates, especially in over-heated rooms. This may serve to explain a certain number of deaths from pyæmia occurring in spite of all the precautions of the antiseptic method.—*La France Medicale, Oct. 30, '80.*

BOOK NOTICES AND REVIEWS.

A Practical Treatise on Tumors of the Mammary Gland: Embracing their Histology, Pathology, Diagnosis, and Treatment.

By Samuel W. Gross, A.M., M.D., Surgeon to, and Lecturer on Clinical Surgery in, the Jefferson Medical College Hospital and the Philadelphia Hospital ; President of the Pathological Society of Philadelphia, etc. Illustrated by 29 engravings. New York : D. Appleton & Co., 1880. Pp. 246.

This work is a valuable contribution to medical literature, alike useful to the busy surgeon, who will glance hastily at the carefully prepared tables of differential diagnosis, and to him who will study at his leisure with a view to mastering a difficult subject.

Though it can hardly be said that the microscope has done as much for this subject as the opthalmoscope has done for ophthalmology, it is plain that modern histological researches have so completely revolutionized our knowledge of the various new formations that there is excellent reason for the appearance of this new and systematic treatise on tumors of the mammary gland. The book is written by the younger Dr. Gross, a man specially qualified for the work by his pathological studies and by his exceptionally fortunate position in being able to command a large amount of clinical material—sixty-five cases of cysts and nine hundred and two neoplasms, the nature of which has been confirmed by the microscope.

The first chapters of the book treat of the classification and relative frequency of the various tumors, of their evolution, transformations and etiology ; then each class is considered in a separate chapter, and, finally, a chapter is given to diagnosis, one to treatment, and one to the tumors of the mammary gland of the male.

"Not the least important part of the work," as the author states in his preface, "is that in which the view is sought to be maintained by an abundant array of facts, that carcinoma may be permanently relieved by thorough operations practiced in the early stage of its evolution." He is "aware that this doctrine will not meet with general acceptance on the part of those purely mechanical surgeons who believe that freedom from recurrence denotes an innocent neoplasm. In every case of final recovery mentioned in this treatise, however, the diagnosis was based upon minute examinations conducted by trustworthy microscopists, whose reports have been utilized to the exclusion of the descriptions of the early writers on carcinoma."

A Treatise on the Practice of Medicine, for the Use of Students and Practitioners.

By Roberts Bartholow, M.A., M.D., LL.D., Professor of Materia Medica and General Therapeutics in the Jefferson Medical College of Philadelphia; formerly Professor of the Theory and Practice of Medicine and of Clinical Medicine in the Medical College of Ohio, etc. New York: D. Appleton & Co. 1880. Pp. 853.

On perusing the preface to this work, the reader unacquainted with Dr. Bartholow's past achievements in medical literature might be somewhat repelled by the intense personality therein displayed. A careful reading of the book itself, however, would convince him that this egotism was turned to a good purpose, and that the author was more than justified in believing there was room for yet another "Practice of Medicine," for his is unique in style, and will occupy a field of usefulness that no present book can successfully compete in. It will doubtless be to the student and younger practitioner of the present what the admirable works of Dr. Tanner were to those of the past, and serve as a healthy protest against the neglect " to act in the living present," and against that proclivity to wrap the therapeutic talent in a napkin, that so many eminent pathologists

display. Dr. Bartholow has combined scientific research into the action of drugs with a large practical experience in their application to disease, and his book abounds, moreover, in many of what Dr. Milner Fothergill happily terms "wrinkles." We must admit, however, that disappointment to obtain *all* the happy results that the author promises for some of his remedies waits upon his followers; but his treatment is always rational, and he gives good reasons for the faith that is in him. The descriptions and diagnoses are concise, to the point, and good epitomes of our present knowledge of the subjects. In the treatment of gastro-intestinal disorders, we especially agree with his advocacy of small doses of arsenic under the various conditions named.

We are glad, also, to see favorable mention of euonymin and iridin as cholagogues; practical experience so far has confirmed the experiments of Rutherford, and they appear free from the drawbacks of most other bile-compelling drugs. We are unable to find any mention of the treatment of simple constipation, or of the proved efficiency of cascara sagrada in overcoming that difficulty. In the article upon pleuritis we are glad to see a protest against the exploded theory of the aplastic power of mercury lately revived under the authority of Ziemssen.

The section devoted to phthisis pulmonalis is an admirable resume of what is known of its pathology, and the treatment embraces many valuable hints; in the etiology, however, we highlanders are at issue with the doctor in the view expressed in the following sentences: "Variability of climate and rapid and extreme atmospherical vicissitudes have a most injurious effect on those having a tubercular diathesis." In the next sentence he writes: "Elevation and dryness are as conspicuously beneficial as the opposite conditions are hurtful to those having a phthisical tendency."

The first sentence is altogether too sweeping; the advantage or disadvantage of variability of climate, and rapid and extreme atmospherical vicissitudes, hinges upon the presence or absence of "humidity." In a damp climate these elements are undoubtedly highly injurious, but in a dry climate such as the Rocky Mountain region, we believe that *even great barometrical and thermometrical changes are markedly beneficial in their results*, for a confirmation of which opinion we refer to Dr. Julius Braun, in his work on "The Curative Effects of Baths and Waters," and, in fact, to most writers upon the subject of altitude and phthisis. With this belief, also, our clinical experience in Colorado is in accord.

Elevation and dryness must invariably be accompanied by extreme range of temperature and occasional high winds, and in or near the mountains sudden and violent storms of rain, hail or snow are common. In a dry climate the absence of watery vapors to retain heat or cold, makes the changes between day and night, sunshine and shadow, extreme. Elevated regions are necessarily cold when the sun is not shining, while, on the contrary, the sun's rays passing through a dry air suffer little absorption or obscurity from vapor, and consequently strike the body with peculiar directness. This property, therefore, exaggerates the natural change from cloud to sunshine. Then, on mountains or elevated plains the movement of air is mostly in extremes, and in the mountain valleys, though sheltered perhaps from most points of the compass, violent gusts of wind are common. In short, on elevated and dry regions the changes are for the most part sudden and extreme, and the sojourner there must be ever on the alert to counteract their effect with a handy wrap; but as the dry air does not rob his body of its electricity, his nerves act promptly to the stimulus of cold, the cutaneous capillaries are emptied, and the larger quantity of blood thus freed, keeps up the heat of the body, and the individual readily reacts from the shock of the storm. The success that has attended the treatment of phthisis by altitude, is, we believe, largely due to the impulse given by these atmospheric changes to the organism, thus causing a vigorous circulation of health-bearing blood in the

place of the Lethean stream that crawls through the veins of the consumptive.

In passing on to the subject of diphtheria, we see the author adopts the view that true croup and diphtheria are non-identical, and shows good reasons for his opinion, and we are inclined to side with him in spite of so much good evidence having been produced to the contrary. The recent experiments of Wood and Formad showing that the diphtheritic membrane can be produced by ammonia, or by non-specific septic matter, tend to make the question still more complicated. The use of lime-water or lactic acid we especially commend.

The whole subject of nervous diseases, as might be anticipated from the author's well known researches in that field, is treated with a masterly hand, though here, also, the process of condensation has been rigorously carried out.

In the treatment of cerebral hyperæmia, occurring from an over-worked brain, we would have liked to have seen the relaxation of wood-chopping recommended, for it has undoubtedly helped many a statesman and many a busy worker in the professions from falling under his mental burden. In the treatment of writer's cramp there are two practical points that we would have wished recorded, viz: that the continued grasping of metal on the pen holder seems to aggravate the symptoms, and that the covering of that portion of the holder with rubber lessens them; then, when the patient cannot give up writing altogether, it is best for him to learn to write with his left hand without delay.

In discussing remittent fever we see the doctor has decided that mountain fever is typhoid with a malarial element; we who witness it here are hardly prepared to accept that theory, though we are still in this matter but "infants crying for the light." We would fain draw attention more fully to the excellencies of this work did space permit. We can, however, most cordially advise its being placed upon the book shelves not only of the student, but of every busy practitioner, to whom its brevity, point and eminently practical character, especially commend it.

MEDICAL NEWS.

SULPHATE OF COPPER FOR THE ERUPTION CAUSED BY RHUS TOXICODENDRON, RHUS RADICANS, ETC.—Dr. A. W..Wiseman, of Jerusalem, N. C., has an article on this subject in the November number of the "Virginia Medical Monthly," in which he recommends the use of a solution slightly colored by the salt, as a local application, three or four times a day, gradually increasing the strength until it produces a slight stinging. A case is mentioned in which a negro slave was cured by this treatment after long suffering from rhus-poisoning, which the usual remedies had failed to relieve.

MEDICATION THROUGH THE UTERUS.— Dr. Robert Barnes, in a recent clinical lecture (Lancet, July 24, 1880), calls attention to medication through the genital mucous membrane. He says: "Of late years a method practiced by Harvey of washing out the uterine cavity when charged with noxious matter has come into vogue. The favorite at present in use is a solution of carbolic acid, and thus we get the twofold good of clearing the uterus of foul matter and of modifying the surface which produced it. But we may, I am persuaded, obtain a third good. We may by throwing carbolic acid, or, better still, iodine or quinine, or sulphite of soda, into the uterus, chase the septic matters through the tissues and vessels by which it entered. We may thus, whilst cutting off further supply of poison to the system, do much to neutralize the action of that which has already entered."—*Med. News and Abstract.*

SULPHIDE OF CALCIUM IN THE TREATMENT OF BUBOES.—Dr. F. N. Otis remarks: I have taught, up to within six months, for many years, that when a gland is enlarged as a result of the chanchroidal secretion, suppuration is inevitable ; a chanchroid is at that moment established in the centre of the gland, which goes on inevitably to suppuration and the formation of an abscess, the secretion of which is chancroidal. But the experiments which were made last winter in the Blackwell's Island Hospital, under my supervision, by the administration of the calcium sulphide in all cases of bubo associated with chanchroid, have led to the belief in the possibility of that suppurative action being arrested, because fifteen out of eighteen buboes so associated with presenting chancre were brought to resolution. The fact that the administration of one-twelfth of a grain of the calcium sulphide every two hours, combined with such other treatment as rest, iodine and pressure, resulted in the resolution of fifteen out of eighteen such cases, is sufficient to warrant its trial.— *Med. and Surg. Reporter.*

PULMONARY SYPHILIS SIMULATING TUBERCULOSIS.—Dr. R. E. Thompson, of London, has recently studied sixty well marked cases of pulmonary syphilis which had occurred in his practice. In all, the signs of pulmonary disease of a peculiar character were present ; they were associated with symptoms of syphilitic cachexia, and were relieved by antisyphilitic remedies. The signs might be distinguished from those of other pulmonary diseases, and were sufficiently peculiar to establish the nature of the disease. Briefly, these physical signs were dullness of percussion, and a peculiar alveolar rustle (resembling the crumpling of thin paper), with bronchial respiration and bronchophony of varying degree. These signs were not to be classed under the signs of phthisis, and the pulmonary condition indicated by them was notable for an absence of signs indicating destruction of lung-tissue. There was marked dyspnœa occurring after exertion, especially in raising the body upstairs or uphill. Hemoptysis of small amount was frequently present, and the expectoration was sometimes abundant. These characteristics of the disease were accompanied with thoracic tenderness and other evidence of syphilitic complications. The pathology of the disease was very obscure, inasmuch as the disease was very chronic, and seldom, if ever, fatal ; only one necropsy having been made by the author, and in this case death was due to other causes. The morbid condition of the lungs in this case was given in detail.—*Brit. Med. Journal.*

POST-PARTUM HEMORRHAGE.—This common and most formidable accident of labor was the subject of a spirited discussion in the Obstetric section of the British Medical Association at the meeting at Cork, in 1879. The subject was introduced in a paper read by Dr. More Madden, of Dublin, in which he strongly recommended as preventive measures during labor when there was reason to expect hemorrhage, rupture of membranes during the first stage ; stimulating enemata of strong infusion of ergot, hypodermic injections of ergotine in the second stage ; constant firm pressure on the fundus uteri from the time the child's head appears at the vulva until the completion of the third stage ; complete avoidance of the least traction on the cord. In actual hemorrhage would not trust to hot water injections ; advocates the use of perchloride of iron by saturated sponge introduced into the uterus ; thinks the method by injection dangerous. Spoke of the necessity for stimulants in collapse, brandy by mouth, hypodermically, or per rectum. Ether hypodermically. Thought he had saved a case in the last extremity by hypodermic ether ; thought that its use would occasionally obviate the necessity for transfusion. Transfusion may be at any time necessary, but we still lack a method which gives better results than the ruder methods of Blundall and others 50 years ago. Dr. Walter, of Manchester, gave the results of treatment of eleven cases by injection of water at a temperature of from 110° to 120°. Concludes that it possesses advantages ; it is ready to hand, cleanly, not disagreeable to the patient, but it cannot be relied on to produce permanent contraction. He recommended using the thermometer to test the temperature of the water, as Dr. Max Runge had by his experiments shown that a temperature of 100° to 104° Fah. was that which seemed to be most useful in producing permanent contraction. Dr. Norman Kerr, London, had found hot water 105° to 110° very valuable in arresting post-partum hemorrhage. Dr. Dill

recommended cold douche to abdomen. Dr. J. Thompson, of Leamington, advocated continuous flapping of the abdomen with a towel dipped in cold water. Had found it very effectual in keeping up contraction, and it did not wet the patient and the bed as in the case of the cold douche. Dr. Cordes, of Geneva, spoke highly of hypodermic ether, and hypodermic ergotine. Said that syncope might be avoided by postural (positional) treatment, and Esmarch bandages to the limbs, so as to retain as much blood as possible in the large blood vessels near the heart. Dr. Atthill, Dublin, said that ergot was unreliable except to anticipate hemorrhage. Cold was perhaps the most effectual agent if used in proper cases at the proper time, while the patient is warm and reaction likely to follow. When reaction ceased to take place the hot water at 100° came in usefully. Did not claim that it would always supersede the use of iron perchloride injections, but that it often would. Perchloride of iron was needed in some cases, he had used it several times, and saved lives by it, but had known one case in which its use was followed by instantaneous death : was not prepared to say from what cause, possibly air in the veins. Dr. Malins, of Birmingham, had used perchloride of iron by means of the sponge—thought was safer than by injection. Mr. Pollard, of Torquay, England, has had excellent results in a good many cases from large doses of turpentine.—*Canada Med. and Surg. Journal.*

MR. SCHAFER'S REPORT TO THE OBSTETRICAL SOCIETY ON TRANSFUSION. — Mr. Schafer's first work was to ascertain, by microscopic examination, the effect on the blood of other fluids than the blood of an animal of the same species.

1. *A weak solution of common salt* (1 oz. to the gallon of water).—This is innocuous to the white corpuscles, but renders the red corpuscles crenate, and prevents to some extent the formation of rouleaux.

2. *Cow's Milk.*—Fresh milk has no immediate ill effects on human blood-corpuscles, but if it is at all sour, it kills the white

corpuscles. Even if fresh, if the preparation has been kept for some hours, the same
effect takes place, and reaction is speedily
produced from commencing fermentation.

3. *Blood or Serum of other Animals.*—The
admixture with human blood of the blood
of many animals, especially the common
domestic animals, exerts a most deleterious
action on the red blood corpuscles. They
become decolorized and swollen, and the
coloring matter is discharged into the serum.
The white corpuscles die, as shown by the
cessation of the amœboid movements and
distinct appearance of nuclei. The blood
of the ox and sheep were found to have the
most rapid action; that of the rabbit and
guinea pig the least. This discharge of the
coloring matter leads to bloody urine, ec-
chymoses, fibrillation and embolism.

Salt solution, although it has been shown
to be innocuous, is nevertheless useless, as it
is not so much deficiency of quantity from
blood lost, but quality, diminution in the
number of oxygen carriers, the red corpuscles. This accounts for the dyspnœa of
these cases.

Effects of Milk Injection.—On rabbits, if
not previously depleted, the injection into
the vessels of even a very small quantity of
ordinary fresh milk (not sour) had a most
injurious effect. London milk almost invariably produced death in twenty-four hours.
The red corpuscles were extensively destroyed, and bacteria were developed in large
numbers. If the milk was boiled just previous to injection, the syringe scrupulously
clean, and with precautions to exclude germs,
milk could be injected in large quantities
without ill effect. The same result occurred
when the milk was allowed to spurt from the
clean teat of the animal into a vessel super-
heated. In depleted animals, milk thus injected never produced any permanent good
effects. For these reasons milk and other
similar fluids must be rejected for transfusion,
and we are reduced to the necessity of using
the blood of some other animal of the same
species, or at least genus. In man, the
blood to be transfused must always be human,

and this may be either in the normal or
defibrinated condition.

As to the best method, Mr. Schafer, after
full consideration of all the methods that
have been proposed, concludes that the
simpler the form of apparatus, the better;
that the simplest and best form of apparatus
is a short flexible tube, with glass canulas at
each end, this tube being used to connect
directly the vein of the giver to the vein of
the receiver, or an artery of the giver to an
artery of the patient; that the amount given
is to be regulated by the duration of flow.
Ordinarily, three or four minutes is enough
in the case of veins; from artery to artery,
half a minute to one minute. As to the
relative advantages of arterial and venous
transfusion, the latter is much the easier
operation; but in the case of a patient *in
extremis* from loss of blood, centripetal arterial transfusion may be expected to yield
the best results—results which, as Mr.
Schafer says, may be truly magical. "Even
if the heart should have ceased to beat while
the operation of inserting the tubes was
being effected, I doubt if resuscitation might
not result from the connection of the two
arteries." Blundell proposed this many
years ago. Failing the possibility of direct
connection of a blood vessel of the donor
with that of the receiver, which is always
more difficult than mediate transfusion, Mr.
Schafer says the latter method may yet be
tried, but it is more dangerous to patient,
by means of an elastic pump or syringe. In
this case the arm of the giver and the interior both of basin and instrument ought
to be washed with carbonate of soda solution in hot water. The blood ought to be
injected quickly, without defibrination, and
with every precaution to prevent coagulation. Injection towards the heart, as recommended by Blundell, ought to be preferred
in very urgent cases. If the injection is
into a vein, a funnel, with rubber-tube and
spring-clip attached, and with the canula
for the vein attached to the end of the tube,
is as simple and effectual an apparatus as can
be devised.

During the discussion which followed the reading of the report, the president, Dr. Playfair, said he had always used defibrinated blood with good results.

Dr. Braxton Hicks had used a mixture of saline solutions with blood to prevent coagulation. He had used phosphate of soda solution, one to three of blood, and, as in defibrinated blood, the whole preparation could be made in the next room, which was a great convenience. In the case of defibrinated blood, it must be very difficult to exclude minute clots from the circulation. Dr. Aveling believed immediate venous transfusion with his apparatus to be the best method, and considered the objections to his apparatus to be groundless. He wished to call the attention of the Fellows to auto-transfusion. By raising the legs and hips of the patient to an angle of forty-five degrees, enough blood might be made to flow towards the heart to preserve life. This method ought always to be practiced before transfusion is resorted to, especially as it conduces to arresting uterine hemorrhage.—*Canada Med. and Surg. Journal.*

MORAL INFLUENCE AND GENERAL MANAGEMENT VERSUS SPECIFICS IN THE TREATMENT OF PULMONARY PHTHISIS.—So much attention has been paid to climate and specific medication for consumptives that there seems a tendency to lose sight of other important features in its management.

As to specific medication, many remedies that are valuable in certain cases at certain times, are decidedly injurious in others. For instance—cod-liver oil, that so often proves beneficial to patients who can assimilate it without injury to digestion, proves prejudicial to those who cannot digest it at all, and who, by its administration, are incapacitated for taking and digesting other articles of food. So often does this occur that one is sometimes led to conclude that there is in the aggregate in the world more harm than good done by the indiscriminate use of this remedy.

So with the various astringent and sedative remedies for the cough ; while they soothe a little and procure temporary rest in most instances, and are highly beneficial and necessary where there is either functional diarrhœa, or that from intra-intestinal complication with tubercle, they prove decidedly injurious to those in whom there is habitual constipation and chronic gastric catarrh.

Neither is it in this connection clearly safe yet to forget everything else for the Salsbury plan.

[To summarize with regard to medication, the better rule would seem to be to avoid too much reliance on specifics and attend almost wholly to restoration of the usually disordered digestion, assimilation and elimination, and to the removal, if possible, of any complicating malady.

As to specific climates, it is not always safe to rely on locality alone ; but also in that connection on a proper adaptation of food, social enjoyment and occupation ; for, in the majority of cases, the benefits derived from change of climate are largely due to the coincident change of scene, to change from in-door to out-door life, from sedentary to more active habits, from brain-work, brain-worry and general nervous exhaustion to rest of the whole nervous system ; and the forced diversion of the mind of the patient from the gravity of the affection.]

Of course, probably no one supposes that all, and not many yet that even a majority, of the cases of pulmonary phthisis are curable ; but every one probably concedes that there is a sufficient number of cases so markedly amenable to treatment, and a sufficient number that are absolutely and permanently curable, to render the subject of advice and treatment for them still one of the worthiest in medicine.

No attempt is intended to be made here to discuss the pathology of phthisis ; but that there is a catarrhal or intra-alveolar, and also a fibrous or inter-alveolar form of invasion, neither of which is tubercular in its origin, seems probable. If so, inasmuch as they are purely local, and dependent on no special dyscrasia, it seems reasonable to suppose that arrest and permanent cicatrization is as likely to take place here as in inflammatory and ulcerative processes elsewhere. With regard to the tubercular variety, if the deposit of tubercle in the lungs be large and rapid, or exist elsewhere too, there is perhaps no hope from any treatment ; but where the deposit is small and the individual of good vitality, arrest is possible and does occur. Hence only in rare cases should the disease be considered and dismissed as incurable, or at least as not greatly amenable to treatment. [And one of the most important rules would seem to be to adopt no treatment or residence that will not be accompanied with the desired impression to be made upon the mind. For whatever may be the importance of the many other features in the management of curable cases, this transcends them all.

In advising people to leave home, it should be remembered that they differ widely in their natures and tastes. Those of romantic natures and more public social tastes are usually much benefited by change from the routine duties and associations of home, wholly apart from the specific quality of the new atmosphere. Of course, if there is no reasonable hope of improvement, it is a cruel thing to send patients away from comfortable homes ; but there are many cases offering a fair prospect for treatment, where the propriety of sending them to a distant health resort is also questionable. It will not usually do well to send patients of very domestic natures and habits to rough it alone among strangers in inferior hotels in strange places, no matter what the climate may be. They never acquire that very essential buoyancy of spirits and interest in life that comes of agreeable surroundings and familiar and congenial associations. Nor is it safe to send them away from cheerful companions to associate with persons, the majority of whom are afflicted with the same disease as themselves ; nor beyond the reach of that intelligent medical advice so essential to regulate the mode of living, and that so frequently assists to give the desired confidence and hope in life.

Whether at home or abroad, then, the associations, amusements and occupation should be so adapted to and impressed upon the over-apprehensive mind as to leave no time for dread, and the patient kept, if possible, as absorbed in surroundings and oblivious of self as a little child.]—*Dr. T. N. Reynolds in Detroit Lancet.*

Note.—The editor takes the liberty of placing in brackets certain portions of the above paper, which he thinks particularly valuable and deserving of emphasis.

VENESECTION IN THE TREATMENT OF HÆMOPHILA.—Mr. Henry Finch reports the following case in the Lancet, Oct. 2, 1880 :

A stout, healthy lady, between fifty and sixty years of age, was suddenly attacked in bed, at night, with profuse bleeding from the nose. I was sent for, and the patient, whose indomitable pluck and steady coolness under a full knowledge of the danger she was incurring, contributed much to the fortunate issue, told me that a brother had died in a few days, from uncontrollable bleeding following some very trivial skin wound, and that an aunt had frequently suffered from copious bleeding consequent on very inadequate exciting causes.

The patient was treated in the routine manner—astringents syringed into the nasal fossæ, ice, iron, and ammonia internally, and finally plugging the nares ; but all to no purpose. The bleeding, if it ceased for a time, soon came on with redoubled energy ; the patient at intervals vomited quantities of dark blood.

I was with the patient two nights, who, notwithstanding the increasing pallor, anæmia, and weakness, preserved her spirits wonderfully. After forty-eight hours of this continuous bleeding, it was evident that unless the hemorrhage could be stayed death would speedily ensue, and as a last resource, in a sort of empirical way, we determined to try the effect of diminishing the blood-pressure by opening a vein in the arm. The state of the patient, while it nerved us to adopt any probable means of staying the flow, made recourse to this particular remedy without previous experience an anxious proceeding. Mr. Sloman opened a vein in the bend of the left arm with some difficulty, owing partly to its small size, and partly to the presence of much superficial fat. The nasal bleeding stopped as soon as a little dark blood had flowed away. After making sure that no further loss was taking place, the vein was closed. The hemorrhage did not return, and from thence forward the patient made a protracted but perfect recovery.

In two other cases which I can call to mind—a gentleman of thirty and a girl of seventeen years, both with well-marked hemorrhagic diathesis, and with hereditary history of the complaint—I have tried venesection, and in both cases success has been immediate and complete.

ACUTE PHTHISIS—THE CURABILITY OF ITS ATTACKS.—-Dr. T. M. McCall Anderson (Brit. Med. Jour., Aug., 1880) says that by acute phthisis he means an acute pulmonary affection, accompanied by high and continued fever, running a rapid course, and leading invariably to more or less destruction of lung tissue, if the patient survived long enough. Three varieties of this disease are : (1) Acute pulmonary tuberculosis. (2) Acute pneumonic phthisis. (3) Acute pneumonic phthisis, complicated secondarily with the development of grey miliary tubercles.— The last two varieties he does not think can be distinguished during life. The first may be suspected when the disease sets in suddenly with high fever, great prostration, profuse perspiration, lividity and great acceleration of breathing, and when these symptoms are out of all proportion to the results obtained from a physical examination of the chest. In these cases, hitherto regarded as very hopeless, he has obtained excellent results from treatment of which the following is an outline : (1) Careful, skilled nursing, with constant feeding and stimulants in small quantities often, from four to ten ounces, daily. (2) Each night a subcutaneous injection of atropine, 1-100 to 1-60 grains. (3) Remedies specially adapted to the removal of fever : (*a*) Ice cloths to the abdomen. (*b*) Ten to thirty grains of quinine, in a single dose, once daily.

(*c*) A pill composed of one grain of quinine, half a grain of digitalis, and from a quarter to three-quarters of a grain of opium, every four hours. In addition to this, special symptoms, diarrhœa, constipation, and the like, must be treated on ordinary principles. Further, the treatment must be adapted to the surroundings of each individual case.

THE TREATMENT OF NEPHRITIC COLIC.— Dr. C. G. Stockton, himself a sufferer, says on this subject, in the Buffalo Medical and Surgical Journal, Nov., 1880 :

Sydenham enthusiastically advocates the use of beer " to cool off the ardent humors that remain in the kidneys and produce stone." Segalas and more recent writers reinforce this statement. I found that beer aggravated these symptoms, as did sour wine and cider. Most writers pronounce beer and all fermented drinks injurious. Sydenham having experienced personal benefit from the frequent use of cathartics, recommends the use of manna and lemon juice. (This acted unfavorably in my own case, where the stone was oxalate of lime.)

But diluents, recommended by the most ancient of medical authorities, still best subserve our purpose at this time. So much for a condition of comparative repose. It is after some unusual exertion that a sensation of uneasiness and qualmishness is experienced, followed soon by intense pain in the lumbar and inguinal regions of the affected side ; this shoots down into the corresponding testicle, which alone, or with its fellow, is violently retracted. Micturition is painful and almost continuous, the urine scanty, highly acid, of a grayish hue and mingled with blood. The gastric apparatus is greatly disturbed ; there is flatulence, colic, frequent dejections, tenesmus.

The pain is best controlled by morphia and atropia, hypodermically, and the application of heat, either by long continued general bath, or by rubber hot-water bags placed under the back and about the scrotum. I cannot speak too highly of the hot-water bag as an epithem. It is more efficacious than the bath, since it can be used continuously without fatigue to the patient. Enormous eating should be proscribed ; the meals should be small, frequently and regularly taken. Hard, particularly lime waters, should be interdicted ; but the drinking in abundance of wholesome waters advised.

Upon drinking Niagara water, I find the urine more acid and turbid. Drinking Bethesda, Apollinaris, or rain waters, increases and clears the urine, rendering it less acid. This is especially true of Bethesda water.

The Gettysburg Katalsyne is an agreeable water, and has a favorable reputation as a remedy in renal concretions, as its constituents would lead us to suppose.

Probably the springs of Vichy provide the best waters for cases of the uric acid diathesis.

The Celestine spring is most frequented by patients with gravel, though without any apparent reason, as the spring is less alkaline than its fellows.

The Malvern water is a favorite in England. It is almost perfectly pure ; so far as I know, the purest spring water on record. There is no end of good waters. I have for some time been drinking " Rosicrucian springs " water ; it comes from Maine, is very agreeable, and is said to be like Apollinaris without the gas.

TREATMENT OF PRURIGO BY PILOCARPINE.—From the observation of the fact that sufferers from prurigo feel relief when the secretion of the sudoriparous glands is active—as, for example, in summer—O. Simon (Allgem. med. Centr. Zeitung) has been led to try the preparations of pilocarpine, and of jaborandi itself, in this distressing condition. A very numerous series of trials have persuaded him of the beneficial action of this means of treatment. In adults he uses a subcutaneous injection of pilocarpine, or prescribes a syrup of jaborandi. The patients, soon after the administration of the medicine, are enveloped in blankets for two or three hours. In patients suffering from psoriasis the perspiration is very scanty, while in pruriginous patients it is very abun-

dant. The effects of this mode of treatment were abatement of the accustomed sense of pruritus, softening of the skin, and diminished tendency to relapse. In general the case did not last longer than a fortnight, and in very severe cases three weeks.—*Dublin Journal of Med. Science, April, 1880, from Lo Sperimentale, Gen. 1880.*

THE EPHELIDES OF PREGNANCY.—Newman recommends (Union Medicale) for these an ointment of chrysophanic acid, one part, to lard, forty parts, well mixed. Gently anoint the part, previously washed with soap and water ; then apply a piece of linen, to prevent staining. Repeat the application three or four times at two days' interval, being careful not to touch the eyelids and not to apply too strong an ointment on persons of delicate skin. The parts to which it is applied become red, then black ; the skin desquamates, and the stain disappears. The same remedy may be used for pigmentary stains occurring independently of pregnancy.

THE DIAPHRAGM : ITS FUNCTIONS.—Dr. W. S. Forbes (Amer. Jour. Med. Siences, July, 1880), from a careful study of the diaphragm, concluded : (1) The vena cava inferior opening, the highest point in the central tendon of the human diaphragm, holds a constant and fixed relation to the right anterior inferior border of the ninth dorsal vertebra. (2) That portion of the central tendon embraced by the base of the fibrous pericardium is prevented from descending in inspiration by the superior tendinous crura of the diaphragm, which are formed by the lateral parts of the fibrous pericardium ascending on either side in two planes, to be attached to the apex of the bony thoracic cone, and through the deep cervical fascia to the processes of the cervical vertebræ and to each stylo-maxillary ligament. (3) These superior tendinous crura of the diaphragm are connected together by transverse and oblique fibrous bands, thus forming a fibrous scaffolding for the support and protection of the heart and the great cardiac vessels. (4) The opening in the fibrous scaffolding for the lodgment of the ductus arteriosus in fœtal life is closed by the contraction of the muscular fibres ascending on the fibrous pericardium from the anterior left side of the diaphragm in the neonatus. (5) The blood in the pulmonary artery of the neonatus is forced into and through the right and left pulmonary arteries and into the lungs by the contraction of the right ventricle, which at this moment has its walls as thick as those of the left ventricle, and by the elasticity of the pulmonary artery, and is not drawn into the pulmonary arterial branches and into the lungs by the expansion of the lungs. (6) The superior fibrous crura and the fibrous scaffolding between them are made tense and open for the lodgment and for the protection of the heart and its great vessels, and for the promotion of the circulation of the blood through them by the contraction of the muscular diaphragm, independent of the descent of its lateral wings, though in the descent of the lateral wings of the diaphragm the vertical area of the thorax is extended. (7) In the contraction of the muscular diaphragm its descent is not necessary, as it contracts on its own planes, which may be supported by the contraction of the strongest abdominal muscles. (8) The diaphragm is rather an appendage of the circulatory apparatus, and not " essentially the chief agent in respiration."

How to Cure a Cold. — One of our readers who has been troubled with a severe cold on the lungs, effected his recovery in the following simple manner : He boiled a little wormwood and horehound together, and drank freely of the tea before going to bed. The next day he took five pills, put one kind of plaister on his breast, another under his arms, and still another on his back. Under advice from an experienced old lady, he took all these off with an oyster knife in the afternoon, and slapped on a mustard poultice instead. Then he put hot bricks to his feet and went to bed. Next morning another old lady came in with a bottle of goose-oil, and gave him a dose of it on a quill, and an aunt arrived about the same time from Eccleshall, with a bundle of sweet fern, which she made into tea, and gave him every half-hour until noon, when he took a big dose of salts. After dinner, his wife, who had seen a fine old lady of great experience on doctoring in High street, gave him two pills of her own make, about the size of a walnut and of similar shape, and two tablespoonfuls of home-made balsam to keep them down. Then he took a half-pint of hot rum, at the suggestion of an old sea-captain visiting in the next house, and steamed his legs with an alcohol bath. At this crisis two of the neighbors arrived, who saw at once that his blood was out of order, and gave him a half-gallon of spearmint tea and a big dose of castor-oil. Before going to bed, he took eight of a new kind of pills, wrapped about his neck a flannel soaked in hot vinegar and salt, and had feathers burnt on a shovel in his room. He is now thoroughly cured and full of gratitude. We advise our readers to cut this out and keep it where it can be readily found when danger threatens.—-*Students Journal.*

Temperature of the Breath.—Mr. R. E. Dudgeon has been trying some experiments on the temperature of the breath, and infers from the results that it is considerably higher than has generally been stated, and that it is variable. First, on rising in the morning, having ascertained the temperature of his body as shown by the thermometer in the axilla and mouth to be normal— about 98½°—he wrapped the thermometer tightly in a silk handkerchief and breathed upon it. In five minutes it indicated 106.2. At 7 p. m., after a brief walking exercise, and when he had eaten nothing but a spoonful of boiled rice, and drunk only half a glass of water and a mouthful of ginger-beer, his breath raised the mercury to 107°. Immediately after a dinner at which only water was drunk, a temperature of 108° was shown. At other times the thermometer would not rise, under apparently the same conditions, higher than 102° to 105°. He can suggest no way of accounting for these indications otherwise than by admitting that they show the actual temperature of the breath as it issues from the lungs. "If so," says Mr. Dudgeon, "it is by the breath that the system gets rid of its superfluous caloric." The experiments seem to show that the temperature obtained from the breath is higher when the surrounding air is warm than when it is cold, indicating possibly that more heat is passed off by the breath when less can escape from the general surface of the body.

THE AMERICAN PUBLIC HEALTH ASSOCIATION.—The eighth annual meeting of this association will be held in New Orleans in December, commencing Tuesday, the 7th, and continuing three days. Papers will be read on various topics relative to the subjects in view, and the following special questions will be discussed:

What are the best means of securing reliable and prompt information as to the presence and location of cases of such diseases?

What are the best means of securing isolation of the first or of single cases of such diseases, and what are the chief difficulties in securing such isolation?

Under what circumstances is it proper to declare such diseases epidemic in a place?

Under what circumstances is it proper to recommend the closure of schools on account of the prevalence of such diseases?

What precautions should be taken at the termination of each case as to : a, Care and disposal of the dead ; b, Disinfection and cleansing of the room and house ; c, Period of time at which it is safe to allow the convalescent to return to school or society?

The meeting will be entirely public, and papers on the questions referred to are solicited from members and others. It is requested that notice of papers to be presented be given beforehand to some of the officers, that the programme may be arranged. Dr. J. S. Billings, U. S. A., Washington, D. C., is president, and Dr. E. H. Jaynes, of New York city, secretary.

CREASOTE AS A THERAPEUTIC AGENT IN CHEST AFFECTIONS.—Dr. Reuss, of Paris, has devoted a good deal of attention to this remedy in phthisis, and has found it very effectual in his dispensary practice, where he was able to give it systematically, and formulate the results. The irritation which it is apt to set up after a time in the air passages, and also in the alimentary canal, he has found, in common with others, to be the great drawback to its general employment. He tried it in alcohol, then with cod-liver oil or glycerine, but in all these media he suspected that it was now and then the cause of a dangerous or even fatal exacerbation of throat or intestinal inflammation. He was led eventually to try balsam of tolu, and found it in all respects a safe and effectual solvent, counteracting, as one would expect, the irritating properties of creasote and its tendency to arrest expectoration. It is essential that the creasote be absolutely pure, the test on which he relies being that it does not coagulate collodion. He prescribes it in lozenges (dragees), the formula for each being :

Pure Balsam of Tolu 20 cent.   (3 grs.)
Pure Beech Creasote, 5 cent.   (¼ gr.)
Excipient,   q. s.

Two of these for a dose ; given at first night and morning, and gradually increased, sometimes up to ten lozenges in the day. He tabulates the results in twenty cases. Five patients in the first and three in the second stage were apparently cured ; three in the first and three in the second stage improved ; three in the second stage were unaffected, and one in the second and two in the third stage died.

More recently he gives a detailed account of a case which was apparently passing into the third stage when the above treatment was begun. The patient was a stone-cutter, æt. 40. He had been losing flesh for a considerable time, perspired profusely, and had severe cough with fetid purulent expectoration. Liquid rales, with other signs of a like import, were heard in both apices. Dr. Reuss gives an exact and detailed report of the patient's progress over a period of ten months till, on 1st April, he is stated to be regularly at work in the country, robust and strong, and with only a slight roughness of the respiratory murmur over the upper part of chest. It must be noted that he took latterly, in addition to the creasote, quinine wine and cod-liver oil.—*Glasgow Medical Journal, Sept., 1880, from Journal de Therapeutique, Aug. 25, 1879, and May 10, 1880.*

PROGRESS IN THE TREATMENT OF STRIC-TURE OF THE URETHRA.—Some remarks were made on this subject by Sir H. Thompson, at the annual meeting of the British Medical Association, in Cambridge, August, 1880. As illustrations of this advance during the last thirty years in England, the doctor mentioned five points :

1. A general recognition of the principle that a delicate and gentle manipulation of any instruments in the urethra is alone trustworthy or permissible, in the place of that which was formerly greatly prevalent, viz., that urethral obstruction might often be overcome mainly by force.

2. The substitution of very pliable and taper instruments for silver and stiff gum-elastic instruments in much of the treatment, both in ordinary and in continuous dilatation.

3. A more general acceptance of the doctrine that, given time, patience, and gentle handling, very few strictures should be met with which cannot be fairly and successfully traversed by an instrument passed through them into the bladder. At the same time, an undoubted improvement is to be noted in the mode of operating for those exceptional cases in which the surgeon fails to accomplish that object.

4. A more general acceptance of the doctrine that dilatation of the urethra, whether with or without incision, may be carried with advantage to a somewhat higher degree than had for some time previously been regarded as desirable.

5. The substitution of internal urethrotomy in some form for the application of caustics, and for external urethrotomy on a guide.

Each of the topics named is then consid-ered somewhat in detail. In connection with the subject of the "calibre," or "diameter," of the urethra, or the amount of its dilatability, he refers to Dr. Otis's revival of the theory of "the large diameter of the urethra." He records his sense of the value of this point, but he adds that "it is a very easy thing to damage irreparably some individuals by over-distending the urethra." Thompson also opposes another doctrine which is associated with the preceding, viz., that stricture of the urethra is permanently cured by complete division of all the diseased tissues affecting the passage. In speaking of the many methods of performing internal urethrotomy, he says that the principles which govern a sound procedure are more essential points for the surgeon to discover and to teach, than a consideration of small details. These principles he briefly states as follows : 1. The necessity for a physical examination before operating, to detect and estimate the narrowed portions of the urethra. This is best accomplished, in his opinion, by means of a series of metal bulbs on slender stems, taking care not to regard as changes of disease those points at which the urethra itself is naturally only slightly dilatable. These bulbous exploring sounds he invariably used, advocating them as essential to diagnosis, in his first work, twenty-six years ago. He still prefers them to any others, as safer, less irritating, and not less efficient than more complex instruments which have been devised. 2. The necessity for accomplishing a complete division of all the morbid tissue constituting the stricture, by an incision carried through it, no matter what part of the urethra, or how much of it, is involved in the disease. As a general rule, he thinks, this is most efficiently done by a slender blade, carried beyond the stricture and made to cut from within outward, this latter proviso being, however, an open question. The important point is that any alleviation of the patient's condition attained by operation will be transitory if any part of the narrowing be left undivided. 3. He regards it as essential, after such division, to place at once a full-sized catheter for some hours in the bladder, to insure a free outlet for the urine, and prevent all possibility of extravasation of urine into and through the incisions thus made. 4. The necessity for passing full-sized bougies subsequently, at occasional intervals, in order to effect free distention of the walls of the urethra, which lie in almost constant apposition, and so to prevent reunion of divided surfaces by the first intention. Finally, he declares that the great desideratum of the present time unquestionably is the discovery of a mode of treatment which shall permanently restore to the strictured passage its original dilatability ; and he adds that a thoughtful consideration of the pathological condition which constitutes organic stricture does not embolden him to hope that such a result can be insured by the application of any principles of action at present known to us.— *The British Medical Journal, August 28, 1880.*

ACTION OF BENZOATE OF SODA IN SCARLET FEVER AND TRUE DIPHTHERIA.—Dr. Demme makes the following statement in the yearly report of the Children's Hospital at Bern (Allegemeine Wiener Medizinische Zeitung, No. 24) : He has treated twenty-seven cases of diphtheria with benzoate of soda internally and externally. Internally, as large a dose as possible was given (5 to 20 grams daily, dissolved in 100 to 125 grams of water, with the addition of 1 to 1.5 grams of liquorice juice). The external application was made by sprinkling the diphtheritic patches with alcoholic solution of benzoate of soda by means of an ordinary laryngeal insufflator. The applications were repeated every two to four hours. If the local disease were spreading rapidly and the lymphatic glands of the throat were swollen, Dr. Demme injected benzoate of soda into the retromaxillary and submaxillary regions, and even into the swollen tonsils. Cold wrapping of the body was employed at the same time for the lowering of the temperature, and cooling baths when

there was severe fever. In septic forms of the disease, he administered cognac (5 to 75 grams daily). Of the twenty-seven cases which were treated in this manner, six, or 22 per cent., died, which must be called very favorable when the severity of the cases is considered. With regard to the special effect of benzoate of soda, Demme arrives at the following conclusions : 1. Benzoate of soda is an effective antimycotic, both as an internal and as an external application. 2. The application of benzoate of soda, in the form of insufflation on the infected spot, favors the section of the mucous membrane and essentially promotes the separation of the diphtheritic deposit. 3. A reduction of temperature is not produced by benzoate of soda. 4. In all his cases, Demme saw under the continued use of benzoate of soda an increase in intensity of the contractions of the heart, generally with diminution of their frequency, and increased discharge of urine. 5. Benzoate of soda had no effect on nephritis, with regard to the secretion of albumen. The doses, which produce a good effect in diphtheria, are, according to Demme, the following : for children from 3 to 6 months old, 2½ grams daily ; from 7 to 12 months, 5 grams ; from 1 to 2 years, 7.5 grams ; from 3 to 7 years, 12 to 15 grams daily. Demme never observed unpleasant symptoms after such doses.—*Lond. Med. Record, August 15, 1880.*

TREATMENT OF DIPHTHERIA BY CARBOLIZED CAMPHOR.—M. Perate has for the last two years used carbolized camphor for the treatment of diphtheria. He paints the surface with a pencil dipped in the following mixture (Bulletin de Therapeutique, July 15): Carbolic acid, 9 grams ; camphor, 25 grams ; alcohol, 1 gram, diluted with equal parts of oil of sweet almonds. The paintings are made every two hours in the day, and every three hours in the evening ; then, after some days, they are divided by periods of three, four, or five hours, according to the improvement of the patient. These paintings are made over the whole extent of the false membranes, and with troublesome children the pencil is plunged as deeply as possible to the bottom of the throat, being, of course, previously drained. The mixture has an extremely disagreeable taste, to which, however, the patient soon becomes accustomed. M. Perate has been very successful with this plan of treatment.—*Lond. Med. Record, August 15, 1880.*

BOOKS AND PAMPHLETS RECEIVED.

"The Journal of Psychological Medicine and Mental Pathology." Edited by Lyttleton S. Forbes Winslow, M.D., D.C.L. Publishers, Bailliere, Tindall & Cox, London. [Placed on our exchange list.]

"The Journal of Nervous and Mental Disease." Edited by J. S. Jewell, M.D., Chicago. Associate editors: W. A. Hammond, M.D., Meredith Clymer, M.D., New York; S. Weir Mitchell, M.D., Philadelphia. [Placed on our exchange list.]

"Annals of the Anatomical and Surgical Society." Edited by Charles Jewett, M.D., associated with E. S. Bunker, M.D., G. R. Fowler, M.D., L. S. Pilcher, M.D., F. W. Rockwell, M.D. Brooklyn, N. Y. [Placed on our exchange list.]

"The Obstetric Gazette." A monthly journal devoted to obstetrics, with diseases of women and children. Edward B. Stevens, A.M., M.D., editor and publisher, Cincinnati. [Placed on our exchange list.]

"Canada Medical Record." A monthly journal of medicine, surgery and pharmacy. Editor, Francis Wayland Campbell, M.A., M.D., L.R.C.P., London. [Placed on our exchange list.]

"The Ohio Medical Recorder." Editors, J. W. Hamilton, M.D., J. F. Baldwin, M.D. Cott & Hann, publishers, Columbus, Ohio. [Placed on our exchange list.]

"The Medical Summary." A monthly journal devoted to practical medicine, new preparations, etc. R. H. Andrews, M.D., editor and proprietor. [Placed on our exchange list.]

"Hygienic and Sanative Measures for Chronic Catarrhal Inflammation of the Nose, Throat and Ears." Part I. By Thos. F.

Rumbold, M.D. Publishers, Geo. O. Rumbold & Co., St. Louis.

"Naso-Pharyngeal Catarrh." By Martin F. Coombs, M.D. Professor of physiology, ophthalmology and otology in the Kentucky School of Medicine, etc. Publishers, Bradley & Gilbert, Louisville, Ky.

"The Brain as an Organ of Mind." By H. Carlton Bastian, M.A., M.D., F.R.S. Professor of pathological anatomy and of clinical medicine in University College, London; physician to University College Hospital and to the National Hospital for the Paralyzed and Epileptic. Published by D. Appleton & Co.

"Journal d'Hygiene." Published by Dr. P. de Pietra Santa, Paris. [Placed on our exchange list.]

Publications of "La Societe Francaise d'Hygiene," viz., "Societe Francaise d'Hygiene, sa raison d'etre, son but, son avenir," "Les Hospices Marins, Les Ecoles de Rachitiques," "L'Annuaire pour 1880."

"Index Catalogue of the Library of the Surgeon-General's office, U. S. Army," Vol. 1. A—Berlinski, Washington: Gov't Printing office, 1880.

"The Proceedings of the Medical Society of the County of Kings." Conducted by the Council; Issued monthly; Brooklyn, New York. [Placed on our exchange list.]

"The Medical Gazette," a weekly Journal of Medicine, Surgery and the Collateral Sciences. Chas. L. Bermingham & Co., publishers, New York. [Placed on our exchange list.]

"Quarterly Epitome of Practical Medicine and Surgery," being an American Supplement to Braithwait's Retrospect. Part II, June, 1880. W. A. Townsend, publisher, New York. [Placed on exchange list.]

The "Philadelphia Medical Times," a bi-weekly Journal of Medical and Surgical Science. Edited by Horatio C. Wood, M.D., Philadelphia. [Placed on exchange list.]

"The Detroit Lancet," a monthly exponent of Rational Medicine. Edited by L. Connor, A.M., M.D., Detroit. [Placed on our exchange list.]

"La France Medicale." Published every Wednesday and Saturday. Edited by Dr. E. Bottentuit, Paris. [Placed on exchange list.] "The Druggists' Circular and Chemical Gazette," New York. [Placed on exchange list.]

"A Short Course in Qualitative Chemical Analysis." By John H. Appleton, A.M. Philadelphia: Copperthwait & Co. 4th edition; 1879. Pp. 112.

"The Young Chemist:" a book of Laboratory Work, for Beginners. By John H. Appleton, A. M. Philadelphia: Copperthwait & Co. 2nd edition; 1878. Pp. 110.

"School and Industrial Hygiene:" one of the American Health Primers. By D. F. Lincoln, M.D., Phila. Presley Blakiston, 1880. Pp. 144.

"On the Production and Reproduction of Sound by Light." By Alexander Graham Bell, Ph. D. Reprint from the American Journal of Science.

"Seven Cases of Retroflexion of the Uterus, with Peritoneal Adhesions of the Fundus, treated by Forcible Separation of Adhesions." By Aug. F. Erich, M.D. Reprinted from the American Journal of Obstetrics and Diseases of Women and Children. New York, 1880. (From the author.)

"A Device to Facilitate the Removal of Deep Wire Sutures in the Operation for Ruptured Perineum." By Aug. F. Erich, M.D. Baltimore, Md. Reprint from Maryland Medical Journal. (From the author.)

"The Cincinnati Lancet and Clinic:" a weekly Journal of Medicine and Surgery. [Placed on our exchange list.]

"The Chicago Medical Journal and Examiner." Editors, N. S. Davis, M.D.,LL.D., James Nevins Hyde, A.M., M.D., Daniel R. Brower, M.D. Published by the Chicago Medical Press Association. [Placed on our exchange list.]

"Saint Louis Clinical Record:" a monthly Journal of Medicine and Surgery. Edited by Wm. B. Hazard, M.D. [Placed on our exchange list.]

"St. Louis Courier of Medicine and Collateral Sciences." Published monthly by the Medical Journal Association of the Mississippi Valley. [Placed on exchange list.] "The Virginia Medical Monthly." Landon B. Edwards, M.D., editor and proprietor. Richmond, Va. [Placed on exchange list.]

"New York Medical Eclectic," devoted to Reformed Medicine, General Science and Literature. Edited by Robert S. Newton, M.D., and R. S. Newton, Jr., M.D. Published monthly by the Eclectic Med. College of the City of New York. [Placed on our exchange list.]

"The American Journal of the Medical Sciences." Edited by I. Minis Hayes, A.M., M.D. Philadelphia: Henry C. Lea's Son & Co. [Placed on our exchange list.]

"A Paper on Some Impurities of Drinking-Water." By Professor W. G. Farlow, M.D., of Harvard University. Reprinted from the Report of the State Board of Health, Lunacy and Charity of Massachusetts. Boston, 1880.

Atresia of the Genital Passages of Women. A paper read before the Chicago Medical Society, July 19th, 1880. By Edward W. Jenks, M.D., LL.D. Pp. 24.

"The New York Medical Journal." Edited by Frank P. Foster, M.D. New York: D. Appleton & Co., publishers. [Placed on our exchange list.]

"The Medical Brief." A monthly journal of practical medicine. J. J. Lawrence, A.M., M.D., editor and proprietor. St. Louis, Mo. [Placed on our exchange list.]

WE cannot refrain from publishing the following extract from a letter received by the editor of the "REVIEW," from Professor Le Conte, the eminent scientist of California:

"Allow me to say that a somewhat careful examination of the copy sent me impresses me with the fact that .your journal is far in advance of any except the very best in the country. I wish we could show anything like it in California. I will take pains to show it to my medical friends in Oakland and San Francisco.

"Very respectfully yours,
"JOSEPH LE CONTE."

ROCKY MOUNTAIN MEDICAL REVIEW.

Vol. I.   COLORADO SPRINGS, DECEMBER, 1880.   No. 4.

ORIGINAL ARTICLES.

OBSTETRICS.

Report of the Committee on this subject read before the Colorado State Medical Society, June 30th, 1880.

By H. K. STEELE, M.D., OF DENVER.

The object of this paper, is to present a few points of personal experience, on a subject not yet exhausted, very old, but still new to the busy practitioner and fully as important as any other we may contemplate.

The first practitioner of medicine seems to have made a specialty of obstetrics, and the reporter of this first case on record, observes that terseness and simplicity of description, that comprehensiveness and candor in detail, which classes it as a model report, that might well be copied by the greater number of those who have succeeded him.

His report is in the following words: "And Adam knew Eve, his wife, and she conceived and bare Cain." To have entered fully into the particulars of this first labor, would have been certainly excusable and exceedingly instructive and entertaining, and especially would it have been interesting to know, whether the curse, "In sorrow shalt thou bring forth children," rested as heavily upon our first parent, as upon those of the succeeding ages, who have followed them. Has civilization with its attendant evils heightened the sorrow? Have the ills that are accumulating upon the human race, as generation entails its diseases upon each succeeding generation, so increased the woes of maternity that the method nature has taken to propagate our species, is wrought out with more sorrow and pain, and difficulty each year?

It is the effort of the obstetrician, to so extend his aid that this grievous sorrow may be alleviated, and the mission of our profession is not accomplished until it attains this end. The idea advanced by many, that great distress is necessary to be borne, in order to give a healthful progeny, and to secure a successful labor, is barbarous, and no more logical than to contend that a man should submit to a great loss of blood and suffer the pain incident to a necessary amputation, in order to make a good recovery from it. The bandage of Esmarch, the anæsthetics, ether and chloroform, give life and health, they do not destroy them.

There are undoubted grounds for the belief that nature intended labor to be attended with great sorrow, for do not physiologists narrate this singular fact, "That the same system of nerves supplies all the hollow viscera of the body, and while the contractions of that hollow viscus, the womb, in its healthy, physiological action, are always attended with pain, pain which is beyond description in its severity, the contractions of the distended bladder, another hollow viscus, are attended generally with sensations of an agreeable nature, and the contractions of the heart which are unceasing and force through it, in an ordinary life-time, half a million tons of blood, are accomplished for the most part without the consciousness of the individual."

Yet with this evidence against us, ought we not to hope that before the grave of the 19th century is filled, some heaven-endowed benefactor of his race, will have arisen to point out the remedies to relieve the pangs of travail? Whether or not we have accomplished much in this line, since Adam first began many thousand years ago, still let it be said of us, we labor on with that end steadfastly in view. And now it becomes us to

enquire, where we stand in the line of progress, and what advances we have made this
year.

FORCEPS.

First and pre-eminent what of the forceps?
From discussions held, since our last meeting, over the land, in which many of the
men eminent in this branch of medicine have
taken a part, we cull the following among
many other important points that seem to be
conceded.   In speaking of the effects of
forceps delivery, it was agreed :

1st. That it had rarely ever any agency in
the production of vesico-vaginal fistula, when
performed with ordinary skill.

2nd. That the direct cause of vesico-vaginal
fistula, was the delay in delivery, after impac-
tion had taken place.

3rd. That in the hands of an experienced
manipulator, many lives would be saved,
much injury to the mother prevented, and
great suffering relieved, by the judicious early
use of the forceps.

4th. That the exact time at which the
forceps should be applied, is when the head
ceases to recede, as well as advance, after
the occurrence of a pain.

5th. The force employed should be mainly
extractive instead of the old rule, two thirds
extractive, one third lateral.

6th. A precautionary step, that should always be adopted, in cases of instrumental
labor, is to introduce the catheter and
empty the bladder before proceeding, as that
is in most cases the indirect cause of vesico-
vaginal fistula.

In passing, we would suggest the necessity
of emptying the bladder, in case of tedious
and protracted labor, as a full bladder is
known to be the cause in such labor, of producing still born children, as well as the other
evil, vesico-vaginal fistula.

We cannot content ourselves with the belief, that we are holding a neutral, or safe
position, when in case of an impacted head,
we withhold the use of the forceps.   Experience teaches that we must expect great
injury to be inflicted upon the maternal parts,
unless we rise to the assistance of nature,

when duty calls us to perform that, which
we would naturally shrink from.   These sentiments are alongside those of the advanced
profession throughout the land, and the forceps have a record of being superior as life
and labor preservers, to ergot, to manual version, to instruments used for the destruction
of the child, and to unassisted natural protracted labor.

When the forceps are brought to that state
of mechanical perfection, in which the force
of traction will not interfere with the acts
of rotation, flexion, and extension, and will
always be in the direction of the axes of the
pelvis, no accoucheur will be justified in not
becoming an expert in their use.

The statistics of labor in Colorado will,
we think, coincide with the assertions above
made, and although we may not all be skilled
in the use of the forceps, yet if we mistake
not, the majority of us are ready to agree
that they are more safe and satisfactory than
any of the other means resorted to to facilitate labor, or to the delay we submit to by
inaction.

ERGOT.

We will not draw a comparison here, between the use of ergot and the forceps, for
that has been thoroughly enough discussed,
but take occasion to say, that when we can
escape the use of ergot, we should do so. The
disagreeable consequences that often follow, to
mother and child, in the wake of large doses
of ergot, do not seem to compensate for its
selection.   In addition to the nausea that
frequently attends its administration, and its
inertness, when nervous prostration is extreme, or physical exhaustion from loss of
blood exists, the production of one of its
recognized physiological effects is very har-
assing ; namely, the difficulty attending micturition after labor, and the frequent calls for
the use of the catheter.

We should not forget to add to this, the
effect of the closure of the os, by the oxytocic
action of the drug, commencing in the lower
segment of the uterus, thus often impeding
the delivery of the placenta and inducing the
very evils we are seeking to combat ; retained

placenta, attached placenta, hour-glass contraction, etc., etc.

As compared with manual version, the opinion of the profession is almost unanimous in giving preference to the forceps, where the choice is offered, and as to embryotomy no controversy exists.

POST-PARTUM HEMORRHAGE.

What have we learned this year in reference to post-partum hemorrhage? It is well to devote a few moments to the consideration of this subject, for the point has been made by practitioners in some of our neighboring cities, and I am forced to concur in it, that there is more tendency to post-partum hemorrhage, in mountainous regions, where we now live than in the lower levels where we have been accustomed to practice. There is no doubt that there are but few of us, who have not at one time, or another, been appalled at the blanched face and sightless eyes of our patient, from whom was pouring in a gurgling stream her life blood, and have been almost as incapable, for the moment, as she, of checking its flow. We turn in anxiety from one remedy to another to afford relief. Those who have lost no patients, are ready to fly to the same remedies that have served them so well before. But it is well to know, that all of these expedients may prove inefficient sometimes. We apply pressure, but the womb obstinately refuses to respond. Ice is resorted to, as it has proved at times one of the most reliable and safest remedies. We make pressure with it on the abdomen, we apply it to the vulva, we push it forward into the womb, place it on the feet, and sometimes our efforts are successful, but it may happen they are not. We therefore crowd our hand into the womb, never for a moment letting go our hold externally on the fundus through the walls of the abdomen, but these means fail. We are compelled now to resort to expedients that are considered more effectual, although not so free from danger. Crouching over our patient and sweating from every pore, we ask ourselves shall we inject the womb with powerful astringents. We have read of the danger, and hesitate ; death

approaches. We remember that water at a temperature of 110° Fahrenheit has been highly recommended. This is certainly safe and we try it, but the remedy fails. In this emergency, Dr. Penrose, of Philadelphia, whose reputation entitles him to the greatest respect, recommends the following, which has been universally successful in his hands ; and in his extensive practice, he reports that he has not lost a case since he began its use. His formula is this : "Saturate a rag with common vinegar, carry it into the cavity of the uterus and squeeze it. It is easily obtained, easily applied, always cures, excites the most sluggish womb to contraction, and yet is not so irritating as to prove injurious, and withal it is antiseptic and astringent. It seems really to possess all the qualities we need, and as such, recommends itself very highly."

Now although Dr. Penrose is not the real author or originator of the vinegar treatment, Paul of Ægina, Leroux, Des Granges and many others, having preceded him in it, yet he may consider himself the author of the manner of application, and as such this remark of Desormeaux seems to apply : " It often happens that men, even those who are otherwise worthy of credence, are more successful with remedies of their own invention than any one else."

Each one's experience, is often, to himself, a better guide than that of others far more learned and skilled than he. This may result from the manner in which such experience is made available, or from the capability of the person to employ it. Notwithstanding the great recommendation of so able a man as Dr. Penrose and the eminent success with which he and others have employed it, it has proven a failure in my hands. In two instances within the last six months, I have almost sacrificed two lives, depending upon vinegar to control the hemorrhage. In both instances, one pint of sharp cider-vinegar was brought in contact with the entire cavity of the womb without producing any appreciable good effect. In one case after its application the radial pulse ceased to be perceptible, the patient felt death fast approaching, committed her children to the care of her husband, and awaited the arrival of death. Cold, ergot, pressure, introduction of hand within uterine cavity, vinegar, had been employed, all unsuccessfully; the flow continued. In this emergency, as a last resort, Monsel's solution of persulphate of iron, 1. part to 4 of water, was thrown up to the fundus of the womb, through a gum catheter; two fingers being retained within the os as a precautionary measure, to permit the reflow from the womb.

The result of this treatment was a complete and entire suppression of all hemorrhage, contraction of the womb, and restoration to life and health without an unfavorable symptom. The second case was similar in most respects, hemorrhage from inertia, relieved in like manner. Monsel's salt was selected because of its more astringent and less caustic property than the other preparations of iron. The solution of the perchloride is not so valuable, because of its known property of requiring about 30 seconds to coagulate the blood after it is applied, which may render the effects entirely nugatory in rapid hemorrhage. It possesses also caustic properties, in greater proportion than the persulphate.

That this treatment is contrary to the views of many prominent obstetricians we are aware, some of whom have gone to extreme lengths in their denunciation; but it is supported by such reputable authorities as Fordyce Barker, Robert Barnes, Outrepont, Kewisch, and others.

Barker says: "Inject very carefully and without force, into the uterine cavity a half ounce of the solution of the persulphate of iron diluted with an equal quantity of water." Barnes gives high commendation of the remedy, using however the perchloride more frequently than the persulphate. He says four ounces of the liq. ferri perchlor. fortior. may be dissolved in twelve ounces of water for the injection, or one half ounce of solid perchloride or presulphate dissolved in ten ounces of water. He summarises his views

in these words: "I have found nothing of equal efficacy to the injection of perchloride of iron into the uterus, after clearing out the cavity of placental remains and clots. I have used it in a large number of cases after labor and abortion and have always had reason to congratulate myself on the result. The perchloride has the further advantage of being antiseptic."

It is a safe principle in medicine to cure your patient of the disease which threatens life, even if you have to resort to the most fearful measures, and take your chances of combatting the evil that may follow, rather than fold your hands in submission to what seems the inevitable. Desperate cases require desperate remedies.

That all cases of post-partum hemorrhage do not arise from inertia of the womb we are well aware, and some that prove the most dangerous are unsuspected, because they occur with a contracted womb. The abnormal location of a blood vessel, that may be ruptured, or laceration of the neck sometimes produces the most fearful flow. In these cases astringents prove most valuable and afford another argument favoring their use.

POSITION.

Much has been said of late in reference to the posture the parturient female should assume, and we believe it is a question of much more importance than generally considered, and we do not hesitate to affirm, that much comfort may be afforded the patient and labor may be facilitated by selecting for her the position she should take. No arbitrary rule can be laid down in regard to this, nor can we say this position or that is the best. A position suitable to a woman, in one confinement, may not be adapted to, her condition in the next, nor need it be the best for another woman. The same position may not be retained throughout the whole course of a labor, with profit either, and in the great majority of labors we find it beneficial to have the position changed. This knowledge is not to be gathered from the woman herself, but the accoucheur with his finger on the child's head should be the magister, who should direct the changes and guide the position as the helmsman does his vessel after consulting his chart.   Does the accoucheur find, for instance, that rotation is performed tardily, let him direct the woman to turn to the side towards which he would have the head proceed, and he will find that he has thus the aid of the body of the child to assist nature in the work she would accomplish. Is the axis of the womb, not in the direction of that of the strait through which the child should pass, let it be corrected by the intelligent direction of the accoucheur. Does a pendulous abdomen retard the progress, the proper position will correct that dystocia.   Does the head come to a stop in passage through the straits by some unaccountable circumstance, let a change in posture be assumed and often the difficulty will be overcome and labor will proceed as usual, and thus we are prepared to say that an occasional, if not a frequent change of position, will often quicken the slowest labor.   Does not the experience of every one teach him, how rapidly, lagging labors have been converted into quick ones, by a fortunate change of position?

When the head is pressing upon the perineum and about to emerge, then we are prepared to say, the lateral position is the best and should be retained until the labor is completed, and this for two reasons :

1st. It will be found that while the patient is in the dorsal decubital position, any effort to raise the hips, or pelvis, or pubic region, is attended with a contraction of the perineal and levator muscles and thus the perineum is made tense and an obstable offered to the passage of the head.

2nd. The bed on which the woman rests is often hollowed out and thus a resistance is offered, which is often quite effective.

This question of posture, is by no means a new one; every age and every people have had their superstitions and views on this subject. We remember that some stood upon their feet, some knelt and bent forward, some stooped as at stool, some suspended themselves by their hands, some were held in the erect position by their husband clasping them around below their breasts, some sat in a chair or in the lap of a friend, some posed as instinct taught them; but we contend that the proper position is that, in which they should be placed, by the intelligent judgment of their physician, who directs each case, as the progress of labor indicates.

LACERATED PERINEUM.

Very many practitioners seem unwilling to admit that they encounter any lacerated perineums in the cases which they have attended, and you hear it spoken by some, and written by others, that out of the many thousands of cases of labor, that have fallen into their hands, scarcely a dozen have suffered with laceration.   Happy practitioners ! fortunate patients !   Were it not for these statements we would be ready to say that a very large number of women in childbirth, suffer from a laceration of a greater or less extent, and that in a majority of all these instances, the parts readily heal and no permanent inconvenience follows.   That very few are lacerated, so as to involve the sphincter ani, does not admit of a doubt, but as to there being no ruptures we are incredulous.

It is gratifying to us to know that some of the best authors recommend delay in the operation to relieve the rupture, until nature has time to show what part she can perform in the reparative process.   When nature fails, let surgical science commence its work. Until the operation is simplified (which we have reason to believe will be done before long) we think the results oppose immediate operation to cure an imperfectly lacerated perineum.   It is time to dissolve the myth, that the numbness of the parts immediately following labor destroys the sensation of pain. We have yet to find that benumbed condition existing, and generally the lacerated parts are very tender.

LAPARO-ELYTROTOMY.

This operation, a substitute for Laparo-Hysterotomy or Cæsarian section, as revised by Dr. Thomas, of New York, has met with much favor, and inasmuch as it aims to avoid the dangers arising from peritonitis, metritis, and incarceration of the intestines in the uterus, is certainly worthy of our consideration and trial, even in those cases where we would shrink from the performance of the Cæsarian section. As practitioners it is our duty to become familiar with all these operations whose specific object is the immediate preservation of life. In our practice we may not be called upon to operate previous to the death of the mother, although such occasions have overtaken some of us, and we may not be called upon to remove an undelivered child from the body of its dead or dying mother, but we should not forget that duty may demand it, and lead us to remember that well authenticated cases exist where the child has been delivered alive, one hour after the mother's death.

THE CORD—THE BINDER—THE WASHING OF THE CHILD.

Attempts have been made at innovations upon the customs of the profession, in reference to tying the cord, washing the child and the application of the binder, indicating that the cord should not be tied, the baby not washed, and the binder not applied. It will not do to dignify these ideas with the title, new, for the Bible records in Ezekiel, 16th chap., 4th verse, in the shape of a curse upon Jerusalem, a great many years before Christ, that "In the day thou wast born, thy naval was not cut, neither wast thou washed in water to supple thee, thou wast not swaddled at all, nor salted at all." Little difference perhaps it makes how the cord is treated, whether it is bruised or bitten in two by the father, or cut with the scissors and tied ; but the application of the binder we do not consider so insignificant a matter. We confess to great surprise that some advocate the omission of the binder, and we cannot recognize as tenable the reasons they advance in advocacy of their neglect. The almost universal expression from the parturient woman, when the binder is carefully, and we may say properly, applied, "Oh how comfortable that

feels, Doctor," is a sufficient warranty to me, that benefit accrues from it. The healthful pressure it exerts on the emptied abdomen, the steadiness that it imparts to the tender womb, are the strongest evidences we can ask that it is almost indispensable. Have we not seen on the removal of large tumors from the abdomen, when the pressure is thus suddenly taken off from the large vessels and viscera, syncope result ? And have we not been compelled to replace the tumor, or apply pressure temporarily, until a gradual distribution of the blood has taken place ? This part, the binder, after parturition, performs. The pressure is removed, when the eight or ten pound child is delivered, the heart and brain feel it, an unaccustomed vacuity exists. Relief is afforded by the application of the binder. Who will deny the woman this compensation ? Apply it properly and she will reward us with thanks. Apply it improperly, too tight, or too loose, too much like a ligature, or too nearly like a rag, and she will feel discomforted, and we can detect no benefit arising from its application. Under such circumstances we may well decry the bandage. Cazeau speaks of the efficacy of the binder in preventing and relieving syncope after rapid labors. Barker says one of the exciting causes of puerperal convulsions after labor, is the sudden change in the circulation following the removal of long continued pressure on the great abdominal vessels. Tyler Smith says he has known cases where he has been obliged to attribute a fatal result from hemorrhage to the neglect of a binder after delivery. Lumley Earle, a distinguished obstetrician of London, also forcibly urges this same opinion. As to the washing of the child we may set it down to the question of taste. If we want a clean child we had better wash it.

We have thus briefly, Mr. President, opened up a few points in the great field of obstetrics, that have particularly attracted our attention the past year. Many other equally important ones have been omitted, that perhaps would have interested the Society more, but in those that have been touched upon, there is abundant room for thought, and the time appropriated to their consideration, simple as they may appear, may not have been misspent.

IS QUINIA AN OXYTOCIC?

By C. BARDILL, M.D., of LONGMONT, COLO.

For years past and even to this day it has been considered by some physicians to be dangerous to administer quinia in large doses to pregnant women, for fear of its causing miscarriage. So far as my experience of its use in obstetrics during ten years is concerned, I have come to the following conclusions:

First, that it acts as an oxytocic in labor at full term.

Second, that its action is the reverse in cases of threatening miscarriage, due to reflex action or other causes not mechanical.

The following cases taken from my note book may have some weight in substantiating the above theory.

In 1870 I was practicing medicine in Missouri in a locality where malaria abounded throughout the year, and where consequently quinia was indicated in every case of sickness. I therefore resolved to try the effect of large doses of the drug in the first case of pregnancy occurring in my practice. I had not to wait long. One evening I was called into the country in great haste to see Mrs. F. According to her statement she was then in the sixth month of her second pregnancy. She had had a chill daily during the last five days. Three hours previous to my arrival, labor pains had commenced. Upon examination I found the os uteri dilated to the size of a 25 cent piece, the pains recurring every fifteen minutes. As she could assign no cause for them, I suspected malarial poisoning. I administered ten grains of quinia at once and watched the effect. In one hour the pains had ceased altogether. I then prescribed four doses of the same drug, of five grains each, to be given three hours apart. The pains did not

return until her confinement at full term, when I was again called to attend her. They were irregular, causing her great distress, without producing any good effect. I gave her ten grains of quinia ; in ten minutes the character of the pains improved, labor progressed rapidly and was completed in an unusually short time.

During my practice in Missouri I was called to several cases similar to the above, and administered quinia with good effect.

After practicing in Colorado for some time, I was called to see Mrs. G., then in labor with her second child. Labor pains had been constant for the last four hours, she said, with scarcely a minute's intermission. She was an anæmic, weakly woman. I made an examination and found the os uteri dilated to the size of a 25 cent piece, but hard and unyielding. It had been my habit to administer a quarter of a grain of morphia hypodermically in similar cases with good result, but this patient claimed that morphia did not agree with her, therefore, supposing that the irregularity of the pains was due to reflex irritability, I decided to try quinia. I gave ten grains at once and in fifteen minutes the pains ceased altogether and the patient went to sleep. After a rest of about fifteen minutes, regular pains come on and she was confined in two hours.

In March, 1879, I was called to see Mrs. C., a stout, healthy woman, pregnant with her sixth child. She informed me that she had had regular labor pains every ten minutes for the last four hours, but had not expected to be confined for at least two weeks. On examination I found the os uteri dilated to the size of a 25 cent piece. Considering the regularity of the pains and the fact that according to her own statement the time of her confinement was near at hand, I concluded that she had miscalculated and that she was at full term. After waiting an hour, during which time no progress was made although the pains were regular, I administered ten grains of quinia. To my astonishment the pains ceased altogether in the course of the next hour, and did not return until two weeks later, when she was confined at full term according to her own calculation.

In May, 1880, I was called to see Mrs. M., then in the eighth month of her first pregnancy. She was of a very nervous temperament and had been threatened with miscarriage two months before. During the last twelve hours she had had strong pains at intervals of fifteen minutes. I gave her ten grains of quinia at once and in an hour and a half the pains subsided. When she had reached full term, I was called to attend her. The pains were weak and irregular for five hours, producing no effect, after the administration of ten grains of quinia, the pains improved in strength and she was confined in two hours.

The above cases seem to show that quinia acts differently on the pregnant uterus under different circumstances, and the question would naturally arise : does quinia act as a tonic to the longitudinal fibres of the fundus uteri, at the same time allaying the irritability or spasmodic condition of the circular fibres of the cervix uteri ?

I have administered quinia in sixty-two cases of labor, and in not one of the cases did post-partum hemorrhage, convulsions or puerperal fever occur, but I must add that in cases of nervous women I combined 20 grains of bromide of potash with the quinia. On several occasions I tried five instead of ten grain doses of quinia, but was disappointed in every instance and was obliged to increase the dose to obtain the desired effect.

In two cases I repeated the dose of ten grains within an hour without causing headache, ringing in the ears, or any of the symptoms of cinchonism. In my opinion quinia is much safer in obstetric practice than ergot, for it can be administered in any stage of labor whether the cervix uteri is dilated or not, without danger to the mother or child. Such is not the case with ergot. I have been disappointed so often in the effects of ergot that I have almost entirely substituted quinia for it in my obstetric practice. Judging from my own experience, I think

that if quinia is not an oxytocic it is certainly the best remedy we have where labor is delayed by inertia of the uterus.

ON WATER ANALYSIS.

By R. S. G. PATON,
Chemist of Chicago Health Department.

Water Analysis, as every one feels and believes, is a matter of intrinsic value to every individual in the community. The question as to whether this branch of chemistry has arrived at the fine point of delicacy we have reached in some other branches, or no, is not yet settled. That is to say, that, if we can to-day make an accurate analysis, do our results entirely bear us out in the statement that this water is pure and healthy, and that is not ? We can now say that a certain water is good and that a certain other is bad, because they belong to the extremes ; but, when it comes to a differentiation—the demarcation of a distinct line—below or above which it shall be good or bad, or vice versa, I think chemists and physicians should hesitate ere they make a final decision. First then, for what ingredients should we examine a water ; and, having found them and estimated their proportions, what knowledge have we gained ? It will, of course, be absolutely essential that our methods of analysis shall be, so far as the present state of the science will allow, absolutely correct. I shall therefore confine myself, at present, to the chemical aspect of the question specially with regard to the estimation of organic matter in water.

Fifteen years ago, when I commenced the study of chemistry, we had only one method for this determination. That consisted in the evaporation of a known quantity of water to dryness and the incineration of the residue. The loss on burning was called organic matter. This is evidently a very rough method and is liable to a great many errors. Chemists differed in the length of time necessary for incineration ; hence their results would not agree on the analysis of the same water. Again, as some silica or silicates are always present, decomposition would take place on heating these with nitrates and carbonates, also liable to be present. The latter error was attempted to be prevented by a final drying-up with ammonic carbonate, but was only partially successful. It would therefore be extremely incorrect to say that the entire loss on incineration was organic matter, (more or less carbon might be left entangled in the residue); yet, even were we certain of the accuracy of the process, what information have we as to the source of this organic matter? None whatever. Waters which had their rise in boggy land would assuredly contain distinct traces of organic matter, but who would say this is deleterious to health? Certainly the determination of the amount of sodic chloride (common salt) present— usually derived from the juices and liquids of the animal system, though it might have its origin from the soil of the district— would have a very considerable influence in reporting on the sample, yet the data derived from this old method of analysis could not in any case be depended upon. There is also the great objection to the evaporation of large quantities of water with its consequent liability to loss—perhaps from decomposition—especially where it has to be transferred, in the process, from one vessel to another.

Frankland and Armstrong, of London, some years ago, brought out a method known as the "organic carbon and nitrogen process." To this we have also the objection of the evaporation of a large quantity and the transference of residues. A certain amount of sulphurous acid is added to each portion previous to evaporation in order to reduce the nitrates. This, on the face of it bears rather a dubious appearance, as no knowledge of the water is had until after the analysis is completed; so that it will not do to add always the same amount, as too little would leave nitrogen as nitrates and too much might have some reducing effect upon the residue. This latter is then subjected to the ordinary combustion analysis and great ideas are supposed to be derived

from the knowledge of the resulting amounts of carbon and nitrogen. The figures would not show the derivatives of the ingredients, hence the data may be considered valueless. The process has, in my opinion, one grand fatality, viz., that the experimental error (even stated by Frankland himself to be as high as 1-22 of the whole amount contained) is sufficient of itself to entirely mislead in making out an opinion of the quality of any single water examined.

The third method of analysis I shall mention is that known as the "Wanklyn or ammonia method." Right here let me state that it is that in use to-day by nearly every chemist in the world, and is employed by Frankland even and some of those who stick to the older analytical process, as, as they say, "a check upon their results." To say that they use it as a check upon their methods would lead one to suspect there was a screw loose somewhere—probably not in the ammonia process. This process employs, as Wanklyn (who with his unfortunate companion, Chapman, brought the method to its present condition) says, the water itself, and not simply the residue. A measured quantity—usually half a litre—is placed in a retort with a capacity of 1½ litre, fitted to a Liebig's condenser. The whole apparatus is first, however, just previous to use, cleaned out by distilling some pure water through it. Fifty cub. cent. of the sample are then distilled over and nesslerised. One hundred and fifty cub. cent. more are distilled off and thrown away; the flame removed from beneath the retort for a moment and fifty cub. cent. of a standard solution of potassic permanganate and caustic potash added. Distillation is again commenced and the distillate collected by the fifty cub. cent. at a time, and nesslerised until one hundred and fifty cub. cent. more have passed over. That which distils over before the addition of the permanganate solution is termed the "free ammonia," representing the nitrogenous organic matter which has already undergone decomposition into ammonia. Urea and similar nitrogenous organic matters rapidly become converted into ammonia, or some volatile salt thereof, so that any water contaminated with sewage will usually contain appreciable quantities of free ammonia. "Albuminoid ammonia" is the name given that which passes over after the addition of the permanganate solution. This is looked upon as representing the nitrogenous organic matter not yet decomposed in the water but ready and liable to become so. The relative proportions of these two kinds of ammonia will greatly assist in the report as to the condition and fitness of the water for domestic use. It will at once be observed that the analysis should be made as early as possible after the drawing of the sample.

Nesslerising has been mentioned above. You will permit me to explain it. The several portions of the distillate are received in colorless glass cylinders having as nearly as possible the same height and diameter, placed upon a white porcelain tile; two cub. cent. of Nessler's reagent (potassio-mercuric-iodide in caustic potash solution) are then added to each, which will strike from a yellow to a deep brown color or precipitate according to the amount of ammonia present. If a precipitate be obtained the examination must be commenced anew, diluting the original sample with a known quantity of pure water. The colored solution of the distillate with the nessler reagent must be imitated with an ascertained amount of a standard solution of ammonic chloride added to pure water and nessler. From the amount of ammonic chloride required one can readily calculate the quantity of ammonia present in the sample. This is an exceedingly pretty process which, in the hands of an ordinarily careful manipulator, will give very accurate results and, besides, give data of the condition in which the nitrogenous organic matter actually exists. No one would object to the nitrogenous organic matter in the shape of a good beefsteak, but they might to the same if it were in a state of decomposition, as we may say a water is when it contains both free and

albuminoid ammonia. It may safely be asserted that a water holding in solution a considerable quantity of sodic chloride and free and albuminoid ammonia is contaminated with sewage matter. A water should contain *no* free ammonia and never more than eight parts of albuminoid ammonia in one hundred millions of water. A deep spring water may often be found to contain as little as one part and, unless mixed with surface water, need not yield more than five parts of the albuminoid ammonia to one hundred millions of water. The presence of free ammonia alone in waters would not militate against their use for culinary purposes, as that would show that decomposition (as in filtration through animal charcoal) had completely taken place, the resulting ammonia compound not being poisonous. All of it would also be dispelled upon thorough boiling. I regret that I have no analyses of any waters in this vicinity. I have made a few analyses for a prominent physician in Cleveland, O., and find the following figures in my note-book:

PARTS PER MILLION.

	Free Ammonia,	Albuminoid Ammonia,
No. 1	0.00	0.03
No. 2	6.40	0.80
No. 3	6.00	1.00

No. 1 was a spring well water. No. 2 was also a well water. The well, being situated immediately *under* the back-kitchen, was the recipient of all the waste from the house. This water, I may mention, likewise contained large quantities of chlorides and nitrates. The owners protested they did not use the well any more, but the bucket and chain were in an exceedingly good working condition when I went to take the sample. No. 3 is the analysis of the sewage of Cleveland as it runs into lake Erie. It will be observed that No. 2 is exceedingly like No. 3.

Nitrates should also be carefully taken into consideration. They are the product of the oxidation of nitrogenous organic matter and are found largely in waters which find their way through or from cemeteries. In themselves they are innocuous but are liable to become extremely dangerous from their action upon lead pipes. No water containing nitrates should be passed through lead pipes, nor stored in leaden cisterns, as a portion of this metal may be taken up, and very serious may be the consequences arising from the constant use of such a water-supply. They might also be derived from such rocks as the granite, etc. When such is the case their action on lead may be questioned.

With regard to the mineral constituents I have little to say excepting in the way of calling attention to the apparent prevalence of gravel amongst those using waters containing much calcic sulphate, and an idea, suggested by Wanklyn, of the desirability for determining the amount of magnesic sulphate in drinking waters.

SELECTIONS FROM JOURNALS.

THE TELEPHONE AND MICROPHONE IN AUSCULTATION.

By C. J. BLAKE, M.D.

Read before the Boston Society for Medical Improvement, November 8, 1880.

It is too soon to have forgotten the enthusiasm which greeted the first publication of the electrical transmission of articulate sounds as made possible by the invention of the telephone, and the interest which was awakened in the medical profession in anticipation of its value in auscultation, and especially for the purposes of clinical demonstration,—expectations which were still further encouraged at a later day by the results of experiments made almost simultaneously by Mr. Berliner, of Boston, and Prof. Eli W. Blake, of Brown University, in this country, and Professor Hughes, in England, and published to the world by the latter in the various forms in which the discovery was demonstrable, as the microphone.

It is now more than four years since the introduction of the telephone, and more than two years since the appreciation of the full value of the "broken circuit" in connection with the telephone for sound trans-

mission ; in addition to the various forms of "transmitters" which are used on telephone lines, and which are in fact microphones, differing only in the mechanical devices for reception of the sound waves and adjustment of the contact surfaces, several forms of microphone have been constructed especially for purposes of auscultation, but none have as yet, even in a slight degree, answered this purpose. Setting careful investigation of their capabilities aside, this is sufficiently evidenced by the fact that none of these instruments have come into general use.

That, with numerous experimenters in the field, nothing in this line of research has been satisfactorily, practically accomplished awakens the question by those who await results as to why it has not been done, and it is an endeavor to answer this question which forms the basis of this paper, in which I shall have to apologize for partially retelling a twice-told tale in explanation of experiments bearing upon this subject made at different times during the past four years, at first with the hope of devising an auscultation telephone and microphone, and finally for the purpose of determining the reasons for the want of success in this attempt.

Having failed at one end, it was necessary to begin afresh at the other; this I am inclined to think has been the experience of other investigators of this subject, and as achievement is so often, finally, the result of the study of apparent impossibilities, it is to be hoped that a practically useful auscultation microphone may yet be devised. When the discoveries of the telephone and microphone were first announced, some very enthusiastic gentleman went so far as to predict that telephonic consultations would be held, and that eminent special practitioners, who "listened to the heart beats of a nation," as a matter of business, would each settle themselves down in the centre of a web of wires and auscult at indefinite distances from the patients ; the general argument being that an instrument which could make audible the foot-fall of a fly or the rustling of a camel's-hair pencil could certainly transmit at their full value the sounds from within the chest cavity which are so loudly heard in the stethoscope. But little of that knowledge of the practical working of the telephone over lines in use for purposes of ordinary communication, which came with the general introduction of the instrument, was necessary to prove the baselessness of this scheme; the very delicacy of the telephone in its susceptibility to interrupted currents and its almost fatal propensity—if such an expression may be used—to pick up sounds that did not belong to it were enough to show that it could not be used for so delicate a purpose as auscultation over any circuit of sufficient length to expose it to the influence of other electric currents. The first experiments, begun in 1877, were therefore made over a private wire, extending a distance of about eight hundred feet from one house to another, overhead, isolated as far as possible from other wires, the nearest overhead wire being six feet distant, but having a ground connection ; the telephones used were the ordinary hard-rubber case hand telephones, Bell. The telephone was placed upon the bared surface of the chest, the mouth-piece of the telephone being pressed upon the surface, the auscultant listening at the second telephone at the other end of the line. The experiment was several times repeated with the telephone upon different portions of the chest, and with varying degrees of pressure ; with exception, in one instance only, of the suspicion of a barely perceptible " thud," no sound which could be referred to the heart as its source was heard, although with the telephone in this position the voice and words of the experimenter could be heard by the auscultant.

There could also be plainly heard, in consequence of the ground connection, the snapping and crackling noises indicative of earth currents, the clicking of the Morse instruments, and the sound of a " fast speed transmitter" on the Western Union lines running along the Providence Railroad, and the ticking of the clock connected with the

Observatory in Cambridge. It was very plain, therefore, that even if the heart sounds could be transmitted they would be drowned by extraneous noises, and the experiment was repeated by making the auscultation in the same room with the patient, the two telephones being connected by the short flexible wires, about three feet long, in common use. Even under these favorable circumstances no sound originating in the chest cavity could be heard.

Disks of postal-card paper, having small disks of iron in the centre, were substituted for the metal disks of both telephones, but with no better results, and the experiments with the telephone alone were abandoned.

The introduction of the microphone awakened new interest in the possibilities of auscultation, and a series of experiments was instituted with various forms of this instrument, the Bell telephone being used as a receiver.

Guided by former experience, a short line without ground connections was always used, and the microphone was placed either directly upon the chest or upon a small box of proportions suitable to its action as a resonator for very low tones. In no instance were sounds heard in the telephone which could be referred to the heart or lungs as their source, but there could be distinctly heard rubbing and rustling noises resulting from the contact of the microphone apparatus with the surface of the skin and the small hairs upon it ; and when the microphone was held in the hand, a short distance above the surface of the chest, noises could be heard which could be referred only to the disturbance of the opposing microphonic surfaces in consequence of the slight involuntary movements of the hand. Auscultation of the tracheal respiration was also attempted, with the same results. Of the various forms of microphone used in these experiments one was the transmitter invented by Mr. Francis Blake and used by the National Bell Telephone Company ; another a microphone having a curved membrane, modeled upon the human membrana tympani ; and a third a common form of microphone, but arranged with an adjustment which seemed to fit it particularly for the transmission of low tones. In order to test the capacity of the latter instrument in this respect, a resonator was tuned as nearly as possible to the pitch of the first beat of the heart (in this case a tone of about 170 v. s.) and placed upon the microphone. On blowing into the resonator forcibly, and thereby giving a tone much louder than that of the heart, there could be heard in the telephone a rushing sound of comparatively high pitch, but no fundamental tone of the resonator.

A large tuning-fork of the same pitch, set in vibration and held over the microphone, or placed upon the table at a short distance from it, was distinctly heard.

It seemed, therefore, that the microphone would not transmit audibly tones as low in pitch as those of the heart beats unless they were of considerble intensity; in other words, the volume of sound from the heart would need to be greatly increased, or its pitch considerably raised, in order to make it audible by microphonic auscultation with any present apparatus. The susceptibility of the microphone, moreover, to movements causing slight disturbances of its contact surfaces and the correspondingly loud noises induced in the telephone present a very serious difficulty in the way of auscultation.

In other words, for successful auscultation a microphone must be constructed which will respond only to very low tones of slight intensity, the contact surfaces of which shall not be subject to slight mechanical disturbance, and the adjustment of which shall be fairly permanent. In view of the work in electrical sound transmission during the past four years, no one would be so rash as to assert that these conditions cannot be fulfilled.

The reasons for the failure in auscultation as above stated are :—

(1.) In the telephones, the very considerable loss of power in sound transmission. In a series of experiments made for another purpose* in 1878 it was found that in response to a tone of 448 v. s. the centre of the disk of the transmitting telephone, without the magnet, moved through a space of 0.2625 millimetre, while the disk of the receiving telephone made a corresponding movement of only 0.0135 milimetre, a loss in motion of 92.9 per cent. between the two telephone disks. With so great a loss by telephonic transmission in a tone of considerable intensity, it is claimed that the proportion of so weak and dull a sound as that of the heart, its overtones damped by transmission through soft tissues, given forth by the receiving telephone, would be inaudible to the human ear.

(2.) While the telephone creates its own current of electricity, a certain proportion of the working force which moves the disk of the transmitting instrument being expended in this operation, the microphone takes a current already provided by a battery, and merely varies the amount of the current passing from one to the other of its opposing contact surfaces. The whole of the motive force, therefore, generally speaking, is expended in varying the resistance. With this great saving the instrument is much more susceptible to very slight sounds, but it is also fatally susceptible to the influence of very slight mechanical movements which vary the relations of the contact surfaces to each other, thereby varying the amount of current passing, and producing corresponding sounds in the telephone. These sounds are usually very loud and sharp, and interfere materially with the hearing as well as with the transmission of regular musical tones. It may be briefly put that for the delicate purpose in question the telephone transmits too little, and the microphone, in any of its present forms, transmits too much.

Several auscultation microphones and sphygmophones have been constructed in France and Germany. I have had no opportunity to experiment with them, but

* British Society of Telegraph Engineers, London. Sound and the Telephone. C. J. Blake.

from the details of their construction they do not differ essentially from those used in these experiments, and I should judge them to be open to the same objections.—*Boston Medical and Surgical Journal.*

THE MANAGEMENT OF THE SECOND STAGE OF NATURAL LABOR.*

BY WILLIAM T. LUSK, M. D.,

Professor of Obstetrics and Diseases of Women and Children, and of Clinical Midwifery, in Bellevue Hospital Medical College.

The management of the second stage of labor calls for considerable tact on the part of the medical attendant. It is incumbent upon him to make frequent examinations, to determine the degree of rapidity with which the descent of the head takes place. So long as the advance is regular, he should abstain from interference. Should the pains slacken, however, he should not allow the duration of the second stage to exceed the physiological limits. It is not easy to define exactly what is implied in the expression, "physiological limits." As a rule, a very rapid second stage is not physiological, as it endangers the integrity of the vagina and perineum, and predisposes to post-partum hemorrhage. Still, now and then labor is ended by a single pain after rupture of the membranes without detriment to the mother. Of course, such cases are extremely uncommon in primiparæ. They require an unusually distensible condition of the soft parts, and an extraordinary degree of resiliency in the uterus. On the other hand, pressure of the head, after its descent into the pelvic cavity, leads, if too long continued, to pathological changes in the tissues of the canal and of the outlet. It is usual, therefore, unless the head is small or the pelvis roomy, to use the resources of art to terminate labor, when the head remains stationary after two hours of effort at the perineal floor. It is desirable, therefore, when the pains are weak and ineffective, to make use of all the simple adju-

vants which experience has shown to possess real efficacy in increasing the activity of labor.

Changes of posture increase the power of the pains temporarily. When head flexion is incomplete, it has been recommended to place the patient upon the side toward which the occiput is turned. Others, again, claim that the descent of the occiput is best effected by placing the mother upon the side toward which the child's forehead is directed. In point of fact, either posture frequently leads to the desired result, simply because the change from the dorsal to the lateral position is apt to be followed by a temporary addition to the uterine force.†

In many women, owing to defective innervation or to insufficient development of the muscular structures of the uterus, it is of great moment that the expulsion of the child be aided by the voluntary pressure of the abdominal walls. To be sure, in most cases the reflex impulse to bear down is imperative, but in others, where the impulse is feeble, or held in abeyance by the dread of the patient lest she increase her sufferings, it becomes the duty of the physician, in tardy labors, to see to it that all the auxiliary forces are brought into play. To this end, he should instruct his patient to fix her pelvis, either by pressing her feet against the foot-board of the bed, or by drawing up her knees and resting them against an assistant, who assumes the position best adapted to furnish the requisite support. Then the nurse or other suitable person should grasp the woman's hands, so as to enable her to fix her thorax and to bring all the expiratory muscles into full exercise. Often, when the agony is intense, the patient can be induced to strain with her pains, if her sufferings are first dulled by small doses of chloroform. When the head is on the perineum, the physician may further expulsion by rubbing the abdomen to excite pains, and by pressing upon the breech through the fundus.

During the second stage, the patient's

*Extracted from the author's forthcoming work, entitled "The Science and Art of Midwifery." New York: D. Appleton & Co.
†LAHS, "Die Theorie der Geburt." Bonn, 1877, p. 237.

posture should be left in general to her own volition. The physician should accustom himself to conduct labor with equal facility, no matter whether the woman lies upon her side or upon her back. The left lateral position, affected by English accoucheurs, is very convenient at the time of delivery, especially when there is occasion to support the perineum, and when, owing to the flatness of the nates, the vulva is scarcely raised in the dorsal posture above the level of the bedding.

THE PRESERVATION OF THE PERINEUM.

By far the most delicate task which the physician has to fulfill toward his patient, in the expulsive stage, consists in so regulating the exit of the child's head as best to avoid perineal lacerations. It is needless to state that such lacerations, unless of slight extent, entail upon women a variable degree of subsequent discomfort and suffering. When the perineum is examined with care after labor, a practice which should be invariable with a conscientious attendant, the frequent occurrence of more or less extensive rupture of its tissues is a matter of easy confirmation. Statistics of their frequency are of little value, much depending upon individual skill in management.

Olshausen* reports, as the result of the preventative measures adopted at the Clinic in Halle during a period of ten years, 21.1 per cent. of perineal injuries in primiparæ, and 4.7 per cent. in multiparæ. These percentages did not include slight tears confined to the frænulum. He regards 15 per cent. as not too high an estimate for the absolutely unavoidable lacerations, due to defective distensibility of the perineum and to the disproportionate size of the child's head.

The aim of prophylactic measures should be to develop the elasticity of the soft parts to the fullest practicable extent, and to cause the head to pass through the distended orifice of the vulva by its smallest diame-ters. Preliminary softening of the perineum is best accomplished by the continuous, but not too rapid, descent of the presenting part. The relaxation, as a rule, begins earlier and is more complete in multiparæ than in primiparæ. In a few cases, the soft parts will already have ceased, by the end of the first stage of labor, to offer any effective barrier to delivery. The distensibility of the soft parts may be fairly inferred from the presence of a copious discharge of glairy mucus. When rupture takes place, the vaginal mucous membrane is the first structure to give way. In the ordinary form, the perineal body tears from the commissure backward to the rectum. In rare cases a central perforation may result, and the child be delivered through a rent situated between the vulva and the anus.

When the head begins to make the perineum bulge, the physician should be on the alert, and inform himself during each contraction of the strain to which the parts are subjected. At first, it is only necessary to rest the hand lightly on the perineum. Direct pressure is to be avoided, except when the perineum is stretched to a membranous thinness, and the danger of central perforation threatens. As the head begins to distend the vulva, the tension at the frænulum should be carefully gauged by a finger introduced between the labia. Measures to avert rupture may be classified under three headings, viz.: 1. Those designed to check the exit of the head before the fullest expansion has been secured, and to prevent expulsion during the acme of a pain, when the borders of the orifice are most rigid; 2. Measures which impart an upward movement to the head, with a view of making all unoccupied space beneath the arch of the pubes available; 3. Measures which favor expulsion during the interval between the pains, or at least after the acme has subsided.

In ordinary cases, Hohl's method, recommended by Olshausen,† has rendered me

---

*OLSHAUSEN, " Ueber Dammverletzung und Dammschutz." Volkmann's " Samml. klin. Vortr.," No. 41, p. 360.
† Loc. cit., p. 366.

excellent service. It consists in applying the support, not to the perineum, but to the presenting part. To this end, the thumb should be applied anteriorly to the occiput, and the index and middle fingers posteriorly upon that portion of the head which lies nearest to the commissure. The unconstrained position of the hand enables the operator to exercise effective pressure in the direction of the vagina, while the posterior fingers favor the rotation of the head under the pubic arch. The patient should at the same time be directed not to hold her breath during the pains, except when they are weak and powerless. Where the impulse to bear down is irresistable, chloroform should be given to annul the excessive reflex irritability. Under the most skillful management, laceration is liable to occur unless the physician is able to control the action of the auxiliary expulsive forces.

So soon as the biparietal diameter passes the tense border of the vulva, the perineum retracts rapidly over the face, and the delivery of the head is completed. It is during this period that laceration is most apt to occur. The danger is, however, greatly lessened, if the head is made to issue through the orifice after the pain has subsided, and when the soft parts are in a relaxed and dilatable condition. To accomplish this, in many instances where the resistance to be overcome is slight, it is sufficient for the woman to hold her breath during an interval between the pains, and voluntarily call into play all the muscles of expiration. In the larger proportion of cases, however, these efforts are futile, because of the comparatively feeble motor force brought into action.

An excellent method of manual delivery we owe to Ritgen,* which consists in lifting the head upward and forward through the vulva between the pains, by pressure made with the tips of the fingers upon the perineum behind the anus, close to the extremity of the coccyx. Of course, the method

is only available after the head has descended sufficiently for the pressure to be exerted upon the frontal region.

Rectal expression has lately found warm advocates in Olshausen † and Ahlfeld ‡. The maneuver consists in passing two fingers into the rectum toward the close of the second stage of labor, and hooking them into the mouth or under the chin of the child through the thin recto-vaginal septum. By pressing the face forward and upward, the normal rotation of the head beneath the pubic arch can be effected, and the delivery can be accomplished between the pains at the will of the operator.

When rupture is felt to be imminent, mock modesty should be discarded, and the parts imperiled should be unhesitatingly exposed to view. If, owing to its elasticity, the occiput, in place of being directed forward to the vulva by the perineum, distends the latter so that central perforation threatens, the hand should be applied in such a way as to give direct support to the stretched tissues, and to guide the head upward to the outlet. If, on the other hand, the danger arises from defective elasticity, the physician, standing to the right of the patient with his face toward the foot of the bed, should pass the left hand between her thighs, and press the head upward and inward during each pain, with the thumb and two fingers, as previously described. At the same time, the movement of extension, should it threaten danger to the parts, should be hindered by pressing backward upon the frontal region through the perineum with the disengaged hand.

Dr. Goodell § recommends hooking two fingers into the anus, and drawing the perineum forward during a pain, to remove strain from the thinned border of the vulva and to promote the elasticity of the tissues.

Fasbender §§ places the patient upon the left side; then, standing behind her, he seizes the head between the index and middle fingers of the right hand, applied to the

*See AHLFELD, " Das Dammschutzverfahren nach Ritgen."   "Archiv f. Gynaekol.," vi, p. 279.   † Loc. cit., p. 369.
‡Loc. cit.   § "Am. Jour. of the Med. Sci.," January, 1871.   §§ " Zeitschr. f. Geburtsh. u. Gynaekol.," ii, 1, p. 58.

occiput, and the thumb thrust as far into the rectum as possible. By this manoeuvre the head is held under complete control, the rectal wall hardly affecting the grip in any appreciable manner. During a pain the progression and extension of the head are readily prevented. During the interval between the pains, by pressure with the thumb through the rectum and the posterior portion of the perineum, the head can be raised forward and upward at the will of the operator.

Between pains, I have been in the habit, in cases of rigidity, of alternately drawing the chin downward through the rectum until the head distends the perineum, and then allowing it to recede. It is astonishing how often apparently the most obstinate resistance can be overcome by the simple repetition of this to-and-fro movement, the parts becoming rapidly soft and distensible. Of course, it should be discontinued the moment contraction begins, and care should be taken to effect delivery after uterine action has subsided.

With judicious management, the number of unavoidable lacerations can be restricted to a small proportion of cases. Still there are individual peculiarities which will now and then render abortive the best of prophylactic measures. In this category I have already alluded to a primitive lack of development of the maternal parts, to unusual size of the child's head, and to the excessive rigidity of the perineum of primiparæ, especially after the thirtieth year. In addition should be mentioned, cases where the pubic arch is diminished by the approximation of the pubic rami, or where the tissues have been rendered friable from chronic œdema, from a varicose condition of the veins, from condylomata, from syphilitic sores, or from inflammatory infiltration consequent upon undue prolongation of the second stage of labor. Lacerations are more frequent in occipito-posterior positions, and in the delivery of the after-coming head, where hasty extraction is demanded in the interest of the child.

When, in the judgment of the physician, rupture of the perineum seems inevitable, he is justified in making lateral incisions through the vulva to relieve the strain upon the recto-vaginal septum. To this operation the term episiotomy is applied. By it not only is the danger of deep laceration through the sphincter ani prevented, but owing to their eligible position, the wounds themselves are capable of closing spontaneously; whereas, when laceration follows the raphe, the retraction of the transversi perinei muscles causes a gaping to take place which interferes with immediate union. As, however, every wounded surface is a source of danger in childbed, episiotomy should never be performed so long as hope exists of otherwise preserving the perineum. It is essentially the operation of young practitioners, the occasion for its employment diminishing in frequency with increasing experience. The chief resistance encountered by the head is not at the thin border of the vulva, but is furnished by a narrow ring, situated half an inch above, and composed of the constrictor cunni, the transversi perinei, and sometimes of the levator ani muscles. Incisions should be made during a pain, when the ring becomes tense and rigid and is easily recognized with the finger. As it is not desirable that the head should be driven suddenly through the vulva during the act of operating, the time selected for performing episiotomy should be at the commencement or close of a contraction. The division of the rigid fibers may be accomplished by means of a blunt-pointed bistoury or a pair of angular scissors. So far as practicable, the incision should be confined to the vagina, and should not exceed three quarters of an inch in length. In cases where the head is on the eve of expulsion, the bistoury may be introduced flat between it and the vagina, half an inch anterior to the commissure, and the section made from within outward. Care, however, should be taken at the same time to avoid severing the external skin, by draw-

ing it as far back as possible.* In central
perforations, it is best to divide the band
left attached to the vulva, as its preservation
is of no after-advantage.

THE EXPULSION OF THE SHOULDERS.

After the delivery of the head, mucus
should be wiped from the mouth and nose,
and cleaned from the throat with the finger,
should laryngeal rales indicate an embarrassment of the respiration. If the cord is
found coiled around the neck, it should be
loosened by drawing upon the placental
end until the shoulders can pass readily
through the loop. Should this be found
impossible, either because the cord is unusually short, or because it is wound several
times around the body, a ligature should be
applied, the cord should be cut between
the ligature and the placenta, and delivery
should be hastened by manual efforts.†

In the majority of cases the shoulders are
expelled spontaneously. Still, it is a good
plan to expedite the descent by pressure
made with the left hand at the fundus of
the uterus. Care must be taken lest the
lower shoulder convert a slight tear in the
perineum into an extensive laceration.
The right hand should therefore be applied
to the perineum in such a way as to lift the
shoulder upward, and, at the same time,
furnish a bridge over which it can glide in
its movement forward. Sometimes, after
the passage of the head, a deep vaginal laceration coexists with an intact condition of
the external parts. The shoulder then tears
through the skin, and a complete rupture
ensues. Olshausen recommends, in cases
where rupture is imminent, to turn the
shoulders so that they clear the vulva one to
the right and the other to the left.

If, after the expulsion of the head, the
child does not breathe, and asphyxia threatens, the physician should rub the uterus
with the hand through the abdominal wall,
to excite a pain, during which he should
urge the patient to press down, and thus aid
expulsion. The most common hindrance

to delivery consists in an arrest of the upper shoulder beneath the pubes. Usually
its release is readily effected by seizing the
sides of the head with the two hands and
drawing directly downward. It is rarely
necessary to raise the head subsequently, or
to hook the finger into the armpit to extract
the posterior shoulder.

TYING THE CORD.

When the cord is torn across, as sometimes happens in street births, no hemorrhage takes place from the lacerated vessels.
Of course this occurrence deprives the physician of the power of choosing the point at
which the division shall be made. As it is
desirable, for the sake of convenience, to
sever the cord about two inches from the
navel, it is the custom in all civilized countries to cut it with scissors, and to prevent
hemorrhage by the application of a ligature. Almost any material may be employed for the latter purpose, though
nothing is so handy as the narrow flat bob-
bin which most nurses keep in readiness.
The ligature should be applied tightly, and
the cut surface should subsequently be examined once or twice by the physician
before leaving, to make sure that the arteries
are sufficiently compressed to prevent oozing
from taking place. The cord should be
held in the hollow of the hand at the time
of its division, to avoid the possibility of including accidentally any portion of the
child between the blades of the scissors.
Commonly two ligatures are applied, and
the cord is severed between them, though
the question of one or two ligatures is, except in twin pregnancies, of trifling importance.

In practice, it is very desirable that the
physician should understand the physiological difference between the effects of the
early and those of the late application of the
ligature. The custom, as regards this point,
has been by no means uniform. The
ancients deferred the ligature until after the
expulsion of the placenta. Mauriceau,

---

*OLSHAUSEN, loc. cit., pp. 372, 373.   †TARNIER recommends dividing the cord, and then compressing the proximal end
between the thumb and the index finger.   The proximal end is distinguished by the spurting of the two umbilical arteries.

Clement, and Deventer followed the same plan, but employed artificial expedients to complete the third stage of labor rapidly.* The common practice of the present day is to tie the cord immediately after the birth of the child. Still, there have not been wanting in recent times warning voices against precipitate action. Naegele advised waiting until the pulsation of the cord had ceased ; Braun† first describes the changes from the fœtal to the post-natal circulation, and then says : "This stupendous process should be taken into consideration in the treatment of every case of labor, and because of it the cord should never be severed or tied so long as pronounced pulsations can be felt near the navel "; Stoltz‡ remarks that, "after the child has respired well, division of the cord is followed by an insignificant loss of blood, while, after immediate section, blood escapes in abundance."

In 1875 Budin, at that time _interne_ at the Maternite of Paris, undertook the following experiments at the suggestion of Professor Tarnier ; in one series the cord was tied immediately after the birth of the child, and the blood which escaped from the placental extremity was measured ; in the other, the quantity of blood was determined in cases where the cord was not tied until several minutes after delivery. By a comparison of the results thus obtained, he found that the average amount of placental blood was three ounces greater in the first than in the second series of experiments. Welcker estimated the entire quantity of blood in the infant at one nineteenth the weight of the body, which would amount, in a child of seven pounds, to six ounces. To tie the cord immediately after birth would therefore be equivalent to robbing the child of three ounces of blood, which would otherwise pass into its circulation. This startling result has in the main been abundantly confirmed by subsequent observers. Two years later (1877) Schuecking, extending Budin's ex-

periments by weighing the child at birth, and then observing the changes that took place up to the time of the cessation of the placental circulation, found that the child gained from one to three ounces in weight by delay. It is certain that these amounts do not represent the entire increase, as a portion necessarily escapes observation in the interval that must elapse before the weight can be ascertained.

There is a difference of opinion as to the mechanism by which the transfer of the blood from the placenta to the child takes place. According to Budin, the principal factor in the accomplishment of the result is thoracic aspiration. With the first breath, the afflux of blood to the lungs developes a " negative pressure" in the vessels of the larger circulation, so that a suction force is exerted upon the placental blood, which continues until the equilibrium is restored. To tie the cord prematurely, therefore, is to cut off from the child a supply of blood, for which the establishment of the pulmonary circulation had created a physiological need.

Schuecking,§ on the contrary, maintains that after the first inspiration thoracic expansion ceases to operate as an active force, and that the main agent which drives the blood from the placenta through the umbilical vein is the compression exerted by the retraction, and, at intervals, by the contractions, of the uterus.

The difference in the theoretical standpoint of these two observers is of practical importance, for, if the movement of blood to the child results from thoracic aspiration, the quantity which enters its circulation will not exceed its requirements; while, if the movement is due to uterine compression, the question arises as to whether the forcible transfusion thus accomplished is compatible with the child's safety and welfare.

The ultimate decision will depend partly upon experimental and partly upon clinical observations. Provisionally, the case stands

* Budin, "A. quel Moment doit-on operer la Ligature du Cordon Ombilical?" " Publications du ' Progres Medical,' " 1876.
† "Lehrbuch der Geburtshuelfe," p. 192. ‡ "Nouveau Dictionnaire," p. 283, art. "Accouchement Naturel."
§ "Zur Physiologie der Nachgeburtsperiode." " Berl. klin. Woch.," Nos. 1 and 2, 1877.

as follows: The manometric observations of Ribemont* show that the pressure in the umbilical arteries is uniformly greater than that in the umbilical vein; during a series of deep inspirations and expirations, the blood in the umbilical vein is subject to marked oscillations; after the pulsations of the cord have ceased, the uterine contractions alone are insufficient to propel the placental blood through the umbilical vein to the infant. Again, Budin (Discussion upon Ribemont's paper), in a breech delivery, compressed the cord at the vulva as far as possible from the navel; at birth, the vein was distended with blood, but with the first inspiration it was instantly emptied. Thoracic aspiration does, therefore, exist as an operative force. On the other hand, Schuecking found that, when the placenta was rapidly expelled by Crede's method, so as to remove it from the influence of uterine retraction, the pressure in the vein was slightly lessened, and the total amount of blood transferred to the infant was greatly restricted.

According to the clinical observations of Budin, Ribemont, and Schuecking, infants which have had the benefit of late ligation of the cord are red, vigorous, and active, whereas those in which the cord is tied early are apt to be pale and apathetic. Hofmeier,† Ribemont, Budin, and Zweifel‡ have shown that the loss of weight which occurs in the first few days following confinement is less in amount and of shorter duration when the cord is not tied until after the pulsations have ceased.

There appear to be no harmful results to the child growing out of the practice of late ligation. Porak, indeed, reports two cases of dark vomiting, two of melæna, and two with sanguineous discharges from the vagina, which he is convinced were the result of the practice; but the extensive trial to which it has since been subjected in the principal lying-in institutions of the continent has sufficiently demonstrated that it is exempt from danger.

In late ligation, the amount of blood retained in the placenta and the increase in the weight of the child differ materially in different cases §—a difference which seems to indicate that, so long as the placental circulation is left undisturbed, the amount of blood passing to the child will be measured by its needs. In a case of Illing's,|| on the other hand, after the placenta had been expressed from the uterus, its contents and that of the cord were forcibly squeezed into the circulation of the child, and death followed from over-distention of the heart. Porak and Georg Violet¶ claim that there is a special predisposition to icterus in children when the cord is tied after the placental circulation has ceased. Violet attributes the discoloration, not to bile pigment, but to a rapid disintegration of the excess of bloodcorpuscles. Helot, he says, found, on the first day after birth, a difference of 900,000 corpuscles to the cubic millimetre between cases of late and those of early ligation, while on the ninth day the difference fell to 300,000. Others have failed to notice any characteristic icteric discoloration peculiar to late ligation. Neither Porak nor Violet attaches any pathological significance to the symptom.

The outcome of the foregoing observations may fairly be stated as follows:

1. The cord should not be tied until the child has breathed vigorously a few times. When there is no occasion for haste arising out of the condition of the mother, it is safer to wait until the pulsations of the cord have ceased altogether.

2. Late ligation is not dangerous to the

---

* "Recherches sur la Tension du Sang dans les Vaisseaux du Fœtus et du Nouveau-ne." "Arch. de Tocologie," Oct., 1879.
† "Der Zeitpunkt der Abnabelung," etc. "Zeitschr. f. Geburtsh. u. Gynaekol," iv, 1, p. 114.
‡ "Gentralbl. f. Gynaekol.," No. 1.
§ "See WIENER, "Ueber die Einfluss der Abnabelungszeit auf den Blutgehalt der Placenta," Arch. f. Gynaekol.," xiv, 1, p. 35; also, MEYER, "Centralbl. f. Gynaekol.," 1878, No. 10.
|| "Inaug. Diss." Kiel, 1877.
¶ GEORG VIOLET, "Ueber die Gelbsucht der Neugeborenen und die Zeit der Abnabelung. "Virchow's "Archiv," lxxx, 2, p. 353.

child. From the excess of blood contained in the fœtal portion of the placenta, the child receives into its system only the amount requisite to supply the needs created by the opening up of the pulmonary circulation.

3. Until further observations have been made, the practice of employing uterine expression previous to tying the cord is questionable.

4. In children born pale and anæmic, suffering at birth from syncope, late ligation furnishes an invaluable means of restoring the equilibrium of the fœtal circulation.— *New York Medical Journal.*

EDITORIAL.

The Guy's Hospital Dispute.

The long and bitter fight between the governors and the staff of Guy's hospital, London, has ended in the complete defeat of the latter.

For those of our readers who may not have followed the case, we shall sum up the facts briefly.

About a year ago the governors saw fit to fill the post of matron to the hospital without consulting the staff. The woman appointed seems to have been in every respect fitted for the position. But it was the manner of the appointment that displeased the staff. In Guy's hospital, as in several others of the London hospitals, the matron has complete control of the nursing, and this one chanced to be a reformer. Many of the old nurses were discharged and new ones, for the most part tainted with the ideas of Miss Lonsdale and her school of lady sentimentalists, were put in their places. In all this, the staff, who surely were vitally interested in the character of the nursing which their patients should receive, had no voice. Remonstrances proving of no avail, the aggrieved physicians turned to the press. This was the beginning of that war which has raged for months in the columns of the London journals, medical and non-medical.

Partisan statements and arguments abounded on both sides. Scarcely a fair-minded, temperate article was written during the whole controversy. Finally, about two months ago, the staff brought affairs to a crisis. They sent a letter to the governors, signed by the senior physician and senior surgeon, and purporting to represent the sentiments of the staff as a body. The letter was very intemperate in language, and the governors were justly indignant and demanded its withdrawal. This being refused, the resignations of the senior officers were demanded. The staff now confessed their weakness, withdrew the letter, and made other concessions to the governors. This practically leaves the governors triumphant.

The question of hospital management in London and elsewhere is still a living one. Let us hope that some day it may find men of broad views and impartial judgment to investigate it.

Pure Drinking Water for Denver.

All interested in the sanitary condition of Denver will be pleased to learn that a company has been formed to supply that city with pure water for drinking and domestic purposes.

The water of the Platte river becomes daily more impure, unfitting it for household use. It can still be used, of course, for irrigation and mechanical purposes. Indeed, the new company undertakes to furnish water only for domestic use, and, in case of necessity, for fires. It is to be brought from the Chicago lakes, which are forty miles distant in the mountains. The undertaking will be attended with great expense, but nevertheless should be encouraged. The lakes are above all source of impurities, and will supply Denver with pure drinking water for all time.

The following vigorous and characteristic language of Dr. Henry I. Bowditch, of Boston, may not be out of place in this connection : "Only one-third of the towns and cities of this nation make any claims, even the most trivial, of endeavoring to procure pure potable water for their inhabitants. The remainder (65.73 per cent.) either confess carelessness or ignorance of the subject. In other words, over one-half of the people of these United States are openly and avowedly living in a senseless disregard as to whether they are drinking pure water or water contaminated by every kind of filth.''

The Regulation of the Practice of Medicine in Colorado.

We reprint this month the bill for the regulation of the practice of medicine in Colorado. With the text of the bill will be found an article commenting upon it, written by a prominent lawyer of Leadville, as well as some editorial remarks, all taken from recent numbers of the Leadville Chronicle.

Since our last issue we have received letters from two physicians practicing in Longmont, both of whom disapprove entirely of section 12 of the proposed bill. In this they agree with the writer of the legal article mentioned above.

We are glad to note that the citizens of Colorado, without regard to profession, are alive to the importance of this subject. We shall be glad to receive and to publish communications bearing upon it from all quarters. It is only by a free and intelligent discussion of the measure by all persons interested that we can arrive at the result desired. Now is the time to make objections against and amendments to the bill, not at the moment of its presentation to the legislature.

The bill for the regulation of the practice of medicine in Massachusetts might not have failed so signally last spring, if the public mind had been better prepared for the measure. Secrecy is not called for in matters which concern the public at large.

THE paper on Water Analysis published on page 124 is written by R. S. G. Paton, of 77 Major Block, Chicago, who is employed as chemist of the Health Department

of that city. He is engaged in making analyses of all foods used in Chicago, but also acts as an analyst for any person who may desire his services.

THE continuation of Dr. Solly's paper on "The Influence of Altitude upon Respiration," which was to have been published in this issue of the REVIEW, has been unavoidably postponed.

TRANSLATIONS.

By JOHN W. BRANNAN, M.D.

The Physiological Action of the Climate of Mentone.

Dr. Cazenave de la Roche, a few months after his arrival at Mentone, noticed a remarkable emaciation in himself and his family, as well as in his servants. They all, however, maintained their usual state of health.

In seeking for an explanation of this fact, he directed his observations, first to the native inhabitants of Mentone, and then to the foreign population. He soon ascertained that the Mentonians, as a rule, were of spare build, muscular, and very seldom disposed to corpulence. Following the Hippocratic precept that man is the faithful mirror of the locality in which he lives, he inferred that this absence of fat was due to climatic influence.

Turning now to the foreign population, he weighed frequently and with regularity a certain number of persons recently arrived in the country. In all cases he found a steady and more or less rapid loss of weight. The individuals under observation were all in good health, some of them but slightly disposed to *embonpoint*, while others were fat and even corpulent.

Observations tending to the same result having been reported from Nice and Hyeres, he is led to conclude that the atmosphere of the Mediteranean coast in general, causes loss of weight.

The writer finds a physiological explanation of the above facts in the emulsionary power of the pancreatic fluid. He would argue that the atmosphere of Mentone acts as an especial excitant to the pancreas, leading to an increase of its secretion. Hence an increased digestion of fat, and a diminution of the body weight. The writer supports his opinion by referring to the analogous effects of the climate of the equatorial and polar regions upon certain organs of the body. The moist heat of the tropics causes the liver to secrete more actively—the low temperature of the frigid zone increases the activity of those organs which form the chyle and hematose. As the hot and cold climates have each their elective affinity for a certain organ of the body, the writer would allot to the temperate climate (in particular, that of the Mediterranean coast) the power of acting especially upon the pancreas. He, however, lays more stress upon the therapeutic value of his observations than upon their etiological significance.—*La France Medicale, Oct. 27, '80.*

A SIMPLE WATER FILTER.

Take a large flower pot and close the hole in the bottom of it with a sponge. Then place in the pot, first a layer of clean fine sand, upon this a layer of charcoal, then another layer of sand. The water, after having passed through these filtering layers, can be drawn off at will, by attaching a stopcock to the bottom of the pot.—*Journal d' Hygiene, Nov. 4, '80.*

ABSINTHISM.

Mr. Lanceraux has been studying the effects in men and women of the excessive use of absinthe. He distinguishes three forms of absinthism, the acute, the chronic and the hereditary. They all present phenomena analogous to those of hysteria. The acute form, that following a single excess, is characterized by convulsions, which Lanceraux compares with those of hysteria, while Magnan and Challand regard them as epileptoid. In the chronic form, following constant long-continued abuse of absinthe, derangements of sensibility were observed, perfectly hysterical in

character. Hereditary absinthism Lanceraux considers to be generally confounded with hysteria. He raises the question whether many of the cases of hysteria reported in men may not really belong to either the chronic or the hereditary form of absinthism.—*Journal d' Hygiene, Nov. 11, 1880.*

BOOK NOTICES AND REVIEWS.

SCHOOL AND INDUSTRIAL HYGIENE.

By D. F. LINCOLN, M. D.,

Chairman Department of Health, Social Science Association. Philadelphia : Presley Blakiston, 1012 Walnut street. 1880.

The task which the author of this volume, the last of the series of American Health Primers, had laid before him, was one beset with many difficulties, and we congratulate him upon the successful accomplishment of his work. On the one hand, there is the vastness of the subject to be treated; on the other, the limited space given for the work. He had to comply with the editor's purpose, "to furnish the general or unscientific reader, in a compact form—reliable guides for the prevention of disease and the preservation of both body and mind in a healthy state;" and yet, he had to write so that his work would stand the brunt of scientific analysis. The book does not claim to be exhaustive or critical, those qualities are not to be expected in such a limited work, intended as it is for the laity; but, nevertheless, the professional reader will find, in condensed form, much that he would have to glean from larger works on hygiene.

Viewing with distrust the tendency to popularize medical knowledge, we feel that the author has been very discreet in imparting truths, which should be valuable in preventing disease, and in avoiding all hobbies in regard to theories and treatment.

Surely the subject is one of much importance, and one upon which we need enlightenment, the professional man, not less than the layman. But we need wise instruction and correct inferences drawn from analysis of the past, and it seems to us that Dr. Lincoln has endeavored to be impartial in his conclusions.

The arrangement of the book commends itself, and the subjects treated of are practical.

There is a tendency on the part of the author, we think, to place too much responsibility upon the teacher, and too little upon the parent. It seems almost too much to expect of a teacher, that she should ascertain if the children's feet are wet and should send a child home, in case its parent has been foolish enough to allow it to go to school poorly shod. It appears to us that just here is one point of error in our public school system, viz: in educational matters the parent has no control over the child. It is true that many parents are only too willing to shirk this responsibility, just as it is also true that many parents are not qualified to decide what is best for their children in this respect; but that large class in every community, those parents, who are both interested in their children's welfare and are also capable of judging what is good for them, this class have to remain perfectly passive, unless they are wealthy enough to provide private instruction.

The large sum of money spent annually in public education, necessarily involves a complex system in its disbursement, and, as in business, method and organism are essential; but we regret exceedingly the refinements that are daily creeping into the system, considering them subversive of true education. The present idea seems to be, to consider our school children as a vast army, a mass, to be disciplined to study and recite as if moved by clock-work. On this account we are glad to see that the author recognizes the individual in his suggestions on education, that he urges that allowances be made for differences of temperament, constitution, age and sex. In this connection, we commend his suggestions in regard to the number of school hours *per diem.* That the number of hours should be diminished, would seem rank heresy to those parents who regard the time

their children spend in school as just so many hours of freedom from care and annoyance for themselves. But we do not mean that our schools shall be mere "lock-ups" for children, places to put them in for a certain number of hours to keep them out of mischief. Dr. Lincoln is an advocate of fewer hours of school, with the idea that much of the time spent in school is wasted, and that "*three hours of good work are better than five hours of poor work.*" Of course there are two main considerations on this point. The health of the children must be considered, and, second to this, the actual number of hours required for a fair amount of work to be expected of the children. Our school system makes no allowance for a child's health, the school day is a fixed quantity, a certain number of hours during which the child of seven or eight, as well as the youth of eighteen; the weak, as well as the strong; the girl, as well as the boy; the sickly, menstruating girl, as well as her sturdy brother; during which one and all have to be at their appointed tasks. Evidently such a system is wrong and needs condemnation.

The number of hours necessary for the accomplishment of the work is an uncertain quantity. It must vary with the abilities of the individuals and with their ages. Indeed the factors bearing on this subject are various, and it is a matter of some nicety to strike a fair average. It seems pitiable to us, however, even if we grant for the sake of argument that the school day is none too long, it seems pitiable that when children get out of school, they should not be through their work. It is a luxury that most men demand, that with the closing of their store-doors they leave their work behind them. How different is the exaction made of our children, who, after a full day's work, bring home their books for several hours more of study! It is obvious that we either give our children more work than they ought to do, or that they do not work as they should in the hours allotted to their tasks. We think that a great deal of unnecessary work is imposed upon children, and we also believe that we foster a tendency on their part to become listless over their work, by having them study when they are physically unequal to the task. We know it to be a physiological fact, that after a meal the blood is drawn to the abdominal organs, for the purposes of digestion ; and that in consequence, for the time being, the brain is not in its best condition to work. It is partly on this account that late dinners are becoming prevalent. It is obvious that late dinners are injurious for children, and that their best development requires that their heartiest meal should be taken in the middle of the day. This being the fact, it seems cruel that physiological laws should be broken in their case, and that close mental application should be required of them, when they are in a poor condition for mental effort. In other words, it seems that if we wish to teach children the value of concentration, of close mental effort, we are taking a very poor way of doing it ; but if we wish to make children laggards, indifferent workers, contented with simply spending a given amount of time over their work, we take the surest way of accomplishing our ends. Dr. Lincoln's denouncement of the afternoon session, for children, at least, seems to us to accord with known physiological facts.

The subject of "exercise" has been ably treated by the author, but the happiest solution of the difficulty, it seems to us, would be to lessen the hours of confinement indoors and to provide suitable play grounds. The gymnasium and calisthenics are but makeshifts for the romping, out-door sports boys and girls should enjoy.

The questions of ventilation, site of building, arrangement of school-room, etc., are of vital interest, but they are subjects for the few, who have such matters in charge, rather than for the people at large. We should have liked a free discussion of the subject of sex in education from the author's pen, and wish that he had devoted a chapter to the subject, even if he had left out the one on the "Care of the Eyes."

This latter has been much more fully treated of in another book in this series. The question of seats for shop-girls is being agitated so much nowadays that the author could have profitably devoted some space to its consideration, and not have contented himself with dismissing it with a single sentence.

The chapter on "Contagious Diseases" is full of valuable hints, and the rules laid down by the author for the prevention of the spread of contagious diseases in school, it seems to us should be enforced by every school-board.

The part of the work devoted to "Industrial Hygiene" is crowded with statistics and facts. The subject evidently interested the author less than did that of the previous part of the work, and he has given us but few of his own ideas on the subject. We are glad to have these facts where we can turn to them, but they interest us less than a pleasant discussion or a frank assertion. In conclusion we welcome this earnest appeal for the "*Mens sana in corpore sano*," and we feel assured that the diffusion of the ideas contained in this book will tend to improve our school system, and also the health of children, and, in consequence, of the people at large.   S. A. FISK.

INDEX-CATALOGUE

Of the Library of the Surgeon General's office, United States Army, Vol. 1, A to Berlinski. Washington: Government Printing Office, 1880.

This first volume of the series opens with the official report of its compiler, Surgeon J. S. Billings, U. S. Army, announcing the completion of the volume and setting forth the scope of the whole work, its general arrangement, etc. Following this report is a short table of abbreviations used in the volume, and then an alphabetical list of abbreviations of titles of medical periodicals which are on the library shelves and have been employed in this work ; the list covering 126 pages and comprising medical journals and transactions of societies throughout the world and as far back as the earliest publications. Finally appears the index-catalogue proper in double-columned pages to the number of 888. Like former publications of this office the volume is well printed and the proof reading is a marvel of accuracy.

The immensity of the labor of preparing such a work is inconceivable. There must of course first be a transcription of the titles of all *original* medical papers in the more than one hundred thousand books and pamphlets in the library. A duplicate transcription has also to be made of author-titles. "The present volume includes 9,090 author-titles, * * * * 9,000 subject-titles of separate books and pamphlets and 34,604 titles of articles in periodicals." Presupposing that there will be issued in all ten volumes of equal size, more than 500,000 titles will be included. Anyone who has ever attempted to arrange a few hundred titles in appropriate groups will be able to form a very dim idea of the difficulties of the task undertaken by the compiler and his assistants. Not only is such a mass of papers difficult to handle, but doubts as to headings will arise all along the line. "Where there is doubt as between two or more subject-headings, cross-references are given." The repeated revision which the proof of such a work requires in passing through the press will be appreciated only by those who have been engaged upon similar technical matter.

It is natural and pardonable that we should all feel a pride in the fact that this work is a production of American enterprise; a work which the nations of the old world with greater resources and facilities at command have so far failed to undertake. Its advantages are already recognized at home and abroad. To every inquirer is presented a condensed statement of previous publications in his particular direction of inquiry, saving the time and expense of and facilitating his own research. It is true that the exhibit is only to the extent furnished by this one library; but this is so complete, that there is hardly a medical subject of a character even the most remote, concerning which there may not be found there satisfac-

tory information. It will doubtless surprise even physicians that a mere list of medical periodicals should cover 126 pages; and yet there are some which have not yet, though much desired, found their way to the library shelves.

From the great mine of citations arranged under appropriate and convenient headings from A to Berlinski, it will suffice to adduce a few illustrations of the completeness and thoroughness of the work.

Within a few months the subject of *abstinence* and *fasting* has acquired an unusual interest. In the volume before us, after referring to other allied headings, *forced alimentation, diet in disease, famine, inanition, starvation* and *temperance*, we are confronted with 21 author-titles on this special subject and more than a hundred journal articles.

Addison's disease, after referring to *discoloration of the skin* and *pathology of suprarenal capsules*, cites 26 author-titles and 300 journal articles.

Disease of old age, after referring to *bloodletting in old age, senile dementia, typhoid fever in old age, senile gangrene, marasmus, pneumonia in old age, enlarged prostrate* and *prurigo*, cites 37 author-titles and 50 journal articles.

Compressed and rarefied air as a remedy, after referring to *haemospasia* and *treatment of phthisis*, gives 20 author-titles and 200 journal articles.

Anatomy inclusive of allied headings covers 80 columns ; *aneurism* in the same manner 120 columns ; *angina pectoris* 7 columns ; *aphasia* 10 columns; and *asthma* 13 columns.

The minute sub-division of subjects facilitates research, and the examples given above show the fulness of information that the library affords.

The index-catalogue has been supplemented by the *Index Medicus*, a monthly publication which has brought out the later medical information continuously to date.

On the whole, the index-catalogue as well as the index-medicus are invaluable as a ready means of research ; and both the inception and execution are alike creditable to the gentlemen, who, without additional emolument, have risen above their office routine to grapple with this stupendous undertaking.   D. S. L.

MEDICAL NEWS.

DIAGNOSIS OF CUTANEOUS SYPHILIS.

Dr. George Henry Fox, in a clinical lecture on the diagnosis of cutaneous syphilis, strongly urges that the diagnosis should be made without seeking aid from the patient. Many patients, even those of intelligence, are ignorant that they ever had an initial lesson or any of the early secondary symptoms. On the other hand, others will tell you that they have had "the disease," when they may have had simply a venereal ulcer or a gonorrhœa.

In discussing the localization of syphilitic eruptions, he says "the earliest hyperemic macules are commonly seen upon the abdomen, upon the lateral aspects of the trunk, and upon the inner portion of the arms and thighs. They are most distinct where the skin is thin and delicate." The papules of early syphilis are often well marked upon the forehead, the back of the neck, the palms, and the soles, as well as scattered over the trunk and extremities. The early syphilitic pustules are somewhat similarly distributed. They are usually abundant on the lower extremities, and are sometimes found upon the glans and the sheath of the penis. The early syphilitic eruptions are always disseminate in character. The eruptions of late syphilis, on the other hand, are no longer disseminated, but are grouped together upon limited portions of the body, and show a marked tendency to a circular or crescentic arrangement.

Fox does not believe in the diagnostic value of the "copper" color of syphilitic eruptions. He thinks that the color of lean ham is better suited for comparison with certain syphilides than that of copper. But he considers the color as the least important of the various characteristic features of

syphilis. The scales occasionally found upon syphilitic papules and tubercles are usually thin and adherent. The crusts of late syphilis are thick, rough and greenish-black as a rule. Syphilitic ulcers are usually circular, except where two or more have coalesced, and their borders are usually sharply cut. The "horseshoe" form of ulcer is the result of the tendency to heal upon one side while the ulcerative process continues upon the other.

The absence of pain and itching is laid down as a very important diagnostic feature of syphilitic eruptions. In any case of skin disease accompanied with severe itching, the eruption is not syphilitic. The presence or absence of itching is to be determined, not by the testimony of the patient, but by the presence or absence of scratch-marks. In certain cases of syphilis, where itching was said to be present, the scratch-marks were found upon the sound portions of skin, not upon the eruption. Syphilis did not protect the patients from lice.

The Vienna school of dermatologists, as represented by Hebra, Kaposi and Neumann, agree with Dr. Fox in depending upon the objective signs rather than upon the testimony of the patient, when making a diagnosis in skin affections.—*New York Medical Journal.*

THERAPEUTICAL USE OF THE MAGNET.

Dr. William A. Hammond, in a paper read before the New York Neurological Society, and published in the New York Medical Record for November, has presented to the profession some very interesting facts in regard to the therapeutic use of the magnet.

He calls attention to the fact, that several years ago Baron von Reichenbach spoke of the sensitiveness of individuals, in perfect health, to the magnetic influence; the sensations varying from those of simple heat or cold, to those of pricking, drawing and creeping, headache even, being produced in some instances. Men, as well as women, were found to be susceptible, and children peculiarly so.

To eliminate, as largely as possible, any errors arising from the "principle of suggestion," Dr. Hammond then performed the following experiment : A gentleman, thirty years of age, of an unimpressionable nature, was blind-folded, his right arm was then laid bare and every effort was made to fix his attention upon his arm as the part to be experimented upon. A strong horse-shoe magnet was then held over the nape of his neck, about an inch from the skin. In thirty-two seconds, he said : "I feel nothing at all in my arm, but I feel a queer, numb sensation at the back of my neck." The magnet being then brought over the top of his head, while his arm was being stroked with a paper-knife: "I feel you rubbing my arm with something," he said, "but the numbness has gone out of my neck, and is just on the crown of my head." This, and other experiments showed that the magnet produces sensations in parts of the body where it's proximity is not suspected.

Reichenbach further claims, that if persons of neurotic temperaments be brought into a dark room, in which are several magnets, they can in a few minutes determine the position of the magnets by the luminous rays given off from their poles.

Dr. Vansant has demonstrated that the conjunctival membrane indicates by feeling, with which pole of the magnet it is touched ; the southward, or positive pole, producing, when applied, a sharp sensation and an involuntary closing of the eyelid ; while the northward pole can be applied without causing pain or winking. He also found that the pain of facial neuralgia was increased by the application of the northward pole ; but was diminished by the application of the southward pole.

Dr. Hammond then gives the results of his own experiments. In two cases, out of nine, of chorea, "complete cures" were produced in a few minutes, by the use of the magnet ; in the seven others, no results followed. In the first case cited of "Paralysis from Cerebral Hemorrhage," where the patient was hemiplegic on the left side and had lost the power of speech, a double horse-shoe magnet was applied on the fifth day after the attack. Sensation was restored in ten minutes, throughout the thigh, leg and foot. After three hours the patient was able to "walk across the floor without difficulty." A ten pound magnet was used.

The second case was one of paralysis of the right side of the face, with cutaneous insensibility, and loss of motion in the right leg and arm. Pinching the skin, so that the nails almost met, produced no sensation; in three minutes, however, after a five pound magnet was placed upon the affected side, its poles being turned upwards, the "restoration of sensibility was thorough and it has remained intact up to the present time." Dr. Hammond cites two cases of a similar nature, reported by Mr. Debore, where recovery followed on the use of the magnet.

In conclusion Dr. Hammond says : "It would be asking too much to claim that the cures in the instances cited, whether of chorea or of paralysis, were due to any specific influence of the magnet. It is possible that the association was a mere coincidence, or that the relief was due to a strong mental impression made on the mind of the patient. At any rate, the cases are interesting as facts, and are worthy of being allowed to exert influence in directing further inquiry."

Two Examples of Purpura Hæmorrhagica from Widely Different Primary Causes.

By WALTER C. STILLWELL, M. D.

Chas. H., æt. 8, while playing in a stable yard was incarcerated in a dung-pit by his playmates for about twenty minutes to half an hour, and a few hours after being liberated he was seized with a chill, followed by the appearance of red spots over the whole body, which gradually became of a purple color, and soon after he began to have hemorrhage from the nose, which was very profuse and lasted for some considerable time ; he became exceedingly weak, his pulse grew feeble and very rapid, and for a time his condition was very alarming. On making an examination of his throat, I found the tonsils so much enlarged that they almost filled up the fauces ; they were of a dark-purple color, and looked as if the blood was about to start from them. The other portion of the mucous membrane of the mouth and throat was very pale and unnatural in color.

Attention was first directed to stopping the hemorrhage, which was done after much difficulty ; the various vegetable and mineral astringents were used without effect, and, as a last resort, I was compelled to plug up the nostrils with cotton saturated with a strong solution of alum.

The tinct. ferri chlor. with potass. chlorat. in large doses was administered internally, in conjunction with strong milk-punch. This treatment was continued for about five days, when the sulphate of quinia was added to the other medicinal agents. A week or ten days passed before I considered him in a fit condition to leave.

The other case was that of a little girl, 9 years of age, who was a typhoid fever patient, the purpuric spots appearing about the tenth day of the attack, accompanied by a profuse bleeding from the nose, as in the other case, and after a day or two, cerebral symptoms set in and the child died.

Here are two cases of purpura hæmorrhagica, with the extravasation of blood in or upon the true skin, appearing in spots over the whole body ; both are caused by a depraved or poisoned state of the blood and the relaxed condition of the capillary vessels.

In the first case, the child inhaled the ammoniacal and other putrescent vapors from the refuse matters in the pit, and the inoculation or absorption of the poison was through the respiratory apparatus ; but in the second case, the typhoid fever patient, it occurred through the medium of the bowels, the matters from the diseased glands of Peyer being absorbed into the blood.

MISCELLANEOUS.

ANTI-VIVISECTION.—"We observe that a memorial recently presented to Mr. Gladstone, urging him to do all in his power for the absolute abolition of vivisection, was signed by "one hundred representative men ;" among them Cardinal Manning, Prince Lucien Bonaparte, John Ruskin, Alfred Tennyson, Robert Browning, James Anthony Froude; the head-masters of Rugby, Harrow, and seven other large schools ; twenty-one physicians and surgeons; and thirty-seven peers, bishops, and members of parliament. The memorialists take the ground that vivisection, even with anaesthetics, should by law no longer be allowed, and they quote the opinions of Sir William Fergusson, Sir Charles Bell, and Mr. Syme, that "it has been of no use at all, and has led to error as often as truth." They add that the utility, if proved, would not, in this case, excuse the immorality of the practice. Of the persons herein named as being in favor of total abolition of vivisection, Sir William Fergusson and Mr. Syme may by accepted as competent to express an opinion with as much authority as belongs to eminent surgeons of the past age, which believed in nothing new and practiced rule of thumb. Of the remaining "one hundred men" probably not one knows anything whatever of the practices which he seeks to abolish, or of the results of those practices. Most of them are essentially "emotionalists" who would sign any earthly manifesto on a humanitarian subject if only their bowels of compassion were sufficiently harrowed up by the falsehoods we know, by experience, anti-vivisection agitators to be capable of. They are all either professed philanthropists and religionists whose signatures to such a document are a "matter-of-course," or else they are well-intentioned people who are simply bored into signing without inquiry or serious thought. As for the "twenty-one physicians and surgeons" we covet further information. We should like to analyse the list to ascertain how many of them are sim-

ply traders on gushing philanthropy as a means of advertisement or of fee-getting ; or how many of them are of the fossil *laudator temporis acti* class, who are still sceptical as to germ-theories, septicæmia, and evolution, and who believe in bleeding for a cold in the head. We wish the anti-vivisectors would offer us the name of even one single modern physician or surgeon of repute for learning and research who will go for total abolition of vivisection. One such signature would be worth the whole hundred of peers, members of parliament, poets, and schoolmasters, or of any number of the class of doctrinarians who are in the habit of making up their minds without thinking, and uttering manifestoes without learning."
—*Medical Press and Circular.*

A BILL

FOR AN ACT TO PROTECT THE PUBLIC HEALTH
AND REGULATE THE PRACTICE OF MEDICINE
IN THE STATE OF COLORADO.

Be it enacted by the General Assembly of the State of Colorado :

SECTION 1. That a Board is hereby established which shall be known under the name and style of the State Board of Medical Examiners, to be composed of nine practicing physicians, of known ability and integrity, who are graduates of medical schools of undoubted respectability, giving each of the three schools of medicine (known as the Regular, Homœopathic, and Eclectic schools) a representation proportionate to the number of graduates of said school within the State.

SEC. 2. The Chief Justice of the Supreme Court of the State, shall, as soon as practicable after this Act shall have become a law, appoint a State Board of Medical Examiners as provided in section 1 of this act, and the members first appointed shall be so designated by the Chief Justice that the term of office of three shall expire in two years from the date of appointment ; and the term of office of three shall expire in four years from the date of appointment ; and the term of office of three shall expire in six years from the date of appointment. Thereafter the Chief Justice shall biennially appoint three members, possessing qualifications as specified in section 1, to serve for the term of six years, and he shall also fill all vacancies that may occur, as soon as practicable ; provided, that in making biennial appointments

or filling vacancies, the ratio of representation of the medical schools in the Board shall not be changed from the original basis as in section 1.

SEC. 3. The Board of Medical Examiners shall as soon after their appointment as practicable, organize by the election of one of their members as president, one as secretary and one as treasurer, and adopt such rules as are necessary for their guidance in the performance of the duties assigned them.

SEC. 4. That every person practicing medicine in any of its departments shall possess the qualifications required by this act. If a graduate in medicine, he shall present his diploma to the State Board of Medical Examiners for verification, or furnish other evidence conclusive of his being a graduate of a legally chartered medical school in good standing. The State Board of Medical Examiners shall issue its certificate to that effect, signed by a majority of the members thereof, and such diploma or evidence and certificate shall be conclusive as to the right of the lawful holder of the same to practice medicine in this State. If not a graduate of a legally chartered medical institution in good standing, the person practicing, or wishing to practice, medicine in this State, shall present himself before said Board of Medical Examiners and submit himself to such examination as defined in section 7 of this act, and if the examination be satisfactory to the examiners, the said Board of Medical Examiners shall issue its certificate in accordance with the facts, and the lawful holder of such certificate shall be entitled to all the rights and privileges herein mentioned. All persons who have made the practice of medicine and surgery their profession or business continuously for the period of ten (10) years within this State, and can furnish satisfactory evidence thereof to the State Board of Medical Examiners, shall receive from said Board a license to continue practice in the State of Colorado.

SEC. 5. The State Board of Medical Examiners, within ninety (90) days after the passage of this act, shall receive through its president applications for certificates and examinations. The president of said Board of Medical Examiners shall have authority to administer oaths, and the said Board of Medical Examiners to take testimony in all matters relating to its duties. It shall issue certificates to all who furnish satisfactory proofs of having received diplomas from some legally chartered medical institution in good standing. It shall prepare two (2) forms of certificates—one for persons in possession of diplomas, the other for candidates examined by its members. It shall furnish to the county clerks of the several counties a list of all persons receiving certificates. Certificates shall be signed by a majority of the members of the said Board of Medical Examiners granting them.

SEC. 6. There shall be paid to the treasurer of the State Board of Medical Examiners a fee of one dollar ($1.00) for each certificate issued to graduates or practicioners of ten (10) years' standing, and no further charge ahall be made to the applicant; candidates for examination shall pay a fee of ten dollars ($10) in advance.

SEC. 7. All examinations of persons not graduates shall be made directly by the State Board of Medical Examiners ; examinations may be in whole or in part in writing, and the subjects of examinations shall be as follows: Anatomy, Physiology, Chemistry, Pathology, Surgery, Obstetrics and Medicine (exclusive of Materia Medica and Therapeutics.)

SEC. 8. Every person holding a certificate from the State Board of Medical Examiners shall have it recorded in the office of the Clerk of the County in which he resides, and the record shall be endorsed thereon. Any person removing to another county to practice shall procure an endorsement to that effect on the certificate from the County Clerk, and shall record the certificate in like manner in the county to which he removes, and the holder of the certificate shall pay to the County Clerk a fee of one dollar ($1.00) for making the record.

SEC. 9. The County Clerk shall keep in a book provided for the purpose a complete list of the certificates recorded by him. If the certificate be based on a diploma, he shall record the name of the Medical Institution conferring it and the date when conferred. This register shall be open to public inspection during business hours.

SEC. 10. The State Board of Medical Examiners may refuse certificates to individuals guilty of unprofessional conduct of a criminal nature, and they may revoke certificates for like causes.

SEC. 11. Any person shall be regarded as practicing medicine within the meaning of this act who shall profess publicly to be a physician and prescribe for the sick, or shall attach to his name the title " M. D.," or "Surgeon," or " Doctor," in a medical sense. But nothing in this act shall be construed to prohibit gratuitous services in cases of emergency.

SEC. 12. Any itinerant vender of any drug, nostrum, ointment, or appliance of any kind intended for the treatment of disease or injury, or who shall by writing or printing, or any other method, publicly profess to cure or treat disease, injury or deformity, by any drug, nostrum, manipulation or other expedient, shall pay to the treasurer of the State Board of Medical Examiners the sum of one hundred dollars ($100) quarterly for a license, to be collected by the treasurer of the State Board of Medical Examiners.

SEC. 13. Any person practicing medicine or surgery, in any of their departments, in this State, without complying with the provisions of this act, shall be punished by a fine of not less than fifty dollars ($50) nor more than three hundred dollars ($300), or by imprisonment in the county jail for not less than ten (10) days nor more than thirty (30) days, or by fine and imprisonment, for each and every offense ; and any person filing, or attempting to file, as his own the diploma or certificate of another, or who shall give false or forged evidence of any kind, shall be guilty of a felony, and upon conviction shall be subject to such fine and imprisonment as are made and provided by the statutes of this State for the crime of forgery.

SEC. 14. All fees received by the treasurer of said Board of Examiners, and all fines collected by any officer of the law under this act, shall be paid into the State treasury ; and all necessary expenses of the Board shall be paid for out of the fund of the State treasury not otherwise appropriated ; but no fee shall be required or accepted by any member of the Board for services.

SEC. 15. The State Board of Medical Examiners shall meet as a Board of Medical Examiners, in the City of Denver, on the first Tuesday of January, April, July and October of each year, and at such other times and places as may be found necessary for the performance of their duties.

SEC. 16. Justices of the peace and all courts of record in the State of Colorado shall have full jurisdiction over and power to enforce the provisions of this act.

To the Editor of the Leadville Chronicle.

At the opening session of the Legislature there is, we understand from that able journal, the *Rocky Mountain Medical Review*, to be presented a bill entitled, "A bill for an act to protect the health and regulate the practice of medicine in the State of Colorado." The general scope and intention of the bill, which may not perhaps be fully conveyed to the public by the title thereof, is to provide for the exclusion of improper persons from holding themselves out as practitioners in either of the accepted schools of medicine. The bill is of such importance to the public at large, as well as the medical profession, that it should receive careful consideration at the hands of our legislators, and if inconsiderately passed, such an event should be deprecated by unprejudiced right thinking men. Without in this article attempting to elaborately criticise the bill, but rather to invite enlightened discussion concerning its merits, we may say that it seems to incorporate several provisions which are ill advised and pernicious.

To begin with Section one. While it is marked with a spirit of progressive fairness, as emanating from the medical profession, in that it provides for the appointment of a board of medical examiners composed of nine persons, "giving each of the three schools of medicine (known as the regular, homœopathic and eclectic schools) a representation proportionate to the number of graduates of said school within the State," it is nevertheless, as we think, open to the objection that it leaves the whole matter entirely in the control of the profession, disregarding the salutary effects of the system of "checks," which has now engrafted itself upon all our legislation, both state and national, to such an extent that we may assert a deviation from it to be against the genius of our institutions. It is a matter of impossibility for a set of men whose prejudices and interests run in the same groove, to sit in the administration of judicial and executive functions at once. In fact, in this State, considering the disparity in the proportionate number of the other two schools to those of the so-called "regular," the control of the administration of the powers sought to be given to the Board of Medical Examiners would practically be with those of the "regular school."

But this aside, we suggest the advisability of the substitution for three of the number mentioned, three persons not members of the profession, leaving a clear majority of three to the medical gentlemen. The ready objection to this, we are aware, will be that none are so competent to pass on the competency of applicants as the members of the profession to which they wish to gain admission. But it will be seen that such extensive powers are given to the Board by sections 3, 4, 5, 8 and 10 of the proposed act, among others the power to "revoke" licenses already granted, as well as to refuse certificates to persons "guilty of unprofessional conduct of a criminal nature," that the rights of the citizens would greatly be jeopardized, were there no check upon the exercise of such almost arbitrary powers by the inclusion of an unprejudiced element.

Section four, it seems to us, is liable to a serious objection, in that without sufficient reason it departs from a custom which is at least time honored, to-wit: That the appointment of State officers should be made by the executive, and what is much more grave, in fact, one of the most salutary of the rules which has as yet not been destroyed by the prentice hand of new fledged legislators, to-wit: That the judiciary should never be invested with political patronage of any kind whatsoever. Never! Space forbids our noticing in detail at present the other specific provisions of sections 3, 4, 5, 6, 7, 8 and 9, which will doubtless hereafter be fully discussed prior to any action being taken on the final passage of the bill.

Section 10, it appears to us, should be absolutely expunged. Such powers as are here sought to be given, are so dangerous as to shock one in contemplation of the injustice which might ensue in their exercise by the tribunal attempted to be constituted. If time permitted we could show how such powers cannot be constitutionally delegated to such a tribunal. Let the law and medical profession by all means ostracise and

eject unworthy members from their voluntary associations, and thus set their brand
upon them under the responsibilities to
which the law properly subjects the libellant.
But to bestow such a power absolutely to
one class of persons, and that an interested
class, is an attempted exercise of power
which must shock the conscience and "make
the judicious grieve."

As to the 12th section of the proposed
Act, we scarcely know what to say. If it
were not that we desire to speak courteously
of a measure evidently intended on the
whole for the benefit of the public at large,
as well as for the honorable members of the
medical profession, we could not refrain
from characterizing it as absurd inherently.
It is full of contradictions, and entirely inconsistent with what has preceded it. It
can not be quoted at length, but suffice it to
say that where in such a wholesale way it
makes a distinction between "itinerant"
venders of drugs, etc., and local venders of
the same merchandise, "itinerant" professors to cure or treat disease and local professors of the same art, by imposing such a
heavy burden upon the "itinerants" as to
amount to a punishment, and we might say
an extinction, it savors not only of injustice,
but illegality. But when we come to consider the latter portion of the section,
wherein it is provided that by paying $100
quarterly to the "Treasurer of the Board of
Medical Examiners" for "license," the
"itinerant" may "work his own wild will"
upon the public, its inconsistency with the
apparent general spirit of the bill is so conspicuous as to lead one to believe almost that
it was the intention of the framers thereof to
provide a means for levying a species of
blackmail upon the class denounced as
"itinerants."

It may perhaps be claimed that section
fourteen relieves the act of the obnoxious
feature above referred to, but all lawyers
know, as well as laymen, that "fees," when
their uses have been designated in the previous portions of the law, can only be

applied to such uses, and that "license" is
a term which refers to something entirely
separate and distinct.

In another column will be found an article
on a subject which has already been agitated
in many quarters, and which is of peculiar
importance to the inhabitants of Leadville.
The subject is that of controlling and super-
vising the practice of medicine in this state,
and protecting the people from the ravages
of unprincipled quacks who, without any
thorough medical education and the necessary practice and experience, undertake to
meddle with health and life itself. The
article is from the pen of Mr. W. J. Sharman, of the law firm of Caspar & Sharman,
and merits careful perusal. It is a legal
opinion given at our request without any
particular study and preparation, but—as
we said before—the position taken by Mr.
Sharman merits careful and deliberate consideration. We print in another portion of
the paper the bill to which it refers, which
was prepared by the State Medical Society
for the purpose of being laid before the
state legislature at its next meeting.—*Leadville Chronicle, Dec. 15, '80.*

An article in our yesterday's issue, from
the pen of one of our prominent lawyers,
commenting on the bill proposed by our
State Medical Society and urged by them
for passage by our state legislature at its
approaching session, has been variously
commented upon. It is important that this
subject should receive early and full consideration. Leadville, particularly, has suffered from the presence of incompetent and
brazen-faced quacks, who ought to be weeded
out of the medical fraternity by the strong
arm of the law. No other power will do it,
for this class of pretenders manage to
ingratiate and keep themselves in the eye of
ignorant and gullible people with a perseverance worthy of a better cause, while the
worthy and reliable physician of good
standing is kept by the ethics of his profes-

sion from thrusting himself before the people. We publish to-day the text of the proposed bill, which, taken in connection with the legal opinion which we published yesterday, will give our readers an insight into the merits of the case.—*Leadville Chronicle, Dec. 16, '80.*

OZONE.—M. P. Hautefeuille communicates to the Academy of Sciences the observation that ozone at a high tension appears of a beautiful blue color, which deepens as the pressure increases.

EFFECTS OF SANITARY IMPROVEMENT.—Two reports have recently been published in Great Britain which illustrate what has been accomplished in lessening the prevalence of disease and prolonging human life by measures of sanitary improvement. The improvement trustees of Glasgow have given out a statement showing that the average death-rate per one thousand persons has been reduced nearly eleven per cent. in twelve years, under the operation of the sanitary measures instituted by them, which included the demolition of unwholesome dwellings, and the provision of ample hospital space for small-pox and fever cases, and for the control and limitation of epidemic disease. They also cite from the report of the Register-General figures showing that a similar improvement in sanitary condition has been wrought in other towns ; in Edinburgh of fourteen, in Dundee of twelve, in Aberdeen (where the death-rate was already very low) of three and one-half per cent. The figures given of a number of English towns show a less striking rate of improvement. Dividing the twelve years into two groups of six years each, it is found that, in twelve leading towns, 61,000 fewer deaths occurred in the second six years (1873–'78) than would have occurred under the higher death-rates of 1867–'72. The sanitary officer of Manchester has reported to the bishop of the diocese that, under the operation of the measures which have been

adopted in that city, "typhus and typhoid fever, though not absolutely extinguished, are of comparatively rare occurrence, and nearly all other infectious diseases have been largely reduced in amount, while the general health has been improved."—*Popular Science Monthly.*

BOOKS AND PAMPHLETS RECEIVED.

" The Symptoms of Sexual Exhaustion," (Sexual Neurasthenia). By George M. Beard, A.M., M.D. Reprinted from The Independent Practitioner. Baltimore : Practitioner Publishing company, 1880. Pp. 19.

" Pocket Therapeutics and Dose Book." Second edition. By Morse Stewart, Jr., B.A., M.D. Detroit, Mich.: George D. Stewart, 1878. Pp. 263.

" The Medical Record ;" a weekly journal of medicine and surgery. George F. Shrady, A.M., M.D., editor. New York : Wm. Wood & Co. [Placed on our exchange list.]

" The Boston Medical and Surgical Journal ;" a journal of medicine, surgery, and allied sciences ; published weekly. Boston : Houghton, Mifflin & Co. [Placed on our exchange list.]

" Ninety-eighth Annual Catalogue of the Medical School (Boston) of Harvard University. 1880–81." Cambridge : Charles W. Sever, University bookstore, 1880.

" The Southern Clinic ;" a monthly journal of medicine, surgery and new remedies. C. A. Bryce, M.D., editor and proprietor. Richmond, Va.

" The Medical and Surgical Reporter ;" a monthly journal. D. G. Brinton, M.D., editor. Philadelphia. [Placed on our exchange list.]

" The Medical Library Journal ;" issued on the first of each month. Boston.

" The Chicago Medical Review ;" a journal of medicine, surgery and the allied sciences. Published semi-monthly. E. C. Dudley, editor. Chandler & Engelhard, publishers, Chicago. [Placed on our exchange list.]

" The Sanitary Journal ;" a journal of hygiene and public health. Edited by James Christie, A.M., M.D. Alex. Macdougall, publisher, Glasgow. [Placed on our exchange list.]

ROCKY MOUNTAIN MEDICAL REVIEW.

Vol. I. COLORADO SPRINGS, JANUARY, 1881. No. 5.

CLINICAL LECTURES.

LECTURE I.—CLINICAL SURGERY.

By David W. Cheever, M. D.,

Professor of Clinical Surgery, Harvard Medical School; Surgeon to the Boston City Hospital.

Reported for the Review by Mr. H. C. Coe, of the Harvard Medical School, Boston.

Case I. Boy of 12 years. This case is one of congenital naevus, appearing at birth as a small pimple spot on the middle of the lower lip, which has gradually developed into the large mass which you see at present. It includes the whole of the lip, and, as I invert it, you see that the growth extends widely beneath the mucous membrane. It is due to a local dilatation of the blood-vessels. These often shrink and atrophy as dentition is accomplished; hence if the naevus is not large, it is better to wait till teething is over and see if the mass does not disappear of itself. When large these naevi resemble cirsoid aneurisms.

Treatment.—When small and occurring in an infant the naevus may be vaccinated, and thus obliterated by the subsequent cicatrization. It may also be frequently painted with contractile collodion. If we wish to employ more radical measures, there are four methods open to us, viz : a. If small and circumscribed, naevi may be excised. b. If not circumscribed, the vessels may be strangulated. c. We may inject astringent solutions (especially of the Fe salts) directly into the growth. d. Electric currents may be passed through needles, introduced into the naevus. There is more or less danger of embolism in the use of the last two methods, and on the whole, rather neater results may be obtained by excision than by ligation.

In this case I make a V shaped incision through the integument, its angle being just above the chin. I now dissect around the entire mass, tie the spurting vessels (as I am obliged to cut the coronary artery) and having pierced the base of the naevus with a needle, carrying a stout double ligature, the vascular supply is easily cut off. The growth is now drawn into the mouth and left there, the external wound being closed by hair-lip pins and sutures. In a few days the strangulated mass will slough off, and we shall have a good lip, with very little, if any, deformity.

Case II. I thought fit to bring this case before you and operate in your presence. The operation will be long and tedious, but it is a comparatively rare one, and I thought that you might be interested in it. This young woman has long been under treatment for bronchocele, but with no apparent benefit. The external use of iodine and the internal use of the iodides and bromides often seem to be efficacious in glandular swellings, but in these cases excision is the only radical cure. I shall not discuss the subject, but proceed at once to the operation, describing the steps as I proceed. A long incision is made over the tumor, parallel with the sterno-mastoid, and another at right angles to it. The superficial and deep layers of fascia, platysma, sterno-hyoid and thyroid are successively divided on a director. It is necessary to proceed slowly and cautiously, as we are in the vicinity of important parts. The thyroid veins prove very troublesome and must be ligatured as they are divided. Having fairly enucleated the tumor, and made sure that no important veins or vessels are implicated, its base is pierced with stout curved needles carrying waxed ligatures, and the vascular supply is completely cut off.

I now excise as much of the tumor as possible above the ligatures, and apply the thermocautery to the stump, carefully drawing the sheath of the great vessels and nerves to one side to avoid injuring it. The external wound is not to be entirely closed, but abundant provision be made for free drainage. Such wounds as these form very ugly receptacals for pus, which I have known to burrow down even as far as the sternum. I shall leave a drainage-tube in the wound and have it syringed out daily. In a few days I hope that the stump will slough off. Hemorrhage is to be carefully watched for. The process of healing is apt to be retarded here, from the anatomy of the parts and the impossibility of securing perfect rest, because of the respiratory movements. On making a section of the tumor it is sure to be very vascular, and to contain a large cavity filled with the products of colloid degeneration.

Case III. This man has a fistula, upon which I shall operate, and, while he is being etherized, I shall say a little about this affection. Fistula arises from a small abscess in the ischio-rectal fossa. Some authorities think that it begins as an ulceration *inside* of the rectum, others that it originates *outside*. The opening usually forms just inside the sphincter, but the external abscess may run up for an inch or two along side of the gut.

Treatment.—This is comprised in the finding and obliteration of the internal opening. Having introduced a fine probe through the external opening, pass the finger up the rectum and try to feel the probe. Be very careful not to make a false passage, for this has often occurred by forcing the probe inward under the mistaken idea that it was already within the rectum, when it was separated from the finger by only a thin layer of mucous membrane. Next introduce a speculum and endeavor to see the end of the probe. Pass a director through the opening and slit up the sinus. If there are several sinuses running along under the skin, slit them all. The raw cavity resulting must be carefully packed with small sponges or

carbolized cotton for several weeks, to prevent the old sinus from bridging over again. The average time of cure varies from three to six weeks. One case I treated for upwards of a year before complete cure was attained, being obliged to pack the cavity nearly every day.

LECTURE II.—CHRONIC LARYNGITIS.

By E. Fletcher Ingals, A.M., M.D.

Clinical Professor of Diseases of the Throat and Chest, Woman's Medical College; Lecturer on Physical Diagnosis and Diseases of the Chest, and on Laryngology, Rush Medical College, Chicago.

C— O——, aet 27, sailor. This patient comes to us complaining of "tightness of the chest and hoarseness, with palpitation of the heart." The tightness seems to have begun about a year ago, when he gave up his ordinary avocation on the sea and began to work ashore. These sensations have some of the time been quite troublesome, but at present he chiefly complains of his throat. For the past two months he has noticed an impairment of the voice, which totally prevents him from singing, and which somewhat interferes with speaking. This hoarseness is very marked in the morning, but it gradually wears away during the day, and by evening his voice is quite clear. For the past five days he has been very hoarse as the result of an acute cold.

An inquiry into the former history of this patient, and into his hereditary history, reveals no predisposing cause for his present difficulty. We can obtain no family history of pulmonary or cardiac disease, and he affirms that he has never suffered from venereal disease. However, considering the patient's occupation, this statement must be taken *cum grano salis*. Besides the sensation of tightness at the lower part of the sternum, and the frequent palpitation of the heart, our patient complains of a tickling sensation in his throat, which causes more or less cough. He tells us that he has lost about ten pounds of flesh during the past year, but he appears well nourished.

We find his skin cool and moist, pulse 84 and respiration normal. His voice is quite hoarse. He tells us that he frequently has severe paroxysms of cough, but that he expectorates only a small quantity of tenacious mucus, which comes up in small pellets after prolonged coughing. Besides this, a small amount of frothy mucus is expectorated. The patient's tongue is large and white ; the mucus membrane of his throat is nearly normal. His appetite is good, bowels constipated, kidneys acting naturally.

The sensations of tightness and the palpitation at once direct our attention to this patient's chest.

A careful physical examination, however, reveals no abnormal sound, and, therefore, yields only negative signs ; but we have the larynx yet to examine.

Upon laryngoscopic examination we discover the cause of hoarseness, for we find that the mucous membrane covering the supra-arytenoid cartilages is slightly congested and swollen, and both vocal cords are of a light red color, and appear thicker than normal ; the cord of the left side being more swollen than its fellow. In respiration and phonation the cords move naturally.

We have here the physical appearance of subacute or chronic laryngitis. Upon glancing down the trachea, its surface is seen to be redder than natural, or, in other words, there is evidence of tracheitis. With these signs before us, we need not hesitate in pronouncing this a case of Laryngo-tracheitis, and from the history, we know that it has become chronic in character, but that an acute exacerbation of the inflammation has recently occurred.

Leaving aside the symptoms of "tightness and palpitation," the statements of this patient point directly to chronic laryngitis. There has been gradual failure of the voice, owing, as it seems from the patient's later remarks, to straining of the voice in singing. This is the most common history of the beginning of chronic laryngitis. The recent acute exacerbation is one of the ordinary sequences. The tickling sensation in the throat and the prolonged and severe attacks of cough, which result in the expulsion of

only a small pellet of mucus, are among the usual accompaniments of this affection.

Loss of flesh is also a frequent symptom in chronic laryngitis, but in the case before us, an uncertainty on the part of the patient should cause us to almost ignore this symptom.

In chronic laryngitis the hoarseness is nearly always worse in the morning, use of the voice and quickened circulation in the parts seeming to remove a part of the congestion and swelling, causing the voice to clear up toward the latter part of the day. This symptom of hoarseness, however, may result from many other causes, among which the most likely to mislead us, are paresis of the laryngeal muscles and chronic œdema of the larynx.

In paresis, the hoarseness is usually less after rest, and it increases with use of the voice. Inspection of the larynx reveals loss of motion in paresis, and in œdema a semitransparent, or a light, pinkish swelling, which has the appearance of being caused by submucous infiltration, neither of which conditions is present in uncomplicated laryngitis. The appearance of this patient's tongue, and the condition of his bowels, are those generally found in chronic laryngitis. Now it still remains for us to ascertain the cause of the palpitation and the "tightness" of the chest. I find upon inquiry that this patient uses a considerable quantity of tobacco, and in the absence of physical signs, I am inclined to atribute his palpitation to this fact. The sense of tightness may be referred to the same source ; but there is another cause operating in this case which may account for the latter symptoms. Inspection of the trachea has shown us that its mucous membrane is inflamed ; now it will often be found that patients affected with chronic inflammation of this tube or of the larger bronchi, complain of a sense of constriction at the lower part of the sternum, which is variously described by them as "tightness," "oppression," "pressure" or "difficulty in breathing."

Finding no hereditary disease or venereal

taint in this case, and no evidence of the gouty, rheumatic, or dartrous diathesis, I conclude that this inflammation is of a simple catarrhal character.

My prognosis is favorable, providing the patient will second the internal and local treatment by strict compliance with the hygienic regulations which I shall direct. The first and most important direction which I shall give him, is that he must not use his voice whenever it can be avoided; that he must not sing or shout, and that when compelled to speak, he must do so with as little effort as possible. Next, I shall interdict the use of tobacco, and, finally, I shall enjoin the greatest care on his part in the avoidance of all causes of acute colds.

If this patient were so situated that I could see him frequently, I should make topical applications to the larynx of strong solutions of the mineral astringents; but as it will be impossible for him to call on me regularly, I must attempt by other means to obtain a favorable result.

In simple catarrhal inflammation of the mucous membrane, no internal remedy is more generally useful than chlorate of potassium; therefore, I shall order it in this case, to be taken in doses of four to eight grains every one or two hours during the day. One of the most satisfactory methods of administering this remedy is in the form of compressed pills, which should be allowed to dissolve slowly in the mouth. After ten or fifteen days, if this remedy does not prove effectual, I shall resort to chloride of ammonium, and later, possibly, to the iodide of potassium with small doses of copaiba.

In these cases much benefit is often derived from the use of vapor inhalations, therefore these will be conjoined with the internal treatment. I shall recommend for this patient the inhalation, night and morning, of vapor from the compound tincture of benzoin. For this remedy, if it is not alone sufficient, may subsequently be substituted oil of white pine or creasote, in the proportion of one-half to one drachm to the ounce of water.

The directions accompanying either of these will be that a pint of water at 150° F., or as hot as the patient can possibly bear his hand in, be placed in the inhaler and that to it be added a teaspoonful of the medicine. From this the patient should take thirty or thirty-five long and deep inspirations, which should occupy six or eight minutes. The patient must not go out of doors for half an hour after the inhalation.

If these vapors are not sufficiently stimulating, camphor may be added, five or ten grains to the ounce of the other medicine. By pursuing this course I hope for a complete cure, but the treatment is likely to be prolonged over several months.

ORIGINAL ARTICLES.

RECENT PROGRESS OF MEDICINE.

By D. S. LAMB, M.D.,
Professor of Anatomy, Howard University, Washington, D.C.

ERYSIPELAS IN SMALL-POX.

Cavarre, in the service of Brouardel (Jour. de Med., Nov. 1880, p. 514), has studied this question and believes the complication to be more common than is generally admitted, especially in epidemics of the benignant and discrete form; he considers that the appearance of erysipelas, the more so if limited to the face, renders the prognosis more favorable. It occurs almost always at the period of desquamation and may be unperceived. Erysipelas of the trunk and limbs makes the prognosis grave.

DIPHTHERIA.

Dr. Day, Victoria, Australia, uses table salt as a gargle, a tablespoonful to a tumbler of water, both as a prophylactic and cure. Children too young to gargle may drink a teaspoonful or two occasionally.

Dr. Zinke, of Cincinnati, Ohio, (St. Louis Med. and Surg. Jour., Nov. 20., p. 585), recommends the inhalation of a solution of quinia, a drachm to the ounce of water, administered every hour through the steam atomizer. He believes that it dissolves the false membrane.

Vidal recommends tartaric acid locally

every three hours ; acid 10 parts, glycerine 15 parts, mint water 25 parts ; followed shortly afterwards by an application of lemon juice.

DANGERS OF USING THE FOOD-PRODUCTS OF ANIMALS AFFECTED BY TUBERCULAR CONSUMPTION.

This important subject was considered at the reunion of the session of the International Congress at Turin (Annals d'Hygiene pub. Nov., 1880, p. 472).

Prof. Brusasco, after giving a complete history of the question, based upon the most recent observations and experiments, expressed the opinion that the milk and flesh of tuberculous animals ought to be excluded as food. Prof. Bassi, though agreeing that preservative measures should be taken, recalled the opinion of Virchow against the identity of the consumption of beeves and the tuberculosis of man. Profs. Bizzozero and Nocard protested against this opinion and cited numerous facts, tending to show the identity of the two affections; they asked that measures be taken to preserve the public health. Considering on the one hand the fewness of the facts tending to show that the piece of raw meat contains the elements of virulence, and on the other hand the evils which would follow too great a diminution of the products of the slaughter-house, Nocard desired that the destruction of meat from tuberculous animals should be required only in the gravest cases. Bocard and Bassi wished to have it spread abroad that milk should be boiled and beef cooked well-done. After a long discussion the section voted that popular instruction should be printed and diffused ; that there ought to be established in the larger cities an inspection service of dairies ; and that the inspectors of slaughter-houses should eliminate from the animals to be slaughtered for food, all those in whom general tuberculosis was appearing.

SYMMETRICAL NEURALGIA IN DIABETES.

Dr. Worms, at the session of the Academy of Medicine, Sep. 28, 1880, (Archives gen. de Med., Nov. 1880, p. 625), read a memoir on this subject ; he concluded that there is a form of neuralgia peculiar to diabetes, in which the inferior dental and sciatic nerves are affected, and which does not yield to the usual treatment for neuralgia.

HEART DISEASE.

M. Potain, at the meeting of the French Association for the Advancement of Science at Rheims, August 1880, recommended milk diet in cardiac affections of renal or gastric origin on the principal of giving rest to those organs ; provided that the milk is digested and assimilated.

M. Barthe (La France Med.) recognized in a woman near the end of her pregnancy a blowing murmur instead of the first sound of the fœtal heart. The child was still-born and on examination was found healthy except that the heart was enormously hypertrophied and the tricuspid valve showed "plastic vegetative endocarditis."

LOCUST BEAN IN THE RIGHT BRONCHUS.

A case is reported (Brit. Med. Jour., Sept. 25, 1880, p. 508) of a boy, age 15, who was admitted to the St. Mary's hospital Dec. 24, 1879 and died Feb. 6, 1880. His symptoms which were perplexing were explained at the post mortem examination. There was destructive inflammation of the lower part of the lung and later an abscess below the diaphragm followed by pyaemic abscesses of the live ; the locust bean was in a dilated bronchus of the right lung.

HEMIGLOSSITIS.

This very rare disease (Jour. de Med., Sept. 1880,) is said to be found only on the left side of the tongue. The case reported was that of a man, age 25. The symptoms were the swelling of the tongue on the left side, dysphonia, nasal twang, abundant salivation, fœtid breath, but no pain in the face, neck or tongue ; the diseased part was hard and thickly coated on the surface. Emollient gargles with chlorate of potash internally. Recovery in ten days.

Dr. Gauthier (same journal, Nov. 1880, p. 499) reports a case of parenchymatous inflammation of the left side of the tongue ending in suppuration. There were pain and swelling of the tongue with dysphagia. The left side of the tongue was thickly furred ; some redness of throat ; abundant salivation ; pulse 100. Leeches were applied to the infra-hyoid region ; superficial scarification of the tongue ; gargles. The swelling became so great that the tongue protruded ; pulse became more frequent. Longitudinal incisions were now made on the left side of the tongue. The local trouble then diminished in intensity but the dyspnœa became agonizing and necessitated the next day that the abscess should be opened. The discharge of pus was attended by the abatement of the disease. Eight days later he had entirely recovered.

Dr. G. gives a curious explanation of the appearance of the disease only on the left side, viz : that on the morning of the attack, the patient, who was a great smoker, had been exposed to a very cold rain and had smoked all the time, and, as was his habit, held the pipe in the left side of his mouth ; that the irritation then produced on the left side of the tongue aided in localizing the inflammation. The reporter recalls the localization of labial carcinoma in smokers.

FLATULENCE, &C.

Dr. Ringer (Lancet) recommends for flatulence, acidity and pyrosis, glycerine in teaspoonful doses either before, with or after meals ; in water, coffee, tea, or lemon or soda water.

BILIARY CALCULI.

Dr. Kennedy (Lancet, Sept. 18, 1880), reports the prompt and painless expulsion of biliary calculi by the use of olive oil. Six ounces of the oil are given at bed time and a purgative dose of castor oil in the morning. The olive oil may be repeated more or less frequently according to the presence or absence of symptoms of obstruction.

SWEATING OF THE FEET.

Dr. Thin (British Med. Jour., Sept. 1880), recommends the following treatment for excessive sweating of the feet with bad odor. The stockings are to be changed twice daily ; they should be placed in a jar containing a saturated solution of boracic acid ;

after being dried they are ready to wear. Cork soles are used in the bottom of the shoes ; these are changed daily ; after using they are soaked over night in the boracic solution, and put aside for one day to dry.

CHRONIC RHEUMATOID ARTHRITIS.

Mr. Fletcher Little at a meeting of the Yorkshire branch of the British Med. Ass. June 16, 1880, reports three cases of this disease which were much improved under oil-rubbing, the galvanic current and Russian vapor bath.

INTERMITTENT HYDARTHROSIS.

(Jour. de Med., Nov. 1880, p. 512). After collecting the case reported by Landrieux in Jour. de Therapeutique and that of Sieligmueller in Deutsch. Med. Woch., and referring to that of Panas in the Jour. de Med. 1878, and the thesis of Ragon, Paris, 1877, the following comment is made: "The points common to all these observations are that in an individual otherwise well, without prodromata, without any appreciable cause, there appears an intermittent swelling of one or both knees, without inflammation and without fever. The swelling increases rapidly until it has attained its maximum ; then it becomes stationary for several hours and gradually disappears. In most cases these phenomena are reproduced with mathematical certainty on fixed days. The intervals last eight days or even a month. The attack lasts oftenest from four to six days and affects the knee or hip joint.

As to the nature of the disease one can say nothing certain. In thirteen cases collected by Sieligmueller, only two had been exposed to malaria, which militates against this hypothesis. S. is disposed to regard the disease as a vaso-motor neurosis. He believes that the sudden dilatation of the blood vessels of the synovial membrane can produce very rapidly a serous effusion in the joint. It remains to explain the periodicity of the affection. In the published cases both local and general treatment have been useless. In two cases only, quinine and arsenic have ameliorated somewhat the condition.

PSORIASIS.

Dr. Charrasse, of Montpellier, applies an ointment of 5 to 15 per cent. of pyrogallic acid; one to four frictions daily. No odor and no pain. Arsenic is given internally at the same time. The acid often produces a brown color which, however, disappears after a few days.

VACCINATION IN CHRONIC SKIN DISEASE.

Dr. Bessey (Canada Med. Record, Nov. 1880, p. 28), reports the successful treatment of eczema, psoriasis and other skin diseases which had become chronic, cured by vaccination.

OPIUM HABIT.

Dr. Irwin (St. Louis Clinical Record, Oct. 1880), adds his testimony to the curative effects of the fluid extract of coca in tablespoonful doses, for the opium habit.

GENERATIVE ABERRATION.

Paul Moreau, of Tours (Jour. de Med., Nov. 1880, p. 509), in a very elaborate work of 300 pages upon the aberration of the genesic sense, describes a very curious state, happily not very frequent, supervening with some individuals the first days following marriage. Examples of insanity determined by the first conjugal approaches are not rare. Esquirol relates having attended a lady who had had a maniacal attack the first night of her nuptials, her modesty having revolted at this critical moment. A young woman was so unhappily affected by the first approaches of her husband that her reason fled immediately. Skae gives to this kind of insanity the name of post-connubial. It is observed among vigorous individuals who have preserved a severe continence up to the epoch of their marriage. The first night of pleasure is followed by attacks of short duration similar to the epileptiform congestions of general progressive paralysis; oftener the symptoms presented by these patients are those of acute dementia. They are stupid, incapable of replying; these symptoms are temporary and the prognosis is favorable. Among women the disorders are more grave; there is notably a change of humor, which makes the woman detest the man whom she had adored. Patients may be dangerous both to themselves and others. The morbid delirium is so intense as to necessitate the greatest surveillance; they may acquire finally an incredible cunning.

Sexual excess which follows marriage, gives place to disorders of another kind, to tabes dorsalis, to acute mania, and according to Dr. Blandfort to general paralysis. Skae gives in corroboration a most curious observation; a man 42 years old, after long years passed in *affaires*, married and presented immediately after his nuptials, the symptoms of post-connubial insanity characterized by a spontaneous and unjustifiable hatred against his wife and he threatened to kill her. The latter had the courage to hide the mental state of her husband for four years, when his sickness changed its form. He suddenly fell in love with his wife and as a result of his ardor he became a father. But from that time he fell into dementia and died shortly afterwards of progressive general paralysis.

DANGER OF COMBINING ANODYNES.

A case is reported in which a young man who was delirious and had taken chloral without beneficial effect, was given a hypodermic injection of $\frac{1}{4}$ or $\frac{1}{6}$ grain of morphia; he soon became quiet, livid, and died in two hours. The cause of death was not explained.

ANTAGONISM OF MEDICINES.

Prof. R. Bartholow (Med. Record, Dec. 11, 1880, p. 645) considers chloral hydrate and strychnia to be antagonistic, but that chloral is more effective in strychnia poisoning than strychnia in chloral poisoning. The chloral should be given in sufficient quantity to stop the strychnia spasms. In chloral poisoning, 1-60 grain of strychnia should be given at first and 1-120 every half hour till the maximum is approached, to overcome the tendency to failure of heart and respiration.

Opium in medicinal doses counteracts the depressing effects of veratrum viride, gelseminum and aconite.

Morphia antagonises tea, coffee and their alkaloids ; and *vice versa.*

Morphia and chloroform. Bassis sa?s : "There may be obtained in man, with a little attention, by the combined action of chloroform and morphia, a state of complete insensibility to pain, with preservation, to a partial extent, of the intelligence, tactile sensibility, auditory and visual, and of the voluntary movements. From a practical point of view, the analgesia obtained by the combined action differs completely from the demi-anæsthesia caused by the employment of chloroform and ether singly in that it is not preceded or accompanied by a period of hyperæsthesia with violent excitement, and the tendency to exaggerated reflex arrests of the heart and after syncope." Prof. Bartholow recommends morphia ¼ and atropia 1-100 grain.

IODOFORM.

Dr. Biermann uses five to eight drops of oil of fennel to fifteen drops of iodoform to cover the odor.

CHIAN TURPENTINE.

Prof. Flueckiger (Pharm. Jour. and Trans. Oct. 16, 1880) suggests Algeria as a sufficient source of supply for this turpentine ; the island of Chio being unable to supply the demand.

—————•◦•—————

ACUTE RHEUMATISM AND THE SALY-
CILATES.*
————
By S. EDWIN SOLLY, M. R. C. S., ENG., Vice-President Colo-
rado State Medical Society.

GENTLEMEN.—Two years ago I had the honor of reading to you a paper upon this subject. The opportunities for studying this disease in this climate are few, and therefore the cases in my own practice that I can relate are scanty. There has, however, during the last two years, been much written on the subject and much difference of opinion expressed. This literature, Dr. Sinclair aptly terms chaotic. My endeavors will be to try and find what is reliable and true, and what are fair assumptions, that can be drawn from

this medley of experience and opinions, so that we who are from time to time called upon to treat this disease, and yet have but meager opportunities of personal study of its course and therapeutics, may learn from these battles of the giants on which side to range ourselves, or to follow the classic maxim, "In medio tutissimus ibis," which in the American vernacular means, it is safest to sit on the fence.

It may be remembered that Drs. Gull and Sutton, who treated a large number of cases simply by mint water, diet and rest, establish-ed the facts that the average mean duration of the pyrexia, pain, and other acute symptoms was from the first day of going to bed, about 12 days, by which time the patient felt well though weak, and that the subsidence began from the 8th or 9th day, and the case was discharged, when there were no relapses or complications, about the 21st day. In the cases which they treated by alkalies, blisters, iron, quinine, etc., the same average du-rations were observed and the same propor-tion of cardiac complications arose. The observations of Drs. Findlay and Lucas, Lancet, Sept. 20, 1879, show the duration of the days of pyrexia to be about the same under treatment by alkalies and quinine, viz: from 10 to 12 days. Other observa-tions I find generally accord with these, and therefore from 10 to 12 days may be taken as the average mean duration of the acute symptoms, in a case put under good hygienic and dietetic conditions and subject to almost any treatment save by the salycilates, or to no treatment at all. As to the question of length of days in the hospital, it is very dif-ficult to strike a fair average, for, as we know, patients are kept a longer or shorter time in the hospital for so many reasons over and above their physical condition, and there is no exact point of convalescence at which they are all discharged. Therefore the lee-way to figures on that point must be very great ; moreover we cannot easily find a similar period in private practice to which to compare the duration.

I have not found any statistics stating the average proportion of relapsing cases in a number treated by the expectant plan. Dr. Sibson reported in the British Medical Journal, August 13, 1870, that he observed that in a fourth of all cases the joint affection lasted over 21 days, and that half of his cases had relapses. Dr. Wade of Birmingham, in 1874, writes that out of 109 cases 25 relapsed.

The average number having heart complications has been variously estimated by Brouillard at 50 per cent, and by Chambers at 5 to 7 per cent. Fuller, who made a careful examination of a very large number of statistics, puts the average at one third. These figures are mostly taken from hospital experience and the complication had generally arisen before admission or treatment; in private practice where our patients are, as a class, better fed and housed, and are usually treated from the outset of the affection, the average is undoubtedly not so high.

Dr. Reginald Southey, who has devoted much time and thought to the subject, speaks very positively of there being two distinct forms of acute rheumatism, which he calls the continued and the relapsing forms. The features of the continued form are absence of cardiac and other complications, the pain and fever remitting generally on the 8th or 9th day, and about the 12th day in bed the patient feels well but weak, and is discharged from the 14th to the 21st day.

The relapsing cases which are as severe at first have generally been ill longer, but have gotten up and then relapsed; they mostly have an endo-cardial murmur, endo-carditis being as much a feature of the disease as the articular inflammations. Almost invariably there is a remission between the 5th and 8th day, and between the 12th and 13th day, a relapse lasting 24 to 48 hours; there may be other relapses or the case enters upon tardy convalescence about the 21st day and the average mean residence in the hospital is 42 days.

Roberts Bartholow in the Maryland Medical Journal, January, 1880, says that rheumatism without treatment hangs on about 42 days and tends to get well from the 14th day; and further that there are three types of rheumatic patients, the feeble and nervous, the florid and flabby, and the vigorous and full-blooded, each requiring different treatment. I need not here recapitulate the various plans of treatment in vogue before the introduction of salycin and the salycilates, but in going through the literature of the subject it is very evident that although some forms of treatment modified the symptoms, the number of days of fever and pain were about the same as when no drugs were used.

I cannot quote any statistics proving it, but there is a very general belief that the alkalies given in large doses greatly lessened the danger of cardiac trouble and were useful in modifying its effects when it arose. Dr. Austin Flint has recorded his opinion to this effect. The blister treatment as carried out by Dr. Davis, of London, undoubtedly relieves pain, and apparently also lessens the chances of cardiac complications, and this it is believed to accomplish not so much by the counter-irritation as by rendering the blood alkaline. The treatment by acids, including Garrod's lemon juice plan, undoubtedly often gave as much relief as that by alkalies. The tincture of perchloride of iron given to the anæmic has yielded good results.

Salicylic acid was first tried in the treatment of rheumatism from theoretical inferences: in 1872 Binz discovered that quinine which was known to be an antipyretic was also an antiseptic. Kolbe then found that salicylic acid was a much more powerful antiseptic than quinine, and therefore argued that it would prove also to be a more powerful antipyretic, which theory he shortly verified. It is now almost universally conceded by those who desire to procure the effects of salicylic acid, that it is best and safest to use one of its salts and the salicylate of soda has been the one generally adopted, it being more soluble and not having the irritant and disagreeable effects of the acid. It is also more rational to use it in preference to the acid, if, as most chemists state, it has to be converted into the soda salt before its action upon the disease begins. Salycilic acid is very rapidly eliminated as shown by its presence in the urine ten minutes after it is taken (E. Prideaux). It increases the excretion of urea and favors the elimination of uric acid.

M. Marret has shown that this increased excretion in acute rheumatism appears before the swelling of the joints subsides and continues after the fever has fallen, "so that not merely a febrile waste but a true crisis is induced, similar to what may always be observed before the termination of rheumatic fever. Contrary to what is seen in other fevers, instead of the dilated vessels allowing a greater discharge of heat the local liberation of salycilic acid checks molecular change, and the production of heat is diminished. Further evidence of this has been given by proof of the greater rapidity with which salycilic acid is disengaged and eliminated during an attack than after it is over; hence, during fever a larger quantity of the drug may be well borne, but its action is also more rapid and requires to be carefully watched" (Dr. W. Squire, Lancet, December 20, 1879). If toxic effects are exhibited it is advisable to examine the urine to ascertain whether the acid is being diminished or accumulating in the system. All observers record certain symptoms when the drug is pushed to extremes, the first of which are usually tinnitus aurium, deafness, vertigo, vomiting, depression of heart's action, and feebleness of vision running on to complete temporary blindness; these symptoms, however, disappear very quickly on discontinuance of the drug. Drs. Wolffberg and Reiss discovered hemorrhagic erosions of the stomach and intestines, which in some cases, however, were undoubtedly due to adulteration with carbolic acid. It is reported that a girl took 340 grains in 6 hours and recovered in 10 days.

It has been repeatedly stated by the opponents to its use, that it directly induced the cardiac complications, but I can find no evidence of this in the data at present collec-

ted. "It slows the pulse, lowers the blood pressure and diminishes the vascular tension" (Squire); if continued too long its depressing action therefore would make a cardiac lesion more dangerous. The mass of evidence however, goes to show that it has no direct influence one way or the other over either peri- or endo-carditis either during the attack or its consequences. Dr. Maclagan, the apostle of salycin, affirms that myo-carditis is common in acute rheumatism but not easily detected, as can be imagined, if this is so. The depressing effect of salycilic acid would much increase the danger, this opinion, however, I do not find corroborated elsewhere.

The records reported of cases of acute rheumatism in which salycilic acid or its salts were administered that I have found available are as follows: The report of Boston city hospital 1877, shows the average mean duration of fever days to be, 2.85 days, average mean duration of residence in hospital 18 days out of 106 cases. Dr. Greenhow reported to the Clinical Society (May 20, 1880) 50 cases treated by salycilate of soda in which the fever days, without counting relapses, were from 2 to 3, the average mean duration in hospital was 57 days. In obtaining this result he excluded 14 light cases, 2 cases of hyperpyrexia and 2 fatal cases. Dr. Hermann Weber reported 44 cases treated by salycilate of soda to the Clinical Society, in which the average number of febrile days, including relapses, was 14, whereas by other treatment it was 19. Drs. Findlay and Lucas (Lancet, Sept. 20, 1879) reported 60 cases treated by salycilate of soda, 60 cases by alkalies, and 38 by alkalies and quinine. Dr. Southey (St. Barth.w's Hosp. rep., vol. xv) does not state the number of febrile days in cases under salycilate of soda, but admits their being lessened though finding relapses frequent. Dr. Reiss, of the Berlin hospital, reported 400 cases treated by this salt, in most of which there were rapid reduction of temperature and pain, and hastened convalescence; though the mean duration of febrile days is not given in the abstract. Dr. Stricker reported 14 cases where the febrile days averaged 2. Professors Traube and Senator give similar evidence. Dr. Cavafy (St. George's Hosp. rep., vol. viii) shows the rapid effect of the drug in reducing pain and fever. Dr. Julius Pollock in reporting 16 selected cases, shows the same effects.

From this evidence, meager as it is, but coming alike from friends and foes of the drug, and it is singular the amount of partizanship that these discussions have revealed, it would appear that the number of febrile days exclusive of relapses, in cases under the salicylic treatment, and in cases under other or no treatment, stands about 2 to 5 days under the former, to about 10 to 12 days under the latter.

With regard to the total length of time under treatment there does not appear from the evidence taken as a whole, to be any marked difference, though Drs. Southey, Greenhow and others think the use of the drug delays the recovery. Cardiac complications appear to occur rather more frequently under the salicylic treatment than under the alkaline. It would seem, however, as if this could be obviated by combining the two plans.

In respect to relapses, the evidence as in the cases of Drs. Findlay and Lucas, rather points to their being more frequent, but most physicians believe this to be due to the sudden discontinuance of the drug, or too early return to full diet, and most agree that the symptoms disappear rapidly again with the renewal of the drug.

Out of this entangled mass of observations I believe the following facts stand out :

That salicylate of soda excels by far all other remedies used in acute rheumatism, in its power to control the pain and fever.

That its action is physiological, is shown by its speedy elimination by the urine, by its increasing the excretion of urea and by the large doses necessary in disease not being borne in health.

That its beneficial results are obtained short of its toxic effects.

That if the drug is discontinued too early or the dose too soon reduced, relapses may occur which will again yield to its administration.

That it has no direct influence over cardiac complications. Indirectly on the one hand by its speedy reduction of the fever, it lessons the chances of their occurrence, but on the other, by its depressing effects when given in large and often repeated doses it increases the risk of the heart's failure where that organ is affected and its action feeble.

As regards the use of salycin I find Dr. Maclagan to be the only authority preferring salycin to salicylic acid, and prescribers of the latter now use the acid combined with a base, and soda is the one at present most in use.

The salycilates of potash (bicarbonate and acetate), of ammonia, of quinine, and of iron, have each found their advocates.

Dr. M. Donelly, of New York (Medical Record, March 6, 1880) speaks highly of the value of the potash salt. " The large doses of salicylic acid or the salicylate of soda required to act in acute rheumatism, is one of the greatest objections to their use, especially when any cardiac disease exists or is feared. In employing the salycilate of potash, large doses of the acid are not required ; 8 to 10 grains of the acid to double the quantity of potash, seems quite sufficient. The rapidity of action of this combination is quite remarkable ; judging from the effects in a dozen cases it is no over-estimate to say that salycilate of potash seems as thoroughly abortive of acute rheumatism as quinine of intermittent fever."

The use of salicylic acid with the carbonate or acetate of ammonia is recommended by Drs. Agle, Barclay, Erskine Stuart and others, and is reported as modifying the depressing effect of the acid upon the heart.

The treatment by the salycilate of quinine as recommended by Dr. Graham Brown, Dr. Hewan and others, appears to have the same advantage. Its expense, however, is a serious drawback to its general use. How far the salycilate of cinchonidia, which is cheaper, can take its place is yet uncertain. Dr. Willis White of Glasgow, reports favorably of the action of salycilate of iron, but the bulk of authority is against the use of iron in the acute stage, except in the form of the tincture of the perchloride in anæmic cases. I hope shortly to present to the society a report of some cases under my own care, treated by these salts, but I may add that so far my personal experience agrees in the main with the conclusions arrived at from the observations recorded. The remedy, powerful as it is for good in the majority of cases, does not evidently control all the phenomena of rheumatism and the exhibition of alkalies, stimulants and tonics is generally necessary to effect a cure.

A MODIFICATION OF THE ANTERIOR SPLINT.

By ROSWELL PARK, A. M., M. D.,

Demonstrator of Anatomy, Chicago Medical College, etc.

This splint, of which an idea may be formed from the accompanying wood-cut, was first shown to the Chicago Medical Society and subsequently at various other medical societies in 1878. Since that time a sufficiently large number of them have been sold, and tests enough have been successfully made, to

always be on the inner side ; the knob resting in the perinæum serving a double purpose, being more comfortable and at the same time keeping bandages from slipping. It can be changed from one side to the other in an instant. The sliding thimbles carrying hooks E E, serving as points of suspension, are movable along the tubes, thus permitting adjustment with reference to the centre of gravity. Thus every feature of the splint can be regulated as desired ; and to this extent it is simply an adjustable style of the good old anterior splint of Smith or Hodgen.

But this splint, unlike any other of its class, can be bent opposite to the knee, thus allowing a certain passive motion to be made at the joint, which materially diminishes the irksomeness of confinement and the stiffness which is the inevitable result of the fixed dressing on the old plan. Provision is also made for obtaining extension in the direct line of the femur when the knee is bent at an angle greater than 12°, as will be readily comprehended after a glance at the cut. The rods L I may be removed when the limb is to be placed in the straight position, and may be easily slipped into their stand-

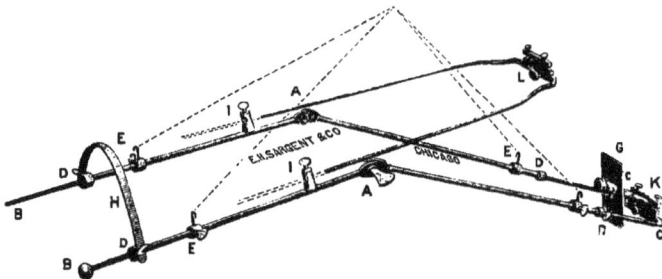

warrant giving the apparatus a wider notice than it has yet had.

Among its marked features are, first of all, its adjustability. Rods at each end sliding into tubes furnished with thumb screws, an easily worked clamp at the lower and an adjustable band at the upper end, permit the splint to be adapted to any size of limb to which it can be desirable to apply it. The rod furnished with a knob, as at B, should

ards at I I, when wanted. A clamp at L similar to that at K permits their distance to be adjusted.

To facilitate extension, the following method, original, so far as I know, in this application, is employed : At G is a thin wooden block around which the adhesive-plaster strips coming down the limb are fastened as usual. To this, through a perforation and by aid of a cork stopper, or by other suitable means, one end of a piece of pure rubber tubing is attached, the other end of which is held between two bars of the clamp K, and may be firmly held at any point. Too much can hardly be said in favor of this elastic extension ; it is found to overcome the tension of the muscular fibre more easily and quickly than any other method. Its general adoption, when possible, in orthopædic apparatus is evidence of this.

To properly use the splint it is only required to put on two extra strips of adhesive plaster which shall leave the limb at the knee and form a loop around another wooden piece like that at G ; the rubber tubing being simply transferred from one clamp to the other, when the extra attachment is to be used. In this way the position of the limb can be changed as often as comfort requires, say once or twice a day, while the site of the lesion, supposing it to be a fracture of the femur, and the adaptation of the fragments are undisturbed, and the extension is always made in the same line.

The suspending cords, represented in the cut as dotted lines, run up to a knot, and, finally, the usual extension is made by a single cord running up to a pulley at the ceiling, the pulley being located so as to cause the cord to pull the splint *away* from the patient, the weight of the limb being ample for the counter-extension.

It is, in the writer's experience, difficult for those who have never used an anterior splint, or even seen it used, to comprehend its *modus operandi* to a sufficient extent to use it with success ; but those who are familiar with it generally use it in preference to all others. To the latter class, I venture a hope, the additional features of the splint described and figured above will commend themselves. The writer, therefore, claims for his apparatus the following especial advantages :

1. Portability ; it can be taken apart and made up in a package the size of a roll of music.

2. Adaptability ; it can be fitted as de-scribed to any size of limb, or to a patient of any age above mere childhood.

3. Applicability to any fracture between hip and ankle except of the patella. While one splint is thus suitable for almost every fracture of the lower extremity, it is not, of course, intruded as a substitute for plaster of Paris dressings, though it may be made an admirable adjuvant, as in cases of compound fractures.

4. The benefit of the principle of elastic extension.

5. It allows the position of the joints and soft parts to be changed often enough to avoid stiffening.

6. For convenience in dressing compound fractures it meets every requirement. The bandages in which the limb is swung can be changed as often as required, while the parts are exposed for the inspection of the attendant, with a minimum of trouble to the surgeon, whenever desirable.

A CASE OF PLEURITIC EFFUSION; ASPIRATION; RECOVERY.

By SAMUEL J. ROSS, M.D., OF LONGMONT.

March 22nd, 1880. I was called to see a little girl aged 8 years, 6 months, and the following history was elicited :

She had been attacked with pneumonia (probably pleuro-pneumonia) of the left lung on December 25, 1879. The disease ran through the first and second stages with some degree of severity. Resolution was finally established, but progress was slow and finally reached a point where it ceased to clear up, leaving a space corresponding to the lower half of the left mammary region that remained dull on percussion, while general health was fast failing. Symptoms of hectic fever were ushered in, there being a daily chill followed by fever. A hard, dry, hacking cough was constant, accompanied with pain in left mammary region ; the pulse was 150 and respiration 32 to the minute.

March 25th. I saw the patient again and found the symptoms worse, chills occurring twice a day followed by fever, the pulse and respiration still more accelerated. The dullness had increased, now corresponding to the left mammary infra-clavicular and axillary regions; the intercostal spaces were full, but not bulging. With the finger I could feel over the intercostal spaces an impulse synchronous with the heart-beats. The urine was suppressed, the amount voided being but 2 oz. in 24 hours. It did not contain albumen, though there were general ana-sarca and ascites.

I recommended thoracentesis. The parents objected and I waited three days before they would give their consent. In the meantime the patient was placed on elaterium, bitartrate of potash and digitalis.

March 28th. I called prepared for the operation. Upon examination I found the symptoms to be as follows: the dullness was increased, being complete from base to apex, respiration was 40, the pulse 160 to the minute. The dropsical condition was somewhat improved and 4 oz. of urine had been voided in the last 24 hours. This improvement was probably due to the above treatment. I selected a point between the fifth and sixth ribs at the angle and inserted a No. 2 aspirating needle connected with a vacuum bottle, but found no fluid, probably because of adhesions. I made a second puncture between the fourth and fifth ribs at the angle and drew off 6 oz. of liquid, containing some coagulable lymph. The liquid contained about 90 per cent. per volume of albumen, and under the microscope showed some pus cells. The patient was relieved by the operation and progressed rapidly to convalescence under the use of Iodoformi et Ferri Pil.

WRITERS for the REVIEW are respectfully requested to make their manuscripts as legible as possible. Especial care should be exercised when using technical medical terms or words from foreign languages. It should be remembered that we are in a new country and that our printers have not yet had time and opportunity to familiarize themselves with the language of science.

* Philadelphia Medical Times, Nov. 6, 1880.

RECENT PROGRESS IN CLIMATOLOGY.

BY JOHN W. BRANNAN, M.D.

MINNESOTA.

WHAT CAN MINNESOTA DO FOR CONSUMPTIVES? *

In this article the writer aims to present certain well-ascertained facts with regard to the climate of Minnesota, and to indicate what class of cases is likely to be benefited by a removal to that State. He very fairly tells us also that certain forms of consumption are aggravated by residence there.

Minnesota lies about nine hundred feet above sea-level. Its soil is composed of drift deposits of blue clay, stratified clay and gravel and sand; above these is a rich silicious loam about two feet in depth. The following is the annual mean temperature for the last four years:

	Maximum.	Minimum.	Difference.
1876	68.5°	14.5°	54.0°
1877	69.9	20.4	49.5
1878	71.8	23.7	48.1
1879	71.7	14.2	57.5

The following is the barometric table for the same period:

	Highest.	Lowest.	Difference.
1876	30.444	29.376	1,068
1877	30.372	29.444	0.928
1878	30.315	29.353	0,962
1879	30.437	29.397	1.041

The mean relative humidity for 1878 was 67.7; for 1879, 65.5.

The next table shows the amount of rain and melted snow in inches:

1876	23.67	1878	22.78
1877	28.81	1879	32.39

The prevailing wind in summer is from the southeast, in the autumn and winter from the north and northwest. In Minnesota the northwest winds are dry, coming as they do from over a large surface of land. Northeast storms are very infrequent. The writer claims that Minnesota has a stimulating climate, as expressed by the following conditions, viz: considerable elevation, a soil admitting of rapid absorption, low mean temperature, somewhat low barometric pressure, and winds which, though very strong at times, bring but little moisture. He therefore would expect to find that pathological processes marked by active hyperæmia and a tendency to rapid extension would be aggravated rather than relieved by such a climate. On the other hand, he would look for improvement in chronic cases, where there is an indurated wall between the diseased pulmonary tissue and the sound portions of lung.

Patients in the incipient, and, if possible, in the pre-tubercular stage of consumption are most benefited. Next to this class, come those in whom the form of the disease is essentially chronic. Hemoptysis is not a contra-indication to a trial of the climate.

Of course the earlier relief be sought the better ; but even when recovery is out of the question, persons have come to Minnesota apparently in the last stages of consumption, and have had their lives prolonged for from five to ten years.

In many cases a permanent residence is a *sine qua non* to relief; a prolonged stay is necessary in almost every instance. It is obvious that invalids should come to Minnesota in the summer or early autumn rather than in winter.

To sum up, phthisis of an acute type is not likely to be benefited. As the climate of Minnesota tends to set up catarrhal affections of the respiratory tract, catarrhal phthisis, especially when the bronchial apparatus is much involved, is not often improved. Pneumonic and tubercular phthisis are both benefited, as a rule.

The following table shows the percentage of deaths from phthisis in Minnesota during the last four years :

1876	10.5	1878	10.5
1877	11.1	1879	9.9

In this table are included the deaths of invalid visitors as well as those of old residents of the State. Probably many of those whose death is attributed to consumption really die of senile bronchitis. In Minnesota, as in most other health-resorts, there are numbers of consumptives who do not seek the climate until the disease is so far advanced that relief cannot be obtained. The deaths of these help to swell the percentage of deaths from phthisis.

FLORIDA.

ON CLIMATE IN THE PREVENTION AND CURE OF PULMONARY CONSUMPTION, WITH SPECIAL REFERENCE TO THE PENINSULA OF FLORIDA. *

Dr. C. J. Kenworthy, of Jacksonville, Florida, writes in answer to Dr. Talbot Jones, of St. Paul, Minnesota. Dr. Jones had pleaded for cold climates in the treatment of consumption, condemning Florida in unqualified terms. He called its climate hot, moist, enervating and depressing, and said that consumption was exceedingly fatal to old residents of that State, as well as to persons visiting it in delicate health. Dr. Kenworthy claims that the air in a large portion of Florida is dry and bracing, that atmospheric changes are infrequent ; rains and cloudy weather are the exception, and sunshine the rule. The State possesses a variety of climates, there being tropical, semi-tropical and cooler sections ; level, rolling and hilly lands ; dry and bracing and somewhat moist localities. Should a case demand a change of climate, a suitable one can easily be found.

Dr. Kenworthy deems the climate, during the five winter months at least, eminently adapted to the treatment and cure of pulmonary affections.

The following is the mean temperature of the peninsula for the cold months :

	No. of years.	Nov.	Dec.	Jan.	Feb.	Mar.	Mean for 5 years.
Key West	5	74.5°	70.5°	70.5°	71.7°	73.8°	72.2°
Punta Rassa	5	69.7	68.4	65.5	65.9	69.8	67.1
Jacksonville	5	62.1	55.8	56.2	56.9	62.7	58.7

During the winter, consumptive patients engage in riding, driving, rowing, sailing, botanizing and other active exercise. Even in summer the heat is not so oppressive as in many portions of the north. Though the thermometer be high, there is a very low relative humidity, with a refreshing breeze. A severe case of sunstroke has not yet occurred in the State. The nights are always cool, refreshing and invigorating.

Dr. Kenworthy combats strongly the popular opinion that the climate of Florida is

---

* New York Medical Journal, Oct. and Nov., 1880.

humid. Mentone, on the Mediterranean, is famed for its low mean relative humidity. Dr. Kenworthy gives the following table of the mean relative humidity of the cold months for a period of five years for stations in Minnesota and Florida, and for three years for Mentone :

	No. Years.	Nov.	Dec.	Jan.	Feb.	Mar.	Mean for 5 mos.	Mean for 3 mos. for States.
Mentone, Mediterranean	3	71.8	74.2	72.0	70.7	73.7	72.4	
Breckinridge, Minnesota	5	76.9	83.2	76.8	81.8	79.5	79.6	
Duluth, Minnesota	5	76.0	72.1	72.7	73.3	71.0	72.6	} 74.3
St. Paul, Minnesota	5	70.3	73.5	75.2	70.7	67.1	71.3	
Jacksonville, Florida	5	71.9	69.3	70.2	68.5	63.9	68.8	
Key West, Florida	5	77.1	78.7	78.9	77.2	72.2	76.8	} 72.7
Punta Rassa, Florida	5	72.7	73.2	74.2	73.7	69.9	72.7	

He thus shows that the mean relative humidity of Mentone exceeds that of Jacksonville by 3.6 per cent.

Dr. Kenworthy admits that malarial diseases exist in some localities during the summer and autumn. But he shows by the aid of statistics from the army medical bureau that these diseases are of a much milder type in Florida than in any other State in the Union. In the Middle, Northern, and Southern divisions of the United States the ratio of deaths to the number of cases of remittent fever is much greater than in Florida. In explanation of this fact, he says that the luxuriant vegetation, which in the Southern and Middle States passes through all the stages of decomposition, is, in Florida, generally dried up before it reaches the putrefactive stage of fermentation ; consequently less malaria is generated than in climates more favorable to decomposition. The soil of Florida is almost everywhere porous and absorbent so that moisture remains but a short time on its surface ; the atmosphere is in constant motion, and there is more clear sunshine than in the more Northern States. During the cold months malarial diseases occur very seldom.

The following table shows the percentage of deaths from phthisis in Florida during the year 1878 :

Percentage of deaths from phthisis, including those of invalid visitors 6.6
Percentage of deaths from phthisis among residents 5.5

COLORADO.

COLORADO FOR INVALIDS.*

This article by Dr. S. E. Solly is addressed to the laity, but it is of value also to the profession as it contains facts of scientific interest and importance with regard to the climate of Colorado.

According to Dr. Solly, the climate is dry, bracing and cool ; there is an abundance of sunlight, the atmosphere is antiseptic and highly charged with electricity, the elevation varies from four to eight thousand feet above sea-level. The following table shows the average annual rain and melted snow fall along the foothills as compared with that of New York, of Boston and of St. Louis :

Colorado .... 15 inches.
New York.. 44   "
Boston.. 45   "
St. Louis.. 42   "

The average annual humidity compares as follows with that of New Orleans, of Santa Barbara and of Philadelphia :

Grains to cubic foot of air.
Colorado.. 1.13
New Orleans... 5.11
Santa Barbara. 3.98
Philadelphia 2.35

The above tables show that the climate of Colorado is emphatically a dry one. The number of clear days is estimated at three hundred and two in the year. The mean annual temperature is 47°.

The presence of ozone in the air has never been demonstrated, but all indirect evidence is in favor of its being present.

In the winter, the cold is at times very severe, though as a rule the days are bright and warm, while the nights are intensely cold. Even in summer, the nights are cool. The snow-fall is mostly in the early spring and is very light. On about a third of the days of winter the noon-day meal can be taken in the open air. The winds are sometimes very strong, especially in the spring. There is no rain except during the summer. In that season thunder-showers are frequent in the afternoon, but are of very short duration. The changes of temperature are sudden and extreme, hence wraps should always be kept at hand.

---

* Reprinted from " New Colorado and the Santa Fe Trail."

In a consideration of the diseases in which a change of climate may be advisable, consumption comes first on the list. The climate of Colorado, owing to its stimulating properties, is pre-eminently fitted to cure the early stages of consumption. In later stages, especially when softening is taking place, the destructive process is apt to be aggravated. But in all cases, it is a question of the amount of healthy lung left, and of the vitality of the patient. When a large portion of the lung is involved and the patient is much reduced in health, he would do better to seek an equable sedative climate rather than the stimulating rarefied air of Colorado. But even where the fever is high and night-sweats are constant, Dr. Solly is in favor of trying this climate, if the other conditions are favorable. On account of the stimulating effect of the climate on the action of the heart, the existence of organic heart disease is a contra-indication to a trial of the climate.

In neuralgic heart affections, the distress is usually increased.

Asthma is always relieved, to a greater or less degree according to the elevation reached.

Cases of nervous exhaustion are almost always relieved.

Acute organic disease of the nervous system is aggravated.

Acute disease of the kidney is made worse, but chronic disease is often benefited.

In rheumatism and in liver derangements the anæmic and debilitated improve, the florid and full-blooded grow worse.

The dry air irritates the mucous membrane in throat affections and nasal catarrh ; but where the condition is mainly owing to debility, treatment may modify the local effect, and the constitution is usually benefited.

Diseases dependent upon a scrofulous taint in the system or upon general debility are benefited, as a rule. Dr. Solly considers the climate of Colorado to be the deadly enemy of scrofula.

Most skin-diseases in the anæmic are improved, owing to the stimulus given to the cutaneous circulation.

The signal service office in Denver gives the following meteorological facts for 1880.

During the year there were 322 clear or partly clear days and only 44 cloudy days. On 60 days rain or snow fell. The precipitation from these storms amounted to 9.49 inches, which is 6.5 inches below the average annual rain-fall. Snow fell on May 9th, and the last frost of spring occurred on May 11th. The first snow of autumn was on September 25th, and the first frost on October 13th.

The mean relative humidity was 47.

The mean annual temperature was 49.3. On June 19th, the highest temperature of the year was recorded, viz., 96° in the shade.

On November 17th, the coldest day of the year, the temperature reached 13° below zero, thus giving a yearly range of 109°.

The temperature was 90° or above, twenty-two times, about equally divided between June, July and August. The thermometer was below zero twenty-three times—four in January, four in February, five in March, nine in November and one in December.

The prevailing direction of the wind was from the south.

The barometric readings are unfortunately not corrected for elevation, hence they are of no value for comparison with those of other climates. The yearly range of the barometer was 1.147 inches.

SOUTH CAROLINA.
AIKEN AS A HEALTH RESORT. *

A residence of ten years at Aiken qualifies Dr. W. H. Geddings to pronounce upon its climate. Long before the civil war Aiken was the chosen summer residence of the South Carolina planters. There they found a dry, bracing air and a climate free from malaria. During the war a commission appointed by the Confederate Government

* Gaillard's Medical Journal, Dec., 1880.

selected Aiken as the most suitable locality for a hospital for soldiers suffering from pulmonary diseases.

Since the war Aiken has been a most popular resort for invalids of all kinds.

SOIL, ALTITUDE AND CLIMATE.—Aiken is situated in the Sand Hill region. The soil is composed of loose sand ; being very porous it readily allows of the passage of water. The easterly wind having to pass over two hundred miles of dry soil loses much of its moisture before reaching Aiken. There is no surface water ; wells must be dug from eighty to two hundred feet deep in order to reach water.

Aiken is six hundred feet above sea-level.

There is no malaria, and Dr. Geddings has seen but seven cases of typhoid fever during a practice of ten years. According to Schoenbein's test, there is an abundance of ozone and peroxide of hydrogen in the atmosphere. Pine forests extend for miles in every direction about Aiken.

TEMPERATURE.—The following table shows the mean temperature for November, December and January at Aiken as compared with that of some noted European health resorts :

Aiken	48.53°
Nice	50
Pau	42
Cannes	50
Mentone	50

In order to show the relative equability of Aiken as compared with that of some other health resorts of the United States, the following table is given : ·

TABLE SHOWING MEAN OF VARIATION IN 24 HOURS.

	Sept.	Oct.	Nov.	Dec.	Jan.	Feb.	Mar.
Key West	10.00	8.87	8.43	8.54	8.95	9.83	9.18
San Diego	12.00	12.77	14.30	17.16	12.93	12.67	15.19
Aiken	10.03	17.03	18.26	18.06	12.45	19.14	17.64
Jacksonville	15.00	19.38	15.76	19.22	14.54	18.60	20.32
San Antonio	14.36	25.33	23.15	21.33	22.22	22.05	16.46
St. Paul	19.00	16.41	14.46	17.61	20.09	20.82	19.00
Colorado Springs	28.50	24.22	27.85	26.98	31.87	24.28	25.22

From the above table it will be seen that Key West and San Diego are the only American health resorts that surpass Aiken in equability. The European health resorts are even more equable than Aiken, partly because of their greater humidity.

The mean relative humidity of Aiken is 59.45 per cent. According to Vivenot's tables, these figures would cause Aiken to rank only as moderately dry.

The following table shows the relative humidity of Aiken, as compared with other places in the United States :

Denver, Colorado	44.7 per cent.
Aiken, S. C.	59.4 "
Cincinnati, Ohio	63.2 "
Los Angeles, Cal.	64.1 "
St. Paul, Minn.	67.6 "
Jacksonville, Fla.	68.6 "
New York, N. Y.	69.7 "
Santa Barbara, Cal.	70.0 "
San Diego, Cal.	72.8 "

Denver is the only place with a less percentage of moisture than Aiken. This dryness is characteristic of all the Rocky Mountain stations.

On comparison with the European resorts Hyeres is found to be the only one surpassing Aiken in dryness :

Hyeres	58.0 per cent.
Aiken	59.4 "
Cannes	62.0 "
San Remo	65.0 "
Mentone	70.0 "
Nice	71.0 "
Palermo	73.0 "
Madeira	73.9 "
Pau	77.0 "

Frosts are not very frequent even in midwinter. Fogs are rare and seldom last more than a few hours.

The average number of rainy days in the year is 89, of which 41 are in the colder half of the year. Only very little snow falls at Aiken.

Dr. Geddings characterizes the climate of Aiken as "cool, sunny, bracing, stimulating and dry."

He considers the climate of Aiken to be indicated in the following diseases :

1. Bronchitis.

2. Consumption in all its stages, except the last, and in all its forms, except acute tuberculosis and laryngeal phthisis. Persons with marked tendency to consumption, but in whom it is as yet undeveloped, are especially likely to be benefited by a residence at Aiken.

3. Malarial Diseases.

4. Dyspepsia.

5. Anæmia.

6. Diseases of Females, excepting cases complicated with severe neuralgia.

7. Diseases resulting from overwork, confinement, &c.

8. Convalescents from Pneumonia and Pleuritis.

9. Convalescents from Typhoid Fever and other exhausting diseases.

10. Syphilis.—Old cases of this disease requiring a change in winter on account of debility and anæmia, do well at Aiken.

11. Children convalescing from scarlatina, measles, and whooping-cough, others with scrofula, and suppurating glands.

Aiken is contra-indicated in the following diseases :

1. Laryngeal Consumption.

2. Laryngitis.

3. Bright's Disease.—Cases of this disease should winter at Nassau or Florida.

4. Eye Diseases, the glaré from the white sand at Aiken being very injurious.

5. Many Nervous Diseases.

As to rheumatism and asthma, no fixed rule can be given. Some cases do well, others need warmer and moister regions.

The finest portion of the year at Aiken is from the 1st of October to Christmas. But the invalid may remain with advantage until the 1st of June. After that time the weather grows too warm to be beneficial.

NEW JERSEY.

WINTER HEALTH-RESORTS—THE CLIMATE OF ATLANTIC CITY, AND ITS EFFECTS ON PULMONARY DISEASES. *

Dr. Boardman Reed, after condemning the climates of Pau, Nice, Mentone, Cairo and other health-resorts in the old world, as well as those of Minnesota, Colorado and Florida in the United States, proceeds to give his reasons for considering Atlantic City a perfect winter health-resort for invalids. The chief recommendation of Atlantic City over and above the claims of other health-resorts is its accessibility. Invalids from New York, Philadelphia and New England need not dread the terrors of exile from their friends.

* Philadelphia Medical Times, Dec. 18, 1880.

Atlantic City faces almost directly southward, and south as well as east winds blow across the Gulf Stream. At this point on the coast the Gulf Stream is four hundred miles broad, and one of its currents approaches within sixty-five miles of Atlantic City.

The only ocean breezes not warmed by the Gulf Stream are those from the northeast. The prevailing winds in winter are from the north and northwest, which are usually dry and bracing. The unpleasant northeast winds usually prevail for several days at the time of the equinoctial storms, but are infrequent during the rest of the year. The east and south winds are warm and moist.

The following table shows the temperature, humidity, barometrical pressure and rain-fall at Atlantic City :

1880	Mean Temp.	Range of Temperature		Mean Humidity.	Mean Barometer	Rainfall in inches.
		Max.	Min.			
January........................	41.1	64	13	79.3	30.189	1.70
February.......................	38.2	71	11	74.4	30.129	2.85
March..........................	40.1	72	18	71.9	30.061	5.97
Mean for 3 mos......	39.8			75.2	30.126	10.52

During December, 1879, there were but five days during which the thermometer fell below the freezing-point. During January, 1880, the thermometer fell below freezing-point during seven days. During the same two months, the thermometer at noon registered below 40° on only three days, and on twenty-nine days it registered 50° or above at the same hour.

The high mean barometer at Atlantic City is considered by Dr. Reed to be a matter of importance. There are no sudden barometrical changes to distress the invalid.

He thinks that nervous, rheumatic, gouty, dyspeptic and most other chronic ailments are usually benefited by a winter residence there. Convalescents from acute disease, or from surgical operations, improve on removal to Atlantic City. As to diseases of the respiratory organs, bronchial and laryngeal cases improve, as a rule. Consumptives in the pre-tubercular or incipient stage, and even those in advanced stages where the progress of the disease is slow, derive great benefit.

Pneumonia and bronchitis are of infrequent origin at Atlantic City, and when they do occur the patients almost invariably recover. Dr. Reed has never known an uncomplicated case of either disease to prove fatal.

Dr. Reed gives extracts from letters of twelve prominent physicians of Philadelphia. These letters were written in answer to his inquiries as to their opinion of the climate of Atlantic City. All but one of them gave testimony favorable to the climate, especially in the relief and cure of pulmonary diseases.

Dr. Reed considers the good accomplished by this climate to be due not to any specific influence of the air upon the lungs, but to its tonic and alterative properties, acting by the improvement of digestion and nutrition, the promotion of sleep, etc.

CLIMATE CURE—NEW JERSEY PINES. *

Dr. Joseph Parrish sets forth the claims of Lakewood, New Jersey, as a health resort in winter, as well as in summer. This village is about midway between New York and Philadelphia, on the New Jersey Southern R. R. It is nine miles from the sea coast. The soil is a deep, dry sand, allowing of the rapid absorption of water. There are extensive pine forests all about, which protect the invalid from the northerly, westerly and easterly winds. There is an abundant supply of pure water. The hotels are unusually good and private cottages may be had.

Various forms of disease of the air passages, including acute laryngeal and pulmonary phthisis as well as the milder forms of bronchial irritation, have improved there. Persons suffering from overwork find a pleasant change of thought and scene. Its accessibility is a great recommendation. Altogether, Lakewood seems to be a desirable resort for those invalids who cannot safely reside in the eastern cities during the winter and spring months.

REPORTS OF SOCIETIES.

Proceedings of the El Paso County Medical Society.

JAS. A. HART, M. D., SECRETARY.

January 11, 1881. Dr. J. W. Collins presided, in the absence of the president, Dr. Kimball.

Dr. S. E. Solly read a paper entitled Rheumatism and the Salycilates.†

After the reading of the paper Dr. J. Reed opened the discussion. Dr. Reed did not favor the use of the salycilates in rheumatism. He considered that salycilic acid was a dangerous remedy when given in large doses and that it was a direct irritant and depressant to the heart. He took issue with Dr. Solly on the point that shortening the days of the fever was a preventive of lesions. Salycilic acid irritates the stomach, hinders nutrition and produces dangerous aftereffects. Even the advocates of the drug admit that relapses occur on discontinuing its use. Beyond cutting short the time of suffering, no benefit is derived from the use of any of the forms of salycilic acid.

He favored the expectant treatment. He was in the habit of wrapping his rheumatic patients warmly in blankets, packing cotton about the joints, placing them on a light diet with much water or lemonade to drink. Under this treatment defervescence took place on the 7th or 9th day, and on the 12th day the patients were able to get out of bed. Many patients got well with no treatment whatever.

Dr. Strickler stated that he had a different view from Dr. Reed, and considered the salycilates to have almost a specific effect in rheumatism. He cited several cases in support of this view. The first case was one of sciatica, in which acute rheumatism developed on the second day. There was intense pain in several joints. He prescribed salycilate of soda in 20 gr. doses every two hours. On the next day there was great improvement, the fever and pain both having

---

disappeared. Under the old treatment the same class of cases lasted from five to six weeks. One patient was troubled with nausea if too long a time passed between the doses, but when the drug was regularly administered, she was all right. In a case of muscular rheumatism, salycilate of soda produced ringing in the ears, but relieved the pain and lowered the temperature. In another case there was an old mitral murmur. Under the administration of 20 gr. doses of salycilate of soda convalescence occurred in 7 days. A relapse occurred on the discontinuance of the drug, but on a return to it the symptoms improved.

He had no fear of bad effects from the use of salycilic acid, and had never been obliged to discontinue it.

Dr. Warren, of Waltham, Mass., said that he had not been in practice for a number of years, hence he had had no experience in the use of salycilic acid. In old times, he had found the lemon-juice treatment very successful. A combination of opium, bicarbonate of soda and the sweet spirits of nitre had given him good results.

Dr. Crocker had used salycilate of soda combined with bicarbonate of soda. Convalescence resulted·without leaving any bad effects on the heart.

Dr. Tucker had found recovery speedy with the use of salycilate of ammonia, dose 10 to 12 grs. every two hours. He remembered only one case in which there were cardiac complications when this salt was used. In that case the murmur was audible on the third day before treatment. In anæmic cases he also gives quinine and the tincture of the chloride of iron. In old times he had pursued the alkaline treatment.

Dr. Goodspeed reported one case in which there was complete recovery after 5 doses of salicin of 10 grs. each every two hours. The fever had been as high as 102°.

Dr. Collins said that his experience had been favorable to the use of salycilic acid. In one case of a girl of 7 years there was swelling of all the joints from the shoulder to the ankle, with a temperature of 104°—

106°. Salycilic acid was given her in 7 gr. doses every 3 hours until noon of the next day, at which time the pain had subsided and the temperature was 97°. During six days' further observation the disease did not return.

In a second case, the temperature was 105°, and the feet and knees were enormously swollen. Salycilate of soda was given in 20 gr. doses every two hours. On the next day the disease appeared in the arm, but was much less severe. In three days there was recovery and in one week the patient was perfectly well. In addition to the internal treatment, the joints had been wrapped in cotton batting soaked in a lotion of aconite and chloral. There was no nausea or depression.

Dr. Strickler said that patients themselves believe in the drug, sending back the old bottles to be refilled in subsequent attacks.

Dr. Brannan, referring to Dr. Reed's statement that salycilic acid only lessened the time of suffering, said that he thought this a very important point. Morphia was given to lessen pain, why should not salycilic acid be used for the same purpose. He had seen cases in which the change from great agony to perfect ease and relief was very striking. The fact that relapses occur on discontinuing the drug was a point in favor of its use rather than against it. All testimony shows that the relapses are perfectly controlled by the drug.

Dr. Hart said that he had treated but few cases of rheumatism in the West. While practicing in the East, before the days of salycilic acid, he had seen cases recover in 2 or 3 weeks under the treatment of cotton batting and oiled silk. .

Dr. Brannan mentioned a case of rheumatic meningitis occurring during his service in the Mass. Gen. Hospital, in Boston. The meningitis was an accidental complication, supervening after an operation on the hand. The patient was wildly delirious and his temperature ranged from 103°—105°. For two days no internal treatment was attempted. Salycilate of soda was then given in doses of 15 grs. every two hours. In less than 24 hours the delirium had disappeared and the temperature was normal.

Dr. Strickler said that he, for his part, was perfectly satisfied as to the anodyne effects of the drug. With regard to relapsing cases, judging from his experience, salycilate of soda always checked the relapses at once.

Dr. Solly, in closing, said that the point of relief of pain had already been pretty well covered.

The effects of salicylic acid are easily watched and controlled. There had been no authentic case of poisoning reported. His feeling was that a man required to be as good a physician in giving this drug as in giving any other. He was not simply to order a certain number of doses and then to go off, feeling that he had done his whole duty.

Delirium is said to have been caused by salycilic acid—it is probable that more cases have been checked by it. Dr. Brannan's case was a very characteristic one.

He did not consider the drug a specific— it was not known to have any effect on plastic exudation. It is best to combine alkaline treatment with it to accomplish this. The point that relapses are controlled by the drug has already been taken. The drug should not be too long continued, and it was well to give tonics. Salycilate of quinine points to our being able to combine tonics with salycilic acid. The salycilate of ammonia is a good form of the drug, as being stimulating and alkaline. Great care must be exercised in anæmic cases. The disease does not always take the same form. In eighteen cases he had found but one in which salycilic acid had not worked well, and that was on account of heart trouble. He thought it best not to press the remedy as he could not watch the case.

The salts of salicylic acid do not cause nausea and vomiting, as does the acid itself. The salycilate of potash is more palatable than salycilate of soda.

EDITORIAL.

THE REVIEW FOR 1881.

The season has arrived at which the journals of the country are expected to come forward and modestly state their aims and prospects for the coming year. Fully aware of our extreme youth, it is with becoming diffidence that we take our place in line. Nevertheless, we have a past, though it be but a short one, a present, and we hope a future.

Our main aim will be to present to our readers as much original matter as possible, having recourse to the "scissors" only when compelled by absolute necessity. The original department will be seen to be largely increased in the present issue. The department of Selections from Journals has been entirely done away with. Under the head of Reports of Societies we hope to present in every number a report of the proceedings of some medical society of good standing. In the absence of hospitals, we of this new country must do without notes of hospital practice, unless some of our more favorably-situated brethren will kindly furnish us from their super-abundant store. As the number of our foreign exchanges increases, the department of Translations will be enlarged. Under the heading of Book Notices and Reviews will be noticed, as heretofore, all medical and scientific publications of importance that are submitted to our consideration. Abstracts of articles from other journals, that would formerly have been given in full as Selections will now appear as Medical News.

It is in the field of Climatology that this journal would prefer to work. Therefore, while we solicit papers on all subjects of scientific interest, we shall be especially pleased to receive articles bearing on the science of Climatology. It is hoped that before the end of the present year something may be done to place the science of Meteorology on a better footing in Colorado.

AMENDMENTS TO THE BILL.

The bill for the regulation of the practice of Medicine in Colorado has been presented to the Legislature which is now in session at Denver. But previous to its presentation it was amended in several important particulars by the committee of the State Medical Society having the bill in charge. Amendments were made to sections 1, 2, 10 and 12.

The amendment to section 1 changes somewhat the representation of the three schools of medicine in the State board of medical examiners. The representation is not to be proportionate to the number of graduates of each school within the State, but the board is to "consist of six physicians of the regular, two of the homœopathic, and one of the eclectic school or system of medicine." This amendment was made in response to the appeal of the committee of the homœopathic medical society of Denver.

Section 2 has not been definitely amended, but it is left to the discretion of the Legislature to designate the State officer who is to appoint the board of medical examiners. It is a question whether the executive or the judiciary shall be invested with this appointing power.

Section 10 now reads as follows: The State board of medical examiners may refuse certificates to individuals *who have been convicted* of conduct of a criminal nature, and they may revoke certificates for like causes. The medical examiners are not to decide the question of the guilt of the individual; that matter is left to the courts, where it belongs.

Section 12 has been amended in a most radical and thorough manner; it has been erased entirely from the bill.

We see every reason to concur in all of the above amendments. Even were it a question as to which of the three schools should rule in the board of medical examiners, a clear majority of three ought to satisfy us. Should the bill become an act, we trust that the members of the board will work harmoniously in their endeavors to accomplish good. The

three schools have surely a common cause in their war against charlatans.

It has been left to the higher powers to decide who shall appoint the board of examiners. As the investment of the judiciary with patronage (*not* political in this case, we take it) shocks the legal mind, pray let us yield that point.

The change in section 10 is a good one. Physicians are not judges and should not be invested with judicial powers.' When the supposed criminal is convicted of his crime by the properly constituted authorities, then let the board of examiners reject or eject him, as the case may be, from the body of legally practising physicians.

We are devoutly thankful that section 12 has been expunged from the bill. It was a stain of so deep a dye that all the good points of the bill could not cover it.

As the bill now stands, it is worthy of the hearty support and commendation not only of the profession, but of all citizens who have at heart the well-being of their fellowmen. We have every reason to believe that the Legislature will pass the bill.

THE FATE OF THE INDEX-MEDICUS.

We regret to hear that the Index Medicus is in great danger of suspending publication. The medical profession of the United States have failed to support it. This great work, "A Monthly Classified Record of the Current Medical Literature of the World," has just closed the second, and we fear the last, year of its existence. That it has lived so long is mainly due to the untiring energy and industry of its chief editor, Dr. John S. Billings, the President of the American Public Health Association. He is noted not only for being a hard worker himself but for his power of getting work out of others. But the personal efforts of the editor cannot alone carry on a journal, especially a publication of such immense scope. The publisher, Mr. F. Leypoldt, of New York, reported a deficit of over $4000 at the end of the first year, and now at the end of the second year this deficit has increased to $5000. Naturally this state of things cannot go on much longer. It is not often that American enterprise meets with such poor recognition. The suspension of the Index Medicus would be an irreparable loss to the medical profession the world over. Nevertheless it will come to pass unless the present subscribers renew their subscriptions for the coming year and also induce others to subscribe.

THE ABUSE OF MEDICAL CHARITY.

We are pleased to see that this important subject is at last beginning to receive proper attention. Spasmodic attempts at reform have been made from time to time both in England and in this country, but no concerted action has yet been taken.

The most important move in the right direction was made last month in Boston. At a recent meeting of the Boston Society for Medical Improvement, Dr. J. H. Whitmore, the Superintendent of the Massachusetts General Hospital, read a paper entitled "Are Free Dispensaries Abused?" His conclusions were that there *was* abuse and that there should be complete co-operation of all the dispensaries and hospitals in order to accomplish the reform needed. He suggested that a competent paid inspector should be appointed at each one of the large dispensaries. This officer should examine all applicants for treatment and should admit *only* the deserving poor. He should also devote a portion of each day to visiting the patients treated, in order to ascertain that they were as represented—both poor and deserving.

In the discussion which followed the reading of the paper, the great majority of the speakers agreed with the opinions expressed by the reader. Not only is medical pauperism encouraged, but great injustice is done to the younger members of the profession. There are many men who must earn their living from the start, and who are willing to take small fees for their services. But un-

fortunately there are also many comparatively well-to-do people, who would disdain the name of paupers, but who are only too ready to accept medical aid gratis.

The educational side of the question must of course be considered. Students must be educated and abundant clinical material is needed for that purpose. But the number of patients at any given dispensary in Boston could be reduced one-half and still leave ample material for clinical instruction.

A plan similar to that pursued by the Associated Charities in Boston, if carried out by the various hospitals and dispensaries, would do much good. The name of every person assisted by the Associated Charities is registered and he is visited at his house in order to see that he is really worthy. The number of paupers in the city has been greatly reduced by this system. We hope that some such plan will soon be put into effect by the hospital and dispensary officials in Boston.

TRANSLATIONS.

By John W. Brannan, M.D.

THE HOANG-NAN.

This is the name of a climbing plant found in certain ranges of mountains in China. It is employed by the natives of those regions as a remedy for hydrophobia, leprosy, serpent-bites, etc. Botanically it is known as *strychnos gautheriana.* The bark of the plant is the part used, and chemical analysis has shown that it contains the alkaloids strychnia and brucia, the latter predominating in quantity.

According to Dr. Simon, of Marseilles, the action of hoang-nan, though somewhat analogous to that of strychnia and brucia upon the nervous system, is less general in its effects. At Tonkin hoang-nan is given in combination with alum and the red sulphate of arsenic. The dose is one pill containing 1½ grains each of hoang-nan and arsenic and 1 grain of alum. In hydrophobia, if violent symptoms have already begun, one or two pills are given in a tea-spoonful of vinegar; the dose is repeated until the appearance of tremblings in the limbs and especially trismus.

If the symptoms of inoculation have not appeared, one pill is taken the first day, two the second, and the number is increased day by day until trismus appears. Spirituous liquors and highly-seasoned foods are to be avoided during treatment. This remedy is said to be absolutely unfailing in its curative action in cases in which violent symptoms have not appeared, and even after the madness has manifested itself a cure is usually obtained.

Hoang-nan serves to relieve the fears of persons who have been bitten, but in whom the poison has not yet manifested itself. If they are not inoculated with the virus, trismus will appear soon after the first dose of the drug. The bites of the most venomous serpents are cured in the same manner. M. Hillairet, of Paris, reports the cure of a case of leprosy by the use of hoang-nan.—*Journal d' Hygiene.*

A REVOLUTION IN THE LIGHTING OF HOUSES.

M. Kordig, of Hungary, recently exhibited in Paris a curious illuminating liquid. It was a very light and volatile hydrocarbon, presenting the following remarkable properties:

1. It is volatile at ordinary temperatures and boils at the temperature of the hand.

2. It burns equally well at a relatively low temperature.

3. It produces a white light which is more beautiful and has greater illuminating power than the same volume of gas.

M. Kordig, in order to show that the new liquid need cause no fear of fire or of explosion, poured a large quantity of it on his hat and then set it on fire. The flame rose to the ceiling; the exhibitor placed the hat on his head and wore it until the flame died out. The hat was found to be uninjured. Handkerchiefs. light-colored gloves

and silk ribbons were then dipped into the liquid and set on fire. They remained unhurt by the flames.

According to the inventor, this interesting substance is an essence of naphtha. It has a slight smell not at all disagreeable, and has a trace of ether in its composition. When placed upon the hand it gives a sensation of cold. This new mineral essence is said to come from the wells of natural oils recently discovered in Hungary.—*Journal d' Hygiene.*

LEPROSY IN THE SANDWICH ISLANDS.

In March, 1880, the hospital for lepers on the island of Molokai contained 684 patients, 424 of whom were men and 260 women.

Besides these, the majority of the lepers on the island are treated at the outpatient department of the hospital and live in the midst of the population. The average mortality in the hospital is in the ratio of 58 to the 1000 yearly.

As the cold and moist season approaches, towards November, the lepers increase in number. Leprosy is contagious, as is shown by the fact that since its introduction into the Sandwich islands in 1856, the number of lepers has steadily increased. It is the opinion of Dr. Emerson, the physician in charge of the hospital, that there is no cure for leprosy.—*Journal d' Hygiene.*

TREATMENT OF SYPHILIS.

M. Martineau gives the following as his method of treating a well-authenticated case of syphilis. During the first year he prescribes mercury for the first four months, then iodide of potash for six months, followed by a month of no treatment at all. During the second year he gives mercury for one month, then iodide of potash for two months, followed by two months of no treatment; then again mercury for one month, iodide of potash for three months, followed by three months during which he gives no mercury or iodide of potash but begins the use of sulphur baths and sulphur internally. The treatment during the third year is the same as during the second. In M. Martineau's opinion, this prolonged treatment is the only means of guarding the patient from the visceral complications which are so frequent.

He always gives mercury before giving iodide of potash, his experience having proved to him that the anti-syphilitic action of iodide of potash cannot show itself until after that of the mercury. Should tertiary symptoms fail to yield to large doses of iodide of potash, he omits the treatment of the iodide for a week, replacing it with mercury. At the end of the week, the iodide is resumed and produces the desired effect.

The sulphur water serves as an important auxiliary to the mercurial treatment. It favors the absorption and the elimination of mercury, and makes the body more tolerant of the drug. As outbreaks of syphilis coincide ordinarily with the changes of the seasons, especially in spring and autumn, M. Martineau chooses those periods for beginning the mercurial treatment or for increasing the doses of mercury or of potash. He does not give mercury on the first appearance of the chancre but waits for the secondary symptoms.

In syphilis in the infant, M. Martineau prefers inunction, using 15 grs. of mercurial ointment daily. The mercurial treatment is followed by that with iodide of potash, 3 to 7 grs. being given daily during two or three months. This treatment is continued for two years, at the end of which, sulphur water is given for two weeks, in the form of Challes water, one or two teaspoonfuls daily.—*La France Medicale.*

PARASITES IN FISH.

Trichinae were recently discovered in the body of a pike caught near Ostend. Dr. Elentin, of that city, examined the fish under the microscope and found it filled with these parasites. The fish had probably devoured the remains of an animal which had died infested with trichinae, and thus trichinosis had developed in it. Until this discovery, this disease had never been found in any animal except the hog. It has long been known that a peculiar kind of solitary worm is found in fish, much resembling the toenia soleum.—*Journal d' Hygiene.*

HYSTERIA AND HYDROTHERAPY.

M. Leroy-Dupre is strongly impressed with the efficacy of hydrotherapy in the treatment of hysteria. The following is his definition of this malady : When a woman is in good health, there exists in her a perfect harmony between the cerebral or voluntary innervation and the spinal or involuntary innervation, with a preponderance of the former over the latter. If the spinal innervation begins to predominate, the result is nervous *anarchy* with all its consequences, especially weakening of the will (cerebral paresis).

He thinks that we should exercise great caution in giving a favorable prognosis when proceeding to treat a case of long confirmed hysteria.—*Journal d' Hygiene.*

HYGIENE OF THE EYES OF THE NEW-BORN.

Dr. Briere, in a popular treatise, recommends the following precautions for preserving new-born infants from purulent ophthalmia :

1. Keep the eyes very clean. As soon as the child is born, wipe (before doing anything else) the eyelids with a dry linen cloth ; then wash the face and the head before the rest of the body.

2. Avoid the cold ; if the infant is taken out during the first days after its birth, it should be warmly dressed and the head well covered.

3. If, two or three days after birth, the eyelids swell and a secretion escapes, first of tears, then of a greenish-yellow matter, do not make use of such unmeaning remedies as washing with the mother's milk etc., remedies which are inactive, useless or even

hurtful. It is not to be thought that it is but a light matter; often much valuable time is lost until it is too late for proper treatment to save the sight.

4. If the secretion of matter and the swelling of the eyelids last more than 24 hours, send at once for a physician, who will understand the gravity of the case and will stop the disease in time.

5. The principal point is to wash the eyes frequently, separating the eyelids so as to clean the interior. Do not use a syringe or a sponge; use a linen cloth and plenty of water.—*Journal d' Hygiene.*

PROLONGED INCUBATION IN A CASE OF HYDROPHOBIA.

M. Colin recently reported a remarkable case of hydrophobia to the Academy of Sciences in Paris. A French officer was bitten by a mad dog in Algeria, in November, 1874, and it was not until August, 1879, that the symptoms of hydrophobia appeared. The period of incubation was therefore nearly five years long. Previous to this case, one year was the longest period of incubation reported.

M. Colin said that some observers, among them Devergie, would call this a case of nervous hydrophobia. He, however, combated this opinion strongly, contending that the various features of an attack of hydrophobia could not be simulated by a person acting under purely nervous excitement. He was of the belief that the progress of science would greatly reduce the number of *fatal* cases reported of *nervous* or *spontaneous* hydrophobia. He considered that an incubation of six months or one year was just as difficult to understand as an incubation of five years.

Brouardel entirely supported Colin in his opinions.

In closing, Colin said that of 26 cases of hydrophobia occurring in the French army, between the years 1862 and 1879, only 8 were among the soldiers stationed in France, while 18 were among those stationed in Algeria.—*Journal d' Hygiene.*

BOOK NOTICES AND REVIEWS.

The Young Chemist: A Book of Laboratory Work, for Beginners.

By John H. Appleton, A. M., Professor of Chemistry in Brown University. Second Edition. Philadelphia: Cowperthwait & Co.

This little book, now in its second edition, seems to us to be admirably adapted to supply a want in our school text-books. There is a growing demand for a knowledge of chemistry, and it is possessing increasing fascinations for our scholars. This may be due to the fact that this study appeals to the senses more than most, that in it there is a chance for manipulation, the hands and eyes are occupied as well as the brain, and the method of instruction is more demonstrative and less didactic than that of most branches.

While we feel that the public perhaps misjudge the practical importance of this study, and while we are convinced that, to be of any use at all, one's knowledge of chemistry must be accurate and somewhat extensive, we nevertheless favor this growing taste for this study, as one of the sciences, the study of which affords so much pleasure as well as imparts useful knowledge.

For this reason we welcome any text-book that helps to make the subject plain and intelligible. Our experience with text-books on chemistry, is, that, in general, they are made as blind and unintelligible as possible; and after studying some of them we have felt sympathy for the student who summed up his knowledge of chemistry with the words, "chlorine is green, chlorine is green."

This book is less objectionable in these respects than most of the books we are acquainted with, but nevertheless, it has its faults.

Exact definition is quite essential to a correct understanding of chemistry, but it has been our experience that the definitions in most text-books are singularly unfortunate and incomprehensible. As illustrations of blind definitions in this book, we noticed especially the definitions of salts, and of anhydrides.

We commend the general arrangement of the book and the fullness with which it is illustrated. The introduction of experiments helps to make the subject interesting and to fasten the subject matter upon the memory. As its title indicates it is intended for the beginner in chemistry. It is far from exhaustive and might be styled a primer, and as such it ought not to be compared with longer works. We commend the introduction of "rational" formulae, and the use of "graphic" symbols, but we deplore the change in terms, and we much prefer the old and familiar terms Hydrochloric acid and Sulphuretted Hydrogen, to Chlorohydric acid and Sulphydric acid.

It is evident that the book would be of but little value to a practitioner, whose needs would demand a more exhaustive work.

A Short Course of Qualitative Chemical Analysis.

By John H. Appleton, A. M., Professor of Chemistry in Brown University. Fourth Edition. Philadelphia: Cowperthwait & Co.

This book is well adapted for an advance from the preceding one, and we should judge that it was the author's intention that it should be so used. Taken independently of the other it seems to us to be open to the objection, that it starts immediately upon a course of analysis, before the scholar is made acquainted with the properties of quantities to be analyzed. For this reason we prefer Clowe's work, which dwells on the chemical properties and the reactions of quantities before taking up the general analysis. We favor the introduction of formulae and reactions, as is done in this work, as helping to familiarize the student with an important but difficult part of chemistry; but our criticisms on the nomenclature of the previous work applies equally to this. The plan of analysis is that which seems now to be generally adopted, and the classification is the common one. For a physician who would like directions in analysis, this book is well adapted to the purpose.   S. A. F.

A Practical Treatise on Nasal Catarrh.

By Beverly Robinson, A. M., M. D., (Paris), Lecturer upon Clinical Medicine at the Bellevue Hospital Medical College, New York; Physician to St. Luke's and Charity Hospitals, etc., etc. New York: William Wood & Co.

This book, evidently the work of a specialist in the subject of which he treats, can be read with advantage by every practitioner of medicine. It gives the results of the personal study and experience of the author and does not attempt to go into the literature of the subject. It is thus enabled to justify most thoroughly its claim of being practical.

In the first chapter an elaborate division of the affections of the nose is given. The author then proceeds to confine himself to that which he has set before him, viz: the study of the catarrhal inflammations of the nose. One chapter is devoted to the anatomy, physiology and pathology of the nasal passages. In the chapter on instruments for the treatment of the nasal cavities, Dr. Robinson strongly condemns the nasal douche, whether used for cleansing purposes or for medicating the nasal cavities. He claims that it not only fails to cleanse or medicate the vaults of the pharynx and the superior meatus and turbinated bones, but that it frequently causes inflammatory diseases of the ear. The same objections apply, though not so strongly, to the different forms of syringes. For cleansing the nasal cavities, Dr. Robinson prefers to use atomizers. For purposes of medication, he uses atomizers, or an inhaling apparatus, or powder-blowers. These last he finds most generally satisfactory.

The chapter on prophylaxis and general remedial treatment of various forms of coryza we consider the most important in the whole book. Under the head of prophylaxis Dr. Robinson considers the following points: 1. care of the feet; 2. cold baths; 3. friction and shampooing; 4. clothing and temperature. He advocates the wearing of thick soles to the shoes in all weathers and seasons; when there is snow or rain upon the ground, over-shoes should be worn. He is in favor of the daily cold bath; the reaction following the bath relieves any tendency to congestion not only of the internal viscera but also of the mucous linings of the nose. Friction or shampooing for a few moments before the bath aids in the production of a good reaction after it. Flannel undergarments should be worn in winter and summer. The outer clothing should be warm and comfortable, but not oppressive. Wear no mufflers around the throat. Breath through the nose when in the open air. If a person is suffering with acute coryza or a severe chronic coryza, it is well to wear a respirator covering both mouth and nose during inclement weather. Never take a warm drink just before going into a cold or damp atmosphere. Dr. Robinson is strongly opposed to the use of furnaces or stoves for heating the living rooms of houses. Only halls and corridors should be so warmed. The open fire-place is the great purifier and ventilator of rooms, large or small.

The above chapter, which we have considered at such length, might be read with profit not only by the profession, but by the laity as well.

Dr. Robinson's book should win a high place ; the subject is most important and he has handled it ably.　　　J. W. B.

Genital Irritation, together with some remarks on the Hygiene of the Genital Organs in Young Children.

By Roswell Park, A. M., M. D., Demonstrator of Anatomy, Chicago Medical College, etc. Chicago: Tucker, Newell & Co.

This interesting article, read before the Chicago Medical Society, in October last, brings to the notice of the profession some points of interest, in addition to those already treated of by Drs. Sayre and Beard, of New York, on this subject. Be the reason what it may, the Hygiene of the Genital Organs has not received that attention from medical writers which the importance of the subject seems to demand. Every practitioner knows, only too well, the dire effects produced by the morbid dwelling on some slight real or supposed affection of the

genital organs. Among our hardest patients to treat is the hypochondriac, but we can all attest the difficulty of treating the hypochondriac, made so by brooding on some slight abnormality or some disease of the genital organs. We remember reading in a life of Mr. Webster, that one of his most important cases, involving the disposal of a large estate, hinged on the supposed imbecility of a young man, who, in truth, was a hypochondriac from some imagined genital trouble. There is no class of cases which furnishes richer spoils for " quacks" and "irregulars " than these. A simple herpes preputialis, or a balanitis, is made by these cheats to appear a terrible disease, and by them the patient is made most wretched.

A certain false delicacy, it seems to us, is common among practitioners in imparting useful knowledge in regard to the cleanliness and care of the genital organs. We are squeamish and over nice on this point, or else we falsely judge the subject unworthy of our attention. The careful and conscientious practitioner does not deem it beneath his notice to recommend cleanliness to his patient. Almost every doctor has his hobby in regard to baths, sponge baths, warm baths, tepid baths, cold baths, but how many of them impress upon their male patients the necessity of cleanliness about the prepuce? The train of symptoms which a balanitis may set up, as pointed out by Dr. Park, is alarming, a urethritis, a cystitis and even a nephritis resulting in some cases.

We justly consider prevention better than cure, and prophylaxis better than treatment. On this ground we should urge, as a prophylactic measure, that boys be instructed to draw back the fore-skin and cleanse the parts of smegma, when taking their morning sponging. In advising this we are simply calling attention to the teaching of Acton and others on this subject.

Every practitioner must recognize the harm resulting from leaving a phymosis untreated, and would agree with Dr. Park that the necessity exists, in such cases, of enlarging the orifice, either by circumcision, or by slitting up the dorsum; and Dr. Beard has called attention to the efficacy of these measures, in at least assisting in the cure of neurasthenic symptoms, where phymosis coexists. We think that Dr. Park lays a little too much stress on the necessity of breaking up adhesions of the preputial mucous membrane, in the case of infants. We are led to think this affection quite common and one which is generally cured with time.

In conclusion, we would call attention to the two cases cited by Dr. Park, where the cure of phymosis acted as a cure of a chronic diarrhoea in two boys aged respectively three and five years.   T. A. F.

Colorado for Invalids.

By S. Edwin Solly, M. R. C. S. Eng., L. S. A. Lond., Fellow of the Royal Medico-Chirurgical Society, Vice-President of the Colorado State Medical Society. Colorado Springs, Colo.: Printed by the Gazette Publishing Company. 1880.

This little brochure of Dr. Solly's is quite timely in its appearance. The literature bearing on the climate of Colorado is very scanty, and we welcome this addition to it. As the pamphlet is addressed to the general public, the style is of course popular, at times perhaps a little too much so to please some tastes. The anecdote of the London alderman is very amusing in its way; but it hardly adds dignity to a treatise intended to convey to the laity, truths which are the results of years of scientific research and observation.

But, setting all this aside, there is much in the little work to commend. It gives the views of an observer who has resided for six years in Colorado, and who has thus had abundant opportunity to study its climate and its effects upon various diseases. Other observers may differ with him in some of his explanations of accepted facts, and in some of his conclusions, but it would be remarkable if such were not the case.

Dr. Solly's first aim was to give good advice to persons seeking a climate better suited to themselves than their own. It seems to us that he has accomplished his

purpose very satisfactorily. He has given, in our opinion, a very fair view of the case. Invalids who follow his directions will not go very far wrong. Too much space is possibly given to descriptions of physiological and pathological processes in the body, considering that he is writing for the unscientific mind. But the average layman is flattered when such food is offered him and enjoys the treat hugely.

In guiding the invalid to a choice of residence, he hits off the characteristics of the various towns of Colorado very well. But we would like to know to what he refers when he speaks of places of entertainment in Denver. Surely he cannot have in his mind the Palace Theatre!

No one will find fault with Dr. Solly's terming the climate of Colorado, "dry, bracing and cool." But we fancy that many will object to his union of the words moisture and consumption. Physicians practicing in Florida and in the health-resorts of southern Europe speak in decided terms of the relief that consumptives experience in those moist climates. And there are many islands in the Southern Ocean where consumption is almost unknown. We incline to think that there are other factors besides moisture which must be taken into account before deciding that a given climate favors the development of consumption.

Dr. Solly thinks that cold aids in curing consumption. But dwellers in warm climates will produce tables showing the marked prevalence of consumption in various cold countries of the world. Here again other factors must be considered.

The great objection urged against Colorado by writers living in equable climates is its sudden and extreme changes of temperature. But Dr. Solly boldly claims that these very changes are beneficial to the consumptive whose disease is not too far advanced. This is spiking the enemy's cannon most assuredly and it remains to be seen how they will return to the charge.

On the whole, we think that this little book will be of great service to the invalid who questions whether or not this is the proper climate for him, and it must also be considered as a valuable contribution to the scientific literature of the climate of Colorado.   J. W. B.

MEDICAL NEWS.

THE AVERAGE AMOUNT OF URINE: ITS IMPORTANCE IN DIAGNOSIS.—This article, read before the Boston Society for Medical Improvement, is worthy of weight and consideration, both on account of the subject and also as coming from the authority that it does.

In opening his article, Dr. Wood calls attention to the fact that this important means in diagnosis is very generally overlooked by practitioners, who are content simply with testing for albumen and searching for casts. He lays emphasis on the fact that "it is not possible in many cases to ascertain, even approximately, how much work the kidneys are capable of doing, without knowing approximately the average amount of urine which the kidneys are eliminating." The total amount of solids is learned approximately by multiplying the last two figures of the specific gravity by 2⅓, giving the number of grains of solid matter in one liter (about a quart) of urine. The amount of urine, which is also a guide for ascertaining the rapidity of metamorphosis going on within the body, is determined by measurement.

In acute parenchymatous nephritis the amount is at first much diminished, about 500 cub. cent. (or a pint); and then gradually increases, with the diminution of the inflammation, up to the normal amount; with convalescence it exceeds the normal, and where complete restoration of the kidney follows, it falls again to normal.

In chronic parenchymatous nephritis the amount is always below normal; diminishing with the increase of the disease, especially if dropsy set in. If the progress is not active the amount is but little below normal.

In both interstitial and amyloid degener-

ation the amount of urine is largely increased, even to three or four times the normal (sometimes as high as five pints), except during a short time previous to death, when the quantity may be less than normal.

In complicated cases, which are far more common than uncomplicated cases, the amount of urine varies according to the nature and extent of the complicating affection. A parenchymatous affection existing as a complication of either an interstitial nephritis or amyloid degeneration, is almost sure to be detected by the increased amount of albumen and the character of the urine, and in these cases the measurement of the amount is of importance.

In passive hyperaemia of the kidneys due to cardiac or other disease, the average daily amount of urine is diminished.

In active hyperaemia of the kidneys, due to the elimination of some virus or drug, the amount of urine is usually diminished.

In typhoid, rheumatic and other fevers we often see in the urine a few hyaline casts and a trace of albumen with a diminished amount of urine. In many cases of acute rheumatism, especially after the exhibition of salicylic acid and the salicylates, a trace of albumen may be found in the urine and hyaline and finely granular casts with blood and renal epithelium in the sediment. This has often led to a mistaken diagnosis of chronic parenchymatous nephritis, a mistake which would not have occurred had due attention been paid to the average daily amount of urine.—*Boston Medical and Surgical Journal.*

PHOSPHORESCENT MEAT.—The French journals tell of some perfectly fresh meat that became phosphorescent. Some cutlets of raw pork shone so brightly in the kitchen that it was possible by the aid of the light to tell the time by the watch. The butcher from whose shop they came said that all the meat of which they were a part of the stock became phosphorescent within a short time after having been put into his cellar. The phosphorescent meat did not otherwise differ in aspect or odor from common meat;

it had not been exposed to a temperature of more than 50°, and entire freshness seemed to be a condition of phosphorescence, so that when the meat began to smell it ceased to be bright. The phosphorescence generally disappeared on the sixth or seventh day.—*Popular Science Monthly.*

RENAL ALBUMINURIA AS A SYMPTOM.—An editorial, in the New York Medical Record for Jan. 1, 1881, in treating of this question, points out that the old notion that albuminuria meant renal disease has been disproved. Dr. Ellis has shown it to be present under 150 conditions and even his list is incomplete. Transient albuminuria may be observed in a great variety of disturbances, and even the prolonged presence of albumen in kidney secretion does not *per se* argue permanent histological change in these organs. Certain articles of diet, notably eggs, produce it—a change in the condition of life may lead to it. The effects of climate may also act as exciting causes. And all this in persons of apparent excellent health. Indigestion appears to be frequently accompanied with a renal excretion of albumen. Stimulation of the spinal cord below the medulla, irritation of the renal nerves and other neurotic influences give rise to it. It may also result from nervous depression ; from vaso-motor disturbances, as in fevers ; in anæmia, especially after exertion.; in cardiac failure. Remeberg explains the transient albuminuria of healthy persons as follows : He asserts that the transudation of semen-albumen takes place in the glomeruli on account of the increased permeability of the walls of the blood-vessels composing the tufts. The epithelial investment of the latter participates in this abnormal permeability. Dr. Nunn has pointed out, that eleven per cent. of persons examined for life insurance, where the individuals appeared, after careful examination, to be otherwise in perfect health, had albumen in their urine.

Albumen, though occasionally an ingredient in the urine of healthy individuals, indicates an abnormal state at the time it is voided ; but "meanwhile we should solace ourselves with the comforting thought that albumen may often be made to disappear by proper attention to the exciting cause."

DISTINCTIONS BETWEEN REAL AND APPARENT DEATH.—Dr. William Fraser, in opening this interesting article, says, "a satisfactory definition of life should involve every stage of vital development, but never be identified with any mode of inanimate existence." A most important postulate, and one to which he most consistently adheres throughout his treatise.

The introduction, though perhaps somewhat prolix, and abounding in terminology of a rather technical cast for the pages of a popular magazine, is mainly devoted to illustrations of the opening statement, after having defined life provisionally, as the continuous individual integration and differentation of material energy.

Man, with his powers unimpaired, manifests his vitality in unmistakable terms, but conditions not incompatible with resuscitation may occur wherein all his functions are so reduced as to be directly imperceptible. In such cases, to prevent premature burial, it is important to discover some sign absolutely diagnostic of real or apparent death.

In some simple forms of life, vital action may be suspended by dessication, and again restored by moisture. In some of the cold-blooded animals the same effect may be produced by congelation, and counteracted by the subsequent absorption of heat. In man, however, the animal functions may be suspended, and even some of the organic processes interrupted, without destroying life ; but certain of his functions cannot be interrupted, even for a limited period, without causing death. These are the functions of circulation, innervation, and respiration— all three so interdependent that the complete interruption of either necessarily leads to arrestment of all, and consequent death.

The blood is the seat of two distinct modes of motion—a sensible circulation through the heart and vessels, and a subtiler change with tissue-elements. The latter process, being invisible, must be largely accounted for on hypothetical grounds; yet there are certain associated phenomena which are quite observable under certain circumstances, and which tend to throw light on the physico-vital relations of the blood. Such are the local variations in the total quantity of its mass, and in the relative proportion of its various constituents. As there are means of exciting a recognizable preternatural activity of the circulation in parts open to observation during the minimum degree of vitality, such a possibility affords a reliable method of deciding as to the existence or non-existence of this vital process.

Heat—the most potent and available form of irritant—when applied to the skin so as to elevate its temperature considerably above the normal point, causes first an efflorescence of surface, diminishing gradually in intensity from a central point. By increasing the heat, or prolonging its action, the color becomes more distinct, till, at the point of its greatest intensity, the cuticle becomes detached by the gradual exudation and accumulation of a fluid which thus forms a true vesicle. In *post-mortem* vesication, the contents are generally gaseous, and even if fluid, from infiltration in an œdematous or dependant part, this is always serum, unlike the vital, fibro-albuminous solution coagulable by heat. The pathognomonic distinction, however, is found in the underlying cutis, which, after death, has an unalterable yellowish-white, crisp, horny appearance, in obvious contrast to the efflorescence of vital active congestion, which can be repeatedly displaced and removed by recurrent pressure.

The retention of carbonic-acid in the anterior openings of the air-passages is evidence of vitality, where its absence, under appropriate tests, is proof of the opposite condition.

Innervation is blended with and controls, all the vital operations, being particularly implicated in muscular contractions, an act primarily concerned in the movements of respiration and circulation. The frequently repeated transmission of intense electric currents is the most powerful stimulus of contractility, and when such a measure fails to excite contractions in muscles essential to life, death must have occurred.

Rigidity and putrefaction may be regarded as infallible *post-mortem* indications. In the former state, the muscles of the lower jaw and neck are generally first involved; those of the lower extremity, last.

In the case of frozen limbs—a condition with which *post-mortem* rigidity might sometimes be confounded—the differential diagnosis may be found in the creaking sound which is yielded by a frozen limb on forcible flexion, from breakage of congealed moisture, and also in the spasmodic contraction which resumes its morbid position on removal of the correcting force.

Putrefaction succeeds rigidity as a bluish-green tint of skin, beginning usually over the abdomen, and spreading over the body. Similar gangrenous appearances may occur during life, but the line of displaceable redness at the confines of the living tissues, is the distinguishing characteristic.

The desideratum, however, is some infallible proof of death, whereby this state can be decided without waiting for the tardy supervention of these positively *post-mortem* phenomena. Cadaveric aspect; coldness and lividity of surface; imperceptibility of the respiratory movements; experiments upon the pupillary muscles, and examinations of the fundus of the eye; the effects produced by tightly ligating a limb; the changes induced in a polished needle inserted deeply into the living tissues;—none of these affords a perfectly satisfactory and reliable solution of the problem. Even the difference between vital and *post-mortem* vesication might be rendered obscure by circumstances.

Consistently with his definition of life, the author holds that the possibility of absolutely deciding, in doubtful cases, as to the presence or absence of vitality, depends on the possession of artificial means which will sensibly demonstrate the minimum activity of each of the essentially vital processes, the utter failure of each of the specific reactions, under its proper tests, being conclusive proof of each.

The respiratory test is made by collecting the carbonic acid, at its point of exit, into a small, transparent vessel, containing clear lime-water ; its merest presence, in contrast to other reagents, changing this fluid at once, on shaking, into an opaque milky solution.

The innervation test is rendered practicable through the inseparable connection of this attribute with muscular contraction. The body is placed before a bright light, and a laryngoscope introduced well back into the pharynx, so as to bring the superior laryngeal aperture into view. After death, the rima glottidis presents the narrow, elongated form, from the close approximation of its chords. If, under the repeated transmission of intense electrical currents, properly directed, there is no responsive contraction so as to sensibly widen the aperture, death is certain.

The circulatory test, or the attempt to excite an actively congested state of the cutaneous capillaries, is pre-eminently the best, as it requires only simple and easily procured appliances, which always yield decisive results either in the living or dead subject. The application of heat, and the act of cupping are both effective topical means for arousing this preternatural activity of the cutaneous circulation, even in the most languid condition of the system compatible with vitality. The entire absence of such distinctive physiological reactions, and the occurrence of merely physical alterations, is undeniable proof of death. Over the heart is the most suitable region whereon to operate, as there the skin retains longest its vital warmth.

Hold the flame of a candle close to (but not in contact with) the skin, sufficiently long to render the cuticle easily detachable ; if the body is dead, the cutis will present a crisp, yellowish-white, horny appearance, unaffected by pressure ; if alive, there will be a readily perceptible vital redness, dis-tinguishable from all *post-mortem* discolorations by its repeated displacement and reappearance under alternating pressure. These phenomena may be rendered more evident by the use of the magnifying glass, the part being viewed by a bright light.

Kindle a piece of paper soaked in any alcoholic liquor, put it into an ordinary drinking-glass, and invert this over a part where all its edges will come into accurate contact with the skin. If there remains a minimum degree of vitality, a state of superficial capillary congestion will be induced, whereas the actual inability to produce such vital reaction, and the production of solely physical effects by such potent agencies are infallible evidences that life is irreparably destroyed.—*Popular Science Monthly, Jan. 1881.*

OPIUM NARCOSIS IN AN INFANT SIX WEEKS OLD.—RECOVERY UNDER THE USE OF BELLADONNA.—Dr. S. C. Chew reports the following case as a contribution to the history of "Physiological Antagonism between Medicines :"

On Nov. 30, 1880, he was called to see a child aged six weeks and two days. On arrival at the house he was informed that six hours previously a powder containing one-quarter of a grain of sulphate of morphia had been given the child by mistake. The child had been given milk immediately afterwards, and had regurgitated a part of it. Possibly some of the poison may have been ejected in this way. At all events, the child was in profound narcosis when Dr. Chew arrived. The pupils were contracted to the size of pin points, the respiration was very feeble and shallow and the pulse at the wrist scarcely perceptible. It was too late to use an emetic or the stomach-pump. Dr. Chew directed that strong coffee be given the child in teaspoonful doses every few minutes, while he went to his office for some tincture of belladonna. On returning he gave a drop and a half of the tincture at once, repeating half a drop every twenty or thirty minutes. At the same time he applied a cloth dipped in ice-water to the back of the neck for a moment at a time with the effect of causing deep gasping inspirations. Gentle shaking and slight pricking with a pin were occasionally employed, but all violent measures were avoided.

After four drops of the tincture of belladonna had been taken, a bright scarlet hue diffused itself over the face, arms, and gradually over the entire body of the child; upon this the interval between the doses was extended to an hour. Synchronously with the redness of the skin other evidences of the physiological action of belladonna were apparent; the pupil began to expand, the pulse became quicker and stronger, and the respiratory acts deeper and more frequent. As the respiratory functions responded so promptly to the influence of the belladonna, Dr. Chew did not consider it necessary to employ the faradic battery which he had in readiness.

Six hours after the doctor had first seen the child, its respiration and circulation were completely re-established, and it was easily kept awake. He therefore felt justified in leaving it, directing simply that it should be awakened every hour. On the following morning it was perfectly well.

This case of Dr. Chew's is especially interesting when taken in connection with the lectures which Dr. Bartholow has been recently delivering in New York. Dr. Chew agrees with Dr. Bartholow that the state of the pupil is not a safe guide when giving belladonna or atropia to counteract the effect of opium. The belladonna should be given not to dilate the pupil, but to maintain the respiration and circulation; this is effected by the stimulant influence of small doses.— *The Medical Record, Jan. 8, 1881.*

METASTASIS OF ACUTE ARTICULAR RHEUMATISM TO THE MENINGES—RECOVERY.— Dr. Ranney reports this rare complication of acute rheumatism. A woman, after suffering for five days from acute articular rheumatism, involving the knee, ankle, and elbow joints, found that the pain had entirely left the joints but that she was troubled with a headache. The headache increased until she became delirious and finally comatose. The pupils were regular, neither dilated or contracted, and responded sluggishly to a strong light. The temperature was 104.5°, pulse 100, very hard and somewhat irregular, respirations twenty to the minute.

It was necessary to draw the urine with a catheter; it was highly acid, of a specific gravity of 1019, and contained no albumen.

Three leeches were immediately applied to each mastoid process, and an ice-cap to the head. A drop of croton-oil was placed on the tongue, and tepid-water baths were given every half hour. The temperature fell gradually to 101° in about three hours, and was kept at this point by occasional baths.

On the next day the patient was conscious, but stupid. The temperature remained at 100.5°, very few baths being now needed to keep it at this point.

On the second day the patient was in her right mind. The rheumatism reappeared in the right elbow and wrist. No heart or lung complication was discovered at any time in the course of the disease. Salicylate of soda was now given, 25 grs. every three hours. Under this treatment the rheumatism disappeared entirely in the course of three days.

Dr. Ranney gives the following causes of marked cerebral symptoms in simple acute rheumatism:

1. Some affection of the heart or, more rarely, of the lungs. Of these complications, endocarditis and pericarditis far outrank all others as regards frequency.

2. Hyperpyrexia; where the abrupt rise of temperature is the striking symptom, and the cerebral disturbance follows this.

3. Meningitis, either as a complication or as a true metastasis, similar to the change of seat of inflammation in the joints.

The following train of symptoms leads Dr. Ranney to regard his case as one of true meningitis: headache, delirium, and coma, the moderate pyrexia, the sluggish pupils, the hard, somewhat irregular and but slightly accelerated pulse as compared with the temperature. That it was also a true metastasis he considers shown by the fact that when the articular swelling and tenderness disappeared the meningeal inflammation commenced, and, on the other hand, with the subsidence of the meningeal affection the articular symptoms reappeared.

The same patient had a second attack of rheumatism about three weeks later; it was speedily cured by the use of salicylate of soda, lasting only ten days.—*The Medical Record, Jan. 1, 1881.*

A CASE OF ACUTE FATTY EMBOLISM OF THE LUNGS.—Dr. Claussen reports the following case: A German, aged 25, was thrown by his horse in such a way as to strike his pelvis and thigh against a tree. The diagnosis of fractured thigh was made. A Hodken's splint was applied, and for three days the patient did well.

On the fourth day the patient complained of headache and dizziness. Soon after, his pulse was 120; temperature, 101.5°, and respiration 30. During the afternoon he became delirious. Examination of heart and lungs revealed nothing abnormal except the rapidity of action.

On the next day the urine required to be drawn with the catheter. The patient appeared rational; the temperature was 102.2°; pulse 135; respiration 36. In the evening he became drowsy and fell asleep. The temperature was 102.5°; pulse 140; respiration 46; heart and lungs normal. At 2 a. m. the next day, he died.

Autopsy: Heart normal; lungs, with the exception of being a little darker than usual, appeared to be healthy. Liver, spleen and kidneys normal. The right os innominatum was broken from the symphysis pubis through the body of the pubes through the acetabulum into the great sciatic notch. The tissues in the neighborhood of the fracture were infiltrated with blood but there were no signs of inflammation. The femur was fractured at the junction of the lower and middle third.

The following was the result of the microscopic examination of the lungs: In every part of the lungs a great number of the small arteries and of the capillaries were found filled with fluid fat, in some parts there was a perfect injection of the capillaries of the inter-alveolar tissues, often enough to make the capillaries protrude into the air cells.— *Chicago Medical Review, Jan. 5, 1881.*

INFLAMMATION: ITS CHEMICAL CAUSE, AND CURE.—Dr. W. Y. Brunton believes that fermentation and its resulting product, an acid, are the prime factors in producing inflammation. A foreign body in the system, whether it is a thorn, or a germ, or the surgeon's knife in the act of cutting, becomes an impurity, and, having heat and moisture in the system, it ferments.

As acid is the active principle of fermentation, Dr. Brunton would employ an alkali as a remedy in all cases. Liquor potassae answers the purpose very well. He makes a solution of one part of liquor potassae to eight or ten parts of water and applies it to all kinds of suppuration or ulceration, modifying the strength of the alkali to suit circumstances. The alkali must produce a slight tingling sensation in order to be of effect in all cases; on the other hand, if it feels hot after its application, it must be discontinued for the time being, and when next applied the strength of the solution must be reduced.

Dr. Brunton instances the following natural illustration of the truth of his theory: If a person were to die on the plains of Colorado, the body, if left exposed to the rays of the sun for months, would not be decomposed, in consequence, he believes, of the alkali contained in the soil preventing fermentation and, therefore, decomposition.

SOME NOTES ON A NEW ANTI-PRURITIC REMEDY.—Dr. Bulkley, after reviewing the list of remedies used internally to relieve itching of the skin, concludes that chloral and bromide of potash are almost the only ones that are of much use. As these drugs cannot safely be long-continued or too oft-repeated, he sought for some other neurotic.

As gelseminum occasionally gives relief in spasmodic asthma, and in certain cases of neuralgia of the fifth pair, he thought that it would probably act as a nervous sedative on the skin. He prescribed it for a considerable number of persons, mainly those suffering from eczema, and found it an efficient adjuvant for the relief of itching in certain cases. In several cases it failed, and he cannot tell under exactly what conditions it will succeed or not. Dr. Bulkley has used the drug only with adults; he would not yet be willing to give it to children or to those who were not able to watch its effects by their personal feelings in other respects than the itching. He uses the tincture in gradually increasing doses, repeated every half hour or every hour, until the pruritus is relieved, or until some of the unpleasant symptoms are experienced. Ten drops is the first dose, then twelve or fifteen and so on, until results are obtained, or until a drachm or so has been taken in two hours. He usually orders the medicine to be taken immediately before going to bed.

One patient, who had suffered greatly from eczema of the genitals for a long time, and on whom chloral and the bromides had lost all soothing effect, obtained perfect relief from gelseminum. She described her sensation as follows: She felt as if a wave were passing over her first, with a thrill, as if something were circulating through the blood to every portion of the body, and then a sense of quiet or ease followed immediately, with an indisposition to move. There was no unpleasant sensation, absolutely no effect upon the mind, but she wanted to sleep from the relief which was obtained, and the sleep was described as "delicious."—*New York Medical Journal, Jan. 1881.*

PHOSPHORUS POISONING.—"In phosphorus poisoning there is one certain antidote, viz: Carbonate of magnesia in one drachm doses until no phosphorescent breath is observed. Phosphate of magnesia is formed; the uncombined magnesia, by its mechanical action, protects the coats of the stomach from any further action of the phosphorus, and any free phosphoric acid is neutralized by it as it is formed."—*Birmingham Medical Review—Canadian Journal of Medical Science.*

BEEF SUPPOSITORIES.—Dr. Jas. J. Tucker uses Johnson's or Liebig's beef extract incorporated with cocoa butter in the form of suppository to support life in chronic gastric disorders, adynamic diseases and all cases where the administration of food by the mouth is impossible. The beef combines easily with the butter, or, if time must be saved, the hollow suppositories may be used. —*The Chicago Med. Jour. and Ex.*

THE HAMMOCK SUSPENSION IN THE APPLICATION OF THE PLASTER JACKET IN THE TREATMENT OF SPINAL CURVATURE BY DAVY'S METHOD.—This article presents a modification in the mode of application of plaster of Paris in the treatment of spinal curvature. The objections to which the application of the plaster-jacket, by the so-called tripod method of extension presented by Dr. Sayre, is open, are several, and chief among them is the fact that the pain suffered by the patient during the suspension necessitates a hurried application of the bandages, and sometimes requires that the patient be taken down before the process is completed, thus changing his position before the jacket has hardened, and running the risk of injuring the mould, if not of rendering it entirely inefficient.

During a recent visit to London, the author had the opportunity of seeing a method of extension which seemed to possess all the advantages and none of the disadvantages of Dr. Sayre's method. It was introduced by Mr. Davy, Surgeon to the Westminster hospital, and is called the hammock suspension, in contradistinction to the tripod suspension of Sayre.

The hammock may be improvised out of canvas, burlap, or strong muslin. When an expense of two or three dollars is not an objection, Traver's American hammock is very convenient for the purpose, its meshes not being fixed by knots, a matter of importance, as will be seen further on. Points of suspension can be easily obtained in a small room, by bars temporarily nailed to a window and opposite door, or by staples driven into the walls. The hammock is hung by pulleys so that it can be raised or lowered at will. It is then lowered until it lies loose upon a table provided with a blanket or comfort.

The patient is prepared by putting on a jacket over which the plaster is to be applied. The author calls attention to a special jacket which he deems much superior to the undershirt usually used for this purpose. It is simply a knit sacque, open at both ends, and may be made of wool or cotton, the latter preferred. The material by its elasticity adapts itself to the size and inequality of the body, and should be long enough to reach from the axillae to the hips, with a margin left which can be tucked under during application, and afterwards turned over so as to protect the soft parts from contact with the hard edge of the plaster.

Thus prepared, the patient is placed, face downwards, upon the hammock which has slits cut in it through which he passes his arms. The bony prominences are protected by pads of cotton, and rolls of the same material are placed on either side of the spinous processes, especial care being devoted to the protection of the prominent points opposite the seat of disease. The time and care which can be devoted to the protection of points likely to be injured by pressure is one of the great advantages of the method. The protecting pads having been arranged, the hammock is drawn closely around the patient, and any of the cords that do not apply themselves closely to the body, are cut with scissors. The plaster bandages are now applied, and must include the hammock. They are prepared as suggested by Dr. Sayre; fine plaster is rubbed into cross-barred crinoline, which has been cut into strips three and a half inches wide and about five yards long. The plaster jacket should encircle the body from under the crest of the ilium, well up

under the axillae. The application can be arrested if necessary to readjust the protecting pads.

In order to secure full expansion of the chest the patient is directed to grasp a rope fixed to a ring of the hammock, beyond his head, and make steady traction upon it, thus putting the pectorial muscles on the stretch.

After the application of the plaster, the patient remains suspended till it is thoroughly hardened, and is enabled to pass the interval without inconvenience by reason of his comfortable posture. The hammock is then lowered to the table, and cut off just beyond the edge of the plaster.

As to the matter of extension which might be urged by the advocate of the other method, the author claims that, owing to the sway of the hammock, the prone position of the body tends to force apart the bodies of the vertebræ , the transverse processes acting as a fulcrum, a tendency which can be increased or diminished at will, respectively by lowering or raising the hammock.

The merits claimed for this method are enumerated as follows :

It is applicable in all cases where the other method can be used.

It is applicable where from advanced disease, or other cause, suspension might be dangerous or inadvisable.

It avoids all pain or discomfort to the patient.

It allows ample time for careful application.

Full time can be given for the plaster to harden, and finally, it renders the physician independent of skilled assistants.

After citing five cases in which his method was applied, the author says that in none of them was there any complaint of pain, or even discomfort while suspended in the hammock ; and all the patients have rapidly improved, which is the best possible proof that the support is efficient.— *Dr. N. P. Dandridge, in the Cincinnati Lancet and Clinic, Jan. 22, 1881.*

DETECTION OF ARSENIC IN WALL PAPERS.—

The nitrate of silver test has for some time been regarded as the "most simple and reliable test for arsenic in wall papers." Dr. W. B. Hills, of the Harvard Medical School, writes to prove that it is one of the *least* reliable tests for the purpose mentioned.

Dr. Hills claims that the cases in which the nitrate of silver test gives fairly reliable results are comparatively few, and are essentially those in which arsenite of copper (Scheele's green) or aceto-arsenite of copper (Schweinfurt green) is used as a pigment, mixed with other substances. These arsenical compounds, however, are commonly used mixed with other substances, by which means various tints are obtained, and in which, as a rule, the arsenical green cannot be detected by any physical appearance. In these cases the above test will not usually show the presence of arsenic, especially if organic pigments are mixed with the arsenical green. These latter pigments, if treated with ammonia, will often give colored liquids. If nitrate of silver is added to such a liquid, the precipitate produced, whatever its true color may be, will usually appear of the color of the liquid. Therefore, in such cases no conclusion can be drawn as to the presence or absence of arsenic. Moreover, various organic solutions give precipitates with ammonio-nitrate of silver, even in the absence of arsenic.

A yellow precipitate is sometimes obtained from wall paper, by nitrate of silver, even when other tests have proved the absence of arsenic. This precipitate, according to Dr. Hills, is probably phosphate of silver, which cannot be distinguished by physical appearances from arsenite of silver.

The yellow sulphide of arsenic is sometimes used in wall paper, and cannot be detected by nitrate of silver.

Dr. Hills states that arsenic acid is extensively employed in making the aniline colors, and that a considerable proportion of arsenic is often retained in them. The shades of red are especially liable to contain arsenic from this cause. In these cases, the ammonio-nitrate of silver test gives colored liquids, which may disguise the color of the precipitate.

Dr. Hills believes that persons employing the nitrate of silver test alone will report most arsenical papers as free from arsenic, and will be in danger of reporting, as arsenical, papers entirely free from arsenic.

Dr. Hills' own method of proceeding is as follows: "Take a sample three or four inches square (less will suffice with plain paper), cut into small pieces, moisten with concentrated sulphuric acid and heat carefully till the paper is thoroughly charred. Let the charred mass cool, add to it about one fluid ounce of water, grind the black mass fine that the water may come in contact with all parts of it, filter and wash. The arsenic will be found in the filtrate, which is examined, by Marsh's test. All chemicals must be free from arsenic. A paper which, treated carefully in this manner, furnishes no arsenical mirror on porcelain does not contain any appreciable amount of arsenic." —*Boston Medical and Surgical Journal, Jan. 13, 1881.*

CASTS AS A SYMPTOM OF BRIGHT'S DISEASE.—

At a recent meeting of the New York Pathological Society, Dr. Amidon presented a specimen of urinary casts, which was interesting as bearing upon the question of the prognosis in kidney disease. The specimen was obtained eighteen months ago from an engineer of a Pacific mail steamer, who was suffering at the time from symptoms of southern malarial fever. A diagnosis was made accordingly. On examining a specimen of urine under the microscope the following day Dr. Amidon was very much surprised to find casts of all varieties except the hyaline. The urine contained five per cent. of albumen and two and eight-tenths of urea. Dr. A. was so much alarmed at the condition of the urine that he made haste to modify his original diagnosis by adding thereto the grave one, of complication of acute nephritis. An unfavorable prognosis was also given. Quinine was administered hypodermically. The next day the temperature had fallen almost to the normal point, the specific gravity of the urine increased, there was no albumen, but few casts, and the urea was increased to three and eight-tenths per cent. The patient continued to improve, so that at the end of a week he was fairly convalescent. The patient at the time of making the report was in perfect health. Here, then, was a case in which all the varieties of casts save one were found in the urine, and yet the patient recovered.

At the same meeting, Dr. Peabody presented a specimen showing the state of the mucous membrane of the uterus during menstruation. The mucous membrane was uniformly thickened throughout the body of the organ. This thickness was equal to six or eight millimetres. Dr. Peabody had not met with a similar condition before, in an apparently healthy uterus, but Prof. Dalton, who had seen the specimen, had informed him that it was the normal condition of the uterine mucous membrane preceding menstruation. A corpus luteum, three weeks old, was found in one ovary. In the opposite ovary was found a cyst large enough to admit the tip of the thumb.—*The Medical Record.*

HYPODERMIC INJECTIONS OF ALCOHOL.— M. Maxime Dumonly has recently reported the experiences gained in the wards of the Germain See, from the use of alcohol administered hypodermically, and arrived at the following conclusions :

1. In small doses, alcohol stimulates the digestive organs, but in large doses disturbs them.

2. Alcohol accelerates the respiration and circulation.

3. In large doses, it depresses the nervous system ; in moderate doses, it is a cerebral stimulant.

4. In very large doses, it reduces the temperature of the body.

5. Alcohol does not act as an antipyretic; it cannot favorably modify the course of fever.—*Lyon Medical—Cincinnati Lancet and Clinic.*

BOOKS AND PAMPHLETS RECEIVED.

"Electricity in Medicine and Surgery" with cases to illustrate, by John J. Caldwell, M.D., Baltimore, Maryland. (Through the Author.)

"The Treatment of the Genito-Urinary Organs," the use of electricity, damiana, etc., etc., by John J. Caldwell, M.D., Baltimore, Maryland. Reprinted from St. Louis Medical and Surgical Journal. (Through the Author.)

"Genital Irritation," together with some remarks on the hygiene of the genital organs in young children. By Roswell Park, A.M., M.D., Demonstrator of Anatomy, Chicago Medical College, etc., etc. Chicago: Tucker, Newell & Co., 1880. (Through the Author.)

"California Medical Journal," a monthly devoted to the advancement of Medicine, Surgery and the Collateral Sciences. Edited by D. Maclean, M.D., J. H. Bundy, M.D., and D. D. Crowley, M.D. Oakland, Cal. : Published by D. Maclean.

"The International Journal of Medicine and Surgery," published weekly. Edited by B. Newton, M.D., A. Rose, M.D., N. Senn, M.D., H. A. Bunker, M.D., and C. H. Ten Eyck, M.D., New York.

"The American Medical Bi-Weekly." E. S. Gaillard, A.M., M.D., LL.D., Editor and Publisher, New York.—This journal, which was formerly published in Louisville, Ky., and which had nearly reached the close of the eleventh volume when the severe and protracted illness of the editor, Dr. E. S. Gaillard, compelled its temporary discontinuance, is now, on his entire recovery, restored to its position among the medical journals of this country. The Bi-Weekly is now published in New York. The first number of volume twelve has been received. It is now a double-column journal, larger than the old Bi-Weekly, and is published at $1.00 a year.

"The Herald of Health," edited by Dr. T. S. Nichols, London, Eng.

ROCKY MOUNTAIN MEDICAL REVIEW.

Vol. 1.   COLORADO SPRINGS, FEBRUARY, 1881.   No. 6.

ORIGINAL ARTICLES.

A Case of Religious Insanity Instantly and Permanently Cured by Falling on the Head from a Window.

By JOHN MULVANY, M.D., ROYAL NAVY.

In March, 1878, when stationed in the Falkland Islands, I was asked to visit a tailor who was said to be loco or insance. I knew the man well, and as I had seen him very recently I was shocked at the intelligence and went immediately. He was an Italian about 35 years of age, married and the father of one child, a daughter. Both mother and daughter were living at Milan in Italy. Two years previously he emigrated to the "Platte," but not finding it the El Dorado it proves to so many of his compatriots, he came to the Falklands under an engagement with a Portuguese tailor, with whom he had been domiciled about ten months. He was of the nervous temperament, well educated, courteous and refined, his manners being much superior to his state in life. His employer considered him to have been a good workman, assiduous and contented, until he received an intimation that his services would soon have to be dispensed with, in consequence of the falling off in business which had for months been experienced in the settlement. This produced depression of spirits, and he began to neglect his work, to stroll about the roads, and pay frequent visits to the R. C. Chapel; on these occasions he was observed to be strange in his manner, he would make such a noise in the passage as to disturb the congregation, or he would slam the door violently on entering or leaving. Next, he began to see visions, and to carry on imaginary dialogues with a few of the saints; when I saw him he had the pictures of two over his heart, and he gave me a long history of his intercommunion with them. On other subjects he was quite rational, but would shortly and abruptly revert to his visions or his prayers *mens illis imaginibus addicta erat.* He had a furtive, suspicious look, conjunctiva somewhat injected, a tongue thickly furred, loaded bowels, no appetite, and had slept very badly for some nights past. Under the employment of stomachic aperients, and narcotics at bed time, a decided improvement took place, but not of a permanent nature, the tongue cleaned, the bowels functioned healthily, and sleep was followed by long periods of sanity. But in about ten days he was as bad as before, sleep became more difficult and was not followed by lucidity; his visits to the chapel became more frequent and were sometimes paid at night. On one occasion he was found by his employer praying outside its door at midnight, with two candles alight to enable him to read. Now, by the laws of the Colony, anyone introducing a foreigner is answerable for all expenses the Government may incur on his account, wherefore the ghostly spectacle afforded by the loco kneeling in the snow and reading by candle light in the open air at the witching hour of night, conjured up such a vista of government interference, asylum, keepers and heavy bills, that the dismayed Lusitanian, yielding to feelings congenial to the circumstances, gave him a good sound drubbing, illustrating thereby how closely impulse may tread upon the heels of science, and how the ignorant tailor, guided only by his feelings, arrived at the same *methodus medendi* as did the learned Celsus, by *majore studio literarum disciplina*, who prescribes *ubi peperam aliquid dixit aut fecit, fame vin-*

culis plagis coercendus est, and who says *subito etiam terreri et expavescere in hoc morbo prodest.* For some time after this there was a marked improvement in him, but gradually the little rift within the lute enlarged, the prayers and raphsodies became more frequent, the eyes wilder and the sleep more difficult and scanty. To prevent nocturnal visits to the chapel he was locked in at night, but on one occasion he forced open a small window at the back of his room and endeavored to make his *point de depart* through it to the chapel, in doing so he fell on his head into the yard ; the noise awoke his employer, who was quickly on his trail and found him stunned, and bleeding very freely from a deep wound in the scalp. I was sent for, but on arrival I found him conscious and quite sane, he said he was cured, that the fall from the window did more for him than all my medicine, and time proved the permanency of his cure.

A case somewhat similar to this, in point of cure, was related to me some years ago by the superintendent of the asylum at Dartmouth, Nova Scotia ; a young lady whom circumstances suddenly reduced from affluence to penury was obliged to become a governess to some boys who were learning latin and mathematics, in both of which they were more proficient than she. To not allow this to become apparent she was obliged to sit up at night and learn the lessons for next day's tuition; the mental strain and loss of sleep produced insanity, and she was sent to the asylum an almost hopeless case. One day she succeeded in hanging herself by a sheet from one of the bars of a window, but was seen and cut down before life was extinct. A jugular vein was immediately opened and blood drawn copiously, with the happiest results, for not only was her life restored, but also her reason.

In these cases we find in common a cerebral shock and a copious bleeding, and one naturally wonders if recovery were a resultant of both factors or one of them singly, and if an indication for treating a similar case might not be furnished by them.

Medical Facts Relating to the Zuni Indians of New Mexico.

By Dr. H. C. Yarrow, U. S. Army.

The following brief paper has been prepared entirely from the notes of a lady of culture and refinement, who last summer accompanied her husband to New Mexico and remained at the town of Zuni for several months.

Living with these Indians in their rude abodes, sharing their domestic cares and pleasures, and sympathizing in their interests, she was enabled to gain their confidence, and in this way obtained a knowledge of many matters of interest relating to their customs, which heretofore they have been unwilling to communicate to white persons.

She deserves great credit for her efforts, since the attempt has been frequently made before to acquire this knowledge, but no one has succeeded in this particular regard save herself. The people of whom we speak are peculiarly interesting as being a part of those tribes known as the Pueblos, an agricultural race dwelling in towns with tolerably well constructed houses, which are built of adobes or sun dried bricks. Ostensibly Roman Catholics they still cling to a religion of their own, the principal features of which appear to be keeping up an eternal fire and · watching for Moctezuma. From these habits and others, many suppose them to be the descendants of the ancient Aztecs, who have gradually worked their way up from the south, although I have been informed by some of the old men that they originally came from the north. The account which now follows is substantially the same as furnished to me by the lady mentioned :

"Among the various studies which I made during my sojourn among the Pueblos of New Mexico, none were of greater interest to me than the domestic system of the strange people of the Pueblo of Zuni. Although allied to the savages of the northwest in their superstitions, rites, ceremonies, and the common practice of scalping, yet in all other respects they are radically different.

Nowhere have I seen such happy homes, where family ties were apparently cemented by the warmest love; in no place have I seen such individual rights accorded each member of a household. During my intimate relations with these people for a lengthened period, I never heard a harsh word spoken to a child, and I believe disobedience to parents is quite unknown. It is quite common to see men caring for the babes while their wives are busy with household duties. I found no surer way to win the favor of a Zuni father than to notice his little ones. To purchase or trade for an article belonging to a child too young to express his or her desire in the matter, was impossible, so equitably are personal rights regarded. The men are kind and tender to their wives, never putting upon them heavy burdens. The children, with few exceptions, are afflicted with a terrible cutaneous eruption from birth, which continues until they are sufficiently grown to play in the open air, when the trouble ceases, and none over four years of age were seen suffering from this malady, although a few seem to have weak eyes. The infants present pitiable pictures with their faces almost an entire scab, and yet the cry of a child is a remarkable sound in Zuni, the little creatures seeming endowed with wonderful powers of endurance. It is not uncommon for children to die at birth, but the mothers with rare exceptions recover from confinement. Their practice of obstetrics is very novel to a civilized woman, and is confined to the hands of female midwives, two of whom are at Zuni and attend when needed. From what I can learn, the Pueblo women are much more careful of themselves during pregnancy than those of the nomadic tribes. When their hour of travail is at hand the nurse is sent for, who at once proceeds to administer a hot tea made from a certain root; for a while the patient lies upon the floor face downwards, a hot stone placed under the abdomen; the labor increasing she changes her position to a sitting one. The nurse kneeling behind pulls her slightly back, constantly pressing downwards

on either side of the abdomen. This is done to help the progress of the birth, and the nurse throws her whole strength into the task. It is during one of these efforts that the child comes suddenly into the world. They suffer but little, they claim, owing to the virtues of the hot draught of which they frequently partake during the progress of the labor, and to the great assistance rendered by the midwife. The heated stone before mentioned is kept constantly applied to the abdomen. The midwife severs the umbilical cord with a common knife, and ties it with a string as we do; then the child is laid upon a heated white earth which is considered most excellent for the newly born, and a blanket thrown over it. The navel is sprinkled with hot water and a cloth laid upon the abdomen. This sprinkling is several times performed, and in three days the navel is entirely healed. The removal of the afterbirth is not attended with any special treatment, but is awkwardly removed by the midwife. Bandages for mother and child are unknown. The mother is dosed with more hot tea and lies down for a few hours, and the following day she and her babe go out. The mother receives no other treatment but washes in warm water to cleanse, taking the tea occasionally to hasten the cessation of the lochia, which disappears entirely in four days. They state that the tea if taken earlier than the day of confinement would color the child black.

Women are always confined in their own homes, houses for this purpose not existing in the village, so far as known, but as they use the Spanish word for house both in the true sense and for a room, and as they always retire to another than their general living and sleeping room to have their children, blunders may occur from one not knowing the word *casa* is used in a double sense. A man is never present at child birth, it being considered most indelicate, and the presence of their present missionary in his wife's room during her delivery, created much talk among the women, they considering her subsequent long illness due to his attendance at that time. I was closely questioned by one of the midwives, if in Washington men were allowed to be present on such occasions, and upon my replying that men usually delivered the women, she was greatly shocked, and denounced the custom. I should mention that during the infancy of a child, ashes mixed with cold water are kept constantly upon the face and portions of the body, and this application seems to prevent most effectually the growth of hair.

The practice of abortion is well known but seldom employed, except by women of ill repute, virtuous females considering it a most disgraceful vice, and I really believe the proportion of good women is equally as great among these people as with those claiming the highest advantages of civilization. The only method followed in producing abortions is a continuous pressure downward upon the abdomen, the same as employed by the midwives during labor. During the menstrual period, the Zuni women do no hard work and walk but little, and are always relieved from the duty of carrying water from the well. They busy themselves with indoor work, usually weaving or grinding (for one of these women is never idle), and sit at the loom or kneel before the mill over heated earth spread thickly upon the floor, their robes brought up around the waist and a blanket fastened over the shoulders falling loosely to the floor covering all traces of the earth. Great care is taken by the young women that this period shall not be detected. I was pleased as well as surprised to learn with what extreme delicacy they regarded this subject, and not until a most intimate acquaintance with a family could I induce them to speak of it. I should add that during menstruation a heated stone is worn in the belt, and a hot tea frequently used. The period continues from two to four days, and pain is never experienced.

In Zuni there are several men who keep vegetable medicines, which are dispensed among the sick in a somewhat charitable manner. Those needing drugs procure them by trading ; the medicine man requir-

ing only what the sick are able to give. These drugs are highly prized, and are kept carefully stored in little bags, which hang suspended to the wall of the druggist's room. It is very difficult for a stranger to procure any of these medicines, the Indians claiming that they are only adapted to Indian maladies.

The medicines used in cases of child birth are kept and furnished by the professional midwives, who receive compensation both for their medicines and services. Medicines for all other needs are furnished, as above, by the medicine men.''

Thus ends the account furnished, and I regret that time will not permit me to review the narrative in detail and call attention to the customs of other tribes of Indians in regard to the obstetrical procedure and menstruation, but, I may say that as a rule the Zuni methods are peculiar to themselves and are followed by none other with whom I am acquainted. Perhaps at some future time a paper on the subject may be presented and prove interesting.

Thoughts in Reference to Choice of Climate as a Therapeutical Agent.

BY J. W. COLLINS, M.D., COLORADO SPRINGS.

The "climatic cure of consumption" is a subject that has interested the profession for many years and in many countries. It is a question that has been, and is now, involved in many doubts and difficulties.

The good resulting therefrom, has been perhaps equalled, if not excelled, by the evil consequences, until to-day there are serious doubts in many minds, in and out of the profession, as to the utility of all climatic health resorts. It shall be the object of this short paper to attempt to point out a few of the leading questions to be determined in the choice of climate. There is no argument needed to justify the assertion that, the indiscriminate sending of the consumptive away from the comforts of home, and the consolations and attentions of friends and relatives, and from the intelligent and discriminating care of his family physician, should be condemned. For there is often more harm than good as the result. But, on the other hand, many who were in the most hopeless condition have been restored to health, home and friends, by a timely and judicious resort to the proper climate. The question may be asked, why are not all more or less benefited? Why are so many valuable lives shortened by change of climate? It is often remarked that Mr. A. was sent to a different climate and was benefited, or had his health completely restored, and that Mr. B. was sent to the same point, and that he soon died, or was injured by the climate. These effects we see almost every day. To the answer of these questions I will now address myself.

It seems but proper that we should divide consumptive diseases into two grand classes — those of a catarrhal nature, and those of a non-catarrhal. I am perfectly aware that this division is arbitrary, and open to many objections, and does not express the true pathology of all cases. But I am, on the other hand, almost sure, that every observer will admit, that very nearly two-thirds of the so-called cases of consumption we are called upon to treat, have their immediate beginning in catarrh—nasal, naso-pharyngeal, pharyngeal, laryngeal, or bronchial, and on the indirect effect of the catarrhal diathesis. This fact has been also noticed by Dr. Beverly Robinson in his late valuable treatise on catarrh, although he failed to give it due prominence. These cases of catarrhal consumption, are almost always, if not always, characterized by hypersecretion, excessive expectoration &c., and to the relief of these symptoms we address our treatment, and in this class of cases lies all our success. We use astringents, tonics &c., and out-door life is of great benefit. It is these cases that improve or get well in a high, dry, and cool climate, like that of Colorado. The climate effects, in many cases, what we attempt to do by our treatment. It acts as a tonic, stimulant and astringent. The secretions are rendered normal in amount, the strength is restored

by the cool bracing atmosphere, and the vital forces are rendered more active by the constant out-door life possible in this land of sunshine, where we have but little snow and no rain for nine months of the year. I am forced to the conclusion, however, that only consumptives, who have hypersecretion, profuse expectoration &c., do well here,—and not even these, if they come here too late, that is, after cavities, abscesses &c., have formed, and the life forces too far exhausted. High altitude, with rarity and dryness of atmosphere, should, then, be chosen for those cases of a catarrhal diathesis, as also broncho-pneumonias &c., before the formation of large cavities and complete exhaustion of strength. The presence of hectic fever should not prevent the resort to this climate, as it is often checked before any other beneficial effect is seen. That this class of consumptives are often cured here, is attested by the fact that perhaps more than half of the pioneer settlers of this state, came here as a last resort and were cured. They now may be found in our counting rooms, our mines, and upon our ranches. They are building up on these plains, in our valleys and among our mountains, a grand civilization. They are pushing the iron horse through our gorges and canons, and over the cloud-capped and snow-covered mountains. They sit as Solons, in our legislative halls. They pursue science amid natures grandest evolutions. They are rearing with their wealth and energies, grand cities and smiling villages, where once the red man trod supreme and desolation reigned. These have been left to us, restored to health by the influences of this climate. But, on the other hand, hundreds have fallen victims to an improper choice of this altitude, with its dry stimulating atmosphere. That certain conditions of disease are incurable ; that no drug, no combination of drugs, no climate, will cure all, is as true as the fiat of our creator at the beginning, that "dust thou art, and unto dust shalt thou return." The climate of Colorado certainly is not perfect, and I believe is not suited to any class of consumptives but those of a catarrhal type, characterized by hypersecretion; and I also think that diseases of this type affecting the nasal, naso-pharyngial and laryngial passages, with the same characteristics, are most surely benefited here. Upon the reverse, I should not advise patients with tuberculosis, acute or chronic, or disease of the respiratory tract, with scanty secretion and difficult expectoration, or those with large cavities, or organic heart, or nervous troubles, to seek this climate. I should in these, advise a lower, moist, warm climate, as offering the best chance of cure, or prolongation of life. These opinions, crudely and hastily expressed, are the result of some experience and much thought, and are in the line of all enlightened treatment. I have seen patients here with solidification of lung tissue more or less extensive, with bronchorrhoea, hectic fever &c., the beginning of whose disease was in the nasal passages, improve rapidly, under proper treatment; and others in first stages of tuberculosis, as rapidly decline and die. Then, as a rough guide, I would say—send your cases here, who have this catarrhal diathesis, profuse expectoration and hypersecretion, with every hope of being benefited. For the other class, be they catarrhal or non-catarrhal, with scanty secretion, difficult expectoration, large cavities &c., I should advise a climate the very opposite of this, for the same reason that I should adopt opposite plans of treatment in these cases.

CLINICAL REPORTS.

Removal of Both Ovaries by Abdominal Section, with History of Case.

By J. H. VAN EMAN, M.D., KANSAS CITY, Mo.

Mrs. R——, aet. 36, native of New York, married, wife of a merchant. First menstruated at the age of 11. Was awakened one night by violent pains in the back and pelvis; has suffered some from dysmenorrhœa all her life. Married when about 20; has never been pregnant; has had catarrhal disease for many years, affecting more particularly the nasal passages and alimentary canal; has had cervical uterine leucorrhœa at times. Five years since had a small uterine polypus, which was removed without difficulty. Her dysmenorrhœa being of an obstructive character, she has been treated by the gradual dilatation method. In the spring of 1879 I divulsed the cervical canal. She has been treated by a number of physicians both east and west. All treatment, however, failed to give relief; instead of getting better she gradually grew worse. She is of a decidedly nervous temperament. Some five years since, while under treatment for uterine disease, some very singular nervous symptoms manifested themselves. She first complained of a queer "trembling twitching feeling" in the pelvic region. This feeling, more disagreeable than painful, grew worse and worse, and was soon followed by a rythmical contraction of the psoas, illiacus internus, and other muscles concerned in flexion of the thigh upon the abdomen. This involuntary flexion of the thighs upon the abdomen would continue with violence and perfect regularity for hours, unless controlled by profound anæsthesia. Large doses of morphia hypodermically controlled these movements, but did not entirely remove them. If, when she felt the first intimation of one of these attacks, she took an enema containing belladonna, or a large dose of asafœtida per orem, and kept perfectly quiet in bed, the attack would be warded off. Morphia, and, in fact, all narcotics were very badly borne. The inhalation of chloroform almost always produced violent vomiting, which would continue for many hours. These convulsive attacks assumed the greatest intensity during her menstrual periods, but were liable to occur at any time, from over-exertion or mental worry. During the last year or two, an ordinary action of the bowels would bring on an attack, unless she laid down and kept perfectly quiet for an hour or so. About the 1st of February, 1880, she decided, as a dernier ressort, to undergo an operation for the removal of both ovaries, having lost all hope of any less radical treatment effecting any good, and life being a burden to her under existing conditions. On the 5th of February I performed double ovariotomy upon her, at the Sisters Hospital, Kansas City, Mo., assisted by the following gentlemen : Drs. Todd, Schauffler, Johnson and Bogie. Ether was chosen as the anæsthetic, on account of the gastric trouble produced, in her case, by chloroform. Abdominal section was chosen as the form of operation, on account of the possibility of extensive adhesions, and, on the further ground, that in the operation by abdominal section (though perhaps somewhat more dangerous), you are much more certain of being able to complete the operation, and, at the same time, if your patient survives, you are much more likely to expect a cure. An incision nearly five inches in length was made in the linea alba down to the peritoneum. After all oozing had ceased, the peritoneum was pinched up, a small opening made, and then enlarged on a grooved director. Some difficulty was experienced in bringing the right ovary to the edge of the incision, on account of abnormal adhesions. The left ovary was easily brought out. Small parovarian cysts were found on both sides. The left ovary appeared to be healthy. In the right there was a cyst, the size of a large hazelnut, which was broken open in getting out the ovary. After the ovaries were separated from the surrounding tissues, carbolized silk ligatures were passed round the ligaments, drawn tight, cut short, and the pedicle severed and returned to the pelvis. The pelvic cavity was next carefully sponged out, and the incision closed by means of silver wire sutures passed through the peritoneum and all overlying tissues. Two superficial silk sutures were also used, one at the lower, and the other at the upper, end of the incision. Adhesive strips were applied, next cotton, and then a compress, the whole secured by a Scultetus bandage. The operation now being completed, the patient was put to bed at 2 p. m.; the operation lasting a little over one hour.

Feb. 5, 3 p. m., pulse 78,—7 p. m., p. 75. temp. 99 1-5.° Has had some pains in abdomen since 3 p. m. and some vomiting. Prescribed McMunns elix. opium, gtts. xl every 2 to 4 hrs. according to pain.

Feb. 6, 1:30 a. m., p. 75, t. 100°, pain still severe, with violent retching and vomiting. 9 a. m., p. 75, t. 100 1-5°. Gave 1 drachm doses of sub. nit. bismuth every hour or two during the latter part of the night, also small quantities of milk and lime water, 3—1. 11:30 a. m. pain much less severe. Had 40 drops elixir opium at 9 a. m. and again at 2 p. m. 1 p. m., p. 75, t. 100 4-5°. 8:30 p. m., p. 80, t. 100½°. Kidneys acting well ; some nausea, but no vomiting since noon. Passed flatus from bowels ; comparatively comfortable.

Feb. 7, 6 a. m., p. 68, t. 99½°. Skin moist. Took opium at 10 p. m. last night ; rested well afterwards all night ; menses appeared during the night. 1 p. m., p. 77, t. 100°, very comfortable. 9:30 p. m., p. 72, t. 99 4-5° ; skin moist ; no pain ; quite comfortable ; menstrual flow very free.

Feb. 8, 8 a. m. Had a good night. P. 70, t. 98 4-5°. 7:30 p. m., p. 68, t. 99½°; complained of feeling tired.

Feb. 9, 8 a. m., p. 70, t. 98 3-5°.

Feb. 10. Complained of headache and general bad feeling, room too warm.

Feb. 13. Removed all the sutures, the wound having healed by first intention, excepting a small portion a little above the center, and about one fourth of an inch of the lower portion. This latter portion of the incision does not look well.

Feb. 14. A little suppuration at lower portion of the wound. Moved bowels with injection ; drew off urine with catheter for the first ten days after the operation ; patient improving gradually.

March 7. Thirty days after the operation the patient went home. From this date she seemed to do well in all respects save one. The incision had healed perfectly, excepting about one inch of the lower third. This portion would appear almost well except an unusual amount of induration in its immediate vicinity. She would have light chills or chilly feelings with pain, heat, redness and swelling at that part of the wound. It would open and pus be thrown out. The inflammatory condition would then to a great extent subside. This occurred again and again. In the latter part of May, every thing was cleared up. While dressing the wound she found something that had a pointed end projecting from an opening nearly an inch from the edge of the incision. On forcibly removing it, it was found to be one of the silver sutures, which in the violent vomiting that had occurred in the first 36 hours after the operation, had been broken off at both sides close to the skin, and had at once disappeared from view. This source of irritation removed, the patient's general condition very rapidly improved.

In a like operation I would advise a careful counting of the sutures ; a thing I neglected to do, but am quite certain I shall not err in that direction again.

Now as to results. I saw this lady Dec. 17, 1880, ten and a half months after the operation. She is doing her own work. Has had no menstrual discharge since immediately after the operation. Has had no return of the convulsive movements, and not the least symptom of them. If she works too hard she has some pain in the pelvic region, at times quite severe. She has improved greatly in personal appearance and flesh, and is once more able to take some enjoyment in life. Although not so strong as she once was, she is yet gradually regaining her health and strength.

SELECTIONS FROM JOURNALS.

Notes on the Anatomy of the Encephalon, Notably of the Great Ganglia.

By Edward C. Spitzka, M.D.

The anatomy of no portion of the brain is so obscure and so imperfectly known as that of the so-called Thalamus opticus. One of the first requisits to a comprehension of its relations is the establishment of a proper nomenclature, and the point to start from is the very name under which the great ganglionic mass is known. Since it is not exclusively or even in the main connected with the optic tracts in any animal or man, and, indeed, is in the lower sauropsidæ and amphibians not connected with them at all, the affix _opticus_ should be dropped, and the first word involving that very uncompromising conception of an elevation at the ventricular floor may be retained : _Thalamus_

The current conception that the Thalamus is an elevation at the floor of the _lateral_ ventrical is incorrect. One of our leading comparative anatomists will shortly review this question, and it will therefore be but necessary for me to refer to the matter.

In the cat's brain it can be clearly seen, that (aside from membranous separations) the great mass of the Thalamus is excluded from the cavity of the lateral ventricle by the fusion of the lateral edge of the fornix with the corpus striatum, or rather with the ependyma of that ganglion. Consequently, the two thalmi are included in the third ventricle, which cavity on cross section resembles an upright T, whose vertical branch descends between the thalami, as a deep ditch, the _vulva cerebri_ of the old anatomists.*

Luys, who was unfortunately wedded to certain physiological prejudices as to the function of the thalamic centres, restricted the term _Thalamus_ to the most external mass. Meynert called all the centres in the aggregate by that term as a collective designation. He excluded, however, that gray mass which lines the sides of the vertical slit of the third ventricle.

Now, the third ventricle, as shown by Hadlich and Wilder, extends over the entire thalami ; it would be, therefore, incorrect to limit the designation "central tubular gray of the third ventricle" to that portion only which lines the vertical slit. Either this latter designation should be extended to the

---

*The corresponding _penis cerebri_ of the same anatomists has, by more fastidious colleagues, been rebaptized _pinus cerebri_ and the later _pineal gland,_ now known as the _epiphysis cerebri._

entire thalamic masses or the term thalamus should be extended to the so-called central tubular gray.

Thus interpreted there would be, strictly speaking, but a single thalamus, consisting of two main masses, and a commissural part. The commissure is double. The thalami are primitively united by the lower of these commissures, which I propose to term "basilar commissure."† Secondarily, and only in animals above marsupials (as far as I am aware), do we find another commissure produced at an advanced period of embryonic developement by apposition of the main masses. This is the so-called middle commissure of the brain, the *commissura grisea, c. mollis.* I should consider the least ambiguous designation, "the *thalamic fusion.*"

In a manner similar to that which separates the caudate and lenticular nuclei from each other, and which divides the latter into subsidiary "articuli," the chief mass of each thalamus is separated into an inner and outer zone. The zones are separated from each other by a white intercalation, and especially the outer zone (also in part the inner) presents a beautiful alternation of gray and white laminæ. ‡

These two gray zones constitute the fundamental demarcation of the thalamus; they may be termed *zona grisea medialis* and *zona grisea lateralis.* In animals above the rank of marsupials we find added a round nodular mass, distinctly prominent at the ventricular floor, which lies anteriorly, while in still higher groups a second nodular prominence develops posteriorly. The latter is known as the posterior tubercle or pulvinarium, the former as the *anterior* or *superior tubercle.* The former designation seems the best to me, for although what I call the undifferentiated parent mass of the thalami is visible

in sections anterior to those in which the anterior tubercle is reached, yet the latter, which I propose to term the *anterior nodule* of the thalamus, is the first differentiated centre reached. In man, the *zona grisea medialis* is faintly seen before the anterior nodule is reached, but the anterior nodule reaches its main developement before the zones do, and is absent where these are most prominent. In the carnivora, generally, the anterior nodule projects far in advance of the zones. In these animals, too, a more complex arrangement of this nodule is found than in man, inasmuch as the anterior part of the internal slope of the thalamus shows several elevations absent in the human thalamus.

The *zona grisea medialis* appears pretty equally diffused, and exhibits its lamination evenly both in front and in the middle of its course. The same applies to the human brain for the *zona grisea lateralis.* In the cat, ‖ however, the anterior part of the external zone appears as a beautiful round compact ganglionic mass, protruding boldly into the internal capsule, and which acquires the characteristic lamination only in posterior planes.

It is interesting to note that the ganglionic matter of the thalamus is continuous with that of the ventricular nucleus of the corpus striatum (nucleus caudatus). Indirectly it is connected with the extra-ventricular nucleus, through that great common basilar gray mass, which is the *rendezvous,* as it were, of all the gray categories of the forebrain.§

In an earlier publication (Architecture and Mechanism of the Brain—Journal of Mental and Nervous Diseases, 1879), I have called attention to the fact that the ventricular nucleus of the corpus striatum is the representative of the primordial cerebral gray, inasmuch as the nerve cells of the em-

---

† Continuous in front with the *loci perforati antici,* behind with the infundibulum. Atrophic over the chiasm, it exhibits a set of transverse fibres and gray substance elsewhere.

‡ And yet the latest pretended description of these Ganglia, admitted, notwithstanding numberless glaring errors, into a journal of the standing of "Brain" (that by Dalton), has the Thalamus "homogenous."

‖ As seen in a series of transverse sections prepared by Dr. Graeme Hammond.

§ Here meet the olfactory gray, the cortex *basis capitis nuclei caudati,* the *nucleus lenticularis,* the *claustrum,* the thalamic axial gray, etc., etc.

bryonic and lower amphibian hemisphere are concentrated immediately subjacent to the ependyma of the latter ventricle. The majority of these cells are crowded away from the ventricular floor by the white substance developed in higher animals, and only a portion of the primitive gray remains subependymal. This is precisely what constitutes the corpus striatum. Now the corpus striatum actually *lines* the ventricle ; it' not only lies at its floor ! Any section transversely to the cerebral axis and striking the forepart of the lateral ventricle in the Hippopotamus, Horse, Dog or Cat, will show that an attenuated part of the corpus striatum is continued around *over* the ventricle, and constitutes a greater part of its roof.

A similar comparative study shows that the nucleus lenticularis is also a subcortical development, that is, it results from the individualization of a gray mass originally continuous with the cortex, by means of an irruption of white masses. These at first separate fasciculi (as in the dog) in higher animals coalesce to constitute the external capsule. The segmentation of the lenticular nucleus into three distinct *articuli* so characteristic of the human brain, is not found in the carnivora ; only the outer articulus is demarcated, and that but imperfectly.

In the carnivora the *laminae medullares* or white streaks of the lenticular nucleus are conspicuously absent in the anterior half of that ganglion ; in its posterior half they appear and they rapidly increase in bulk as we proceed backwards, so that in planes where the human lenticular nucleus is still quite massive, we have in the dog only slight ganglionic masses intercalated between the fibre tracts. The *claustrum* is, in the carnivora, not the thin expanded lamina found in man, but a low and massive accumulation, hardly separated from the cortex of the Island of Reil. This fact strengthens Meynert's view that the claustrum is but an individualized cortical layer.

In conclusion I would mention as an isolated fact, and disconnected from the

main subjects dealt with in these notes, that the anterior pyramids of the brain of the large Ceylon fruit bat (Pteropus fuliginosus) undergo a superficial decussation, as patent, and.more so, as that of the optic chiasm. The pyramidal tract after decussating is continued as a distinct fasciculus on the lateral aspect of the medulla oblongata. In the same brain the fibres of the fornix can be clearly seen to terminate in the thalamus without descending to the base of the brain. Whether this applies to the whole of that tract, I am not able to say.

I would also note that in the brain of a large Ara (*Ara ararauna*) obtained from the superintendent of the Central Park Zoological Gardens, Mr. W. A. Conklin, I found what appeared to be a thin commissure uniting the two cerebral hemispheres in their posterior half. This (commissure ! if the observation was correct) was not, like the Corpus callosum, a connection between the internal white matter of both hemispheres, but merely a union of the superficial white, which in lower animals is well developed outside of the cortical gray.

In the *carnivora* the *Ganglion* of Soemmering (the *Substantia nigra* in the human brain) is continuous with the innermost part of the lenticular nucleus. This fact strengthens Meynert's proposition, that the Ganglion of Soemmering, like the caudate and lenticular nuclei, should be considered as parts of one system, whose ganglia are connected with the fibres of the *pes pedunculi*.

In the elephant, whose brain, both in its mass, the preponderence of the hemispheres, and the concealment from view of the so-called "trapezium," takes a high rank as regards the grade of development, I had the opportunity to make and examine transverse microscopic sections from the Pons Varolii. The remarkable discovery was made that the descending (longitudinal) fibres of the Pons are wanting. Nothing but transverse fasciculi are seen in the field.' Since the former fibres constitute part of the pyramidal tract, it follows that the tract of the voluntary impulses, the "will-tract," must take another course in the elephant, one which may be considered aberrant; for in all other placental animals so far examined by myself, the pyramidal tract runs through the Pons Varolii, as in man.

A Malformed Heart.

By C. H. A. KLEINSCHMIDT, M.D.,

Prof. of Physiology, Medical Depart., Georgetown University.

The specimen presented before the Medical Society of the District of Columbia, October 6, 1880, was obtained through the kindness of Dr. Gieseking of this city. The patient, female, was 64 years of age, and had, up to a few years before her death, enjoyed excellent health, but for the last six years suffered from repeated and severe attacks of cholera morbus, supervening upon constipa-

denly on the morning of February 23rd, after having breathed laboriously and rapidly for about six hours. During this period of rapid breathing the pulse at times was so rapid that it could not be counted.

The autopsy showed rupture of the left ventricle about midway between apex and base, a little to the left of, and at an acute angle with the anterior interventricular groove. The rupture is rectangular, the angle opening to the right, its descending limb being 1.50 Cm., and its transverse limb 1 Cm. in length. Viewed from the outside, the organ, from which, unfortunately, the auricles and great vessels were in great part removed by a hasty autopsy, presented nothing abnormal. Upon longitudinal section through the walls of the supposed right ventricle, however, the following abnormal conditions were discovered : The left ven-

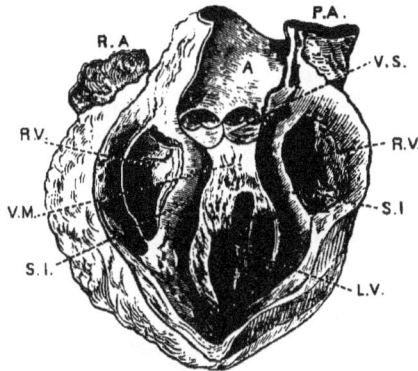

A. Aorta.
P. A. Pulmonary Artery.
L. V. Left Ventricle.
R. V. Right Ventricle.
S. I. Interventricular Septum.
V. S. Semilunar Valves
V. M. Mitral Valve.
R. A. Right Auricle.

tion and flatulence. During one of these attacks her pulse presented a peculiar wavering action, and for the last three years its rate was about 90 per minute. This, as far as can be learned, was the only sign pointing to an abnormal condition of the heart. The last attack of cholera morbus commenced February 17, 1880, and she died sud-

tricle occupies the space normally allotted to both ventricles. The interventricular groove marking no interventricular septum. The aorta arises directly from the center of the base, being normal in size and with valves apparently competent. The mitral valves are also normal, but are attached to papillary muscles arising from the posterior and right lateral portions of the apical muscular tissue, leaving a large cavity to the left, which opens into the aorta. The left auricle is situated upon the right posterior half of the organ. The right auricle is placed immediately in front of the left and opens through what appears to be normal tricuspid valves into a totally misplaced right ventricle. This latter, instead of occupying a position parallel with its fellow, is placed across and in front of it, passing along the upper portion of the ventricular walls ; it is of a semilunar form, its convex border being below. The left horn of this ventricle, the apex, turns suddenly upward as it reaches the interventricular groove, and here gives origin to the pulmonary artery, which thus arises immediately to the left of the aorta, and, like it, possesses competent valves.

The misplaced interventricular septum is thus created by a portion of the anterior wall of the left ventricle, and forms the posterior wall of the right ventricle, while the latter may be described as having been hollowed out of the anterior and antero-lateral muscular wall of the left ventricle, which virtually occupies the ventricular portion of the heart. The rupture of the heart is, in part at least, acccounted for by the fatty degeneration, shown to exist by Dr. E. M. Schaeffer, who examined a section of the cardiac tissue under the microscope. The malformation can easily be explained by supposing that the original tubular canal, forming the first rudiments of the heart, became bent upon itself in such a manner that the venous entrance, instead of being behind and somewhat below the arterial opening, was placed in front of the latter, while the partition, longitudinal in the normally developed heart, grew transversely, and thus separated the anterior (right) from the posterior (left) ventricle. The peculiarity of the pulse was quite likely due to the enfeebled condition of the ventricular walls, induced by the fatty degeneration.

The following are measurements of the several parts of the heart :

Length of organ from base to apex, 10 Cm.

Length of left ventricle from apex to aorta, 7.5 Cm.

Length of right ventricle from auricle to apex, 6.5 Cm.

Diameter of right ventricle at section, 4 Cm.

Diameter of external wall of right ventricle, 8 Mm.

Diameter of internal wall of right ventricle (interventricular septum), 12 Mm.

Diameter of wall of left ventricle, 2.5 Cm.

From these data it is evident that a heart thus malformed, was fully able to perform its functions in an efficient manner, and that the abnormal condition could have escaped notice during life.—*Walsh's Retrospect, Original Department.*

On the Influence of Altitude with Reference to the Treatment of Pulmonary Disease.*

BY WILLIAM MARCET, M.D., F.R.S.

The following communication may be divided under three heads.

1. The effects usually produced on a person in good health on ascending high mountains.

2. The influence of altitude in disease, with especial reference to pulmonary affections.

3. The influence of altitude upon respiration as accounting, in a great measure, for the beneficial effects of high stations in some cases of phthisis.

When, from a low altitude, approaching that of the sea-level, a walk is taken over high cliffs or a neighboring hill, a sensation of easy breathing is experienced, the body appears to feel lighter, and a state of general vigor and comfort ensues. If a mountain a few thousand feet high be ascended, the sensations experienced on low hills are intensified, the breathing becomes fuller, and a pleasurable excitement from the pure state of the atmosphere, united to Alpine scenery, well repays the fatigue of the climb. Seven or eight thousand feet in the Alps appear to be the limit which those accustomed to mountaineering usually attain

* Read in the Section of Medicine at the Annual Meeting of the British Medical Association in Cambridge, August, 1880.

with perfect comfort; above that height, the respiration, in the sitting posture, begins to indicate infrequency, although in an unconscious way; the breathing also becomes deeper, and an unpleasant nervous discomfort is often induced, productive of fatigue, and interfering with sleep at night. At ten or eleven thousand feet, the cold from the low temperature of the atmosphere, and increased evaporation from the skin and lungs, begins to be seriously felt, mostly at night, when no amount of blankets can keep the body warm. After a chilly, restless night, an attack of mountain sickness is not unlikely to set in at three or four o'clock in the morning; and, although attended with considerable distress, resembling very much that of sea-sickness, the attack may have passed off altogether in time for a fairly early start for some higher altitude. I need not add that those who are in good training for mountain excursions can rise to a very great height without inconvenience, or any degree of ill-health.

A sojourn of eight days in succession, in 1875, on the well known pass of St. Theodule (10,899 feet), another of three days, this last summer, at 10,050 feet (Col Sen Geant), and excursions to much greater heights, have afforded me many excellent opportunities of observing the influence of altitude on health. It is singular that the power of digesting food should often fail at a considerable elevation above the sea; as a rule, tourists eat little; their guides have a large share of their provisions to fall back upon, and to that they usually do not fail to do justice. It is not that the appetite is at fault, but, after partaking of very little food, there is no desire for more; and the food taken is often digested but slowly and with much difficulty. I conclude, from my own experience, that milk, coffee, tea, bread and cheese, chocolate, and dried fruit, are about the best articles of food at great altitudes, and a great deal of physical work can be done upon them; it is remarkable that such a diet is precisely that which Dr. Edward Smith has found, experimentally, to

yield most carbonic acid in the body. After living some days on such a short allowance of food, and returning into the plains or valleys, where good hotels are to be found, it is wonderful how keen an appetite is felt, and how much can be eaten and digested with the greatest ease. A stay of a few days at an altitude of about 11,000 feet in the Alps, is productive of fatigue and debility, with an undefinable state of discomfort; there is a peculiar harshness of the air breathed, and a great desire to return into lower altitudes. This I fully experienced at St. Theodule, and I believe my guide, although more accustomed than I was to such a kind of life, was not sorry when the time came to move downwards. The feeling of comfort after descending only a thousand feet was most marked, and the relief to the respiration very positive. A similar grateful sensation was experienced after a stay of three weeks on the Peak of Teneriffe; but it was due in this case not only to the increased density of the air as we progressed downwards, but also to the greater atmospheric humidity. After living under a nearly tropical sun, we entered the zone of trade-wind clouds which surrounds the peak up to between five and six thousand feet, and has a thickness of about 2,300 feet; the sensation on losing sight of the sun, and breathing the damp fog in a denser atmosphere, was exceedingly refreshing.*

On rising above the sea, less and less oxygen is held in a given bulk of air, although remaining the same relatively to the nitrogen; and the atmosphere becomes, as a rule, colder and drier, the dryness being, of course, especially conspicuous in fine weather.

The air also contains a smaller proportion of carbonic acid, and is freer from

* The editor thinks it of interest in this connection to cite the experience of a gentleman, a physician, who, suffering from tuberculosis, visited Lookout Mountain, in company with a healthy companion. During his ascent to an altitude of 2,000 feet, his sensations were those of great relief, but on reaching the above point he became enveloped in a cloud, which extended to the summit of the mountain (2,700 feet), and which had the effect of changing his sensations of relief into those of great depression; while his healthy companion was greatly refreshed, and enabled to accomplish the remainder of the ascent with much more ease.

organic germs productive of fermentation and putrefaction. On the other hand, the sun becomes more and more powerful, as it passes through a smaller thickness of the atmosphere before reaching the earth, and a thinner layer of atmospheric moisture, the latter absorbing the sun's heat very readily.

The amount of carbonic acid in the atmosphere at increasing altitudes is an important circumstance to notice. Not long ago, it was thought that there was rather more carbonic acid in the atmosphere at an altitude of six or seven thousand feet than at the sea-level; recent experiments, however, by M. Truchot, have shown that such is not the case, and that the carbonic acid in the air undergoes a somewhat rapid reduction at increasing altitudes; thus, at an altitude of 1,296 feet, he found 3.13 parts of carbonic acid in 10,000 of air; and at 2,056 feet, its proportion had fallen to 1.72 in 10,000. I have myself been surprised on various occasions on noticing to what a comparatively small extent air precipitates a solution of hydrate of barium at ten or eleven thousand feet high. While I was engaged in experimenting on the Col St. Theodule (10,899 feet), a pool of glacier-water freely exposed to the atmosphere was found to be absolutely free from carbonic acid, but perhaps the temperature of the water may have had something to do with the circumstance. The presence of a comparatively small proportion of carbonic acid in the air breathed on high mountains must be beneficial to diseased lungs. . The air breathed is certainly the more pure on that account, although, with healthy people living at sea-level, and in the midst of a luxuriant vegetation, the amount of that gas which they inhale is perfectly consistent with health. If an excess of carbonic acid in the atmosphere, beyond its normal proportion, is known to be injurious where a disposition exists to consumption, it is but natural to conclude that, the smaller the amount of that gas in the air breathed, the more likely will it be to agree with such patients. The fact that the entire vegetation at Davos is

concealed under a thick coating of snow throughout the winter, must assist in keeping the proportion of carbonic acid in the air breathed very low indeed.

As to the organic germs in the atmosphere, from Dr. Tyndall's experiments, they may be considered absent, or nearly so, in the atmosphere of such a place as Davos, except, of course, inside or in the immediate proximity of a dwelling.

The heat of the solar rays, as previously stated, is much greater on the mountains than at the sea-level; thus the sun may act with great power through an atmosphere which is nearly freezing. The air itself absorbs but little heat; and at Davos, as soon as the direct rays of the sun are hidden, either by a passing cloud or the shade of some other object, an immediate transition from great heat to a piercing cold is often experienced. On the Peak of Teneriffe, under the twenty-eighth degree of north latitude, Mr. Piazzi Smyth, the well known astronomer, has reckoned that the temperature of the direct sun-rays at 9,000 feet amounts to no less than 212°, or the boiling point of water at the sea-side. I have seen, on that very mountain, water left in a plate on the hot sand in the evening, to be frozen next morning into a solid mass of ice.

The intensity of the sun's light on stations situated at about 5,000 feet above the sea is a circumstance also well worth taking into account. The brightness of the sun's light is certainly one of the great attractions of the Riviera, and no doubt this is an important feature of a climate for consumptive invalids. There are direct experiments bearing on this subject. Moleschott, in 1855, observed that light had a distinct influence on the production of carbonic acid in animals, and he concludes from his inquiry: 1. That frogs of similar weights, and in equal periods of time, exhale from one-twelfth to one-quarter more carbonic acid when breathing under the influence of light than while in the dark, so long as the temperature remains the same or varies but slightly; 2. That the production of carbonic acid is greater and greater in a direct ratio with the increase of the light to which animals are submitted ; 3. That the influence of light towards an increase of the amount of carbonic acid expired acts partly through the eyes, partly through the skin. The excess, however, of heat and light at great altitudes, especially when the earth is covered with snow, may be productive of ill-health ; and people wintering on mountain stations should be warned against the glare of the sun, lest their eyes should become affected from this cause. Snow-blindness in the high Alps is not uncommon, and I had an opportunity of seeing a severe case of this painful affection at Chamounix not long ago.

THE INFLUENCE OF ALTITUDE ON HEALTH.

It is well known that a moderate altitude exerts a bracing or tonic influence on health ; and this applies, not only to the air of mountains, but to that of places raised but a few hundred feet above a patient's dwelling. I have had many opportunities of testing the beneficial influence of the hills skirting the Mediterranean coast, and have obtained excellent results by sending cases of phthisis with acute symptoms from the sea-level either to the hills at the back of Nice, or to the picturesque slope rising near Cannes. These stations, if on the top of a hill or a raised plateau, are rather colder than the seaside, but they are somewhat warmer if situated on a slope facing the south. It is quite remarkable what benefit intractable cases of hæmoptysis, for instance, often derive from such a change. The continued influence throughout a whole winter of a high station, such as Davos, must be considered to some extent as different from that resulting from a stay of a few days only in such a place. Experience certainly shows that any prolonged strain of the functions of the body must prove objectionable. Those who are born in the mountains, and have never left them, are in their normal condition under light mountain air, and it will be a strain on their functions to take up a residence near the sea-level. A long-continued stay on high mountains is well known

to be productive, for the inhabitants of the plains, of exhaustion, debility, loss of energy, and want of reaction against external influences. Dr. Lombard, Dr. Jourdanet, and M. Paul Bert, who are well qualified to express an opinion on the influence of mountain climate, bear out this view. I have been told by the monks of the Great St. Bernard (8,115 feet) that, although they are all young, strong, and healthy, and indeed men selected for the work, most of them are obliged to leave, after a residence of one or two years, from ill health and exhaustion, and most, if not all of them, lose their energy more or less.

THE INFLUENCE OF ALTITUDE ON DISEASE, WITH ESPECIAL REFERENCE TO PULMONARY AFFECTIONS.

There is abundant evidence to show that phthisis is nearly unknown amongst the inhabitants of altitudes of 5,000 or 6,000 feet, and higher—a fact, the explanation of which will be found, I think, in the last portion of the present communication. They are subject, however, to other diseases, and can hardly be considered, I should say, as living so long and healthy an existence as the inhabitants of the plains, who, moreover, enjoy more comforts, and have, as a rule, better food.

Inflammatory diseases are more frequent amongst the inhabitants of the mountains than those of the plains; and Lombard remarks that, in the highest towns in Europe, nobody can be found who has not suffered repeatedly from what Dr. Albert of Briancon calls "inflammatory fever without any local affection" (*sans localisation*), ending with perspiration, loaded urine, and a slight eruption at the lips. Bronchitis, asthma, emphysema, pneumonia, pleuropneumonia, and pleurisy are amongst the most frequent diseases met with in mountain towns and villages, such as those situated on the high plateau of the Engadine in Switzerland. At Chamounix, about one-fifth of the death-rate is said to be owing to pneumonia.

Rheumatic affections are very common amongst mountaineers in northern latitudes, although nearly unknown, according to Dr. Tschudi, on the high plateaus of Peru. Dr. Lombard considers rheumatism as mostly prevalent under a climate free from extremes of cold or heat. My own experience is quite in accordance with this observation ; and I well recollect, when engaged with the out-patients of the Westminster Hospital, that, in the spring of the year, cases of pulmonary affections, frequent in winter, fell off rapidly, and were replaced by rheumatic affections.

The stomach is not unlikely to suffer after a prolonged stay above six or seven thousand feet ; the St. Bernard monks complained much to me of gastric symptoms. Under the tropics, diarrhœa and dysentery prevail in mountainous regions ; and, according to Lombard, they form, together with affections of the liver, about one-sixth of the total number of deaths.

Hæmoptysis certainly appears checked by removal from the plains into the mountains. It is remarkable that inflammations of the lungs and air-passages should be so prevalent in high regions, although the reverse be the case with phthisis ; a circumstance appearing to show that there is a greater difference between tubercular disease and inflammation of the lungs than is usually supposed. Are we also to conclude that, above 5,000 feet, pneumonia, although common, does not run the same course as it often does in the plains, and yields no caseous deposit likely to soften down and give rise to a cavern ?

If pulmonary inflammation be common in mountainous districts, it is not so on the Riviera and Mediterranean coast ; people live there under the full atmospheric pressure, and winter in a comparatively warm genial climate ; breathing is carried on quietly and regularly, and every precaution is taken to avoid the cold air, which is usually limited to the early morning and after sunset. Hence there is no excitement of the pulmonary circulation, and no increase of cold in the pulmonary organs, from evaporation of moisture from the air-pass-

ages. The consequence is, that those who suffer from bronchitis in England, when the cold weather and fogs begin in autumn, experience great relief on the Riviera and Mediterranean coast. So far, the difference between the influence of mountain air and the Mediterranean climate is most clearly defined. We now come to the main point of the present communication, namely :

The Influence of Altitude upon Respiration as accounting in a great measure for the Beneficial Effects of High Stations in some Cases of Phthisis.—On first undertaking, in 1875, an experimental inquiry on the influence of altitude upon respiration, I little anticipated obtaining a result which was likely to throw much light on the influence of high winter stations on the progress of phthisis. My first paper on the subject was published in the *Proceedings of the Royal Society* for 1878, and a second in 1879. My experimental stations were situated, at the following altitudes, in Switzerland : Yvoie (near Geneva), 1,230 feet ; the Great St. Bernard, 8,115 feet ; the Reffel (Zermatt), 8,428 feet ; the St. Theodule Pass, 10,899 feet ; the summit of the Breithorn, 13,685 feet ; and in the Island of Teneriffe, 7,090 feet ; on the Peak of Teneriffe, 10,720 feet. My present object is to limit myself to those results which bear on the subject of the present communication.

The first point is the increased bulk of air breathed, under low barometrical pressure, in order to supply the necessary amount of oxygen to the lungs. The following were the results obtained :

AIR EXPIRED PER MINUTE.

ALPS. Altitudes.	Air actually Expired per Minute. Litres.	Air Expired per Minute Reduced to Freezing-point and Seaside Pressure. Litres.	Temp.
1,230 feet	5.66	5.14	57.8
8,115 "	6.05	4.42	43.7
8,428 "	6.50	4.64	52.4
10,899 "	5.03	4.67	39.2
13,685 "	7.94	4.85	34.9
TENERIFFE.			
Seaside	6.44	5.84	—
7,090 feet	7.62	5.47	—
10,700 "	8.07	5.14	—
11,745 "	8.04	4.99	—

It is seen, therefore, that the volume of air breathed does not increase exactly in the same ratio as the atmospheric pressure falls, although there is an approximation to such a change ; but the amount or weight of air or oxygen inhaled shows a decided tendency to diminish as altitude increases. This is seen in the Teneriffe experiments more distinctly than in those undertaken in the Alps ; but, in the latter series, the mean of the experiments at the high stations yields 4.65 *litres* (reduced) of air expired per minute ; while 5.14 *litres* are expired at the lowest station, the difference amounting to 9.5 per cent. In the case of the Teneriffe experiments, a similar calculation gives, oddly enough, exactly the same reduction. Now, 9.5 per cent., or nearly one-tenth, less oxygen taken into the lungs constitutes a very great change in the function of respiration. It would appear, at first sight, that a fall in the amount of carbonic acid expired at increasing altitudes might be expected, as less oxygen is taken in. I have shown, however, that, far from this being the case, there is critically an increased amount of carbonic acid expired by a person as he rises above the sea in the Alps, the extreme extent in my case amounting to 15 per cent., at an altitude of 13,685 feet. My experiments on the island of Teneriffe show that near the tropics, and probably also between them, the amount of carbonic acid expired cannot be considered as increasing as the atmospheric pressure falls : the difference between my results, under the two latitudes, being obviously due to the temperature of the air, which is high and nearly uniform on the island and Peak of Teneriffe, while it falls considerably at high altitudes in the Alps. It may therefore be concluded that, as a tourist rises on the Alps (by tourist is meant an inhabitant of the plains who finds himself under new physiological conditions in the mountains), and while in the sitting posture and remaining perfectly quiet, he will take in a smaller weight of air than he usually does, and give out more carbonic acid. If such be the case, the same weight of carbonic acid will require the inhalation of less oxygen high up in the Alps than at the sea-

level. On the island of Teneriffe, although there is no particularly marked increase of carbonic acid expired high up on the Peak, still, as the air breathed undergoes a positive reduction, the same remark will hold good—namely, that on the higher stations the carbonic acid expired will require, comparatively, the inhalation of a smaller quantity of oxygen. The following is a table showing the proportions of air expired corresponding to the expiration of one *gramme* of carbonic acid, both in the Alps and at Teneriffe :

Altitude.	Temperature of the Air	Carbonic Acid Expired.	*Litres* of Air Expired Reduced.
THE ALPS.			
	Deg. Fahr.	*Grammes*	
1,230 feet	57.8	1	1.36
8,115 "	43.7	1	1.04
8,428 "	52.4	1	1.14 } Mean at and
10,899 "	39.2	1	1.16 } above 8,115
13,685 "	34.9	1	1.08 } feet, 1.105.
TENERIFFE.			
Seaside.	75	1	1.24
7,090 feet	69.6	1	1.19 } Mean at and
10,700 "	64.2	1	1.18 } above 7,090
11,745 "	64.0	1	1.06 } feet, 1.143.

Nothing can be clearer than the result I have obtained. In the Alps, the experiments made, at altitudes varying from 8,111 feet to 13.685 feet, shows that the mean amount of air expired at such altitudes corresponding to one *gramme* of carbonic acid amounts to 1.105 *litres*; while at 1,230 feet, in the neighborhood of the Alps, the volume of air expired for one *gramme* of carbonic acid amounts to 1.36 *litres*, the difference being, therefore, 19 per cent. It may therefore be said that 19 per cent. less air (in weight) was required to make one *gramme* of carbonic acid in the body at these altitudes than at 1,230 feet.

On the Peak of Teneriffe, the expiration of 1,143 *litres* of air (reduced), at altitudes varying from 7,090 to 11,745 feet, corresponds to one *gramme* of carbonic acid ; while, at the seaside, the volume of air expired, for the same weight of carbonic acid, is 1.24 *litres*, giving 8.8 per cent. less air for one *gramme* of carbonic acid at high stations than at the seaside.

These results, obtained from a great number of experiments, at altitudes varying considerably, and under different latitudes, certainly show that the inhalation of a smaller weight of air is required for a similar degree of animal combustion on the mountains than near the sea-level ; or in other words—

The air breathed finds its way more readily through the pulmonary tissue into the blood at a certain altitude than nearer to the sea-level ; and this law applies equally to various latitudes, although less marked in the South.— This phenomenon must be intimately connected with the influence of mountain stations on consumption ; the increased readiness with which the oxygen of the air finds its way through the pulmonary tissue into the blood being clearly opposed to the progress of tubercular disease. It is also apparently this same circumstance which makes consumption so rare amongst the inhabitants of places situated at certain altitudes above the sea.—*British Med. Jour.*

RECENT MEDICAL PROGRESS.

Quarterly Report on General Medicine and Therapeutics.

By D. S. LAMB, A.M., M.D.,
Professor of Anatomy, Howard University, Washington, D.C.

CAUSES OF SUDDEN DEATH.

The causes are cardiac or pulmonary. Cardiac oftenest by *asthenia*, which is usually due to fatty degeneration, as in fevers, diphtheria, delirium tremens ; also by asthenia in pericardial and pleural effusions and valvular disease. By *anæmia*, as in ruptured aneurism. Pulmonary cause of death is by *asphyxia* as in hæmoptysis, spasm of glottis and embolism.—*Birmingham Medical Review, April, 1880. p. 513.*

SLEEP PRODUCED BY THE APPLICATION OF CHLOROFORM TO THE SKIN.

Brown-Sequard (Gaz. Hebd. Nov. 19, 1880, p. 755) communicated to the Societe de Biologie an account of the production of chloroform anæsthesia lasting for hours, sometimes proving fatal, produced in cats, dogs, etc., by simply dropping the liquid,

drop by drop, upon the shoulder or neck of the animal.

PREMATURE BURIAL (?).

Dr. Job reports (L'Union Med., Nov. 13, 1880, p. 785) the case of a consumptive, age 26, who had been under his care for a year. The doctor left his patient one evening in the agony of death, and next morning, finding him dead, gave a certificate to that effect. The deceased was buried the following morning. That evening the doctor was summoned by the authorities to go see the dead (?) man. The grave-digger, while digging a grave next to where the man had been buried, had heard a noise as of pounding in the coffin ; the coffin was therefore opened, but the body was cold and still. The doctor having arrived, investigated the source of the noise, and found that it was due to the swelling of the dry fir wood of the coffin in the wet soil. The *working* of the wood gave a cracking sound quite similar to the cracking of a new piece of furniture.

HYDROPHOBIA.

M. Colin (Jour. de Med., Dec., 1880, p. 530) reported a fatal case of hydrophobia in an army officer four years and a half after he had been bitten by a mad dog ; a comrade had been bitten at the same time, and had died in forty days. M. Bouley claimed that the case was not proved to be one of hydrophobia, no autopsy having been made.

TURPENTINE IN WHOOPING-COUGH.

Dr. Barety (L'Union Med., Nov. 4, 1880, p. 739) ascertained, by accident, that a child having the disease, appeared to be benefited by the vapors of turpentine. The doctor then employed the remedy, and with success. He had a plate containing it placed under the bed and another in the corner of the room. In this room the child passed a part of each day ; the air of the room being twice renewed during the day. The case steadily improved and in a month was well.

SCARLET FEVER AND DIGITALIS.

Dr. Atkinson (Med. and Surg. Reporter, Dec. 4, 1880, p. 485) quotes authorities and adds his testimony to the efficacy of the drug in this disease. It may be given to the child in one to two drop doses every hour or two ; or an infusion may be made of sixty grains of the powdered leaves to twelve tablespoonfuls of boiling water ; after cooling, it may be given in teaspoonful doses every hour or two.

CEREBRAL RHEUMATISM.

Dr. Woillez (Bull. gen. de ther., Nov. 15, 1880, p. 397) reports cases in which he has used cold baths, and says he has never had to regret doing so, but, on the contrary, has twice seen patients die for the want of them.

CHOREA : BROMIDE OF CAMPHOR.

Dr. Larmande (Jour. de Med., Dec., 1880, p. 540) reports the case of a woman, age 20, attacked with hemichorea. She had been anæmic for two years, and was also hysterical. Baths of sulphate of potassium, with the bromide internally twice daily for five months, produced no positive benefit. Bromide of *camphor* in twelve grain doses was then given daily, and she was recommended to travel. In a month she was well. Several other successful cases are related.

PECULIAR CAUSE OF DEATH.

A case of death under peculiar circumstances is reported (Med. Times and Gaz., Oct. 9, 1880, p. 438) as having occurred in a lunatic asylum. A man was found dead in bed, lying on his back obliquely across the bed, his head hanging over one side upon a pillow lying on the floor. The bed clothes were arranged in an orderly manner. The air passages were free ; the face much congested. An autopsy discovered congestion of the brain with chronic disease of the same, and some fluid effusion. It was concluded that the man laid down in the position described and died from its results.

TETANUS.

Case reported in the Cincinnati Lancet and Clinic, Nov. 6, 1880, p. 412, from St. Petersburg Med. Woch., No. 36, 1880. A young man, age 21, a traveling acrobat, well developed, began to have sudden

spasms of the muscles when undertaking any violent exertion. The muscles affected were those employed in the exertion. After resting from his occupation, these contractions disappeared. Two months later, on attempting to lift two heavy dumb-bells, the spasms recurred. All the muscles were now involved and the spasms occurred on slight effort. Bromide of potassium was given, absolute rest enjoined for a time, and he gradually improved, so that he could walk and do light manual labor.

ANOMALY OF THE HEART.

A case of anomalous inter-communication between the auricles is reported (L'Union Med., Oct. 28, 1880, p. 698). The foramen ovale was dilated, the septum thinned, and in part of its extent resembled network of wire. The lesion was not congenital ; it was slowly developed. The signs resembled those of narrowing of the pulmonary orifice. The patient died at the age of 20, having been in hospital at intervals for the greater part of the previous six years. He never had cyanosis. At the autopsy, the lungs were found adherent, the right especially ; the apex was honeycombed by cavities, and there were tubercles throughout the rest of its extent. The left lung was less involved. The heart was small ; no trace of pericarditis ; ventricles appeared normal ; aortic orifice normal ; a very slight thickening of mitral edges ; a gelatinous plate was found in the aorta at the orifice of the subclavian ; the right side of the heart was as large as the left ; tricuspid and pulmonary valves normal. The inter-auricular septum as above described. Fatty liver ; small spleen with many accessory spleens. Congestion of kidneys.

CARDIAC AND RENAL AFFECTIONS.

Dr. Debove (L'Union Med., Oct. 3, 1880, p. 558) believes that in hypertrophy of the heart with Bright's disease, the heart and kidneys are affected by a common morbific cause, which, however, may in some cases affect every one of the organs.

EMBOLISM OF SUPERIOR MESENTERIC ARTERY.

A case is reported in Glasgow Med. Jour., Dec. 1880, p. 485. A man, age 39, was admitted to hospital Dec. 6, 1879. He had been sick eighteen weeks with severe cough, occasional hæmoptysis, dyspnœa, palpitation, weakness and dropsy of the lower limbs. There was bulging in the præcordia; pulsation and tenderness on pressure in the epigastrium. Sounds of heart irregular; presystolic murmur, most distinct at apex. Respiration hurried and shallow, with fits of coughing. Expectoration mucous and frothy. Chest barrel-shaped and hyper-resonant on percussion in front; comparative dulness behind at base of right lung. Appetite fair; bowels regular; liver enlarged. Urine scanty. No history of rheumatism. He improved till about January 25th, when he had pains across the abdomen, just below the borders of the ribs, and slight pain in back in region of kidneys, especially the left. The pain grew worse, became violent and griping, occurring in paroxysms every half hour. Pain relieved by pressure. Slight vomiting of food. 26th, diarrhœa; 27th, diarrhœa worse, vomitus black and very fluid; 28th, vomiting altered blood; no blood in stools. He died in a paroxysm of vomiting. At the autopsy, the heart weighed 17¼ ounces; there was mitral contraction and a rough, ragged surface, as if a piece had been carried away. The left auricle was distended and contained vegetations (clots?). Right ventricle enlarged; tricuspid valve dilated; right auricle enlarged. There was recent peritonitis; the folds of intestine were dark red, the redness stopping abruptly near the beginning of the transverse colon; a soft reddish material in the ileum and colon. Superior mesenteric artery plugged and distented below where the colica dextra is given off. There were infarctions in the spleen and left kidney.

HÆMOPHILIA.

A case reported (in Med. Press and Circular, Nov. 10, 1880, p. 391), at St. Bartholomew's hospital. A man, age 32, waiter, was subject to severe hemorrhages from trifling causes, from early childhood. At 22 he had hematuria, with pain in the right iliac region, relieved on the discharge of blood. Admitted to hospital Sept. 3, 1880. There was extravasation of blood in posterior part of left thigh and front of right tibia. The external condyles of the elbow joints became thickened; thickening took place also in the knee joints; it was believed that hemorrhage had occurred into the joints, the blood coagulating and remaining unabsorbed. 13th, there appeared extravasation in the right upper eyelid. 17th, extravasation in the outer side of the iris of the right eye. He was given a half meat diet, greens and two lemons daily, and the perchloride of iron. Improved, and October 8th was discharged. There was no family history of hæmophilia.

NUTSHELL IN THE LUNG.

A young man, age 17, while eating hickory nuts, started suddenly to run, and inhaled a piece of shell. A continuous cough supervened with discharge of mucus and occasionally a show of blood; and later, pus appeared. After several years, in which, in spite of cough and expectoration he kept in good health, he suddenly had profuse hæmoptysis, which nearly proved fatal. A profuse discharge of pus mixed more or less with blood now began, and continued eleven months. One night he had a severe paroxysm of coughing, and raised a large quantity of pus with blood and a piece of shell with the kernel in it. The shell was nearly perfect and about half an inch in diameter. The patient steadily improved.—*Nashville Jour., Dec., 1880, p. 250.*

HEMIGLOSSITIS.

Med. Inspector Cleborne, U. S. Navy (Med. Record, Nov. 27, 1880, p. 612), reports a case of this rare disease affecting the right side of the tongue. The man, a marine, age 27, 1879, had sore throat and erysipelas; five days later, right hemiglossitis. The tongue was enormously enlarged; there was dysphagia and dyspnœa. Tracheotomy was performed ; incisions were made in the tongue ; stimulants and nutritive enemata were given. He died on the seventh day.

STOMATITIS.

Case reported of pultaceous stomatitis (Gaz. Med. chir. de Toulouse, Nov. 1, 1880, p. 241), in a boy of 14, who had been many times sick, especially with subnasal impetigo. October 27 there was swelling of the upper lip and right side of cheek ; mouth red and covered with whitish, pulpy matter, especially on the side of the frænum, where there was the chief pain. Gums swollen, red, covered with tartar. Bad odor of mouth. The pure *tincture* of *iodine* was applied. In three days he was well.

ASCARIS LUMBRICOIDES.

A case reported (Jour. des Sciences Med. de Lille, July, 1880) in a girl of 11. Her intelligence weakened, irritable, perverse, memory failing ; insomnia, restlessness, vision impaired ; deafness ; aphasia. Opium, chloral, bromide of potassium and baths failed to cure. A vermifuge was tried ; thirty-seven worms were passed : she improved at once. In twelve days she passed eighty worms. In two months she recovered entirely, and there was no return of the symptoms within two years.

TAPEWORM.

A case reported (Bull. gen. de Therap., Nov. 30, 1880, p. 463) in a girl of five and a half years, expelled after the administration of six centigrammes (about one grain) of sulphate of pelletierine. A laxative was given in the morning, a light repast in the evening, and the medicine the next morning, followed in an hour by a purgative. Two hours later the worm, including the head, was expelled. The reporter takes exception to the advice of some, not to give this medicine to children under ten or twelve years.

PROLAPSUS ANI.

A case reported (Western Lancet, Dec., 1880, p. 441) of a boy of five years who had had prolapsus for two years ; a protrusion of two and a half inches at each stool. Ergotin, two grains in suppository, was given after each passage. After the use of a few suppositories, the prolapsus did not recur.

Dr. Weber (Med. Record, Dec. 18, 1880, p. 682) used hypodermic injections of strychnia as far back as 1868. He inserts the needle about three-fourths of an inch from the anus, directing it upwards and parallel with the gut. After a number of injections, four or eight, cures are effected even in cases of several years' standing. No pain nor unpleasant symptoms follow the operation.

ABSCESS OF LEFT OVARY OPENING INTO THE RECTUM.

The case is reported in the Med. Times and Gaz., Oct. 30, 1880, p. 509. A woman of 33, the mother of one child, had had good health till 1876. She then had low fever and diarrhœa, from which she recovered. Admitted to Westminster hospital. February, 1880, she had obstinate diarrhœa, night sweats ; her legs were swollen and œdematous. Examination per vaginum showed a swelling in the right (?) side of the pelvis. At the autopsy, old adhesions of the lungs with emphysema were found ; frothy exudation in the right lung ; in left pleura were two pints of purulent fluid, and patches of yellow lymph. In the peritoneum were two pints of flaky lymph ; in the pelvis the small intestine, ovaries and fundus of the uterus were adherent, the adhesions being old. The left ovary was the size of an orange and contained an asbcess cavity, communicating with the rectum by an opening four inches above the anus ; the rectum below this opening was congested and ulcerated.

SWEATING OF THE FEET.

A case is reported (New Remedies, Nov.. 1880, p. 326, from La Ruche Pharm., 1880, p. 90) of a strong and vigorous factory hand who had offensive sweating of the feet : the workmen refused to work near him ; the skin of the soles of the feet was white from maceration. He washed in a one per cent.

aqueous solution of chloral, and his feet were enveloped in towels saturated with the same. The offensive sweating disappeared.

Dr. Ambrook, of Boulder, Colorado, (Med. Record, Nov. 20, 1880, p. 585) uses a one per cent. solution of permanganate of potassium, bathing the feet night and morning, and letting them dry afterwards.

EDITORIAL.

FLAGRANT ABUSES.

Along with the many beneficial effects an increased liberty of thought and action has enabled us to enjoy, there have grown up abuses that call for serious attention. While greater opportunities have been afforded for enlightened discussion and development of opinion, in like ratio has been the freedom extended to malice, ignorance and peculation. With improved intercommunication and increased national wealth, the opportunities for ill-doing have been proportionately augmented; as communities have widened, and attainable knowledge of our neighbors' character diminished, the chances for speculation, swindling and traducing have increased. A century since, the people were divided, with few exceptions, into small communities, where the position and character of every citizen was well defined; now, we are compelled to rely almost entirely upon our newspapers and exchanges for information as to such matters. Not only are we so dependant upon them, but the majority are led by their representations (through advertisements and notices) to place trust in schemes they could otherwise have no knowledge of. It, therefore, devolves upon editors to exercise a careful surveillance over all that appears under their authority. In arrogating to themselves the functions of educators and reformers, they should be quite sure that their pretensions bear no unfortunate disproportion to their actions; and, as abuses naturally arise in this as in all other departments of every day life, the law should, when possible, afford

ample and thorough redress, and not leave too much to the editorial "conscience."

Unfortunately, however, our legislators seldom have had sufficient time, for many years past, to make laws—being deeply occupied with other serious matters; and in consequence of this, there have arisen many deep and widespread abuses. To attest this, it is only necessary to refer to the provisions in reference to libel and malpractice, and the total lack of all penalty attached to the sale of noxious quack medicines. This state of affairs is one of the most shameful anachronisms of the age. In the first case, any malicious scoundrel can, with impunity, blast the reputation of the purest woman or most Christian man that breathes. With a stroke of the pen he can fill hearts with wretchedness and turn happiness into bitterness and misery. However much the innocent victim may writhe, no redress is attainable—unless actual money loss be proven, so say our wise decrees, there can be no recovery. Thus, the social life of every defenceless woman and sick or timid man, is at the mercy of any malicious reporter or narrow and puny-minded editor that ekes out existence by pandering to the pruriency of his readers. That this species of cowardly assassination is not of unfrequent occurrence, the columns of many of our daily papers only too often show. That where personal retaliation is impossible, ruined lives are the consequence, numberless lost women and broken-spirited men, bear witness. Nor is there any less need of reform in the matter of malpractice and the sale of quack medicines. All the evil effects arising from these causes are far from conceivable at first glance. In a country like ours, where constant movement is the rule among a large portion of the population, the masses are peculiarly liable to become victims of medical ignorance and inefficiency. Where, as in many States, it is within the power of any one to profess a knowledge of medicine without the slightest education therefor, the chances are largely in favor of invalids falling into the hands of some incompetent pretender. Add to this the fact that this class of humbugs arrest the attention of the masses by every conceivable method of advertising, and it is no longer difficult to appreciate the flourishing condition of this class of swindlers. With a medical attendant, ignorance is tantamount to guilt ; for through ignorance, not only is the patient afforded no relief, but by the careless administration of drugs a train of ill effects is laid that too often result in death or permanent disarrangement. How large a proportion of legitimate practice is derived from the after effects of malpractice, it would be hard to say, but that it is far beyond general conception, may safely be assumed. Here, too, the law is lamentably inefficient. Unless death follows quick upon the administration of some deadly poison, there can be no conviction. This may be, in part at least, remedied by the introduction of some such statute as that recently adopted in our own and a few other States, requiring all practitioners to procure a license, by exhibiting a certificate from some well-known and reputable medical college, paying fees, etc.; and inflicting heavy penalties on all those practicing without such license.

Finally, with reference to the sale of quack medicines. These compounds when least harmful, are merely stimulants. They always contain a large proportion of alcohol and anodynes thinly disguised by the addition of other ingredients. Their agreeable effects are owing entirely to the former qualities. They foster a desire for and reliance upon alcohol and anodynes in all those addicted to their use. Even the so-called temperance substitutes invariably contain a percentage of liquor. In short, outside of simple cathartics, the whole rank and file of patent medicines, are either of no effect, or result evilly. They merely furnish a poor substitute for plain whiskey at an exorbitant expense. Moreover, by their ordinary mode of advertising, *i. e.* furnishing a long list of "symptoms" common to numberless affections ; the imaginative, the igno-

rant and the foolish are imposed upon. Glancing at the list of "symptoms" there is hardly one man in a thousand but thinks himself subject to some of them. That, therefore, the ordinary run of men, without the slightest acquaintance with the subject, are seduced into using these noxious and expensive compositions, is but natural.

What we would most complain of, is the encouragement and assistance lent to these nefarious projects by our newspapers. As we have already stated, we expect no immediate revolution in the ordinary editorial conscience. As long as erring humanity pruriently looks first to the scandal column ; as long as imaginative humanity insists upon possessing a disease, just so long will newspapers contain reports of moral leprosy ; and furnish hypochondriacs with the latest scheme for slow suicide. All that we dare ask is that, in the first case, the tale be not, as it too often is, one long, infamous lie ; and that in the latter, the editor have some assurance he does not, by his advertisements, offer disease or death to his subscribers. We know that we now trench upon uncertain ground. We are quite aware that " taking one consideration with another, the editor's life is not a happy one." Nor would we have ventured upon offering advice to the concentrated wisdom of the land, had we not precedent for so doing in the action of some of our most influential and widely known journals. These have long since taken the initiative ; have banished from their columns unfounded calumny ; and carefully abstained from negative, participation in the thousand nefarious schemes that float and eddy on the surface of business life. Even in reference to the advertisement of quack medicines we find so well known a journal as the " American Agriculturalist" inveighing against and excluding religiously such advertisements from its columns. In reference to them, it says, " No journal assuming to be a director of public opinion, or a conservator of the public good, should be so ignorant as to have any part in this nefarious business." This opinion being held by a paper having an immense circulation, of course will hardly apply to the small fry that form the rank and file of journalism. Like skirmishers and sharpshooters, the ordinary laws of war do not apply to them. Each is "a law unto himself," and that law is mainly dependant on his opportunities, and the state of his exchequer. Still, despite these facts, we are of opinion that a little more care, and perhaps a little more truth, might not in the end bring them to ruin.

Bureau of Ethnology—Smithsonian Institution: Its Past and Present Work—Rich Fields for Investigation Open to the Diligent and Observing.

Some years back, when Maj. J. W. Powell was about to carry on a geographical and geological survey of the Colorado River, under the auspices of the Smithsonian Institution at Washington, Professor Henry, who was then in charge of that Institution, recommended a simultaneous study of the Indians of that country—their language, manners and customs. From this nucleus there has grown up what is now known as the Bureau of Ethnology, under the directions of Maj. J. W. Powell, the object of which is to carry on a thorough and exhaustive study of the Anthropology of the North American Indians. Thus far, the subject has been divided into six branches—"Indian Language," "Sign-Language," "Mortuary Customs," "Medical Practices," "Mythology" and "Sociology." The first three are under the personal supervision respectively of Maj. J. W. Powell, Col. Garrick Mallery and Dr. H. C. Yarrow.

These gentlemen, owing to their extensive travels and scientific attainments in general, are especially fitted for the work set before them; and that they may make their individual works the more comprehensive and accurate, forming complete text-books for the student of Archæology and Ethnology, they have invited the co-operation of the cultured generally; and to further the facilities and

systematize the efforts of such co-operators, they have issued three large introductory pamphlets upon their special branches.

These books are exceedingly interesting, alike to the mere curious, the professional man, and the special student. They are intended as a guide and help to further investigation, and will be furnished gratuitously to all those manifesting a desire to prosecute the work, or who think they can at present or from time to time in the future furnish information relative to or bearing directly upon the investigations in question.

This department of research cannot be too much lauded, nor its practical value and application to the various departments of every-day life too highly estimated; it forms, in fact, the foundation for the evolution and construction of such works as Spencer's Philosophy.

No doubt there are many, who, capable of seeing no further than their noses, and thinking only for the present, will contend that all such work is foreign to physics proper, or to common life, and a needless waste of time and energy on the part of those other than *special* students. Such, however, is far from being the case. All knowledge relating to the laws of life and to the government of society, however abstract, are exceedingly interesting and of immense *practical* value alike to the learned and reliable physician and those following the more ordinary paths of life, who are anxious to take advantage of the recorded experiences of others, and the general principles of life as collated from diversified sources. The accurate observer, trained in scientific methods, can hardly read a paragraph of such matter without being able to draw valuable deductions therefrom.

The laws of life include not only all bodily and mental processes, but by implication all the transactions of the house, the street, the commerce, the politics, the morals of men; and to be properly understood must be studied in its most complex forms. "If men are to be mere cits, mere porers over ledgers, with no ideas beyond their trades—if it is well that they should be as the cockney whose conception of rural pleasures extends no farther than sitting in a tea-garden smoking pipes and drinking porter ; or as the squire who thinks of woods as places for shooting in, of cultivated plants as nothing but weeds, and who classifies animals into game, vermin and stock—then indeed is it needless for men to learn anything that does not directly help to replenish the till and fill the larder. But if there is a more worthy aim for us than to be mere drudges, if there are other uses in the things around us than their power to bring money—if there are higher faculties to be exercised than acquisitive and sensual ones—if the pleasures which poetry and art and science and philosophy can bring are of any moment" (as a wise philosopher has it)—then such investigations are of value, and should be assiduously cultivated.

The educated generally, and particularly the physicians, throughout this whole western country, possess unequaled opportunities for rendering assistance in this direction, and we would, therefore, most earnestly appeal to them for such aid as is within their power ; and to those so inclined, we would offer ourselves as a medium of communication. It should be remembered that due credit will be given all investigators for their work ; and to give a more definite and authentic idea of the scope of this work, we here append the concluding paragraph of Maj. Powell's preface to "Introduction to the Study of Indian Languages :"

"This field of research is vast ; the materials are abundant and easily collected ; reward for scientific labor is prompt and generous. Under these circumstances American students are rapidly entering the field. But the area to be covered is so great that many more persons can advantageously work therein. Hundreds of languages are to be studied; hundreds of governments exist, the characteristics of which are to be investigated and recorded. All these peoples have, to a great extent, diverse arts, diverse mythologies, as well as diverse languages and governments; and while the people are not becoming extinct but absorbed, languages are changing, governments are

being overthrown, institutions are replaced, and arts are becoming obsolete. The time for pursuing these investigations will soon end. *The assistance of American scholars is most earnestly invoked.*"

WHILE THERE'S LIFE, THERE'S HOPE !

The Index Medicus must not go down ! The profession of America *must* support it ! This " Monthly Classified Record of the Current Medical Literature of the World" was started two years ago by Drs. Billings and Fletcher, of the Surgeon-General's Office, at the suggestion and solicitations of Mr. F. Leypold, the medical publisher, New York. It is purely an American enterprise, and beautifully exemplifies American industry and promptness in recognizing and providing for the needs of science. When this publication was projected, it was thought that the profession throughout the world would be quick to appreciate its great value, and lend their pecuniary aid to its support ; the former anticipation has been realized, but the latter, as regards the profession abroad, has not, for reasons, perhaps, best known to themselves. And now that foreigners have failed to contribute any material assistance toward the support of this unprecedented work, which is alike valuable to the profession in New York and London, and in the most remote portions of Asia, and in which we may well take a National pride, it behooves us to come forward with our wonted patriotic energy and magnanimity, and nurse this child of American birth to maturity ; and this we can do.

Not only is the Index Medicus of incalculable value to every physician whose compositions occasionally grace and adorn the literature to which it is the Index, but, no cultivated physician, desirous of keeping apace with the times, should be without it ; and if he himself is not capable of appreciating its intrinsic worth, and profiting thereby, he should remember that his subscription constitutes an aid to the common cause of medicine, and that he is therefore indirectly benefited.

The great army of American medical editors should alone give sustenance to a work which they well know to be indispensable to themselves, and instead of expecting to receive it in exchange, so foreign and costly is it, they should gladly subscribe, and, if necessary, contribute a few dollars over and above the subscription price. Their journals should be sent to the Index Medicus solely for the benefit they derive thereby. We might better pay twenty dollars a year than be compelled to dispense with this publication. In accordance with these views, the ROCKY MOUNTAIN MEDICAL REVIEW has this day sent in its subscription, and would respectfully suggest to its *confreres* that they "follow suit," with the hope that the Index may yet be saved from being a "thing of the past," as well it may, if a good proportion of the Medical Journals in the United States will but follow our example; for it must be remembered that Leypold has simply sent out "final" circulars intended to test the sense of the profession, and that "while there's life, there's hope."

COLORADO ABREAST WITH THE BEST.

The people in general, and the medical profession in particular, are to be congratulated upon the passage of the "Act to protect the public health and regulate the practice of medicine in the State of Colorado," for which the better educated portion of the public have so long been clamoring. The bill as it finally passed the Legislature, although far from dealing full justice to the present state of medical science or the demands of general science, was perhaps as good as could be expected in this era—public sentiment always occupying a position considerably in the rear of scientific progress.

Perhaps it would be well to remind those who have given but little thought to the subject, that such laws are not intended for the benefit of the medical profession, but for the protection of the people, inasmuch as they are as liable to be swindled in this as in any other department of life ; and that the active part the physician takes in such movements is purely from philanthropic motives, and the fact of its falling within his sphere ; just as it falls within the province of the "Society for the prevention of cruelty to animals" to prevent dumb beasts from being abused, and to the lot of Science to expose Keeley motors and the like. Such laws, then, are passed to protect the people against this particular class of *infamous swindlers* and *swindling schemes*, just as similar enactments are made to suppress and protect the people against other nefarious projects and general practices calculated to be derogatory to their interests, such as gambling, postal confidence games, lotteries, and, in some places, the sale of intoxicating beverages, opium, and poisons in general. Are such laws considered to be unconstitutional ? By no means ! And this great hue and cry about "trammeling the liberty of the people" and "restricting their rights," originates with that very class of scoundrels the law is intended to effect, who are too lazy to work and earn their livelihood in a legitimate manner, and too indolent to enter the profession through the proper channel (which is open to all), but who rather prefer to ingraft themselves, human parasites that they are, upon the public, and feast and grow fat upon their credulity.

The bill in question having become a law, it devolves upon the Governor to appoint a Board as therein provided. The committee having in charge the draughting of this bill was of the opinion that it would be better to have the appointing power reside in the Judiciary rather than in the Executive, in order to prevent intrigue or political influence from exerting any control over the appointments ; but such a provision was found to be unconstitutional. And now that the duty devolves upon the Governor, we would earnestly appeal to his good judgment and integrity, and pray that, in the present case as in the past, he lay aside all personal and political considerations, and appoint *purely* upon merit. Manifestly, if this bill is to be of any practical value whatever, it is of paramount importance that the appointing power should in no way be hampered by politicians or the recommendations of any person or persons outside of representative men in the medical profession, or, preferably, the profession as a body. This contingency should be guarded against most carefully, otherwise the whole matter had better been left unagitated ; for it is plain to be seen that in this way the entire meaning of this act might be subverted.

At a meeting of the Denver Medical association held at the office of Dr. Warn, on the 12th inst., at which twenty-five members were present, an informal ballot was taken expressive of the sense of the society in the matter of presenting six names to the Governor with a hope that he would choose Denver's quota from among them. The ballot resulted in the selection of the following named gentlemen : Drs. H. K. Steele, H. A Lemen, A. Stedman, Wm. H. Williams, C. M. Parker and J. Culver Davis. Drs. Steele and Lemen begged that they be excused as they were already serving on the State Board of Health, and their request was granted.

On the second ballot Drs. W. H. Whitehead and Charles Denison were chosen to fill the places of Drs. Steele and Lemen, and on motion of Dr. Buckingham, the names as selected at the informal ballot were made the unanimous choice of the meeting.

ANNOUNCEMENT !

We have just completed arrangements for an able corps of regular correspondents and specialists in charge of special subjects upon which quarterly reports will be made. The list is not quite complete yet, but will be made so as rapidly as the proper men can be found. The men we have already secured are widely known, and peculiarly and admirably fitted for the departments allotted them. We have taken particular pains and are very fortunate in having been able to make many of the appointments from Washington, where the facilities for the prosecution of such work are superior to those of any other place in the world. They have there the "Surgeon-General's Library" (the most complete medical library in the world), the "Patent Office Library," "Congressional Library," "The Library of the Smithsonian Institution," "The Library of the Agricultural Department," and many others.

The Surgeon-General's Library receives regularly all the domestic and foreign medical and scientific periodicals that can be had for " love or money." Manifestly then, these gentlemen, all being familiar with several modern languages, have superior advantages at their command for making masterly reports, filled with rare; interesting and valuable material, that it is impossible for other journals to come in contact with.

These quarterly reports upon the different departments of medicine will be superior to those found in any other journal in the United States.

With the next number we expect to introduce a new department in the REVIEW, to be known as "Scientific Summary," which will comprise monthly reports of from one-half to one page each on Anthropology, Astronomy, Botany, Chemistry, Geology, Mineralogy, Physics, Zoology or Palæontology, and Biology. These will be written in an easy, flowing style, and in a way easily to be comprehended by even the moderately cultivated. It affords us great pleasure in being able to announce that Dr. H. C. Yarrow, of the United States Army has essayed to take charge of the department of Anthropology. A better man could not be found— a physician and scientist combined, who has devoted special attention to Anthropology in its broadest sense, and who is at present detailed upon the work of the Index-Catalogue, and located in Washington.

Owing to the large editions of the REVIEW we have been getting out, and the recent blockades, we have run short of the paper our journal was originally printed on. With the next number, however, we will return to the original paper.

At present, arrangements are being made by which we will be enabled to tender all our subscribers unusual facilities and advantages ; notice of which when completed will appear in the "Standing Notices."

We will improve this opportunity to thank all our friends for the material assistance they have rendered us, and for their manifest appreciation of our "daring enterprise," and in conclusion will call attention to the original article from England, in this number, and its *significance.*

TRANSLATIONS.

The Influence of Wounds on the Development of Constitutional Affections.

BY DR. J. CORNILLON.

Translated for the Review by Dr. R. Tilley, Chicago.

Fifteen years ago Dr. Verneuil called attention to the fact that wounds are frequently the cause of constitutional affections. Since that time the instances cited in the various published works have fully confirmed the assertion, and a wide gap in the plan of general pathology has thus been filled up. The following case is one in point:

D., age 36, formerly army surgeon, time of observation September, 1880.

Family history : Father's age at birth of D. 50 years. Suffered for an indefinite and long period from chronic general rheumatism. Died at the age of 85. Mother died of capillary bronchitis.

Personal history : At the age of 10 years, acute gastritis ; at 14, scarlatina, accompanied with rheumatic pains ; at 28 years, syphilis.

March 25, 1879, then 35 years, the patient received a violent kick on the external aspect of the left leg near the knee. An arthritis was rapidly developed and necessitated a dressing, rendering the knee immovable.

Scarcely recovered from the effects of this wound, he was attacked, at the end of May following, with the gout in the left foot,

with the formation of a small tophus at the metatarso-phalangeal articulation of the big toe: The gout became more and more obstinate, and seemed to receive fresh violence about every three weeks. It was confined entirely to the left foot. The big toe became swollen and developed an erythema which extended rapidly over the dorsum of the foot. The tibio-tarsal articulation became swollen and red, especially about the external maleolus. Pressure was painful over all the inflamed parts ; movement of the foot or big toe was very painful ; all the other joints were free.

This case is worthy of note relative to two points :

1st. A contusion of left knee developed in a short time an attack of gout, which was repeated at short intervals.

2nd. The pathological condition developed itself only in the small joints of the contused leg, the other leg remaining perfectly free.

Hitherto, wounds seemed only to have exercised an influence over the rheumatic diathesis. Dr. Verneuil has several times observed acute poly-articular rheumatism develop from a wound, but has never observed the gout to develop for the first time as a result of traumatism. He has observed the gout in a previously gouty person, to arise from an insignificant wound.

In the case in question *the first attack of gout* made its appearance immediately after a violent bruise, and this is the point to which attention is especially directed. It is true that he was naturally predisposed to this affection, being the son of a rheumatic subject ; but the blow which he received facilitated its developement, as Dr. Verneuil puts it, *"absconditos morbos vulnera actegunt"* — wounds reveal hidden diseases.

To those who are acquainted with the usual development of gout, it will seem strange that in the present case only the joints of the bruised leg were affected. Especially in this case, as it usually makes its appearance in the big toe of the right foot ; but if we call to mind what has been said about the *loci minoris resistentiæ* (points of least resistance) we see that in this case the general rule has been illustrated.

Morbid localizations in the organs of least resistance are frequently observed as a consequent of traumatism. We constantly see among those suffering from heart affections that a small wound in the pelvic region will establish a considerable swelling on the same side, while on the opposite side the swelling is quite insignificant. It was then quite to be expected that in the present case the gout should develop itself in that member which was weakened by a traumatic arthritis. Of course it cannot be said that in the future the other parts of the organism will remain free.—*Free translation from Le Progres Medical, Jan. 15, 1881.*

BOOK NOTICES AND REVIEWS.

Hygiene and Treatment of Catarrh; Hygienic and Sanative Measures for Chronic Catarrhal Inflammation of the Nose, Throat and Ears.

By Thomas F. Rumbold, M.D. St. Louis, 1880. George O. Rumbold & Co., Publishers.

During the past few years a number of popular medical works and journal articles upon " Vocal Physiology and Hygiene," " The Throat and the Voice," "Treatment of Nasal Catarrh," etc., have sprung into existence, flooding the literature of the hygiene and treatment of diseases of the nose, pharynx and larynx. One fails to observe a material difference between some of these works, and several are simply repetitions of generally accepted and practiced teachings. We do not wish to be understood as opposing any reasonable and necessary addition to medical literature, and we will always welcome a work whose *raison d'etre* is apparent as another brace to the foundation of medical science.

Many of the articles on " Nasal Catarrh" appearing during 1880, are apparently reproductions of each other, and lack the practical points brought out in the ordinary text-book.

The work before us has received thorough perusal, and, our opinion of it upon the whole is favorable.

" Hygiene of Catarrh," the name which greets us not only on the back of the book, but also at the top of the title page, is, we must confess, a somewhat ambiguous title for a work addressed to the medical man, but one which will be readily understood by the general public, who are wont to refer catarrh at once to the nasal passages.

Chapter I explains in brief the manner in which the mucous membrane becomes more and more susceptible to the inroads of disease under repeated attacks.

Chapter II, among other things, advocates the use of the much neglected " nightcap" of by-gone days. The employment of scarfs and neck-wrappings is now nearly discarded, upon the principle that the more you " coddle" a particular organ, the more delicate or tender does it become. It is questionable whether harm does not often result from the use and subsequent neglect to use not only scarfs, but even " nightcaps," by patients suffering from catarrhal rhinitis, pharyngitis or laryngitis.

Chapter XI describes the influence of illtemper upon nasal catarrh, there being sound argument in the facts alluded to.

Chapter XII is devoted to the consideration of the cleansing of the nasal passages by patients. The general usefulness of snuffing properly medicated solutions into the nostrils cannot be denied, but the explanation here given of its *modus operandi* in reaching the various portions of the nares is not clear. It is probable that the direction taken by the air-current on entering the anterior nares is not that of a straight line at an angle of 45° to the plane of the forehead, but rather that of a curve. The normal nares it is understood are here referred to, for the thickening and connective tissue hyperplasia render it impossible to locate the line traversed by the air-current in disease.

Chapter XIII. The author here enters his protest against Thudichum's nasal douche; and the application of the "starch test" and glass plate over the posterior nares in order to demonstrate that a solution reaches only the inferior portions of the nares, and not the seat of trouble, is certainly ingenious.

Chapter XIV contains a thoroughly practical and concise article on cleansing the ears, and the details therein given should be more generally appreciated and practiced.

In Chapter XV the author says, "I have observed for many years that diseased teeth and gums tend to maintain the catarrhal inflammation of the mucus membrane of the nasal and pharyngo-nasal cavities, the throat and the ears." This is an important statement and should be borne in mind in intractable cases. A special chapter (XVIII), denoting close observation and familiarity with the subject, is given to tobacco and its ill effects.

In conclusion, whilst there is in this work an abundance of matter that is "True and not new," but little that is "New and not true," the accuracy with which our author deals with his subject is worthy of much praise, and the book merits better typography, a large sale and careful reading.

ETHELBERT C. MORGAN.

Washington, D. C.

MEDICAL NEWS.

POSITION IN PALPITATION OF THE HEART. —In the Lyon Medical, Dr. Bouchut states that by the assumption of what he terms the "congestive attitude," nervous palpitations, not dependent on organic disease of the heart, may be instantly arrested. His directions are these: The patient stands erect, fixes his lower limbs, and then stoops over rapidly in such a way as to touch his toes with the tips of his fingers. The head thus falls forward, and its vessels are at once rendered turgid. If the hand be now placed on the cardiac region, it will be found that the palpitation has ceased, the disordered impulse being replaced by a regular and rhythmical beat, which indicates that the organ has re-

sumed its normal action. It is obvious that this treatment is not applicable to the case of the aged or those who indulge in alcohol, or, in short, those in whom the integrity of the arterial or venous system is doubtful.— *Med. and Surg. Reporter.*

PARIS ABOUT TO ADOPT THE NEW YORK AMBULANCE SYSTEM.—Dr. Henri Nachtel, to whom is due the night medical service in New York city, has just addressed the Academy of Medicine, in Paris, in regard to our system of ambulances. Dr. Nachtel, who has thoroughly studied the method of transporting those suffering from accidents in New York to the hospitals, is loud in his praises of the rapidity and efficiency with which the system is carried out. In France, the method of transporting the wounded through the streets seems to be of the most primitive kind. In Paris, when a wounded or ailing man is found, policemen take charge of the person, and he is carried to the nearest apothecary. Though police stations abound in Paris, a person suffering from an accident is never carried there. A litter has to be hunted up, and then porters found, who transport the person either to his house or to the hospital. Before all this can be accomplished, minor accidents, for want of immediate attention, become of the most serious character. Dr. Nachtel thinks even . that deaths frequently occur in Paris from want of celerity in the transporting of patients. In dwelling on the many advantages of the New York ambulance system, Dr. Nachtel expressed the hope that our method might be adopted in Paris. A commission, composed of MM. Larry, Legouest, Vulpian, and Chereau, was appointed by the Academie de Medicine to study the whole subject. It will be pleasant to know that Dr. Nachtel has been the means of introducing into Paris the New York ambulance system.—*Med. and Surg. Reporter.*

ANÆSTHETICS IN 1651.—In a recent work, entitled "Histoirie de la Medecine a Troyes," Dr. Guichet relates that the College of Physicians of that town brought an action against a certain Nicolas Bailli for administering internal remedies to his patients and *putting them to sleep.* In defence, Bailli declared that having observed that in great operations, amputations, incisions, actual and potential cauterizations, many patients slipped through his hands for want of sleep, he had studied the secrets of nature, and had at last found a cordial, or marvelous essence, which put them to sleep softly, and appeased their sensibility to pain.—*Medical Gazette.*

A NEW METHOD OF CONTINUOUS INHALATION.—Dr. Feldbausch (International Jour. of Med. and Surgery) describes a new apparatus for the inhalation of volatile substances. It differs from other apparatuses, in that it is introduced into the anterior nares and is self-retaining, for this reason the inhalation is easier and can be kept up for any desired length of time. The instrument consists of two small tubes connected together, each containing a piece of blotting-paper or flannel, which can be saturated with any desired medicinal agent. It is made in three patterns, the second being small, and the third smaller and of two separate tubes. These of course will not admit of as large an exposed surface, but are more hidden from view. It is most useful in recent colds, etc., but will be of value in some cases of chronic catarrh, irritating cough of phthisis, and, the writer claims, a preventative of infectious diseases, such as diphtheria, scarlet fever, etc., and, on this account, will be a protection for the physician who is treating this class of cases.

PERITONEAL TRANSFUSION.— Dr. Kaczorowski (Dent. Med. Wochenschrift) has reported five cases of transfusion of blood, by injections into the peritoneal cavity. In four cases it was well borne, in the other, tenderness at the point of injection was noticed, which lasted for several days ; in all cases there was marked improvement. The method employed was that of simply plunging a curved trocar into the abdominal cavity in the median line, connecting a rubber tube with the canula, and injecting defibrinated blood. This operation will be of great value in cases of chronic

anæmia and prolonged fevers where there is danger of cardiac failure. It cannot, however, take the place of the old method of vascular transfusion in acute cases.

INDIGESTION A CAUSE OF NERVOUS DEPRESSION.—In the Practitioner, October, 1880, we find an article upon the subject of nervetive overwork and depression which is especially interesting to us, as it is upon a disease or condition very commonly met with by physicians. Of late years its importance has become better appreciated, and it receives the careful attention of all good writers. The author of the paper referred to is T. Lauder Brunton, M.D., F.R.S.

The even-tempered man becomes irritable ; the clear-headed man muddled ; the active, lazy ; the sober, perhaps a tippler ; and the cheerful and buoyant, depressed and melancholy. The brain performs all its functions with difficulty, and the mind is so altered that it does not seem to be that of the same individual. He takes butcher's meat three times a day, and perhaps also strong soups, to say nothing of the wine or brandy and soda to pick himself up. He says: "I take all sorts of strengthening things, and yet I feel so weak." Dr. Brunton thinks if, instead of using these words, he would say: "Because I take all sorts of strengthening things I feel so weak," he would express at least a part of the truth. A want of oxygen, and not of fuel, he thinks, more frequently causes the fire of mentally hard-worked men to languish. If the fire is dull, consumption of the fuel will not be complete, and much refuse matter will remain. It would seem that vital processes are much more readily arrested by the accumulation of waste products within the organs of the body than by the want of nutriment to the organs themselves. The writer reminds us of the fact that we are now completely alive to the important results produced by the absorption from the intestinal canal of poisonous matter, such as typhoid germs, arsenic or strychnine, introduced into it from without. But perhaps we are not yet sufficiently alive to the important results produced by the

absorption from the intestinal canal of substances generated in it by fermentation or
imperfect digestion. We recognize the
danger of breathing gas from a sewer, but
probably we do not sufficiently realize that
noxious gases may be produced in the intestine, and being absorbed from it into
the circulation, may produce symptoms of
poisoning. One, at least, of the chief com-
ponents of sewer-gas (sulphuretted hydrogen), may be formed in the intestine. This
gas, which is so readily recognized by its
smell resembling that of rotten eggs, was
found by Dumarquay to be very quickly absorbed when injected into the rectum, and
to be quickly excreted from the lungs, sometimes appearing to produce, during its elimination, some inflammation of the trachea
and bronchi. It seems not improbable that
the production of this gas in the intestines
may have something to do with bronchitis,
which is not unfrequently observed in connection with digestive disturbance. In cases
of indigestion, this gas seems to be not unfrequently formed, because persons often
complain of the taste of rotten eggs in the
mouth or in the eructations. Even in such
small quantities, it is not improbable that it
may exert a deleterious influence both upon
the nervous system and upon the blood ; for
it is a powerful poison, in its action somewhat resembling hydrocyanic acid, though
not so strong. It destroys ferments, and
robs the blood corpuscles and the seeds
and roots of plants of their power to
decompose peroxide of hydrogen. This
gas is rarely generated in the intestine in
such a quantity as to give rise to symptoms of
acute poisoning, but it sometimes has this
effect. It seems probable, however, that the
substances both gaseous and solid, formed
in the stomach and absorbed from it, are,
upon the whole, less poisonous in cases of
indigestion than those which are produced
lower down in the intestinal canal. The
mere omission to evacuate the contents of
the bowels at the usual time will lead to
headache in the course of the day. No
doubt such headache may, in part, be due to

the nervous irritation caused by the fæces,
but it is also in part due to absorption of
fæcal matter or some product thereof.

NOVEL TOXIC EFFECT OF SALICYLATE OF
SODA.—Dr. Gatti, in Lancet (Nov. 13, 1880)
reports a case of salicylate of soda poisoning
in a girl sixteen years of age, who had taken
two drachms of the drug in the course of ten
hours. The disturbance was not confined to
the hearing, but also involved the sight,
rendering the patient totally blind for a
period of ten hours. The influence seems
to be upon the *centres* rather than upon the
organs of special sense. The drug seems
to have been retained within the system instead of being eliminated.

THE INTERIOR OF THE BLADDER—ITS EX-
AMINATION BY MEANS OF AN ELECTRIC
ENDOSCOPE.—Not only is the ear, nose and
throat, now examined by means of the electric
light, but also the bladder. This latter is
accomplished by means of an instrument
known as the *Nietze-Lietner Endoscope*, described by Sir Henry Thompson in the Lancet. The idea originated with Dr. Nietz, of
Vienna, and the instrument was perfected
and is manufactured by Leitner. This instrument will be of great value in the diagnosis of obscure diseases of the bladder, and
will be of value in determining the position,
size and shape of a stone, or other foreign
body.

WHAT HAS LONG BEEN ANTICIPATED AT
LAST ACCOMPLISHED.—The Chicago Medical
Review reports an operation of Dr. Fenger's
(Pathologist to Cook County Hospital), of
exceeding novelty and interest. It was an
attempt to save the life of a man suffering
from a large gangrenous cavity of the right
lung, "extending from the second to the fifth
rib, and from the sternum to the posterior
axillary line." The operation consisted of
a "transverse incision between the third and
fourth ribs, and one and a half inches to the
right of the sternum, dissecting carefully
through the overlying soft tissues, down towards the pleura. When the intercostal
muscles were laid bare, an aspirating needle

was introduced to ascertain whether or not the subjacent lung was adherent." It was found to be adherent and the cavity was opened, about a pint and a half of fetid pus escaping. A counter opening was made in the axillary space and the cavity washed out with a carbolized solution. A "coherent, yellowish white, gelatinous mass" was removed, which on microscopical examination was found to be the cyst of an echinococcus. A drainage tube was introduced connecting the two openings, and the wound dressed antiseptically. At the time this was reported, some ten weeks after the operation, the patient was doing well, and it was thought would be discharged in a few days.

OYSTERS AS A POSSIBLE SOURCE OF TYPHOID FEVER.—In the British Medical Journal, Sep., 1880, p. 471, Dr. C. A. Cameron draws attention to the danger of oyster-beds being laid down in the vicinity of the mouths of sewers, seeing that oysters so placed are found with their intestinal canals full of the sewage matter.—*Med. Times.*

JABORANDI IN ASTHMA.—Dr. H. J. Thomas (Chicago Med. Jour. and Exam.) reports his experience in some twenty cases, in all of which he obtained good results. The fluid extract was used in four drop doses in the forenoon, afternoon and evening, and an eight or ten drop dose at bed time. In case the secretion of saliva was excessive the dose was lessened.

FOREIGNERS APPRECIATE THE THERAPEUTICAL VALUE OF AMERICA'S CLIMATE.—From the following editorial in the British Medical Journal, entitled "America for Phthisis," it will be seen that foreigners are inclined to think Americans belittle the therapeutic value of their own country :

AMERICA FOR PHTHISIS.—A large yearly emigration of American consumptives takes place to European winter stations, although their own magnificent continent with its mountain-ranges and extensive sea-board, offers a far greater choice of climates, inland and marine, high level and low level, warm and cold, than is to be found in Europe.

To judge, however, by the number of pamphlets and books on American climate, there is no lack of energy on the part of the resident medical men in making known their respective health resorts.

Commencing with low level stations, and starting from the west, there are, on the Pacific Coast of California, San Diego, Los Angelos, San Jose, and Santa Barbara, enjoying a mild, equable, and somewhat moist climate, due to the latitude and the Pacific influence.

Santa Barbara has, according to Dr. Dimmick, a mean annual temperature of 61° Fahr., and a winter mean of 54° Fahr., with a difference between summer and winter of only 13° The rainfall is 16 inches, the rainy days occurring during the winter season. and the average humidity percentage is 69. According to Mr. Culvertson, a resident of four year's standing, Santa Barbara is accessible by steamer from San Francisco in thirty-two hours; and here, taking the rainy season and the dry season, the foggy days and the windy days, there were not more than 20 out of the 365 in which an invalid may not enjoy a walk in the sunshine some part of the day. The summer season is, if anything, more pleasant than the winter, for the extreme heat is tempered by breezes from the Pacific. More bracing and less equable climates are to be found at Aikin in South Carolina. and at Ashville in North Carolina, 2.200 feet above sea-level, or on Walden's Ridge in Eastern Tennessee, or among the Pine Forests of Georgia. In Florida, there are Palatka and St. Augustine, where the moist and sedative climate has been compared to that of Madeira, though meteorological observations show the latter to be far more equable. St. Paul's and Brainherd, Minnesota, are resorted to as types of a cold inland climate, with but little elevation (1,200 feet). According to Dr. Talbot Jones's figures, the cold is extreme in winter, and there appears to be a good deal of wind; while the amount of rainfall and the number of rainy or snowy days are considerable, though the climate has the character of being dry and stimulating.

America is, however, chiefly famed for its high altitude climates, and the practice of treating consumption in high level stations originated in South America; for, long before the existence of Gorbersdorf and Davos, it was the custom of the Peruvians to send consumptives from the coast-line to heights of 8,000 to 10,000 feet in the Andes, generally with signal relief to the sufferers.

North America, by reason of its being traversed from north to south by the Sierra Nevada and the Rocky Mountains, abounds in lofty plateaux of different altitudes, and situated in various latitudes; and here height has so much influence on climate, that it is found that to the east of the Rocky Mountains, and from Central Wyoming to Old Mexico, differences in temperature are due in a greater measure to altitude than to latitude, the isothermal lines running rather north and south than east and west.

As we ascend from the south, we come to the plain of the Anahuac, varying in height from 6,000 to 8,000 feet, where is situated the city of Mexico; northward lie the dry and arid plains of New Mexico and Arizona, where the cacti grow to the size of trees, and where Dr. Denison predicts, will be found the extreme climatic remedy for American consumptives in winter, who are now without much hope anywhere.

It is, however, to the States of Colorado and Wyoming that we wish to draw special attention, as within their confines the Rocky Mountains reach their greatest heights in Mount Lincoln and Pike's Peak, and many of their valleys form the so-called beautiful parks, North, Middle, Estes, and San Louis, in which are to be found numerous and rich mineral springs. The principal towns are Cheyenne, Denver, Colorado Springs, Santa Fe, and Pueblo, all of which are fast becoming resorts for consumptives.

The elevation of this region varies from 5,000 to 6,000 feet, and its climate possesses the characteristics of mountain districts. Dr. Denison's little work, though written with, perhaps, too strong a flavor of local enthusiasm, furnishes many important details, and more carefully discusses the factors of mountain climates than any other work with which we are acquainted. He holds the peculiar climate to be due (1) to diminished humidity depending on rarefaction, the atmosphere containing a third of the humidity present at New Orleans, and half the percentage of Santa Barbara, the latter place being situated in much the same latitude; (2) "to diathermancy of the air, *i. e.*, the heating power of the sun's rays, or the difference in temperature between the sunshine and the shade;" this amounts at Denver to an average of 50° Fahr. Dr. Denison tried the experiment of placing three thermometers in pasteboard boxes with glass covers, and lined with black velvet, at three different elevations; and found that at Denver (5,200 feet) the difference was 72° Fahr.; at Alma (8,800 feet), it was 86° Fahr. and at Mount Bross (13,400 feet) it amounted to 106° Fahr. He deduced the important law that " there is one degree greater difference between the temperature in sun and shade for each rise of 235 feet," and proved it by a table of comparison between the temperatures in sun and shade observed at a large number of heights.

This quality of diathermancy arises partly from the thinning of the atmosphere, and partly from the diminution of moisture in suspense; and to it may be referred some of the phenomena of radiation witnessed in the mountains, such as the rapid rise of temperature immediately after sunrise and the equally rapid fall after sunset.

It must be admitted that these observations, as far as they go, are of great interest; and, taken in com-

bination with similar ones made at Davos, they suggest that specific climatic conditions exist at high altitudes apart from mere cold ; and that, when invalids winter at Davos, " they probably do more than thereby reproduce, as regards cold and sunshine, the conditions of Archangel," as Dr. Henry Bennet strongly affirms. Further confirmation is, however, desirable from other mountain stations, such as the Andes or Himalayas, to make the theory well proven. Of the third condition, viz., increased electricity, which Dr. Denison claims for high altitudes, we cannot see that he has succeeded in proving its existence, and we much doubt whether the effect ascribed to this cause may not be found at low as well as high levels. The physiological results of the Colorado climate appear to be similar to those of other mountainous regions, and consist of quickening of the circulation and respiration, as has been noted by Coindet in Mexico, where the French soldiers respired 19.6 times per minute, instead of 16 times as in France. The Mexicans themselves respire 30 per minute.

Statistics are furnished of 202 consumptives who passed 350 winters in Colorado; and, without going into tedious details, we may state that, compared with those of many other health-resorts, the figures are decidedly favorable ; 69 per cent, were greatly or slightly improved; 22 per cent showed favorable resistance to the disease ; and in 20 per cent. " advance" and " extension" were noticed. The consumptive cases which appeared to do best were those of inflammatory origin, while for the catarrhal and chronic tubercular the results were unfavorable ; and Dr. Denison considers that, for these forms, Colorado is contraindicated. Cases of pulmonary hemorrhage prosper, provided cavities have not formed.

To insure the full advantages of the climate, a somewhat prolonged residence is necessary; and an open-air life, passed to a great extent in camping out and visiting the many beautiful districts of this region, considerably improves the chances of recovery.

It might be worth the while of many English consumptives, especially in cases where the disease has an inflammatory origin, who have undertaken long sea-voyages, to try the Colorado district of North America.

PERIPHERAL NEUROGANGLIOMA.—Dr. Axel Key, of Stockholm, describes in *Hygeria* for 1879 (Nordiskt Medecin. Arkiv, Band xii.), the microscopic structure of a nerve-tumor removed from a journeyman tailor, aged 31, who was discharged cured after a stay of 14 days in hospital. The tumor had commenced as a small knob in the soft parts in the neighborhood of the left ala nasi, and had grown in the course of a year to the size of a plum. The extirpated tumor was encapsuled; it was not adherent either to the skin or the subjacent bone; it was grayish-red in color, homogeneous, and rather soft. After hardening in Muller's fluid, it appeared to be of tolerably firm consistence; it was sharply defined, with smooth, round projections, and was of somewhat irregular flattened shape. To one of the projections were attached some shreds, like connective tissue. Macroscopically, it was like a sarcoma, and had been diagnosed as such in the hospital. The microscope, however, revealed an entirely different structure; namely, very large cells, which completely resembled ganglion-cells, and were enclosed in perfectly developed capsules, the interior of which had the same appearance as the capsules of ganglion-cells. The cells were apolar; and not only one, but two or three, or even more, were contained in the same capsule. In the shreds connected with the tumor was found a nerve broken up into fine fibres of unequal size, probably a portion of the infra-orbital. On examining these branches of nerves, it was clearly seen that the large ganglion elements of the tumor were developed from the nerve-fibres. Thus there was in this case a true ganglioma, which throws light on the hitherto unsolved question, whether a tumor of this kind can be developed on a peripheral nerve. Only one case of ganglioma is described in medical literature, viz., by Lorentz in Virchow's Archiv.; but this tumor, which was of the size of an egg, had proceeded from a pre-existing ganglion, and was thus a simple hyperplastic new growth. The case reported by Professor Key renders it certain that a ganglioma may be developed from a peripheral nerve, altogether independently of preceding ganglionic formations; and hence, in order to avoid all misunderstanding, Dr. Key calls this new growth, " neuroganglioma verum periphericum."— *British Med. Jour.*

ON THE INHALATION OF CARBOLIC ACID IN DISEASES OF THE RESPIRATORY ORGANS. —Dr. Monroe has found that, in his experi-

ence, the inhalation of carbolic acid has proved beneficial; and he is certain that it has helped to cure some cases which would otherwise have died. He commenced to use it first on the supposition that it acted as a disinfectant and stimulating lotion to the ulcerated and diseased tissues of the lungs, and he has in consequence only tried it in the second or suppurating stage of phthisis; the real benefit may, therefore, have been due to the destructive influence of the acid on the bacterium germs. Dr. Monroe has observed that little good followed the inhalation unless it was used frequently, and he has latterly been in the habit of seeing that the patient performed the act of inhalation properly, as many were apt to draw the steam into the mouth without actually inspiring any of it. Dr. Schuller recommends solution of benzoate of soda ($2°-5°$), to be inhaled by means of a spray producer two to four times a day, for about half an hour each time; whilst Dr. Mackenzie practices the inhalation of carbolic acid or creasote by means of a respirator containing a sponge saturated with a strong solution of the drug. The author always employs the inhalation of steam from a mixture of carbolic acid and hot water (1 drachm to the pint) by means of an inhaler or even common jug.— *The Glasgow Med. Jour.*

A FŒTUS OF 148 DAYS LIVES EIGHTEEN HOURS.—Dr. F. B. Watkins describes this most interesting case in a letter to Dr. Gaillard. Its form was perfect; motions natural; rectal and vesical evacuations plentiful; all functions performed normally; during the eighteen hours it lived it consumed three ounces of hot milk. Dr. Watkins says " the catamenia appeared on the 26th of June, and disappeared, according to her habit, in five days. The child was born on the 26th of November, 153 days after the first day, and 148 days after the last day of the menstruation.

Now, it is established beyond the shadow of a doubt that this lady was not pregnant before the 26th of June:

Query—Was that fœtus a legal person,

and in the event of title, or property, could she inherit ? The French law establishes personality of the fœtus if alive at 180 days, for no child is known to have matured at less age.

Remember, this diminutive child (for it weighed barely one pound) gave all the evidences of vitality for eighteen hours, performing every function."

PROF. POLITZER, in the Wiener Med. Woch. (No. 31, 1880), recommends the use of *spiritus vini rectificatus* for the removal of polypous growths and granulations from the middle ear ; for the destruction of vegetable parasites of the meatus ; the treatment of *otitis mycosa ;* and likewise in chronic otorrhœa, where there exists abnormal conditions of the mucous membrane of the middle ear. The head being held on one side, the alcohol, moderately warmed, is poured into the ear by means of a teaspoon, and allowed to remain from ten to fifteen minutes. When productive of much distress, the alcohol should be diluted with equal parts of distilled water. Prof. Politzer thus sums up : The indications for the employment of alcohol may be said to be—1. For the removal of the remains of polypi from the meatus, the membrana tympani, and especially the cavity of the tympanum, in cases in which these are not removable by operative procedures. 2. In multiple granulations of the meatus or membrane. 3. In diffused and excessive growths from the mucous membrane of the cavity of the tympanum. 4. In cases in which, on account of some mechanical obstacle in the meatus, the removal of the polypus cannot be accomplished by instruments. 5. In timid persons and children, in whom operative procedures meet with great obstacles, and often can only be executed under anæsthetics.

ANTHROPOLOGICAL SOCIETY OF WASHINGTON.—At a recent meeting of this Society, Maj. J. W. Powell was re-elected President ; and Colonel Garrick Mallery, Dr. George A. Otis, Professor O. T. Mason, and Dr. H. C. Yerrow were elected Vice Presidents.

MISCELLANEOUS.

A BIOLOGICAL SOCIETY has recently been organized in Washington with the following officers : President, Theodore Gill ; Vice-Presidents, C. V. Riley, J, W. Chickering, Henry Ulke, Lester F. Ward ; Secretaries, G. Browne Goode, Richard Rathburn ; Treasurer, Robert Ridgway ; Council, Geo. Vasey, O. T. Mason, J. H. Comstock, and Drs. Schafer and A. F. A. King.

NEW YORK is agitating the necessity for the erection of a public library, and we hope her legislators will show their good judgment and appreciation of the requirements for advance by liberally responding.

THE CLERGY AND THE DOCTORS.—Perhaps no laymen have given the clergy more trouble than the Doctors. The old reproach against physicians, that where there were three of them together there were two atheists, had a real significance, but not that which was intended by the sharp-tongued ecclesiastic who first uttered it. Undoubtedly there is a strong tendency in the pursuits of the medical profession to produce disbelief in that figment of tradition and diseased human imagination which has been installed in the seat of divinity by the priesthood of cruel and ignorant ages. It is impossible, or at least very difficult, for a physician who has seen the perpetual efforts of Nature—whose diary is the book he reads oftenest—to heal wounds, to expel poisons, to do the best that can be done under the given conditions—it is very difficult for him to believe in a world where wounds can not heal, where opiates can not give a respite from pain, where sleep never comes with its sweet oblivion of suffering, where the art of torture is the only science cultivated, and the capacity for being tormented is the only faculty which remains to the children of that same Father who cares for the falling sparrow. The Deity has frequently been pictured as Moloch, and the physician has no doubt often repudiated him as a monstrosity.

Upon the other hand, the physician has often been renowned for piety as well as for his peculiarly professional virtue of charity— led upward by what he sees to the source of all the daily marvels wrought before his own eyes. So it was that Galen gave utterance to that psalm of praise which the sweet singer of Israel need not have been ashamed of; and if this "heathen" could be lifted into such a strain of devotion, we need not be surprised to find so many devout Christian worshipers among the crowd of medical " atheists.''

No two professions should come into such intimate and cordial relations as those to which belong the healers of the body and the healers of the mind. There can be no more fatal mistake than that which brings them into hostile attitudes with reference to each other, both having in view the welfare of their fellow-creatures. But there is a territory always liable to be differed about between them. There are patients who never tell their physicians the grief that lies at the bottom of their ailments. He goes through his accustomed routine with them, and thinks he has all the elements needed for his diagnosis. But he has seen no deeper into the breast than the tongue, and got no nearer the heart than the wrist. A wise and experienced clergyman coming to the patient's bedside—not with a professional look on his face which suggests the undertaker and the sexton, but with a serene countenance and a sympathetic voice, with tact, with patience, waiting for the right moment—will surprise the shy spirit into a confession of the doubt, the sorrow, the shame, the remorse, the terror which underlies all the bodily symptoms, and the unburdening of which into a loving and pitying soul is a more potent anodyne than all the drowsy syrups of the East. And, on the other hand, there are many nervous and over-sensitive natures which have been wrought up by self-torturing spiritual exercises until their best confessor would be a sagacious and wholesome-minded physician.

Suppose a person to have become so excited by religious stimulants that he is subject to what are known to the records of in-

sanity as hallucinations; that he hears voices whispering blasphemy in his ears, and sees devils coming to meet him, and thinks he is going to be torn to pieces or trodden into the mire. Suppose that his mental conflicts after plunging him into the depths of despondency, at last reduce him to a state of *despair*; so that he now contemplates taking his own life, and debates with himself whether it shall be by knife, halter, or poison, and after much questioning is apparently making up his mind to commit suicide. Is not this a manifest case of insanity in the form known as *melancholia?* Would not any prudent physician keep such a person under the eye of constant watchers, as in a dangerous state of at least partial mental alienation? Yet this is an exact transcript of the mental condition of "Christian" in "Pilgrim's Progress," and its counterpart has been found in thousands of wretched lives terminated by the act of self-destruction, which was so nearly taking place in the hero of the allegory. Now the wonderful book from which this example is taken is, next to the Bible and the Treatise of Thomas a Kempis, the best-known religious work of Christendom. If Bunyan and his contemporary, Sydenham, had met in consultation over the case of "Christian," at the time when he was meditating self-murder, it is very possible that there might have been a difference of judgment. The physician would have one advantage in such a consultation. He would pretty certainly have received a Christian education, while the clergyman would probably know next to nothing of the laws or manifestations of mental or bodily disease. It does not seem as if any theological student was really prepared for his practical duties until he had learned something of the effects of bodily derangements, and, above all, had become familiar with the gamut of mental discord in the wards of an insane asylum.

It is a very thoughtless thing to say that the physician stands to the divine in the same light as the divine stands to the physician, so far as each may attempt to handle subjects belonging especially to the other's profession. Many physicians know a great deal more about religious matters than they do about medicine. They have read the Bible ten times as much as they ever read any medical author. They have heard scores of sermons for one medical lecture to which they have listened. They often hear much better preaching than the average minister, for he hears himself chiefly, and they hear abler men and a variety of them. They have now and then been distinguished in theology as well as in their own profession. The name of Servetus might call up unpleasant recollections, but that of another medical practitioner may be safely mentioned. " It was not till the middle of the last century that the question as to the authorship of the Pentateuch was handled with anything like a discerning criticism. The first attempt was made by a layman, whose studies we might have supposed would scarcely have led him to such an investigation." This layman was " Astruc, doctor and professor of medicine in the Royal College of Paris, and court physician to Louis XIV." The quotation is from the article " Pentateuch," in Smith's " Dictionary of the Bible," which of course lies upon the table of the least instructed clergyman. The sacred profession has, it is true, returned the favor by giving the practitioner of medicine Bishop Berkeley's " Treatise on Tar-water," and the invaluable prescription of that "aged clergyman whose sands of life"—but let us be fair, if not generous, and remember that Cotton Mather shares with Zabdiel Boylston the credit of introducing the practice of inoculation into America. The professions should be cordial allies, but the church-going, Bible-reading physician ought to know a great deal more of the subjects included under the general name of theology than the clergyman can be expected to know of medicine. To say, as was said not long since, that a young divinity student is as competent to deal with the latter as an old physician is to meddle with the former, suggests the idea that wis-

dom is not an heirloom in the family of the one who says it. What a set of idiots our clerical teachers must have been and be, if, after quarter or half a century of their instruction, a person of fair intelligence is utterly incompetent to form any opinion about the subjects which they have been teaching, or trying to teach, so long !— *Oliver Wendell Holmes, in North American Review for February.*

PROCEEDINGS OF CHEMICAL SOCIETIES.— The January *Conversazione* of the American Chemical Society was held at the rooms of the Society on Monday evening, January 17. The Vice President, Dr. Albert R. Leeds, of the Stevens Institute, exhibited a new modification of Dinitro-orcine and certain of its salts. These salts were originally prepared by Professor Leeds at his own laboratory in the course of his investigations of Hyponitric Anhydride in organic substances.

Specimens of Dibenzole and Diphenyle were also exhibited by the same gentleman. Several of the members took advantage of the occasion to visit the laboratory and see the recently patented electrical inventions of Dr. O. Lugo.

The Chemical Society of Paris announces that among the vice-presidents, according to the constitution, the president shall be chosen from the following gentlemen : MM. Grimaux, Salet and Berthelot, and that the Council nominates MM. Grimaux and Salet ; therefore M. Berthelot will remain as vice-president during 1881, and in consequence of the regretted decease of M. Personne, M. Berthelot will be the only occupant of that office.—*Science.*

CONTRIBUTION TO ELECTROLYSIS. — L. Schucht describes the electrolytic determination of uranium, thallium, indium, vanadium, palladium, molybdenum, selenium, and tellurium. For qualitative analysis he uses a strong test-glass, 10 to 12 c.m. high, and 1.5 c.m. wide, fitted with a cork coated with paraffin. Two platinum wires, 1½ m.m. in thickness, pass through the cork down to the bottom, and are connected above the cork with the polar wires of the battery by means of small binding screws. This decomposition tube may be held in a wooden clamp. After the current has passed through the solution to be analyzed for ten to fifteen minutes, the stopper with the wires is drawn out, without interrupting the current, and the deposited metal is determined by its color, lustre, solubility in acids, &c. The manner of decomposition and the slight or strong evolution of gas is noticed. The solution is completely precipitated, rendered alkaline, and again electrolysed, after the wires have been cleansed. Copper is recognised by its color; mercury by the precipitated globules; nickel and cobalt by their lustre and sparing solubility in acids; zinc and cadmium by their color and solubility in potassa. The formation of peroxides is characteristic for lead, silver, bismuth, thallium, manganese. Bismuthic acid is gradually formed, whilst the peroxides of lead, silver, and thallium are deposited at the beginning of the precipitation. Silver peroxide dissolves in ammonia with liberation of nitrogen. The decomposition of the alkalies and alkaline earths is best effected in a U-tube. The hydroxides of the latter are separated in a voluminous form; those of calcium and magnesium in white crusts. The hydroxides of barium, strontium, and the alkalies dissolved on the negative wire. —*Berg-und Hutten Zeitung, 39, 121.*

BERGH AND VIVISECTION.—Dr. Pratt, in his New York letter to the Chicago Med. Jour. and Examiner, thus concludes: "In closing, let me say a word or two regarding a lecture I attended a few evenings since. Mr. Henry Bergh addressed the citizens of this city on the subject of 'Vivisection.' That Mr. Bergh has done a good work in the prevention of cruelty to animals, no one will gainsay, but when he comes before an enlightened audience with such twaddle as he presented on this occasion, one not only perceives the gross ignorance of the man, but is led to doubt his soundness of mind. Hear him : ' Of what use is vivisection ? A friend of mine had a beautiful little daughter lying

dangerously sick with diphtheria. The attending physician called in consultation a well-known physiologist of this city, who advised making an opening into the windpipe. The dear little girl screamed aloud at the horrible suggestion, and said : " 'Father, they will kill me!' " The operation was done ; the child died, and her infantile prophecy was · fulfilled. Had vivisection taught this man anything ? ' The argument is unanswerable. Comment is useless.

Recall to mind those eminent physiologists, whose arduous labors and patient researches have added so much to the sum of earthly happiness; whose whole lives were spent in the interest of their fellow men, and who went to their rest at last, recompensed by the thought of the good they had accomplished. And now listen to the tribute of respect paid them and their followers, by this champion of the brute creation. 'If there is a hell, and hell-fires are burning, I am sure these men will feel their torments to the utmost extent.' The long continued hissing that greeted this expression, must have startled somewhat this noble-hearted man, though we would not expect one capable of conceiving such a fiendish thought to be much disturbed. These extracts will suffice to illustrate the method by which Mr. Bergh hopes to convince the legislature of this State of the uselessness and wickedness of vivisection.''

BOOKS AND PAMPHLETS RECEIVED.

"Announcement of the Thirty-Second Annual Session of the Medical Department of the University of Georgetown," District of Columbia. Collegiate year 1880–'81.

"Walsh's Physicians' Combined Call-Book and Tablet." Ralph Walsh, M.D., publisher, 332 C street, Washington, D. C.

"The Study of Mortuary Customs Among the North American Indians." By H. C. Yarrow, M.D., Act. Asst. Surg. U. S. A., Smithsonian Institution, Bureau of Ethnology, Washington, D. C.

"The Abuses of Medical Charities." By

M. P. Hatfield, A.M., M.D., Professor of Chemistry, Chicago Medical College, and Roswell Park, A.M., M.D., Demonstrator of Anatomy, Chicago Medical College. Reprint from the Chicago Medical Gazette. " Dipthonia Paralytica." By Ethelbert C. Morgan, A.B., M.D. Late assistant to Prof. Johannes Schnitzler, in the Department of Diseases of the Throat and Lungs in the Poliklinik, Vienna, Austria. Reprint from the National Medical Review. Also by the same author " A Contribution to the study of Laryngeal Syphilis." Reprinted from Va. Med. Monthly.

" How shall the Degree of Doctor of Medicine be Conferred?" By E. Fletcher Ingals, M.D., Lecturer on Diseases of the Chest and Physical Diagnosis, and on Laryngology in the Post Graduate Course, Rush Medical College, etc., etc. Reprinted from the Chicago Medical Journal and Examiner. Also by the same author " Treatment of Diseases of the Larynx." Reprinted from the same.

" Microscopical Studies on Abscess of the Liver." By J. C. Davis, M.D., Denver. Reprint from " Archives of Medicine."

" First Biennial Report of the Department of Diseases of the Eye" of the Central Free Dispensary of the District of Columbia, Washington, D. C. By Swan M. Burnett, M.D., Surgeon in Charge.

" Yellow Fever: Its Ship Origin and Prevention." By Robert B. S. Hargis, M. D., Pensacola, Florida. D. G. Brinton, M.D., Publisher, Philadelphia.

" Chronic Bright's Disease in Children Caused by Malaria." By Samuel C. Busey, M.D., Washington, D. C. From Transactions of the American Medical Association, Philadelphia, 1880.

" The Medical News and Abstract." A consolidation of the "Medical News and Library" and the "Monthly Abstract of Medical Science." Edited by I. Minis Hays, A.M., M.D. Henry C. Lea's Son & Co., publishers.

" A Conspectus of the Different Forms of Phthisis, Intended as an aid to Differential Diagnosis." By Roswell Park, A.M., M.D. Demonstrator of Anatomy, Chicago Medical College; Surgeon to the South Side Dispensary. Reprint from the Chicago

Medical Journal and Examiner. Also by the same author—"A Conspectus of three Different Forms of Acute Inflammatory Cardiac Disorder." Reprint from the Chicago Medical Journal and Examiner. Also "Maternal Impressions; Mother's Marks. An Expose of a Popular Fallacy." Reprinted from " The Southern Clinic." Also "Dermatitis Venenata ; or, Rhus Toxicodendron and its Action." Reprint from Archives of Dermatology.

" Yellow Fever a Nautical Disease ; Its Origin and Prevention." By Prof. John Gamgee. Published by D. Appleton & Co., New York. 1879. Also by the same author, "On Artificial Refrigeration." Also " On the Sanitary Urgency of the Florida Ship Canal." Also, " The Inter-Oceanic Ship Canal and the Yellow Fever Zone." Washington, D. C.

" Interpretations of the Structure and Function of the Kidney." By Andrew W. Smyth, M.D., of New Orleans, La. Reported by Prof. John Gamgee. From the N. O. Medical and Surgical Journal.

" Smithsonian Report." Washington : Government Printing Office.

" The Kansas City Review of Science and Industry." Edited by Theo. S. Case. Press of Ramsey, Millett & Hudson.

" Rocky Mountain Health Resorts." By Chas. Denison, A.M. M.D. Second edition.

" The Alienist and Neurologist." A quarterly Journal of Scientific, Clinical and Forensic, Psychiatry and Neurology. Edited by C. H. Hughes, M.D. Ev. E. Carrerar, publisher, St. Louis, Mo.

" Atresia of the Genital Passages of Women." A paper read before the Chicago Medical Society, July 19th, 1880. By Edward W. Jenks, M.D., LL.D., Professor of Medical and Surgical Diseases of Women and Clinical Gynecology in Chicago Medical College.

" Colorado for Invalids." By S. Edwin Solly, M.R.C.S., Eng. Reprint from "New Colorado and the Santa Fe Trail."

" Introduction to the Study of Sign Language Among the North American Indians, as Illustrating the Gesture Speech of Mankind." By Garrick Mallery, Brevet Lieut. Col. U. S. A. Smithsonian Institution—Bureau of Ethnology, Washington, D. C.

"The Clinical News." A National Weekly Journal of Clinical Medicine, Surgery and Gynecology. Edited and published by Samuel M. Miller, M.D., Philadelphia.

ROCKY MOUNTAIN MEDICAL REVIEW.

Vol. I. COLORADO SPRINGS, MARCH, 1881. No. 7.

ORIGINAL ARTICLES.

TO THAT WHICH HATH VIRTUE, CREDIT IS DUE.

By J. H. KIMBALL, M.D.

In the MEDICAL REVIEW for January, I notice that reports on climatology have received special attention. Contributions on this subject are valuable, and should be welcomed by all readers, provided that the writer shows himself free from the perhaps natural prejudice which is apt to exist in the professional mind when writing up a particular locality. The fact is—and it is a fact, too, which should be conceded at the outset —that there is no *perfect* climate for phthisis. What we should seek is the climate which shall benefit individual cases. In studying the effects of climate cure, we must at the same time critically examine the case in hand. A physician at all familiar with health resorts will, at each of them, meet invalids who have received benefit.

The physician then who advises this or that resort, should not presume to do so unless he is qualified to select one by honest research and inquiry, or better still, by personal observation. As a matter of fact, Minnesota, Florida, Colorado, South Carolina, and other States, can point to hundreds who, according to popular phraseology, " came here to die," and who now enjoy a new lease of life.

But while giving statistical tables of temperature and humidity, but few have noted the characteristics of the cases benefited.

Dr. Boardman Reed utters a sweeping condemnation of health resorts generally, alone excepting Atlantic City. One is apt to conclude that his observation has been circumscribed and his inferences drawn from limited sources. It is very true that the reputation of Nice, Mentone and Pau has waned, and with reason ; but Malaga, Algiers, and both Upper and Lower Egypt, still maintain their standing as among the best of health resorts, especially for cases of phthisis of catarrhal origin. The resorts of this country corresponding in general character to the above mentioned, are Aiken, So. Carolina, Northern Georgia, and the neighborhoods of Placerville, California, and Tallahassee and Gainesville, Florida. All of them coming under the class of climate termed " bracing," though strictly speaking, Gainesville might with equal propriety be classed among the " sedative" climates. The health resorts of Florida most frequented are Jacksonville, St. Augustine and Palatka ; but a long acquaintance with these places has convinced me that better results can be obtained at Aiken and Northern Georgia.

Along the St. John's river one often sees laryngeal and bronchial irritations allayed at the expense of the general health and strength, and I have not infrequently met with cases of hemorrhage in systems thus relaxed, where they had previously been unknown.

The statistics of deaths from phthisis, as given by Dr. Kenworthy, whom I know well and esteem, and than whom no one is more reliable, are correct for 1878, and yet, from May 1, 1879, to Nov. 1, of the same year, there occurred in St. Augustine, five deaths from phthisis among the native residents—all colored—out of a total mortality of thirty-two. Dr. Solly's claim that sudden and extreme changes of temperature are beneficial has before been urged. Southgate, in "Army Medical Statistics," wrote " Equability can hardly be considered as the most vital element of climate; the highest degree of physical vigor being attained in strikingly variable climates, the human constitution being adapted to such mutations; and its powers would languish under the monotonous impressions of a uniform temperature for a long time." Hirsch says, "Dry air, either with continuous heat or sudden changes, does not favor phthisis."

It is difficult with our present knowledge to assume that temperature has much to do with the prevalence of phthisis, as there are low as well as high altitudes, warm as well as cold countries, which claim exemption from its ravages. Dr. John C. Thorowgood, of London hospital for diseases of the chest, in a paper recently published, recognizes two classes of phthisis—the "Catarrhal" and "True Tubercular," or as Prof. A. L. Loomis, of New York, terms it, "Fibrous Phthisis." The former mentioned class he has seen relieved in mild climates, while the latter requires a cold one. In his opinion as regards the non-catarrhal type, Prof. Loomis coincides, while for the catarrhal class he is strongly in favor of the Adirondack region, though the objections to this region are many and the advantages claimed for it can be obtained in other localities of moderate altitude, as for example, Ashville, North Carolina, and the pine district of Georgia, north of Atlanta. The same altitude (1800 feet) can be found both in North Carolina and Ga., while the excessive annual rainfall—55 inches—and heavy dews will be avoided.

The question whether individual temperament should influence us in deciding upon the most proper climate, is an interesting one, and I do not think it has usually received attention. It has been observed in Florida that phthisical patients with a nervous or sanguine temperament would improve by a prolonged stay, while those with the melancholic temperament would not; but the observations have been limited and are of little value excepting as they may direct future attention to the matter. It is acknowledged, I think, by the profession, that in the high altitudes there are often de-

veloped diseases of a functional nervous character, which have never appeared on a lower level; and where they have previously existed, they are aggravated on first moving to them. We may, from our present knowledge, conclude that among the bracing climates will be found one suitable for nearly every case of incipient phthisis that may present itself—the high altitudes for true fibrous phthisis, especially when uncomplicated with nervous derangement; the lower altitudes and milder temperatures for the catarrhal type. Of these, Aiken and Northern Georgia offer greatest advantages in the winter months, while to the former class, Colorado and New Mexico appear preferable to any others; but while we of Colorado feel that we possess a climate pre-eminent among those belonging to the high altitudes, we should not, in urging its claims, forget that other resorts are entitled to consideration. Climate cure is a science in itself, and is worthy of attentive study from medical men.

King Block, Denver, Col.

RECENT MEDICAL PROGRESS.

QUARTERLY REPORT ON LARYNGOLOGY AND RHINOLOGY.*

By ETHELBERT C. MORGAN, A.B., M.D.,

Formerly Assistant to Prof. Johannes Schnitzler in the Department of Diseases of the Throat and Lungs, in the Poliklinik, Vienna ; Surgeon in charge of Diseases of the Throat and Ear in Providence Hospital, Washington, D. C.

SECONDARY SYPHILITIC LARYNGITIS—PARALYSIS OF LEFT VOCAL CORD.

M. Jovel (Rev. Mens. de Laryngol., etc., Feb. '81,) reviews the accepted teachings on the above subject, and mentions a case of paralysis of the left vocal cord resulting from secondary laryngeal lesions. Laryngeal paralysis of syphilitic origin is not necessarily associated with the presence of a syphilitic tumor or gomme upon the nerve tracts, but is attributable to the same cause as neuralgias, cardiac palpitation, anæsthesia, facial

* To insure more careful attention in cases of books, pamphlets and reprints coming under this head and intended for review, two copies should be sent—one to the *Review*, the duplicate, to the gentleman in charge of the department.

paralysis, and the disturbance which manifests itself in the second period of variola. A stout, healthy man of forty, complained of great hoarseness with loss of voice towards evening. Complete paralysis of left and partial paralysis of right vocal cord were detected by the laryngoscope, but there was neither redness nor swelling of the laryngeal mucous membrane, the lungs, heart and blood vessels being normal. His vocal troubles had lasted eight days, and denying ever having had a chancre, no diagnosis of syphilitic paralysis could be made. Local applications of zinc chloride were therefore made to the larynx and purgatives administered. At the end of eight days the patient returned with cutaneous syphilides, and then admitted having had a pimple on his penis six months previous. Mixed treatment (mercury and iodides) was used and in thirteen days the voice was restored.

THE METHOD OF REMOVING FOREIGN BODIES ENGAGED AT THE RIMA GLOTTIDIS.

M. Krishaber (France Med. 16-19, Dec. '80) relates four cases in which the foreign body rested on the superior surface of the vocal cords. In two cases a 50 centime piece was retained by the laryngeal ventricles in a horizontal position. One of these patients was placed upon his abdomen in recumbant posture, the index finger of the left hand passed as far as the edge of the epiglottis, and a pair of laryngeal forceps introduced, but it was impossible to seize the piece, which slipped from the grasp of the instrument. The patient was then placed in front of the operator, a pair of closed laryngeal forceps introduced below the glottic orifice, then opened, being careful to exercise pressure from below, upwards. The piece was dislodged and accidentally swallowed by the patient.

In the second case the piece was pushed out by means of a sound introduced through a tracheal opening made to save life, which was threatened by acute œdema. A bone was removed from the glottic orifice in the third case by means of the forceps.

In the last of the four cases a piece of cop-

per was extracted by the same procedure which failed in the first case mentioned.

Krishaber advises that except in the case of infants in whom the larynx is easily reached by the finger of the operator, the performance of tracheotomy, the tamponing of the trachea and the extraction of the foreign body *per vias naturales.*

MUCOUS PATCHES OF THE LARYNX.

Dr. Gougenheim (Lyon Med. No. 41. 1880) contends that "Mucous patches" can even invade the larynx, a fact which has long been denied ; but in the face of so many accurate observations it is impossible to escape discovering them. These patches, according to Dr. G., appear in three forms, viz., the erosive, the circular and the excavated. The epiglottis is their most frequent seat ; after it, the vocal cords. The laryngoscopic examination should be made by the aid of sunlight, and they should be treated by applications of a solution of silver nitrate, 10-25 grs. to the fluid ounce of water.

APHONIA SPASTICA.

Fritsche describes (Centralb. f. d. Med. Wissensch, No. 36, 1880) this interesting affection in a most complete and satisfactory manner, citing the history and results of six cases. The affection was described by Schnitzler in 1875, and consists, according to him, in a functional spasm (Krampf) of the muscles of the vocal cords, analogous to the want of co-ordination in writer's cramp. Objective symptoms do not enable us to distinguish Aphonia Spastica from ordinary paralytic Aphonia. The voice frequently fails on account of the spasm or cramp during utterance of a word, and on laryngoscopic examination, during attempted phonation, we observe that the vocal cords are pressed tightly together, leaving a mere line between them. The best therapeutic results are to be expected from the use of the Galvanic (constant) current. Five of the above cases were thus benefited. One case resisted the use of Faradization and the insufflation of potassium iodide. The articles close by describing a new laryngeal electrode, differing from those in general use by being articulated in the manner of closing the circuit, and in the facility with which it is grasped by the hand, like a pistol or pen.

HYPODERMICS OF MORPHIA IN LARYNGEAL HYPERÆSTHESIA.

In laryngeal Hyperæsthesia, Rossbach (Lyon Med. No. 40, 1880) also (Wien. Med. Presse, No. 40, 1880) advises the hypodermic injection of six milligrams (1·9 gr.) of muriate of morphia on either side of the neck at the point where the superior laryngeal nerve enters the larynx (under greater horn of hyoid bone.)

NEW OPERATION FOR THE REMOVAL OF LARYNGEAL POLYPI.

Rossbach of Wurtzburg (Berl. Klin., Wchnschr. No. 5, 1880) describes a new operation for the destruction of intra-laryngeal polypi situated so deeply that they cannot be reached by the laryngeal forceps. In the subcutaneous procedure, as it is called, a straight spear-pointed bistoury is introduced into the larynx in the median line, through the crico-thyroid membrane. The knife is brought into view by means of the laryngoscope, and the small tumors on the laryngeal wall, cut off. There is no scar after the operation, but care must be taken to enter the bistoury only in the median line. otherwise the arterial branches may be injured.

COMPARATIVE RESULTS OF THE FARADIC AND GALVANIC CURRENTS IN TREATING PARALYSIS OF THE LARYNGEAL MUSCLES.

Dr. Massie of Naples (Lyon Med. No. 41, 1880) compares the results obtained in treating paralysis of the larynx by the Faradic and Galvanic currents. In his opinion, Faradization is insufficient; it may stimulate in a measure the nerve centres, but it is by the galvanic current that we arouse contraction of the paralyzed laryngeal muscles.

HEMORRHAGE FROM AND LACERATION OF THE VOCAL CORDS.

Professor J. Schnitzler (Wien. Med. Presse, Nos. 38-41, 1880) in a clinical lecture reports a few of the ten cases of this rare accident which have come under his observation. A woman thirty-nine years of age had

long been treated at the clinic for ordinary pharyngo-laryngeal catarrh, the vocal cords had become nearly normal, her voice was suddenly lost, and an examination with the laryngoscope revealed the fact that the entire length of the left vocal cord was infiltrated by blood, the cord itself flapping loosely backward and forward during respiration and phonation. Violent vocal exertion in conversation was the direct cause of the accident.

One of the most popular singers of the Vienna Opera was required to sustain a high note very long and to throw herself with force upon the stage ; a moment later, her voice refused to act, though it was perfect an instant before. The next morning Prof. S. examined her larynx and discovered that the whole left vocal cord was infiltrated by blood, whilst its fellow was normal. Astringent inhalations and insufflations of lead acetate succeeded in producing absorption of the extravasation, so that in fourteen days the singer appeared again in public and the press said her voice possessed its usual charms.

LARYNGEAL WOUNDS—THEIR GRAVITY AND TREATMENT.

M. Raoul (Paris Thesis, No. 394, 1880) fully discusses this subject, of which the following is a very brief summary :

Ancient authors make no mention of wounds of the larynx. Hippocrates describes tracheal wounds, and indicates in a clear manner the operative procedure for tracheotomy in saying, "When a patient is suffocating, it is necessary to open the trachea and insert a quill (tube)." Suicide is the most frequent source of these wounds, which are more common in men than in women, the latter selecting methods of death unaccompanied by pain or disfigurement. In longitudinal and oblique laryngeal wounds, there is not much disposition to retraction. Of two hundred and eleven cases, one hundred and forty-five were wounds of the larynx, and sixty-six of the trachea. Malgaigne contends that the young, cut oftener at the superior part of the neck, (larynx) whilst old people, on account of the difficulty experienced in lifting the head, cut low (trachea). The carotid artery is *seldom* wounded by suicides ; the voice remains unchanged if the wound is below the vocal cords (p. 21). Deglutition is only interfered withwhenthe thyro-hyoid membrane is divided. Fistula, strictures and atresia of the larynx may result from these wounds.

Small wounds of the larynx are grave, large ones much less so.

The suture is hurtful in all large laryngeal wounds, it augments irritation and inflammation and favors accidents. The union by suture of the skin over a laryngeal wound also united by suture, should be rejected. It is permissible to apply a suture to a laryngeal wound only to unite the soft parts.

The union of the skin above a non-united tracheal wound is not less objectionable, as it allows the blood and pus to flow into the air passages.

Those wounds situated in the thyro-hyoid membrane are more serious than all other laryngeal wounds.

FIRST EUDO-LARYNGEAL OPERATION DURING NARCOSIS—EXTIRPATION OF A LARYNGEAL POLYPUS.

Schnitzler (Wien. Med. Presse, Nov. 28, 1880) describes this unique operation, which was performed on a very excitable child only eight years of age. The tumor, a papilloma, was attached to the laryngeal ventricle, was large, and it was impossible to even make an attempt, or approach the nervous child, for its removal. He was put under the influence of sulphuric æther, and the growth removed.

THE VALUE OF EPISTAXIS AS AN ELEMENT IN DIAGNOSIS AND PROGNOSIS.

M. Viennot, in Paris Thesis, No. 434, 1880, says—Like all morbid phenomena, epistaxis has, *per se*, a certain significance, as its appearance in the commencement of *certain* affections greatly assists in establishing a diagnosis. Epistaxis has no absolute value as a symptom, for it is sometimes absent in the diseases in which it is most frequently seen. It is occasionally the sole symptom,

enabling us to make a differential diagnosis, and whilst denoting little danger at the beginning of eruptive fevers, merits serious attention during the eruptive period, as a prodrome of internal hemorrhage.

Epistaxis in whooping cough is indicative of serious danger. It appears to be favorable in the early stage of typhoid fever, but in the period of decline it is most grave. When profuse and repeated in hepatic trouble (cirrhosis or ectirus), it is a bad sign.

Epistaxis occurring in the course of diphtheria is of serious importance, as indicating an extension of the disease to the nasal fossæ.

AN IMPROVISED PHARYNGOSCOPE.

Paul Landowsky (Lyon Med. No. 41, 1880) recommends that an ordinary bright spoon be fastened to a candle in such a manner that the concave surface of the spoon will act as a reflector. This rough device may render valuable aid at times in illuminating the fauces.

PROLAPSUS UVULÆ AND ITS INFLUENCE UPON THE VOICE.

Dr. Moua, at the International Laryngol. Cong. Milan, spoke of the above affection and said he regarded it as of frequent occurrence among singers, and advised excision of the prolapsed portion by means of the uvula scissors. He said he never feared serious hemorrhage after the operation.

Dr. Blanc (of Lyons) cited a case in which, after the operation, his patient was seized with sudden hemorrhage during sleep. Dr. Labus called attention to the pain attending and resulting from the operation, and Dr. Lennox Browne said he always advised patients to take a hearty meal prior to the excision, in order that the parts might afterwards remain quiet.

TURPENTINE IN WHOOPING COUGH.

M. Barety (L'Union Med. Nov. 4, 1880) noticed some four years since that of four cases of whooping cough he was attending in the same house, the most violent, who was removed to a recently painted room, became suddenly better, the spasm lessening. M. B. has used spts. of turpentine often since, in the following manner: Two plates are kept filled with the fluid, one being placed under the bed, the other in a corner of the room, which is ventilated twice daily; the spasm will quickly diminish, the disease assumes a milder type, and seldom lasts over a month.

LARYNGEAL POLYPUS—CONCOMITANT LARYNGEAL PHTHISIS—REMOVAL OF POLYPUS—CURE.

M. Coupard (Rev. Mens. de Laryngol, etc., Jan. 1881) relates the history of a man twenty-seven years of age, with a papilloma on right vocal cord and concomitant laryngeal phthisis, in whom not only the voice, but the general health, returned permanently after the removal of the tumor. The larynx meantime became normal and presented no signs of previous disease.

MYXOMATA OF LARYNX.

Dr. Clinton Wagner (Archives of Laryngology, Vol. 1, No. 4) reports a case of myxomata of the left laryngeal ventricle resulting in spontaneous expulsion. Such cases Dr. W. says are very rare, and quotes Mackenzie as saying he has never seen an example of a pure myxomata of the larynx. In strange contrast of the above is the statement of Fauvel (Maladies du Larynx), a French author, who reports forty-eight such cases among three hundred tumors which he has removed. Fauvel, remarkable as it may appear, whilst giving the above number of myxomata, neglects to mention how many have been submitted to a crucial microscopic test. Bruns (Polypen des Kel Kopfes p. 17) reports a case of Myxoma Hyalinum.

CLINICAL CONSIDERATIONS' ON ACCESS TO BENIGN INTRA-LARYNGEAL NEOPLASMS THROUGH EXTERNAL INCISIONS.

Dr. Cohen (Reprint from Archives of Laryngology, Vol. 1, No. 2) remarks that by far the greater number of growths are accessible to laryngoscopic procedure, and that no external operation should be resorted to until the failure of intra-laryngeal resources has been amply demonstrated. Section of both the cricoid and thyroid cartilages are rather to be avoided, the former destroying the solidity of the laryngeal skeleton, and in rare instances being followed by necrosis, the latter causing agglutination of the anterior portions of the vocal bands and destruction of the quality of the voice. In the first observation, the middle crico-thyroid ligament was incised and the sub-glottic neoplasm destroyed by the galvano-cautery, with the result of obtaining a fairly good voice and a slight scar on the neck. In the second, section of the thyroid cartilage for evulsion of infra-glottic papillomata, terminated well. The third, describes the removal of a sub-glottic fibroma through the wound of a tracheotomy, which appears to have been likewise successful. The fourth, was a sub-hyoidean pharyngotomy upon a child five years of age for direct access to a benign intra-laryngeal neoplasm, with failure of satisfactory access to the growth.

NASAL POLYPS AND THEIR OPERATION.

The second edition of Voltolinis brochure "Ueber Nasenpolypen und deren operation," does not contain much that is new. He reiterates his claim as having been the first to substitute the wire snare for the forceps in removing nasal polyps, and gives the following propositions: The removal of nasal polypi with forceps is a barbarous method; is painful, unnatural, incomplete, etc., and should be superseded by the snare. We obtain better results with the snare. When the ordinary snare fails we may have recourse to the galvanic wire. Port-caustics are useless.

SYPHILITIC STRICTURE OF LARYNX—A NEW CUTTING DILATOR.

Dr. Whistler (Archives of Laryngology, Vol. 1, No. 4) gives the history of two cases of cicatricial adhesion of the larynx operated upon by means of an almond-shaped instrument, which he has devised, combining the properties of a knife and dilator. The cases appear to have resisted all forms of medication, and both were tracheotomatized previous to using the Laryngotome. Respiration and phonation became almost normal after division and dilatation of the strictures.

TRIANGULAR STONE IN RIGHT BRONCHUS ONE MONTH.

M. Laugier (Proges. Med., No. 2) reports the expulsion of a triangular stone from the right bronchus, where it had remained at least one month. The stone was accidentally swallowed in some thick soup, and occasioned violent suffocative attacks, bronchitis, fever, purulent and bloody expectoration. The pain over the middle of the right bronchus· lasted two months after the stone was expelled.

Dr. Catte, of Fiume (Lyon Med., No. 41, 1880), calls attention to a generally committed error—the diagnosis of ankylosis of the crico-arytenoid articulation, as paralysis of the posterior crico-arytenoids. Ankylosis occurs more frequently than paralysis, and the signs which would lead us to diagnose the former in preference to the latter, are as follows : The previous existence of ulcerations, absence of lesions of the nerve centres, the result of the laryngoscopic examination, which demonstrates perichondritis.

MALFORMATIONS OF THE ŒSOPHAGUS.

Dr. Mackenzie (Archives of Laryngology, Vol. I, No. 4) states that there are but fifty-seven recorded cases· of the above trouble, only *one* occurring in his own practice. Of complete deficiency, there are five; of blind termination, thirty-seven ; nine in which there was no inter-communication between the œsophagus and air passages, with deficiency of the former; three cases in which there was communication with a bronchus; two of an otherwise normal œsophagus communicating with the trachea ; two of complete blocking-up of œsophagus from membranous obstruction ; one of congenital pouch, and one of longitudinal diversion of the œsophagus. The cause of congenital malformations of the œsophagus must arise from abnormal conditions either in the spermatozoon, in the ovum, or in the embryo. That the first is sufficient to produce malformation is proved by the same male occasion-

ally producing a similar deformity in the offspring of different women. The symptoms of congenital malformation of the œsophagus are characteristic ; the infant may appear healthy while at rest, but the moment it attempts to swallow, the most distressing attacks of suffocation supervene, and there is great danger of one of them proving fatal. A case of a child who died from milk passing into the air passages is cited, but as a rule, death takes place from inanition, a few days after birth.

916 E St., N. W.

Quarterly Report on General Medicine and Therapeutics.—CONTINUED.

By D. S. LAMB, A.M., M.D.,

Professor of Anatomy, Howard University, Washington, D. C.

LEPROSY.

Dr. Harang (Med. and Surg. Reporter, Nov. 6. 1880, p. 417), in regard to the leprosy of Louisiana, states that in 1869 there were only four or five lepers in his parish, whereas now there are over thirty well defined cases.

METALLO-THERAPIC.

Dr. Bricquet (Bull. zen. de Therapic, Nov. 30. 1880, p. 433), gives a resume. 1st, the restoration of sensibility is nearly always temporary. 2nd, the restoration of sensibility may be compensated by an equivalent loss (as deafness occurring in the left ear after restoration of hearing in the right). 3rd, there is no rule to guide in the choice of metals. 4th, a metal may be chosen which is not appropriate, and then the restoration may instantly lapse. 5th, transfers of sensibility may take place unforeseen and perhaps inconvenient. He presents a table of numerous cures of anæsthesia, promptly and surely, in a total of 120 cases —anæsthesia of all the members and hemi-anæsthesia. He regards the cures as due rather to faradization, which was also employed ; the amelioration being rapid, immediate, quite often instantaneous ; at other times occurring the next day. He often used concurrently, when the improvement did not take place fast enough to suit him, frictions with croton oil, ammonia liniment and sinapisms.

ANTAGONISM OF MEDICINES.

Prof. Bartholow (Med. Record, Nov. 27, 1880, p. 593) considers morphia and atropia antagonistic in their effects on the cerebrum, and usually so on the pupil. Also on the heart, but the effect of atropia is more powerful and prolonged. Also on respiration ; morphia slows it and diminishes the excretion of carbonic acid gas ; atropia increases both. Also on the arterial tension ; opium slows the heart and paralyzes the arterioles. Atropia presents, however, to a large extent, the depression, coldness of surface, cold sweating and cerebral nausea of morphia. Antagonistic also in their effects on the kidneys ; morphia diminishes the discharge of urine. Therapeutically one may utilize these actions. In poisoning by either, give its antagonist in small doses frequently repeated.

IRON.

Dr. Bayard (L'Union Med., Oct. 20, 1880, p. 713) believes that iron is given in too large doses, and quotes Quinquand as saying that it should be given "in small doses, and never in concentrated preparations." Bayard recommends to give it in soluble form and organic combination, and diluted in an appropriate vehicle ; the citrate, tartrate, albuminate with syrup of orange peel, in doses not exceeding three-fourths of a grain of the metal, after the two principal meals ; this is the most assimilable form, best tolerated by the stomach, and furnishes the best results.

ELECTRICITY.

Prof. Bartholow (Med. News and Abstract, Jan. 1881, p. 5), in regard to the use of electricity in disease, makes the following observations. The Faradic current should not be used in ordinary hemiplegia, unless there should be wasting, degeneration and impaired electro-contractility, and also late rigidity. Galvanize the contracted parts and Faradize the relaxed or weak parts.

Paralyzed members receiving their innervation from a diseased part of the spinal cord, lose their electro-contractility to the Faradic current ; but preserve it when that part of the cord whence the nerves are given off is healthy, though the cord elsewhere is diseased.

If the motor trunks are diseased, the contractility declines, the muscles degenerate and fail to respond to Faradization, but yet for a time respond to Galvanism ; finally they are insensible to that. If the nerve recovers, it is found that the response to the will takes place sooner than to electrical stimulation.

When paralyzed muscles respond to Galvanism but not Faradization, the former is used until the time comes when the latter elicits response.

The Faradic current is of little service in loss of sensibility.

The property of relieving pain belongs to the Galvanic current.

In internal maladies Galvanism is used because it penetrates to the deep organs, and Faradization does not ; the latter tetanizes the bloodvessels. Galvanism stimulates the peristaltic action of the intestines.

The tonic and reconstituent effects which follow the application of Galvanism to the cervical sympathetic, pneumogastric and spinal cord, are doubtless due to increased action of the vessels and stimulation of the nervous apparatus which presides over the movements of the chylopoietic viscera. Also in intra-cranial disorders of circulation, due to weakness of vessels, the current should be weak and only applied for a few minutes.

In applying electrodes, he says that, in Faradization. well-moistened, sponge electrodes are used when it is desired to reach the muscles ; for a single muscle, the olive pointed electrode. To Faradize the skin thoroughly, dry it and dust with powder. To Galvanize : for single muscles and separate nerve trunks, use small electrodes ; for large groups and pain in many nerve filaments, use large, well-moistened sponge electrodes.

Salt is to be added to the water only in Galvanization of face and head. In neuralgias of the extremities, use powerful currents. In Galvanization of the head, the *seance* should not exceed five minutes ; in neuralgia, a longer time ; in sciatica, about fifteen minutes repeated every four hours ; in Faradization, five to fifteen minutes twice daily.

TOBACCO POISONING.

A case reported (Pacific Med. and Surg. Jour., Dec. 1880, p. 308), of a man, age 49, manager of a cigar factory, a great smoker, and in the habit of testing the cigars made by the workmen. He frequently complained of weakness, particularly of the legs. He died suddenly. The heart was found contracted and empty, except a small white clot on the left side, somewhat choking the aortic orifice; lungs emphysematous; blood dark; kidneys hyperæmic ; two ounces of serum at the base of the brain. The reporter considers the death due to nicotine poisoning.

CARBOLIC ACID POISONING.

A case reported (Cin. Lancet and Clinic, Nov. 6, 1880, p. 419, from Archiv. f. kinderheilk, 12. 1. 1880), of an infant of ten weeks ; a necrotic spot about an inch in diameter appeared in right axilla. A two, and then five per cent. aqueous solution of carbolic acid was applied. In six hours after the second solution was used, collapse took place ; urine dark, on examination, was found to contain carbolic acid. The child died in 24 hours in convulsions, with contracted pupils and suppression of urine. Lungs, liver, spleen and brain hyperæmic ; spleen three times its normal size ; some coagula in the vessels.

OPIUM HABIT.

A case reported (Virginia Med. Monthly, Dec. 1880, p. 701) of opium habit of long standing successfully treated with coca. A married woman, age 24. The opium was stopped Sept. 28, 1880, and twenty grains of extract of coca given four times daily ; 30th, there was anorexia, nausea, diarrhœa, restlessness, muscular twitchings, pain in back

and joints, double vision. Three-eights grain of morphia concealed in wine, bismuth and catechu, was then given. Oct. 2. she was suffering less severely. One-fourth grain of morphia given as before. Oct. 9. diarrhœa, menorrhagia, formication, pains in the back and limbs. One-eighth grain of morphia given. The coca treatment had been steadily kept up and no more morphia was given. November 19. the desire for opium had ceased.

ANTAGONISM BETWEEN MEDICINES AND DISEASES.

Prof. Bartholow (Med. Record, Dec. 18, 1880, p. 673) considers strychnia antagonistic to paralysis not dependent on organic changes, especially the reflex forms, which are probably often due to anæmia of the motor centre. Strychnia may also be used in enfeebled heart from degeneration, with low tension of the vascular system.

Woorara in hydrophobia has recorded a few cures.

Chloroform by inhalation, chloral, tobacco or nicotine, bromide of potassium, physostigma and gelseminum are used with success in tetanus and strychnia spasm. They agree in diminishing or suspending the reflex function of the cord.

Picrotoxin and bromide of potassium in epileptiform convulsions ; the former when there is anæmia, the latter, in plethora. Nitrite of amyl aborts the convulsions by dilating the contracted arterioles.

In paroxysmal cough, neurotics or chloral, bromide of potassium and nauseants relieve the spasm. Hiccough may also be relieved by the same and also by a Faradic current.

Nitrite of amyl in angina pectoris.

Pain may be relieved either by interrupting its transmission to the conscious centre, or suspending the function of the centre ; for the former, aconite and gelseminum ; for the latter, anæsthetics. The deep hypodermic injection of chloroform is particularly recommended. The hypodermic injection of morphia and atrophia combined, joins the anodyne qualities of each, and obviates, in great measure, their disadvantages.

Insomnia and mania, treated with chloral. High excitement with illusions and great motor activity, by gelseminum, duboisia, hyoscyamia, conium, etc. Melancholia, by morphia. Acute active cerebral congestion, by aconite, veratrum viride and bromide of potassium; the passive form, by ergot, digitalis, etc. Cerebral anæmia by strychnia, brucia, atropia, quinia.

Cardiac disease : Excessive action with diminished inhibition, as in exophthalmic goitre, treated with Galvanism, digitalis, ergot. Aconite in excess of inhibition. Atropia in paralysis of the accelerator apparatus. Palpitation, treated by bromide of potassium or digitalis, according to cause. Digitalis is used with advanatge in mitral lesions with rapid pulse, low arterial tension and high nervous tension. In aneurism, ergot is used. In hemorrhage we have increased action of the heart and relaxation of the vessel walls ; antagonized by ergot, digitalis, bromide of potassium, veratrum viride, etc.

Respiratory diseases : Strychnia is a respiratory stimulant, as in the acute respiratory failure of acute pulmonary affections, some cases of emphysema and chronic bronchitis, in the reflex nausea and vomiting of phthisis. Possibly also in capillary bronchitis, when if the respiration is depressed, atropia increases it. Atropia stimulates the respiratory centre and lessens the irritability of the sensory nerves of the lungs, and increases the circulation through them.

Internal diseases : Belladonna in serous diarrhœa. Opium in diarrhœa and dysentery. Constipation from muscular paralysis by the Faradic current. When secretion is also deficient, a combination of nuxvomica, belladonna and physostigma.

Skin diseases : Night sweats treated by atropia, duboisia, hyoscyamia. Dry skin, by pilocarpin or picrotoxin. Pilocarpin is a galactagogue ; atropia, a galactifuge.

Kidneys: Medicines which stimulate the skin are antagonistic to those which stimulate the kidneys. In diminished activity of the kidneys, squill and digitalis are used.

Incontinence of urine treated by cantharides. Nocturnal incontinence of urine, according to the cause ; if the sphincter is weak and relaxed, give belladonna or ergot ; if the mucous membrane is intolerant, bromide of potassium and alkalies ; if the muscular coat of the bladder is irritable, gelseminum, chloral or conium.

SCIENTIFIC SUMMARY.*

ANTHROPOLOGICAL NOTES.†

By Dr. H. C. Yarrow, U. S. Army,

Member of Acad. Nat. Science, Phila. ; Philosophical, Biological and Anthropological Societies of Washington ; American Asso'n for the Advancement of Science ; French Asso'n for Adv. of Science ; Corresp. Member of Lyceum of Nat. Hist., New York ; Zoological Soc. of London, etc., etc.

SALUTATORY.

At the request of the editor of the ROCKY MOUNTAIN MEDICAL REVIEW, the undersigned, with considerable diffidence, assumes editorial control of the special section devoted to Anthropology and such subjects relating thereto as may be of probable interest to medical men.

Owing to the limited space at his disposal, it will not be possible to furnish more than mere notes or remarks upon certain papers or articles, but great care will be taken to select for notice from those only which appear to possess the most value to the greatest number of readers. At the same time, items of intelligence with reference to anthropological work in this country and abroad will form a prominent feature of the department.

It is hoped that the labor of compilation will be somewhat lightened by the good-will and interest which may induce the patrons of this journal to send contributions from time to time for this department, and favors in this regard are desired and solicited.

With an earnest determination to do his best in the direction indicated, the writer

* Reports appearing in this department will be written with special adaptability and reference to the requirements of the scientific physician.

† Books, pamphlets or papers on anthropological subjects, may be sent the editor of this department, in the care of the Rocky Mountain Medical Review, Colorado Springs, Colo.

invokes a kind indulgence for any shortcomings.   H. C. YARROW.
Washington, D. C., March 21, 1881.

LIST OF RECENT PAPERS ON ANTHROPOLOG-
ICAL SUBJECTS.

Allen, Grant.—Aesthetic evolution in
man. Mind. xx
Bacon, A. T.—The ruins of the Colorado
Valley. Lippencott's Magazine, Nov.
Baillarger et Krishaber.—Cretin. Dict.
Encycl. d. sc. Med., Par., 1879, I. s., xxiii,
126-146.
Bellen, H. W.—The races of Afghanistan.
Calcutta, Tacker, Spink & Co. Rev. in
August Athenæum.
Benedikt, Moriz.—Weitere methodische
Studien zur Kranio, und Kephalometrie.
Mitth. d. Anthrop. Gesellsch. in Wien.,
1880, ix, 348-371.
Blache, R.—Developpement physique de
l'enfant depries la naissance jusqui au
sevrage. Union Med., Par., 1880, 3. s.,
xxix, 386; 433; 441.
Blake, C. J.—On the occurrence of ex-
ostoses within the external auditory canal in
pre-historic man. Am. J. Otol., N. Y.,
1880, ii, 81-91.
Broca, Paul.—Instructions generales pour
les recherches anthropologiques a faise sur
les vivants. 2d ed., Paris, 1879, G. Masson.
301 pp. 4 pl. 8°.
Chevens.—A plea for the study of archæ-
ology. Med. Times & Gaz., Lond. Sept.
DeCosta, B. F.—Glacial man in America.
Pop. Sci. Month., Nov.
Dunbar, J. B.—The decrease of the North
American Indians. Kansas City Review,
Sep. 1880.
Flower, W. H.—The American Races.
Brit. M. J., Lond., 1880, 1, 549; 577; 616.
Holmes, N.—Geological and Geograph-
ical distribution of the human race. Tr. St.
Louis Acad., iv, 1.
Kant, E.—Anthropology. J. of Specula-
tive Philosophy, July.
Kneeland, S.—Traces of the Mediter-
ranean nations. Proc. Bost. Soc. Nat. Hist.
xx, 11

Lanman, C. R.—Recent publications in
the field of American antiquities. Am. J. of
Philosophy, Oct.
Lewis, H. C.—Antiquity of man in Amer-
ica geologically considered. Science, Oct. 16.
Powell, J. W.—The Wyandottes. Sci-
ence, Sept. 11.
Primavera, G.—Ersendo l'uomo carnivoro
di sua natura, poteva egli col tempo farsi
ounivoro impunemente? e fino a qual
punto? Morgagni, Napoli,1879, xx, 569-666.
Rechus, E.—Studies of primitive peoples.
Inter. Rev., N. Y., 1880, n. s., viii, 457-471.
Zabarowski.—Tertiary man. Kansas City
Rev., Oct.

DEATH OF DR. OTIS.

In the death of Surgeon Geo. A. Otis,U.
S. A. (Curator Army Med. Museum), which
occurred in Washington, Feb. 23d, anthropology has suffered a loss almost irreparable.
It was largely due to the interest manifested
by Dr. Otis in craniology that the present
large collection of crania in the Army
Medical Museum owes its existence, a collection hardly equalled in any museum of
the country. It is peculiarly rich in crania
of North American Indians, and it had long
been a cherished scheme with Dr. Otis to
carefully study it and publish the results of his
investigation as a sequel to the valuable
work of Dr. Thos. George Martin, entitled
"Crania Americana;" in fact, a vast amount
of material had been gone over, and if Congress had seen fit to make an appropriation,
the work could have been finished in a short
time. Within the last twelve months Dr.
Otis had prepared a second edition of his
catalogue of the crania in the Army Med.
Museum, and such was the demand for it
that but few copies now remain for distribution.

Those who have been intimately connected with Dr. Otis, whether by social ties
or official intercourse, will not soon forget
his largeness and generosity of heart,
breadth of mind, and devotion to his favorite science. Death has cut the thread of
life, but imperishable monuments of his
genius remain with us.

INDIAN MEDICAL PRACTICES.

Major Powell, Director of the Bureau of Ethnology, Smithsonian Institute, has in course of preparation a volume relating to the medical practice, of the North American Indians. This will not only embrace the subjects of medicine, surgery, obstetrics, etc., but will also contain accounts of conjuring, shamanism, witchery and demonology, which enter largely into the treatment of the sick. The work will be edited by Dr. G. J. Engleman, H. W. Henshen, and Dr. H. C. Yarrow. To induce contributions, a circular has been prepared, which is to be sent to all interested in the subject, and in the forthcoming annual report of the Bureau attention will be called to it.

ABNORMALITIES IN ANCIENT INDIAN SKEL-
ETONS.

Dr. C. L. Metz, of Madisonville, Ohio, reports as follows regarding his explorations of an ancient cemetery near his town : " In taking up the many skeletons (586 in number) from our ancient cemetery, I have secured many interesting pathological specimens of bone, principally of the bones of the skull, showing extensive injuries with subsequent repair. A tibia, that in my opinion shows unmistakable evidence of syphilitic caries ; an adult male cranium in which complete anchylosis of the atlas to the condyles has occurred, the posterior arch remaining free. One of the most interesting and remarkable of the specimens is that of an adult female dwarf skeleton, in which the spinous and articular processes of all the dorsal and lumbar vertebrae are anchylosed, the bodies remaining free, with the exception of two in the lumbar region, which are united by a thin band of osseous tissue. The lumbar vertebrae are solidly united with the sacrum, and the sacrum with the ilia. Several of the carpal and meta-carpal bones were united into a solid bony mass, and the atlas is anchylosed to the skull. * * * I have also some bones showing fractures and repair, which had either been left to nature or the incantations of the medicine man."

Dr. Metz has been exceptionally fortunate in discovering such interesting remains, as they are by no means common. The writer has exhumed many thousands of Indian skeletons and has seen but three cases of pathological interest, one, a case of caries of the vertebrae, a second, a repaired fracture of the neck of the thigh, and a third, repaired fracture of two ribs.

ANTIQUITY OF MAN.

Professor Mudge has presented some interesting evidence relating to the antiquity of man in the Kansas City Review of Science. He starts by assuming the correctness of the generally-accepted opinion among geologists that man was on the earth at the close of the Glacial epoch, and offers evidence to prove that the antiquity of the race cannot be taken at less than 200,000 years. After the Glacial epoch, geologists have 'recognized, by their effects, three others, namely, the Champlain, the Terrace and the Delta, all supposed to be of nearly equal length. His argument for estimating the duration of these epochs is as follows : He takes the case of the Delta of the Mississippi, and notes the fact that, for a distance of about 200 miles of this deposit, there are to be observed buried forests of large trees, one over the other, with interspaces of sand. Ten distinct forest growths of this nature have been observed, which must have succeeded one another. " These trees are the bald cypress of the Southern States. Some have been observed over 25 feet in diameter, and one contained 5,700 annual rings. In some instances these huge trees have grown over the stumps of others equally large, and such instances occur in all, or nearly all, the ten forest beds." From these facts it is not assuming too much to estimate the antiquity of each of these forest growths at 10,000 years, or 100,000 years for the 10 forests. This estimate would not take into account the interval of of time—which doubtless was considerable—that elapsed between the ending of one forest and the beginning of another. "Such evidence," concludes Professor Mudge, "would be received in any court of law as sound and satisfactory. We do not see how such proof is to be discarded when applied to the antiquity of our race. There is satisfactory evidence that man lived in the Champlain epoch. But the Terrace epoch, or the greater part of it, intervenes between the Champlain and Delta epochs, thus adding to my 100,000 years. If only as much time is given to both these epochs as to the Delta epoch, 200,000 years is the total result."

PECULIARITIES OF RACE.

The report of Dr. Smith, stationed with the troops at Fort Clark, Texas, shows that there is quite a difference in taste between the white and negro troops at that place. The troops are furnished with certain rations. If they do not care for these they can sell them and purchase other groceries and supplies in their stead. The doctor finds that the negroes sold most of their coffee and salt, whence it might be inferred that they did not appreciate these articles. On the other hand, the white troops sold their pork and bacon and pepper. A curious circumstance was the fact that the negro soldier consumed twice as much sugar as a white one ; a still stranger fact, the declaration that the darkies use, on an average, three times as much soap as their white brethren.—*Med. and Surg. Rep., Phila.*

ANTHROPOLOGICAL SOCIETY OF WASHINGTON.

At the regular meeting of the Society, held March 1st, two papers of interest were read, one on the " Amphibian Aborigines of Alaska," by Mr. Ivan Petroff, a Russian gentleman connected with the U. S. Census Bureau, the other by Dr. A. F. A. King, on the " Evolution of Marriage Ceremonies."

A RECENT paper which concerns us somewhat as a military nation, is published by Dr. J. Moursu, a surgeon in the French Navy, in the Arch. de Med., March, 1881. The title is as follows : "Recherches Anthropometriques sur les Apprentis Canoniers." The article is not finished in the number of the volume given, but it appears from M. Moursu's studies that a considerable num-

ber of those men who undergo drill with heavy guns (the exercise being very fatigueing and violent), diminish in stature. The author states that this fact at first appeared to him incredible, but after careful measurements he was forced to believe his conclusions correct. At the same time the expansibility of the thorax increases. Generally the weight at first increases, but in some individuals no changes of nutrition take place. Towards the end of the instruction most of the men are found to have lost all they gained previously. It is at this time that they are most subject to pulmonary disorders, but rest from drill soon improves their condition. It is also stated that at first the violent exercise diminishes the capacity of the lungs. The paper contains lists of elaborate measurements.

It might be worth while to institute a similar series of experimental studies among our miners, with the view of determining what effect their hard labor has upon vital longevity.

PROF. F. W. Putnam, of the Peabody Museum of Archæology, Cambridge, Mass., recently read before the Essex Institute of Salem an interesting paper entitled " The former Indians of southern California, as bearing on the origin of the Red Man in America." After giving an account of the discovery of the Peninsula of California in 1534, he called attention to the facts relating to the antiquity of man on the Pacific coast, and to the importance of the discovery in California of human remains, and of the works of man in the gravel, under beds of volcanic material, where they were associated with the remains of extinct animals, and to the necessity of looking to this early race for much that it seems otherwise impossible to account. He thought that what is called the " Eskimo element," in the physical characters and arts of the southern Californians, was very likely due to the impress from a primitive American stock, which is probably to be found now in its purest continuation in the Innuit. In this connection he dwelt upon the probability of more than one type of man. In following out this argument, he called attention to the distinctive characters in different tribes of Indians on the Pacific coast, and stated his belief that they had resulted from an admixture of the descendants of different stocks.

The Californians of three hundred years ago, he thought, were the result of development by contact of tribe with tribe through an immense period of time, and that the primitive aace of America, which was as likely autochthonous as of Asiatic origin, had stamped its impress on the people of California.

The early men of America he believed were dolichocephali, and the short-headed people he thought were made up of a succession of intrusive tribes in a higher stage of development, which in time overran the greater part of both North and South America, conquering and absorbing the long-headed people, or driving them to the least desirable parts of the continent. He thought that the evidence was conclusive that California had been the meeting ground of several distinct branches of the widely spread Mongoloid stock ; for in no other way could he account for the remarkable commingling of customs, arts and languages, and the formation of the large number of tribes that existed in both Upper and Lower California when first known to the Spaniards.

The speaker then gave a review of the arts of the Californians and the physical characters and customs of the people, showing that, notwithstanding the absence of pottery, the tribes, when first known, had passed through the several stages of savagery and had reached the lower status of barbarism• of the "ethnical periods" given by Morgan.

Mr. Putnam concluded by calling attention to the recent explorations of the coast of Southern California and the adjacent islands, by the expedition under Lieut. Geo. M. Wheeler, of the U. S. engineers, in charge of the survey west of the 100th meridian, and the extended explorations of the Santa Barbara Islands, which had been con-

ducted by the Peabody Museum of Archæology at Cambridge. The results of these explorations, he stated, were now embodied in the seventh volume of the Reports of the Survey under the charge of Lieut. Wheeler, and published, by authority of Congress, under the direction of the chief of engineers, U. S. A.

Mr. A. H. Keene commenced in "Nature," Dec., 1880, a series of papers on the Indo-Chinese and Oceanic Races, their types and affinities. The following is the scheme he proposes to follow in the discussion :

A—DARK TYPES.

I. *Negritos :* Aetas ; Andamanese ; Samangs ; Kalangs ; Karons.

II. *Papmans :* Central branch, Papmans proper. Western branch, Sub-Papmans, west. (Alfuros.) Eastern branch, Sub-Papmans, east. (Melanesians.)

III. *Austral :* Australians, Tasmanians.

B—CAUCASIAN TYPE (fair and brown).

IV. *Continental branch :* Khmer or Cambojan Group.

V. *Oceanic branch :* Indonesian and Sawaion, or Eastern Polynesian Groups.

C—MONGOLIAN TYPE (yellow and olive brown).

VI. *Continental branch :* Indo-Chinese Group.

VII. *Oceanic branch :* Malayan Groups.

ASIATIC CULTURE IN AMERICA.

Prof. John Campbell, of Montreal, in No. VI (Vol. IX) of the Canadian Naturalist, attempts to connect the Basques, of Europe, the Nubians, of Africa, the Circassians, on the border of Europe and Asia, the Koriens, the Japanese and the peninsular people of Asia, the Aleutians, Kadiagmuts, Dakotas, Iroquois, Cherokee-Choctaws, Muyscas, Peruvians and Chilenos of America. The work will repay perusal.

PAWNEE INDIANS.

A very interesting paper on the Pawnee Indians has been published by Mr. John B. Dunbar in the November number of the "Magazine of American History," This describes their food, hunting, war, trade, feasts and medicine. The list of plants used for food and the discussion of their medical practices are good. Prof. Mason, in reviewing this paper in the American Naturalist for March, states as follows, and we reiterate his statement: "It has been asserted in very high quarters that the Indian of this continent had primarily no knowledge of the medicinal properties of herbs, aside from incantations. It might be well for Mr. Dunbar [as well as others] to give this question a little attention."

THE NEW JOURNAL—"SCIENCE."

This periodical, which is rapidly achieving the circulation and position it is entitled to, by reason of merit, is edited by Mr. John Michels. A number of valuable anthropological papers have already apppeared, which we commend to our readers.

SKIN FURROWS OF THE HAND.

Prof. Mason, in the Feb. Naturalist, mentions that new anatomical characters are being brought constantly within the anthrological arena. Only a few months ago the relative length of the ring-finger and forefinger, was added to the list of marks for observation. Mr. Henry Faulds, of Tsukipi Hospital, Tokio, Japan, has commenced in Nature of Oct. 28th, a series of pâpers on the ethnological value of careful observations relating to the finger-marks in ancient pottery, to those of criminals, and of the anthropoid apes.

SELECTIONS FROM JOURNALS.

ORGANIC HEALING POWERS.

A Lecture by Rudolf Virchow.

(Translated from the German by the Marchioness Clara Lanza.)

Andrew Jackson Davis, who is called the "Great Prophet" by his German adherents, thus begins a chapter in his "Harmony"* entitled "The Philosophy of Disease:"

"The improvements and progress which have been made in pathological science, are

not by any means in keeping with its actual value and antiquity." And then he adds the following :

"The age of a science or doctrine has but little to do with its reliability, importance or progress. Indeed, the great maturity of any doctrine is almost a positive proof that it originated in ignorance, superstition and error."

The "Great Prophet," who conceives all his ideas without the aid of study, and who, moreover, by a peculiar direction of his will, turns from the confining influences of the material world in order that he may enter the "highest state," has entirely overlooked the fact that the ancient science which he disdains, proceeded from precisely similar revelations as those which he produces with so much pride.

Welcker, in his magnificent work upon the "Art of Healing Among thé Ancient Greeks,"† has given very impressive descriptions of the Epiphania which occurred more than 2000 years ago in the Temple of Æsculapius, and they now possess a double interest in regard to American Spiritualism, or spiritualism of any kind (if we consider for a moment how philologists a quarter of a century ago investigated the question), as to whether the so-called incubation of the Æsculapians was identical with modern clairvoyance. Those seeking to be cured from disease obtained revelations while sleeping or dreaming in the sanctuary of God. Hence medical literature arose, for the afflicted wrote a description of their curses upon the pillars of the Temple or else upon certain consecrated tablets, and from them the forefather of medicine, Hippocrates, collected in the Temple of Kos those memorable "Predictions" which can be considered one of the principal sources of our scientific knowledge.‡

Did all this spring from "ignorance, superstition and error?" The point perhaps cannot be contested, but it contains, never-

---

* Andrew Jackson Davis, M.D.—Harmonious Philosophy Concerning the Origin and Destiny of Man—His health, disease and recovery. Leipzig, 1873, p.93.

† F. G. Welcker—The Art of Healing Among the Ancient Greeks. Bonn, 1850, p. 95, 112. 151.

‡ Magni Hippocratis Opera Omnia—Edit. Kuhn. Leipzig, 1825, Vol. 1, p. 234.

theless, a large portion of veritable experience, and Hippocrates, notwithstanding
his direct descent from heathens, was a too
critical and (remarkable as it may seem) a
too worldly person not to expose everything
which partook merely of a sacerdotal or
superstitious character.

In his writings and in those of his followers, there is nothing supernatural to be found.
The gods no longer heal the sick. Nature
does it, and nature, moreover, does not act
in accordance with instantaneous inspiration.
On the contrary, it is subject to "divine necessity," or rather we should say to eternal
and also divine laws.

Since the remote period before referred to,
opposition has been openly declared between
science and superstitious therapeutics. The
latter even now has certainly not died out.
The countrymen of the "Great Prophet,"
that is to say, the medicine men among the
North American Indians, still boast of their
immediate intercourse with the Great Spirit,
and perhaps it is the proximity of these people which promotes the increase of spiritualism throughout the United States. One
of the nations of North Asia§ beats a magic
drum, while a certain people in South Africa
blow an enchanted trumpet in order that the
evil spirits of disease may be dispelled.
However, we do not need to go so far for
examples of this kind. In our immediate
neighborhood the traditions of heathenism
rise up secretly and flourish, while superstition concerning mystical healing powers is
capable of continually bringing forth fresh
fruit.

Conjuring, however, during the past century has rapidly declined. I, myself, remember that during my childhood many
people of the middle class where I lived believed in fire conjuring. Even at the present day you will scarcely find one German
city where the worth of a fire brigade is not
undervalued on account of the possible termination a conflagration may have in consequence of conjurations.

In one of those old Greek writings, which,

on account of its age, has been attributed to
Hippocrates, and has for its subject Epilepsy, or the divine disease*, which, at that
time was treated by magic, the author says
that those who conferred divine names upon
diseases were merely magicians, purifiers,
pious beggars, and coxcombs, who gave
themselves the airs of God-fearing individuals, but who, in reality, knew no better how
to conceal their perplexity than by taking
refuge behind the deities.

How many years have passed since then !
The Olympian Gods have been shattered for
ages ; even Christianity has by degrees become an old religion, and yet with it all,
epilepsy has not ceased to be the subject of
conjuration and magic.

Superstition, no matter how degraded,
will always outlive faith. The fathers of the
church belonging to the first Christian century, fought and struggled in vain against
the traditions of heathenism. Chrysostom
said that a Christian had better far endure
sickness and death than have his health restored and his life lengthened by means of
amulets and exorcisms. But the Christians
would not listen to this voice, and in the
end the Church was forced to make amends.
When it erected its places of worship upon
the very ground where formerly were temples and sacrifices, and changed the heathen
festivals into Christian ones, new methods
of supernatural cure were instantly put into
practice. Even the kings by God's favor
did not hesitate to adopt this sort of accomplishment—not only the most Christian
kings of France, but also those of England,
until the first representative of the House of
Hanover mounted the throne, Catholic and
Protestant alike cured scrofula by discourses
and sundry calming influences. At that
time the disease was called "Kings' Evil,"
just as epilepsy was termed the "divine
disease."

Such obtuseness in regard to traditional
superstition may seem astonishing, not to
say alarming. It lies, however, deeply imbedded within the human mind. How long

§ O. Peschel. Knowledge of Nations. Leipzig, 1874, p. 274   *Hippocrates. De Morbi Sacro. Welcker, p. 587..

has the fear of ghosts at night been kept up, while scarcely any one dreads spirits in broad daylight? According to the testimony of Signoria Coronedi, people in Bologna burn daily the combings of their hair, to the end that no witchcraft can be perpetrated upon them, and I remember distinctly that when, as a boy, my hair was cut, the clippings were carefully thrown into the stove.

The inhabitants of some of the Malay Islands fear that a magician will have their lives in his power should he take the remnants of their meals and burn them in a peculiar sort of ashes called *Nahak*. Everywhere we find the same childish tricks performed by men in the lower orders of life that they may create fictitious personalities, endow living or inanimate bodies with imaginary powers and trace out the superior force of spirits in purely natural incidents. This is nowhere to be seen so plainly as in the origin and cure of disease, and if the source of various maladies is referred to enchantment, possession or dispensation, it arises mainly in regard to the cure to be effected.

The reason can easily be comprehended. While we are familiar with the natural causes of maladies, we are still in want of a well organized acquaintance with their natural preceding incidents. By taking an unprejudiced view of the case, we can easily see that even Hippocrates had recourse to nature in curing diseases. Physics he designated the basis upon which the healing incidents rested, and there can be no doubt that this term was the same to him as is to us the tautological epithet of the " physical nature of man." If you read attentively the part in which he mentions this, you can no longer doubt that he had the whole question of man's bodily formation foremost in his mind. Taken in this sense, the healing powers belonging to the body itself must consequently be natural or physical organic forces.

The idea, however, was in a certain measure a prophetic one. Knowledge at that time was not sufficiently extensive to admit of, or to supply any explanation of it. Even

the most favorable and clear sighted observations relating to natural incidents in healing, led to nothing more than a superficial, and to a certain extent, brief conception of the events. This sufficed certainly to establish their situation and also furnished abundant cause for application of remedies at certain times and on particular parts of the body, remedies which seemed adapted to facilitate the natural course of events, to favor it, or in case it remained concealed, to bring it forward.

There have been numerous attempts to explain all this. One school after the other produced its doctrines, but each one of them was based upon imperfect or voluntary suppositions. Each new step of progress in the knowledge of various occurrences which take place in the human organization overthrew the opinion under consideration and produced another. Of course this did not conduce to strengthen faith in regard to scientific medicine.

It was only during the period of spiritual inactivity when nature's perceptions remained for a long time unchanged as in the early portion of the middle ages, and the Church as well as Medicine adopted natural science in its system of teaching, that medical doctrines gained for themselves the recognized character of stability. It was then that the physician attained aristocratic honors. However, secondary schools then arose and dilettanteism pushed forward into existence. So it was at the time of the German revolution, the French revolution and the formation of a new German kingdom.

At no period whatever has mysticism been wanting. A peculiar form of it deserves to be especially mentioned. It is called mystical calculation ; its origin lies buried in the most remote practical teachings. Hippocrates himself, observing a country which up to this day is shunned on account of its malarial influence, has established with minute exactness the duration of the feverish maladies which arose from the marshy districts with peculiar regularity. He not only ascertained the precise duration of the fever, but also the days when a decided crisis would appear. The numbers acquired served to denote when the treatment should be discontinued as the critical days, the 7th, the 11th, etc., designated the proper time for the administration of remedies. In this way the calculating system became celebrated, and as it was made a subject of universal contemplation before the days of Hippocrates by the various philosophical schools, we cannot be surprised that those who succeeded them thought to recognize in the theory more than mere expressions concerning the legitimate relation of things to each other.

During the Middle Ages astrology formed a close alliance with medicine, and the constellations occupied the places of the ancient oracles. But even subsequent ages have repeatedly had recourse to conceptions which nearly approach those of the Pythagoreans. Particularly near the close of the preceding century, discoveries in the departments of electricity and magnetism caused the biological sciences to adopt the theory of polar attraction, a doctrine in which the heterodoxy of animal magnetism, and its companion spiritualism, is firmly rooted. In the Pythagorean philosophy, a two-fold existence was supposed to be at the root of everything, and the circulation of this doctrine has resolved itself, so to speak, into the "Great Prophet" of America, according to whose conception Providence is a moving substance formed of positive and negative proportions, and which acts upon matter in different ways through the agency of the number 7.

Among all these attempts to grasp the phenomena in a determined manner, an effort comes to light which is in every way worthy of recognition. It has been shown that the human intellect has no more a universal and spiritual form which can establish the relation and conception of things, than it has a material one. Calculation produces the definite value by which we are enabled to assign things to their proper places. It is for this reason that intricate natural sciences, physics and chemistry partake

every day of a more mathematical character. The descriptive natural sciences follow timidly in their footsteps, and even physiology and psychology have already been made to travel over the same road. How, then, could medicine escape?

However, the numbers 2, 3, 4, 7 and 10, do not suffice to explain the infinite multiplicity of things, even if the combination of ten numbers serves to account for each calculation. Every reckoning about actual things rests upon observation and not upon inspiration. The more difficult the calculation, the more complex must have been the preceding observation which went to supply the elements of the reckoning. This is true, earnest work, such as no one individual is capable of producing. One workman assists the other, and one generation helps another, not only in transmitting results, but also their aim and object.

It will be a difficult task, nevertheless, for any generation to recognize self-acting forces in numbers. If two objects attract each other it is not owing to the things themselves. And there is no number in existence which possesses healing powers, and no talisman compounded of numbers which possesses active force. The numbers supposed to play an important part in disease only serve to give those versed in art the means by which they may discover the time and duration of the malady and arrange their mode of action accordingly.

But just as Astronomy is incapable of moving the moon or planets by means of numbers, so is the physician unable to produce any effect upon the course of disease or recovery by the same process. Numbers are not remedies, for remedies are actual things, which stand at the disposal of medical art; are actually applicable, and which possess in a certain sense real powers of healing. When we consider them, however, we come to a lengthy and apparently increasing contention which is embodied in medical history in the names of physiologists and technologists. Physiologists are those who seek healing powers within the physical organization itself, while technologists think to recognize them in such means or influences which exist independently of the patient and are directed toward him.

It is true that the physiologist does not altogether despise remedies, but they only serve, in his opinion, to set the organic powers at large. The technologist, on the contrary, intrenches upon the organism. He forces life into artificial conditions. He "orders" and "prescribes" where the physiologist is satisfied with existing circumstances and comes forward as Nature's servant.

Of course a long time has elapsed since the controversy between these two schools was at its height, but in some recent accounts it appears again, not only in specified cases of treatment, but also in a general sense.

Not many years ago blood-letting was a daily occurrence in every hospital, and indeed in almost every private practice. Now it has become so rare that young physicians are scarcely acquainted with it. When I was a young hospital assistant I was frequently forced to perform cupping four or five times in one morning. Singularly enough the change came at a time when we were the least prepared for it, In cases of inflammation of the lungs, where the most audacious blood-letting was considered an almost irrefragable means of restoring the patient, they began in the Universal Hospital at Prague to observe the natural course of the disease without the application of any remedies. They contented themselves with giving the patients plenty of fresh air, good attendance, greater cleanliness than they were in the habit of getting, and strict dietetic surveillance. In the way of medicine, they got nothing, and yet very favorable statistics were obtained. In this way physiology gained a victory over technology, and at the first step reached the highest form—nihilism.*

Since then a certain reconciliation has taken place. A firm conviction arose that

hospital practice could not merely be influential to private practice—that the hospital, with its manifold contrivances, its order and regimen, possessed provisos and remedies which in a private family, even a wealthy one, could only be imperfectly established, or else not at all—and finally, that the nihilism of the hospital physician could not be transmitted to families.

Of course, both physiology and technology will continually enlarge in the future, the more so as experience gains new perceptions and increased power. This, we all know, is inevitable, and the public, which might justly reprove medicine for its scientific changeability, should bear in mind constantly that it is the fate of humanity to be fickle, not only in regard to science, but also every other matter from the State to the Church. We can only hope that changes everywhere will be made with as much honest intention as they generally are in regard to science.

It would, perhaps, be possible to check trivial fluctuations if people could only agree better as to proper healing objects. This is precisely the point over which scientific men find it so difficult to attain to a uniformity of opinion. When a physician is called upon to cure, he has the case before him, represented by the patient—a unity so to speak. And yet the malady itself gives the impression of another unity. It has the appearance of some strange being which has implanted itself in the individual. It has been not improperly termed a parasitic organism, which lives in or upon the system of the patient. Numbers of times it has been asserted that a strange existence has penetrated into the sick man and "possessed" him. All these ideas unite in the practical task of expelling the disease by driving it forcibly from the body. Is it not perfectly evident that a double existence takes the place of the former unity? Can any conclusion be drawn from such premises, except that the "case" must be regarded as dualistic? If the physician has the patient *and* the disease before him; if he is

* Archives of Pathological Anatomy, 189, Vol. II, p. 14.

to separate the one from the other ; if the practical endeavor is to act *against* the disease and *for* the individual, can it be a question of a unitarian conception ? Truthfully speaking, such an idea has never properly existed. Even in cases of sickness which were termed rather figuratively universal, it was always understood that a more or less large portion of health should remain undisturbed. It was this remainder that caused " reaction" according to some schools, and led the battle against strange intruders. Paracelsus, in the Middle Ages, expressed these thoughts in the most worthy manner. Let us take up the point and imagine a defensive battle whose seat of action is the human frame. Who are the combatants ? On one side we have the disease, on the other, the healthy portion. The latter, of course, can go forth with no other weapons of defense and attack than those previously possessed. Where can new ones be found ? The means of resistance must necessarily spring from the physical system itself. Thus far the ideas are simple enough. But if we see that the struggle is carried on according to a military principle, that it has a tendency to cure, and that the means of reaching this end are apparently, purposely and systematically chosen and put into action, what power shall we consider the decisive one ? What is the leading principle, and where are we to look for it ? The generality of physicians say with Hippocrates, it is Nature. But do we not, so to speak, run around in a circle when we first of all call all the legitimate formation of the body nature, and then again have recourse to the same term when we wish to explain how this arrangement resolves itself into a systematic unitarian course of action ? Have we not a substance to deal with in the first case, and a force in the second—and an organized force too, a force with designs and purposes—a species of spirit in fact ? Paracelsus was firmly convinced on this latter point. He designated the decisive power the *Archæus maximus*, which corresponded to *spiritus rector*, or leading spirit.

Georg Ernst Stahl, the celebrated clinical lecturer, in the beginning of the past century, went a step further. He set up the soul itself, the *anima*, as the decisive principle. But at that time the philosophy of the unknown was not yet invented, and it was difficult to demonstrate that the hitherto thinking and conscious soul could here work in an entirely unconscious manner, and yet be systematic withal. It was also extremely hard to trace the diseases of cattle, the *morbi brutorum*, or the maladies of plants to a soul, if we did not wish to run the risk of losing the conception of the term by this extensive generalization.

Toward the close of the past century we became more and more inclined to admit the existance of an organic force secondary to the soul—some called it vitality, others natural healing power. Those inclined to the former opinion endeavored to unite a given relation of the healthy organism with an effort directed upon itself. Those who adhered to the latter idea were firmly convinced that a peculiar regulating force existed.

At all events, the much sought for *unity* was driven further and further into the background by the sudden appearance of these new forces. There was no longer merely a *dyas*, but a *trias*. The disease, the remaining healthy portion of the body, and the particular force which ruled it. And no matter what special term was employed to designate the latter, it always partook of a distinctly spiritual character. Many attempts were made to reduce it to a scientific quality ; to construct it according to a physical dynamic system ; to interpret it as a particular form of electricity or magnetism.

However, as soon as the matter was entered upon seriously, and all the systematic plans and workings investigated, natural science became instantly transformed into a spirit.

Nevertheless, assistance was frequently deemed necessary. The course of the struggle was observed more minutely, and if it was found to be too weakly conducted either by vitality or natural healing force, endeavors were made to strengthen both, or at least to incite them to greater activity. But if the battle was found to be sustained with more force than necessity required, pains were taken to moderate and reduce the action. Thus arose a classification of conditions pertaining to disease—asthenic, sthenic and hypersthenic, names derived from *sthenos*, signifying strength.

It would lead us entirely too far from our course, should we attempt to expound the history of the various healing systems. It may suffice to say that every one of them, to use a common expression, has left its traces behind, and that an acute eye can easily detect them. According to our present ideas all these systems rest upon an erroneous conception of life and disease, inasmuch as they endeavor to attribute a more or less personal significance to each of these terms. The perception thus becomes figurative and typical.

Modern medical science has utterly renounced this tendency to personification, where the pre-supposed force does not correspond with an actual demonstrable body. It further separates simple forms from compound ones, although, according to the mode of observation they may possibly produce the impression of unity. For instance, the human organism appears to be a compound form, although we may correctly apply to it a personal expression. Each particular cell can be interpreted as a personality, for they are all self-existing and self-acting, and their power emanates from their own construction —their *physics*. In this sense the human body is not a unity in the strict *material* meaning of the word, but on the contrary, a plurality, a collective form, and in a certain degree a state. There likewise exists no one force which rules it and establishes its action, but on the contrary, a co-operation of many forces which are inseparable from the living element. Even the greatest penomenon in human life, the spiritual I, is therefore no steady, immovable capacity, but a very changeable one.

If the human organic structure appears to us a unity, it is chiefly due to three circumstances : First, in the construction of the vascular system and in the blood circulating through it, there is another perfectly accorded system which pervades the entire body, effects the material intercourse of the various substances, and constitutes a certain dependence of the parts upon the blood. For a long time, therefore, people looked for the source of life merely in the blood, and endeavored to explain all the incidents pertaining to disease and cure by means of the blood alone. When it appeared to be impure it was refined with inappropriate substances. When there was apparently too much or too little, it was drawn off, or attempts were made to produce it. In the second place, in the formation of the nervous system to which man's highest powers are attached, namely, the intellectual, we find an organization extending throughout the entire body, converging to the brain and spinal marrow, and which on one side is qualified to adopt outward impressions and conduct them to the great centre, while on the other side it possesses the capacity to eject any impulse directed upon other portions of the body by causing them to make particular assertions of activity or else to limit them.

Diseases such as fever, for instance, can only become intelligible by referring the great number of collected phenomena which come under this category, to the nervous system.* What wonder, then, that there is continually a fresh attempt to explain disease and cure by means of the nervous system ?

But there is still a third point. This is the enormous mass of tissues of which the body is built up. The compound construction of countless numbers of cellular elements which are organized in the most varied manner, and are capable of producing the greatest diversity of results. Many of them, such as the muscles, appear in a high

*Virchow. Fever. Four Discourses upon Life and Disease. Berlin, 1862, p. 129.

degree to be simple bearers of strength. The blood would be an immovable mass if the muscles of the heart and vessels did not circulate it mechanically. Other tissue formations, as the glands, superintend various things, the act of secretion, for instance, which represents a no less declaration of force. But each of these regulations, every one of these so-called organs, is again a plurality compounded from endless elementary organisms, the cells. And when we see that the nervous system is just as complex, that the vessels, the heart and the blood, are likewise compound combinations, it is well proved that every observation which does not apply to a compound element must be external and superficial.

If such a conception upon first sight results in a detachment of the body, a total breaking up of the perception, a further contemplation will show that these innumerable elements do not exist in juxtaposition. Accidentally or indifferently, they belong to each other on account of their common descent from a simple element which insures a certain original resemblance and relation among themselves, just as there is among the descendents of one family.

This is the "divine necessity" of Hippocrates in its modern form. It does not merely assume the material of all elements to be one organism, but it also concludes that it must form certain combinations by means of which the effect of the different elements through each other produces a legitimate arrangement of the general principles.

Such organizations undoubtedly occur in the vascular and nervous systems, and they exist also in the great masses of superfluous tissues. For even as the vessels and nerves influence these latter, so on their side they influence them. Thus arises a reciprocity of effect which can be beneficial or otherwise, according to circumstances.

As long as the effect is beneficial, so long will the organization appear to be in harmony. And we can experience it in our consciousness as a sensation of well-being.

If the effect should be injurious on the contrary, we say disease has entered the system, and we experience a feeling of discomfort. These sensations do not relate solely to bodily conditions, but to those of the mind, also. There is moral as well as physical indisposition.

In a figurative sense, we might say *equilibrium* instead of harmony, and *loss of balance* instead of discord. In many cases such designations would have an actual significance. The distribution of the blood is arranged to a certain extent, according to simple hydro-dynamic principles. An increase in one part necessitates a decrease in another. The electricity existing in the nerves can be interpreted in a purely physical sense. Here are tensions and accumulations; there, evacuations and discharges of electricity. Even the usual incidents pertaining to the growth of the tissues provide us with numerous examples. If one part increases in strength, another diminishes. A suitable instance of these antagonistic phenomena is given in the difference of incidents pertaining to growth between the male and female sexes.

From these remarks we already see that any disturbance of the harmony or equilibrium does not merely affect the common sensations, and therefore the nervous system, but also other parts of the body, and it can be readily understood that one disturbance will act upon this portion, and another upon that, etc.

All the parts do not stand in equal relation to each other, and those whose mutual dependence is the closest, will, of course, be the soonest affected, while the others will be influenced in a lesser degree or else not at all. We designate the closer relationship as *sympathy*.

All these connections exist uniformly in sound, healthy bodies, and in order to explain them, we have no need to refer to the soul, vitality, or any other special spiritual force. When a diseased disturbance of the equilibrium occurs, they represent what we call *organic healing power*.

In order to obtain a full comprehension of this it is not actually necessary to say much concerning the healing itself. The theoretical discussions which have taken place in regard to this point, and the practical inferences derived from them, have often become very much confused, inasmuch as entirely opposite relations have been drawn together by means of them.

The old word medicine, which is almost synonymous with our modern term therapeutics, led to the misunderstanding that the entire practical energy of the physician should be directed to one particular point of the bodily condition, inasmuch as his chief task is to cure. A closer reflection will show, nevertheless, that this is by no means the case.

Only a certain portion of medical power, although it may be the greater part, has reference to the curing of disease. Important branches of medicine allude to circumstances of sound health supervised by the physician in order to prevent disease. Every year our activity in this respect increases.

Besides the removal of the various causes of disease, there is another cure which we designate as the *curatio causalis*. A foreign body such as a bullet, a glass splinter, etc., penetrates into the organism and remains there. Frequently, if not always, the removal of this body is the proviso of a cure. This of itself, however, is not sufficient, for the cohesion through which the foreign body passed must first be united, and the natural connection re-established, before the actual restoration can be acknowledged.

Very often restoration is spoken of when the case in question consists merely of a disturbance or a simple deficiency. If a person breaks his leg he is not ill. He cannot walk, of course, and an actual malady can proceed from the fracture if the surrounding parts become inflamed and the nerves excited. But the fracture itself is no illness, although it may become the cause of one. In spite of this, however, the sufferer always hopes to be "cured" by the physician.

Now it is unquestionably true that the same principle of observation cannot be applied to all such cases, otherwise we should become hopelessly embarrassed. A broken knee will never set itself; therefore the physician is not to rely at all upon nature, but simply upon his own skill; but he does not occupy himself with the phenomena by means of which the fracture will be re-united. That happens by itself. The medical influence in question is certainly technological. It is by means of force that the physician brings the pieces together in a position which as nearly as possible corresponds to the natural one. It is by means of force that he holds them thus. But all that is not a cure, but merely the stipulation for one. The broken part finally grows together in a very bad shape, and the re-establishment of the connecting portions occurs only with a very unfavorable position of the fracture. Nature in this case works most powerfully.

Every restoration of a broken bone is also physiological, and the physician only endeavors to let it occur undisturbed and under the most propitious circumstances. This "only" is of very great importance to the patient, for a fractured bone which heals crosswise or crookedly can infringe upon the use of a limb for life. But when we come to investigate all the theories of healing we must remain firm in stating that recovery from fracture is not caused by the physician. *The cause of the cure is due to the surrounding tissues.* They produce a new tissue, which forms over the scar.

We now come to *actual diseases.* They are not mere disturbances or yet definite conditions. An actual disease is an incident, also a succession of conditions, one proceeding from the other and affecting vital parts. No lifeless object, no dead body ever becomes subject to disease. An animal or a plant can become diseased, but only while they are alive and only in such parts as are endowed with life. Therefore, every disease is a demolition to sound health, for the same part cannot at once be sick and well. Disease is also an incident pertaining to life.

We call those incidents disease which deviate from the typical form of life and which are at the same time affected by the danger to which they are exposed, for disease strives towards death, be it local or general, and, consequently, it struggles against health. If disease is incidental to life, it must be allied to certain living portions. Therefore we say the disease is "seated," and it is frequently one of the physician's most difficult tasks to discover precisely where this seat may be. But I must correct myself. In many cases the disease is located in several places. If a person has inflammation of the lungs, he usually has a violent fever in addition. In this case the inflammation is situated in the lungs and the fever in the centre of the nervous system—two entirely different places. Is all this one disease? Even at the beginning of the present century inflammation of the lungs was put under the category of fevers. Now it is considered as local inflammation. Still, it is the fever principally that is treated, while the inflammation is left to Nature. I will not enter into the fact that among many people who suffer from inflammation of the lungs, the stomach and kidneys also become diseased. What I have already said will suffice to show that the mere investigation made to discover the location of the disease leads us from the idea that it can be a unity. Unity only exists in so-called imaginary maladies. It is entirely figurative, a simple fancy, an abstract. In reality, most diseases are distinct pluralities, some existing in which the number of locations is countless.

It remains further to be said that in reference to diseases the word "cure" has many significations. If the term in plain language means wholeness without injury, it should designate the entire and complete re-establishment of the condition. Such an interpretation as this speaks badly for technology. If one has a tumor on the knee and the leg is amputated, curing denotes none the less a complete re-establishment. But it does not always agree with physiology either.

There is scarcely a single form of inflammation of the kidneys which admits of complete restoration ; hardly one example of inflammation of the brain which does not always leave certain defects. These diseases therefore are cured but imperfectly, and yet we may say the patients are quite restored, because, in spite of the deficiencies, new relations and connections take place in the body which cause the equilibrium of the actions performed.

As an example of the most perfect cure that we know of, I might mention inflammation of the lungs. Although it happens that in the course of a few days five, eight, or even ten pounds of matter are deposited in the lungs through which the air inhaled should penetrate, we see, nevertheless, that again within a short time the entire mass is loosened and gradually disappears. This is the consequence of mere natural circumstances. But it requires only trivial aggravations, insignificant want of foresight, slight renewal of deteriorating causes, to interrupt this natural incident ; then no relief can occur. On the contrary, the masses of matter remain firm like dead material ; they break in pieces ; the tissue surrounding them becomes impaired, and thus the first step is taken toward that insidious occurrence called consumption. Therefore, the timely advice of a careful physician is very important even if he does not cure, and consequently no one should confidently imagine that all can be satisfactorily arranged independently of him.

Every incident of disease arises either from a defective nutrition or formation, or else from some disturbance of the local actions. A compound disease frequently includes all these reasons at once. Defects of nutrition and formation are generally classed under the category of *organic imperfections*, because in both cases local alterations take place in the organism. For this reason the equalization of the disturbances occurs generally very slowly. The defects can only be removed gradually, and the normal condition established by degrees. Functional imperfections on the other hand can often be removed in a moment, because the inward construction does not change and the local action is altered merely by unusual excitation or oppression. The more the disease is confined to functional blemishes, the quicker it can be removed.

In any case whatsoever, the cure is obtained by complete restoration of the bodily harmony. It consists of a balancing and regulation of the disturbed relations, and indeed, an equalization through inward bodily resources. The healing powers are situated in the vital portions of the organism. These parts nourish themselves, and produce adequate conditions. They bring forth actions which serve to direct, relieve, and repair certain defects of the equilibrium. Even when the physician's utmost power is exerted, when the part in question is cut off or destroyed, then, also, restoration of the bodily equilibrium is necessary before any tolerable result can be produced. Also, when the healing powers remove certain imperfections, when an acid is neutralized by an alkali, or when a dormant faculty is roused into fresh activity by any excitation, the cure can only be perfect if the natural relations return again, or else if new ones are formed. Every outward effect is only a means by which to lead the inward formation of the body to free and regular action.

No physician can trust wholly to nature, but neither can he produce by art that which takes place naturally in the body. That is the work of the organic healing powers. Every medical man must rely upon their efficiency, but at the same time he has no right to sit idle with his hands in his lap in consequence, On the contrary, it is frequently necessary to employ the most forcible interference in order to regulate the action properly. In particular diseases, how much nature is able to perform, and how much the physician is compelled to do, can only be ascertained by personal experience, and can be determined *a priori* by no theory. On the other hand, how far, in certain cases, medical treatment must extend, and how far the natural course is to be

influenced by the physician, is not merely a question of experience, but frequently one of scientific value, which only an educated and cultured physician is capable of undertaking. Experience alone, in the medical world, produces only adventurers, who perhaps may succeed now and then, but for whom self-reliance is always a risk. Such experience as is led and regulated by science alone is capable of removing all barriers, and able to designate the realm in which nature and the physical organic forces have the supreme command.—*Science.*

CEREBRAL LOCALIZATION; OR, THE NEW PHRENOLOGY.*

By Henry De Varigny.

" When the one who listens does not comprehend, and the one who speaks understands as little, you have metaphysics," says Voltaire. Taking this as a true definition, we may say that there has been, and yet remains, much metaphysics in the treatment of the functions of the brain. But the difficulties in cerebral physiology are great. There is divergence of hypotheses, the facts themselves are not settled, and contradictions abound. The foundations of the science are yet deficient. But it does not follow that experiments are useless. For half a century important researches have been carried on ; and more recently facts have been discovered to which we would here draw attention.

From an anatomical point of view the brain is composed of two symmetrical halves, right and left, united by a voluminous commissure, which probably puts into communication the homologous parts of the two hemispheres. Each hemisphere consists of a central mass, with its envelope of convolutions. The central mass, partially separated from its outer covering by the lateral ventricles, is composed of two round bodies, formed of gray nerve-cells—the active part of the nervous system· The office of these rounded ganglia seems to be to

---

* Translated and abridged from the " Revue des Deux Mondes," by Miss E. A. Youmans.

strengthen the impressions that come from without, or from stimulated parts of the brain itself, and they may take part in automatic actions. They are in relation, on the one hand, with the spinal cord, and perhaps, more or less directly, with most of the motor and sensitive fibers of the body ; on the other hand, they are connected by fibers with the gray matter which is spread out in layers over the convolutions of the brain. In other words, the nerve-cells forming the periphery of the convolutions give out white fibers, which penetrate the central ganglia, probably connecting themselves with their cells. From these cells other fibers proceed toward the cord and extremities of the body. The central masses are on the line of the cerebral fibers between their origin in the gray cells of the convolutions and their termination in the cord and body generally.

This anatomical arrangement seems to indicate that the peripheral substance, and not the central mass, is the point of motor stimulation and the seat of sensitive impressions. It is everywhere admitted that the brain is the organ of thought and will ; but for a long time it was believed that the central mass was the important portion, and the convolutions were disregarded. Hippocrates thought they were only a gland. Malpighi and Vieussens thought the same. Ruysch, from their vascularity, considered them as a simple sanguinous network, and Boerhaave and Haller adopted this conclusion. Vic d'Azyr was the first to examine their structure ; then came Baillarger, Ehrenberg, Purkinje, Meynert, Luys, Betz, and Charcot, who revealed their precise anatomy. As to their physiology, Gall taught that intelligence is a function of the convolutions ; Desmoulins added that the degree of intelligence is in proportion to their number and depth ; and Broca, taking the ideas and facts of Dax and Bouillaud, announced the first discovered localization—that of articulate language in the third left frontal convolution.

In 1870 two German scientists, Fritsch and Hitzig, passed a current of electricity across the head and behind the ears of a living subject, which caused movements of the eyes. They referred these movements to stimulation of the gray matter of the convolutions, and set about the verification of their hypothesis. From their experiments they drew the three following fundamental propositions : 1. There are in the head convolutions that may be excited by electricity, and this excitation is followed by the production of determinate movements depending upon the point excited ; other parts being excited without producing movements. 2. The points, which under stimulation induce action in such and such muscular groups, occupy a very limited portion of the cerebral surface. 3. The extirpation of such a point of the surface, known to be a center of distinct movements, paralyzes these movements.

The theory of cerebral localization assumes as proved that there is in the brain a peripheral portion devoted to the production of movements—a motor region ; and another where stimulation is not followed by movements—a non-motor region. It further assumes that the motor region may be subdivided into a certain number of small tracts, definitely circumscribed, each of which presides over the movements of a certain muscular group and this group alone. Ferrier, starting with these conclusions, proceeded with the research. and seems to have established—that the convolutions, in man asw ell as in the lower animals, may be separated into three regions : the anterior, devoted to intellectual functions ; the middle portion, charged with the motor innervation of the body ; and a sensitive region where are received the impressions made upon the sense-organs by the external world. To show how Ferrier reaches and justifies these conclusions is the object of this article.

Two methods of investigation are open to the physiologist—the experimental and the clinical. In the order of time, the first and the most practicable is the experimental method, which consists in observing the brain through openings in the skull and exciting the convolutions or removing them, according to the end proposed. This method is applicable only to animals ; and the monkey, from the likeness of his brain to that of man, is best suited to these investigations. But experiments upon other animals are very useful. They show the homologous regions in different cerebral types ; and the inequality in relative importance of the central masses, the region of automatic functions, and the convolutions, the region of the will. The great advantage of this method is that the operator can repeat the experiments indefinitely upon all sorts of animals, varying the conditions, and choosing his own time and place. Lesions can also be better circumscribed, and the autopsy can be made at will. On the other hand, animals can not tell their sensations, and we have to judge as best we can as to affections of the intellectual and sensational regions. And the preliminary operations may cause general symptoms that mask the phenomena to be studied. Still, this method is in merited favor, and it will yet yield answers to many questions if we know how to put them.

All the processes of experimenters may be reduced to two categories, according to their effects upon the functional activity of the convolutions. These are irritating lesions and paralyzing lesions. Irritating lesions provoke spasms when applied to the motor region, subjective sensations when applied to the sensitive region, and delirium when they effect the intellectual region. Paralyzing lesions in the motor region paralyze the normal action, and in the sensitive and intellectual regions bring on anæsthesia and mental depression. But this division applies only to immediate results, for it is not rare to see a lesion, whether experimental or in clinic, provoke symptoms of irritation in the beginning, followed by paralytic symptoms, and reciprocally.

Experiments that produce prompt paralyzing effects are more numerous than those which excite the functions of the convolutions. But the most preferable one, and that adopted by Ferrier, is the method of limited ablation. By it lesions can be more

circumscribed, it affords a sure means of control, and hence it has come largely into use. Electricity is the only process for exciting the functions to greater activity ; and since it has been much opposed, and is the sole support of the theory of cerebral localization, we must defend its legitimacy.

It may be remarked that the convolutions of the brain are in the form of long, round, gray swellings, separated by furrows. Each of these swellings has a direction, relations, and a situation peculiar to itself, and identical for all animals of the same species ; and the principal ones have their homologues in all animals. They have also their special names. When the excitant is moved about over the different convolutions, the existence of the motor zone is made manifest by the successive movements of the animal ; the sensitive zone is known by signs of sensation, while the indications of an intellectual zone are not thus positive. Observe what passes when we irritate the so-called motor region. We exite a certain point of the convolutions, and a movement is produced. Groping about with our electrodes, we find the effect limited to a small tract where at all the points the same movement is provoked, and that alone. Next to this zone, we can in the same way find the limits of others presiding over other movements. The influence appears greatest at the center of the zone : exciting the periphery sometimes produces a light supplementary movement belonging to a neighboring zone. The special centers or zones being very near together, our means of electrizing them are not perfect enough to prevent a slight diffusion of the current into neighboring centers and faintly exciting them. To get precise separate effects we must excite the center of the special zone.

As the same movement invariably follows the stimulation of a special center (the name of a special zone), we conclude that the center in question has charge of this movement, and, as a great number of these centers, placed side by side, preside over most of the movements of the body, we infer the existence of a motor region of the brain, the seat of voluntary control of the body. But, since, as the galvanometer shows, the electricity is diffused over neighboring centers, what right have we, it may be asked, to assert that the movement observed, when a certain center is excited, is due to that excitation, when neighboring centers are also excited? We reply that to these facts we can oppose other facts. On removing the electrical conductors scarcely a centimetre, we excite perfectly definite unlike movements, although the current is diffused as before. Why are not the movements last produced the same as those observed at the first position? Why such distinct localization in spite of diffusion? Although the occurrence of diffusion is proved by means of sensitive apparatus, it is physiologically insufficient. Besides, it can be prevented by proper precaution.

Again, it is asked if there may not be such a thing as diffusion beneath the surface? This is a grave question, for, under the motor region, there is a large ganglion containing motor fibers going to the muscles. If the current diffuses downward as far as this ganglion, we can not assume that the convolution alone is excited. There are three answers to this objection: 1. The excitation of the convolutions nearest to this ganglion gives the least results; sometimes no movement at all. Unless we admit that the effects are directly as the resistance, which is absurd, downward conduction can not be affirmed. 2. Braun has demonstrated that a section of the white fibers below the excited points, interrupting the physiological continuity, does not arrest electrical conduction sufficiently to prevent movements. 3. Ferrier has shown that the direct excitation of this ganglion produces a general muscular contraction of the other side of the body, and not special isolated movements. So that we again conclude that, though electrical diffusion toward the corpus striatum (the name of the ganglion in question) may exist, it is physiologically insufficent.

But, it may still be asked if electrical diffusion does not excite the white fibers inter-

posed between these convolutions and the ganglion beneath. This objection borrows a character of probability from the slight thickness of the layer of gray matter of the convolutions, and also from the alleged unexcitability of the gray substance, which has been proved in the case of the spinal cord, but conclusive proof in regard to the brain is yet wanting. Many physiologists claim that these cells are only excitable by the will. For the theory of motor-centers they substitute that of psycho-motors. Reserving this discussion for another time and place, we may say that, whether we excite the cells or the fibers that arise from them, the result is the same.

All stimulation, whatever its nature or origin, acts upon a nerve according to its functions. Excite a motor nerve at any point of its course, and you produce movement; excite a sensory nerve, and the subject will feel a sensation which will very with the nature of the nerve. Compress the eyeball, you excite the optic nerve and get the sensation of light; auditive nerves give sensations of sound, and so for all the nerves of special or general sensibility. If, then, in the brain we excite the motor region, the origin of the motor nerves is irritated, and we get movements; excite the sensitive region, and in place of movement we have sensation owing to the connection of these nerves with sensitive cells. So that the electrization of the gray matter of the convolutions acts in the same way as the electrization of a nerve on any point of its track; the only difference is that in one case we excite them at a point near their origin. From the anatomical relations that exist between the white fibers and gray cells, we infer that the cells play the role of center to the nerves.

We come now to the details of Ferrier's experiments. Ferrier operated chiefly on monkeys, because in them the will is in the ascendant, while in lower animals automatism preponderates. As the result of his experiments, he affirms the existence of three zones—the intellectual, motor, and sensitive—into which the surface of the brain can be divided. The one best known, and to which least objection has been made, is the motor zone. In passing the electrodes over its surface, we soon find the little, well-defined centers that preside over particular groups of muscles. Under the microscope there is seen at these points a limited mass of large cells, called *nests* by Betz. The functions of these centers are best shown when they are electrized at the central point, the current being then less likely to spread, and so produce more complicated movements that mask the true function of the center under experiment. Here electricity provokes movements of the leg on the opposite side, owing to the cross-action of the hemispheres. It moves as if to go forward; or the movement may be limited to the foot, or even the great-toe. Sometimes the motions are still more complex, and involve many muscles, as if the animal would scratch its breast or press against it some object which it had taken from the ground. Again, it is the arm, forearm, or hand that moves in various ways, and to different ends; sometimes there is combination of movements, like those required for swimming or prehension; the fingers may lock together with force, as if to retain an object, or extend themselves with a lively movement, as if to scatter something. It is probable that, if our instruments were more perfect, we should find in each center a number of subordinate centers for the execution of single movements, or the moving of a single muscle.

The general idea of the relations of the brain and spinal cord is that the brain commands and the cord obeys. The brain requires such or such a movement, and the cord, working unconsciously, coordinates the elementary and individual movements required to produce the desired effect. Electrizing the centers is the same as issing an order to the cord, not of directly exciting movement.

Monkeys being difficult to obtain in our climate, to satisfy the constant needs of experiment, Ferrier operated upon dogs, jackals, Indian pigs, rats, pigeons, frogs, fishes,

anything that was going. These experiments fully confirmed the results otherwise obtained, and showed, besides, that the action of the hemispheres is of less importance as we descend in the animal scale, while automatism rises.

Ablation of a limited portion of gray matter of the brain leads to the same conclusions as electrization. It produces paralysis, and monkeys rarely or imperfectly recover from these lesions, while inferior animals, as the rabbit and Indian pig, recover; thus showing that the voluntary centers are more important in monkeys than in lower animals.

Such are the facts on which we base the existance of a motor region in the brain. Putting aside all questions of interpretation, it is undeniable that there is a region in the brain where stimulation produces movements varying with the zone or center excited.

Behind this region we find the sensitive centers which receive impressions made upon the sensory nerves, and form perceptions from them. As the terminations of the nerves of sensation are specialized, in exciting the centers we produce subjective sensations, like those caused by cerebral maladies, in which the intellect refers to the outward world the origin of sensations, of which the cause is in the brain; and, whether voluntarily or by reflex action, the animal operated on by electricity shows by evident signs the nature of the sensations he experiences. By means of counter-verification we determine the existence and topography of a certain number of centers which Ferrier has studied with great care.

Take, for example, the visual center: its stimulation caused disordered reflex actions, indicating unpleasant visual perceptions. This alone is insufficient proof; but we can control the region in question by ablation, which brings on unilateral or bilateral blindness, according as we operate on one or both of the visual centers. Stimulation of the auditive centers, in the same part of the brain, provokes movements of the ears, eyes, and head, showing astonishment or terror, just like those caused by a violent and unexpected noise. Ablation causes deafness; the animal remains indifferent to all sounds. Excite the centers of touch, and the signs indicate disagreeable or painful tactile impressions. Ablation brings on complete anæsthesia of the same parts; you may prick, cut, and bruise the animal, and he remains insensible.

Some experiments seemed to indicate the existence of centers of taste and odor, but it is difficult to trace their limits. They are intermingled like the gustative and olfactive sensations. Electricity causes movements indicating unpleasant tastes and odors, and extirpation of the parts ends these sensations. The animal will respire odors, or taste savors that in the normal state would make him fly about the laboratory, and it all passes unperceived. Still more hypothetical are the centers of the organic needs of hunger and thirst, and more yet those of sex, but Ferrier's arguments are strong in favor of their existence.

The presence of a third, or intellectual region, is proved, as far as it can be, by experiments on the lower animals. It is difficult to understand the mental action of a dog, Indian pig, or even of a monkey. Ferrier observed numerous facts tending to establish the intellectual function of the anterior region of the brain. Electricity could hardly be employed in these researches; but ablations, when performed with caution, brought on notable changes in the habits of animals. The monkeys chosen by Ferrier were remarkable for their vivacity and intelligence, prying about right and left, and observing everything. After the operation they became stupid and apathetic. But these indications are not convincing. The clinic alone can decide whether physiology sustains the doctrine of cerebral localization.

We know in what the clinical method consists. Applied most often to man, it amounts to this: to observe the symptoms of cerebral disease, and at the autopsy to connect the lesions, discovered by the naked eye or the microscope, with the symptoms,

as cause and effect. It is true that in cerebral pathology there is great difficulty in separating the essential from the accidental, and distinguishing cause and effect among a plurality of causes. Besides, it frequently happens in cases of cerebral disease that at the autopsy no appreciable lesions can be found. The question is still further complicated by the solidarity that binds together the different parts of the cerebro-spinal system, and which makes it probable that a simple local trouble will produce general functional perturbation. The brain is like a complex machine, in which, if a screw loosens, or a nut gives way, or a rod bends or breaks, at once all goes wrong. It is not that the screw, nut, or rod in question, is the immediate cause of the movements of the machine, but that the failure of these accessories may, for the moment, produce accidents as grave as would be caused by disturbance in much more important parts. Again, cerebral lesions tend to spread and become general. And yet, we have to accept the lesions caused by disease, for we can not produce them at will.

With these reservations, the clinical method is still of the first importance. By means of it we verify in man the hypotheses of experiment, and assure ourselves of the existence of the intellectual and sensitive regions of the brain. Neither medicine nor physiolog opposes the use of the clinical method in cerebral localization. But only circumscribed lesions that have little or no tendency to become general, or to act at a distance by compressing the brain, or otherwise, can come to the aid of our theory. When there is a lesion of the cortical region of the brain which fulfills these conditions, the resulting symptoms may be of two orders— either stimulative or paralytic of the true function. These are the two opposed symptoms that we produce experimentally by electrization and ablation of the substance of the convolutions. It goes without saying that the symptoms vary with the locality of the lesion; the intellectual region gives delirium; the motor region, spasms; the sensitive region, subjective sensations. The symptoms of functional paralysis are also diversely represented by mental feebleness, motor paralysis, and anæsthesia limited to one sense. A lesion frequently presents both orders of symptoms, which succeed each other, or alternate, according to its nature. This fact is as important as the division of the symptoms into two great classes. We will now consider the facts in the same order as before.

The middle region of the superior face of the brain appears to be the motor region. In fact, limited lesions of this region bring on marked troubles in the motor innervation of the body, such as monoplegia, or limited paralysis, or equally limited spasms. Putting aside those cases where the lesions cause general trouble, and regarding those where the symptoms are limited, we come at a constant relation between certain lesions and certain troubles. In ocular monoplegia, the eye can not be controlled by the will. Brachial and crural monoplegia are more frequent; sometimes a single member, arm or leg, sometimes both; but successively, because of the extension of the lesion to both centers, which are near together. In this case the lesion advances slowly and invasively; at the autopsy we can often appreciate the differences of age of the extreme points of the diseased spot. Not far from the brachial and crural centers is the center that presides over the muscles of the face. As before, this center may be affected alone or simultaneously with the other. It depends upon the nature of the lesion whether the invasion is sudden, or slow and progressive, ending in feebleness rather than in paralysis. The proximity of the brachial and facial centers explains their apparent solidarity in the normal state, as shown by the grimaces that often accompany vigorous use of the arms. It seems as if the second center were stimulated by the activity of the first. We close our enumeration of the centers which the clinic has proved to exist by reference to the center of articulate language, discovered long ago by Broca, which attends to

the co-ordination of the phonetic movements.

We have seen that lesions of the motor region of the brain may be manifested by spasms as well as by paralysis. These monospasms have been long known, but it was Hughlings Jackson who first attributed them to lesions of the motor region of the brain. Prior to this they had been described by Bravais, but he did not seek for their origin or signification. They are localized convulsions, or partial epilepsy, and Hughlings Jackson thinks they are due to nervous tension. Any new excitment added to those already stored up will produce discharge or spasm. Like monoplegia, they may be limited to an arm or leg, or even the face; but these phenomena are seldom noticed, and we have few observations upon them. When the spasm involves several parts of the body, it always begins at the same point and follows the same order. Dr. Maragliano has made a very interesting study of partial epilepsy, and explained its causes and signification. Both monospasms and monoplegia indicate the same localization of power.

The sensitive region is found, by experiment, behind the motor centers. While limited lesions of this region often manifest themselves externally as circumscribed anæsthesia, it sometimes happens that they remain latent when they are seated on only one hemisphere. There is no sign of pathological perturbation, and in this case we seem forced to admit functional substitution, or the possibility of the regular action of two sensitive, homologous regions, notwithstanding the absence of one of the two corresponding cerebral hemispheres. What does this signify? Must we abandon the doctrine of localization as regards the sensitive centers? A single center suffice for the two parts of the body? This anomaly is probably due to insufficient observation. Disease of the cerebral centers may give no further symptom than enfeebled sensibility, which might pass unperceived. Lesions of the motor region often result not in total paralysis, but in slight paralysis—a feebleness and not an abolition of the functions. But there are cases where lesions of the sensitive region are accompanied by less equivocal symptoms, and it is from these that we affirm localization.

Symptoms may be of two orders, according to the nature and the phase of the disease ; symptoms of excitation, which produce subjective sensations responding to nothing external, and symptoms of anæsthesia manifested by an abolition of the perceptions belonging to the affected part. These two classes of symptoms may also alternate in the same disease, as in the motor and intellectual regions.

A characteristic case, confirming this theory, is that of a child who fell on his head and buried a portion of the parietal in the surface of the brain. He became blind in the eye of the opposite side. He was trepanned, and the blindness ceased immediately. Soon inflammation arose at the wounded part ; blindness returned and lasted till the inflammation disappeared. Compression of the visual center, first by the bone and then by the products of inflammation, was the evident cause of this intermittent blindness. Other cases establish the possibility of abnormal stimulation of the visual center.

The centers of hearing, taste, odor, and touch, are localized in the same way, and, if the observations are not very numerous, they make a strong presumption in favor of the localization actually adopted. It sometimes happens that several centers are affected at the same time. In these cases, if the lesion is an irritating one, there occurs from time to time a simultaneous discharge producing a singular mixture of sensations. One such patient, observed by Ferrier, said that he had the sensation of a horrible odor and green thunder. We admit that the clinical arguments in favor of the localization of the sensitive centers are not so numerous or conclusive as could be wished. But only lately have they been sought, and each day brings its contribution, which, considering the rarity of limited lesions of the brain, can not be very considerable.

The localization of intelligence in the frontal region of the brain was thought of long before our day. Gratiolet used the expression *frontal races* for intelligent races ; and those of least intelligence have been called *occipital races*. The frontal region is greatest in man along with the predominance of reason and logic, while in women, who are dominated by their sensibilities, the occipital region prevails. We may cite to the same point the researches of Bordier on the skulls of assassins, of Luys on the brains of fools and idiots, of Benedikt on the brains of criminals, of Lombroso on the characters of habitual criminals. Their conclusions are analogous, and favor more than they oppose the popular idea.

But these arguments are not precise and and positive. Happily there are others more scientific and more conclusive. Take the celebrated crowbar case, where a young man, who was blasting, had a pointed bar of iron about three-quarters of an inch in diameter and weighing three pounds, driven, by a sudden explosion, upward through his head. It entered at the angle of the under jaw, passed behind the nose and eyes, penetrated the skull, and cutting the cerebral substance of the frontal region passed out at the top of the head, above the forehead and to the right of the median line. The wound was frightful. All one part of the brain was disorganized, without counting the multiple fractures of the skull and face. He was alone, but in less than an hour after the accident, without help, he walked to the surgeon's, went up the steps, and related the circumstance clearly and intelligibly. He recovered, but died of epilepsy some twelve years later. His physicians and friends observed that his character and intelligence changed notably after the accident. From being intelligent and active he became capricious and unsteady, and had to retire from his post of overseer. Analogous cases to the same purpose might be cited, but we have no space. We need new facts, but those we possess strongly favor the theory of Fritsch, Hitzig and Ferrier.

We have passed in review the experimental and clinical arguments in support of this theory; and there are others of not less importance drawn from pathological phenomena. But we have no space for their consideration.

M. Brown-Sequard has shown the greatest hostility to this theory. His chief argument is that lesions and symptoms are not coextensive. An insignificent lesion causes general. trouble ; a considerable lesion remains latent in the matter of symptoms. This is true, but it is the exception ; whereas he ought to show that it most frequently happens. The real question is, does the seat of a lesion signify nothing, and may we have identical symptoms with two very different lesions ? And we have demonstrated that this can not occur. The facts cited in opposition to the theory of Ferrier may be embarressing, and at present inexplicable ; but such facts would be far more abundant if we admitted the theory of M. Brown-Sequard.

Again, there is the theory of Vulpian, who thinks that the stimulation of the gray cells by electricity is not possible. For motor centers he substitutes psycho-motor centers. In his view, stimulation of the convolution acts, not on the cells, but on the white fibers which proceed from them. But his mode of interpretation does not alter results, nor set aside the centers.

Goltz made a curious experiment, showing clearly the office of the centers. He took two dogs of the same species, one having the education common to all dogs, and the other some supplementary accomplishments, and among them that of giving the paw. In both dogs he removed the center which presides over the movements of the forepaw of one side—the one given by the knowing dog. They soon recovered, and could run about. Running is a reflex act, that does not require the intervention of the centers. But, while the learned dog could use his legs, and go and come, he could not give his paw. This was a superior, voluntary act, which could not be performed in the absence of the corresponding center. It is in this

differentiation of the organs of voluntary activity from those of automatic activity that we find the explanation of so many singular facts which at first sight seem to contradict the theory of localization.

It is undeniable that there is yet much to be done in this domain. Bnt the results obtained by Ferrier are so encouraging that we hope this new way or studying cerebral physiology will be followed and explored with more care than ever.

CORRESPONDENCE.

CHICAGO LETTER.

CHICAGO, March 25, 1881.

Editor of the "Review"—Dear Sir : The physicians of this city have had a busy winter. Sickness of all kinds has prevailed. We have had more scarlatina and diphtheria than usual, and considerable anxiety has been rife in lay circles concerning the unusual amount of small-pox in the community. Vaccination has, in consequence, become quite fashionable. Most of our physicians prefer bovine virus, and the demand for ivory points and for quills has been quite brisk. One of our dealers has averaged three thousand points a day, though largely for country orders, for some time. It is the common experience here that the period of incubation of vaccinia induced by animal matter is considerably longer than when humanized virus is used ; it being not infrequently two weeks from the vaccination to the first symptom of a "take." We have also had to treat a larger number than usual of cases of rheumatism and diseases of the respiratory passages ; and a great many have suffered from what is called here "winter cholera," a species of serous diarrhœa with more or less pain, and accompanied or followed by a considerable amount of prostration. It usually yields well enough to treatment, but relapses from very slight indiscretions in diet are very frequent.

Nearly all the medical colleges have closed. Rush Medical turned out a batch of one hundred and seventy, the Woman's College a larger class than ever ; while the Chicago College, having the longest term, does not close till the end of this month. All these colleges, like those at the East, are gathering about them the enterprising and intelligent young men who are willing and content to do most of the hard and telling work, while the elder men look on approvingly, waft messages of encouragement and—reap the harvest. But no matter how patriotic or enthusiastic in the matter of "reform in medical education" these old gentlemen may be, it is seldom or never noticed that they step down and out to give a chance for genuine reform.

During the winter one expects most of interest to emanate from the medical societies. The leading society—and most respectable on account of age—is the Chicago Medical. Its membership is large, its ranks include nearly all of the profession here of any standing, and some of no standing ; in fact it is too much of an omnibus. Its meetings are held semi-monthly. Once in a while its papers and discussions rise above mediocrity, but as a rule it hardly pays a man to attend. In fact, I recall nothing for some months back that is worth mention in your Journal. This is not because its members are not capable of good work, for it contains men the peers of any in the land, but it results from lack of interest. The society has too much "lumber" in it. It was once the writer's lot to hear a member refer several times to Tarrant's "*Excelsior*" Aperient, with other blunders to match.

The medical gentlemen of the West Division of the City are largely interested in the West Chicago Medical Society, in which much better work is done, The gynæcologists of the city, too, have banded themselves together into a Gynæcological Society, which meets once a month at the Grand Pacific Hotel, to discuss quadrivalves, etc., first, and then an appropriate lunch of *bivalves* and Roman punch before going home. Here there is a truce to professional rivalry, and rival aspirants for professorial

chairs shake hands with apparently extreme cordiality. Little of their work, however, is made known to the public.

But the society which has done the best and most genuine work since its foundation two years ago, is the Biological Society. It is composed entirely of young men who are known to be workers and industrious students. Most of its members are physicians, consequently the papers contributed are mostly on medical topics. At the January meeting the retiring president, Dr. Lester Curtis, gave his address, in which he took up the matter of the histology of the white blood corpuscles. Klein, of London, following Heitzman, of New York, had figured and described an elegant and subtile network in this cell, only visible under high powers and by careful manipulation. Dr. Curtis proved conclusively that this whole appearance was an optical delusion—one of light and shade from an irregular surface, and that the figures so beautifully drawn and lithographed were wholly inaccurate. He also read some letters from Dr. Heitzman, which were as amusing as they were characteristically assuming. Dr. Curtis has the corroborating testimony, in this matter, of some of the most expert microscopists in the country.

At the same meeting a discussion on the germ theory was begun. It was opened by Dr. Henry Gradle, who is Professor of Physiology in the Chicago Medical College. He is an enthusiastic advocate of the theory, and his arguments, logical and easily grasped, made an excellent impression on his hearers. At the next meeting the discussion was continued, and Dr. Gradle was answered by Dr. Lyman, Professor, at one time of physiology, in Rush Medical College. The latter only accepts the germ theory with a great deal of allowance, and he spoke at length of the chemical nature of fermentative processes and the possibility of bacteria *et id genus omnes* being factors in the causation of fermentation only, as they were bearers or producers of the chemical ferments. The whole discussion was one of great interest, and one of the most notable of its kind ever held here.

At the March meeting Dr. Curtis contributed the outlines of some recent researches into the histology of the blood, and exhibited drawings of peculiar, very fine granules which he had found and believed to result from the degeneration of white blood corpuscles. Similar granules, identical, in fact, in appearance, he had also found in vaccine virus, adipose tissue and semen. He had also observed very finely-branching threads of fibrin, each thread connecting with a white blood corpuscle. These observations were original.

So much for society work; in my next I will give you some items concerning our clinical advantages, and other matters of general interest.

EDITORIAL.

THE LITERARY REVOLUTION.

"Chambers's Encyclopædia," which is fully described below, ought certainly to be in the hands of every reader of this Journal. No work ever published should more properly be considered a *necessity* to any person aspiring to even ordinary intelligence, than a cyclopædia. It should be ranked even before a dictionary. Before the days of the "Literary Revolution" encyclopædias were an impossible possesion to the majority of readers, because of their expensiveness. But a few years ago Chambers's could not be had for less than $50.00. Now you can get it for but a fraction of that amount, or WITHOUT COST in money by terms given below.

AN AMAZING OFFER.

By special arrangements with the publishers (The American Book Exchange), we are enabled to offer our subscribers the following very extraordinary opportunities to secure this great encyclopædia.

FREE. For a club of 3 paid subscribers to this Journal at $5.00 each, we will supply "Chambers's Encyclopædia," 15 volumes bound in cloth, *free.*

For $8.00 we will supply "Chambers's Encyclopædia" as above, and the *Review* for one year.

This edition of *Chambers's new and complete Encyclopædia* is an exact reprint of the very latest (1880) London edition. It is in FIFTEEN VOLUMES, well bound in cloth, and each volume contains nearly NINE HUNDRED PAGES.

We are perfectly well aware of the fact that this is a most astonishing and incredulous offer, advertising as we do, to furnish fifteen bound volumes of the best Encyclopædia in the English language, in connection with the *Review*, for the mere nominal sum of eight dollars. And yet there is no trickery or deception about this thing; we mean and will do just what we say. The explanation is very simple: The publishers are willing to pay thus handsomely for the advertisement of their work in this Journal. That is to say—for each copy of the Encyclopædia sold to the *Review*, they receive so much in money and so much by advertisement. As the result, the Profession has the opportunity of obtaining Chambers's Encyclopædia at a price marvelously low and wholly unprecedented.

FURTHER INDUCEMENTS.

Walsh's Call Book (price $1.50) and *Walsh's Retrospect* for one year ($2.50) will be furnished in connection with the *Review* for one year at $5.50. *Walsh's Retrospect* is an exquisite and invaluable publication, and is the *only* quarterly compendium of American medicine and surgery extant. His "Combined Call Book and Tablet" is too well known and appreciated to require our lauding.

The above inducements are only held out to those subscribing for volume two of the *Review*, which commences in September, 1881. And all those sending in their money before May 1st, will receive the remainder of the present volume *gratis.* All communications and money should be addressed to the Editor and Publisher, Dr. A. Wellington Adams, Denver, Colorado.

DO HYBRIDS BREED ?

A private letter from an officer of the army serving in Texas has been shown the editor of this journal, from which the following is excerpted :

" By the way, we had a curious incident to occur this afternoon. A mare mule, one of K Co.'s pack mules, gave birth to a mare mule colt—honest Injun, no sell, on my honor. One of the sergeants saw the birth and immediately called the company commander. Woodward, Wasson, Gen. Mc-Laughlin, Col. Carpenter, all of us, in fact, are just back from inspecting the wifit. We out here think it as singular a thing as there is on record. The colt was born dead. No hair on it, save about the eyes and on the hoofs ; perfectly formed ; long mule ears. The mother is a full-blooded mule— no jackass about her. The father of the monstrosity is supposed to be an Indian pony, a very pretty little stallion, which Lebo captured in the Staked Plains some years ago. What do you think of this ? Do you know of many such occurrences ?"

REMOVAL AND CONSEQUENT DELAY.

The delay in issuing this number of the REVIEW has been unavoidably caused by our removal to Denver. Hereafter, everything intended for the REVIEW should be addressed Denver, Colorado.

OBITUARY.

WAR DEPARTMENT,
SURGEON GENERAL'S OFFICE,
Washington, February 25, 1881.

It is with profound regret and a sense of loss, not only to his Corps, but to the medical profession, that the death of George Alexander Otis, Surgeon and Brevet Lieutenant Colonel, U. S. Army, is announced to the Medical Corps of the Army.

Born at Boston, Massachusetts, November 12, 1830, he graduated with the degrees of A.B. and A.M. from Princeton College ; entered the Medical Department of the Uni-versity of Pennsylvania, and received his degree of M.D. from that Institution in 1850 ; visited Europe, and prosecuted his studies in London and Paris, and returning to this country he established himself at Springfield, Massachusetts ; appointed Surgeon 27th Massachusetts Volunteers, September, 1861, he held this position until appointed Surgeon, U. S. Volunteers, August 30, 1864. After the close of the war he entered the Medical Corps, U. S. Army, as Assistant Surgeon, February 28, 1866 ; became Captain and Assistant Surgeon, July 28, 1866 ; Major and Surgeon, March 17, 1880, having received the four brevets of Lieutenant Colonel of Volunteers, Captain, Major and Lieutenant Colonel, U. S. Army, for meritorious services during the war. While Surgeon of the 27th Massachusetts Volunteers he served in Virginia, North and South Carolina, and was on special duty in charge of the Hospital Steamer " Cosmopolitan" in the Department of the South. Assigned to duty in this office July 22, 1864, he was Curator of the Army Medical Museum, and in charge of the Division of Surgical Records until his death.

He was editor of the Richmond Medical Journal for three years, member of the leading medical societies of America and corresponding member of various similar societies in Europe, and a contributor to prominent medical journals. Surgeon Otis, with his personal observations of the surgical collections abroad, brought indefatigable industry and untiring energy to the development of the surgical and anatomical collections of the Army Medical Museum, which he has made the most valuable of their kind in the world. The compilation of the Surgical Volumes of the Medical and Surgical History of the War has placed Surgeon Otis confessedly among the most prominent contributors to surgical history.

While on duty in this office, Surgeon Otis wrote for publication no less than ten reports on subjects connected with Military Surgery, &c. ; among which are his most valuable and exhaustive reports on "Excision of the head of the femur for gunshot injury," and on "Amputation at the hip-joint in military surgery." Of great culture, retentive memory, and with a remarkable facility of expression, he was, as a compiler and writer, conscientious in his analyses, giving his deductions from the facts before him with modesty, but decision. With such a record it is needless to speak of his zeal, his ambition or his devotion to his profession and especially to the reputation of the Corps of which he was so bright an ornament. While devoting himself to the preparation of the Third and last Surgical Volume (now more than half completed) of the Medical and Surgical History of the War, he died in this city February 23, 1881. His untimely death will be deeply deplored, not only by the Medical Corps of the Army, but by the whole Medical Profession at home and abroad.

J. K. BARNES, Surgeon General.

TRANSLATIONS.

By DR. R. TILLBY, CHICAGO.

NERVE STRETCHING FOR EPILEPSY.

Speaking of the value of nerve stretching for the treatment of different nervous affections, the Progress Medical, Feb. 5, 1881, says : M. Gilette stretched the median and ulnar nerve in the upper third of the arm of a patient who had been afflicted with congenital epilepsy. The operation was performed the 31st December, 1880. Although the short time that has elapsed does not permit us to speak of the final result, yet it must be stated that so far it has afforded some relief. The attacks, which were about ninety a month, have been reduced since the 1st of January to eighteen. Moreover, these attacks have diminished in length and intensity. The aura, which showed itself by severe pains in the arms and by a sense of oppression, has completely disappeared, and there is very much less hebetude after the attack. The inconvenience associated with the operation was not worthy of notice. The wound healed by first intension.

Resection of Small Intestine—Recovery.

M. Koeberle performed laparotomy on a young girl who had several times exhibited symptoms of obstruction in the intestines. After opening the abdominal wall four strictures of the small intestines were visible, extending over a space of two metres. These two metres were removed. The two ends stretched together, about twenty mesenteric vessels tied, and the wound closed. The operation lasted three hours and succeeded admirably. No Listerism and yet no fever.—*Le Progres Medical, Jan. 29, 1881.*

Eczema and Ear Piercing.

M. Constantin Paul, having pierced the ears of a woman in good health, witnessed the development of a chronic eczema attributable to the piercing in question. The eczema appeared annually. Led by this circumstance to investigate the matter, he was able in one year to collect 120 cases demonstrating the development of eczema in scrofulous women as a result of piercing the ears.—*Le Progres Medical, Jau. 29, '81.*

(NOTE.—The translator has seen two cases which seemed to develop after vaccination..)

BOOK NOTICES AND REVIEWS.

Archives d'Ophtalmologie is the title of a new publication lately issued in Paris. It is to appear bi-monthly and the first number was issued on the first of November, 1880. The literary editors are F. Panas, Professeur de clinique ophtalmologique a la Faculte de medecine, etc. ; E. Landolt, Cherurgien Oculist consultant a l'Institution des Jeunes Avengles ; F. Poncet (de Cluny) Medecin principal d'Armee, Professeur agrege au Val-de Grace, etc. The publishers are Messrs. Delahaye & Lecrosnier.

The object of the journal is to afford an opportunity for the publication of original research on the subject of ophthalmology. A section will be reserved for bibliography, but discussions and isolated observations are to be excluded, excepting such cases as offer a genuine scientific interest.

The names of the above mentioned ophthalmologists will be a sufficient guarantee to the medical public for the character of the work, but to furnish a more definite idea of the nature of the publication, we append a translation of the table of contents of the first number.

Paralysis of the Motor Oculi, consequent on Traumatism of the Cranium, by Prof. F. Panas.

New Method of Blephoroplasty, by Dr. Landolt.

A Telemetre, by Dr. Landolt.

Pterygium, by Dr. F. Roncet.

A contribution to the Clinical and Anatomical study of Ocular Tuberculosis, by Menfredi and Cofler.

The Sense of Light and the Sense of Color, by Professor A. Charpentier, of Nancy.

Reflexions in Physiological Optics, by Dr. Badal.

Treatment of Detachment of the Retina by subcutaneous injection of Nitrate of Pilocarpine, by Dr. Dianoux.

Sudoriparous Cysts of the free border of the Lids, by Dr. L. Defosses.

Bibliography. By Dr. Thomas.

Almost all the articles are illustrated.

MEDICAL NEWS.

EXCISION OF A PORTION OF THE STOMACH. —This is the second time the operation has been performed. The case in question was operated on by Professor Billroth, for an infilliating carcinoma. About six inches of the greater curvature, including the pylorus, was removed. At last accounts (a week after the operation) the patient was doing well in all respects. (Since reported dead).

A CASE OF LUXATION OF THE HEAD OF THE FIBULA.—Dr. Oldright (Canadian Jour. of Med. Science) reports a case of luxation of the head of the fibula in a boy two years of age. The head of the bone was freely movable and easily replaced, but as easily slipped back. The limb was incased in a plaster of Paris bandage with a pad, secured by surgical plaster, to keep the bone in place. Absolute immobility was maintained for forty-one days, and at the end of that time the head of the bone slipped back quite freely.

JUGLAUS NIGRA, A REMEDY FOR DIPHTHERIA.—Dr. C. R. S. Curtis (in Boston Med. and Surg. Jour.) reports some thirty cases of diphtheria, in all of which this remedy was used. A strong decoction of the leaves was used, and in the more malignant cases the hulls of the green walnuts were added. The decoction was used as a gargle, in the form of spray in the steam atomizer, and in young children by means of the swab. In cases with œdema of the neck and swelling of the glands, a poultice of the leaves was employed. In one or two instances the remedy was given internally, and with advantage. No complaint was made of the distastefulness of the remedy, as the throats of young children were swabbed out with little difficulty. In most of the cases iodine was used locally in conjunction with the black walnut ; and in all cases a general tonic treatment was employed. The results were most favorable, as *all of the cases recovered.*

HEADACHES DEPENDANT ON DISEASE OF THE DURA MATER.—Dr. J. S. Jewell, in an article on the nature and treatment of headaches (*Journal of Nervous and Mental Diseases,* vol. vi., 1881, p. 64), says that headaches depending upon acute, but much more frequently subacute, forms of disease of the dura mater are much more common than is ordinarily supposed. Affections of the dura due to blows, etc., and which are accompanied by pain, may occur suddenly, or after months, or even years. Tubercular deposits, rheumatic and syphilitic influence, may also give rise to pain in the dura. It sometimes follows in the wake of sunstroke or severe exposure to cold,or arises from the extension of disease from the nasal to the cranial cavity through the cribriform plate of the ethmoid bone, or from the middle ear, as in otitis media, also from unknown causes in cerebro-spinal meningitis. In addition to these, diseases of the bones, or tumors or growths, may lead to painful affections of the dura. Also, diseases of the Pacchionian bodies, or of the brain itself, may lead to affections of the dura of a painful nature. The inflammations of this membrane are ordinarily localized rather than general. Painful affections of the dura may occur at any age, but are commoner during later childhood, youth, and the middle period of life.

With regard to the symptoms, the pain is usually definitely localized. It does not shift from place to place, as it does in many of the circumscribed pains of neuralgia; it is persistent, seldom entirely ceasing long at a time so long as the meningitis continues; it commonly begins gradually and disappears slowly; it is aggravated by anything which increases the activity of the intercranial circulation; it is aggravated by shocks to the head; it is not relieved in assuming the lying-down position. As a rule, it is made worse by increased barometric pressure and by the occurrence of cold weather, or by exposures of the surface to cold by which the cutaneous vessels are contracted, or by any other means by which vascular tension is increased, or by any means by which the cutaneous circulation is diminished in activity or repressed. It occasionally throbs when the heart's activity is increased. It is accompanied generally by more or less mental depression and by nervous irritability, discouragement, and disinclination for mental and physical labor. It is rarely, though sometimes, accompanied by nausea. It is rarely accompanied by increase in temperature of the head, though local rise is usual.

The treatment of these affections involves keeping the patient as quiet as possible and away from bright lights and exciting circumstances of all kinds. All exercise, whether physical or mental, except the most moderate, should be avoided. The diet should be simple and unstimulating. If any acidity should appear in the stomach, it should immediately be neutralized. The bowels should be kept entirely free. If there is irritation of the bladder, measures should be taken to allay it. All sexual indulgence or excitement should be avoided. All the sleep that can be secured should be had. The patient should sleep upon a gently-inclined plane, formed by putting blocks of wood from four to eight inches in height under the head-posts of the bed. By this means the blood is made to gravitate away from the head, and relief is obtained. If possible, a warm climate and a locality high above the sea-level, where the barometric pressure is low, should be secured. The surface should be thoroughly protected from exposure to cold air. Alcoholic stimulants and strong coffee and tea, as a rule, should be avoided in this form of headache. Protracted hot foot-baths are in order.

Among medicines, opium is the most generally useful, either in the watery pilular extract or in the deodorized tincture. It should be given in doses of such size and frequency as to subdue the pain, and should be continued in conjunction with other measures until the pain subsides, when the use of the anodyne may be gradually withdrawn. Side by side with this it is necessary to employ large doses of the iodide of potassium. For an adult, ten grains may be given three times a day to begin with. Each day the dose may be augmented by five grains, until decided evidences are given that the remedy has produced results. If duly diluted with water, from fifty to one hundred grains may be given three times a day if necessary. If the disease is of syphilitic origin, inunctions of mercury may be employed, For this purpose the oleate is to be preferred in the strength of ten grains of the stronger oleate to an ounce of cosmoline perfumed with a little oil of roses. Inunction should be practiced twice daily until the effect of the mercury is produced. Counter-irritation behind the ears and along the back of the neck by means of the actual cautery or by blistering collodion, Dr. Jewell has found useful. Under this treatment, in the course of a few days, or, at most, of a few weeks, the pain abates. Tonics, as acid solutions of strychnia and quinine, may be given at this stage as required.

SEPTIC POISONING IN AN INFANT.—In the New York Medical Journal for March, 1881, Dr. J. Foster Bush, of Boston, relates a fatal case of septic poisoning in a new-born child and remarks, as an interesting fact concerning the communication of septic diseases in the puerperal state, that the mother's convalescence was not interrupted, and that the child's grandmother, a midwife, spent her time alternately in attending to the child and in following her avocation, and that, so far as he could ascertain, no harm resulted. If a physician should do this, he pertinently asks, what would the professional verdict be?

SYPHILITIC ULCERS.—Salycilate of lime, one part to fifty of water, is highly recommended as an application to syphilitic ulcers. Its effects are described as almost magical in certain phagedenic cases.—*Canadian. Jour. Med. Science.*.

PEPTONE IN PUS.—Hofmeister has lately determined the existence of peptone in pus. The quantity varied from .367 to 1.275 grammes in 100 centimeters, and was found to be greater in proportion to the thickness of the pus. The corpuscles contained it in abundance, whilst the serum was free from it.

BOOKS AND PAMPHLETS RECEIVED.

"Syphilis and Marriage." Lectures delivered at the St. Louis Hospital, Paris. By Alfred Fournier, Professeur a la Faculte de Medecine de Paris, Medecin de L'Hospital St. Louis, Membre L'Academie de Medecine. Translated by P. Albert Morrow, M.D., Physician to the Skin and Venereal Department, New York Dispensary ; Member of the New York Dermatological Society ; Member of the New York Academy of Medicine. Published by D. Appleton & Co. New York. Price $2.00.

"The Heart and Its Functions." One of the Health Primers. Published by D. Appleton & Co. New York. Price 40 cents.

"Introduction to the Study of Indian Languages," with words, phrases and sentences to be collected. By J. W. Powell. Second edition. With charts.

"Mississippi Valley Medical Monthly." Editors and proprietors J. J. Jones, M.D., Julius Wise, M.D. Memphis, Tenn.

"Annual Reports of the State Board of Health, of Colorado." For the years A.D. 1879 and 1880.

"The Indiana Medical Reporter." Edited by A. M. Owen, M.D., J. W. Compton, M.D., J. Harker, M.D., Arch. Dixon, M.D., J. Gardner, M.D. Published by J. W. Compton, M.D.

"Osservatorio Di Moncalieri." Osservazioni Meteorologiche ; Fatte Nelle Stazioni Della Corrispondenza Meteorologica Italiana Alpino Appennina. E. Publicate Per Cura Del Club Alpino Italiano.

"Le Progres Medical." Journal de Medicine, de Chirurgie et de Pharmacie. Paraissant le Samedi, Redacteur en chef: Bourneville.

"Proceedings of the Alumni Association of Rush Medical College, Chicago, 1880."

"The Monthly Index"—To current periodical literature, proceedings of learned societies and Government publications. Published at the office of the American Publisher, New York.

"The American Naturalist." Devoted to the Natural Sciences in their widest sense. Editors, A. S. Packard Jr., and E. D. Cope. Publishers, McCalla & Stavely.

"La Gáceta de Sanidad Militar," Periodico Cientifico Y Ofcial del Cuerpo de Sanidad del Ejercito Espanol. Madrid. Imprenta de A. Gomez Fuentenebro, Bordadores, 10.

"The Romance of Astronomy." No. 20 of the Humboldt Library. By R. K. Miller, M.A., Fellow and assistant Tutor of St. Peter's College, Cambridge, England. With Appendix, by Richard A. Proctor.

"Anæmia in Infancy and Early Childhood. By A. Jacobi, M.D., Clinical Professor of Diseases of Children in the College of Physicians and Surgeons, New York. Reprinted from the Archives of Medicine.

"A Conspectus of three Different Forms of Cardiac Trophic Disease." By Roswell Park, A.M., M.D., Demonstrator of Anatomy, Chicago Medical College, etc., etc. Reprint from the Southern Clinic.

"Lectures on the Diseases of the Rectum and the Surgery of the Lower Bowel." Delivered at the Bellevue Hospital Medical College. By W. H. Van Buren, M.D., LL.D. (Yalen), Professor of the Principles and Practise of Surgery in the Bellevue Hospital Medical College, etc., etc. Published by D. Appleton & Co., New York.

"The Transactions of the American Medical Association." Instituted 1847. Vol. XXXI.

"Is Scondary Syphilis communicable as such?" A paper read before the St. Louis Medical Society, by G. M. B. Manghs, M.D.

ROCKY MOUNTAIN MEDICAL REVIEW.

Vol. 1.   DENVER, APRIL–MAY, 1881.   Nos. 8-9.

ORIGINAL ARTICLES.

On Microscopical Measurements.

By LESTER CURTIS, A. B., M. D.,
Professor in the Chicago Medical College.

The attempt to determine from what sort of an animal a given sample of blood has come, has lately become a matter of great practical importance in a medico-legal way. The blood corpuscles of most mammals resemble one another in shape and general appearance, the main difference between them being their size. The corpuscles in individuals of the same species average nearly the same, while the corpuscles of different species differ, more or less, widely.

But the corpuscles themselves are microscopic and the differences between them are, of course, exceedingly minute. In order to determine this question, therefore, it is necessary to resort to very careful measurements under the microscope. It consequently becomes a matter of popular, as well as scientific, interest to know something of the methods used in making these measurements, the difficulties attending them, and the degree of success which has attended the efforts to overcome the difficulties. We must remember that the use of any instrument designed to attain great accuracy, is difficult just in proportion to the degree of accuracy aimed at.

An astronomer in pursuing some of his investigations wished a table quite level. He employed a carpenter and furnished him with a sensitive astronomical level to test the accuracy of his work, the level tipped about in all sorts of irregular ways and the carpenter was unable to bring it to the centre. He finally went to the astronomer and told him that the level was good for nothing, he had a level at home that he could make work every time. The man was not used to such an accurate instrument and did not know how to use it. The microscope is no

exception to this rule that the use of an instrument of precision requires care and skill. It must be correctly focussed and adjusted or the image will be blurred.

In the examination of coarse objects slight defects in these adjustments are not noticed, just as slight defects were not noted by the carpenter's own level, but let one try to make out the structure of some finely marked object, such as a difficult test object, or the ultimate structure of a muscular fibre, and he will begin to appreciate the difficulty of using the instrument in really exact work. This difficulty of focussing and adjusting the microscope increases with every increase in the magnifying power. The use of a low power is easy enough—a power of 2,000 diameters is only fit for an expert.

When we come to the use of the microscope for purposes of measurement we add to what is already no small load, the further difficulty of measuring. Carrying a surveyor's chain cannot be entrusted to every bungler, only careful and experienced men are fit for the work; and when extreme exactness is required, as in measuring the base line used in the coast survey, the difficulties are tremendous, and are only to be overcome by the greatest care and skill of trained mathematicians. In measuring with the microscope, then, we have not only the difficulty of focussing and adjusting the instrument, but also the difficulty of applying the measure.

And, first, as to the microscope. The microscope must be exactly focussed or the edge of the object to be measured, instead of being a sharp line, will shade off each way and it will be impossible to tell where it begins or where it ends, and there will be no certain point from which to measure. Just here is a chance for a considerable error. The accurate focussing of all glasses and especially of high powers is a matter of such nicety that it is necessary to have a somewhat complex apparatus for its accomplishment, capable of slow and steady motion which is under the perfect control of the observer. This is accomplished in most of the older patterns of English and American microscopes by means of a tube sliding within the tube of the microscope, like a telescope tube.

This tube holds the objective and is moved to and fro by a lever which bears against a spring and is moved by a screw. The motion of this screw moves the lower combination of glasses—the objective, while the upper combination, the eye-piece, remains stationary. But this operation of lengthening and shortening the tube alters the magnifying power of the instrument, and the higher the power used the greater the change by the action of the screw.

Here, then, every change of focus alters the magnifying power of the instrument, and to secure accurate results it would be necessary to compute the magnifying power for every position of the screw, of course, an impossibility.

In many of the continental models of microscopes and in some of the most improved patterns made in this country, this difficulty is avoided by a contrivance which moves the whole body of the instrument without changing the relative distances of the eye-piece and objective.

Having got rid of this difficulty of the alteration of the magnifying power by the focussing of the instrument, there remains a still greater one to contend with. Nearly all objects which are to be examined by the microscopist are placed upon a slip of glass and covered with a smaller and thinner piece. Now, the light passing through this cover is turned slightly out of its course. In finely corrected glasses this turning of the light causes a distortion of the image, like that produced by objects being out of focus, with, of course, the same blurring of the edge. In order to do away with this blurring, the best dry glasses are provided with an arrangement called "cover correction" to adjust the glass to the different thicknesses of cover. The cover correction is constantly called into use in examining the finer structures of objects with dry glasses, for no two covers are of the same thickness, even the two sides of the same cover, often vary enough to produce a noticeable blurring of the image. But a slight movement of this adjustment alters the magnifying power considerably. This difficulty is so troublesome that with dry glasses it is, by no means, easy to get correct

measurements. There has lately, however, been introduced a sort of objective, called "homogeneous immersion," in using which a drop of fluid of nearly the same refractive power as the cover and the front lens of the objective is used. A drop of this fluid is put upon the cover of the object to be examined and the end of the objective is immersed in the drop while the object is studied. The drop of fluid is supposed to unite the cover and the front glass of the objective into one homogeneous mass, and light passing through the mass suffers no change in its direction, consequently a cover correction is not needed.

Even here, however, we are not entirely free from danger of error. A fluid of exactly the same refractive power as glass has, up to the present time, not been discovered, consequently even in these glasses it is necessary to have some sort of a cover cor-

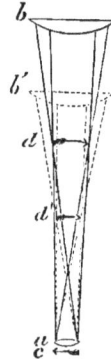

FIG. 1. *a*. Combination of lenses forming the objective. *b*. Combination of lenses forming eye-piece. *a*. *b*. Length of the tube of the microscope. *c*. Object. *d*. Image formed by rays converged by the objective, this image is again magnified by the eye-piece.

The dotted lines show the same combination of lenses with a shortened tube. *a'*. The diminished image.

rection. This is sometimes accomplished by lengthening and shortening the tube of the microscope which brings back our old error again. We can, however, by careful attention to details and by selecting covering glasses of measured and uniform thickness reduce this error to a minimum.

So much for the microscope. Now, for the application of the measure. The first thing necessary for this is the obtaining of an accurate standard of comparison. A standard divided into comparatively long spaces, as inches, halves, quarters, and so forth, or into similar divisions of the metre. We also need finer divisions of the same standard, as into hundredths and thousandths of an inch, or some small division of a metre, as tenths and hundredths of a milimeter, which can be placed upon the stage and viewed through the microscope. The former of these we have in the ordinary foot-rule, the latter in that ruled slip of glass called the stage micrometer.

The ordinary foot-rule is made by hand, the workman has a copy divided into the same sort of divisions that he wishes in the measure that he is to make. He lays several of the bits of wood, which he wishes to divide, side by side parallel with his standard. He then takes a T square, lays it upon his standard and moves it from one division to another making each division in succession upon the bits of wood with a knife adopted to the purpose. It needs but little reflection to see that such a method can only produce results approximately accurate, and a little examination of almost any foot-rule will show a difference in the lengths of the divisions so large as to be easily detected with an ordinary pair of compasses. Besides, even if they were accurate the lines are so coarse and so uneven in thickness as to be not well adapted to delicate measurements. Such a rule is commonly used as a standard to determine the magnifying power of the microscope. I need hardly refer to the errors into which it may lead.

But stage micrometers are also inaccurate. Professor Rogers, of Cambridge, has collected great numbers of these micrometers and has compared them with a carefully corrected copy of the standard yard and the standard metre. This comparison has shown that the micrometer of no single maker is accurate. The errors are of two sorts:

1st. An error in the length of the whole space divided. 2d. An irregularity in the sub-divisions of this space. If, for instance, the space is called a tenth of an inch, it is nearly always greater or less than the tenth of an inch. This error, of course, makes the sub-divisions all wrong. But besides this error the sub-divisions are unequal. Mr. Beck, the prominent instrument maker of London, has said that this spacing can be done so as to be practically accurate. I took the trouble to go over my micrometer—a Beck's—and find in it a very considerable

irregularity. My memorandum has been mislaid and I cannot put my hands upon it just now, but I can confidently say that the difference between the longest and smallest of what should be equal divisions exceeds ten per cent*. He has made a great number

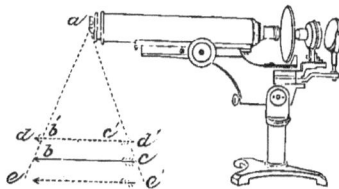

Fig. 2. Microscope arranged to draw with camera lucida. *a.b.*and *a.c.* The direction of the rays to apparent image on the paper. This direction of the rays is constant. If the distance of the paper from the camera is lessened the size of the image is diminished, if the distance is increased the size of the image is increased.

of experiments in ruling micrometers and has come to the conclusion that it is impossible to get one absolutely correct. The only way to avoid error is to carefully compare each micrometer with a standard, and note the errors, which can then be allowed for in measuring.

And now as to the actual practice of measuring. A method sometimes used is to turn the instrument into a horizontal position, project the image of the stage micrometer on a screen, and compare it with an ordinary foot rule. This will give the magnifying power, then project the image of the object upon the screen, measure with a pair of compasses and apply to a foot-rule. The result divided by the magnifying power will give the size.

In this case the screen must be rigid like a wall, or a piece of ground glass, or its vibrations will make the result valueless. But however carefully the proceeding is done it is open to the errors of focussing and adjustment, Every change of focus alters the magnifying power, and it is practically impossible to keep it constant.

The object, if very thin and transparent, may be placed upon a stage micrometer that has been ruled upon the surface of the slide, not on the cover. This may be covered with a thin glass and the image of the lines on the micrometer and of the object all thrown

---

*Since writing this I have found the memorandum, it agrees substantially with the above.

together upon the screen. The object must be very thin and lie flat down upon the surface of the micrometer or they will not both be in focus at the same time, the room also must be dark, and it is difficult to make accurate measures in such a room, indeed, it is not at all easy to measure accurately any image projected upon a screen.

Another method is to photograph the object and measure the photograph. The photographic prints have been used for this purpose, but they are so liable to be more or less distorted in the process of washing, drying and mounting, that the process is extremely inaccurate.

The photographic negatives themselves have been used for the same purpose, these are pretty accurate, but the method is impracticable to one who has never learned photography—it has been also claimed by some that the collodion used to fix the image to the glass is liable to contract and produce minute changes in the size of the objects photographed upon them. Some processes are said to be peculiarly liable to this accident—some are claimed to be very accurate. I am unable to form any opinion of my own from personal experience.

The next instrument for measuring is the cob-web micrometer. This is an eye-piece containing two fine wires or cob-webs in its focus. These wires can be moved towards or away from each other by means of a screw attached to one of them. The head of this screw is large and graduated upon its rim. Within the eye-piece there is also a row of notches to indicate how many turns the screw has made, one complete revolution moving the wire one notch. By bringing one of these wires to one edge of the object, the distance between the two sides, can be easily measured by turning the screw until the other wire rests on the other edge of the object, and noting the number of divisions on the head of the screw which pass the fixed index in doing so. Then, by comparing this distance with the spaces ruled on the stage micrometer we get the distance.

In using this method we are, of course, liable to the errors of focussing and adjustment, and also to the personal errors of observation, which are considerable. Besides these, we have a source of error in the screw. Professor Rogers has shown that a screw cannot be made so true that turning the

head a certain definite distance will always move the nut a uniform distance. This error exists in even the most carefully made screws. If the screw is not carefully made and the nut is not properly set the error is increased. But if the instrument is carefully made and the same part of the screw is always used, this method is, by far, the most accurate known. It is claimed by competent observers that the error can be reduced to less than 1-1,000,000 of an inch by this method.

The next method of measuring is by the Jackson eye-piece micrometer. This consists of a piece of glass upon which fine lines are ruled, similar to those on an ordinary stage micrometer. The instrument is placed in an opening in the eye-piece of the microscope when its lines can be seen over the object in the microscope. By determining the value of one of these divisions by comparing it with the stage micrometer we can then use it in very much the same way as we would an ordinary foot-rule. It is, of course, subject to all the errors of the

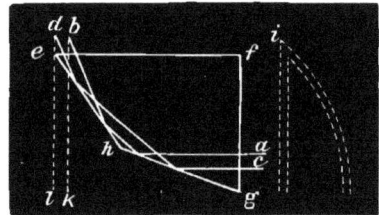

FIG. 3. Course of the rays through camera lucida. *e,f,g*. *h*. Camera. *a.b*. Direction of rays to form image, which appears to be on the paper at *k*, when the eye is at *b*. If the eye is moved to *d* the course of the rays will be changed to *c.d*, and the image will be seen at *l*.

adjustment of the instrument, as are all measuring appliances, together with the irregularity in the divisions—though these are usually so small as to be of no practical importance, besides, the lines are fixed and consequently bodies the borders of which fall between the lines, can only be measured approximately. It is therefore impossible with it to measure any body with extreme accuracy. Also, the glass of the micrometer interferes, to some extent, with the definition of the object.

Another method is by the use of some form of the camera lucida. This is an instrument placed over the eye-piece, usually with the microscope turned so that the tube is horizontal. In this position the image of the object in the microscope is reflected into the eye-piece but appears to be thrown down upon the table, where its outlines can be traced by a pencil, or other means, upon a paper.

The method of measuring by means of this instrument is to determine the magnifying power by putting the stage micrometer upon the stage and tracing its divisions upon paper and comparing them with a foot-rule. If we have some definite division as, for instance, a 1-1,000 of an inch of the stage micrometer traced upon paper by aid of the camera lucida we can easily tell by measuring the tracing how much the division is magnified—if, for instance, it is an inch long it is magnified one thousand times in length. Having determined our magnifying power, if we place any other object upon the stage of our microscope, trace its outlines and measure them with a pair of compasses and apply them to a foot-rule we shall have the size of the object; magnified, of course, a thousand times; dividing by 1,000 we have the size. If, for instance, we have a blood corpuscle under our microscope, and after tracing its outlines upon paper we find it to be one-third of an inch in diameter we know of course, that it is 1-3,000 of an inch across. This method is very convenient in some respects, and is sufficiently accurate for ordinary work, but where great accuracy is required, as in medico-legal cases, it is open to many serious objections.

In the first place it is not the easiest thing in the world to trace quite accurately. Let any one who has never tried it lay down a piece of paper over a drawing and trace it. If he then compares this tracing with the drawing he will find that while he has preserved the general shape and size of the original there will be a quite measurable difference between the two. Now this difficulty is much greater while tracing an image seen with the camera lucida. However sharp and clear the image may appear when examining it with the instrument, the instant we begin to trace, it assumes a certain shadowy vagueness, and not a little concentration of vision and mind is required to finish the work. In drawing with the camera lucida, the whole of the drawing must be made without changing the focus of the microscope, not because the magnifying power

would be changed but because the focussing of the microscope moves the object, and if this were to be done while part of the object was drawn, the image would be distorted on the paper. But for an object to be all in focus at the same time it must be small and thin. Even then it is often the case that with high powers, objects cannot all be seen at once. Indeed a good observer when he is studying an object with the microscope is always in the habit of toying with the fine adjustment. But with all object glasses of high power, the whole field is not in focus at the same time; in other words, the field in no objective is perfectly flat, and the attempt to make it so would be at the expense of other and more valuable qualities. But if this is the case, it is impossible to have all objects in the field in focus at the same time, consequently if we wish to make the drawing accurate enough for purposes of measurement we must change the focus of the microscope, which would distort the image drawn on the paper. We are then between the two horns of a dilemma, either to draw our object out of focus or else make a distorted picture. This is a serious trouble very often, in ordinary drawings where exact accuracy in the proportions of the drawing is not essential, but for purposes of measurement it is fatal. The camera lucida is a mirror, and from every point of its reflecting surface the image is reflected; any slight movement of the eye then would not cause any disturbance in the image any more than a slight movement of the eye of one looking in a mirror would cause a disturbance in that image. But such a movement of the eye would cause a movement of the image on the paper and would, of course, lead to distortion. Now in tracing an object the eye instinctively follows the point of the pencil, and of course must move, and the image must move with it. With one who has experience this movement is not very great, but it is always enough to destroy the accuracy of delicate measurements.

But even supposing that the eye could be kept perfectly still, there is yet another difficulty with camera lucida drawings. The ray of light which passes to the eye from the point directly under the eye, is, of course, a perpendicular line, but all other rays of light are more or less oblique. Now, it is impossible to see accurately any object that is not seen in this perpendicular line. In looking at objects with the naked eye in the ordinary way we are accustomed to keep the head and the eyes in slight motion, and we make up our opinion of the form of the object by a combination of various images. If we are deprived of this power of motion of the head and eyes, our view of any object is necessarily imperfect. What person that has ever been a boy and has experienced the pangs of unsatisfied curiosity while peeping through a crack in the fence, or a hole in a circus tent, can not, with his whole heart, testify to the truth of this statement. In ordinary observations with the microscope, we are accustomed almost, perhaps quite, unconsciously to take advantage of the same movement of the eye in order to get a better view of the object. In drawing with the camera lucida we are shut up to one point of view. As soon as we begin to trace with this instrument we begin to experience another difficulty, that is, tracing along this oblique line must necessarily be erroneous. This error may not be specially noticeable in small objects which are brought to the center of the field of the microscope, but in large objects, or in those at the outside of the field, the error is very considerable, as any one may see by experimenting a little.

Again, the apparent size of the object varies with the distance of the paper from the eye—if anything moves the paper from the eye the image enlarges, if anything brings it near the eye it diminishes. Now, in drawing, it is impossible to get paper to lie perfectly flat, it is continually springing, and with every spring the size of the object changes. Such drawings, therefore, however skillfully performed, can only be approximate and of small value as matters of scientific accuracy.

In conclusion, I would say that the only method of measuring which can be relied upon for absolute accuracy is the use of the cobweb micrometer of first-class construction compared with a corrected stage micrometer and in the hands of a practiced and skilled observer—or possibly the measurement of a photographic negative as recommended by Dr. Woodward.

Trichina.

BY PROF. D. S. LAMB, A. M., M. D.

This entozoon has been known for nearly fifty years. It has been found in the intestines, or muscles, or both, of many animals; the dog, cat, rabbit, rat, mouse, guinea-pig, mole, hedgehog, marten, pole-cat, fox, chicken, pig, sheep, hippopotamus; and according to some, in the horse and ox and in fishes.

If trichinous flesh be fed to animals, the larval trichinae may attain sexual maturity in less than forty-eight hours. Six days later the female will contain perfectly developed free embyos, which pass out of her genital parts into the intestine of the subject, and set to work to penetrate directly through the wall of the intestinal canal in all directions towards the voluntary muscles; where they assume the spiral form, become encysted, and are called *trichinæ spirales*.

According to Cobbold, the young trichinæ, as found in human muscles, present the form of spirally coiled worms in the interior of small, globular, oval or lemon-shaped cysts, which latter appear to the naked eye as scarcely visible minute specks. These specks sometimes resemble little particles of lime, and, indeed, the cyst-walls are more or less calcareous, according to the amount of degeneration. The cysts average 1-78 of an inch in length, and 1-130 of an inch in diameter; the uncoiled worm 1-125 of an inch in length, and 1-630 of an inch in diameter; the cysts are often wanting. The worm has a well marked digestive apparatus and reproductive organs. The number of the larvæ may amount to millions in the same subject; they have great vitality; Davaine kept them alive in water for a month. The adult worm perishes rapidly.

The presence of trichinæ in the alimentary canal gives rise to diarrhœa; this flow may or may not expel the worms. Their passage through the intestinal walls may give rise to peritonitis. Their presence in the muscles is usually announced, about two weeks after ingestion of the food, by rheumatic pain increased by attempt to use the muscles. They prefer the muscles, and the voluntary ones at that, to all other parts, and are naturally most numerous in the abdominal muscles and diaphragm, as being nearer the intestines. The muscles degenerate and their functions appear to be interfered with by the presence of the cysts; so that an animal which has withstood the diarrhœa, the fever, and the other symptoms, typhoid in character, which attend the presence of the entozoa in the tissues, may die eventually from debility.

The progress, duration and severity of the disease, depend upon the number of trichinæ in the subject; the disease may last from one to three months. The diagnosis may sometimes be made by removing a portion of muscle and examining it microscopically. Protection consists in avoiding uncooked meat. The treatment is directed first to the removal of the worms from the intestines. After their arrival in the muscles, it is not probable that any treatment would directly affect them.

In the *Practitioner*, XXIV 1880, pp. 388 and 466, is a report of an outbreak of trichiniasis or trichinosis, among the boys of the reform ship, Cornwall, in September, 1879, and which was at first thought to be typhoid fever. Reference is there made to a similar attack in 1875, affecting nearly fifty per cent. of the boys, and which was also thought to be "enteric fever."

A long article on "La trichine et la trichinose," by Ed. Dele, *Annales de la soc. de med d'Anvers*, 1879, p. 473, discusses very fully the whole subject.

There is also a critical review in *Schmidt's Jahrbucher*, 1878, p. 955, which is reproduced in the *Annales d'Hygiene*, 1879, tome (vol.) III, p. 497.

Some Cases of Toxhæmia Causing Temporary Insanity.

By Dr. F. P. Blake, Canon City.

CASE I.

Sometime in September, 1879, I was invited, by Dr. James, of Silver Cliff, to go over to the McClure House and see a patient that he had brought down, or rather, had accompanied so far on his way east to his friends. The case had been declared typhoid fever. I regret that I am unable to give a complete history. The symptoms had not been marked, except one, that had been so from beginning to end, and that was insanity. The patient was violently maniacal and required, at times, the united efforts of two strong men to restrain him. There had been insomnia. No amount of opium, chloral, hyoscyamus or whiskey had the least effect. This condition amounted, in fact, to coma vigil. A business partner had undertaken

to see him home, and no argument could dissuade him. After resting at the hotel over Sunday they continued their journey. The patient died on the train east of Pueblo—duration of illness between three and five weeks.

CASE II.

One evening in the summer of 1880 a man called at my office, accompanied by a younger companion who was complaining of feeling confused. They hailed from the New England sea coast. While in the office the young man betrayed, through his random talk, that his mind had gone wrong. Thought if he was married and had some children it would cure him. Pulse was quick. Tongue dry. Nights restless. Bowels costive. I ordered his friend to keep a close watch of him. Gave a hydrargogue cathartic, and followed the dose with sedatives. The medication was responded to very happily, but by the second day, the patient awoke from a nap (his friend was enjoying the shade on the sidewalk in front of the hotel) and finding himself alone, and in a strange place, I suppose he became alarmed, for, rushing from his room with a cry, he entered another room where was another patient, whom he attacked by might and main, biting and scratching and otherwise acting on the offensive. The outcries of the two soon brought a mob, among which were three or four men who had the courage and tact to overcome the poor fellow. He was taken to the county jail where, after a two or three weeks' illness, he recovered in body and mind and went to Denver to work at his trade.

CASE III.

An old gentleman, who was boarding at one of the hotels, became mentally deranged. He would promenade in front on the sidewalk for hours at a time, refusing to speak to any one. Then he would take long walks through the heat and dust, returning and ordering his meals at untimely hours, and then he would go to bed assuming an attitude of prayer. He would lie on his back for a whole day—eyes fast closed, the flies swarming over and on his face and in his eyes, but he made no effort to drive them away, for his hands were fast clasped above his head. During these devotional exercises he would refuse to open his eyes or answer a question, though he did not resent any ordinary physical examination. Pulse about 100. Temperature not taken. Tongue dry and coated. Bowels constipated. Patient refused to take any medicine of any kind. As he had no friends he was allowed his own way. The county paid his board, entered into correspondence with his friends in the east and were about to send him off. In about three weeks' time the patient got up one morning, went down to breakfast and called for his bill. The three weeks' time was all a blank to him. He refused to settle with the landlord only for four weeks. He had been in the house seven weeks. During all this time the man denied being ill, refused all medicine, recovered and went right out in search of work. He was a stonecutter by trade.

CASE IV.

Sometime during the past summer I was invited by Dr. Craven to see a patient with him. If I remember rightly the pulse never was above a hundred. The temperature never very high. Tenderness in right iliac region. Patient able to sit up, and would sit up if allowed to do so. For a week or ten days before taking his room this patient had been taking pedestrian tours up the Grand Canon and to other points of interest, making sketches, for he imagined himself an artist, while his sketches, I was told, were rude and betrayed his ignorance of art entirely. He would talk of his wife, though he had none. Under a supportive treatment for about three weeks he rallied, and mentally was as strong and sound as any patient sick for some length of time. His father came to him and when he recovered accompanied him home.

CASE V.

A man was arrested at the depot for noisy demonstrations. He had inflicted injuries upon himself, had thrust sticks down his throat trying to reach the poison that was there. Had delusions; thought the newspapers were all talking about him and a famous wedding that was to come off, himself one of the happy participants of the coming nuptials. He was exceeding wealthy, and so was the prospective bride. Pulse 110. Tongue dry and coated. Bowels constipated and bad—been so for some time. Gave hydragogue cathartic, diuretic and febrifuge sedatives to give sleep at night, which he needed sadly. The acute symptoms subsided as did the offensive conduct of the man, when the jailor turned him loose. As he had no money, one hotel after another turned him away, when, late at night, he got into a quarrel with the clerk and threw a stone through the window. This caused his re-arrest, when he was again placed in jail. A jury was summoned who declared him insane and he was sent to the asylum. I have no doubt that he recovered or died in a month's time, and that he was no more insane than the others and would have rallied under continued proper treatment, or, like Case 3, with none, if he had had a good boarding place, with his bill paid for him. These are five of some ten or twelve cases I have treated, or been invited to see, during the past two years. There has been this similarity in all of them:

(1.) The duration of illness about the same.

(2.) Recovery in most of the cases as good as in any other case of typho malarial fever.

(3.) Absence of other marked symptoms in most of the cases.

I have no doubt that the cause in all was the malarial poisoning of the brain through the blood.

Asphyxia of the New-Born Child.

BY DR. HOUZEL, in the Review Medicale.

Translated by A. Labrie, M. D., Denver.

When a child is born seemingly in a state of death, I immediately remove the mucus from the mouth and the glottis, and at the same time order two pails of water to be brought, one at the temperature of about 50° (cent.) and the other as cold as possible. If insufflation does not succeed, which is nearly always the case, I take the child with one hand on the buttocks and the other on the shoulders and nape of the neck, and plunge it in the warm water to its neck, and hold it there for about half an hour, meanwhile practicing artificial respiration by pressing the thorax with both hands. I then raise it suddenly and put it at once in the cold water for a few moments and replunge it again in the warm water, keeping up this process until the respiration is perfectly established. Frequently at the third or fourth immersion, a deep inspiration is observed. By persevering with this process the inspirations follow one another at shorter intervals, till finally the normal function is completely established.

I have now employed this simple process for over twelve years with nearly always perfect success. From my observations, whatever is the degree of asphyxiation of the child at birth, provided the sound of the fœtal heart was heard just before the labor began, we must always endeavor to recall it to life.

It has frequently cost me more than two hours of persistent effort, but I have often been rewarded by saving the lives of children which the assistants and midwife herself had given up for dead. In no case has the cold water immersion been followed by the slightest accident. I have been led to employ this method by the observation which is observed in the cold bath.

Plunge an adult in cold water, and at once a sudden impression is produced on the skin, which is followed by an irrisistible reflex action of all the muscles of inspiration. Involuntarily a deep inspiration is produced, and frequently even a cry. That is what I am after and what I am nearly always able to obtain in all cases, in throwing, so to speak, the new-born child in cold water.

I put it in cold water first in order to avoid any sudden refrigeration which might prove fatal, and at the same time to establish a still greater transition of temperature, and be able to replunge it in cold water with impunity as often and as long as may be necessary.

What I seek above all is the energetic and sudden impression of the skin, and from my observations it is nearly always at the moment that the asphyxiated child touches the cold water that it takes its first inspiration, which is often so sudden that it is accompanied by a cry.

Whatever may be the interpretation of the fact, the result alone is important. I have employed, I repeat, this means a number of times without ever noticing the least accident.

Lacerated Cervix—Report of Three Cases—Operation—Recovery.

By Thomas H. Hawkins, M. D., Denver, Late attending Physician and Surgeon, Out-Door Department, Bellevue Hospital.

CASE I.

During the month of September, 1880, I was consulted by Mrs. S., age fourty-two,

married twenty-seven years, the mother of seven children, was perfectly well up to the birth of her last child, six years ago. Her last labor was a tedious one, and instruments were used; since which time she has been an invalid, spending most of her time in bed. Walking or even standing caused her much pain. She has suffered greatly in consequence of menorrhagia and dismenorrhœa during the past five years, also from obstinate constipation and vesical tenesmus, urinating five to ten times in a single night, sleeping when under the influence of an opiate. An examination revealed a rectum filled with hardened fæces. The anterior wall of the vagina relaxed, and a well marked cystocele; excessive tenderness about the region of the rectum and bladder. The uterus engorged and very much enlarged, slightly prolapsed and completely anteverted. A bilateral laceration of the cervix extending to the vaginal junction, and the lips hypertrophied, the anterior one so much so that it was mistaken by one physician for a fibroid tumor. The uterine canal measured three and three-fourths of an inch in length.

On January 20, 1881, I closed up the lacerated cervix. The patient being brought thoroughly under the influence of ether and placed in Sims' position, the uterus was readily drawn down by means of tenacula. The edges were pared, a V shaped piece removed from the angle on either side and the denuded surface brought together by silver-wire sutures. The probe was then introduced and the canal was found to measure three inches, being three-fourths of an inch less than before the operation.

The patient was kept quiet and a full dose of opium administered. She suffered but little pain, and there was only a slight elevation of the temperature at any time. Vaginal injections of hot water were resorted to every day. At the termination of ten days the sutures were removed and union was found to be perfect, except on the left side, where, between two of the sutures, a probe was readily passed into the uterus. The edges of this small aperture were pared and another suture introduced. This united kindly. At the end of another ten days this suture was removed and union was found to be perfect. The patient was put on small doses of the fl. extract of ergot, tonics administered, and the bowels kept open.

The uterus rapidly reduced in size. Hot vaginal injections were used once a day, and after each injection, the vagina was packed with absorbent cotton. The cystitis, in a short time, subsided, and Gehrung's pessary for cystocele and anteversion was inserted. At first the patient experienced much difficulty in wearing a pessary. By alternating, the cotton one day and the pessary the next, she was enabled to tolerate it and gradually could wear the pessary two days and the cotton one, etc.

At the present writing (April 17th) she is free from pain, wears a Smith's anteversion pessary with perfect ease. This keeps the uterus in place. The cystocele has nearly disappeared, there is no cystitis, the cervix is solid and reduced to nearly its natural size. No pain during menstruation, has taken no opiates for more than a month, and sleeps moderately well. Walks a mile twice daily. The constipation, though not entirely relieved, is much improved. It might be interesting to mention that five homœopaths and seventeen doctors had treated her, but none had ever made a correct diagnosis, one calling it a "high inflammation of the uterus," another, "cancer," and so on.

CASE II.

Was called to see Mrs. M—— December, 1880, and recorded in my note-book the following: Aged twenty-eight years, has three children, the youngest are twins, has been married five years, been perfectly well up to the time of her last labor—fourteen months ago.

She is a typical blonde, weighs twelve pounds less than her normal weight. Menstruated eight months after the birth of the twins, and every two weeks since. The flow is profuse and usually lasts from three to five days.

Present condition : She has been flowing eleven days, is very weak, unable to sit up, and her face is blanched.

Examination reveals an anteverted and subinvoluted uterus, a laceration on the left side of the cervix, extending very nearly up to the vaginal junction. Fluid extract of ergot in twenty-five drop doses every two hours, and hot water injections per vaginum was ordered.

December 7.—Still flowing, but not quite so profuse; replaced the uterus and tamponned the vagina with absorbent cotton.

December 8.—No flow, cotton removed and vagina syringed out with hot water and small amount of carbolic acid.

This patient, after being prepared in the way which I shall speak of later on in this paper, had the lacerated cervix closed. The uterus was easily drawn down, in fact, the cervix was pulled out even with the vulva.

The entire operation, including the administration of ether, only lasted thirty-five minutes. On the tenth day the sutures were removed and union was perfect.

During this ten days she did not have a particle of pain. Hot water injections per vaginum were used every second day. The uterus put in position and supported with cotton saturated with glycerine, small doses of ergot given internally. At the end of one month the uterus was much reduced in size and remained in position without support.

May 19.—The patient is well, menstruates normally.

CASE III.

Miss C., came under my care November, 1880, and I take the following history of her case from my record book:

Miss C., aged 28, has had one child, being barely seventeen when it was born. Her health was good up to the birth of this child, eleven years ago. For the past five or six years she has suffered from severe pain in her back, (lower part of spine, as she calls it) frequent desire to urinate, suffers from menorrhagia and metrorrhagia. By an examination the uterus was found to be enlarged, tilted forward, the fundus resting on the bladder and the cervix on the rectum. The os uteri was surrounded by extensive erosions, the cervix soft, enlarged and lacerated on both sides, extending half way up to the vaginal junction. In this case the flow was so frequent and excessive that I thought it best to make an attempt to arrest this before performing hystero-trachelorrhaphy. By the use of hot water injections every day, and cotton saturated with the glycerite of tannin, packed around the cervix, thus lifting up the fundus and improving the circulation of the blood in the pelvic viscera, this difficulty, to a certain extent, was relieved. Small doses of ergot were also administered during this time.

January 29.—The patient being etherized and placed in Sims' position, I proceeded to close up the lacerated cervix. In paring the edges it was found that the laceration had, in the first place, extended up to the vaginal junction, but partial repair had taken place by the filling in of a large amount of cicatricial tissue. A V shaped piece was excised from both sides and extending up to the vaginal junction, great care being taken to remove every particle of cicatricial tissue. The cervix was then closed with thirteen silver-wire sutures, a light tampon of dry cotton was placed around the cervix, a full dose of morphia administered, the hips elevated and the patient kept perfectly quiet. During the ten days following the operation, hot water injections were used and she was kept on a light, but nutritious diet. No opium was required, as she suffered little or no pain. On the tenth day the sutures were removed, union being perfect, but the cervix was still spongy and soft. The os was somewhat dilated and bled on the slightest touch. On the fifteenth day after the operation she menstruated. The flow was very profuse and lasted seven days. After this the uterus was replaced and kept in position by means of cotton saturated with glycerine. She was given internally a half drachm of the extr. ergot three times a day, and general tonic treatment. Hot water injections per vaginum every second day.

February 25.—Patient's general health much improved. Uterus somewhat reduced in size, but she menstruates every fifteenth day as before. At this examination it occurred to me that I had not used the curette. Suspecting that the uterine cavity was lined with fungoid growths, this instrument was introduced and an unusually large number of these growths removed. This operation gave rise to but little pain, and the hemorrhage was slight. After this she was kept quiet for a few days and the ergot continued. The uterus being put in position, a Smith's pessary kept it there.

May 19.—The uterus is in position and she still wears the pessary. The pain in the back has disappeared. The cervix is reduced in size and feels solid and firm to the touch. She menstruates once a month, and the flow is in every way normal.*

I will close this paper by offering a few suggestions with reference to certain precautions necessary to be observed in order to arrive at a correct diagnosis, as well as

to ascertain what the indications are for the operation. Also how to prepare a patient for hystero-trachelorrhaphy. To arrive at a correct diagnosis the physician should first of all gain the entire and complete confidence of his patient. By so doing he will save her much unnecessary embarrassment, to say nothing of the nervous sensitiveness that would otherwise exist, and thus prevent a thorough examination. If, in making a digital examination, the finger detects what seems to the touch a lacerated cervix, note carefully the os and also whether the lips of the cervix appear to be everted or not.

To ascertain whether this is a laceration or an erosion of the cervix, examine the patient standing. If it is an erosion (or ulceration) there will be no change in the os or cervix, and but little in the position of the uterus. But if a lacerated cervix, and especially if a bilateral one, the uterus will be found to occupy a lower plane in the vagina. The os as it appeared in the first examination will have disappeared, and the finger will pass, if it be an extensive laceration, very nearly if not quite up to the internal os. The lips will be much more everted and drawn upwards.

If there has been an attempt at repair, and there is a large amount of cicatricial tissue, the above condition will not be so well marked. To decide more positively the actual condition in a case of this kind, use a bivalve speculum, or, better still, a Sims', exposing the os and cervix clearly to view in good light; with two small tenacula approximate the lips. If, when the lips of the cervix are drawn together, the deformity disappears and the probe being introduced shows the canal to be shorter than when they are not approximated, you have unquestionably a lacerated cervix. There is another form of laceration of the cervix, that of the internal os, but of this I shall have more to say at some future time. The indications for the operation on a lacerated cervix are numerous, some of the more prominent ones are the laceration of the cervix itself, which weakens the support of the uterus and interferes with the circulation of the blood in the pelvic viscera in general. This condition, in turn, causing sub-involution, partial prolapsus, some one of the displacements, and fungoid growths in the cavity of the uterus. This last condition usually giving rise to metrorrhagia and menorr-

*I was assisted in these operations by Drs. Cox, Kimball, Edmondson, Cole, Skinner, Smith and Kline.

hagia. The above indications are sufficient
to justify the closing up of a lacerated cervix, providing there are no contra-indica-
tions, such as pelvic cellulitis and peritonitis,
ovarian cyst, large fibroid tumors, softening
of the cervix, old age, etc., etc.

The carrying out of certain preparatory
measures which are absolutely necessary if
any degree of success is to be obtained in operating on a lacerated cervix, is of the greatest
importance. First, you have the debilitated
condition of the patient to contend with.
This may be, in part, remedied by tonics,
fresh air, putting the uterus in position and
maintaining it there by the use of cotton and
glycerine for a few days. Hot water per
vaginum should be used every night at bedtime. To obtain the best results from this
remedy, the patient should be placed on her
back, hips elevated and a bed-pan used. The
patient should never be allowed to sit up on
a chamber while using these injections, for,
by this means, a valuable remedy will be
brought into disrepute and she will derive
no benefit therefrom. A "Davidson's" syringe, or one constructed on the same principle, is better than the "Fountain," for reason of the interruptible or jetting stream. A
hard rubber nozzle is preferable to a metalic
one. The water should be as hot as the patient can possibly endure, and the temperature gradually increased. If the patient has
metrorrhagia, menorrhagia or dismenorrhœa,
the os should be thoroughly dilated, while
the patient is kept in bed, and the cavity of
the uterus examined for polypi and fungi.
If the former are found, remove them by
tortion or otherwise. If the latter, Sims' curette, should be freely used, thoroughly scra-
ping every part of the uterine cavity. After
using this instrument, administer an opiate
and enjoin rest for at least three days. During this time injections of hot water should
be used with great care, the os being dilated.
If the cervix is spongy and soft, paint it
with Churchill's tinct. of iodine every third
day, and continue with the hot water; also
with the cotton and glycerine, with the addition of tannin, (glycerite of tannin) for another ten days, the patient will then be in a
fair condition for the operation. During
this time the bowels must be particularly
looked after and special attention paid to
diet.

The operation should never be performed
immediately before or during the menstrual
flow. The fifth day after its cessation I consider the best time.

The Doctrine of Enucleation.

By Dr. Litton Forbs, L. R. C. P.,
Fellow of the Ophthalmological Society of Great Britain.

There is perhaps no operation in the whole
range of surgery, with the exception possibly of craniotomy, which throws so much
responsibility upon the surgeon, or about
which it is more needful for him to have
clear and precise notions, than that of the
enucleation of an eye. There is no operation which at times makes greater demands
on the faith of a patient in his medical
attendant, and there is no operation in which
the consequences of an error of judgment
on the side either of the patient or surgeon,
is more likely to prove disastrous. Partial
or total blindness is not the greatest evil
which can result, for more than one case is
on record in which a patient has actually lost
his life from surgical aid having been inju-
diciously withheld or administered.[*]

In the present paper I would venture to
discuss from a purely practical point of view
the Doctrine of Enucleation as at present
received. The subject is perhaps one of
quite as much importance to the general
surgeon as to the specialist, for cases requiring enucleation occur as often in the practice
of the one as in the other.

I shall here endeavor to lay down a few
general propositions which may guide us in
the performance of this very important operation in doubtful cases. My excuse for so
doing is that, for some years past I have devoted myself specially to this subject, and
have enjoyed, what I may call, exceptional
opportunities for studying it. At the Royal
London Ophthalmic Hospital, with which I
have been connected for the last five years,
as many as 100,000 patients have been
relieved in one year, of which an even larger than average number were traumatic
cases. This is explained by the proximity
of the Hospital to the docks and other great
industrial establishments of East London,
which, as is well known, are always prolific
of surgical injuries. Subsequently in Paris,
at the Clinique of DeWecker, and at the

*See Mackenzie's Maladies des yeux, Vol. II., p. 127.
Translated by Warlomont and Testelin.

Krankenhaus, in Vienna, under that veteran surgeon and ablest of Ophthalmologists— Prof. Arlt, I was enabled to compare the practice of the leading ophthalmic surgeons of England with that of their confreres in Europe.

This subsequent experience has led me to modify a good deal in my own practice the theoretical doctrines which, like heir-looms, have been handed down to British ophthalmology, and which may almost be said to be traditional in the London school. Some of these doctrines were essentially and primarily developed amid the pauper surroundings of the large English hospitals, and received, as it were, a great increment of support by the Scotch adaptation of the American discovery of anaesthetics. Sympathetic ophthalmia was first described as a distinct disease by Mackenzie,* of Glasgow, an ophthalmologist of immense clinical experience and great natural acumen. Not long afterward, Sir J. Simpson announced the discovery of chloroform, with this result, among many others, that probably many eyes were enucleated which would previously have been left in the patient's head.

Of late years there is no doubt but that a reaction has set in somewhat against the heroic doctrines of the English school, as laid down by Lawrence, Critchett and other writers. The protest has come most clearly and earnestly from Prof. Mauthner, of Vienna, who, in his admirable essay,† on "Sympathetic injuries of the eye," makes a forcible appeal to the pathologist, not to enucleate an eye merely "in order to examine it under his microscope." I would say, in passing, that in no place was this advice more needed than in Vienna itself, where pathology, rather than therapeutics, is the dominant passion. Mauthner's protest has, however, I think, found supporters among the more thoughtful of the younger ophthalmologists, both in Europe and America. This is exemplified by the earnestness with which attempts to find some substitute for enucleation are being pursued. We are all familiar with Critchett's abscission, with Weeker's and Knapp's modifications of that radically

bad operation; with Meyer's ciliary neurotomy; with optico-ciliary neurectomy, and other operations having in view the saving more or less completely of the outer semblance of an eye hopelessly damaged, while, at the same time, shielding its fellow from future risks.

In the following observation I will venture to give the results of my own clinical experience, from which I have deduced a code of rules, which, though necessarily imperfect and subject to modification, may prove useful in the conduct of doubtful cases.

I think it will conduce to clearness if we consider the indications for enucleation under three heads, viz.:

1. Those cases in which enucleation may be considered as demanded.

2. Those in which it is of doubtful utility.

3. Those in which it is contra-indicated.

As regards the first, I think it may be taken as an axiom that every eye hopelessly damaged, and at the same time painful, should be removed. This for several reasons: First, because it is functionally useless and may eventually become a source of danger. Second, because, should it subsequently endanger the other eye, its removal, then, might be of no avail. Third, because it is unsightly and ill adapted for carrying any artificial appliance. Hence, by an immediate removal, the patient's safety, comfort and personal appearance may be considered as best provided for. These are self-evident propositions, which probably no one, except a refractory patient, will care to call in question. In London, an eye is considered irretrievably lost when wounded in the ciliary region, and is generally enucleated at once, say an hour after the receipt of the injury. In the practice of other surgeons, an injured eye will be perhaps left unmolested for a year, and though completely atrophied will not give rise to any trouble in its fellow. I believe one of the longest, if not the very longest, period on record between the destruction of an eye and the supervention of sympathetic symptoms in the other, occurred last year in my own practice. I enucleated a stump fifty-six years after the receipt of the original injury, with excellent results, in a woman aged sixty-four. The whole question, of course, resolves itself into what constitutes an irreparable injury. My answer,

---

*See DeWecker's Ocular Therapeutics—translated and edited from the French, by Dr. Litton Forbes, p. 237.

†Die sympatische storungen des Anges—one of the series of Professor Ludwig Mauthners' excellent vortraege aus dem gesammtgebiete der augenbeilkunde—of J. Bergmann Wiesbaden. 1879.

in general terms, would be—whatever of necessity destroys the functions of the eye. Many injuries do this instantaneously, others only at a later period ; if, however, it is evident from the first what the result will be, I think the eye should be enucleated without useless delay. Of course the great argument in favor of early enucleation is that it lessens the risk of sympathetic inflammation. Granted that it lessens it, but there is no ophthalmic surgeon of any experience but is aware that it does not do away with it altogether. Instances of sympathetic irritation or even of sympathetic ophthalmia are unfortunately too common, even after immediate enucleation, to enable us to neglect them in our calculations. I have fortunately not met with any such in my own practice, but three have lately been reported by Mr. Nettleship, in the last number of the British Medical Journal (April 16, 1881),and a case is mentioned of sympathetic iritis following five days after enucleation. Such examples should make us pause before assuring a patient that after immediate enucleation he may consider his remaining eye as absolutely safe. It, however, assumes still greater force when we come to consider the second part of the subject, that is, those cases in which enucleation may be considered as of doubtful necessity.

These certainly form a most difficult class of cases, concerning which a great deal has been said and written. Every surgeon must, however, frame rules of action in such cases for himself, and not be too ready *jurare in verba magistri*. It has been laid down by several authorities that the question of immediate or deferred enucleation should, in a great measure, be decided by the social condition of the patient. If he is a laboring man, unable to afford either the cost or the delay, consequent on a long course of special treatment, enucleation had better be performed without loss of time. If, however, his circumstances are such as to enable him to command prompt surgical advice, then enucleation may be delayed until the course which the injury will eventually take has been manifest. Suppose, for instance, a case where a man receives a pellet of shot in the eye, which, having perforated the sclerotic, vitreous and retina, has lodged in the posterior tunics of the globe. It may be asked, should such a case be submitted to immediate enucleation if occurring in no matter

what condition or calling of life? I would answer, No! with considerable confidence, because there is a chance that the shot may become encysted, and that eventually the eye may retain useful vision. If, however, in spite of the shot having become incysted, the eye should continue tender to the touch after a reasonable time has elapsed, I should then recommend immediate enucleation, and that for two chief reasons ; first of all, because an eye containing a foreign body and being itself in a state of irritation, is a standing menace to its fellow, and, secondly, because the spontaneous intra-ocular inflammation, of which pain is the symptom and proof, must eventually destroy sight in the wounded eye. Apart from the mere traumatic lesion, the subsequent contraction of cicatricial tissue will be enough to effect this. This contraction may be powerful enough to eventually cause a detachment of the most anterior portion of the ciliary body. In such a case, any attempt later on to form an artificial pupil may insure the extension of the irritation to the other eye, which, perhaps, till then had remained unaffected. Should the pellet of shot have traversed in its course the ciliary region in addition to the vitreous and retina, there can I think be no hesitation in recommending immediate enucleation.

In the case of foreign bodies within the eye, there is, of course, very frequently the chance that they may be successfully removed by surgical intervention. Hence, no surgeon, I take it, would be justified *prima facie* in removing an eye because he found a foreign body in the iris, lens or vitrious. He should, in all such cases, attempt its removal, reserving to himself, with the patient's consent, the right of immediate enucleation then and there, should the operation not prove feasible. The question of enucleation, in cases of foreign bodies in an eye which may still preserve some useful vision, is, by no means, an easy one to decide upon. I do not think any generally applicable rule can be laid down on the subject. Such cases must be treated on their own individual merits and on broad surgical principles. I will merely add that no class of cases presents such a wide field·for differences of opinion among specialists. In the multitude of counsels which such cases call forth, the patient, at least, will be strongly in favor of retaining his damaged eye, till all the doctors, without exception, shall agree that it must be removed. Unhappily, when this occurs, enucleation often comes too late to be of much benefit.

There is another large class in which the advisability of enucleation at once presents itself. I refer to cancerous growths. Most of us have probably watched with the ophthalmoscope, the slow and stealthy onset of an intra-ocular tumor. Appearing, at first, as a dark, perhaps jet-black, ill defined mass somewhere behind the ciliary margin of the iris, it is seen with difficulty, and diagnosed with still more difficulty. It causes the patient but little annoyance and does not interfere with vision. Removal of the eye is, however, sooner or later inevitable, but I am, myself, opposed to too early a removal.

It is a great thing to have your patient thoroughly on your side in every case of enucleation, and I have, more than once, seen patients change their medical adviser at the bare mention of the remote possibility of such an operation. Hence, I would venture to recommend delay until sight has been completely, or in part, destroyed, and until there can be no possibility of doubt as to the diagnosis. The operation should not, however, be deferred until the cancerous mass perforates the ocular tunics and appears in the orbit. Enucleation is then useless; nothing but exenteration of the orbit will be of the slightest avail, if even this has not come too late. In the case of melano-sarcoma commencing on the cornea, the patient will, probably, not at once consent to enucleation,and, therefore, the growth may be treated for some time, locally, without much danger. When the summit of the cornea has, however, become involved, sight is usually irretrievably lost, and, therefore, no further delay is desirable. It might seem that no doctrine could be more irrefragable than that of the necessity of enucleation in true cancer of the eye-ball, or its surroundings, and such, no doubt, is the case. But, probably, more eyes have been unnecessarily sacrificed for so-called glioma retinae than for any other disease. Glioma is a most malignant form of cancer, and probably always recurrent. When it does not recur after enucleation, it is hardly too much to say that it was not present in the first instance. In its earlier stages it is, by no means, easy to distinguish from retinitis with effusion of lymph or post

retinal formation of pus. This detachment of the retina has led to many eyes in children being unnecessarily removed. My attention was first called to this "false glioma" by Dr. Brailey, curator of the Royal London Ophthalmic Hospital. The condition is not described with sufficient fullness or emphasis in any of the text-books with which I am acquainted. The chief points of diagnosis are that, in glioma, the ocular tension is increased, whereas, in detachment, it is diminished; glioma is highly vascular, detachment of the retina, much less so; glioma generally exhibits small cauliflower excrescences, detachment of the retina never; in glioma the surface of the tumor is generally more or less polished, with, perhaps, nodules on it; the surface of a detachment is generally somewhat dull and of a subdued yellow color. Of course, the Ophthalmoscope, especially when used for the direct image, will give the most certain information. Assuredly it would be a most gross and grave error for a specialist to enucleate an eye for glioma, where there existed only detachment of the retina. He would not merely expose his patient to the risks of a severe operation at a very tender age, but would also destroy all symmetry of face by effectually checking the development of the orbit and surrounding parts.

I pass on now to the consideration of the third part of my subject, viz. : Those doubtful cases in which the balance of evidence is against enucleation as a method of treatment. Perhaps the most important group of such cases is that in which sympathetic ophthalmia has already set in. Suppose that an eye A has been wounded in the ciliary region,and developed unmistakable symptoms of chorido-cystitis ; and that eye B has sympathized with it, and vision is now considerably reduced, but on the whole is somewhat better in eye A, that is, the one first injured, than in B. In such a case what treatment would give the patient the best chance? Would it be to enucleate A, and treat B energetically, hoping later on to perform iridectomy or iritomy, and so to restore some useful vision ; or would it be to abstain from all physical interference? The latter would, I think, be the best plan to pursue. It is evident that the removal of either eye would not at this stage materially affect the prognosis or course of the disease. A great deal in such a case might be done by energetic

medical treatment with mercury, as recommended by Stelwag and DeWecker, and later
on, after all the inflammatory symptoms had
disappeared, an iritomy performed with a
DeWecker's scissors, might restore a
considerable amount of vision. Similarly
the question of enucleation as a means of
treatment, either *per se* or in preference to
some other operation, will frequently arise.
Thus, in buphthalmus, a choice will frequently
lie between iridectomy and enucleation, and
in such cases the balance is nearly always in
favor of the latter; the same may be said of
staphyloma with enlarged blood-vessels,
where any operation short of extirpation
may produce a most dangerous hemorrhage.
Enucleation, however, is most strongly
contra-indicated in all inflammations of the
eye or its appendages during the acute stage.
In panophthalmitis an heroic incision
should be made across the front of the eye
or in the sclerotic between the recti muscles,
and the whole contents of the globe evacuated. Enucleation in such cases is fraught
with the greatest danger. Many cases are
on record in which, after enucleation, the
sheath of the optic nerve has been
invaded by pus, and where fatal meningitis
has quickly supervened. DeWecker relates
such a case in which thirty hours after the
operation the patient was seized with hemiplegia of the left side, and died six days later
with all the symptoms of basilar meningitis.

In conclusion I should like to formulate the
following propositions, which I think will
include the modern doctrine of enucleation.

1. Every eye hopelessly blind, if also tender and painful, should be enucleated.

2. An eye should not necessarily be enucleated, especially in young persons, merely
because it is inapt for vision.

3. Enucleation is probably the best prophylaxis we possess against sympathetic ophthalmia.

4. Enucleation of an actively inflamed
eye may be an operation dangerous to life.

It should be remembered that the delay
which has necessitated the issuing of the
April and May numbers together, was unavoidably caused by our removal to Denver,
and consequent change of business arrangements, place of printing, etc. Hereafter, re-
mittances and all business communications
should be addressed—Rocky Mountain Medical Review, Denver, Colorado.

RECENT MEDICAL PROGRESS.

Quarterly Report on General Medi-
cine and Therapeutics.

By D. S. LAMB, A.M., M.D.,
Professor of Anatomy, Howard University, Washington, D. C.

SMALL-POX.

Dr. Schwimmer recommends (Gaz. des
Hop, Dec. 18, 1880, p. 1171; from Nouv. jour.
Med.) to make a mask of soft linen, with
openings for the eyes, nose and mouth; and
spread on the side toward the face, one of
the following formulae. The mask is removed every twelve hours. The ointment may
be applied to the parts of the face not touched
by the mask, and to the hands, on compresses:

Form. 1.—Carbolic acid, one to two and a
half drachms; olive oil, ten drachms; powdered chalk, two ounces. Form. 2.—Carbolic
acid, 75 grains; olive oil, ten drachms; very
pure starch, ten drachms. Form. 3.—Thy-
mol, thirty grains; linseed oil, ten drachms;
powdered chalk, two ounces.

DIPHTHERIA.

Dr. Misrachi (Gaz. des Hop., Oct. 7, 1880,
p. 932) adds his testimony to that of Letz-
erich and of Klebs, in regard to the efficacy
of benzoate of soda; which, he thinks, is
not due to the solvent action of the drug on
the diphtheritic membrane; nor to disinfectant properties. It was used by pulverizing
a ten per cent. solution; at the same time it
was given internally in the dose of forty-five
grains in water every three hours. Ice, soup
and wine were also used.

THE PLAGUE.

Sir Joseph Fayrer (Practitioner, Dec. 1880,
p. 470) in an address before the Epidemeo-
logical Society of London, discussed the
plague of 1878-9 on the banks of the Volga.
The commissioners appointed by the English and German governments respectively,
differed as to the origin of the disease. The
former regarded it as most likely endemic;
the latter thought it was due to the importation of infected articles from the Asiatic seat
of the Russo-Turkish war.

DIABETES.

Hanocque (Societe de Biologie, Jan. 15,
1881, in Le Prog. Med. Jan. 22, 1881, p. 62)

says that uro-chloralic acid found in the urine of certain chloralized animals can give analogous chemical reactions to glucose, and the two cannot be distinguished by the polarimeter.

CARBOLIC ACID IN FEVER.

A further paper on this subject was read before the Academy of Medicine (*Bull. Gen. de Therap.*, Dec. 30, 1880, p. 550). The conclusions are that the acid is a sure and prompt antipyretic, but the action is brief. It can be employed in all fevers. It should be used boldly, although its effects, except in the beginning, should be watched. Its intermittent administration in large doses, gives better results than its continuous administration. During a lengthened employment of the acid, it is well to watch the heart and kidneys, although, so far, there is nothing to prove a resulting degeneration of these organs.

SCROFULA AND TUBERCULOSIS.

Constantin Paul (*Gaz. Hebdom.*, Jan. 21, 1881, p. 39) gives a new sign of the scrofulous diathesis, viz.: After piercing the lobule of the ear, to insert an earring, a scrofulous ulcer appears, which endures for months and sometimes years. Besnier and Fereol confirmed this sign.

Dr. Thaon, in a note to the same society, Dec. 24, 1880, (*L'Union Med.* Jan. 8, 1881,) states the results of his anatomical researches upon tuberculosis and scrofulosis. The scrofulous ganglion has all the histological characters of tubercular lesions. If, then, ganglionary scrofula is tuberculous, phthisis is a scrofulous manifestation—scrofula of the lung, which may be independent of scrofula elsewhere. Geographically, lung scrofula does not appear in the same zone as external. The former is almost unknown in the Alps, where external scrofula is very common. The reason is that, in · mountainous countries, the lung is preserved because the nutrition is perfect; the air is always pure. In the lower lands, cities and shops, where the air is bad, the lung suffers, and scrofula localizes itself in the feeble point. The skin, however, feels the intemperateness of the mountain climate and is, therefore, prone to show scrofula. Tubercle is an inflammation, perhaps specific if ever its contagiousness is demonstrated in the existence of a germ. Like all inflammations it appears as miliary foci or extended infiltration; it un-

dergoes a special evolution which leads to caseification or fibrous transformation. It is always accompanied by fever.

Dr. Heitler (*Jour. de Med.* Jan. 1881, p. 26) examined the lungs of over 16,000 cadavers and found 780 cases dead from non-tubercular disease which had evidences of the cure of *phthisis pulmonalis*. All the subjects belonged to the working classes. In 651 cases the lesion was bilateral, nearly always limited to the summit. There were cicatricial nuclei, strongly pigmented, very black and surrounded by grey or yellowish nodules, and cicatrized caverns, varying in size from that of a nut to an egg. The foregoing statements, which are extracted from the *Med. Jahrbucher* and *Lyon Med.*, are supplemented by the editor recalling the researches of Bean at the Salpetriere, which were to the same effect.

ABSCESS OF BRAIN.

Ribbert (*Berlin Klin. Woch.*, 43, quoted by *Amer. Jour. Obstet.* Jan. 1881, p. 260) reports an autopsy on a twelve-day-old child, whose mother had puerperal ulcers. The throat, mouth and larynx of the child were covered with *thrush* membrane; there was an abscess of the cerebrum containing the spores and threads of the thrush fungus. A second similar case is referred to.

ATHETOSIS.

Du Cazal, at the *Soc. Med. des Hop.*, Paris, (*Gaz. Hebdom.*, Dec. 2, 1880, p. 793) reported the case of a young soldier, who, four months previously, during convalescence from typhoid fever, was attacked with incomplete right hemiplegia, with hemi-anaesthesia and aphasia. Twenty days previously he had a unique epileptiform attack. To-day, the hemiplegia is diminished; aphasia total; he nevertheless replied in an intelligent manner by yes and no to questions addressed to him, and could write his name and age. When at rest, the right hand was quiet, but when walking, athetoxic movements affected it. The aphasia and hemiplegia could be explained by a cortical lesion of the third left frontal convolution and ascending frontal and parietal convolutions; but the athetosis could be explained only by a lesion of the third posterior portion of the internal capsule. Quinquand has observed an analogous case of athetosis, with right brachial monoplegia and aphasia, in which the autopsy had revealed only a cortical lesion of the foot of the convolution of Broca.

TETANUS.

Boon (*Practitioner*, Dec. 1880, p. 438) objects to purgatives, on the ground that they irritate,and the patient should be spared irritation; and instances his experience that his cases recover more promptly and certainly without than with them.

CHOREA.

Dr. Gauthier (*Jour. de Med.*, Feb., 1881, p. 66,) reports a case in a woman seventy-five years old, as well in health and intelligence as is usual at that age; without antecedent rheumatism or heart trouble. One morning on rising she noticed an uncertainty and notable exaggeration in the movements of the left lower extremity. The next day inco-ordination was complete in the entire left side. Five days later the chorea invaded the right side. She could not eat; walking was uncertain and stumbling; she grimaced; speech was intact. The jactitation diminished when she lay down ; but still the leg would by contraction suddenly be raised and thrown in different directions. This state lasted seventy-four days; the general sensibility, organs of sense and intellectual faculties remained intact. The choreic movements ceased during sleep, which was difficult; they were always more marked on the left side. Bromide of camphor was given and also an arsenical solution. A month later she recovered.

The cause seems to be that she had a left molar tooth extracted, and an hour or two fterwards there began to be involuntary spasms of the lower jaw; it stopped only during sleep and did not reappear.

ANÆSTHESIA.

Dr. Crombie (*Practicioner*, Dec., 1880, p. 401) adds testimony to the advantages of morphia and chloroform combined. After a hypodermic injection of morphia,one-sixth grain, it requires but a small amount of chloroform to anaesthetize, a half drachm to a dram ; and this quantity is usually enough to keep up anesthesia for a half to three-quarters of an hour. Vomiting is rare and asphyxia *does not occur*. He mentions the advantage of shoving the lower jaw forward in front of the upper, to advance the tongue and avoid the use of the tongue forceps.

MEMBRANOUS CROUP.

Dr. Taylor (*Med. and Surg. Rep.*, Feb. 5, 1881, p. 151) recommends iodide of potassium in fifteen grain doses every two hours.

EPISTAXIS.

A case is reported (*Gaz. des Hop.*, Oct. 16, 1880, p. 964, from *Jour. de Med. de l'Ouest*) of a man, aged sixty-six, with epistaxis. The anterior nostrils were tramponned. The same day he was admitted to the Hotel Dieu of Nantes, where the tamponning was renewed, and a solution of twenty-five drops of perchloride of iron in about four ounces of *eau sucre* was applied. This solution was replaced the next day by an aromatic tonic lotion. Four days afterwards he was well enough to leave the hospital. Five days afterwards he was readmitted with an intense angina accompanied by aphonia, dyspnœa, difficult and painful deglutition. During his absence from the hospital he had lost much blood and had been tamponned by a mid-wife with dry charpie,after having injected a yellow liquid of very bad taste and of which he knew not the name. Four days after readmission, a progressive dyspnœa supervened, with great general feebleness, and he died the next day asphyxiated.

At the autopsy there was found a light pseudo-membranous exudation upon the vault of the palate, the epiglottis was injected, mucous-membrane swollen and indurated; aryteno-epiglottidean fold thickened and indurated; larynx, trachea and bronchi covered with friable pseudo-membrane, not adherent to the mucous-membrane; the latter was redish and foul. In the bronchial ramifications the mucous-membrane was violet colored; the tubes contained a grayish liquid in which floated detached pseudo-membranous clots. There was a gangrenous focus the size of an orange in the middle of the posterior borders of the right lung; it was badly circumscribed, lungs emphysematous in front; hypostatic engorgment behind. Aorta, its valves and the mitral valves atheromatous; heart fatty; cerebral arteries atheromatous; much fluid in ventricle; liver large; cartilaginous granulations in spleen.

LIPOMA OF TONGUE.

A case reported (*Le Prog. Med.*, Dec. 11, 1880, p. 1014) in a man, age forty-eight, in whom the tumor appeared at twenty-three. It was removed and the microscopic examination showed it to be a lipoma.

PSEUDO-POLYPI OF COLON.

Surg. Woodward, U. S. Army, (*Amer. Jour. Med. Sci.*, Jan., 1881, p. 142) reports a specimen received at the Army Med. Museum, Washington, illustrating this rare condition. They belong to the results of follicular ulceration of the colon, and were contributed by Prof. John T. Hodgen. The patient suffered for seven months from a chronic alvine flux, and finally died with the usual symptoms of follicular ulceration of the large intestine. The minute structure of these polypi, from the resemblance it bears to what has been described as characteristic of *carcinoma*, might cause the question to be raised as to the true nature of the lesion. Dr. Woodward seems to question whether there be any specific histological peculiarities by which cancer may be distinguished from chronic inflammation.

ANTAGONISM OF MEDICINES AND DISEASES.

Prof. Bartholow (*Med. Rec.* N. Y., Jan. 1, 1881, p. 1) continues his lectures on this subject. To the first stage of *inflammation*, *morphia* and *quinia* are antagonistic; the former tends to remove congestion by increasing the vascular tension and checking the vital processes; the latter, in large doses, arrests the migration and multiplication of white corpuscles, diminishes the oxidizing function of the blood and diminishes the excretion of urea and uric acid. Digitalis, aconite and veratrum viride may also be used in the first stage; they all lower the circulation. *Digitalis* slows but energizes the heart and increases the vascular tension; the great objection to its use, is the slowness of its action. *Aconite* lowers the arterial tension but also diminishes the flow of blood towards a part. It lessens oxidation by diminishing the work of the lungs; and reduces temperature. *Veratrum viride*, like aconite, diminishes the amount of blood flowing to a part and also diminishes the oxidation.

In the second stage of inflammation the antagonists are those which prevent or remove exudation. *Chloral* is available; it reduces temperature, dissolves exudation and quiets restlessness and delirium. In order not to paralyze the heart the doses should be small and at intervals of at least two hours. If

atropia be added to it the cardiac depression will be avoided. The *alkalies*, especially the potassium, sodium and lithium salts, by increasing the alkalinity of the blood, check the formation of exudation and partially cause their solution after formation. Perhaps the best is ammonia, and in the form of the carbonate, dissolved in liquor ammoniae acetatis.

In the third stage, digitalis and quinia again come into use. They seem to give tone to the partially paralyzed vessels and favor the removal of the inflammatory products. Digitalis is, perhaps, most serviceable, as especially giving tone to the heart.

Fever heat is reduced by agents acting on the heat producing sources which facilitate radiation. In the first group are remedies which diminish or suspend muscular activity; also quinia, salicylic acid, resorcin, chloral, digitalis, aconite, veratrum viride; and all agents which depress the respiration and circulation. *Quinia* in large doses, twenty grains, is the best. In malarial fevers it has also a specific effect and may be given in smaller doses. Salicylic acid is of less value, except in acute rheumatism. *Resorcin* is a new remedy given in doses of about a drachm. At first it produces a quickened action of the heart, flushed face, warmth and praecordial depression; later, profuse perspiration. *Digitalis* is useful in exanthematous fevers; *aconite* and *veratrum viride* in high fever due to sthenic inflammation. *Cold* is the most efficient in reducing high temperature in essential fevers; by cooling the surface-blood, the whole quantity of blood is lowered in temperature. The cold bath, cold pack, ice water enemata, ice bags. In sun stroke (or thermic fever) and certain cases of acute rheumatism, where the high temperature may prove rapidly fatal, these are useful.

Some favorable reports have been made regarding the anti-convulsive effects of *woorara* and *pilocarpine* in *hydrophobia*.

In *diphtheria*, pilocarpine promotes free salivary discharge, and is said to soften and detach the false membrane. It may, however, paralyze the heart.

In *constitutional syphilis*, use *mercury*. *Iodide of potassium* is a chemical antidote.

POISONING BY A LEECH.

A man having applied a leech to his gums for toothache (*Gaz. des Hop.*, Dec. 11, 1880,

p. 1148) remarked at the end of two hours, upon his lips, a slight swelling, which soon extended to the jaw, neck and chest. Next morning his head was swollen; the tumors formed by the leech bites were joined together; respiration was difficult and he had a high fever. Some hours later delirium set in, accompanied by tremblings and convulsions. He died on the evening of the second day. The autopsy showed that he had died from poisoning. The wound of the leech was large and had black, gangrenous borders.

VULCANIZED RUBBER A POISON.

Mr. Haskell (*Chicago Med. Jour. and Exam.*, Jan., 1881, p. 29) objects to vulcanite in dentistry; first, on account of the loss of bony substance from undue absorption caused by the retention of heat under the plate; second, the poisonous effect of the coloring material which constitutes a third of the whole plate. He objects to celluloid on the same grounds, but states that there is less coloring matter in it. He recommends *continuous gum.*

Report on Mental and Nervous Diseases.

BY H. M. BANNISTER, M. D., CHICAGO,

One of the editors of the Journal of Nervous and Mental Diseases.

The number of articles in the medical periodical press has not been less than usual during the past three months. A complete abstract of this literature would, of course, be impracticable, and I have given below the substance of only a few of the more notable memoirs.

DELUSIONS OF THE INSANE.

Dr. E. C. Spitzka, in an interesting paper, (*Jour. Nerv.* and *Ment. Dis.*, Jan. 1881) discusses the pathology of insane delusions and points out their value in a diagnostic point of view. He divides them into two great classes, the systematized and the unsystematized. The former, which are connected with anomalies or lesions in the associating mechanism of the brain, are characteristic of chronic forms of insanity,and afford a bad prognosis and suggest hereditary or other taint. The subjects of these delusions often reason logically on their false premises and

defend their often absurd conclusions. To this class belong the monomaniacs, erotomaniacs, certain cases of so-called religious insanity, etc. The unsystematized delusion is, on the other hand, characteristic of the more acute forms of insanity and of general paresis,and are found as relics of the primary insanity in chronic secondary mania. The value of these points in relation to diagnosis and prognosis will be at once appreciated.

THE PATHOLOGY OF HALLUCINATIONS.

Luys, in a series of lectures, delivered at the Salpetriere and published in recent numbers of the *Gazette des Hopitaux*, makes some rather remarkable (No. 142, 1880, *et seg.*) statements in regard to the pathological changes connected with hallucinations. He claims that he has observed in a very large number of autopsies of insane persons, who, during life, had been markedly subject to hallucinations, certain conditions of the cortex and optic thalami not met with in other individuals. The cortical change consisted in a gibbosity of the paracentral lobule, so that, viewing the hemisphere on its internal face, its upper margin, instead of the usual regular curve, presented a decided prominence at that point. Examining the convex surface of the hemispheres the two marginal convolutions are also observed to be swollen and tinged in these cases, indicating a previous functional erathism of these parts. These conditions are generally found only on one hemisphere, a fact that Mr. Luys interprets as one of the best evidences of the functional independence of the two halves of the brain.

The thalamic lesions in these cases consist of hemorrhagic nuclei, patches of softening, sclerosis, and atrophy and other changes of the nerve elements, all indicating serious circulatory disturbances, which must have been connected with very decided functional derangements of the thalami.

Mr. Luys statements, it verified by future observers, are of the highest importance, and go far to establish some very important points in cerebral physiology. We hope to hear of further observations in this direction from other sources.

HYDROPHOBIA.

Several interesting communications and discussions on the subject of hydrophobia have been published in recent French journals. At the session of the Academy de Medecine, Paris, of Nov. 2, M. Colin reported the case of (*Gaz. des Hopitaux*, 1880, No. 128) a young officer of artillery who was bitten by a rabid dog while attempting to rescue a comrade in 1874,and who succumbed with all the symptoms of the disease four years and a half later. As the cause of death was important in relation to the allotment of a pension to the family of the deceased, the bite having been received in the line of his duty, M. Colin had carefully investigated the matter,and came to the conclusion that the death was really due to hydrophobia, and no other infection than the injury from an admittedly rabid dog five years before could be found. He, therefore, gave a certificate to that effect. The long period of incubation in this case naturally aroused some skepticism, but apart from its apparent inherent improbability there is no reason to doubt it. It certainly leaves the the question as to the duration of time required for the development of this disease in greater obscurity than before.

At the meetings of the same society, on the 18th and 25th of January, a very interesting discussion occurred on (1 *Bull. General de Terapeutique*, Feb. 15,) the same general subject. M. Raynaud communicated for himself and M. Lamnelongue the results of certain experiments on the transmission of hydrophobia to rabbits. M. Galtier had shown that these animals are impressible by the infection, and that the period of incubation in them was short, rarely passing seventeen days. MM. Raynaud and Lamnelongue made three series of experiments, using the virus obtained from a child that died, unmistakably hydrophobia, in the hospital St. Eugenia on the 11th of last December. They were, first, experiments with the fluids from the child while still living, second, those with the virus from the cadaver; and third, those in which one rabbit was inoculated from another. The first series redemonstrated the fact already known that the disease is transmitted by the saliva, and not by the blood. The rabbits succumbed in from seventeen to twenty-four hours after their inoculation. In the second series of experiments,a rabbit inoculated with the bronchial mucus died within forty-eight hours, another recovered. Out of six others inoculated with portions of the salivary glands, only one succumbed,while others inoculated with portions of nerve tissue from the dead child, died within four days. In the third series of experiments, every inoculation but one caused death.

M. Raynaud thought that there could be no doubt that his rabbits died of hydrophobia; similar experiments on guinea pigs were not followed by the same results.

In the following discussion M. Colin held that the cause of death in these cases was not hydrophobia, but septicæmia, and he was supported by M. Dujardin Beaumetz, who declared that, whatever it might be,the cause of death in these cases was not rabies.

M. Pasteur had made autopsies of the animals experimented upon by MM. Raynaud and Lamnelongue and had found a peculiar organism in the blood, a short, rod-like body, constricted slightly in the centre and surrounded by a pale aureola of gelatinous substance. When cultivated in veal broth these bodies lost the aureola, became thicker, more constricted, and arranged themselves in little wreaths of 150 or more. He considered these bodies the cause of the disease, but whether this was a form of hydrophobia or not he could not say; it was to him, in its symptoms, a new malady.

M. Colin reiterated his opinion as to the disorder being septicæmia, and claimed that the organisms found by M. Pasteur were also met with in certain cases of septicæmia. The discussion was partaken in by other members, MM. Bergeron, Gosselin, etc. The question seems still unsettled as to whether this disease is really connected with hydrophobia; it certainly is not typical hydrophobia.

LOCOMOTOR ATAXIA.

Dr. George Fischer has noticed (*Deutsch. Arch.f.Klin.Med.*,XXVI, p.83) some peculiar features in locomotor ataxia that have hitherto escaped observation,and that are worthy of mention. He found in four cases out of nineteen examined, normal conduction of pain impressions with retained patellar reflex and intact bladder. In one curious case this state of things was unilateral. He believes, on physiological grounds, that in these cases the posterior columns are alone involved in the morbid process, without participation of the grey substance.

He also observed some peculiar conditions of the reflexes in this disease. Layden and Remak have noted that in some cases of tabes the conduction of tact sensation is normal, or nearly so, while that of pain is retarded. Dr. Fischer found this to be the case in eight of his cases. But he also found, testing the reflexes, all of the following variations in his fifteen cases of retarded pain conduction:

1. Two cases of retarded pain conduction without double sensation and without reflex. One of these was hyperalgesic, the other, paretic. In the one, the lack of reflex was due to trouble in the central portion of the reflex arch, in the other, to peripheral disease.

2. Three cases of retarded pain conduction, without double sensation, gave reflexes combined with voluntary reaction, indicating cerebral transmission.

3. One case of retarded pain sense, without Remak's symptom, gave the normal reflex synchronous with the tactile sensation.

4. Two cases of retarded pain conduction, with Remak's symptom, gave reflexes synchronous with the consciousness of pain. In these the action was of cerebral origin.

5. In one case, with double sensation, the reflex occurred synchronous with the prick, and before the pain, indicating that the normal reflex centre was in good functional activity.

6. In some cases, with double sensation, there was a double reflex corresponding with both tact and pain sensations. The first of these was the normal spinal; the second, the cerebral reflex.

It is easy to see that this method of examination gives a certain new light on the condition of the cord, and it is worthy of trial in all cases of this disease.

MM. Debove and Baudet have (*Archives de Neurologie*, I, 42) experimented upon the muscular tone in tabes, using, for the purpose, a new instrument, the myophone. They found a decided difference in the tonicity of different groups of muscles, and conclude that "the incoordination of tabetics, is due to an unequal tonicity of their muscles, the effects of which are diminished by the maximum contraction of their muscles." They do not appear to consider the loss of muscular sensation as having anything to do in the production of this symptom.

NERVE STRETCHING IN TABES.

The operation of nerve stretching in this disease, though not yet established fully as an advisable therapeutic measure in tabes, is now undergoing trial in various quarters. Langenbuch's first case has died, though not from the operation, which gave him much relief. Esmarch has employed it at least once with benefit, and more or less successful cases are reported by Debove and Gillette (*Progres Medical*, 1880, No. 50).

A curious feature of the effects of this operation is the contralateral action observed in some cases. Thus, in one of Debove's cases, stretching of the median and radial nerves of one arm caused improvement only in the symptoms on the side operated upon, while complete relief followed on the opposite side. This points to a central action in the cord in the relief of the symptoms, and is suggestive as regards the indications and the methods of operating in certain cases where the symptoms are altogether unilateral.

At the session of the Societe de Biologie, February 5th, M. Laborde stated that in experiments on guinea pigs, he had found that stretching the nerves, abolished or weakened sensibility, while leaving the motor conduction of the nerves unimpaired, Also that he had found in the first patient, operated upon by M.Debove,a marked decrease of sensibility and of the reflexes, a fact that is notable as agreeing with the results of experimentation on animals.

ABSCESS OF THE LIVER AND HYPO-CHONDRIA.

In 1878 Dr. W. A. Hammond published an account of his success in relieving (*St. Louis Clinical Record*, June, 1878) melancholic and hypochondriac symptoms of obscure origin, by aspiration of abscess of the liver, which had revealed itself by no other symptoms. As an advance in diagnosis and therapeutics, this is important, and further confirmation and experience was highly desirable. A paper on the same subject, based on one of the cases operated on by Dr. Hammond, was read before the Virginia State Medical Society in 1879, by Dr. Marion Sims, and now we have in the *St. Louis Clinical Record*, January, 1881, a second paper by Dr. Hammond, giving the results of his full experience up to the present year. He has operated in altogether forty-three cases. In fifteen of these, pus was found and evacuated, and the symptoms relieved. In one case,there was evacuated a large hydated cyst, with similar relief of symptoms, and in twenty-seven cases, the puncture which is always, of course, largely for diagnosis, revealed no abscess. In no case was the operation followed by any untoward symptoms whatever; under proper precautions it appears perfectly inocuous. Besides Dr. Hammond, Dr. Marion Sims and Dr. C. C. Lee, have performed the operation with success, but these cases both died; Dr. Sims' case from cancer, and neither of them was, in Dr. Hammond's opinion, properly comparable to those operated upon by himself.

His paper also contains two clinical histories of analogous cases, by Drs. J. C. Shaw, of Brooklyn, and W. H. DeWitt of Cincinnati, confirmatory of the views here enunciated. In Dr. Shaw's case there were very obscure symptoms, ending fatally, and abscess of the liver was discovered at the autopsy. Dr. DeWitt's patient recovered from a spontaneous evacuation of the abscess into the intestine.

LOCAL FUNCTIONAL ISCHEMIA OF THE BRAIN.

Prof. Ball, of Paris, in a paper read before the British Medical Association, and published in the *British Medical Journal*, October 30th, describes several, with very serious appearing symptoms of organic lesions of the brain, aphasia, hemiplegia, hemianaesthesia, etc., appearing suddenly, but disappearing so rapidly and completely in a short time as to exclude the possibility of severe structural disease having existed. They all appeared after intense emotional excitement or severe exposure; the possibility of hysteria was excluded by the analysis of the cases, and the following conclusions were deduced:

1. Spasmodic contraction of the brain-vessels may be produced by emotional impressions, fear, anger, or grief, and also by prolonged action of severe cold.

2. All the symptoms of organic injury of the brain may be created by functional ischaemia.

3. Mental disturbances of a peculiar kind, and especially lowering of the intellectual power, as apart from positive insanity, may be the result of this process.

4. Spasmodic contraction of the brain-vessels, when once induced, may persist for a considerable length of time without producing structural changes in the nervous centres.

5. This morbid condition may, in certain cases, suddenly disappear, while it is not unreasonable to suppose that the converse may be equally true, and that the symptoms may culminate in rapid or even sudden death.

Report on the Progress of Ophthalmology.

By Dr. Lifton Forbes, S. R., C. P.

It is generally held that enucleation of a wounded eye before any symptoms of sympathetic irritation have appeared in its fellow, is a certain safeguard against subsequent sympathetic disturbances. This dictum, though true for a vast number of cases, is not so for all. Many cases are on record in which, after enucleation of a wounded eye, the other eye, which had hitherto remained sound, has become affected. Such a case has lately been under the care of Dr. Lloyd Owen, at the Birmingham and Midland Eye Hospital (*British Medical Journal*, April 10th). The patient was a girl, aged 16, whose left eye had been removed on account of an injury. There was nothing remarkable about the operation, but five days later the hitherto sound right eye began to develop symptoms of iritis. Vision was reduced to perception of fingers at three feet. The treatment ordered consisted of careful bandaging of the eye, hourly instillation of a strong solution of atropia, three leeches to the temple, liberal inunction of unguentum hydrargyri night and morning, and quinine internally. In ten days the pain and iritis had quite disappeared, some photophobia only remaining. The eye was kept in modified light for five weeks, when no further evidence of mischief being apparent, it was gradually exposed to full daylight. The removal of foreign bodies from within the eyeball is always a matter of considerable difficulty and uncertainty. Recently, the experiments and operations of Knapp, Cooley, and other ophthalmologists, have shown that, in these cases, where the foreign body is a piece of iron or steel, the electro magnet may be of considerable service. Drs. Brouner and Appleyard (*British Medical Journal*, April 16, 1881,) record a case in which a fragment of metal was thus removed after other means had failed. The foreign body was seen lying on the surface of the lens, the capsule of which it had pierced. A large iridectomy was performed, but the hemorrhage obscuring the position of the foreign body, recourse was had to an electro-magnet, with the result of removing a piece of iron one-tenth of an inch long by one-fifth broad.

At the last meeting of the Ophthalmological Society of Great Britain, (April 7, 1881) Dr. Hughlings Jackson exhibited a patient who had slight changes in the fundus oculæ ten years after optic neuritis. His case had been recorded in the *Medical Times and Gazette*, of December 7, 1872, as one of double optic neuritis with good vision. There was clear evidence of syphils. Dr. Hughlings Jackson remarked that if this patient had not been treated with iodide of potassium he would be blind now. On the same evening Mr. Henry Powers exhibited a case in which a clot of blood of a bright arterial color could be seen with oblique illumination, lying apparently behind the lens. The interest in the case arose from the fact that the blood had remained for a long period unaltered.

Dr. Brailey, of the Royal London Ophthalmic Hospital, has recently presented the report of a committee formed to consider defects of sight in relation to the public safety. The accumulation of the statistics had occupied five months, and the examinaton had embraced 18,088 persons drawn from all classes of society. Of these,615 were pronouncedly color blind, that is, could not distinguish red from green. There were also a few cases in which blue or violet were the defective colors, and three in which there appeared to be no appreciation of any color; these being only seen as shades of varying brightness. Red blindness appeared to be, in the United Kingdom, a little more common than green blindness. Total color blindness was excessively rare. Among 16,-431 males of various ages, and from all ordinary positions of life, in that country, the average percentage of color defects, both slight and serious, was 4.76. The corresponding average among females was .4. Not only were females nearly thirteen times as exempt from color defects, but the forms encountered among them were nearly always slight. Among 1,679 children of Jews exam-

ined, not only was color blindness more frequent than among others of the same age and station, but it was principally of the form known as red blindness. As a general rule color defects were more frequent among the uneducated classes than the educated, and as abundant in adult life as in children. An interesting point was that there were but one or two cases of color-blindness in one eye on record, to these Dr. Brailey now added a third.

At the annual meeting of the American Medical Association held in Richmond, Va., May 3-6, 1880, Dr. Chisholm, of Baltimore, read a paper on the use of the actual cautery as a means of inducing shrinking of the cornea in kerato-conus. A heated needle was thrust suddenly through the cornea, the anterior chamber was at once emptied, and atropia and cold water used. The result was satisfactory. At the same meeting, Dr. Eugene Smith, of Detroit, described a successful operation for blepharoplasty, in which the graft was taken from the arm without any pedicle. A piece of skin, one and a half by two inches, was taken from the arm, and the cellular tissue and a portion of the true skin shaved off. This graft was then applied to the cut surfaces of the lid,being about one-fourth larger than the wound. The case terminated successfully, the motions of opening and closing the eye being perfect.

Dr. J. C. Dalton has lately (February 17, 1881,) read a paper before the New York Academy of Medicine, on the centre of vision in the cerebral hemispheres. His paper is chiefly physiological and experimental. He concludes with the following propositions:

1. Extirpation of the angular convolution causes loss of visual perception on the opposite side.

2. This operation is not followed by any disturbance of the attitude, intelligence, power of locomotion or general sensibility.

3. It does not interfere with the local sensibility of the retina or conjunctiva, the reaction of the pupil to light, or with the normal consentaneous movements of winking. Its effects are therefore confined to the exercise of visual sensibility.

In the discussion which followed, Dr. Jacobi called attention to the difference in results which might follow from the mode in which the experimental operations had been performed; for instance, the use of the knife was more likely to be followed by secondary inflammation than either chromic acid or the actual cautery.

Dr. Loring (*Medical Record*, April 9, '81,) calls attention to the great prevalence of a mild form of conjunctivitis in New York, dating back during the last eighteen months. The affection is, properly speaking, hyperæmia of the conjunctiva, and though a mild affection, is extraordinarily rebellious to all treatment. He has found extremely weak collyria of arg. nit. of from one-eighth to one-sixteenth of a grain to the ounce of distilled water, useful. He recommends the same mild application in certain forms of subacute keratitis, accompanied by muco-purulent secretion.

Dr. David Webster (*Medical Record*, March 5, 1881,) records two cases of that extremely rare affection—sympathetic neuro-retinitis. In each, the eye first injured was removed, when the affection gradually subsided.

Dr. Rockwell (*Medical Record*, March 5, 1881,) reports two cases illustrating the views of Dr. Hughlings Jackson, of London, as to the frequent connection between atrophy of the optic nerve and locomotor ataxy. In one, the patient was aged 45, and had well marked symptoms of spinal sclerosis, together with atrophy of the disc; in the other, the connection was not so marked, inasmuch as altho' the usual disturbance was pronounced progressive, the spinal symptoms had, as yet, scarcely made themselves manifest. The prognosis is, however, most unfavorable in such cases, and it is on record that slight, darting pains in the legs, associated with impairment of vision, have preceded, by a considerable period, any observable symptoms of inco-ordinate movements.

In the *British Medical Journal*, of March 26, 1881, appears a comprehensive article on "Optic neuritis in Intercranial disease," from the well known pen of Dr. Hughlings Jackson. The paper is too long to make an abstract of, but will well repay perusal. The subject is considered under five different heads, viz.: 1. Optic neuritis from the ophthalmoscopic point of view. 2. Clinical facts in connection with optic neuritis. 3. Association of optic neuritis with other symptoms. 4. On the diagnostic and non-diagnostic value of optic neuritis. 5. Various

hypotheses as to the mode of production of changes in the optic discs by intercranial adventitious products.

In the March and May numbers of the *New York Medical Journal*, Dr. William Ayres discourses, at length, about the nature and physiological significance of visual purple. Though originally discovered by Ball, all we really know about it is due to the researches and experiments of Kuhne. Dr. Ayres, after considerable study of this substance, concludes that it is an albuminoid compound, and a secretion of the pigment epithelial cells of the retina, but that this secretion is not controlled by any one of the larger nerve trunks which have a part to play in the functions of the eye. We know of no drug which can diminish its secretion; but, on the other hand, pilocarpine and muscarine greatly increase it. As regards its use, it is probably a conservative compound developed in the eye as a matter of protection, and which enables it to perform its duty under the most varied circumstances. Lastly, according to Dr. Ayres, the idea of utilizing it for medico-legal purposes, is inadmissible; no thoroughly defined or durable retinal picture being obtainable by its aid.

Most of the modern text books on ophthalmology contain some reference to quinine amaurosis. The reported cases are, however, very few. Gruening, (Archiv. of ophthalmology, X, 1,) could collect but eleven. He reports the case of a lady who within thirty hours took eighty grains of sulphate of quinine in ten grain doses. Shortly after the last dose there was a convulsive attack with twitching of facial muscles, but not complete loss of consciousness. When the attack had passed she was totally deaf and blind. The pupils widely dilated did not respond to light, but contracted with strong convergence of the eyes. The media very clear, the optic discs very pale but transparent and well defined; the retinal vessels, both arteries and veins, excessively attenuated. Gruening holds that in view of the constancy of the symptoms, and the uniformity of the ophthalmoscopic picture, we are entitled to consider quinine amaurosis as a distinct affection.

Dr. Wilh. Goldzieber, in the *Wiener Medizinische Wochenschrift*, April 16 and 23, contributes a thoughtful article on gun shot wounds of the orbit, and their ultimate influence on sight. He quotes several cases with clinical commentaries.

Dr. Merklen (*La France Medicale*, 10th March) records a curious case of the occurrence of acute symptoms in the course of exophthalmic goitre of six years' standing. The patient attributed the first appearance of the goitre to suppression of the menses, and its existence was certainly cotemporancous with that of amenorrhœa. During her stay at the Beaujon Hospital she suffered successively from attacks of fever, diarrhœa and general hyperæsthesia, with prolonged intermittance of the heart's action, and epileptiform attacks. A clinical commentary accompanies the case. Under treatment, by Digitalis, the patient recovered.

Report on Otology.

By A. Wellington Adams, M. D., Denver.

Truly, we are becoming a nervously sensitive people, when inflation of the middle ear, by Politzers method, produces syncope; and loss of hearing is sustained by a kiss upon the ear, and yet, we learn from the *Archives of Otology,* that such accidents have actually happened.

Perhaps no more acceptable news could be here heralded than that which comes to us from Dr. Edward C. Mann, of New York, to the effect that both the blindness and deafness, resulting from cerebro-spinal meningitis, can be treated successfully with the galvanic current. He reiterates the statement, made by Niemeyer, in 1870, that "in the constant current we have a means more powerful than any other of modifying the nutritive conditions of parts that are deeply situated." Starting out with the proposition that, in such cases, the blindness and deafness is the result of inflammatory products in the corpora quadrigemina, and in the auditory nerves, resulting in degeneration and mal-nutrition, the beneficial action of the galvanic current is explained by the fact that the catalytic effect of this form of electricity is such as to remove organizable deposits and modify nutrition. To insure success, the doctor enjoins patience and perseverence. In a case of total blindness reported to have been cured by this method, the results were purely negative until after

eight weeks of persistent treatment, when they became positive, resulting in complete cure. It is to be hoped many physicians may find this treatment as successful in their own practice, as does Dr. Mann, for there is, up to the present time, no form of deafness more trying to treat, or more unsatisfactory as regards results; in fact, it has so far been considered hardly amenable to treatment.

Dr. Laurence Turnbull, however, and a few others, have frequently reported very satisfactory results from the use of the galvanic current in those forms of deafness arising from disease or injury of either the ultimate filements of the auditory nerve, or of the nerve centre itself. By the way, speaking of Dr. Turnbull, reminds us of his recently published brochure, on "Imperfect Hearing and the Hygiene of the Ear." Dr. Turnbull has long been known as an able writer and an exceedingly *practical* aural surgeon. The work referred to is a series of monographs on various subjects in otology. The work is calculated to interest and instruct both the specialist and general practitioner. It embraces articles on "the limit of perception of musical tones by the human ear;" "tinnitus aurium, and observations on aural or auditory vertigo, with diagnosis and treatment;" "the importance of the treatment of the naso-pharyngeal space, tonsils and uvula in acute and chronic catarrh of the middle ear;" "artificial perforation of the membrana tympani;" "the mastoid region and its diseases, with illustrative cases;" "the hygiene of the apparatus of hearing, with the prevention of deafness;" "on the method of educating the deaf-mute at home, and on the selection of proper schools for the deaf and dumb;" "a comparison between the audiphone, dentaphone, etc., and the various forms of ear-trumpets." An "introduction" presents the recent progress of otology.

KELOID OF THE LOBULE OF THE EAR.

Dr. Axel Key (*Hygieia; British Medical Journal,* vol. II, 1880, p. 808,) gives a case where keloid growths had arisen from the scars of perforation made for the purpose of introducing ear-rings. Eight of these growths had appeared and been extirpated within a period of twenty years, the last measuring a few tenths of an inch in breadth and one-half an inch in length. In the treatment of such growths, nothing could be better than the method recommended by Dr. Vidal (_Bull. Gen. de Therap._, Vol. I, 1881, p. 136), viz.: That of scarification in parallel and transverse lines.

GALVANO-CAUTERY IN MASTOIDAL TREPHINING.

Dr. Bagroff (_Annales des Malad. de l'Oreille_, 1880,) recommends the use of the galvano-cautery in perforating the mastoid process. After incising the integument and laying bare the bone, the cautery is applied to the point intended to be perforated, until there appears a blackish eschar. This procedure renders the bone more friable, and thus facilitates the use of the gouge. After removing a layer of osseous tissue with the gouge, the cautery is again resorted to, and so on alternately until the opening is sufficiently deep. In this way the danger from lesions of the venous sinuses is avoided.

Bagroff also believes this method to be applicable to the removal of osteomata of the external auditory canal. After preliminary local anæsthesia, the cautery is applied over the most accessible portion of the tumor. After the production of an eschar the chisel is used, and the operation continued as before.

THE NASAL DOUCHE AS A CAUSE OF DEAFNESS?

We are somewhat surprised to find that, in spite of all that has been said and demonstrated in proof of the danger to hearing connected with the use of the nasal douche, Dr. Franke H. Bosworth, in his recent work on diseases of the the throat and nose, should adhere to that pernicious practice. While alluding to the possible danger to the ear from this method, he seems to doubt whether the douche is responsible for deafness occuring in those who have used it, inasmuch as a very large number of persons with impaired hearing, as the result of a catarrhal extension from the naso-pharyngeal cavity through the Eustachian tubes to the ear, have never used the douche; certainly a very illogical conclusion. He thinks there is fair ground for regarding it an open question whether the use of the douche or the original catarrh is responsible for the impairment.

We think after such reasoning the doctor might with equal propriety add as a corol-lary, that inasmuch as many subjects suffer from an aggravated form of naso-pharyngeal catarrh without ever experiencing deafness, that, therefore, the former is not dependent upon the the latter, of course, a most absurd conclusion.

The author has undoubtedly failed to comprehend the force of the accusations our most noted aurists have hurled at the practice in question. A great many cases of acute and chronic purulent inflammation of the middle ear have unmistakably originated from the use of the nasal douche, the fluid thus used having entered the tympanum through the Eustachian tube. This procedure should positively be set down as pernicious, for the accident cannot be avoided, even though the physician himself uses the douche. Besides, the object aimed at, thorough cleansing of the nasal cavity, is not accomplished, as Dr. Rumbold has very ingeniously demonstrated in his recent work (_Hygiene and Treatment of Catarrh_). This author also most emphatically enters his protest against the nasal douche.

SOME POINTS ON CLEANSING THE EAR.

Dr. W. H. Bennett, of Brooklyn, writes (_Medical Record_, March 26, 1881): "Books on aural diseases, and ear specialists generally, advocate for cleansing the auditory canal, the use of a syringe having a short, blunt, or swollen nozzle, and advise further, simply that the canal should be thoroughly straightened while using the instrument. But experience has taught me that this is not the proper syringe, and that a somewhat different method should be pursued. We want free vent for the outflow, or return current, of water (except in the comparatively few cases where it is desired to force the fluid through a perforated drum-head into the parts beyond,) in order that it may carry out the foreign body, wax, or pus; and then we should be able to _see_ what we are about, and to direct the stream with exactness and nicety above, below, or on either side of the substance to be removed, if it be impacted in the canal, and so cause the fluid to accumulate behind it without forcing it farther in. To accomplish these ends we must use the head-mirror, the speculum, and a syringe with a long, slender nozzle. I use the ordinary uterine syringe. By means of these instruments, one can syringe an ear neatly, intelligently, and effectually, and there is no more danger to the patients, nor as much, as by the old method. To my mind it is as unphilosophical to syringe an ear in the dark as it would be to probe the auditory canal blindfolded; and should the army of ear-syringers object to the method here recommended, I can only assure them that when any one of their number has once tried it he will adopt it. The stream of water should not be thrown with quite so much force as is generally employed, for it impinges more directly upon the drum-head or walls of the middle ear, and, of course, the smaller the stream, with the same amount of power, the swifter the current and the more violent its action. The speculum should be as large as it is practicable to use. During the syringing it will frequently fill with water and obstruct the view, but this may be sucked out and the parts again exposed to the eye. If it be a foreign body or a lump of wax we are endeavoring to remove, as it makes its way out, the speculum may gradually be withdrawn until the substance reaches a point where it may be seized by the forceps, or admit of the introduction of a probe behind it. It is understood, of course, that the syringe, and method of using it, here mentioned, is meant to be intrusted only to professional men, as it would be a dangerous instrument in the hands of a layman. A few words regarding the drying out of the canal after syringing, which ought always to be done, except in acute inflammations. For mopping up the residual fluid, a camel's-hair brush, of good size, previously moistened and squeezed out, answers the purpose admirably. It can be sufficiently dried by pinching it between the folds of a towel, and may be repeatedly used. Afterward, for removing every trace of moisture, ordinary blotting-paper, cut into narrow strips, a line or two wide, is the most convenient and by far the least dangerous or irritating material I have found." We perfectly agree with the writer in all that he states, particularly in regard to removing every trace of moisture, but in place of the camel's-hair brush and blotting-paper for drying purposes, we would recommend our practice—that of thoroughly wiping out the external auditory canal with pledgets of absorbent cotton.

AURAL EFFECTS OF NASAL STENOSIS.

Dr. J. O. Roe, of Rochester, in a valuable paper (*Medical Record*, April 30, 1881,) read before the New York State Medical Society,

thus speaks of the effect of nasal stenosis upon the ear and audition: "Toynbee first demonstrated, by a series of experiments,the altered condition of atmospheric pressure in the fauces and ears when swallowing with closed nostrils.

"Lucae repeated these experiments, and also observed that, with obstructed nostrils, these changes in atmospheric pressure produced abnormal tension of the membrana tympani, which gradually produced indistinctness in hearing.[*]

"Let us briefly consider the manner in which these changes are produced. With the completion of the first stage, and during the second stage, in the act of swallowing, the nasal passages and the upper pharyngeal space are almost completely shut off from the pharynx by the soft palate being closed firmly against the posterior pharyngeal wall.

"In the third stage of this act,the pharyngeal constrictors close by reflex action on the substance swallowed and force it onward in its course to the stomach, and at the same time air is exhausted from the naso-pharyngeal space and middle ear, by the suction naturally following the descent of the bolus. This tendency to produce a partial vacuum is prevented by air entering freely through the nasal passages, and the normal air-pressure in the posterior nares and middle ear is maintained.

"If the anterior nasal passages become closed, it is readily perceived that, during each act of swallowing, a corresponding degree of disturbance in air-pressure will take place in the nasal cavity, and in the ear also, because of the direct communication through the Eustachian tube.

"This aural pressure one can very easily illustrate on himself by closing the nostrils while swallowing, when a marked sensation of pressure will be felt in both ears, supposing the Eustachian tubes to be unobstructed, and with the aural speculum this movement inward of the drum-head can readily be seen.

"I have found it to be a rule that when the nostrils are not free enough to permit one to breathe entirely through them, even during a brisk walk, they are not sufficiently

---

*Verhandlg. der Berliner med. Ges., 1867-68, S. 133; Ziemssen's Cyclopædia, vol. iv., p. 111.

free to maintain the aural equilibrium during continued acts of swallowing.

"It is an undoubted fact that, even in a state of repose, air continually permeates the Eustachian tubes,† and that more or less of the aerial‡ conduction of the sounds of the voice in autophony is through the Eustachian tubes.

"Were this not the case, and did air enter the ears only during the act of deglutition, as stated by most authorities, an uncomfortable aural pressure and a slight impairment of the hearing would not take place almost immediately upon the stoppage of the Eustachian tubes by a plug of mucus, or from any other cause, whereas one may remain for hours, awake or asleep, without swallowing, and yet the ears and hearing remain perfectly normal.

"Thus we see that a continually free communication of the external air with the middle ear is necessary to perfect hearing, and, as the air cannot be supplied to the tympanic cavity by any other route than through the nasal passages and Eustachian tubes, so the aural pressure is lessened in proportion to the degree of nasal obstruction.

"If the obstruction is great, aural changes take place rapidly; if it is only slight, they go on more slowly, sometimes imperceptibly, and, sooner or later, the most serious functional and structural changes take place.

"The continuous external pressure increases the concavity, and causes a rigidity of the membrana tympani.

From this, results an inactivity of the ossicular chain, and, from this inactivity, the delicate articulations become stiffened, impacted, and finally immovable. Besides, the tensor tympani muscle and ligament become relaxed and ultimately rigid from disuse, so that, as remarks Cassells,¶ 'were it possible, which it seldom is, to remove the other consequences of altered tension, this contracted tendon and ligament mars the best efforts of the practitioner to effect an improvement.'

"The characteristic symptoms are gradually increasing deafness, giddiness, distressing tinnitus, which diminish or altogether pass away as the deafness deepens.

"If this condition is still allowed to go unrelieved, another, and sometimes more serious, set of changes supervene.

"In consequence of the catarrh and thickening of the mucous membrane of the naso-pharynx, the Eustachian tubes become invaded, and concentrically closed, 1st, by the collapse of their flaccid walls by the suction or negative pressure; 2d, by the catarrh and thickening of the mucous membrane of the naso-pharynx invariably attending nasal stenosis.

The air thus shut up in the tympanic cavities is speedily disposed of, and, as a result of the diminished pressure, engorgement of the lining membrane of the cavity follows, and free serous transudation takes place sufficiently to fill the tympanum, and from the pressure of the imprisoned fluid the membrana tympani gives way, and an otorrhœa is established. Thus, when 'unrelieved by art, nature attempts, although in a rude way, to perform a natural cure,' by establishing an opening to the middle ear."

Report on Orthopædic Surgery.

By Thomas H. Hawkins, M. D.,

Prof. of Principles and Practice of General Surgery in the "Columbia Veterinary College and School of Comparative Medicine and Surgery," N. Y. City.

PHYSIOLOGICAL METHOD OF EXTENSION IN MORBUS COXARIUS.

Dr. Hutchison claims (*Philadelphia Medical Times*, May 7, 1881, p. 481,) many advantages for the "Physiological Method," over the ordinary methods of extension by means of the various apparatus.

His method consists in the wearing of a high shoe on the foot of the sound side; thus raising the body sufficiently, when the patient is standing upright on a pair of crutches, to prevent the toes of the other side from touching the ground.

This method, according to the author, meets all the indications necessary for successful treatment of Hip-Joint Disease, and prevents deformity. The reflex contraction of the peri-articular muscles, intracapsular effusion, and the efforts of the patient to

†This view is quite elaborately advocated in an article on "The Method of Air-Supply to the Middle Ear," by Dr. Thomas F. Rumbold, St. Louis Medical and Surgical Journal, July 20, 1880.

‡"Hearing by the Aid of Tissue-Conduction, the Mouth-Trumpet, and the Audiphone," by Samuel Sexton, M. D., American Journal of Otology, April, 1880.

¶In a masterly article, with the striking title, "Shut Your Mouth and Save Your Life." J. P. Cassells. Edinburgh Medical Journal, February, 1877, p. 740.

keep the diseased joint quiet, will secure sufficient immobility. The weight of the diseased limb (equal to one-fifth of the entire body) will make sufficient extension to overcome all muscular spasm. Muscular rigidity may exist, if there is no muscular spasm, and yet the patient be free from pain. It is claimed for this method, that it is more grateful to the patient, less constraining, and that it incites less "reflex resistance" than any of the other methods.

When the patient has learned that he can be free from pain by suspending his limb, there will be no difficulty in persuading him to walk or stand on his crutches three or four hours every day, and this will sufficiently relax the muscular rigidity to prevent the reflex pain at night.

The absence of the iron apparatus and the perineal band, which so often worries the patient and irritates the muscles, is certainly a desideratum; but the "Physiological Method," until it is better known and understood by the medical profession, must remain *sub judice.*

THE PHYSIOLOGICAL METHOD WITH FLEXION AND FIXATION OF THE LEG AT RIGHT ANGLES WITH THE THIGH.

Dr. R. J. Levis (*Philadelphia Medical Times,* May 7, 1881, p. 505,) does away with the necessity of the "thigh shoe" on the sound side, by lifting the diseased leg from the ground and flexing it at right angles with the thigh, and maintaining it there by a silicate of sodium splint, which, when it is cut open in front, is removable, and can easily be applied by bandaging, etc. This splint is removed at night, and, if necessary, weight is used. When the weight of the limb fails to overcome muscular spasm, he fastens, in the folds of the bandage near the knee, sheets of lead. When fixation of the hip is necessary, the splint is extended up over the pelvis.

MOTION IN THE TREATMENT OF JOINT TROUBLES.

A. J. Steele, M. D., of St. Louis, (*St. Louis Courier of Medicine,* p. I. 1881,) believes that the idea of motion, in the treatment of joint diseases, is an erroneous one, and that splints, which allow of motion at the diseased joint, should not be used so long as there is inflammation in joint structures; but that quiet is the all important therapeutic

principle in the treatment. Passive or active motion to prevent anchylosis should not be resorted to until all signs of inflammation have subsided. The inflammation should first be completely controlled and then motion restored. Chronic joint disease, according to the author, rarely terminates in anchylosis.

SPINA BIFIDA.

H. E. Leach, M. D., (*Walsh's Retrospect,* April, 1881, p. 319,) reports an exceedingly interesting case of spina bifida with talipes equino-varus and anterior flexion of the legs.

POTTS' DISEASE OF THE SPINE OF SYPHILITIC ORIGIN.

Fournier (*Annales de Dermatologie et de Syphilographie,* II, No., 1881, p. 19,) cites an exceedingly interesting case of Potts' disease, which he believes to be of syphilitic origin. The patient was a man 59 years of age; scarcely able to walk. Pain, of dull character, was complained of about the lumbar region. Lesions, undoubtedly syphilitic in their nature, were found in different parts of the body. He was put on antisyphilitic treatment. The external symptoms rapidly improved, while his general condition grew worse, new symptoms developed, he became extremely emaciated and fell into a state of cachexia and died. Post-mortem examination showed syphilitic lesions of the liver, kidney, a gummatous deposit around the fourth left lumbar nerve at its exit. Lesions of Potts' disease in the region of the lumbar spine—third, fourth and fifth vertebrae—also lesions of the bones, etc. The important question is, whether this case of Potts' disease was itself of syphilitic origin, or was it a mere coincidence. Prof. Fournier decides in favor of syphilitic origin because of the advanced age, (tubercular disease of the spine being rare after youth)—the previous history of the patient, unmistakable syphilitic lesions, syphilitic appearance of the diseased vertebrae, and lastly, on account of a gummatous tumor arising from the periosteum of one of the lumbar vertebrae.

ACUTE PRIMARY SYNOVITIS OF THE HIP.

V. P. Gibney, M. D., (*N. Y. Med. Journal,* April, 1881,) reports several cases of this affection which, he believes, occurs oftener than is generally supposed. This affection, the writer avers, usually runs a comparatively brief course, and makes its appearance in from eight or ten to fifteen years of age, and nothing more than the synovial membrane is involved, while a chronic ostitis of the epiphysis, often mistaken for synovitis of the hip-joint, usually occurs in children between two and seven years of age, and runs a much slower course. Ostitis, the author claims, differs in its history from that of synovitis.

In the former, the beginning is imperceptible, the latter is acute and well defined at its onset. In ostitis there is no joint tenderness and the pain does not prevent the patient from going about. Joint tenderness in synovitis is well marked and the child, after a few hours, is unable to walk. Ostitis, as a rule, is chronic. Synovitis is acute. One lasting from two to five years; the other varies from four to ten weeks. In joint tenderness the pain is referred to obturator when the joint surfaces are jammed together. In synovitis the history is ordinarily very clear, the mother naming the very day, and frequently the hour, when the child was first attacked. These cases, it is claimed, make a good recovery and walk without a limp, if the proper treatment be used, viz.: Rest and counter-irritation.

A DEVICE FOR RETAINING DISLOCATIONS OF THE CLAVICLE AT ITS DISTAL END.

Dr. Stiles (*Med. Record*, March 19, 1881, p. 316,) describes a rather ingenious device for retaining the end of the clavicle in place when dislocated at its acromial end. The bone is put in position and a plaster cast is taken of the parts, measuring four by two and seven-eighths inches. A hard rubber plate is made from the cast, in the same way that a plate for a set of false teeth is prepared by a dentist. This rubber splint is lined with lint and applied over the clavicle at the point of dislocation. This is kept in position by two strips of adhesive plaster attached to the outer and upper surface of the splint, and made fast to the posterior part of the chest. Two other strips were applied over these, extending in an opposite direction and somewhat diverging. To these strips were fastened two rubber bands three inches long and three-quarters of an inch wide. To the lower ends of these bands were attached pieces of adhesive plaster, with broad bases. The strips, after making sufficient traction to hold the bones in posi-

tion, were made fast to the anterior part of the chest. A small roll of bandage was placed in the depression of the splint and a strip of adhesive plaster firmly applied from before backwards over the roller bandage. The forearm is placed across the chest and retained there by adhesive strips and a few turns of the roller.

A CASE OF POTTS' DISEASE IN THE ADULT TREATED BY THE PLASTER BANDAGE.

J. F. Walsh, M. D., (*Philadelphia Med. Times*, March 12, 1881,) reports a case of Potts' disease with a large abscess occupying the right iliac, one-half of the hypogastric and a portion of the right lumbar and epigastric regions. The plaster bandage was applied, and seven days later the abcess was aspirated and twenty-eight fluid ounces of pus was withdrawn. Thirty ounces, a few days before the application of the bandage, was removed. The patient made a rapid recovery. The psoas abscess pointed above instead of below Pouparts ligament; somewhat unusual.

"TRIPOLITH."

"Tripolith" was discovered by B. Von Schenke, of Heidelberg, during the summer of 1880, and was recently recommended by Prof. Langenbeck, as a substitute for plaster of Paris in fixation dressing. Tripolith bandages are applied in the same manner as plaster of Paris. The advantages of tripolith are:

It absorbs moisture from the atmosphere less freely than plaster of Paris, and it retains its setting power after exposure to the atmosphere. Bandages made from tripolith are lighter. Tripolith is fourteen per cent. lighter than plaster of Paris. Tripolith dressing sets completely in from three to five minutes; the plaster requiring a much longer time. Plaster of Paris gives off moisture for hours; not so with tripolith. When once dry and hard it undergoes no change when laid in water. A patient could, therefore, bathe in his tripolith dressing, provided measures be taken to prevent the water from getting up inside the dressing.
425 Champa St., Denver.

From 1847 to 1853, syphilis destroyed, in England and Wales, an annual average of 565 infants under one year old in each million of births ; in 1878, the deaths under this head had arisen to 1851 per million.

SELECTIONS FROM JOURNALS.

⋯⊷⊙⊷⋯

The Human Face; Its Modifications in Health and Disease, And its Value as a Guide in Diagnosis.

By AMBROSE L. RANNEY, M. D.,

Adjunct Professor of Anatomy in the Medical Department of the University of the City of New York.

The extent to which the anatomy of the head, as studied from the standpoint of physiognomy, may suggest points of practical value to the physician or surgeon, has not, in my opinion, received sufficient consideration in the popular text-books of the day. From the "British and Foreign Medical Review," of 1841, I quote the following sentence: "Medical physiognomy is, in many instances, a source of diagnosis which seldom fails the practitioner who is himself well versed in it; and we believe that much of the exquisite tact in discrimination of disease, which distinguishes some practitioners and which others can never attain, depends upon the vivid perception of an eye and ear habitually familiar with the lineaments, the tone, and the gestures of disease." Among the earlier authors, who were ignorant of many of the present methods of determining the condition, size, and position of organs, since the art of auscultation and percussion is a growth of later date, the study of the human countenance formed a very important part of the preparatory drill. The followers of Hippocrates and Galen were rendered perfect in their perceptive faculties. The former gave to us, in his masterly work, descriptions of the symptoms of disease, which are still considered classic, while the latter, in his essays on the "Temperaments,"[*] is equally careful to note the most trivial alteration either of the face or of the posture.

There seems to be a growing tendency of late to regard the rational symptoms of disease as subordinate to the results of a physical examination, and as of but little value in themselves, except as confirmatory evidence. Authors frequently render the description of the symptoms of disease so terse and indefinite, that but few of the readers

of the later medical or surgical works could precisely picture to themselves the appearance of a sufferer from any of the maladies, with the pathology and physical symptoms of which they may be thoroughly familiar. It is not infrequently the experience of the most erudite of the profession to be amazed at the gift, which is possessed by some less scholarly brother, of making a diagnosis, which seldom errs, without the aid of the thermometer or the stethoscope; and many an old nurse, long accustomed to spend weary nights in watching the sick, can often render a prognosis which seems little short of inspiration when her utter ignorance of all medical knowledge is considered.

Despite the fact that some of our best authors have denounced the attempts of De-Salle, Jadelot, and Seibert, to establish certain facial lines and wrinkles as of positive value in diagnosis, and have pronounced all such statements as a mere fantasy, still no one of large experience can deny that the face may at times afford most positive and valuable information.

In 1806, Lavater[†] published his work upon this subject, in which he discusses at great length the diagnostic value of general physiognomy. Subsequently, Sir Charles Bell wrote upon the subject from a purely anatomical point of view, and, in 1824, published his "Essays upon Expression." Baumgaertner[‡] added his contribution to the subject in 1839, and Laycock,[§] in 1862, published his course of lectures, with illustrations, which were designed to show the various types of diathesis and their bearing upon the general development. Corfe, in 1867, published a series of contributions to the "Medical Times and Gazette," in which the subject was studied from a clinical point of view, and in which not only the entire field of facial expression, but also that of general physiognomy, was pointed out to the student, so far as the cases under consideration illustrated any points of special interest.

Darwin's great work upon the expression of the emotions in animals, and the contributions of Connelly[*] upon the typical shades of expression peculiar to the insane may

[*] Kuhn's edition.

† "L'Art de connaître les Hommes par Physiognomie." Paris, 1806–'7.

‡ "Atlas," 1839.

§ Med. Times and Gaz., 1862.

* "Med. Times and Gaz.," 1862.

well be read by those who question the utility of this much-neglected department of science. The careful study of the expressions of the face and the modifications which age produces in it is at least very advantageous in furnishing a normal standard by which deviations in disease may be studied. I quote from the most excellent treatise of Blandin† the following sentence: "Those who neglect or seek to ridicule this mode of investigation prove only one thing, that they study pathology without a proper knowledge of anatomy and physiology, upon which the former is founded. The morbid expressions of the face are an extremely useful and often the only guide of the medical practitioner in the case of a very young child, that can tell nothing in regard to its sufferings."

dering; and may sometimes show a series of changes in rapid succession.

These various conditions of the countenance may not only be the direct result of the influence of the ever-varying passions upon the muscles of the face, as is the case in health, but they may also be classed as morbid phenomena, each of which possesses some special significance. Chomel lays great stress upon these variations of countenance, and endeavors to point out the special diagnostic value of each.

FACIAL LINES AND WRINKLES.

The theories of De Salle, Jadelot, and Seibert.‡ as to the diagnostic value of facial lines and wrinkles have had their share of support from time to time; while they have

FIGURE 1.—Transverse Rugæ.

It is with a view to systematize and arrange the collected investigations of the authors previously named and to bring within the compass of a single article such practical information as the anatomy of the face may afford the practitioner, that I am led to draw professional attention to this subject once more.

The physiognomy of the sick presents innumerable shades of expression. It may assume the various conditions expressive of sadness, dejection, attentiveness, indifference, uneasiness, or terror; it may, at times, be smiling; occasionally menacing or wan-

also been considered by some authors as speculative and destitute of any value. The existence of these marks may be attributable to one of two conditions, viz., a disappearance of the fat from the subcutaneous tissues of the face, or the abnormal contraction of certain facial muscles, dependent upon some apparent irritation of the motor nerves supplying the affected muscles. It is important, in using these lines and wrinkles as guides in diagnosis, that the discrimination be made between those lines which are natural to the face of the sufferer and those which are developed as a result of the

---

† "Anatomie Topographique."   ‡Williams. "Principles of Medicine."

disease. For the reason that the face of the adult is always more or less marked by lines,[*] it must be evident that these lines are a more reliable guide in the infant than in later life, if their diagnostic value remains unquestioned. Without entering into a discussion as to the merits of the question, I give the theories advanced for whatever interest or value they may possess to the reader. The wrinkles of the face may be classified into six groups, as follows :

FIG. I.—THE TRANSVERSE RUGÆ.

These are situated upon the forehead, and are formed by the action of the occipito-frontalis muscle. They are thought to be ex-

of acute diseases, an impending efflorescence and sometimes a fatal termination may be indicated by their occurrence. In those types of headache where the pain is very excessive, these rugæ may exist simultaneously with the ones previously described. It is stated that when the former rugæ meet the latter abruptly, during the course of an acute disease, some serious lesion of the brain, or its coverings, is developing.

FIG. III.—THE LINEA OCULO-ZYGOMATICA.

This line (the line of Jadelot) extends from the inner angle of the eye downward and outward, passing across the face of the malar bone. It is said to indicate, in children, a

FIGURE 2.—The Oculo-frontal Rugæ.

pressive of an extreme amount of pain, arising from causes outside of the cavities of the body.

FIG. II.—THE OCULO-FRONTAL RUGÆ.

These extend vertically from the forehead to the root of the nose, and are formed by the corrugator supercilii. They are thought to express distress, anxiety, anguish, and *excessive pain from some internal cause.* It is said that they furthermore indicate an imperfect or false crisis; and that, in attacks

cerebral or nervous affection;[*] and, in adult life, some disease of the genital organs, masturbation, or venereal excess.

FIG. IV.—THE LINEA NASALIS (LINE OF DESALLE).

This line extends from the upper border of the ala nasi downward, in a direction more or less curved, to the outer edge of the orbicularis muscle. This line is said to be strongly marked in phthisis and in atrophy. Its upper half (the linea nasalis proper) is

---

*Blandin, op. cit.

*Vogel, "On Diseases of Children." New York: D. Appleton & Co.

thought to be a reliable indication of *intestinal disease*, if extensively developed and prominent; the lower half (the linea buccalis) is supposed to indicate the existence of some *disease affecting the stomach*. It is claimed by Peiper that, when this line appears conjointly with the line of Jadelot, it may be regarded as a positive indication of worms in children, if a peculiar fixed condition of the eye exists and a pallor of the face is present.

FIG. V.—THE LINEA LABIALIS.

This line extends downward from the angle of the mouth till it becomes lost in the

to be a reliable guide to diseases of the *thoracic and abdominal viscera.**

COLOR OF THE FACE.

The color of the face is subject to variations which to the eye of the medical adviser afford unquestioned aid in diagnosis. *Flushing* of the face, as evidenced by a diffused redness which is of a transient character, is very common in women suffering from irregularity of the menstrual periods and during the menopause. In plethora, especially after exertion or excitement, an unnatural redness of the face may occur, associated with symptoms indicative of cere-

FIGURE 3.—The Line of Jadelot.

lower portion of the face. It is usually developed in connection with those *diseases which render breathing laborious or painful*, and is more common in children than in the adult as a sign of diagnostic value.

FIG. V.—THE LINEA COLLATERALIS NASI.

This line extends from the nose downward to the chin, in a semicircular direction. It lies outside of the linea buccalis, the linea nasalis, and the linea labialis. It is thought

bral hyperæmia. Pressure of tumors, either of the neck or of the thorax, upon the sympathetic nerve may create an abnormal dilatation of the capillaries, thus resulting in a redness of the skin, with an increase of the temperature of the affected region; while section of the sympathetic nerve, although a rare form of accident, would result in a like condition.† *Red patches* occur on the cheek

*Corfe, "Med. Times and Gaz.," 1867.
†M. Foster.

during an attack of croupous pneumonia. In wasting affections of a chronic character, especially of the lungs, such as phthisis, cancer, etc., a circumscribed redness over the malar bones, known as the "hectic flush," is usually present. It may occasionally affect only one cheek,* where only one lung is diseased. *Pallor* of the face is the rule during convalescence from any severe disease, and in patients long deprived of sunlight.‡ A *waxy pallor* exists in chronic Bright's disease, which renders the skin almost transparent. In the chill of fevers and malarial attacks, a *dusky paleness* is usually perceived; while in cases of haemorrhage, where the loss of blood has been sufficient to

it may occasionally result in the deep yellow of jaundice.¶ In the early stages of jaundice, the sclerotic coat of the eye and the corners of the mouth first show the *yellow color*, although the discoloration soon tends to become diffused over the entire face. A *blue tinge* exists in those cases where the venous return to the right heart is obstructed, or where, from any cause, the oxygenation of the blood is imperfectly performed. It occurs therefore in cyanosis, asphyxia, the fevers, certain diseases of the pulmonary organs which interfere with the circulation, and in diseases of the heart which render its action weak or imperfect. In cases of poisoning from the nitrate of silver,

FIGURE 4.—The Line of De Salle.

produce constitutional effects, the pallor of the face assumes a peculiar *leaden color.*§ A *greenish tint* is present in profound attacks of anemia and during chlorosis,‖ giving to the face an appearance similar to that of imperfectly bleached wax.

Malaria and cancer are often manifested by a *light straw color* of the face, although

the skin assumes a still deeper blue tint than in those cases above mentioned, and the staining is permanent. In Addison's disease of the supra-renal capsules, *a dark-brown* color of the skin results, which may be either uniform or in insolated spots, and which may, in severe cases, almost rival the pigmentation of the negro. The redness of erysipelas is usually accompanied by an œdema which renders the face tense and shining, and which often causes a markedly altered expression of the countenance.

The face is the seat of many of the erup-

*Stille.

‡Williams, *op. cit.*

§Sir Charles Bell. "Treatise on Surgery."

‖Niemeyer,,'Text-book of Practical Medicine." New York: D. Appleton & Co.

¶Reynolds, "System of Medicine."

tions, some of which are confined almost exclusively to it, while others are usually found in that region before they appear elsewhere. It would exceed the limits of this article to enter into the description of the characters which stamp each of the various eruptions, since they can be easily learned by reference to any of the special treatises.

Corfe suggests as a guide to the student in physiognomy the following table, which designates the prevailing changes in the

Marshall Hall[*] thus describes a countenance which he considers typical of the acute form of dyspepsia: "This affection is accompanied by some paleness or sallowness, and a dark hue about the eye. The lips are slightly pale and livid. The cutaneous vessels exude a little oily perspiration, and the muscles of the face, and especially of the chin and lips, are affected with a degree of tremor, particularly on any hurry or surprise, or on speaking."

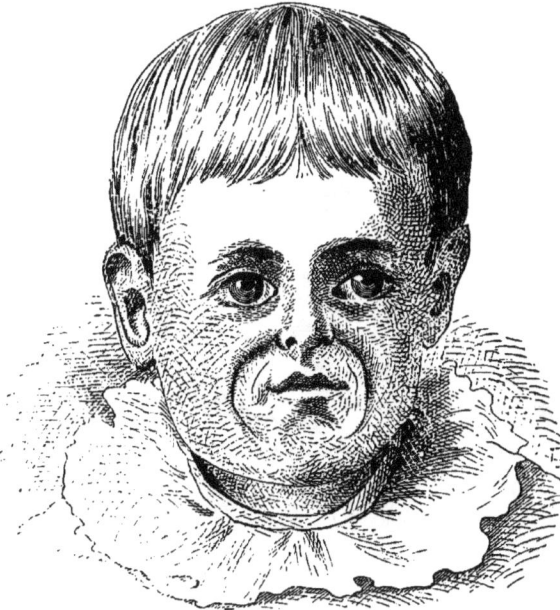

Figure 5.—The Linea Collateralis Nasi.

complexion of the face in the course of the more common disorders. While it is not possible to construct any table which shall give all the information desired upon so important a subject, still this one may prove of some value as a means of aiding the memory:

In cerebral disease	.... the countenance is	lethargic.
In emphysema	" "	livid.
In pulmonary œdema.	" "	dusky, distressed.
In pneumonia	" "	dusky and flushed.
In pleurisy	" '	pale and anxious.
In phthisis	" "	pale and thin.
In malignant disease	" "	sallow and thin.
In icterus	" "	yellow and thin.
In renal disease	" "	thin, puffy, anæmic.
In peritonitis	" "	anxious, dragged.
In uterine disease	" "	sallow and haggard.

THE NOSE.

The nostrils are of some practical interest from a medical point of view. They *dilate* forcibly and rapidly in difficult respiration when produced by disease;[†] and *itching* of the nostril is regarded by many authors as a valuable diagnostic sign of intestinal worms.[‡] The nose seldom points directly forward, being, as a rule, slightly inclined toward the right side. This fact is explained by Bee-

* "On Diagnosis." London. 1817.
† Sir Charles Bell.
‡ Peiper.

lard as the result of the habit of wiping the nose with the right hand, since, in left-handed people, the opposite deflection exists. The nose of a face perfect in its outlines should be one-third of the length of the distance from the root of the hair to the chin; but, in certain races, the variation from this rule affords a special physiognomy. The integument which covers the nose is very firmly attached to the muscles underneath it by a cellulo-fatty layer. Blandin ‡ lays

organ, even if completely detached, with a hope of obtaining union. Among the ancients, amputation of the nose was practiced upon the the criminal classes, and the operation of rhinoplasty was first suggested as a means of relief for those so disfigured.

The redness of the nose after an attack of crying indicates a connection between the sympathetic supply of the capillary vessels of the nose and that of the capillaries of the

FIGURE 6.—"Bell's Paralysis." (Modified for Corfe.

great stress upon this fact as explaining the infrequency of œdema of this region, and as an effort on the part of Nature to preserve the uniformity of contour of the nose, which would be seriously impaired by any local swelling of the face, were the skin over the nose loosely attached. The nose is extremely vascular; hence the custom of surgeons to replace severed portions of the

lachrymal apparatus; hence any form of irritation of either of these localities is liable to be accompanied by symptoms referable to the other.* Injury to the nose resulting in fracture, often leaves a permanent facial deformity, and, even when no evidences of serious injury can be ascertained by external examination, cerebral symptoms are liable to follow, as fracture of the skull may result,

‡ *op. cit.*

*Blandin, *op. cit.*

from a transmission of the force through the perpendicular plate of the ethmoid bone.* Vascular tumors of the region of the nose are not uncommon, while a prominence of the capillary vessels of the nose is met with in the aged as the result of defect in the contractile power of their coats.†

Marked elevation of the nostril is regarded by some authorities§ as an indicator of pain within the cavity of the thorax.

THE EYE.

"It may appear to many a superfluous task to attempt to judge of the character of an forehead, with the small, well-formed mouth, of the philosopher, down to the shallow front and protruded muzzle of the negro, whose habits are more bestial than those of the animals he chases for the support of his life."‖

The intimate communications between the fifth, the seventh, and the sympathetic nerves, through the media of the ciliary, optic, and Meckel's ganglia, would lead us to expect that the eye should exhibit, in its altered appearance, the derangement of internal structures. "When a glance of this

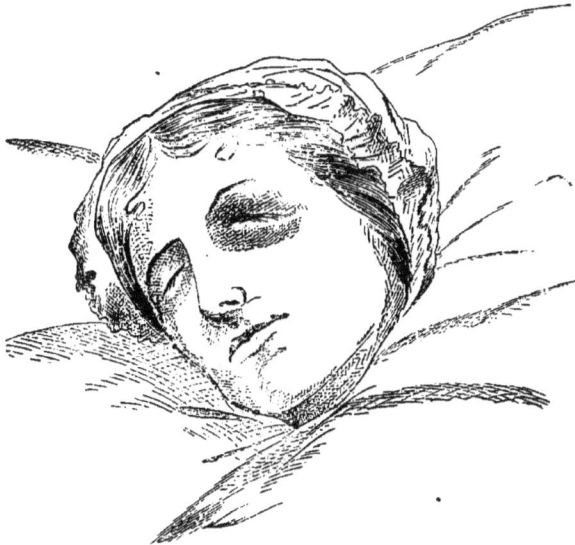

FIGURE 7.—Face after Hemorrhage. (Modified from Corfe.)

individual by a glance at his face, but, whatever may be thought of the possibility of laying down strict rules for such judgment, it is a fact of every-day occurrence that we are, almost without reflection on our part, impressed favorably or unfavorably with the temper and talents of others by the expression of their countenance. The face acquires its expression also from bodily habits and from intellectual or sensual pursuits, so that we may pass from the lofty and expanded organ is caught, what a field of mute expression is open to the mind! This silent and instructive index of the whole man may be bright or dull, heavy or clear, half shut or unnaturally open, sunken or protruded, fixed or oscillating, straight or distorted, staring or twinkling, fiery or lethargic, anxious or distressed; again, it may be watery or dry, of a pale blue, or its white turned to yellow.‡"

The pupils may be contracted or widely

---

*Beclard.

†Holden, "Human Osteology." London, 1855.

§Marshall Hall, *op. ci'.*

‖Corfe, *op. cit.*

‡Corfe, *op. cit.*

dilated, insensible to or intolerant of light, oscillating or otherwise, unequal in size, or changed from their natural clearness of outline. The noble arch of the brow speaks its varied language in every face of suffering humanity. It may be overhanging or corrugated, raised or depressed; while the lid of the eye, an important part of this vault, exhibits alterations of puffiness or hollowness, of smoothness or uneveness, of darkness or paleness, of sallowness or brown discoloration,

centers or upon the optic nerve itself.[†] In adynamic fevers the eyes are heavy and extremely sluggish, and are, as a rule, partially covered by the drooping eyelid; while in certain forms of mania they are seldom motionless.[‡] This latter peculiarity is also often noticed in idiocy.

In the so-called "Bell's paralysis," due to failure of the facial nerve, the eyelids stand wide open and can not be voluntarily closed, since the orbicularis palpebrarum muscle is

FIGURE 8.—Countenance of Emphysema. (Modified from Corfe.)

coloration, of white or purple. Lines intersect this region, and the varied tints are perpetually giving new color, new feature, new expression, by their shadows. If the frontal muscle acts in connection with the corrugator supercilii, an acute deflection upward is given to the inner part of the eyebrow, very different from the general action of the muscle, and decidedly expressive of debilitating pain, or of discontent, according to the prevailing cast of the rest of the countenance. An irregularity of the pupils of the two eyes indicates, as a rule, pressure upon nerve

paralyzed. This condition may be further recognized, if unilateral, by a smoothness of the affected side, since the antagonistic muscles tend to draw the face toward the side opposite to the one in which the muscular movement is impaired; an inability to place the mouth in the position of whistling, since for this act the two sides of the face must act in unison; loss of control of saliva, which dribbles from the corner of the mouth;

[†]Ferrier, "The Localization of Cerebral Disease." New York: G. P. Putnam's Sons, 1879.

[‡]Connelly, "Med. Times and Gaz.," 1861-'2.

and a tendency to accumulation of food in the cheek, since the buccinator muscle no longer acts.

When the third pair of nerves are affected upon either side, the upper eyelid can not be voluntarily raised, for the levator palpebræ muscle fails to act; and the eye is caused to diverge outward, since the external rectus muscle, not being supplied by the third pair, and having no counterbalancing muscle, draws the eye from its line of parallelism with its fellow. In photophobia, attempts to open the eye create resistance on the part

itself tend to greatly alter the normal expression of the face. Prominently among these may be mentioned cataract, glaucoma, cancer, staphyloma, exopthalmus, iritis, conjunctivitis, amaurosis, etc., but the special peculiarities of each need not be here described.

Abnormalities of the pupils may afford the practicioner material aid in diagnosis. The pupils are found to be dilated during attacks of dyspnœa and after excessive muscular exertion,[‡] in the latter stages of anæsthesia and in cases of poisoning from belladonna

Figure 9.—Cardiac Dyspnœa. (Modified from Corfe.)

of the patient, since the entrance of light causes pain; while, as death approaches, or in a state of coma (save in a few exceptions), the eyes are usually open. In cardiac hypertrophy an unusual brilliancy of the eye is perceived,[*] since the arterial system is overfilled from the additional power of the heart. A peculiar glistening stare exists during the course of scarlet fever, which is in marked contrast with the liquid, tender and watery eye of measles.[†] Many diseases of the eye

and other drugs of similar action. A contracted state of the pupils exists during alcoholic excitement, in the early stages of anæsthesia from chloroform, and in poisoning by morphia and other preparations of opium, physostigmin, chloral, and some other drugs. Paralysis of the third cranial nerve creates a dilated condition of the pupil of the same side, since that nerve controls the circular fibers of the iris.

Growths within the deeper portions of the orbit tend to create a displacement of the eye forward, and thus to cause an apparent in-

*Loomis, "Lectures on Diseases of the Respiratory Organs, Heart and Kidneys." New York: William Wood & Co., 1874.

†J. Duggan, quoted by Haviland Hall: "Differential Diagnosis." Philadelphia, 1879.

‡M. Foster, "Text-Book of Physiology," 3d ed. London: Macmillan & Co., 1879.

crease of that organ in size. A similar condition may also result from abscesses or the growth of tumors within the cavity of the antrum. In the so-called Basedow's disease,§ an abnormal prominence of the eyes accompanies a simultaneous enlargement of the thyroid gland. The eye-lashes, if abnormal, not only in themselves create deformity, but also by causing irritation of the conjunctiva, produce an alteration in the normal expression of the eye.

THE CHEEK.

The cheek is capable of a great variety of movement. During the reception of liquid or solid food into the mouth, it is of the

passion, in which the malar region is markedly in sympathy with a general excitation of the whole respiratory apparatus.

The cheek may become the mirror of the soul. When the feelings are gay, it is drawn outward and upward; but, when the mind is depressed or saddened, it is drawn obliquely downward. If these movements be carefully noted, it will be perceived that the movable point of the cheek is situated in the immediate vicinity of the naso-labial groove;* since the attachments of several of the small facial muscles at about this point tend to draw the anterior part of the cheek outward from the line of this groove. It may be noticed as a matter of interest, that,

FIGURE 10.—Face of a Patient with Obstruction at the Pyloric Orifice

greatest assistance, since by its movements the two acts are greatly facilitated; during mastication, the buccinator muscle helps to force the food between the jaws, which are brought into apposition and rubbed together; and, finally, the cheek can act as an important factor in producing the peculiar type of countenance which is so strongly indicative of the desire of taking nourishment. The respiratory motions of the cheek are manifested in the acts of gaping and blowing, and in the exhibition of intense

when the mental impressions are slight and trivial, no traces of their effect upon the face are left upon the cheek; but, when they are of a serious or prolonged character, deep and permanent grooves are formed, which are of interest to the physiognomist as an indication of the temperament, and to the medical adviser as often of positive value in diagnosis. In the young child, the cheek, which is at nearly the same instant alternately moistened with a tear or decked with a smile, preserves in the healthy state the roundness which marks that happy age; but

§F. von Niemeyer, "Text-Book of Practical Medicine." Translated by Hackley and Humphrey. New York: D. Appleton & Co., 1869

*Blandin, *op. cit.*

in the adult, the cheek, on the contrary, presents numerous lines and wrinkles, and this appearance becomes still more apparent as old age approaches. There are, however, lines in the cheek of the aged which should not be mistaken for evidences either of the temperament or of disease, since they are produced simply by the approximation of the jaws. Lavater,* in his work upon physiognomy, locates most of the sentiment of the face in the cheek, and draws comparisons between the base and jealous face and that which is generous and noble, as a support to his theory.

fluenced except simultaneously. The changes in the cheek which effect expression, like the respiratory motions, depend chiefly upon the influence of the facial nerve; and thus it is that children and females, in whom the nervous system is generally more susceptible to impressions, also present, to the greatest degree, more or less transient modifications of the cheek. The cheek suffers a diminution in its fat as age advances, and when the teeth have been lost the approximation of the jaws forces the redundant cheek outward; and its flaccidity, from the loss of fatty tissue, throws it into folds,

FIGURE 11.—Cancer of the Abdominal Cavity. (Modified from Corfe.)

The color of the cheek varies much, both as a direct result of the passions and from special diseased conditions, which have been mentioned previously in this article. In fear and envy, the cheek is usually pale and colorless, while in love, embarrassment, or anger it is often uncommonly red. To the physiologist, these changes are a beautiful exhibition of the sympathy which exists between the mind and the circulatory and respiratory systems, which are seldom in-

which are not present in the face of the infant.

The cheek approaches a triangular form in the infant, but it becomes quadrilateral when the teeth are developed; and in the old man, as the teeth are lost, it again returns to the triangular form as in infancy. The fact that the maxillary sinus is very imperfectly developed in the child, and gradually increases as age advances, explains to a great extent why the triangular form tends to become quadrilateral; and the frequency of abnormal protrusions of this region is explain-

*Op. cit. Hunter's edition.

ed by growths or the accumulation of fluid within this cavity. The changes in the cheek produced by advancing years are also illustrated in its color. In the child, the bright rose tint, which accompanies exertion and frequently the hours of sleep, bespeaks health and general activity; but in adult age this coloring tends to disappear, and in old age the cheek often assumes a striated redness, which is due to an abnormal dilatation of the capillary vessels, especially the veins. The vascularity of the cheek renders the occurrence of erectile tumors common in this region; and the elasticity of

may be associated with that of fissure of the hard palate, and often with imperfect development of the soft palate; and thus not only is the countenance impaired, but the power of sucking, natural to the infant, is destroyed, and the articulation of words is subsequently rendered imperfect. The vascularity of the lips renders the development of erectile tumors of this region not infrequent; while hypertrophy of the tissues forming the lips may occur as one of the types of facial deformity.

The lips of the young child are very much longer in proportion to the face than those

FIGURE 12.—Partial Paralysis of the Facial Nerve from Disease near the Pons Varolii. (Modified from Corfe.)

the tissues affords an anatomical explanation of the little disfigurement which follows the removal of large portions of the cheek, in case surgical interference is demanded from any cause.

THE LIPS.

Certain deformities of the face are common in the region of the lips and mouth. Among these may be mentioned the condition of deficient closure, which is the normal condition of the hare, and to which the term "hare-lip" is applied. This deformity

of the adult, and their increased length renders the act of sucking easier to the infant than if the teeth were present, since the lips can be made almost to cross each other and thus closely embrace the nipple. When the teeth are formed, the excessive length of the lips diminishes, and the expression of the face is thus greatly altered; while, in the old man, as the teeth are lost, the lips again become very long, which accounts for their projection forward when the mouth is closed, and which gives to the face of those advanced in years the peculiar pouting expres-

sion so often seen.* The excessive length of
the lips in the aged furthermore acts as a
hindrance to mastication, and often renders
the articulation of words extremely indistinct.

In sickness, if the angles of the mouth be
depressed, pain and langor may be read;
and, when the corrugator supercilii muscle
co-operates with the depressor muscles of the
mouth, acute suffering is proclaimed.†

Extreme pallor of the lips is observed in
excessive haemorrhage, in purpura, in
chlorosis, etc.; deep lividity denotes a defective oxygenation of the blood, and occurs
chiefly in diseases of the lungs, heart, and
larynx; while pale lividity occurs in cases
where the circulation of the surface is languid or imperfect.‡ In painful affections of
the abdominal organs, the upper lip is
usually raised and stretched over the gums or
teeth, so as to give a diagnostic expression to
the countenance, which is considered by
some as of great value. In anasarca of the
face, the lips, eyes and cheeks are most
affected, since the subcutaneous cellular
tissue in these regions admits of distention
more readily than in those regions where it
is not so loose.

DEFORMITIES OF THE FACE.

Among the extraordinary deformities of
the orbital region, may be casually mentioned those rare cases of absence of the
eyes, and the union of the two orbits, as
reported by Tenon and Bartholine. The
eyelids may also be found deficient or united
at birth; and occasionally turned in or out,
when the skin and the conjunctiva are of
unequal length. The last type of deformity
is most frequently the result of cicatrization
of the tissues of the face, following an injury;
while adhesions of the eyelids to the globe
of the eye may be either a congenital defect
or the result of inflammatory processes. The
pupils may be absent at birth, or may be
partially incomplete,§ while deformities of
this aperture may also be acquired as the
result of adhesions between the iris and the
cornea or the crystalline lens, or as the
result of an operation in which portions of
the iris are excised for the relief of glaucoma.

*Blandin, *op. cit.*
†Corfe.
‡Marshall Hall, *op. cit*.
§Blandin, *op. cit*.

The entire absence of the face at the time
of birth has been recorded by Lecart, Cur-
tius, and Beclard; while in numerous instances the median portions of the face have
been absent, or the existence of deep central
fissures in the face has been detected. Cases
are on record where all evidences of the existence of the nostrils are absent, termed
"anarina"; those where the mouth has been
found absent, termed "astomia"; and those
where a double nose has existed, as recorded
by Beclard. In these abnormalities, as in
those where the cranium has been partially
or totally wanting, an arrest of the process
of development at an early stage of foetal
life must have occurred, the date of which
in pregnancy may be roughly estimated by
the extent and situation of the deformity.
In cases of senile atrophy of the forehead,
the bones are sometimes completely absorbed
and hernia of the encephalon may thus
spontaneously be produced.

Tumors of the face always create a deformity, which is confined to the anatomical
region affected; some of which have already
been referred to in this article in the treatment of certain of the special features.
Many conditions of the face, which may
properly be spoken of as deformities, are dependent upon disease. Some of those which
affect the eye and its appendages, and others
which are due to injury of nerves or to disease of nerve centers, will be described
later on, among the special types of physiognomy which are of interest in their bearing upon general diagnosis. Severe types of
ulceration, as it occurs in lupus and carcinoma, often create so extensive a destruction
of tissue as to give rise to hideous deformities, but they have no special bearing upon
the diagnosis of the existing disease.

SPECIAL TYPES OF FACE.

Many of the specific forms of disease have
their special physiognomy. As examples of
this fact, scrofulous children inherit either a
velvety skin, dark-brown complexion, dark
hair, dark brilliant eyes, and long lashes,
with the lineaments of a face finely drawn
and expressive; or a fair complexion, thick
and swollen nose, broad chin, teeth irregular and developed late, inflammation of the
Meibomian glands, scrofulous ophthalmia,
eruptions of the head, nose, and lips, and
enlarged cervical glands.*

*Williams, *op. cit*.

Hippocrates† describes a characteristic expression, which has been called from him the "facies Hippocratica," in which the eyebrows are knitted, the eyes are hollow and sunken, the nose is very sharp, the ears are cold, thin, and contracted, with marked shriveling of the lobules; the face is pale and of a greenish, livid, or leaden hue; and the skin about the forehead is tense, dry, and hard. This tpye of countenance is a most frequent indicator of impending death from chronic disease, or in an acute form of disease which has been unusually prolonged.

The "facies stupida" is distinguished by a dullness of expression, which is its chief characteristic. A peculiarity exists as regards the eyes, which are extremely dull, and resemble those seen in alcoholic stupor. This type of countenance is identical with the so-called "typhoid face," since it is most frequently met with either in connection with typhoid fever or with the typhoid condition associated with some other disease.*

Another type of countenance to which attention is frequently drawn is called the "pinched countenance." It can be produced artificially by exposure to cold, and is characterized by an apparent decrease in the size of the face, with a contracted and drawn expression of the features, and pallor or livid color of the skin. It is said to exist most frequently in the course of acute peritoneal inflammation.

In the long list of diseases which tend to shut off the supply of air to the lungs more or less suddenly, and in those accidents, such as choking, strangulation, smothering, drowning, etc., where the same effect is accomplished, the symptoms of apnœa are manifested in the face by flushing and turgidity, at first, and, later on, by a livid and purplish color. The veins of the neck become markedly swollen, and the eyes seem to protrude from their sockets. A loss of consciousness, and possibly convulsions, precedes death.†

The countenance of extreme anæmia is seen in those cases where, from sudden or gradual hæmorrhage, the prognosis is rendered alarming. The phenomena which attend this mode of dying are pallor of the face, with a peculiar leaden or clay-like hue,‡ cold sweats, dimness of vision, dilated pupils, a slow, weak, irregular pulse, and speedy insensibility. With these symptoms are frequently conjoined nausea, restlessness and tossing of the limbs, transient delirium; a breathing which is irregular, sighing, and, at last, gasping ; and convulsions before the scene closes.

The expression of the countenance is typically marked in certain of the inflammatory diseases of the eye.* In strumous ophthalmia, the child's brow is knit and contracted while the ala nasi and the upper lip are drawn upward. Those muscles which tend to exclude the light from the inflamed organ, without shutting out the perception of external objects, are called into action ; thus producing a peculiar and distinctive grin. In severe cases, the child will sulk all day in dark corners, or, if compelled to stay in bed, will bury the face in the pillow, since the exclusion of all light tends greatly to diminish the suffering. If brought to the window, the eyes are shaded with the hands or the arms; and, if the eye be opened, a profusion of hot, scalding tears will enter the nose and give rise to sneezing, or flow over the face and cause excoriation of the adjoining parts. This special intolerance of light seems to be a chief characteristic of this type of trouble, since it is often greatly out of proportion to the redness which indicates the extent of the inflammation present. In catarrhal ophthalmia, the inflammation seems to be confined to the conjunctiva and the Meibomian follicles. The eyelids are glued together by the lashes, which are bathed in the excessive secretion of the conjunctiva or of the inflamed follicles; and a redness of the surface of the eye, with some pain and uneasiness, is the only other symptom of special diagnostic value.

The deformity of iritis is characterized by a redness of the sclerotic; a change in the color of the iris, and in its general appearance, as compared with the healthy eye; an irregularity in the pupil, produced by adhesion of the iris to the adjacent structures ; possibly immobility of the pupil, as the result of such adhesions; and a visible deposit of coagulable lymph. The pupil, in

† "Prognostics." (Adam's translation).

*Finlayson, "Clinical Diagnosis." Philadelphia : H. C. Lea, 1878.

†Watson, "Practice of Physic, (Condie's edition).

†Sir Charles Bell, *op. cit.*

*Haynes Walton. "Operative Ophthalmic Surgery." Philadelphia. 1853.

acute iritis, seldom dilates in the dark, on account of the intense congestion which exists;* and it is usually smaller than that of the unaffected eye. Some pain and excessive photophobia are usually also present in attacks of acute iritis. There is something very peculiar in the expression of the countenance of a person suffering from amaurosis, by which alone the physician may almost recognize the disease. Such a patient enters a room with an air of great uncertainty as to movement; the eyes are not directed toward surrounding objects; the eyelids are wide open ; and the patient seems gazing into vacancy. This unmeaning stare of the face is due, in a great measure, to an absence of that harmony of movement and expression which results largely from the information obtained by the exercise of vision.† This seeming stare at nothing is not observed in patients who are blind in consequence of opacity of the crystalline lens or of its capsule, i. e., in consequence of cataract. They, on the contrary, while they can not see, still seem to look about them, as if they were conscious that the power of sight still remained in the retina, although the perception of objects was shut out from it. Patients, afflicted with cataract, who can not detect the existence of a gas jet or a candle in a dark room, are not fit subjects for operation, as the existence of trouble behind the lens may safely be surmised ; since the periphery of the lens seldom becomes opaque to such an extent as to prevent the perception of light by the retina, even if the outline of objects cannot be perceived.

The countenance of chronic hydrocephalus is perhaps the most typical of any of the conditions to which the attention of the physician or surgeon is directed. In it, the frontal bone is tilted forward, so that the forehead, instead of slanting a little backward, rises perpendicularly, or even juts out at its upper part, and overhangs the brow. The parietal bones bulge, above, toward the sides; the occiput is pushed backward, and the head becomes long, broad and deep, but flattened on the top. This, at least, is the most ordinary result. In some instances, however, the skull rises up in a conical form like a sugar-loaf. Not unfrequently the whole head is irregularly deformed, the two sides being unsymmetrical. Some of these

rarer varieties of form are fixed and connate; others are owing, probably, to the kind of external pressure to which the head has been subjected. While the skull may be rapidly enlarging, the bones of the face grow no faster than usual, perhaps not even so fast; and the disproportion that results gives an odd and peculiar physiognomy to the unhappy subjects of this calamity. They have not the usual round or oval face of childhood. The forehead is broad, and the outline of the face is triangular. The visage is triangular. The great disproportion in size between the head and the face is diagnostic of the disease, and would serve to distinguish the skull of the hydrocephalic child from that of a giant. In acute cerebral diseases, the countenance is either wild and excited, or lethargic and expressionless.*

Thoracic affections are all accompanied by more or less change in the color of the face; whereas the alteration in the natural hue of the features is so slight in abdominal diseases, that both the intellect and complexion remain unaltered up to the final struggle, though the pinched and dragged features express the acute sufferings of the patient. In pneumonia, the countenance is inanimate; the cheek, of a dusky hue, with a tinge of red; the eyelid droops over the globe; the brow is overhanging; the lips are dry, herpetic, and of a faint claret color; the chest is comparatively motionless, but the abdomen exhibits evidences of activity; the skin is hot; and the respiratory acts are usually about double the normal number, while the pulse is markedly accelerated. In cases where the dyspnœa is extreme, the patient, entirely regardless of what is going on about him, seems wholly occupied in respiring; is unable to lie down, and can scarcely speak; and the face becomes expressive of the greatest anxiety, while the expanded nostrils and their incessant movement indicate pulmonary distress.

In emphysema, the face is not only dusky but anæmic; the eyes are wide open, as the patient gazes at you; the dusky redness of the lips bespeaks the lack of proper oxygenation of the blood; the neck is thrown backward, and the mouth is slightly open; while the cheek is puffed out during the expiratory act; the distended nostril and the

---

*See the experiments of Mosso, quoted by Michael Foster.

†Watson, *op. cit*

*Sir Charles Bell, *op. cit*.

elevated brow stamp the case as one of dyspnœa; while the coldness of the skin shows that no acute inflammatory condition is present. If we see, in addition to these facial evidences of the disease, the deformity of the chest which has been termed the "barrel-shaped" thorax, the shrugged shoulders, and the absence of that expansive movement so well marked in normal respiration, auscultation and percussion can hardly make the diagnosis more positive.

There are certain facial conditions, which so clearly tell, to the student of physiognomy, of the existence of that most prominent sign of many pulmonary and cardiac diseases, dyspnœa, that it may be well to enumerate the alterations from the normal countenance which chiefly indicate this condition. In all cases where dyspnœa is present, the brows will usually be found to be raised; the eyes will be full, staring, and clear; the nostril will be dilated, and often it may be seen to move with each respiratory act;[*] the mouth will commonly stand partly open, while its angles will be drawn outward and upward; the upper lip will be elevated, so as to show the margins of the teeth; and the utterance of the patient will be monosyllabic, as the rapidity of breathing renders the utterance of long sentences a matter of extreme difficulty. When we add to these symptoms those of imperfect oxygenation of blood, as is met with in all conditions where the free entrance of air is in any way interfered with, we can better understand how the clear eye becomes stupid, as coma approaches, from the carbonic-acid poisoning, and the face cyanotic from the venous tinge of the blood. It thus becomes possible for the student to picture to himself the countenance which must exist in such conditions as acute laryngitis, spasmodic and true croup, thoracic tumors which cause pressure upon the lungs or the trachea, and the various conditions of the lung itself, which impede the entrance of air to the organ, but which are not of inflammatory origin, and which have, for that reason, no distinctive physiognomy.

In cases where renal dropsy has stamped its characteristic marks upon the countenance, we may perceive the signs of dyspnœa, due to the accompanying œdema of the lungs, in the corrugated forehead, the raised eye-

brow, the dilated and waving nostrils; the corners of the mouth will be found to be drawn downward and outward, expressive of some disease of the abdominal cavity; the eye will be full and anxious, indicative of suffering long continued and borne with patient calmness; the conjunctiva may present that pellucid and bleb-like condition, so often seen in this type of disease, and an œdema of the eyelid may greatly alter its appearance; finally, the waxy pallor of the complexion and the pasty and bloated cheeks show the profound anæmia of the patient.

Chronic diseases of the abdominal cavity are usually characterized by a languor of the eye and by an absence of that flash of alarm so peculiar to the acute forms of abdominal trouble;[*] and, if attended with steadily increasing danger to life, the corrugated brow and eyelid, the retraction of the cheek, the dragged and elongated nostrils, the depressed angles of the mouth, the protruded chin, and the parted lips, with the teeth firmly clinched behind them, still further proclaim the seat of the disease.[†]

The pale face, stamped with the signs of anxiety and distress; the head raised upon two or three pillows, and the trunk similarly supported; the knitted brow, which bespeaks the cerebral disturbance; the nostrils, waving to and fro with each breath; and the jugulars, which, as they lie exposed in the throat, show that the valves of the heart are acting imperfectly, by their pulsation or unusual distention; all may be found in endocardal inflammations, or in the conditions of the heart dependent upon chronic valvular disease.[‡]

The countenance of continued fevers is liable to receive a modification from their complication with some morbid affection of the head, viscera of the thorax, or of the abdomen; the dejection produced by the latter of which is among the most important objects in the clinical study of these diseases.[§] In scurvy, the dirty ashy hue of the skin and its characteristic dryness; the blue and bleeding gums; the emaciation and the frequent indurations of the inter-muscular tissue of the cheek; the sunken eyes, surrounded by a blue ring; and the livid

---

*Lavater, *op. cit.*; Sir Charles Bell, "Anatomy of Expression."

*Corfe, *op. cit.*

†M. Louis, quoted by Marshall Hall, *op. cit.*

‡Corvisart, "Disease of the Heart"; Gate's Translation. Boston, 1812.

§Marshall Hall, *op. cit.*

tinge of the lips, makes the diagnosis positive at once.

In Graves's or Basedow's, disease, a peculiarity of the eye is produced, due to its partial protrusion from the orbit, probably from an increase of the intra-orbital fat, which stamps the disease beyond the possibility of error in the diagnosis. In many cases, the inability to approximate the lids, and an absence of power to move the eye, on account of paralysis of the muscles from the stretching which they have undergone, furnish evidence also of disease of that organ which enhances the facial deformity.

In Asiatic cholera, and in children during attacks of profuse diarrhœa, the eyeballs sink into the orbit, a dark ecchymosis appears in the region of the eyes, the lower eyelid forms a prominent fold in the region of its attachment to the cheek, the nose is pointed and sharp, and the lips, normally ruddy and full, become thin and sharply outlined. These changes are chiefly dependent upon a rapid emaciation, which follows the withdrawal of a large proportion of the water from the tissues.* In chronic atrophy, the entire absence of the adipose tissue in the subcutaneous structures causes the skin to become loose and corrugated; while various muscles become prominent from contraction (chiefly the frontalis, the corrugator supercilii, and the levator labii superioris).† Thus the so-called "senile face" or "Voltairean countenance" is produced, which is seldom to be mistaken in the child.‡

Among the diseases of the nervous system, there are certain types of physiognomy which are so characteristic as to be of the most positive value in diagnosis. Thus, in the attacks of epilepsy, the neck at first becomes twisted, the chin raised, and brought round by a series of jerks toward one shoulder. The features are greatly distorted. The brow is knit; the eyes are sometimes fixed and staring, and at other times rolling about in the orbit, and again turned up beneath the eyelid, so that the cornea is covered and only the white sclerotic is to be seen; the mouth is twisted to one side and distorted; the tongue is thrust between the teeth, and, caught by the violent closure of the jaws, is

bitten, often severely; and the foam which issues from the mouth is reddened with blood. The turgescence of the face indicates obstruction of the venous circulation; the cheeks become purplish and livid, and the veins of the neck are visibly distended.

The expressions of the countenance which are produced by paralysis of any of the special nerves of the face have striking peculiarities which enable the skillful anatomist to easily detect the nerve affected. It is important to remember that, if paralysis of any nerve be the result of any form of external injury, a danger is presented in the form of tetanus, which should be guarded against by a quick comprehension of the existing malady and by all known precautions, applied with judgment based on the anatomical course and relations of the nerve affected. It is also well to bear in mind the fact, that any form of severe external violence about the face may, by causing a fracture of the bones through transmission of the force applied, cause injury to some special nerve whose course may lie far distant from the apparent seat of injury. It is not infrequent to find a fracture of the superior maxillary bone followed by symptoms indicative of a foreign body within the cavity of the antrum; and symptoms of irritation of the nasal mucous membrane, or of neuralgia of some of the principal nerve trunks distributed to the face, may likewise follow such an accident. Violence to the vault of the skull may produce not only cerebral lesions and their subsequent evidences in the face and body, but also types of local paralysis,* produced by injury to some of the more important nerve trunks at their point of escape from the skull, in case the base of the skull has been injured.

"A slight tremor of the lips; a hesitation of utterance; a partial loss of power over the lips and tongue, which seem to have lost their grip, as it were, over the consonants; a characteristic stillness of the muscles of expression; and a slight disparity in the pupils are the predominant features of the early stage of development of the general paralysis of the insane."† In those rare cases where the facial nerve of both sides is impaired, symptoms similar to those mentioned

*Vogel, *op. cit.*
†Marshall Hall, *op. cit.*
‡Vogel, *op. cit.*

*Holden, *op. cit.*
†W. H. Gairdner, Article on "Medical Physiognomy," in Finlayson's "Clinical Diagnosis."

above exist, except that the tongue has its normal capabilities of movement, save in the perfect articulation of the labial consonants only, and that a complete absence of facial expression is present. An open mouth; a loss of control over the saliva, which constantly dribbles; an awkwardly moving or motionless tongue; and an indistinct articulation render the labio-glosso-laryngeal paralysis of Trosseau and Duchenne easy of detection.† In the so-called Bell's paralysis,‡ which has been described in previous pages of this article, the patient can not laugh, weep, or frown, or express any feeling or emotion with one side of the face; while the features of the other may be in full play. One-half of the aspect is that of a sleeping or dead person; while the other half is alive and merry. This incongruity would be ludicrously droll, were it not so frightful and distressing.

During the fit of exacerbation, in an attack of tetanus, the aspect of the sufferer is sometimes frightful. The forehead is corrugated and the brow knit, thus expressing the most severe type of bodily suffering; the orbicularis muscle of the eye is rigid, and the eye itself staring and motionless; the nostril is widely dilated, indicating the extreme dyspnœa; the corners of the mouth are drawn back, exposing the teeth, which are firmly clinched together; and the features, as a whole, have a fixed and ghastly grin—the so-called "risus sardonicus." During such paroxysms, as in those of epilepsy, the tongue is liable to become protruded between the teeth and to be severely bitten.

In chorea, the facial muscles participate in the general eccentricity of movement. Watson§ thus describes the peculiarties of this strange affection: "The voluntary muscles are moved in that capricious and fantastic way in which we might fancy they would be moved, if some invisible or mischievous being, some Puck or Robin Goodfellow, were behind the patient and prompted the discordant gestures. With all this, the articulation is impeded; there is the same perverse interference with the muscles concerned in the utterance of the voice. By a strong figure of speech, the disorder might be called 'insanity of muscles'."

†Finlayson, *op. cit.*
‡Sir Charles Bell, *op. cit.*
§*Op. cit.*

In catalepsy, the patient lies often with eyes open and staring, yet without expression indicative of life ; more like a wax figure or a corpse than like a living subject. The features may be made to assume any expression, no matter how absurd, as the tissues have their normal pliability; and they will remain so placed until again mechanically altered. This same peculiarity is also present in the muscles of the extremities, and forms one of the distinguishing tests of the disease. The mental faculties are in abeyance, and all power of voluntary motion is lost. The sensibility of the body seems also to be lost.

The deformities of face and intellect which seem to be the result of residence in special atmospheric conditions, or of certain well-defined localities, are illustrated in that race of people found in Valais and the adjoining cantons of Switzerland, called "cretins". Many of these wretches are incapable of articulate speech; some are blind, some are deaf, and some suffer from all of these privations. They are mostly dwarfish in stature; with large heads, wide vacant features, goggle eyes, short crooked limbs, and swollen bellies. The worst of them are insensible to the decencies of Nature, and in no class of mortals is the impress of humanity so pitiably defaced. They are usually the decendants of parents afflicted with goitre.

In that long list of pathological conditions in which the brain may be subjected to more or less compression of its substance, there are certain signs of positive value in diagnosis which may often assist the medical practitioner to locate the disease. Thus, in depressed fracture of the inner table of the skull, where the signs of external injury are absent ; in abscess within the cranial cavity, during the course of meningeal inflammations ; in apoplexy ; in the development of intra-cranial tumors, etc., the eyelids will usually be closed and immovable ; the pupils generally dilated or irregular, and always sluggish and less sensible to light than in health ; the breathing will be slow and stertorous if coma exists ; the special senses will be in abeyance ; and the temperature will be either normal or increased. The evidences of a paralyzed condition of certain of the cranial nerves may also exist, and thus afford an additional means of determining the exact seat of the disease. A regidity of certain muscles, if present, denotes some special irritation of the nerves which supply them, and it is, therefore, seldom present in cerebral softening, but frequently so in those cases where paralysis is produced by pressure upon nerve centers. In cases where the fifth cranial nerve has been impaired by pressure, injury, or disease, the prominent symptoms are a redness of the conjunctiva on the side of the face supplied by the affected nerve; insensibility of the cornea, nostril, and tongue on the same side; a dullness of hearing; a partial or complete loss of smell, sight, and occasionally of taste also in the anterior two-thirds of the lateral half of the tongue; and a diseased state of the gums, similar to that observed in scurvy.

While many typical varieties of countenance, which are of value to the diagnostician, have been ommitted, since the limits of a single article have possibly been already overstepped, still, it is to be hoped that the facts mentioned, although they are but fragmentary jottings, may tend to kindle among the medical profession a renewed interest in a subject which is rapidly being lost sight of, and the value of which is often ignored. It is not to be expected that sight alone can guide the medical attendant to unerring diagnosis; but that it may prove of the greatest value *as an aid*, can not, I think, be disputed. It is to be remembered, however, that a direct perceptive faculty, like that of touch, hearing, or smell, *grows with use*, and is capable of unlimited development. As with the musician, an instrument which at first produces discord becomes, under skillful hands, one of melody; so the enlightened and accomplished practitioner may often see at a glance what, to one unaccustomed to note facial changes or to interpret their meaning, would escape detection, unless a special effort was made to note and record systematically the peculiarities of each particular feature and anatomical region of the face, and the records afterwards studied, as the mariner studies his chart before he attempts to direct his vessel through channels with which he is not perfectly familiar.

Dr. James J. Hale, of Anna, Ill., strongly recommends an infusion of celery seeds (Apium Graveolens) given almost ad libitum to irritable children. He much prefers it to the stronger narcotics, and says: "It is astonishing what good babies can often be made of the most fretful and restless."

American Medical Association.

The Thirty-second Annual Meeting of the Association was well attended, and the proceedings were harmonious. The President's address and the addresses of the chairmen of Sections were carefully prepared, were worthy of the occasion, and were well received. In the Sections the papers were meagre, notwithstanding the efforts of the chairmen to procure valuable contributions, and elicited no notable discussions. In the business sessions, interest centred around the amendment of the Code of Ethics, and the proposition to substitute a weekly medical journal for the annual volume of Transactions. Dr. Billing's substitute for the amendment to the Code was a happy conception, which to a large extent met the arguments of both sides, and at the same time filled the requirements of the case as fully as was probably practicable under any proposition. The action of the Association in relegating for further consideration the question of publishing a weekly journal was at least prudent. The publication of such a periodical involves a much larger annual outlay than it was apparent the Association had at its command, and therefore, with more wisdom than such large bodies usually display in business matters, it wisely deferred action until it had received more accurate information upon which to form a sound judgment than the committee had been able to lay before it. The selection of a President for the ensuing year was happy, and was a proper recognition of the value of Dr. Woodward's labors, as well as those of the Army Medical Staff, in behalf of medical science.

During the sessions of the Association the *Virginia Medical Monthly* and the *Southern Clinic* issued daily edition, giving full and accurate reports of the proceedings of the day before. To these we are indebted for aid in the preparation of our report of the proceedings of the Association.

To the local committee of arrangements much credit is due for the careful and painstaking way in which they successfully managed the details of the business and pleasure of the Association. The well-known reputation for hospitality which Richmond has always enjoyed was more than maintained, and this year's meeting will long be memorable to those who participated in it for the unbounded hospitality, both public and private, which was extended to the delegates by the physicians and citizens of Richmond.

Proceedings of the American Medical Association, Richmond, Va., May 3, 4, 5, and 6, 1881:

The American Medical Association convened in thirty-second annual session at Mozart Hall, in the city of Richmond, Va., on Tuesday, May 3d, at 11 A .M., John T. Hodgen, M. D., of St. Louis, President, in the chair. The session was opened with prayer by the Rt. Rev. Bishop Keane, and the Association was welcomed to Richmond by Dr. Frank Cunningham, of Richmond, Chairman of the Committee of Arrangements, and by the Hon. F. W. M. Holliday, Governor of Virginia. The President then delivered his Annual Address, which was an earnest and able advocacy of conservative surgery, and was received by the Association with marked evidence of approval.

SECTION I. (*Practice of Medicine*, Dr. William Pepper, of Philadelphia, Chairman). In a paper upon "Bloodletting as a Therapeutic Measure in Pneumonia," Dr. W. C. Wile, of Sand Hook, Conn., advocated the free use of the lancet, in which he was supported by Drs. N. S. Davis, of Chicago; Post, of New York; Quimby, of Jersey City, Whitney, of New York; and S. D. Gross, of Philadelphia. Drs. J. J. Lynch, of Baltimore; Whittaker, of Cincinnati; Oeterlony, of Louisville; and Lester, of Detroit, took opposite ground.

SEC. II. (*Obstetrics and Diseases of Women*, Dr. James R. Chadwick, of Boston, Chairman). Dr. Paul F. Munde, of New York, made some "Practical Remarks on the Use of Pessaries," and the subject was discussed by Drs. Beverly Cole, of San Francisco; Albert H. Smith, of Philadelphia; H. P. C. Wilson, of Baltimore; Maughs, of St. Louis; and Quimby, of Jersey City. Dr. Wilson dwelt upon the importance of ascertaining whether the uterus can be lifted up before introducing any pessary, and expressed his disapproval of stem pessaries.

SEC. III. (*Surgery*, Dr. Hunter McGuire, of Richmond, Chairman). Dr. J. H. War-

ren, of Boston, read a paper "On the Use of Various New Surgical Instruments;" and Dr. Wm. A. Byrd, of Quincy, Ill., reported "A Case of Ulceration and Perforation of the Appendix Vermiformis, with Remarks upon Abdominal Sections in cases of Perforation of the Bowel," in which he advocated operative interference in inflammatory affections around the cæcum. Dr. J. E. Reeves, of Wheeling, presented from Dr. B. W. Allen, of Wheeling, a report of a case of nephritic calculus weighing 480 grains, which had caused a pyo-nephrosis with a large fluctuating tumor on the hypochondrium. By aspiration eighteen pounds of sero-purulent fluid were evacuated, with temporary relief. The patient died, and on autopsy the left kidney was found to have been converted into a sac, fifteen inches long, twelve inches wide, and six inches deep, containing ten pounds of purulent fluid, together with the above-mentioned calculus.

SEC. IV. (*State Medicine*). In the absence of the Chairman, Dr. J. S. Reeve, of Wisconsin, this section adjourned.

SEC. V. (*Ophthalmology, Otology, and Laryngology*, Dr. D. S. Reynolds, of Kentucky, Chairman). Dr. G. T. Stevens, of Albany, described a "Registering Perimeter." Dr. W. C. Jarvis read a paper on "Nasal Catarrh, with Hypertrophy," and advised the removal of the hypertrophied tissue by the ecraseur, and in cases of sessile growths with the aid of transfixion needles. Dr. J. J. Chisolm, of Baltimore, read a paper on the "Treatment of Conical Cornea," in which he advocates puncture of the cornea with a red-hot needle, as advised by some French surgeons.

SEC. VI. (*Diseases of Children*, Dr. A. Jacobi, of New York, Chairman). "The Importance of Physical Measurements in Children" was the title of a paper by Dr. H. P. Bowditch, of Boston. He said that it seemed probable that the accurate determination of the normal rate of growth in children will not only throw light upon the nature of the diseases to which childhood is subject, but will also guide us in the application of therapeutic measures. Dr. Bowditch exhibited a chart representing the case of a child, between two and three years old, in which careful and systematic weekly weighing showed, first, the approach, by some weeks, of a chronic disturbance of nutrition, represented by enlarged cervical glands and clay-colored stools; and, second, after recovery, the approach of an attack of measles, the "danger signal" of progressive loss of weight preceding the eruption by at least a week.

Dr. Lee, of Baltimore, remarked that he had paid especial attention to this subject, and had noticed that a female child could lose more in proportion to its weight, without detriment to its health, than a male child of the same age. He also said that if the loss of weight preceding an eruptive disease was excessive, the case was so much more grave in its prognosis, and that this loss of weight preceded this eruption by from four to five days.

Dr. Busey, of Washington, read a paper on "The Relation of Meteorological Conditions to the Diarrhœal Diseases of children;" and Dr. James C. White, of Boston, forwarded a paper entitled "Some of the Causes of Infantile Eczema, and the Importance of Mechanical Restraint in its Treatment." He described the many and varied external influences which immediately effect the skin of the new-born, as being a common cause of eczema, and laid especial stress on the fact that heat was the more usual cause of the disease than cold. These external influences, however, furnish but a small proportion of all the cases of the disease which occur at this period of life, altho.t, n by m the great part of those concerning the etiology of which we have any positive knowledge. During the last twelve years he had treated at the Massachusetts General Hospital five thousand cases of eczema, of which one thousand seven hundred and seventy occurred in children of ten years of age and under, and of which the largest proportion, viz., 569 cases, was in the first year of life. Eliminating the operation of the causes directly acting upon the skin from without, above mentioned, and a few other extraneous agencies, the parasitic chiefly, Dr. White did not hesitate to say that he knew nothing whatever of the causes of the disease in the remainder; also, that as far as his experience went, eczema affected all classes of society alike, occurred at all seasons of the year, came in children of all degrees of health, in the perfectly sound as frequently as in the feeble; that it had no necessary connection with any other disease of child-

hood; that it showed itself in an equal proportion in bottle babies and those reared at the breast, and was independent of diet; also, that if there were other assigned causes he would here say that his observation gave him no justification for believing any of them.

After speaking of the extreme suffering which the little patients undergo, Dr. White said the prime factor of the treatment was the prevention of scratching, and he advocated controlling the child's movements by a system of swathing in a pillow-case, by which the same chances of success in the therapeutics of infantile eczema is given, as exist in the adult. He said that when the strait-jacket treatment is carried out, the child soon becomes used to the confinement, and a wonderful improvement takes place in the disease.

Dr. L. Duncan Bulkley, of New York, believed that Dr. White had laid far too much stress upon local causes, and had ignored entirely the influence of internal, general, dietary, and hygienic causes. Internal treatment he believed to be to a certain degree absolutely necessary, and without it, physical restraint would be comparatively ineffective. He advocated the internal use of small purgative doses of calomel every other day, and a mild alkali, as acetate of potassa in liquor ammoniæ acetatis, with a little nitre and perhaps aconite. As regards mechanical restraint, if the itching is relieved, it is not required; and if it is not relieved, such confinement is torture beyond description. Diachylon ointment Dr. Bulkley believed to be very inefficient in arresting itching. Tar in some form was far more efficacious; indeed, he had little to desire in the way of an application to infantile eczema beyond the following ointment: R. Unguenti picis, one ounce; zinc oxid, two drachms; unguenti aquæ rosæ, three ounces. Mix. This should be very carefully prepared and very thoroughly and abundantly applied. If it appears stimulating, less of the tar ointment may be used. He laid great stress upon employing the rose ointment and not simple cerate, or lard, or vaseline, or petroleum. The ointment should be made of a consistency to spread easily and yet not to all melt away after application.

Dr. Bulkley was very positive in the directions given in regard to the use of water to eczematous surfaces in children: they were only to be washed according to direction, and that very rarely, often only at intervals of several days; moreover, it was all important that the protective ointment should be replaced *immediately* after the surface is dried, and renewed sufficiently often to keep the parts completely protected, even twenty or more times the first day. On covered parts the ointment may be thickly spread on the woolly side of the sheet lint and bound on. Among the hundreds of cases he had never covered the face with a mask, and had rarely been obliged to restrain the infant much after the first day or so. The only restraint he had ever practiced, was putting on muslin mittens, tied around the waist, and then tapes from these passed behind the back or beneath the leg.

Dr. D. H. Goodwillie, of New York, read a paper on "Thumb-sucking."

Second day. General meeting. The Committee on Nominations was announced, and the Association then proceeded to the consideration of the proposed Amendment to the Code of Ethics, Aar. I., paragraph 1st, by adding " and hence it is considered derogatory to the interests of the public and honor of the profession, for any physician or teacher to aid in any way the medical teaching or graduation of persons knowing them to be supporters and intended practitioners of some irregular and exclusive system of medicine."

Dr. Marcy, of Massachusetts, moved to lay the amendment on the table. Lost. Dr. E. S. Dunster, of the University of Michigan, then addressed the Association in a speech carefully prepared, in opposition to the amendment. Dr. N. S. Davis, of Chicago, moved that the amendment be made the special order after the addresses on Thursday. Dr. Howard, of Maryland, moved as a substitute, that the matter be indefinitely postponed. The substitute was lost, and the motion of Dr. Davis adopted.

Dr. William Pepper, of Philadelphia, delivered the Address in Medicine, which was devoted to the consideration of the great importance of local lesions, and especial catarrhal inflammation of mucous membranes as forming the essential cause of many apparently obscure diseases, and also as adding greatly to the danger of many diseases which are now regarded as due exclusively to the

presence of some specific poison in the blood.

Dr. James R. Chadwick, of Boston, delivered the address on Obstetrics and Gynaecology, in which he directed attention to a statistical consideration of the literature of his specialty during the past five years.

Dr. John H. Packard, of Philadelphia, from the Committee on Journalizing Transactions, presented an elaborate and carefully prepared report upon the subject which closed with the proposal of the following resolution:

Resolved, That the President be authorized to appoint a committee of five to to digest and report in detail as soon as practicable upon the time, place, and terms of the publication of such a journal, to elect an editor, fix his salary, and to arrange all other necessary details.

Dr. N. S. Davis, of Chicago, moved to strike out so much of the resolution as related to the election of an editor. Adopted.

Dr. Marcy, of Massachusetts, moved that as many of the previous committee as were present at the meeting, be members of the new committee. Adopted.

Dr. Toner, of Washington, moved to add the Secretary and Treasurer to the committee. Adopted.

Dr. Toner, of Washington, offered a resolution instructing the Secretary to publish with the forthcoming report of the Transactions of the Association an index of all the previous volumes. Adopted.

Dr. Toner presented the report of the Committee on Necrology.

SECTION I. (*Practice of Medicine.*) Dr. D. W. Prentiss, of Washington, presented a paper entitled "Is croupous pneumonia a zymotic disease?" and Dr. W. C. Dabney, of Charlottesville, Va., a paper on the "Nature and Treatment of Pneumonia." Dr. L. Duncan Bulkley, of New York, read an elaborate paper on "The Diet and Hygiene of Eczema."

SEC. II. (*Obstetrics and Diseases of Women*). Dr. H. P. C. Wilson, of Baltimore, exhibited some instruments, which gave rise to a spirited discussion upon instrumental interference. He showed some dilators by the use of which he is enabled to dispense with the use of tents. Dr. Albert H. Smith, of Philadelphia, expressed a preference for the tents: he said, the instruments make small lesions about the cervical canal, hence the dangers of septicaemia in careless hands. The tents, he thinks, are safer, if not withdrawn before the end of forty-eight hours instead of twenty-four, as is usually done. The tents ought to be cylindrical (not conical), and should be rubbed over with moist soap and dipped in a solution of salicylic acid before being introduced.

SEC. III. (*Surgery and Anatomy*). Dr. C. F. Stillman, of New York, described some "new mechanical appliances," and Dr. Alfred C. Post, of New York, read a paper on "Plastic Operations on the Face." Dr. D. H. Goodwillie, of New York, also read a paper on the "Treatment of Arthritis of the Temporo-maxillary Articulation," by means of an apparatus to relieve the joint of pressure on the inflamed articular surfaces. It is made as follows: an impression of the teeth of either jaw is taken and an interdential splint made, the posterior part of which is raised a little for the purpose of a fulcrum, on which the back tooth of the opposite jaw rests. Another impression is taken of the chin, and a rubber splint is made to fit it. A skull-cap is next made to fit the head closely, with elastic bands on each side passing down from it and fastened to the chin-splint. The interdental splint is placed in position in the mouth, and the back teeth of the jaw closed on the fulcrum of the interdential splint; then, when pressure is made on the chin by tightening the elastic bands connecting the skull-cap with the chin-splint, the joint is relieved from pressure. Dr. Goodwillie reported some successful cases thus treated.

Dr. B. A. Watson, of Jersey City, N. J., read an elaborate paper, the title of which was "An Experimental and Clinical Inquiry into the Etiology and Distinctive Peculiarities of Traumatic Fever."

SEC. IV. (*State Medicine*). In the absence of the Chairman, Dr. J. S. Billings, U. S. A., was elected chairman *pro tem*.

Dr. J. L. Cabell, of the University of Virginia, read an interesting paper on "The National Board of Health and the International Sanitary Conference of 1881." He reported that there was good reason for hoping that an international agreement may be arrived at between the States most frequently threatened with epidemic invasions. And, aside from this, the degree of attention which

as a result of the deliberations of the Conference has been given to the subject of maritime sanitary police, cannot be without fruit in securing greater cleanliness, better ventilation of ships sailing on the high seas; and in general, an improved sanitary condition of these important instruments of commerce, which become so often the carriers of the most deadly contagion, from the failure to use such precautions as sanitary sience suggested, and as it is hoped will now be enforced among the maritime powers of the world.

Dr. C. F. Folsom, of Boston, read a paper on the "relation of the State to the insane." He argued in favor of the establishment of State Lunacy Boards. Among the points made were, first of all, that a lunacy board should embrace men with a thorough knowledge of insanity and its treatment. The chief duties of this board should be to secure proper care for the insane in private dwellings, where they are very much liable to neglect. Secondly, they should examine the commitment papers, and otherwise looking into the cases, so as to be able to tell whether the lunatic should be retained for care or be discharged.

SEC. V. (*Ophthalmology, Otology, and Laryngology*). Dr. Carl Seiler, of Philadelphia, read a paper upon "Syphilitic Laryngitis," and Dr. Chisolm one on a form of tinnitus induced by a rhythmical contraction of the tensor tympani muscle. Dr. Eugene Smith, of Detroit, presented a report of a successful operation for blepharoplasty, in which the graft was taken from the arm without any pedicle.

SEC. VI. (*Diseases of Children*). Dr. R. J. Nunn, of Savannah, forwarded a paper on the "Treatment of Diphtheria." Dr. E. H. Bradford presented an article on "resection of the tarsus in severe congenital clubfoot."

Third Day. General meeting. On motion of Dr. S. D. Gross, of Philadelphia, the By-laws were amended so as to establish an additional section to be known as the Section of Dentistry. A motion to suspend the rules so that the section might be organized immediately was objected to, and therefore could not be considered.

The President announced as the Committee on Journalizing the Transactions, Drs. J. H. Packard, N. S. Davis, J. S. Billings,

L. A. Sayre, and R. B. Cole, with the Treasurer and Secretary.

Dr. Hunter McGuire, of Richmond, delivered the address in surgery, and chose for his subject "operative interference in gunshot wounds of the peritoneum," which he advocated as exchanging an almost certain prospect of death for at least a good chance of recovery. He urged operative interference in gunshot penetrating wounds of the peritoneum with intestinal injury, in penetrating wounds of the peritoneum with any visceral lesion, and in similar cases without visceral injury. The wounds in the abdominal walls should be enlarged, or the linea alba opened freely enough to allow a thorough inspection of the injured parts. Hemorrhage should be arrested. If intestinal wounds exist, they should be closed with animal ligatures, trimming their edges first if they are lacerated and ragged. Blood and all other extraneous matter should be carefully removed, and then provision made for drainage. If the wound of entrance is dependent, drainage may be secured by keeping this open. If the wound is a perforating one, and the aperture of exit dependent, the patency of this should be maintained, and, if necessary, a drainage tube of glass or other material introduced. If there is no wound of exit, and the wound of entrance not dependent, then a dependent counter-opening should be made and kept open with a drainage tube. If it is urged that the means suggested be desperate, it can be said in reply that the evil is desperate enough to justify the means.

In the absence of the address of the chairman of the Section on State Medicine, Dr. J. S. Billings, U. S. A., occupied the allotted time with some interesting remarks on some of the results of the tenth census as regards mortality statistics.

Dr. N. S. Davis, of Chicago, from the Committee on Clinical Observations and Records, presented an elaborate report in which the appointment of a standing committee was recommended. The report was accepted and the recommendation adopted.

The Committee on Nominations presented the following report:

For President, Dr. J. J. Woodward, U. S. A.

For Vice-Presidents, D. P. O. Hooper, Arkansas; L. Conner, Michigan; Eugene Gresson, North Carolina; and Hunter McGuire, Virginia.

Secretary, Dr. Wm. B. Atkinson, Pennsylvania.

Treasurer, Dr. R. J. Dunglison, Pennsylvania.

Librarian, Dr. Wm. Lee, Washington, D. C.

To fill vacancies in Judicial Council, Drs. S. N. Benham, Pennsylvania; J. M. Toner, Washington; D. A. Linthecum, Arkansas; William Brodie, Michigan; H. D. Holton, Vermont; A. B. Sloan, Missouri; and R. B. Cole, California.

St. Paul, Minnesota, was selected as the place for the next meeting; and Dr. Stone was appointed chairman of the Local Committee of Arrangements.

Dr. Billings, U. S. A., moved the adoption of the report of the committee; and paid a high compliment to Dr. Woodward, the nominee for President, and thanked the committee, in behalf of the Medical Staff of the Army, for the high compliment paid it in the selection of one of its most honored members for the highest position in the gift of the Association.

The report of the committee was unanimously adopted.

The next order of business was the further consideration of the amendment to the Code of Ethics. Dr. N. S. Davis, of Chicago, made an able and elaborate argument in reply to Dr. Dunster and in favor of the amendment, and Drs. Martin, of Massachusetts, and Dunster, of Michigan, spoke against it. Dr. Marcy, of Massachusetts, moved to lay the amendment on the table, defeated; and the further consideration of the amendment was postponed until Friday.

The next business was the consideration of the following amendment to By-laws, offered by Dr. J. M. Keller, of Arkansas:

"In the election of officers and appointment of committees by this Association and its President, they shall be confined to members and delegates present at the meeting, except in the Committees of Arrangements, Climatology, and Credentials." The amendment was adopted.

SECTION I. (*Practice of Medicine*). Dr. Whittaker, of Cincinnati, read a paper on the "treatment of dyphtheria," and advanced the view that diphtheria was first a local and afterwards a general disease, and that it is only when the epithelial barrier is broken down that the blood and the body become infected. He contended that, although he could not kill the germs of the disease in the throat, he could so condense its mucous membrane as to make it a dam to the influx of disease, and to this end he recommended the persulphate of iron in full strength, applied well up behind the velum palati.

Dr. I. E. Atkinson, of Baltimore, read an exceedingly valuable and interesting paper on the "production of albuminuria by use of iodide of potassium in syphilitic disease," and Dr. Henry A. Martin, of Boston, a paper on "variola vaccine and variola equinae in Massachusetts."

Sec. II. (*Obstetrics and Diseases of Women*). Dr. Jos. Tabor Johnson, of Washington, read a paper on the "diagnosis of pregnancy in early months," and Dr. Munde, of New York, exhibited a curette for removal of adherent placenta after abortion.

Sec. II. (*Surgery*). Dr. Charles A. Leale, of New York, presented a paper on "carbuncle, and its treatment," in which he recommended free incision and the thorough application of pure nitric acid.

Sec. IV. (*State Medicine*). No meeting.

Sec. V. (*Ophthalmology, Otology, and Laryngology*). The afternoon was occupied by a general discussion of the subject of astigmatism.

Sec. VI. (*Diseases of Children*). Dr. Clarence J. Blake, of Boston, presented a paper on "middle ear disease in children in the course of the acute exanthemata." Dr. A. Jacobi, of New York, Chairman of the Section, then delivered his address.

Fourth Day. General Meeting. The President announced the following as the "Committee on Clinical Observations and Records," Drs. N. S. Davis, of Chicago; J. M. Toner, of Washington; H. O. Marcy, of Boston; W. H. Geddings, of Aiken; and S. M. Bemiss, of New Orleans.

The Association resumed the consideration of the amendment to the Code of Ethics, and Dr. J. S. Billings, U. S. A., offered the following substitute:

"It is not in accord with the interests of the public or the honor of the profession that any physician or medical teacher should examine or sign diplomas or certificates of proficiency for, or otherwise be especially concerned with, the graduation of persons whom they have good reasons to believe intend to support and practice any exclusive and irregular system of medicine." The previous question was moved, and the substitute was adopted.

The Committee on Nominations presented the following additional report:

Section on Practice of Medicine—Chairman, Dr. J. A. Octerlony, Kentucky.

Section on Surgery and Anatomy—Chairman, Dr. J. C. Hughes, Iowa.

Section on Obstetrics—Chairman, Dr. H.O. Marcy, Massachusetts.

Section on Medical Jurisprudence and State Medicine—Chairman, A. L. Gihon, Washington, D. C.

Ophthalmology. Otology, and Laryngology—Chairman, Dr. D. B. St. John Roosa, New York.

Diseases of Children—Chairman, Dr. S. C. Busey, of Washington.

Dentistry—Chairman, Dr. D. H. Goodwillie, of New York.

On motion of Dr. Grissom, of North Carolina, a honorarium of $1,000 was voted to the permanent Secretary.

Dr. Goodwillie, of New York, offered an amendment to the Constitution, enabling permanent members to vote. Laid over for one year.

Dr. D. S. Reynolds delivered the address on Ophthalmology, etc.

The report of the Treasurer showed a balance of $2,208.40.

The report of the Committee on Publication was presented.

The Librarian reported that 273 titles had been added to the library during the year, and asked for an appropriation of $200 for current expenses, and that an appropriation of $50 be made in aid of the publication of the index medicus. Accepted, and appropriations passed.

The Convention unanimously adopted the following:

Resolved, That the thanks of this Association are hereby tendered to the Committee of Arrangements of this Association, for the faithful attention they have given to their duties and requirements; to the medical profession and citizens of Richmond for their hospitality and endeavors to make the time spent by us while here pleasant and agreeable; to Drs. McCaw and McGuire for the elegant special entertainment given by them at the Westmoreland Club; to Mr. McClure, Superintendent of the Telephone Company, for special facilities given the Committee of Arrangements and the Association; to Vice-President Parsons, of the Richmond & Allegheny Railroad, for his kind invitation for a free ride on his road to show us the interior of the State of Virginia; to Mr. Powell, manager of the Richmond Theatre; to the managers of the Mozart Association, and all others who have contributed to our pleasure and comfort; to the press, and especially their reporters, in giving such a full *resume* of the proceedings in the daily papers; to the railroad companies generally who have so liberally reduced the rates of transportation for our benefit and any other modes of conveyance that have so contributed; to Mr. Valentine for his kind invitation to his studio.

Be it Especially Resolved, That our thanks are particularly due to the ladies of Richmond for attention and kind interest in making our sojourn so pleasant and agreeable.

The Association then adjourned to meet at St. Paul on the first Tuesday in June, 1882.

The Association was entertained on Tuesday evening by Dr. and Mrs. McCaw and Dr. and Mrs. Hunter McGuire at a very handsome reception at the Westmoreland Club; on Wednesday evening by an operatic performance (The Doctor of Alcantara), tendered by the physicians of Richmond; on Thursday evening by the citizens of Richmond at a promenade concert at the Richmond Theatre; and on Friday afternoon by an excursion along the valley of the James River, by the Richmond & Allegheny Railroad Company.

THE AMERICAN SURGICAL ASSOCIATION.

The American Surgical Association met at Richmond, May 5th, at the Hall of the House of Delegates. Dr. S. D. Gross, of Philadelphia, presiding, and J. R. Weist, of Indiana, acted as Secretary.

At the request of the President, Dr. Packard, of Philadelphia, read the proposed new Constitution and By-Laws. On motion of Dr. Davis, the Constitution and By-Laws, as read, were unanimously adopted. Under

this Constitution, surgeons known as "specialists" are allowed to become members of the Association. Under the old Constitution, that class of surgeons were excluded.

Drs. Marcy and Martin, of Boston, and Watson, of Jersey City, were recommended for membership.

Dr. Packard, the Treasurer, reported that he had received $925 from members, which, with the interest ($9.25), was deposited to the credit of the Association.

An amendment to the Constitution was adopted, to the effect that all nominations for membership shall lie over until the next regular session.

The Oriental Hotel, Coney Island, New York, was selected as the place for the next annual meeting. The 13th, 14th, and 15th of September was selected as the time.

The Association then adjourned until the above time and place.

The following is the list of officers of the society:—

President—Samuel D. Gross, M. D.

Vice Presidents—Drs. L. A. Dugas, James R. Wood.

Recording Secretary—J. R. Weist.

Corresponding Secretary—W. D. Briggs.

Treasurer—John H. Packard.

Council—Drs. Moses Gunn, John T. Hodgen, Hunter McGuire, J. C. Hutchinson, Samuel D. Gross, J. R. Weist, and R. A. Binloch.

CORRESPONDENCE.

····→≈☼≈←····

TRINIDAD, COLORADO.

Editor ROCKY MOUNTAIN MEDICAL REVIEW.

DEAR SIR:—To the question in Dr. Haws' circular, "Do you regard the so-called mountain fever as a malarial fever?" I answer, "I regard the so-called mountain fever as a so-called malarial fever"; and to the sequent, "what are your reasons?" I replied that I do not believe that a specific aerial poison possessing the peculiar property of causing periodical fevers and commonly called "malaria" exists, because I have no reason to believe it.

Feeling inclined to answer the question in the hearing of all my learned brethern, I have dashed off the following—not doubting but that there are others as incredulous on the subject as myself.

I do not believe that any specific aerial poison as the cause of periodical fevers, and commonly called "malaria", has any real existence. I do not believe it because I have no reason to believe it; I have never seen, felt, smelt or tasted it. Chemistry and microscopy have alike failed to reveal it. I have no evidence whatever of its existence. I have heard it suggested that we are made conscious of its presence by its effect upon our health. I have no more reason to attribute those effects to a specific malaria than I have to attribute them to the diabolical operations of evil spirits, and would regard the latter as about as scientific a conclusion as the former.

It seems to me the time has arrived when we ought to be able to trace the so-called malarial diseases to causes other than a specific *bad air*—non-specific bad air, bad changes of temperature, bad rest, bad exercise, bad diet, bad drink—two or more sanatary evils combined.

As for periodicity, I believe everything in nature is regularly periodical—the earth in its revolutions, and the celestial bodies—the universe is one grand system of regular periodical changes, I believe that everything on the earth, in the sea, and under the earth even down to, and excluding only the ultimate action of inorganic matter, are regularly periodical. I consequently believe that man, in health and in disease, is a periodical being independent of *malaria*, and that it is not wonderful that his natural periodicity should, under the influence of certain morbid conditions, become more manifest than in health. Diurnal periodicity of disease generally, sometimes more, sometimes less distinct, has been observed from time immemorial, and that when the exacerbations are shorter it is evidence that the disease is less grave, and *visa versa*. I take it that the more distinctly periodical a fever may be, the greater is the evidence of the triumph of innate periodicity over the operations of the morbific cause, and should not be mistaken for the playful freak of *malaria*.

Show me the peculiar malaria that produces the periodical fever. Show it to me in a test-tube, under the microscope, or to the naked eye. Let me have just a little of it in my hand so that I may feel it. Let me have

a particle of it on my tongue that I may taste its peculiar flavor. Let me smell its strange aroma, or let me hear its omenous sound. Make me cognizant of its existence in any reasonable way, or let me alone in my belief that it is a myth, a hallucination, a humbug.

M. BESHOAR, M. D.

EDITORIAL.

→≈◊≈←

The Secularization of Hospital Nursing.

The experiment which, for some time past, has been in progress at two of the large Paris Hospitals, of replacing the Sisters of Charity by a lay nursing staff, has excited a good deal of controversy. Some of the best known names in the medical world are to be found ranged on one side or the other, and the discussion of a question, which, after all, should be a mere matter of business, has assumed the dignity of a discussion on first principles. On the one hand, it has been assumed to be a question identical with the interests of religion and morality, and on the other, that the presence of Sisters of Charity in a hospital ward is inimical to perfect civil and religious liberty. When such is the temper of the disputants, their arguments are apt to be acrimonious rather than logical.

To us, the question in dispute appears simple enough. Narrowed down to its ultimate elements, it merely amounts to this—whether or not the nursing staff of a great hospital should be practically independent of the medical staff and lay government?

Now, this latter is precisely the position which sisters of any religious nursing corporation occupy, whether in Paris or elsewhere. They owe allegiance to none but their own theological superiors; they form essentially and emphatically an *imperium in imperio*. They occupy a position of their own, independent of and apart from the general body of employees. In return for certain services rendered, they claim immunity from the ordinary discipline and regulations of the institution in which they serve. Hence, their position is anomalous and illogical, unless, indeed, they can show some paramount claims to consideration and indulgence beyond ordinary mortals.

No one, of course, will deny that in the past these nursing guilds did good work. At some time they, in conjunction with the higher clergy, may be said to have had a monopoly of all tender offices among the poor, the sick, and the dying. But such is no longer the case; the distribution of charity is dependent on mere spontaneous impulse.

With the growth of an elaborate civilization, charity has become an organized department of the State. No one pays poor rates from any sentimental or religious motives, but simply because, as a citizen, he must bare his share of the ordinary burdens of the State, one of which is, the support of paupers. So with the great Hospitals. They are essentially State, or at least public, institutions, and should be managed on the strictest and most prosaic business principles. In every department they should be under one central authority, who should rule with no uncertain sway. Once let there be the semblance of divided authority, and straightway abuses will creep in, and the public interest suffer.

Nursing is no longer the semi-religious, semi-romantic business it was in the middle ages. It is, at the present day, a department of skilled labor, and, moreover, a very honorable and important one. Like every other business, it requires to be learned, and it possesses a market value of its own. Should any guild or body of persons desire to undertake the very arduous and responsible duties of hospital nursing, on principles other than those we have enunciated, it would probably be found cheaper and better, in the long run, to decline their offer of assistance, however well-meant such proposals might be.

America in Advance — Facts and Fancies Concerning Our National Medical Library.

Every American physician has great reason for the exercise of National pride in the work which for some years back has been assiduously prosecuted at our National

Capital, and which has but recently assumed definite shape. We refer to the work of preparing a subject catalogue of the books, pamphlets and periodicals in the possession of the National Medical Library.

The first volume of this stupendous work made its appearance a few months since, and was reviewed in the columns of this Journal, as well as it was possible to review and do justice to so extensive and valuable a production. The work in question is the immediate outgrowth of the indefatigable energy and perseverence of Dr. Billings and his collaborators—Drs. Fletcher, Yarrow and Wise, of the Surgeon General's Office. It not only embraces the titles of all books lodged in the library, but also the titles of all articles appearing in the pamphlets and journals therein deposited; the two combined constituting over 500,000 distinct subject headings.

Although this Library ranks but third in *size*, the work here referred to renders it the most *valuable* in the World; for without such a catalogue, the most extensive libraries are comparatively speaking of no practical value; the treasures stored away in their archives being as good as buried. Such a catalogue emanating from so extensive a library, might almost be considered a complete bibliography of the medical literature of the world. Indeed it is applicable as an index to the contents of all the medical libraries throughout the land, great or small; its value in any one case being proportionate to the size and completeness of the library.

The following incident may, perhaps, in a small way serve to illustrate the amount of prestige such a library bestows upon a country : But a short time back, the British Government appointed a Board to investigate the ravages of diphtheria, its cause and prevention, which at that time was carrying off a large number of the Nobility. Of course, the first step necessary for successfully carrying out such an investigation, was a thorough study of the literature of the subject up to that time. This, however, was impossible without America's aid, owing to the absence of a subject catalogue. Aware of the fact that such a catalogue was in course of preparation at Washington, under the supervision of the Surgeon General's Office, the Board in question recognized and availed itself of our advance, by sending over to this country for an abstract of the literature bearing upon the subject to be investigated; offering, at the same time, to pay any amount required in return for the same. The Surgeon General's Office furnished the desired information gratuitously, and the Board was enabled to carry out its otherwise imperfect work satisfactorily. The literature bearing upon diphtheria was found to be something immense, and far beyond our English brethren's anticipations.

That America should be the first to undertake and carry out such an enterprise is greatly to her credit, and very justly places her in the foremost ranks.

The size of this library, and the value of its catalogue, may be crudely imagined when we consider that the first volume of the latter contains nearly 1,000 pages, and yet only embraces A and a part of B; and that the entire work will consist of no less than ten volumes, with over 500,000 separate titles.

We are sorry to have to say Congress has failed to appropriate a sufficient amount of money to complete even the second volume of this work—a work which not only is invaluable to a fraternity numbering in this country upwards of 65,000, but which also redounds to America's credit, and which will prove indirectly a boon to suffering humanity throughout the length and breadth of the land. Congress might better at once appropriate a sufficient amount to complete this work than squander $120,000 for a projected local scheme—in the form of another hospital for Washington, which place is already amply supplied with such commodities, and needs another only that certain recently dislodged surgeons may find congenial occupation again.

Not content short of perfection, the projectors of the *Index Catalogue*, Drs. Billings and Fletcher, have started a monthly supplement to the catalogue, known as the *Index Medicus*, a monthly classified record of the current medical literature of the World. In this matter again America takes the lead, such an enterprise never before having been undertaken.

The National Medical Library already contains upwards of 100,000 books and pamphlets, and with the additions rapidly being made through the liberal appropriations from Congress, and the two works above referred to, it will soon be the *largest* as well

as the *best* medical library in the World. It should be remembered that all books, journals and pamphlets sent to the *Index Medicus* are lodged in this library, through the liberality of its Editors, and are not used for private purposes.

The library is, at present, located in the old theatre building in which President Lincoln was assassinated. This building is very much dilapidated, and the walls are at any time liable to fall under the excessive weight imposed by so many books. It is therefore to be hoped Congress will make provision for this library in the projected Congressional Library Building, or else appropriate a sufficient sum to erect a separate building for its accommodation, preferably the latter.

With the view of rendering this great medical repository of greater practical value to isolated members of the profession in distant parts of the country, who are unable to visit Washington, but who frequently wish to consult the literature bearing upon some particular point brought up by a society debate, the preparation of a paper or report, we would present our "fancies" and suggest the advisability of inaugurating what might be appropriately termed an abstract bureau; composed of students and young graduates in medicine, under a competent chief familiar with the contents and arrangement of this library. The object of which should be to examine into and report upon all questions involving the literature of a particular point or subject referred to the library by any medical man in good standing, whose application shall have been endorsed by the President of his county society. Charges to be made and the work paid for by the applicant. Any quantity of medical students and young graduates could be found who would be only too glad to perform such work for the mere schooling it would give them in addition to a small pecuniary compensation; and the profession would, without doubt, gladly take advantage of the system.

In this way the usefulness of this library might be immeasurably extended, a greater interest in it provoked among the members of the profession in distant parts of the country, and a fruitful source of revenue created for the enlargement of the library.

THE Colorado State Medical Society will meet in Leadville, Tuesday, September 13, 1881, at 10 a. m. An unusually large attendance is anticipated.

Denver as a Medical and Scientific Centre.

The City of Denver having secured to itself a great amount of prestige, by its central location, numerous railroad centerings and termini, and the vast wealth and productive powers still remaining dormant throughout its environs, has long since established itself as the commercial centre of the "New West." But from present appearances she is not going to stop short here and rest content with fame from a purely utilitarian and business stand point. On the contrary, she aspires to renown as a great scientific, educational and social metropolis. She already has numerous general educational institutions of a superior character. The most prominent of these institutions is the Denver University, which sails under the leadership of the Rev. Dr. Moore, a gentleman of superior talents and unusual liberality, and which was founded principally through the beneficence and cultivated scientific tastes of our ex-Governor, Dr. John Evans, who, wherever he has been, has figured conspicuously in medical and scientific circles. This University, greatly to her credit may it be said, though young, is many, many years in advance of all other similar institutions west of the Mississippi; particularly as regards her Scientific Department, which receives its "spiritual" impetus from that very promising chemist and physicist, Prof. Sidney R. Short, the discoverer of a method of reducing chromium from its oxide to its metalic state. The chemical and physical apparatus of this college far surpasses anything to be found anywhere else in the west.

The most recent advanced step on the part of this institution which we here wish to chronicle, is the establishment of a Medical Department, the faculty of which has just been organized. There seems to be a positive need for a medical college in this part of the country, for the accommodation of broken down students, who come out here to recuperate; as also for the convenience of those desirous of studying medicine, but who are unable to do so from the fact that they find it impracticable to travel thousands of miles to accomplish their object. Although there are many men in Denver not connected with this faculty, who are emin-ently fitted for positions therein, and whom we would very much wish to see recognized, yet, necessarily, there was not room for all. Those who have been selected and appointed are from among the pre-eminently representative men in the medical profession here, and give evidence of being able to conduct very successfully a medical school of high standard. In fact, we are assured from the prospectus, and gleanings through association with these gentlemen, that from the very start a high standard will be established and adhered to strictly, which seems to be very practical, from the fact that there is presented no competition.

The site for the college building has been determined upon, and that very judiciously, and lectures will begin on the first of November, 1881.

The official announcement will appear in our next issue.

An Injustice Rectified.

We are extremely gratified at receiving, from an outside and disinterested physician, a review of the second edition of Dr. Denison's valuable work on "Rocky Mountain Health Resorts," inasmuch as the opportunity is thus presented to expose the motive which prompted the severe criticism the first edition received at the hands of one Dr. Reed, of Colorado Springs. This latter party, it will be remembered, was for a very short time connected with the REVIEW. Dr. Reed, it must be acknowledged, was placed upon the REVIEW without our having sufficient knowledge of his character or principles, and we soon found him out to be not in the least way calculated for a representation upon a reputable medical journal—one of those "combined homœopathists and alopathists," etc., you know.

At the time the review of the first edition of Dr. Denison's book was written, the editor of this Journal had not read the same, but relied solely upon Dr. Reed's representations, which we have since found out to be unjust to an extreme; to say nothing of the palpable ignorance displayed by the reviewer in his criticism.

We have delayed and had hoped to avoid the recital of these facts, but justice to an injured party compels their publication.

Dr. Denison's work has received unanimous commendation, both in this country and abroad, and is generally conceded, by those capable of judging, to be a valuable addition to the scientific literature of the subject it treats. Such men, for instance, as the editor of the *British Medical Journal*, Prof. Austin Flint, Sir. James Paget and Dr. John Ordronaux, have bestowed upon it deservedly the highest terms of praise.

Ourselves.

It will be observed that we have moved the REVIEW to Denver, Colorado; and this fact must account for our delay in issuing the April and May numbers, and finally bringing both out together. We found it very difficult to secure reliable and satisfactory parties, with proper facilities, to do our printing. Everything now, however, is propitious for still greater perfection and success than we have attained in the past; owing, first, to the fact that we have secured a valuable acquisition in Dr. J. H. Kimball as associate editor; and, second, obtained a first-class printer in Mr. C. J. Kelly. Hereafter, our journal will be brought out regularly on the 25th of every month. We feel assured our subscribers will appreciate the obstacles we have had to overcome, and will deal leniently and charitably with us.

IT is through the courtesy of the *New York Medical Journal* that we are enabled to furnish our subscribers with Dr. Ramey's very practical and admirably written article on the "Human Face in Health and Disease," they having supplied us with the cuts illustrating the same.

OBSTINATE VOMITING IN PREGNANCY.—Bailly has recently (*Archives de Tocologie*, January, 1881), found good results from the application of a blister to the epigastrium, and ice along the spine in cases of obstinate vomiting from pregnancy. The remedy is rather heroic, and if used at all should be used only as a last resort in cases where abortion is threatened, and which fail to yield to milder treatment.—*Chicago Medical Review.*

A BILL to regulate the practice of medicine in Arkansas has passed the Assembly.

SCIENTIFIC SUMMARY.*

···➤✹✺◄···

Anthropological Notes.†

By DR. H. C. YARROW, U. S. ARMY,

Member of Acad. Nat. Science, Phila.; Philosophical, Biological and Anthropological Societies of Washington; American Asso'n for the Advancement of Science; French Asso'n for Advancement of Science; Corresp. Member of Lyceum of Nat. Hist., New York; Zoological Society of London, Etc., Etc.

INDIAN OBSTETRICS.

In the *American Journal* of Obstetrics, for April, 1881, Dr. Geo. J. Engelmann, of St. Louis, Missouri, publishes a paper entitled, "The Third Stage of Labor an Ethnological Study," which is a most valuable and interesting contribution to our knowledge of Indian obstetrics. The author divides the third stage of labor and its management into different sections:

1,

those methods which are adopted for the purpose of expelling the placenta, in which the patient retains the same position which she occupied in the delivery of the child, and of these, the one ordinarily practised is by the employment of a *vis a tergo*, most commonly by a force applied externally from above downwards, by manual expression; and, secondly, by action of the diaphragm, by the use of emetics. Much less frequent is the employment of the *vis a fronti*, that fatal traction upon the cord which forms the third group. Somewhat different from these methods is a fourth which I have classified under

II,

comprising the customs of all those tribes who look upon a change of posture as the important element for the purpose of accomplishing the expulsion of the after-birth. A change of position is made immediately after the birth of the child, and the patient assumes a different posture from the one occupied during the earlier stages of labor. This is not frequent in ordinary cases, but is

*Reports appearing in this department will be written with special adaptability and reference to the requirements of the scientific physician.

†Books, pamphlets or papers on anthropological subjects, may be sent the editor of this department, in care of the Rocky Mountain Medical Review, Denver, Colorado.

a usual resort in cases where some difficulty is experienced in the removal of the placenta.

Manual expression of the placenta, Dr. Englemann states, is practiced by the Commanches, Klamachs, Crows, Nez-Perces, Peorias, Shawnees, Kiowas, Caddoes, Delawares, Wyandottes, Ottawas, and Senecas, the patient being generally in a kneeling, or squatting position, the assistant kneeling behind and pressing with her hands upon the fundus of the uterus. The Clatsops place a bandage about the patient's body "to prevent the placenta from going back further into the body." Some of the divisions of the great Sioux nation permit the patient to lie down in the last stage of labor, and judiciously compress the uterus until the after-birth comes forth. Many of the tribes are said to use what is called the "squaw beet," a broad leather band which is strapped, gradually tightening, about the abdomen after the child has been born. In some cases, traction is made upon the cord while gentle manipulation is made over the abdomen.

A number of different examples are given of the manual expression common to many tribes, and the author closes the first section by stating that the Cheyennes, Arrapahoes, and Chippewas never wait for the natural expulsive efforts of the uterus, but drag away the placenta at once by the cord.

In some of the tribes the patient rises to her feet and forces out the placenta after the child is born by use of the "squaw beet;" the Utes place a folded cloth over the abdomen and lean heavily on a stout stick, until the placenta comes away. Emesis is frequently resorted to when other means fail.

A very curious method of placental extraction consisted in attaching to the funis a buckskin cord, the other end being fastened to the great toe; the woman was then directed to stretch herself out from time to time, and it was expected that in this way the after birth would come forth. A steam bath is sometimes used.

A number of examples of peculiar superstition and observances regarding the cord and placenta are given, one being that they must be eaten if this can be done secretly. This, however, does not obtain among our Indians, but those of Brazil.

Taken, as a whole, the paper is most valuable, and it is to be regretted that our limited space will not admit of a further notice. One matter we cannot refrain from alluding to, in regard to the illustrations, which, while they convey admirably the idea of certain processes, the faces and figures of the women depicted are so hideous that we are inclined to wonder how men could ever be courageous enough to approach for the purpose of affording Dr. Engelmann an opportunity of writing on *their* third stage of labor.

THE MILITARY ARCHITECTURE OF THE EMBLEMATIC MOUND BUILDERS.

This is the title of a recent paper by Rev. Stephen W. Pert, in the *American Antiquarium* for January, 1881, which contains quite an elaborate account of different mounds in certain states which Mr. Pert believes to have served for warlike purposes. Some considered to have been signal mounds for the purpose of transmitting information, on the appearance of an enemy, others to have been used for astronomical or religious purposes. The paper is well worth the attention of students of archery, and is embellished with a number of wood-cuts and diagrams.

The anthropological society of Washington met March 15th, Prof. J. W. Powell in the chair.

After reading minutes of the last meeting, the following gentlemen were elected to membership in the society:

Dr. Frank Baker, David Hutchison and Henry M. Baker. A paper entitled "Politics—Social Function," was then read by Prof. Lester F. Ward, and while bearing more directly on sociology than anthropology, was a most careful and interesting study. Considerable discussion followed. Prof. Otis T. Mason then read a paper entitled, "The Savage Mind in the Presence of Civilization." This paper, owing to the lateness of the hour, was not concluded. A valuable resume and history of the progress of education among the Indians was a prominent feature of the discourse.

THE ALASKA INDIAN DOCTOR.

J. G. Brady, in *The Herald and Presbyter*, published the following account of the *Shamam*, or the conjuring medicine man of Alaska.

The Indian doctor is called *Ischt* in his own tongue, and *Shaman* in the Russian. When a male child is born with a curl-lock of hair it is a sign that he is to be a doctor. He is carefully fostered by his parents and friends. His hair is not cut or combed, nor is he allowed to eat clams, crabs, or any beach food. It is seldom that an infant is born with the desired curly lock.

Years ago the credulous were deceived by designing relatives, who would present the child with a curl made by hand. It is seldom that a boy, with a genuine curl, makes his appearance. There are others who aspire to the position and influence of doctor.

When one dies, an Indian will go upon the roof and call for the *Yake*, or demon, who dwelt in the body, which is now lying in state and surrounded by mourners. If he comes he will be apt to enter into one of the young men who are standing around the corpse. He falls as if he were shot dead. This is a sign that the old doctor's demon has entered into the man. He is taken off to one part of the house and covered with a blanket. He pretends to be wholly unconscious. There is a tacit understanding between the man who calls the *Yake* and the one into whom he is supposed to enter.

Sometimes an Indian falls so violently as to injure his head. This has its proper effect upon the bystanders. Others who have neither curly hair nor are possessed, become doctors. Often a nephew of the doctor's sister's son, is the favored one. All candidates must endure the test. When the proper time arrives, the person who is to be initiated, goes to the tomb of a doctor whom he chooses as a sort of partner. He is attended by two watches of relatives—four in each. The test is an absolute fast for eight days. He sleeps one or two nights in the dead house, and the watches are to see that he does not break his fast. He is allowed the use of tobacco while he is fasting, and he makes up his songs which he will sing when called upon to cure a sick person. His guards learn the same songs, for they are to be his attendants in the future when he practices his rites.

In former years, the business of a physician among the northwest coast tribes, while it was a most profitable one, was by no means pleasant, for, if the *Shaman* failed to cure his patient and death resulted, in most cases he was summarily murdered. Of late years, however, the anger of the relatives, in many instances, is appeased by presents judiciously administered by the medicine man. The supposition power of song in healing the sick, as noted by Mr. Brady, finds its analogy in the account of a case reported in one of the eastern medical journals, where the patient was cured by banjo playing.

SYPHILITIC BONES.

Some bones have recently been received at the Army Medical Museum, exhumed from mounds in the vicinity of Alexanderville, Ohio, by Dr. Bicking. The collection consists of two femora, two tibiae and fragments of three crania, the latter being in very bad condition. Several medical men who have examined the specimens are of the opinion that they clearly show evidence of being syphilis, others believe that the exostoses which are vissible in the tibiae only may be the result of periostitis or rheumatic disorder; of course it is impossible to say positively which theory is correct in the absence of corroborative evidence of the disease in the other bones, writer is inclined to believe them non-syphilitic. They will be submitted to the eminent syphilographer, Dr. F. N. Otis, of New York, for his opinion. A good deal of importance is attached to a positive diagnosis regarding these bones, for if it can be found that the mound builders suffered from syphilis, a valuable link will be added to the chain of facts adduced by Dr. Bruhl in his paper on Pre-Columbian Syphilis, published a short time since.

SURGERY AS PRACTICED BY THE INDIANS.

Dr. H. M. M'Clanahan, of Montana, contributes to the *Medical and Surgical Reporter*, of Philadelphia, for April, 1881, a paper with the above title which appears as a continuation of a former article. In this, the manner of treating wounds, fractures, dislocations, etc., is given. For hemorrhage, compresses are used over the seat of injury, and cobwebs are employed. The Indian knowing full well that uncontrolled bleeding means death. First dressings are allowed to remain undisturbed until removed by ulcerative action. In rifle wounds, the ball is not removed unless in sight. Favorite applications to ordinary wounds are marrow fat and prairie dog oil, after a careful washing with cold water. The writer states that erysipelas, gangreen and pyæmia are very rare, and he believes one reason for this is that the air of the lodge is constantly filled with smoke which acts as a disenfectant. Compound fractures are treated by resection with extension and counter-extension. All openings being filled with certain extracts, and over all is smeared a coating of the scrapings of rawhide boiled in water, this forming a gelatinous layer. Over this and around the entire limb is placed a covering of fine bark, and over top of this, splints of willow or some other light wood. In simple fractures, the splints are applied next to the skin. In dislocation, attempts are made at reduction, and the member is then splinted. Stiff joints frequently result from this plan of treatment. Searing with a flint or sharp instrument is a favorite form of counter-irritation for many ills, especially swellings, tumors and deep seated pain.

The article is fairly well written, and is really a contribution to our knowledge of surgical practices among the Indians.

THE PRACTICE OF MEDICINE AMONG THE INDIANS.

This is the title of a paper which appeared in the *Medical and Surgical Reporter*, of Philadelphia, March 26, 1881, by Dr. H. M. McClanahan, and is worthy a perusal. The writer states that his observations have been among the Gros, Ventres and Assinniboines, but that from an examination of his subject he is convinced that all of the tribes have more or less the same practices. We give a few extracts from the paper:

"To know why the Indian believes in his mode of treatment * * * it will be necessary to ascertain his views as to the cause and nature of disease." "He does not consider himself in any way responsible for an attack of illness. He may expose himself to inclement weather and contract a severe cold, but the two do not stand in the relation of cause and effect."

He knows certain diseases are contagious, but he does not think the disease, *per se*, is catching, but that the person taking it is under some peculiar spell. Habits of life, hereditary influence or changes of weather, he does not believe bears any relation to disease, "but that it is governed by some higher power and entirely beyond the influence of man, so far as preventing it is concerned;but that an enemy can, by making bad medicine, cause it to produce sickness. All the manifestations of disease are, therefore, due to evil spirits."

According to Dr. McClenahan, they have no distinct idea of what they mean by the ten spirits, some giving them the form of animals, or insects; others considering them imponderable entities. "Whatever may be the peculiar form of the evil spirit imagined by the individual, all believe this much to be a fact, namely, that it takes possession of the body causing sickness and death. It is a matter of observation among them, that there are many forms of disease, as whooping cough, measles, diarrhœa, etc., but they do not look upon the different forms as distinct types of disease due to specific causes, but simply believe the spirit to be inhabiting different parts of the body." The spirit is omnipotent, therefore, can produce convulsions in one and diarrhœa in another. It will thus be seen that they give no consideration to rational causes, but assign everything to a higher power. I have asked them why it is that the spirit will select one instead of another when the natural conditions are the same.

"In case a child is sick the father believes he has, in some way, offended the spirit and it punishes him through his child, he being too strong for it. Again, they say an enemy can influence the spirit. This is called making bad medicine. I have known of several cases where a man, having several wives, is taken sick, when some one of the wives gets the blame for making bad medicine.

"It is not necessary to multiply examples, for they would only prove what has already been asserted, that disease is caused by spirits. It is to them an all-powerful, always present *something*, capable of anything an imagination, uncontrolled by reason, can conceive of."

It will readily be imagined that the treatment will be essentially the same in all forms of disease. Such is the fact. Disease is an evil spirit, and the object, in all cases, is to drive it out of the body. Still, they do employ other modes of treatment, but simply to relieve some local symptom, and not to counteract the disease itself. We can hardly find fault with the savage for this belief, regarding the spirit form of disease, for such a belief has been wide spread throughout the world, and even up to the present day among communities said to be civilized and enlightened, we find that many give credence to the idea of witchery producing disease.

The author describes the appearance of the medicine man, his antics and frightful noises, which are supposed to frighten the disease away, and explains that he, like many of his civilized philosophical brethren, follow this pursuit for lucre and not for philanthropy. In one respect the Indian doctor is far ahead of his civilized brother, so far as reputation is concerned, for if his patient dies or gets no better, he states that his fee was not large enough, or that some enemy had made bad medicine to counteract all his efforts. As a last resort he may admit that he was not strong enough to scare the spirit. Dr. Mc-Clanahan describes several of the curative, processes employed by the Indian, particularly the sweat-bath, and promises forthcoming articles on surgery and obstetrics. We shall anxiously look for these, feeling sure from the doctor's long experience among Indians, that many valuable facts will be brought to view.

SIAMESE TWINS OUTDONE.

The *Medical and Surgical Reporter*, of Philadelphia, March, 26, 1881, contains the following account of a singular lusus.

An Italian couple, Jocci by name, are at present exhibiting at Vienna, a most remarkable specimen of their progeny, a pair of twins named Jacob and Baptiste. These boys are grown together from the sixth rib downward, have but one abdomen and two feet. The upper part of the body is completely developed in each; their intellectual faculties are of a normal character. Each child thinks, speaks sleeps, eats and drinks independently of the other. This independence goes so far as to admit of an indisposition of the one without in the least affecting the other. They are over three years old, in perfect health and seemingly in excellent spirits.

INDIAN MEDICINE.

It is hoped that the brief notice of Dr. McClanahan's paper on Indian medicine may attract the attention of medical men, particularly those living on the frontier or contiguous to Indian tribes. The Bureau of Ethnology, Smithsonian Institute, have in contemplation a volume on this special subject, and will shortly issue a circular requesting information, and if a little interest is shown by the profession, much valuable material can be secured with very little trouble. The peculiar Indian practice of medicine is rapidly becoming a theory of the past before the march of civilization, and in a few years all

of it will be lost, and known only by tradition. Let us then harvest our facts while we may.

TRANSLATIONS.

Translated for the REVIEW, *by Dr. Liton Forbs.*

WICKERSHEIMER'S FLUID.

The formula of this well-known preservative fluid is given in the Deutscher Reichs-Anzeiger, No. 251, as follows:

Boiling water, . . .	3,000 parts.
Alum, ...	100 "
Calcic chloride, . .	25 "
Potassic nitrate, ' '	12 "
Potash, ...	60 "
Arsenious acid, . . .	10 "

Mix, allow to cool, and then filter. The resulting liquid should be neutral, colorless, and inodorous to every ten litres thus prepared. Add:

Glycerine, ...	4 litres.
Methylic alcohol, . .	1 "

There are various methods of using this fluid according to the object in view. Animal and vegetable tissues and generally speaking objects whose colors it is wished to preserve may simply be immersed in it. Specimens to be dried should be kept in the fluid from six to twelve days, according to their size, and subsequently dried in air. Ligaments, muscles, insects, shell-fish, etc., when thus prepared, preserve their natural elasticity and can be made to perform their natural movements. Hollow viscera, such as the lungs, intestines, etc., should, first of all, be filled with the fluid and then immersed in it and subsequently dried.

For purposes of embalming, simple injection of the fluid into the blood vessels is sufficient. The quantity generally required for a child two years old, is one litre and a half; for an adult about five litres. Bodies thus treated preserve their form, color and flexibility for an indefinite period. After the lapse of years they can be subjected to the most delicate anatomical, medico-legal, and even histological researches; neither putrefaction nor any unpleasant odor is ever generated, while both animal and vegetable colors are preserved unimpaired. Lowenthal has seen such delicate structures as the epithelium, the striae on muscles and the tissues of the eye, months after treatment with this fluid, react in all respects as fresh tissues would have done in presence of coloring re-agents. This valuable invention is due to Herr Wickersheimer, of the Anatomical Museum of Berlin, from whom it has recently been purchased by the German Government.

POISONING BY LEECH BITES.

Professor Kocher, of Berne, Switzerland, reports the following case :

A gentleman suffering from toothache consulted a dentist who recommended the application of leeches to the gums. Two hours after this advice had been complied with, the patient who had up to that time been in excellent health suddenly became seriously ill. The lip rapidly inflamed, a tumor formed in the cheek accompanied by dyspnœa, which proved fatal in three days. The autopsy revealed no adequate cause of death.—*Journal d'Hygiene*, March 24, 1881.

[This would appear to be one of those very rapidly fatal cases of septicemia, in which the whole mass of the blood is altered and rendered incapable of sustaining nutrition.

The dyspnœa may be explained by the fact that the poison attacked primarily the respiratory centres located in the medulla ablongata. Rep.]

ON THE RELATIVE INFLUENCE OF ATROPINE, DUBOISINE AND HOMATROPINE ON THE EYES.

The researches of Schafer on the relative influence of atropine, duboisine and homatropine go to show that as regards dilatation of the pupil, the action of atropine, though slower, is more permanent than that of duboisine, the effects of which latter while sooner manifested, also sooner pass off. Homatropine exhibits its effects quicker than either of the preceding, but they pass away quicker, nor does it dilate the pupil so thoroughly.

Accommodation is most readily paralized by duboisine and homatropine, the latter perhaps, being in this respect the more active of the two. Its effects pass off in about twenty-four hours, while those of duboisine continue from three to four days. The impairment caused by atropine is of much

longer duration, hence in cases where dilatation of the pupil is required merely for opthalmoscopic purposes, or when relaxation of the accommodation is desired in order to test refraction, homatropine should be used ; but where a therapeutic action, such as the breaking down of synechia is to be obtained, homatropine is distinctly inferior to either of the other mydriatics.—*Archiv. für Augenheil Kunde.*

ON A NEW METHOD OF COMBATTING INSOMNIA.

Professor Hoppe, in the *Journal D'Hygiene*, (April 7, 1881,) mentions a plan of treatment in insomnia, which he has frequently found successful.

It consists simply in rapidly closing the eyelids, twenty or thirty times in succession, and thereby causing fatigue of the depressor palpebrae muscles; this will be followed in a few seconds by an irresistable desire for sleep. The professor recommends this treatment more especially in that form of insomnia which accompanies nervous diseases.

The proportion of military surgeons in the principal European armies, in time of peace, is thus given in the *Militaerartz*, (March, 1881).

In the Swedish army	one surgeon to every	223	men.
" " German	" " " "	230	"
" " Swiss	" " " "	235	"
" " Spanish	" " " "	250	"
" " Austrian	" " " "	290	"
" " Belgian	" " " "	304	"
" " Russian	" " " "	307	"
" " Italian	" " " "	362	"
" " English	" " " "	369	"
" " French	" " " "	385	"
" " Turkish	" " " "	400	"
" " Dutch	" " " "	484	"

BROMIDE OF AMMONIUM IN WHOOPING-COUGH.

This remedy which has already been extolled by Harley, Gibb and Ritchie, is the subject of a communication from the pen of Dr. Kormann, of Dresden, in the last number of the *Wiener Medical Wochenschrift*. This observer has lately had a series of cases under his care, in which the value of bromide of Ammonium in whooping-cough, was very conclusively shown.

He gives it in doses of 0.15 grammes (two grains) to infants and 0.25 to 0.40 for children. A few doses are generally quickly followed by relief.

They are best given in the intervals between the paroxysms. The only contra-indication is the existence of catarrh of the bladder. Even young children can take considerable doses of bromide of ammonium; it rarely causes drowsiness or stupor, but should it do so, the symptom will quickly disappear on the withdrawal of the drug. Its unpleasant taste is best concealed by strong coffee or syrup. Owing to the high price of the drug, it is generally preferable in hospital practice to prescribe it in the form of powder.— *Wiener Medical Wochenschrift.*

LONG STANDING GASTRIC ULCER SUCCESSFULLY TREATED BY WASHING OUT THE STOMACH.

The patient, Carl Joseph, was admitted into Bicetre Hospital, as incurable, on January 8, 1881. He stated that ten years previously, during the Franco-German war, he had received a severe fright and had lost consciousness. Recovering from this state he had vomited large quantities of blood, which symptom had since returned on several occasions, alternating with mucous and bilious ejections. When admitted into hospital there was absolute intolerance on the part of the stomach to all food, and the patient was generally reduced to a state of such extreme cachexia and weakness that death seemed inevitable. Under the circumstances it was determined to wash out the stomach, a treatment which had already given good results in similar cases.

The *modus operandi* was the one already recommended by Faucher, viz., the use of a simple india-rubber tube, fitted with a glass funnel.

The superiority of this method is beyond question, as, on the one hand, it obviates all danger of making a false passage, and, on the other, admits of the removal of the whole contents of the stomach; as owing to the flexibility of the tube, it can be forced boldly down into the organ. The only drawback to the method is the extreme difficulty which, in some patients, attends the introduction of the tube. Treatment commenced on January 13th. At each sitting eight litres of plain water and two of Vichy were introduced. The improvement was so great that, two days afterward, the carrying on of the treatment was resigned to the patient himself. Milk diet was ordered, together with two hundred grammes of meat essence prepared by boiling in an air-tight vessel. Six weeks after the commencement of the above treatment a most remarkable improvement had taken place. All pain had disappeared; the vomiting had completely ceased, while strength had, in a great measure, returned.

The stomach had so far recovered as to be able to digest substances generally considered indigestible; such as vegetables, salads, etc.

The patient, while under treatment, increased from 125 pounds to 142 pounds, or at the rate of 125 grammes daily for the first fortnight, and one hundred grammes daily for the four following weeks.—*Progres Medical.*

ON AURAL SYMPTOMS IN BRIGHT'S DISEASE.

As is well known, both in the parenchymatous and interstitial forms of Bright's disease, aural symptoms may develop themselves. These consist generally, either of tinnitus or hissing sounds, or partial deafness; all of which may be either isolated, simultaneous or intermittent. These symptoms may sometimes be traced to a catarrhal condition or an extravasation of blood in the middle ear—sometimes no actual lesion is apparent. In this latter case, the symptoms may perhaps be dependent on œdema of the auditory nerve.

In twenty-seven patients suffering from Bright's disease, M. Domegue found fourteen in whom there were well-marked aural symptoms. In thirty-seven, M. Dieulafoy found fifteen; in eight, Dr. Alibert found five. The predominant symptom in all was tinnitus.— *Revue Medical.*

SOCIETE DE BIOLOGIE, March 27, 1881.—M. Robin exhibited a specimen of blue urine. The liquid was a greenish-blue, and on standing, deposited numerous crystals of the same color. It had been voided by a patient suffering from interstitial nephritis.

Hence, typhoid fever is not the only disease which may give rise to this singular pathological phenomenon, which is, according to the author, directly due to the intra-vesical decomposition of indican.—*Progres Medical.*

Cerebro-Spinal Fever is becoming epidemic in New York City.

DR. DESPRES has recently reported two cases of the communication of syphilis by the use of the razor.

BOOK NOTICES AND REVIEWS.

···→≫◐≪←···

Rocky Mountain Health Resorts.
An Analytical Study of High Altitudes in Relation to the Arrest of Chronic Pulmonary Diseases.

By CHARLES DENISON, A. M., M. D., Denver. Second edition. pp. 192. Price $1.00. Boston: Houghton, Osgood & Co., 1880.

The call for a second edition of Dr. Denison's laborious work on "Health Resorts" reminds us of the reception with which the first publication was favored by the only medical journal in the author's state. There are few scientific works in which the position taken by the author may not be open to dispute. This is especially true of works that are speculative; in such there must be questionable theories, therefore, dogmatism, either in assertion or counter-assertion, is inadmissable. Dr. Denison's volume presents an array of important facts, whose value has been recognized in a manner most flattering to the author by leading men in the profession all over the world. Had these been aware of the fact that Dr. Reed, of Colorado Springs, had condemned the book as a "common and mischievous advertisement," it may be presumed that they would not have fallen into the error of commending a thing so atrocious and "mischievous." Dr. Reed finds the the work a "flagrant instance of a common abuse." Its "*raison d'etre* evident," and yet though common, it is in another place characterized as the first *effort* at a scientific consideration of the subject treated. He writes "*effort*" advisedly, because there can only be an *effort* at *conclusion*. Ergo, the work is only an effort at *scientific consideration* in other words, the treatment of the subject cannot be scientific because the conclusions cannot be final. This amiable and logical statement prepares us for the charming consistency which pervades the review, and which can only be accounted for by supposing that the writer's natural amiability is constantly asserting itself as against the painful severity which his sense of justice requires.

He "regrets" this severity, but sees no alternative; indeed, he has our sympathy and commiseration. He might, however, without sacrifice of taste, have left his regrets to the imagination of his readers, as they are, evidently, purely a matter of imagination. He says: "If it would be pleasant to follow the author carefully throug his different tabulations of cases; they show a patient, careful industry which is highly creditable, and the result can but be useful to the profession, to whom we reccomend careful study. Although we find that this effect, wherein deduction must be but tentative," reaches the point at which the conclusions (which cannot be formed) "are such as will be endorsed by those who have had experience," we find the kind admission that the "pathological and physical speculations are singularly crude and marvellously incorrect." "The work is full of these curious phenomena"—odd, that phenomena should be "curious"—and the author's pathology "perfectly inaccurate" his physics "still worse." The passages which follow, illustrating Dr. Denison's "worse physics" provokes the exclamation, "we feel sorry for these poor blood-vessels, first cupped and then squeezed twenty times a minute," and one instinctively sypathizes with the feeling, by pity, suggesting as it does, a dyspeptic reviewer in whose bowels such horrors are being enacted as serve to deprive them of all compassion.

On reaching page 110 the paroxysm is renewed, and he finds no alternative, but to send the work to the devil, as being "filled with false physics," resounding with resonant phrases calculated to catch the popular ear." He "regrets to use such language," but he ought to be forgiven, the alliteration is so pretty. He "*writes strongly* because he feels keenly the mischief," etc., and yet he believes "this to be the climate," etc. Dryness, purity of air, sunshine and scenery are, he says, *sufficient to account for the beneficial influence.*

Indeed! thus, then, the great question is disposed of with a word.

We trust that no one will be so unkind as to regard this conclusion on the part of the reviewer as a "hasty generalization," a "crude assumption," especially as he adds, "if there are other causes"— and so on. Why, after finding a *sufficient* explanation, he should suppose it necessary to look for a *more* than sufficient, does not appear, but if this reviewer's experience and knowledge should equal his logic an amiability, one can imagine how remarkable would be the work he might furnish on the subject of climate. The samples adduced as false *physics* and false *physiology*, unfortunately for Dr. Ree l's assumptions, happen to be the very instances in which Dr. Denison's theories are supported by the best writers on physiology. For example: The blood-vessels "cupped and squeezed twenty times a minute," for which the doctor expresses such lugubrious pity, (how tender must the heart be that can sympathize with a blood-vessel) according to Foster's physiology, *are* so cupped and squeezed, and it is *nature* that permits that absurdity. Again, the points that are purely theoretical and which are cited as absurd, are presented by such eminent authorities, that we are forced to the conclusion that they are rediculous and obscure in Dr. Reed's eyes, only because his eyes are not accustomed to the light which shines in certain regions of science.

There are errors in the work conspicuous enough, perhaps, and it is a pity, for the reviewer's reputation, that his discerning mind had not been able to discover something that might be damned with discretion. "'Tis an easy thing to praise or blame, the hard task, and the virtue, to do both."

It were a pity to object to an effort at sincere and honest criticism, if one were satisfied of the "*effort*," but *abuse* is not to be dignified by the name of criticism, and satire should seek for itself a better theme than a scientific work, and it should be less clumsy in form before being presented to public view, otherwise it may be like throwing the boomerang.   J. M.

Lectures on the Diseases of the Rectum and the Surgery of the Lower Bowel; Delivered at the Bellevue Hospital Medical College.

By W. H. VanBuren, M. D., L. L. D , pp. 412. Second edition, New York: D. Appleton & Co.

These lectures, twelve in number, describe the affections treated of in an easy, practical manner, and convince the reader that the author has not only mastered his subject, but

has also been able to record the results of his observation so clearly that the practitioner can see that they are obtained by actual demonstration and experience, and is thus the better prepared to accept them as established rules of practice. The sixth lecture on Fistula in Ano and the seventh on Anal Fissure are especially interesting, though there is not a chapter in the book but that is both interesting and valuable.

Syphilis and Marriage. Lectures Delivered at the St. Louis Hospital, Paris.

By Alfred Fournier, Professor a la Faculte de Medicine de Paris, etc. Translated by P. Albert Morrow, M. D. 8 vo., pp. 257. Price $2.00. New York : D. Appleton & Co., 1881.

This superbly written and eminently practical work from the pen of that noted French syphilographer, Dr. Fournier, has for its *raison d'etre* the establishment of the proposition that syphilis is but a temporary bar to marriage; while at the same time pointing out vividly the consequences, both immediate and remote, attending marriage under improper conditions; involving questions not only of pathology, but of morality and domestic happiness on the one side, and professional responsibility on the other. The author certainly possesses great discriminating powers, and throughout the entire work deals with the most complex and delicate problems in a remarkably shrewd manner. Dr. Fournier writes charmingly, and although his style is at times excessively dramatic and forcible, one would never accuse him of exaggeration. The subject is handled in an unusually systematic manner; and the array of clinical evidence is exhaustive.

The author's aim, as tersely expressed in the introduction, is as follows: (1st). A syphilitic subject wishes to marry, and comes to consult us in relation thereto. What conditions ought he to fulfill, medically, in order that we may be justified in permitting him to marry? Or, conversely, in what conditions will it be our duty to defer or absolutely interdict the marriage? 2d. The marriage is consumated, and syphilis introduced into the conjugal bed. What medical indications are then to be fulfilled in order to lessen or avert the danger of such a situation?

In other words, what is, what should be, in this case, the role of the physician either *before* or *after* marriage? Such is the twofold question the author has undertaken to discuss.

Medical Chemistry, Including the Outlines of Organic and Physiological Chemistry.

By C. Gilbert Wheeler, Professor of Chemistry in the University of Chicago, and formerly Professor of Organic Chemistry in the Chicago Medical College. At present, Professor in the University of Denver. Second and revised edition. Philadelphia: Lindsay & Blakeston.

Prof. Wheeler has certainly supplied a very great deficiency in presenting to the medical student a chemistry specially and admirably adapted to his peculiar requirements. Although the work before us is based, in part, upon Riche's Manuel de Chimie, it is decidedly original in its arrangement and general style. It pre-supposes a knowledge of the general principles of modern chemistry, as also an acquaintance with inorganic chemistry.

The metric system of weights and measures has been adopted throughout the entire work. If its merits can only be brought to the notice of professors of chemistry in medical colleges, it will undoubtedly meet with the commendation it so richly deserves.

MEDICAL NEWS.

Resume of Rules for the Use of Pessaries.—At the meeting of the American Medical Association, May 3, 1881, Dr. Paul F. Munde, of New York, made a brief recapitulation of the rules governing the introduction and supervision of vaginal pessaries, (including vagino-abdominal).

1. Always be sure of the diagnosis, of the nature, and degree of the displacement before resorting to a pessary.

2. Always replace the uterus before applying a pessary. This applies particularly to retro-displacements. It is well to replace the uterus repeatedly, every day or twice daily, for several days before introducing a pessary. The replaced organ may be supported by cotton tampon in the interval, if it is desired to distend and toughen the vagi-

nal pouch: or the object of relaxing the abnormally stretched uterine ligaments may have been obtained by the mere repeated replacement. In flexions, chiefly ante-flexions, the frequent straightening of the uterus, or conversion into the opposite flexion by the sound, will often prove beneficial before introducing a pessary.

3. Never insert a pessary when there is evidence, by the touch, of acute or recent inflammation of the uterus or when pressure by the finger on the parametrium (where the pessary is to rest) gives decided pain.

4. When the uterus is not replaceable, that is, when adhesions bind the fundus down, use great caution and discrimination in deciding whether an attempt should be made and is justified by the symptoms, to elevate the fundus by manual or instrumental means, or whether the elevation should first be tried by the gradual elevation of a pessary (this applies only to retro and latero-versions). If neither is to be recommended, do not introduce a pessary until local alterative and absorbent measures have affected a resolution of the adhesions.

5. Always choose an indestructable instrument, if possible. This does not apply to prolapsus uteri.

6. Always measure and estimate the vagina carefully before choosing a pessary, and be careful to adjust the pessary in every particular (size, curve, width), to that particular case. No two vaginae are exactly alike.

7. If the vaginal pouch is not sufficiently deep to accommodate a pessary, (anterior pouch for ante-displacements posterior pouch for posterior displacements), defer the attempt to fit it until the pouch has been deepened by daily tamponning with cotton, or by the upward pressure of a cutter or Thomas' vagino-abdominal supporter. Or the pouch may be gradually depened by using first a small (slightly curved in retro-displacement) instrument, and gradually increaseing its size (or curve) until the desired size and shape for permanency is reached.

8. Never leave a pessary in the vagina which puts the vaginal walls to the stretch, and which does not permit the passage of a finger between it and the wall of the vagina. This does not apply to prolapsus uteri.

9. A vaginal pessary which projects from the vulva, is displaced.

10. A pessary which gives pain must at once be replaced by one that is painless.

11. A well-fitting, properly-chosen pessary should not only give no pain, but should be a direct source of comfort to the patient.

12. Always examine a patient on her feet after introducing a pessary, or when it is desired, at her return, to ascertain its efficiency in sustaining the uterus during walking and exertion.

13. Always tell the patient that she has a pessary in her vagina, or she may not return in spite of your directions, and the pessary may remain for years, to her ultimate great discomfort and danger.

14. Always tell the patient to return within a week after the first introduction, in order that the position and working of the pessary may be looked after, and that, if it does not suit, it can be removed and a better, one inserted. Tell her that several trials and various instruments may be required before one is found which she can wear permanently. Also let her return for inspection once every four to eight weeks, as the case may require. Tell her that if she fails to do so the pessary may cause ulceration, for which treatment will be needed.

15. Tell the patient that she will need to wear the pessary for months, perhaps years, before a recovery can be expected.

16. Never introduce a pessary which the patient cannot remove herself.

17. Tell the patient to remove the pessary herself if it gives pain, and show her how to do it. When she has removed it let her present herself at once for examination.

18. Tell the patient to use daily vaginal injections for cleansing purposes. If she notice profuse discharge, add astringents; if the discharge is sanguinous or purulent, let her come at once, as the pessary has probably caused abrasion.

19. Tell her on removing a pessary to test the result; that the permanence of the benefit obtained therefrom cannot be determined for several days or weeks.

20. Always direct your patients to relieve all superincumbent pressure on the pessary by a proper support of their skirts; and if the displacement be anterior, aid the internal supporter by an abdominal (supra-pubic) pad.

All pessaries may be introduced in the knee-chest position when it is desirable or possible to replace the uterus only in that position.

A Sims' speculum elevates the perineum, the air enters and expands the vagina, and the pessary (chiefly in retroversion and prolapsus) is introduced by touch and sight, and the patient laid over on her left side. For aggravated retroversion and for prolapsus of ovaries or uterus, this position offers many advantages over the left semiprone decubitus. Care must be taken to remember that the position of the patient is reversed, and that the pessary must be introduced accordingly. —*N. Y. Medical Gazette.*

EUROPEAN INSANE ASYLUMS.—In a paper read by Geo. M. Beard, A. M. M. D., before a meeting of the National Association for the protection of the insane, at New York, relating to a visit to several European institutions for the treatment of the insane, he states that the conclusions arrived at are as follows:

1. In the methods of supervision, and in the general care of the insane in public and private asylums, Great Britain has been, easily, first of all nations, Germany next, and following Germany in order of merit, France.

2. Both in Great Britain and on the Continent, the insane are under the supervision of the government. In England, the institutions, both public and private, are visited by the commissioners, and not only the institutions, but each insane person is made the subject of careful inquiry.

3. In the best asylums of Europe, mechanical restraint is reduced to a very small percentage, and instead of restraint, labor is employed as a therapeutic agent.

He found no patient in England or Scotland who was under restraint, and rarely found a padded room occupied.

The washing, cleaning, cooking form work, and various trades were often all carried on by patients of the asylums.

4. In the best asylums of Europe the insane are treated much like children.

They are free to come and go, although constantly watched, to prevent injury to themselves and others.

5. The best asylums of Europe are not enormous or imposing buildings, but a series or collection of small or moderate-sized, unimposing cottages or houses.

The "East House" and "Craig House," belonging to the Royal Edinburgh Asylum, are nearly a mile apart.

The writer concludes by declaring that the method adopted in Europe will become the recognized methods in this country, although in the actual treatment of the insane in the United States, outside of asylums, there has been as much progress as abroad.—*Jour-Psychological Med.*

INFANTILE HERNIA.—Mr. S. Osborne, F. R. C. S. E., has repeatedly met with cases of infantile hernia connected with phimosis, and which he is certain were caused by the phimosis.

The vaginal process of the peritoneum through which the testicles descend, will easily give way under the strain used to overcome the obstruction to the flow of the urine.

He even thinks that the hernia usually occurs on that side on which the testicle was the last to descend; which would account for its greater frequency on the right side.

If a single truss is used and the phimosis not relieved the rupture will often appear on the other side, thus forming a double ruptue.—*London Lancet.*

TRACHEOTOMY WITHOUT TUBES.—Mougeot (de Troyes) in a paper read before Paris Academy of Medicine (*La France Medicale,* April 7, 1881), advances the opinion that the majority of tracheotomised children do not reach their majority, as the Paris military board of examiners have not found one instance of tracheal cicatrix in the recruits examined. This opinion is expressed, not for the purpose of discouraging the operation, but to lead to a modification of it. The operation without tubes, as first performed and and strongly advocated by Dr. Martin, of Boston, is, according to Dr. Mougeot, the one most indicated, as it prevents the production of pulmonary emphysema and laryngeal phthisis, so often consequent on tracheotomy.—*Chicago Medical Review.*

MORPHINE AND CHLOROFORM IN ANÆSTHESIA.—Francois Franck (*Journal de Medecine de Bourdeaux,* April 17), calls attention to certain advantages derived from administering morphine as a preliminary to anæsthesia by chloroform. He has ascertained by experiment, that the preliminary administration of morphine suppresses or markedly diminishes the initial respiratory and circulatory complications of chloroformisation, and that the great perturbations often observed in the circulatory function, when a considerable quantity of chloroform is rapidly absorbed into the lungs, are very considerably deminished by the preliminary administration of morphine. He calls attention, at the same time, to the fact that a pespiratory syncope may be more easily produced by morphine and chloroform than by chloroform alone, and the respiration requires, therefore, in case this method of anaesthesia be used, constant watching. The patient, as a rule, however, could readily be relieved from this syncope in many cases by artificial respiration.—*Chicago Medical Review.*

COLLODION IN SPRAINS.—Dr. A. N. Blodgett (*Boston Medical and Surgical Journal,* April 2, 1881,) claims very good results from the use of collodion in the treatment of strains and sprains. He employs the "contractile" collodion, several coats being successively employed. The advantages of its application are, that it seldom causes any irritation, does not interfere with the circulation of the skin, acts in the same way as an elastic bandage and cooling application. The collodion is first applied over the most swollen parts, and continued outward, even to the uninjured portion. The compression, if applied over the foot, does not produce œdema of the toes, the pressure being equable at all points. The effect was not due to the ether evaporation. The application of collodion permits cooling lotions to be applied, which thereby produce the effects of dry cold. The chief advantages, therefore, are prolonged elastic compression in parts difficult to bandage properly; water-proof protection to the skin from external irritants or applications; hermetical sealing up of wounds in the region of the strain or sprain; constant access to the affected part, without the necessity of the removal of dressings; an uninterrupted view of every part of the injured limb; reduction of heat in the tissues; great acceleration of the process of healing, with great restoration of function; a great degree of immunity from relapse; an absolute simplicity of application. Collodion might, if these claims of Dr. Blodgett be correct, be used with advantage, not only in sprains, but also in incised wounds.—*Chicago Med. Review.*

QUININE AMAUROSIS.—In the March number of Knapp's Archives, Dr. E. Gruening publishes a very well observed case of this affection following the administration of eighty grains of quinine (within thirty hours) in puerperal fever. No adequate cause of the sudden blindness could be found but the drug. The patient recovered the sight, but after the lapse of six months the peripheral vision and the color-perception were not yet normal. The author has compared the histories of eleven other cases, quoted by different authors, and makes the following valuable deductions: "On reviewing the unequivocal cases of quinine poisoning with amaurosis, we find a remarkable congruence in their essential features. The patient after the ingestion of a single dose, or of repeated doses of quinine in varying quantities, suddenly becomes totally blind and deaf. While the deafness disappears in twenty-four hours, the blindness remains permanent as regards peripheric vision; central vision gradually returning to the normal after some days, weeks or months. The ophthalmoscope reveals an ischæmia of the retinal arteries and veins without any inflammatory changes. In view of the constancy of these symptoms and the uniformity of the ophthalmoscopic picture, we are entitled to demand for this distinct type of amaurosis a recognized position in the pathology of the optic nerve and the retina." Another incident of the same accident was reported by Dr. Michel to the St. Louis Medico-Chirurgical Society (St. Louis Courier of Med., November, 1880). No recovery occurred within seven months. The ophthalmoscope showed an almost bloodless condition of the arteries and thread-like veins.—*Chicago Med. Review.*

TREATMENT OF DIARRHŒA AND CHRONIC DYSENTERY by means of a milk diet, and a mixed graduated diet, has been found very successful by Maural (Bulletin General de Therapeutique Medicale et Chirurgicale, March 15, 1881). He concludes respecting it: First, that the same treatment may be used in both these diseases, whatever their origin. Second, that the treatment may be begun by repeated purgation, with advantage. Third, that the treatment by purgation should be succeeded by a diet of milk only. Fourth, that the quantity of milk may vary from one and a half to three litres a day. Fifth, that only when several days have elapsed since the passage of a soft stool is the mixed diet to be begun. Sixth, this last diet should be graduated into six periods, passing gradually from first, eggs; second, roasted meats; third, bread and wine; fourth, dry and green beans; fifth, ragouts; sixth, boiled beef and cabbage soup; extreme care being taken to pass from each only when several days have expired since the excretion of a soft stool. Seventh, that under the influence of this treatment almost all diseases are slowly but surely conducted to complete recovery. Eighth, those not cured are much improved. Ninth, under the influence of this treatment the digestive functions not only improve but the patient's general health also. Tenth, this improvement is shown by the improvement in the patient's appearance, the return of the liver to its normal size and the gradual increase in weight.—*Chicago Med. Review.*

TYPHLITIS.—Dr. V. P. Gibney (American Journal Medical Sciences, January, 1881,) believes that some cases diagnosticated as hip disease and treated as such with gratifying results, have really been cases of subfascial iliac abscess, or, perchance, primary perityphlitis which, as a rule, terminates in resolution. He comes to the following conclusion respecting primary perityphlitis: First, that it presents many of the signs, especially if subacute, of hip disease in the first and second stages. Second, resolution may be secured and promoted by the judicious use of vesication. There have been numerous recoveries under this treatment.—*Chicago Med. Review.*

IODINE AS A SPECIFIC IN CROUPOUS PNEUMONIA.—The treatment of croupous pneumonia has since some time lost its former active character; since we could but recognize that the course of the disease could not be influenced by any of the present plans of medication. A startling statement now appears by Dr. F. Schwartz, in the *Dutsche Medicinische Wochenschrift*, (No. 2, 1881), who claims that the disease may be aborted by iodine or iodide of potassium. He quotes statistics in the first place from the most reliable German sources, according to which the crisis occurred in 0.6 per cent. during the second day, in 4.7 per cent. during the third, in 7.4 per cent. during the fourth, in 15.9 per cent. during the fifth, in 13.6 per cent. during the sixth, in 22.7 per cent. during the seventh, in 13 per cent. during the seventh, in 13 per cent. during the eighth, and in 11.8 per cent. during the ninth day. These statistics refer to 933 cases of croupous pneumonia treated expectantly. In opposition to these figures Schwartz gives his results as the proof of the efficacy of his treatment. He has had altogether ninety-eight cases, in ten of which the abortive treatment succeeded. This treatment is successful only when begun as the disease starts. If instituted later it seems to be powerless. The inference to which the author leads us is that he saw none but the ten cases quoted at a sufficiently early period, but he does not state this point in so many words. By reproducing the temperature curves he proves that the aborted cases commenced in the usual characteristic manner. In all of these the crisis was completed during the second day. The defervescence took from six to eighteen hours, on an average about twelve hours. The quantities given were very small, one-sixth of a drop of tincture of iodine, or about one grain of the iodide being taken every hour. After the fever had ceased the local symptoms began to recede rapidly.—*Chicago Medical Review.*

EARLY MENSTRUATION.—Dr. Cortejarena in *Le Rev. Medical,* a case of menstruation at the age of seven months. The child menstruated regularly, and when twenty-eight months old, was so developed as to resemble a little woman.—*Chicago Medical Review.*

GASTROTOMY IN STRICTURE OF THE ŒSOPHAGUS.—Dr. T. F. Prewitt, Professor of Clinical Surgery in the Missouri Medical College, records (*St. Louis Courier of Medicine,* March, 1881) a case of stricture of œsophagus, (either cancerous or syphilitic) in which he performed gastrotomy when *in extremis,* with an unfavorable result. Dr. Prewitt tabulates fifty-nine cases of the operation, and from their study deduces the following propositions:—

1st. All attempts at dilatation fail in a large proportion of cases of stricture of the œsophagus.

2d. In very few of the cases does dilatation prove beneficial, and in a still smaller number curative.

3d. In malignant and ulcerative conditions catheterization is fraught with danger, and is *absolutely contra-indicated.*

4th. In cicatricial stricture it is permissi-

ble to attempt dilatation with soft, flexible instruments, incapable of perforating the œsophageal walls.

5th. In cases of cicatricial stricture which have failed to yield to reasonable efforts at dilatation, and in which the emaciation is progressive and starvation threatens, and in all cases of malignant and ulcerative stricture, as soon as solids cannot be swallowed, gastrotomy should be performed.—*News and Abstract.*

NON-MORTAL FRACTURES OF THE BASE OF THE SKULL.—Dr. John A. Lidell (*American Journal of the Medical Sciences,* April, 1881), calls attention to the fact, too often lost sight of, that many cases of recovery from fractures of the skull-base subsequently suffer from many serious neurotic symptoms, often ending in the more serious cerebral neuroses. Dr. Lidell rightly claims that this result should not be lost sight of in prognosis, and that it entails judicious treatment of this class of serious fractures.—*Chicago Medical Review.*

SULPHATE OF COPPER IN CROUP.—M. Thibon, in the *Lancette Belge,* becomes quite enthusiastic over the therapeutic effects of this medicament. He considers it to be not only a very efficient emetic, but also a powerful parasiticide, and therefore of especial value in croup, first causing the expulsion of the false membranes and then preventing their reproduction, through its destructive action on the vegetable organisms.

Practically, he asserts this medicament has given him favorable results in very desperate cases; it should be given in sufficient doses; children of two years of age have taken over one gram (gr. xv.) in twenty-four hours without any toxic effect. M. Thibon employs the following portions:

R. Cupri. Sulph., . viij grains.
Aquæ, . . . ij. ounces. M.

A teaspoonful should be given every ten minutes until vomiting is induced; afterwards the same dose may be repeated every hour, and later every second hour.—*Philadelphia Medical and Surgical Reporter.*

SLEEPLESSNESS.—The Courier Medical says that for sleeplessness, when caused by grief, morphine, narceine, and codeine are best suited; when form nervousness and arterial excitement, bromide of potassium is called for, but is contra-indicated if there is anaemia. In purely nervous sleeplessness, chloform in small quantities answers best. Hydrate of chloral suits in all cases, except in dyspeptic and heart disease, and those who suffer great debility. The sleeplessness of aged persons and of debilitated constitutions should be treated with tonic medications—as wine, bitters, etc.—*Dr. Cullen, in Virginia Med. Monthly.*

TREATMENT OF PELVIC ABSCESS BY ABDOMINAL SECTION AND DRAINAGE.—Dr. Gardner, in *Canada Medical and Surgical Journal*, reports that at a meeting some months ago of the Royal Medico-Chirurgical Society, of London, Mr. Lawson Tait advocated performing abdominal section for the purpose of opening and draining the various conditions of suppuration classed as pelvic abscess. In support, he pointed out that many of these abscesses open into the rectum, bladder, vagina, or amongst the muscles of the abdominal cavity. When such natural openings were established, the patients often died; when recovery took place it was very tedious. Dr. Tait narrated six cases of abscess, all of which seemed to originate in extra-peritoneal haematocele. In none of them could vaginal tapping have been effectual in emptying the abscess and removing the debris of the clots. They were opened through the abdominal cavity, the opening in the abscess cavity being carefully stitched to the opening in the abdominal cavity, except in one case, where adhesions had already taken place. Wide glass drainage tubes were first inserted, and then smaller ones of glass, or rubber with wire. In all six cases the abscess closed, and the patients were restored to health in thirty days. From these cases, Mr. Tait concludes that such an operation is neither difficult nor dangerous, and that, by it, recovery is rendered rapid and certain. Mr. Knowsley Thornton said the operation was a great advance in surgery, as other cases of pelvic suppuration had long been unsatisfactory in their results. He had treated two cases on this plan, and was pleased with the results.— *Virginia Medical Monthly.*

DANGERS OF TENTS.—At a meeting of the New York Obstetrical Society, Dr. T. A. Emmett said (*New York Medical Journal*) that in his experience dangerous consequences were especially liable to follow the use of tents in nervous and hysterical subjects. He referred to a case that he had reported last winter, in which trouble did not occur until the seventh day. The patient should never be allowed to get out of bed until the next day after the removal of the tent. In spite of all precautions, he always felt, when about to use a tent, that he was endangering his patient's life.—*Canadial Jour. Med.Science.*

TREATMENT OF THE VOMITING OF PHTHISIS.—HANOT.—In phthisis, in the gastralgic forms of vomiting, the application to the pit of the stomach of a fly blister, or the hypodermic injection of morphia in the same region, often produces very favorable results. Prof. Peter administers before each meal a drop of laudanum in a small spoonful of water, in order to diminish the susceptibility of stomachal mucous membrane, without determining general effects. Dr. N. Gueneau de Mussy also recommends a short while before meals the use of a pill containing one centigramme of ext. belladonna. Dr. Pidoux combats the vomiting of the tubercular by means of nux vomica, which has the advantage of stimulating the stomachal tonicity in place of stupifying it, and of remedying the anorexia so common in the course of pulmonary phthisis.—*Canadian Jour. Med. Science.*

THERAPEUTICS OF HEADACHE.—Massini (*Deutsche Med. Wochen.*, 1881, p. 101; from *Correspondenz-Blatt f. Schweitz. Arzt.*) recommends bromide of potassium, particularly in uraemic headache. Ergotin is useful in paralytic conditions, nitrite of amyl in spasm. Then follow quinine, caffein; but all these medicines fail sooner or later, and then recourse is had to narcotics. There is always fear, however, of morphiomania. Recently, croton chloral, in doses of five to eight grains every three hours until thirty grains have been taken, has been recommended in uraemic headache. Monobromide of camphor in gelatin capsules, in doses of three-fourths of a grain to six grains (in gastralgic forms), is also of use. Aconitia (the English preparation) in doses of 1-60 to 1-30 grain is highly recommended by Massini, who also suggests that the effect of thess remedies may be increased by the external employment of ointments of aconitia and opium. Tincture of gelseminum in thirty to sixty drop doses in neuralgia of fifth pair is, in Massini's opinion, an excellent remedy. —*Philadelphia Medical Times.*

EXCISION OF STOMACH.—On the 29th of

January, Billroth, of Vienna, excised (*London Lancet*) six inches of greater curvature of stomach, including pylorus, for infiltrating carcinoma. Incessant and uncontrollable vomiting determined Billroth to operate. The operation lasted one hour and a half. There were extensive adhesions to omentum and colon. Fifty silk sutures were used to unite the duodenum to the remaining portion of the stomach. In a week sutures were removed from external wound which had united without reaction. The patient was able to take tea, coffee and light nourishment. In 1879, Pean performed the same operation; catgut sutures were employed, and the patient died on the fourth day. --*Canadian Jour. Med. Science.*

STRYCHNIA AN ANTIDOTE TO ALCOHOL. —According to a recent statement in the *Bulletin de Therapeutique*, by Prof. Luton, he has proven, by experiments, that strychnia is the best physiological remedy for chronic alcoholism. He has used hypodermic injections of strichnia sulphate in delirium tremens with excellent results, relieving the tetanic rigidity and relieving the delirium. — *Virginia Med. Monthly.*

REMEDY FOR ASTHMATIC PAROXYSMS.— According to the *Medical Gazette*, Prof. Roberts Bartholow, Philadelphia, says the following prescription gives "relief in a few minutes, and sometimes the relief is permanent:"

R.   Tinct. lobeliæ . . . drms ij.
      Ammon. iodid . . ozs iij.
      Ammon. bromide . ozs iv.
      Syrup tolu . . . . drms iv.

M. S.   Teaspoonful every one to four hours.— *Virginia Med. Monthly.*

IODOFORM FOR NASAL CATARRH.—Dr. H. A. Eberle, of Webster City, Iowa, contributes the following prescription to a late number of the *Michigan Med. News:*

R.   Iodoform (finely powd'd), grs. lx
      Ext. geranium (solid),   grs. x
      Carbolic acid, . . . gtt. xv
      Vaseline (or cosmoline), q.s.drms ij

M.   Make bougies of absorbent cotton, saturated with this ointment, and introduce up the nose at bed time. Leave in all night and blow out in the morning. Repeat for ten days, when the most obstinate case of catarrh will yield. Scarcely any other treatment is necessary except the occasional

use of the posterior nasal douche, with some cleansing fluid, such as a weak tepid solution of chloride of sodium (common table salt), before introducing the "iodoforminized tent."— *Virginia Med. Monthly.*

QUINIA AS AN ANTIPYRETIC.—Dr. Roberts Bartholow, of Philadelphia, in *New York Medical Journal*, thinks that quinia unquestionably holds the first position as an antipyretic. After an exhaustive examination of quinia, salicylic acid, resorcin, chloral, digitalis, aconite, veratrum viride, cold baths, and all methods of hydrotherapy, Liedermeister holds that quinia is entitled to the first place as an antipyretic, and that if he was restricted to one agent he would choose quinia. Although this is the testimony of but one clinician, a representative of the German school, his opinion is but an echo of the general sentiment among the more enlightened thinkers. The utility of quinia consists in its remarkable power to reduce temperature, conjoined with a minimum of evil effect. It reduces temperature by its influence over the vital activity of protoplasm and over the so-called ozonizing action of the blood. The diminution in the ozonizing processes is shown in the great reduction of urea formation. The quantity of quinia necessary to effect any considerable reduction of temperature has been pretty closely ascertained—not less than twenty grains can have any antipyretic effect. It is true in malarial diseases much smaller doses may diminish fever, but here another element enters the problem. Our German *confreres* give twenty, thirty, forty, even sixty grains for the antipyretic effect, and repeat it as may be necessary, to keep the temperature down at the proper level, and withhold when the result is attained, until required again. The popular, and, to some small extent, the professional opinion, that large doses of quinia effect the ears unfavorably, has no support in my experience. I have used large doses with excellent results in inflammation of the middle ear. That it has any other injurious effect on the human constitution, in proper medicinal doses, seems to me not at all probable. That quinia exercises the same curative influence over fevers—typhoid for example—that it does over malarial diseases, cannot be entertained for one moment. The effect it has on the course of fever is due to its antipyretic property; on malarial diseases, the action is specific and particular. It is effective, then, in the treatment of fever according to the degree in which it reduces the temperature, and the value of this is determined by the importance of the febrile element in the morbid states.— *Virginia Med. Monthly.*

A CHARACTERISTIC SYMPTOM OF HEREDITARY SYPHILIS.—Prof. Parrot (*Le Progres Med.*, 1881, p. 125), in a recently delivered lecture on infantile syphilis, speaks of a particular condition of the lingual mucous membrane, first observed by Gubler and Bergeron, and described by Bridoux in his thesis (1872), and which the lecturer considers to be connected with and chacteristic of hereditary syphilis.

The tongue displays desquamation, beginning at the point and borders, and passing over the surface, the process reaching the central raphe (*V. lingual*) by the time reparation is beginning at the border. The circinate form of the desquamation, and its clinical character of proceeding by rapid and successive assaults, recommencing, for example, at the point of the tongue in the newly-formed epithelium, before it has fairly begun to disappear at the base, serve to make this peculiar affection of the tongue in hereditary syphilis an entity not likely to be mistaken for any other affection of this organ.

Histological examination shows, in perpendicular section of the lingual mucous membrane, that there is a tumefaction of the cellular elements of the papillary layer, with superficial shedding of epithelium and proliferation of embryonal elements, as in the cutaneous syphilides. Another proof of the syphilitic character of this affection is found in the fact that its maximum of frequency is at the same age as that of the other syphilitic manifestations. Prof. Parrot does not regard this lesion as contagious, nor does he consider it amenable to treatment.— *Cincinnati Lancet and Clinic.*

SALIVA AND GUTTER MUD AS VEHICLES OF SEPTIC DISEASE.—Dr. Geo. M. Sternberg, Surgeon U. S. A., details in the National Board of Health Bulletin, April 30, 1881, a remarkable series of original observation and experiments with the following conclusions:

In the light of what we already know, it seems very proper that puerperal fever, hospital gangrene, and the various forms of septicæmia known to physicians and surgeons result from the development of pathogenic varities of harmless and widely-distributed species of micrococci, as the result of especially favorable surroundings, such as are found in the lochial discharges of a puerperal woman or in the secretions from the surface of wounds in a crowded and illy ventilated hospital ward.

Just as differences in resisting power to experimental septicæmia are exhibited by different species of animals, so doubtless individual differences exist in man, especially as the result of lowered vitality; and this want of resisting power, from whatever cause resulting, must be counted as one of the conditions favorable to the development and propagation of a pathogenic bacterium. Thus we find that in experimental septicæmia the micrococcus does not invade the blood until the vital powers are at a low ebb, and death is near at hand.*

In the dog the vital resistance is competent to withstand the assaults of a micrococcus—injected subcutaneously—having the potency of those found in my saliva, and the result of such an injection is simply a circumscribed abscess. But the increased power (which is perhaps simply a more vigorous and rapid development) gained by cultivation in the body of the rabbit, enables these organisms to overcome the resistance of the dog, and a diffuse cellulitis results of fatal character.

The fact, observed by myself, that during the summer months the mud in the gutters of New Orleans possesses an extraordinary degree of virulence,† shows that pathogenic varieties of bacteria are not alone bred in the bodies of living animals. The more I study this subject the more probable it seems to me that in this direction lies the explanation of many problems which have puzzled epidemiologists, and that the sanitarians are right in fighting against filth as a prime factor in the production of epidemics—a factor of which the *role* is easily understood, if this view is correct.

*By virtue of some property or mechanism at present unknown, blood, which external to the body is a favorable medium for the development of many species of bacteria, resists their entrance or gets rid of them when they effect an entrance, *e. g.*, by injection, so long as it is circulating in the vessels of a healthy individual.

†There is no reason to suppose that this is peculiar to New Orleans, but I have not yet had the opportunity to extend my experiments to other places.

The presence of septic organisms, possessing different degrees of virulence depending upon the abundance and kind of pabulum furnished them, and upon meteorological conditions more or less favorable, constitutes, in my opinion, the *epidemic constitution of the atmosphere*, which wise men were wont to speak of not many years ago as a cloak for ignorance. It must be remembered that the gutter mud of to-day, with its deadly septic organisms, is the dust of to-morrow, which in respiration is deposited upon the mucous membrane of the respiratory passages of those who breathe the air loaded with it. Whether the peculiar poison of each specific disease is of the same nature or not—a question which can only be settled by extended experimental investigations in the future—it is altogether probable that this factor often gives a malignant character to epidemics of diseases which uncomplicated are of a comparatively trivial nature.—*Cincinnati Lancet and Clinic.*

ANÆSTHESIA BY CHLORAL.—M. Bouchut publishes, in the *Paris Medical*, a case of thoracentesis, in a child six years and a half old, with anæsthesia by chloral. M. Bouchut gives chloral in doses of from two to three grammes, according to the age of the patient, and in a single dose. He asserts that it is a perfect anæsthetic, without any disagreeable result; and that he has administered it in this way in more than ten thousand cases. Anæsthesia by chloral renders operations very easy in children, who move about, struggle, and incline the vertebral column towards the side which is to be operated on. The anæsthetic sleep overcomes this resistance, sometimes so difficult to conquer, especially in children on whom the same operation has been performed more than once. When the little patient awakes, at the end of three hours, he is ignorant of what has been done to him, and finds himself relieved without having experienced any unpleasant sensations—*Cincinnati Lancet and Clinic.*

ABSORBENT COTTON.—"Absorbent cotton," says Mr. Frank L. Slocum, in *Druggist's Circular*, "has in the last few years taken such an important place among the necessities and conveniences of the pharmaceutist and physician that it is very desirable for the pharmaceutist to be able to procure or himself manufacture it. The manner in which

it is prepared is kept secret, and there is no literature on the subject up to the present time, it is believed.

"At the request of Professor Maisch, some experiments have been made, the following process yielded the most satisfactory results:

"Take of the best quantity of carded cotton batting any desired quantity, and boil it with a five per cent. solution of caustic potassa or soda for one-half hour, or until the cotton is entirely saturated with the solution and the alkali has saponified all oily matter; then wash thoroughly to remove all soap and nearly all alkali, press out the excess of the water, and immerse in a five per cent. solution of chlorinated lime for fifteen or twenty minutes; again wash, first with a little water, then dip in water acidulated with hydrochloric acid; and thoroughly wash with water; press out the excess of water, and again boil for fifteen or twenty minutes in a five per cent. solution of caustic potassa or soda; now wash well, dipping in the acidulated water and washing thoroughly with pure water; afterwards press out and dry quickly."—*Philadelphia Medical Times.*

A Society of Public Hygiene has been formed at Bordeaux, France.

Pirogoff will reach the fiftieth anniversary of his professorship on June 5th.

The Toledo Medical and Surgical Journal has suspended publication.

A new French journal assumes the form of a review of military medicine.

Dr. Edmund M. Landis, a prominent physician of Chicago, died May 6, 1881.

Dr. Salvatore Caro, a prominent New York physician, died April 30, 1881, at the age of fifty years.

A Bohemian University is to be established at Prague, in addition to the German University at present there.

The East India Government have an annual profit of $21,500,000 from the Bengal opium imported into China.

Dr. John Day, a prominent physician of Geelong, Australia, died January 10, 1881, in the sixty-fifth year of his age.

Mr. George J. Seney, of Brooklyn, has given two hundred thousand dollars in cash, and seventy thousand dollars in land, for the erection of a cottage hospital.

ROCKY MOUNTAIN MEDICAL REVIEW.

Vol. 1.　　　　　　DENVER, JUNE-JULY, 1881.　　　　　　Nos. 10-11

ORIGINAL ARTICLES.

·····≥≈≤···

Our Eyes—Their Use and Abuse.

By Swan M. Burnett, M. D.,
Secretary Otological, Ophthalmological and Laryngological
Section of Am. Med. Assoc.; Professor of Otology and
Ophthalmology, Med. Dept, Georgetown University,
Washington, D. C.

The eye, when considered from any standpoint, certainly becomes an object of great
interest, and, aside from the brain, assumes first rank among the organs of the
human economy. Whether looked at as a
piece of delicately adapted mechanism, as an
organ of sense, ranking the highest among
the five, or as " the window of the soul,"
through which we watch the varied and
swiftly changing emotions of the heart, it is a
theme of which the physicist, the physiologist
and the poet never tires.

As with most of the wonders of nature, a
daily and hourly experience with its marvelous
phenomena has blinded us, to a certain
extent, to the beauties of its construction and
the exalted position it occupies as an organ
of sense. We take things we are accustomed
to so much as a matter of course that it is
only when we remove ourselves to such a
distance from them as to be able to view
them entirely objectively, that we can see
them as they really are, and correctly understand their relations to their surroundings.
In what we shall here have to say in
regard to the human eye, we will assume such
an objective position, and consider it and its
function in much the same manner as an
ichthyologist would study a fish.

The first peculiarity of the eye which will
arrest our attention when we come to view it
in this light, is that it is an optical instrument
as well as an organ of sense, and when we

come to consider it further we shall find that
it is by no means an optical apparatus of the
simplest character. It should furthermore be
borne in mind, as an important fact having a
practical bearing, that these two functions are
quite separate and distinct. The eye may be
perfectly normal as an optical instrument, but
have its function as an organ of sense totally
abolished; while as an organ of sense it
may be perfect, and its optical part so widely
altered from what it should be as to render
the eye almost useless.

It is the province of the physician to discover
in any individual case of diminished acute-
ness of vision, which of these two functions is at
fault. Fortunately for the human race, the
studies of physiological opticians within recent
years have enabled us, in the large majority
of cases, to perfectly differentiate between the
defects in vision arising from faults in the
optical part and those in the sensory part.
The former class of defects we are able, in
the large majority of cases, to correct by
means of properly adapted glasses, and the
number of eyes that have thus been restored
to useful and comfortable vision are a noble
monument to exact science.

We shall first give a passing study, then,
to the optical apparatus of the eye, and in
doing so we shall regard it just as we would
any other instrument whose office it is to act
upon rays of light and bend them in such a
manner as to form images of objects from
which that light emanates.

I would remark in passing that the higher
degree of perfection of this optical part of the
eye is quite a characteristic of the more
advanced forms of animal life. In the lower
forms of life this apparatus is imperfectly
developed, and in fact, in very low forms can
hardly be said to exist at all. These animals

merely receive the impression of light coming from certain directions or certain objects, but do not have any clearly defined images of these objects. It seems sufficient for their self-preservation that they be aware of something being there, with no definite impression of its exact nature.

As we ascend higher in the scale, however, the necessity for a closer differentiation between objects is established, and as in nature we always find an attempt on the part of all living things to adapt themselves to their surroundings and their most pressing needs, so we find the eye, as an optical apparatus, approaching more and more nearly to a perfect instrument as we ascend higher in the scale of being. How far it falls short of an attainment to perfection we will consider further along.

The object of the eye as an optical instrument is to form clearly defined images of external objects upon the expansion of the optic nerve. It is what is called a collective apparatus, and resembles the instrument known as the camera obscura. The most familiar form of this instrument is the photographers camera. Here the rays of light coming from an object are united into a well defined image, by the convex lens in the front of the square box, on the ground glass at the back. The eye is in all essential particulars like this oblong box, and necessarily so, because the design of both is to obtain distinct images of objects on the posterior portion of the instrument or eye. If we examine both instruments part by part, we shall find that both are constructed upon the principles of optical science which are best calculated to promote the end in view.

The eye, like the camera, is a hollow box, the only difference being that the eye is more in the form of a globe, while that of the camera is more nearly that of a cube, and the eye is filled with a transparent liquid instead of air. Both boxes are lined on their interior with a dead black coating to absorb any stray rays of light that fall on their sides, which being reflected might fall on the image and mar its distinctness of outline. Both have a

screen at the back in the focus of the refracting surface, on which the image is to be formed, and both have a convex lens in front, whose office it is to so bend rays of light coming from any object toward which the instrument may be directed as to form a distinct image of it upon the screen. In both, there is a diaphragm in front of the lens to cut off any unnecessary or dazzling light. In the eye, this function is performed by the iris.

There is this distinction, however, between the eye and the camera. The eye has two lenses, while the camera, as a rule, has but one. Both the cornea and the lens refract the rays of light, and the eye would be very incomplete without either one of them. In connection with this physical fact there is a physiological difference, the importance of which I wish you to bear in mind, as it has a valuable practical significance. You must have noticed when sitting for your photograph, that the operator was at great pains to get the instrument properly focussed, and he did this by turning a screw head which moved the lens backward and forward as he desired. You must also have observed that when he moved the instrument nearer to you or farther away from you it was necessary for him to alter the focus. In other words, when the distance between you and the camera was changed, the focus of the instrument had to be altered. This is called the adaptation or accommodation of the instrument to the required distance. This accommodation is effected in the case of the eye in a very different manner. It would not be possible here to move the lens backward and forward every time we change from vision near at hand to vision at a distance, when we raise our eyes from our reading, for example, to look at an object across the street, and yet, in order to have a clear and distinct image, we must alter the focus of our eyes with every change in the distance of the object looked at. We are, as a rule, not conscious of this accommodation, but it takes place all the same, and is an important factor in the production of some very unpleasant symptoms, as we shall see by and by.

Now, where the screen is immoveable, as in the case of the eye, and the position of the object is changed, there are two methods by which the lens may yet produce a clear image on the screen. The lens itself may be moved, or the curvature of its surface may be changed. Of the two refracting surfaces of the eye, the cornea and lens, it is the lens which brings about this accommodation for vision at different distances, the cornea remaining perfectly passive; and it does it, not by moving backward or forward, but by altering the curvature of its surface.

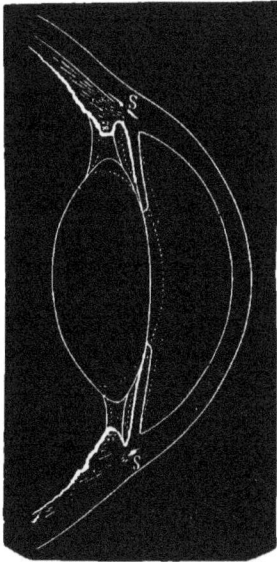

Showing the change which takes place in the crystalline lens during the act of accommodation. The dotted lines show the increase in the curvature of the anterior surface of the lens when objects close at hand are looked at.

The eye when in a state of rest (I mean a perfectly normal eye) is adapted to bring parallel rays to a focus on the retina. That is to say, all objects situated twenty feet from the eye and beyond, even up to the distance of the fixed stars, have their images formed clearly on the retina without any effort on the part of the eye. When objects nearer than twenty feet are seen clearly, there must be an effort made by the eye in order to accommodate it to this distance, the curva-

ture of the lens surface must be increased. This is done by means of a muscle—called the ciliary muscle—situated on the interior of the eye. When this muscle contracts, the curvature of the surface of the lens is increased, and the stronger the contraction of the muscle, the more the curvature is increased and the greater the amount of "accommodation" produced.

It may be well to state, however, in passing, that as an optical instrument the eye is by no means perfect. It is not, as was at one time supposed, an instrument whose perfection man could never equal. It is a very hard thing to have to give up our ideals, but I am afraid we shall have to acknowledge that while the eye is built on strictly optical principles, it has glaring defects in almost every part, such as no optician would permit in an instrument which made any pretense of being ordinarily good. Moreover, these defects are not, as some have supposed, such as may be of service to the eye itself or the individual. They are faults of construction, the result of the peculiar method in which the organ is developed, and are as much the result of the action of law as the growth of a world or the development of a solar system. There is no such thing as blind chance in nature. The Universe is under a reign of law, absolute and unvarying, and the wide departures from the established order of things which we sometimes see, do not depend upon the momentary setting aside of any law or laws, but are the result of the interference with the action of the ordinary laws, by some other law, which at that time is, from some circumstance, the more powerful. We have not the time to enter into a consideration of the causes which have led to these defects of the eye as an optical instrument. They are various, but they are all referable to the laws which control the development of tissue, and it is no wild assertion, or chimerical fancy, to say, that when these laws shall be better understood we will be able to do away with some of these imperfections—however much it may look like "by thought adding one cubit to our stature."

As is the case with all optical instruments of its class, the image formed by the refracting surfaces of the eye on the retina is inverted. The image of a man walking on the street is formed with his head downward as regards the individual on whose eye the image is formed. As we do not see men walking on their heads it was quite puzzling to the ancient philosophers to account for the facts in the case. They tried to explain away the paradox in a number of ways. Some thought that the image was turned right side up in the vitreous humor before it reached the retina, while others supposed that in some mysterious way it was "righted" by the optic nerve before it reached the brain. As is always the case with even the seemingly most mysterious phenomena, the explanation, when

shape, while the other end has a ring, and by a number of experiences it learns to always associate like images formed on its retina as globes and rings. When the rattle is held vertically with the ring downward, it finds, by the sense of touch, that the ring end is towards its feet, and the globular end upward, although the image of the ring is formed on the upper part of the retina and above that of the globe, and so on throughout the whole of its experiences in after life. It finds that the upper part of the image *always* corresponds to the lower part of the object or the part towards its feet, and as all of the outer world maintains the same relative position, there can be no difficulty or confusion in properly placing, in the minds eye, all objects in space.

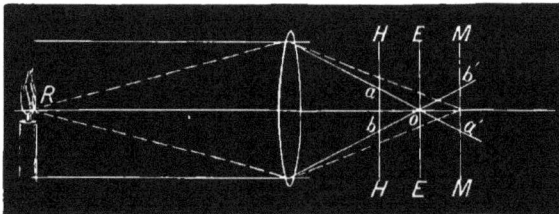

Representing the position of the retina in the three conditions, of far-sightedness (*H*), The normal vision (*E*) and short-sightedness (*M*). Parallel rays are united in the normal eye and form a clear image at *O*. Whereas they form circles of diffusion and indistinct images at *a*, *b*, and *a'*, *b'*, on *H*, and *M*.

we at last find it, is exceedingly simple. The image itself is not turned right side up either in the vitreous humor or on its way to the brain; it is formed inverted on the retina just as your image is formed upside down in the photographers camera. How then is it that we see things in their proper relations? Scientific men explain it in a very simple manner by what they call the "law of projection," which being interpreted means this: The sense of sight is an educated one, that is, every child must learn how to see for itself. The new-born child, even if it had sufficient intellectual capacity, could not form any proper conception of objects simply by their images formed on the retina. It must verify, so to speak, the sense of sight by that of touch. When a rattle, for example, is given it, it finds that the image of one end is globular in

We must call attention to two other points in the anatomy and physiology of the eye, and that is, that, unlike the eyes of some animals, it is movable, and that there are two of them, and that we look out of both at the same time and see only a single object with the two. All of these phenomena hinge on the fact that the eye has six muscles whose office is to turn it in all possible directions. The more important of these muscles for us are the four straight muscles, as they are called. The office of these muscles is to turn the eye upward and downward, outward and inward. Their origin and insertion are such as have an important bearing on the production of certain diseased conditions which we shall consider by and by.

There are many other most interesting points in connection with the physiology of

the eye that we should like to call your attention to, but we must hasten on to the second
and by far more important part, practically,
of the subject—that is those conditions in
which the eye is hindered in the proper performance of its full function. We do not
mean to here consider the *diseased* conditions
of the eye, the results of ordinary inflammation, injuries, etc., but those departures from
the normal condition which are to a certain
degree, preventable, and these form a much
larger part of the cases coming to the oculist
for treatment than is generally supposed.

There is one condition of the eye which
though not a disease, is yet attended with a very
considerable curtailment of its function. To
an audience composed almost entirely of young
men it would seem superfluous to speak of "old

It is true we may view life from a higher level,
and may have managed to extract a considerable amount of philosophy from the hand
to hand struggle with capricious Fortune; we
may have loftier ideas and more expanded
conceptions, and it may be have stamped the
impress of our own individuality upon a resisting world, and made it acknowledge our
power and importance, but when a man
crosses his nose, for the first time, with a pair
of spectacles, he must be made of very
stern stuff if he does not heave an inward
sigh to think that the roseate period of life is
behind him. In youth all things are possible,
in middle age, we find that uncertainty is the
only certain thing in life, and that Life's
"promises to pay" are often but unredeemed
pieces of paper in our hands. Age has its

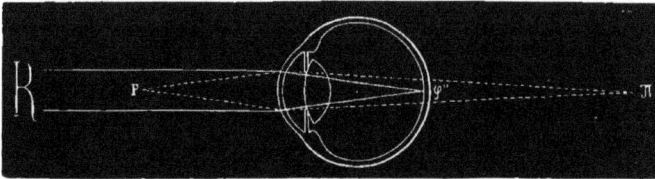

Showing the need of an accommodating power in the eye. Parallel rays coming from a
distant object R are united and form a distinct image on the retina at I''. If there were no change
in the refraction of the eye, when the object was brought nearer to the eye (P), a clearly defined
image would *not* be formed on the retina, but *behind* it at n.

eyes," but whether fortunately or unfortunately, man has not succeeded in finding the
elixir of youth, and it yet remains the order
of nature, as it has been since time began,
that man shall grow old. If your lives are
prolonged sufficiently, the time will inevitably
come to every one of you when, as it has
been facetiously remarked, you will have to
get a pair of spectacles or a pair of tongs
when you read your paper. There must be
something sad about the feelings of a man
when he puts on his first pair of spectacles.
Poets and painters have given us pictures to
convey the immeasurable joy of the boy with
his first pair of boots, but the pathos of the
first pair of glasses is yet to be embodied in
art. They bring the feeling that youth and all
the hopes, anticipations, and longings of that
buoyant period are now things of the past.

compensation in the experience and wisdom,
and wider knowledge which it brings, and no
wise and thoughtful man would, for an instant,
desire to turn Time backward in its onward
flight, but the opportunities and pleasures of
early manhood, like those of youth and childhood, can be known but once. Be wise in
your time, and make the most of them while
you may.

This condition of the eyes, which we are
considering, is dependent upon two causes.
First, the stiffening of the muscle of accommodation, and, secondly, the hardening of the
lens substance. The effect of these two
causes, is that the eye accommodates itself
less readily to objects near at hand.

You will remember I told you that when
we looked at objects close by, it was necessary
to alter the focus of the eyes, and that this

was done by the ciliary muscle acting on the lens in such a manner as to to render its surface more convex.

As the lens becomes harder and the muscle stiffer, it becomes more and more difficult to increase the convexity of the lens, and of course more difficult to see fine objects close at hand, they become bleared, and in order to be seen distinctly, they must be removed farther from the eye.

You must all have noticed how an old man instinctively removes objects from him when he wishes to see them distinctly. He holds his paper at arms length, and throws his head back to still further increase the distance. The first symptoms of which a man complains when he begins to lose this power, is difficulty in reading his paper in the even-

time to time as the power fails with age. The time when this artificial aid begins to be needed varies, but it comes generally between the 43d and 50th year, just when man is in the prime of his vigor. In woman it usually comes earlier, and is earlier in those who are accustomed to doing close work, and is felt first at evening work.

When shall I begin to wear glasses? is a question often asked me. There is a very erroneous idea prevalent among people that the use of the glasses should be put off as long as possible, and that if they are used too early the necessity for them becomes more imperative, and damage is likely to be done to the eye. Much useful and comfortable work is lost through the wide acceptance of these two errors. The time to begin the use

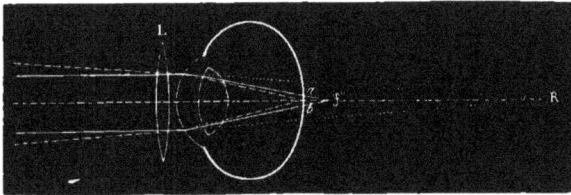

Correction of a far-sighted eye by a convex lens. Parallel rays coming from a distant object would be united and form a distinct image *behind* the retina at *R*. By the intervention of a convex lens, however, the refractory power is increased, and a clearly defined image is formed on the retina at *a*, *b*.

ing. He is not conscious that the fault is in him, and not in the paper, and it is only when some friend, to whom he has made his complaints, suggests that he may need glasses, that the truth flashes upon him that time has laid his inexorable finger on his eyes. He most likely tries the glasses of his friend, and finds that he has no difficulty now in reading as fine print as he ever did, and that too at a reasonable distance from the eyes.

It may be asked how is it that glasses relieve the eye of this short coming. It is done by adding that power to the lens which it is necessary for it to have for vision close at hand, and which it cannot give itself now, by placing a lens outside of the eye, of such strength as shall supply the deficiency. At first the deficiency is small and the lens need not be strong, but must be increased from

of glasses is when they are needed, and this can easily be told. When the time comes when you cannot any longer read your paper in the evening at eighteen inches from the eye, you may be sure that nature wants some help, and you should not be slow to give it to her. As I said, the glasses to be used at first should be weak, generally a glass No. 65, as it is called in the shops, suffices for evening reading for a year or more. This, however, will have to be laid aside for a No. 40 soon, and, as a rule, increased every three or four years. However, no general law can be laid down. It is by no means a bad plan to consult some physician who understands the eye and its functions, when you feel the need of glasses, and have your eyes thoroughly examined and your glasses properly fitted. Much inconvenience and often much money is saved by this. A few words now about the spectacles themselves. Remember that pebbles have no essential superiority over good clear glass, except that they are harder. Opticians are not always honest men any more than other men engaged in business, and they often draw upon their imagination for their facts. More depends upon the accurate grinding of a glass than upon the material of which it is made, providing it is clear of flaws and perfectly transparent. Of course you must remember that you can't get a pair of glasses that will enable you to see at a distance and close at hand too.

There is a condition called far-sightedness, in which no object can be seen, even the most distant one, without an effort. You will remember that I told you the perfectly nor-

cannot be to widely disseminated. We allude to the condition of near-sightedness or myopia as it is called.

You are well aware that the characteristic of this condition is that objects in order to be seen clearly must be brought close to the eye. All objects at a distance are bleared and indistinct. These two phenomena depend upon the fact that in order for the image to be formed clearly and distinctly on the retina, the object must be close to the eye, and the higher the degree of near-sightedness the closer must be the object. This fact again depends upon another physical fact that the eye is too long.

This faulty optical condition is corrected by placing in front of the eye a concave lens of such a strength as will cause parallel rays, after

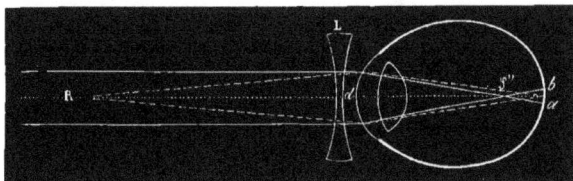

Correction of Myopia by a concave lens. Parallel rays coming from distant objects would be united and form a distinct image *in front* of the retina at **Y**''. If, however, they are compelled to pass through the concave lens (*L*) before reaching the eye, they will assume such a direction as if they came from *R*, and will be united and form a distinct image on the retina.

mal eye saw all objects beyond twenty feet distinctly without an effort. The far-sighted eye, on the contrary, can see nothing without effort. We have not time to go into a full consideration of this class of eye affections, and, besides, it is a matter which belongs more properly to the ophthalmic surgeon, and he ought to deal with it. Suffice it to say that most of the cases of painful vision come from this habitual strain to which the far-sighted eye is subjected. The condition is readily relieved by means of proper glasses. The individual is generally born far-sighted. It depends upon the fact that the eye is too short.

There is another condition, however, which deserves our attentive consideration because it belongs to the class of preventable affections, and knowledge of preventable diseases and the circumstances leading to their production

they have passed through it, to assume such a direction as if they came from the farthest point at which the myopic eye, without any glass, can see distinctly.

If the eye, for example, can see objects distinctly at only ten inches, the lens must be of such strength as will cause parallel rays to diverge as if they came from a point ten inches in front of it. This will put the myopic eye on a par with the normal eye, because, as you will remember, that in the normal eye rays coming from a distance are parallel, and are brought to a focus on the retina, and distant objects are seen without any effort on the part of the eye.

You will observe that I have in this case assumed that the refracting surfaces, the lens and the cornea, are of the same strength as in the perfectly normal eye. This is of course an assumption, but the examination of a large number of short-sighted, and far-sighted as well as normal eyes has demonstrated the fact that these refracting surfaces are in general the same in all kinds of eyes, and that where we have any departure from the normal condition it is through a change in the length of the eye. The eye can be too short, as it is in the case of far-sightedness, or it may be too long as in the case of near-sightedness.

The far-sighted eye, I have already told you, is congenital, that is, it is born so. It is due to an arrest of development. It is true the inconveniences of a far-sighted eye are not felt until late in life, but this is owing to the fact that young eyes possess a surplus of accommodative power, which serves to overcome the defect.

As I have already pointed out to you, the power decreases with age, and we have painful and indistinct vision, and other symptoms into the details of which we cannot here enter. This condition, therefore, is clearly not preventable by any means yet within our knowledge. It can only be corrected by the proper adaptation of glasses.

The condition of near-sightedness, on the contrary, is not, as a rule, congenital. We do not often find near-sightedness in children under 6 or 7 years of age. It is only when the eyes begin to be used for near work that we find myopia making its appearance. Moreover, we seldom find the condition in savage races or among the illiterate classes. I have found but few cases among the negroes, and these have been in individuals who used their eyes much in reading and writing. The ordinary plantation slave was exempted from this curtailment of function. The bright sky, the beautiful landscape, all the glories of nature were open to him without the intervention of a bit of glass.

These facts force upon us the conviction that short-sightedness is one of the prices we have to pay for our civilization and culture. But is it a necessary accompaniment of our advance in enlightenment? Is it not possible that we can attain to the highest degree of culture without having to pay the forfeit of a

curtailment of one of our most important functions? I am glad to be able to state to you that recent investigations have shown that myopia can be arrested in its onward progress, and we may accept it as a demonstrated fact that if proper measures are used, the disease (for such it really is) can be prevented.

An important fact must be borne in mind in this connection. While myopia is not a congenital disease, that is, persons are seldom born with it, the tendency of the disease is in a large majority of cases hereditary. Patients often tell us that one or both of their parents are near-sighted, but that up to a certain time their eye-sight was very good, and it is no uncommon thing for parents and children to become near-sighted at about the same age.

It is very evident that if we wish to prevent a disease, the first thing we must know is how it is produced, what are the causes and conditions leading to its development.

Happily we are in possession of very complete statistics on this subject, for which we are indebted principally to German investigators. Prof. Cohn, of Breslau, examined over 10,000 children in the schools of that city, with special reference to the development of myopia during school life, and other examinations not quite so extensive have been made by others, so that we are now in possession of sufficient data on which to base a very positive opinion.

The result of these examinations has been the discovery of the fact that short-sightedness while very rare in young children, increases with each year of school life, and the highest percentage is found in the most advanced classes, where the eyes are used most continuously. It is, then, a demonstrated fact that the use of the eyes must, at least under some circumstances, lead to the development of myopia. Now, what are these circumstances? They are two: First, a stooping posture, and second, a strong convergence of the eyes. As when we look at objects close at hand.

How is it that these two causes can produce such an alteration in the organic structure of eyes? for as you will remember, a myopic eye is larger than it should be, and something must act upon it to increase its length before it can become near-sighted.

The generally accepted explanation is this: When we look at an object close at hand, both eyes must be brought to bear on the object, and in order to do this, the eyes must be converged, or turned inward, so that the straight lines passing through each of their centres would meet in the object looked at. The office of one pair of the muscles of which I spoke to you is, to turn the eyes inward.

Such are the origin and insertion of these muscles that in contracting to turn the ball they at the same time tend to bring themselves into a straight line, and in doing so make pressure on the equator of the eye, and tend to flatten it, and when it is flattened at its equator it must of course be pushed out at its posterior part where there is little resistance. If the tissues of the eye are healthy and firm this pushing out is resisted and the ball does not alter its shape, but when from any cause the tissues at the back become soft, their power of resistance is lessened, and it can very readily take place. Now what are the circumstances which tend to lessen the resistence of the tissues at the back part of the eye? In the first place, the natural power of resistance may be very weak. This is probably the most important factor. As I before remarked, the *tendency* to short-sightedness is most often hereditary, and this tendency consists in the transmission from parent to child of a softened condition at the back part of the eye. If the eyes are properly used, and the child is kept in the best possible condition, the simple convergence of the eyes is not sufficient to bring about a myopic condition, but if the child is placed under circumstances which tend to increase this softened condition of the tissues, the depelopment of near-sightedness is almost inevitable.

The circumstances which have been found to conduce most readily to the production of this, are those causes which tend to produce a congestion of the parts about the eye, and the most prominent among these is a stooping posture, particularly where the head is bent forward on the neck. In this position, the blood, on account of compression of the veins in the neck, finds a less free exit from the parts about the eyes, and a sort of congestion is the result, which leads, in time, to a softening of the tissues.

This, in brief, is the method in which near-sightedness is generally produced. There are, however, yet other conditions, leading to these, which are of the utmost importance. The question of illumination has been found to be a most important factor. It was discovered in the examination of which I have spoken, that in those schools where the light was bad, or improperly arranged, myopia was more extensively developed than in others where these conditions were better. When the light is not good, we instinctively bring the head nearer the book, and thus bring into play the very two factors which most readily lead to the development of near-sightedness —a strong turning inward of the eyes and a bending forward of the head. The same thing occurs where we sit facing the light; the head is either bent forward to cut off the dazzling light, or the book is brought up close in front of the eyes for the same purpose. In both cases we have a stooping posture and a strong convergence of the eyes.

Now the important question comes—how can we best obviate these harmful conditions? The answer to this must be apparent to you all. We must simply do away with the two conditions of stooping posture and excessive convergence of the eyes, and in doing so must, of course, avoid all those circumstances leading to these conditions.

In the first place, the light must be managed properly. It must be strong enough to enable the finest print to be seen at least eighteen inches from the eye clearly and distinctly. It should not fall in the face, but come from behind and from the left. A direct side-light is almost as dazzling as a front-light. The idea is to have the light on the object and be reflected from this into the eyes. Any light coming into the eyes from any other direction than from the object is dazzling, and mars the distinctness of the image on the retina. The next best position of the light is above. The position of the work or book is another point that must be carefully attended to. The book must be held in such a manner that it can be read without bending the head or body forward. The faulty construction of the desks in the majority of our schools has been a most fruitful cause of the development of near-sightedness. In the large majority of cases they are much too low. The pupil in writing, ciphering and in reading, when the book is laid on the desk, is compelled to bend slightly forward, and this stooping position taken in connection with the bad light which has so often prevailed, has conduced to the ruin of many eyes. And we have here, again, as we often do, a circle of pernicious influences. The bad light and improper position produce the myopia, and this in its turn causes the individual to bring the work still closer in order to be seen distinctly. Hence it is that short-sightedness is, as a rule, progressive, that is, it has a tendency always to get worse. It is a common error that short-sightedness is an advantage to us when we come to be old. It is true we do not then need the glasses that old persons usually do, but for distinct vision at a distance glasses are always required, so that as regards the matter of glasses there is no gain. It must be remembered that the myopic eye is a diseased eye, and requires always the greatest care, because some of the most melancholy cases of blindness are distinctly traceable to that condition.

It is absolutely necessary that the myope be put into such a condition as shall enable him to do his work at a reasonable distance, say eighteen inches or so, and for this purpose he must be provided with glasses which will enable him to do this.

Near-sighted persons often object strongly to wearing glasses. They do not care to see any better at a distance than they do, and they can see well enough near at hand, but the important thing, which they overlook, is that they must *not* bring objects so near as to require a too strong convergence of the eyes.

A few general instructions now in regard to the use of the eyes. The light should come by preference from behind and to the left. The next best position is above. The best form of artificial light is the German student lamp. The next best is gas with an Argand burner; both should have a milk glass shade, and the work should lie where the light can fall directly on it and come from thence into the eye. Desks on which writing or drawing or other work is done, should be high and sloping, so as to obviate the necessity of bending the body or head forward. The work should be at least eighteen inches from the eyes, and when people are so near-sighted that they cannot see their work at this distance, they should wear glasses which will enable them to do so.

By following out these rules, much valuable work will be saved that would otherwise have been lost to the world. The near-sightedness, when once it has been established, can never be *cured*, it can only be corrected by glasses. How important, then, a knowledge of the laws which govern its production, and it is with pride that I announce that the great minds in our profession are now turning their attention so largely to the study of the causes and development of disease. In this study the scientific mind has its highest function, and humanity the broadest field for its Christlike exercise.

LEMON JUICE IN DIPHTHERIA—Dr. J. R. Page, of Baltimore, in the New York *Medical Record*, May 7, 1881, invites the attention of the profession to the topical use of fresh lemon juice as a most efficient means for the removal of membrane from the throat, tonsils, etc., in diphtheria. In his hands (and he has heard several of his professional brethren say the same) it has proved by far the best agent he has yet tried for the purpose. He applies the juice of the lemon, by means of a camel's hair probang, to the affected parts, every two or three hours, and in eighteen cases on which he has used it the effect has been all he could wish.—*Medical and Surgical Reporter.*

Lymphadenoma. Synonyms—Hodgkins' Disease,* Maligant Lymphona, Lympho - Sarcoma,† Pseudo - Lukemia,‡ Pseudo-Leucocythemia.‖

By Thos. H. Hawkins, M. D., Denver,

Lymphadenoma is a disease concerning the etiology of which there is but little known, and the literature of the subject is limited and vague; and as to medicinal treatment, no satisfactory results have as yet been attained. The pathological changes accompanying it, however, are very well understood. The lesions in this disease are similar in some respects to those of true leucocythemia; but in the latter the white blood corpuscles are enormously increased, while in the disease under discussion their proportion may remain normal. In both diseases the changes peculiar to them are mainly found in the lymphatic glands and in the spleen. But in splenic glandular leucocythemia we have first, the blood changes and the anæmia, the glandular enlargements being secondary to these, while in lymphadenoma or Hodgkins' disease, the enlargement of the glands may exist, and usually does, for some time without any marked anæmia, or blood changes.

I have had under my care and observation an unusually interesting case of Hodgkins' disease, a description of which, together with its previous history may not fail to interest.

Mrs. ——, aged 42 years, married fourteen years, sterile, was in perfect health, as herself and friends both claim, until about thirteen years ago, when, after having been exposed during her menstrual period in a very severe snow-storm on her way home from Denver to Golden. She suffered from "suppression", and never "felt right" after that time.

Nine years previous to her death, which occurred June 10, 1881, she noticed a number of small, hard, rounded knots or tumors on the outside of her left breast. They were

not painful and she discovered them only by accident. These growths were movable and were not painfully sensitive to the touch or even firm pressure.

During the next six years these growths slowly enlarged, and, in the meanwhile, others of a similar nature and character made their appearance in different parts of the body in the following order: Left axillary, cervical and inguinal. There was but little pain suffered at any time during the six years, and no great amount of failing in general health, but there occurred a gradual change in her complexion. Her bright, rosy cheeks gradually faded, and in their stead came a pale waxy or tallowish appearance of the skin, the "splenica et lymphatica cachexia", so called. During the latter part of this six years, some choreic movements of the left side were observed. Jerking or twitching of the head to the left. About three years ago, suddenly, the growths rapidly increased in size, and this was soon followed by severe pain in the left arm. The hand and arm became very much swollen, due, doubtless, to the pressure of the enlarged glands in the axilla. Later, intense pain appeared in the left side, extending from the region of the spleen to the ovary of the same side. About this time she experienced an excruciating pain of the entire left side of the head, worse from three to ten P. M. She suffered from occasional nausea and vomiting, which grew worse every day, also from obstinate constipation. She became very irritable and jealous, having no rest nor sleep day nor night excepting when very large doses of narcotics were administered. This state of affairs continued, and gradually grew worse until she became a helpless invalid, unable to walk about or even to sit up for any length of time.

I was called to see this lady, June, 1880, and made the following note of her condition at that time:

The patient is anæmic and emaciated, the pulse feeble, beating 105 per minute. Dyspnœa upon the slightest exertion. A hyperæsthetic condition of the skin. Temperature, under the tongue, 101 3-4. Tongue dry and

---

*Hodgkins first described this disease in 1872. †Vichow, ‡Cohnheim, ‖Bennett.

parched in centre, with red, sensitive edges. The bowels are constipated. Micturition painful, and the urine dark and scanty, emitting a strong ammoniacal odor. A subsequent examination showed a high specific gravity, but no albumen. Almost constant nausea and frequent vomiting. Food, whether solid or liquid, not retained by stomach unless taken in very small quantities. Small cheesy masses are frequently ejected with the vomited substance. She is nervous and hysterical, and suffers from nervous dyspnœa. There is a constant pain in the head, due to the pressure on the carotids and jugulars by the enlarged cervical glands. A disagreeable sensation accompanied by severe pain in the abdominal region is complained of. Every gland in the body, so far as I can ascertain, is enlarged, sized from that of a filbert to a goose egg. The smaller glands are separate and movable, on the neck and axilla they are united, forming a conglomerate mass. The spleen and liver are enlarged and are very tender and sensative to pressure. Much pain is experienced when the ovaries are pressed on. I called in Dr. M. A. Wilson, who thought there might be a syphilitic taint in the case. After controlling the nausea and vomiting, to a certain extent, with a mixture containing carbolic acid, sub-carbonate of bismuth and cherry laural water, regulating the bowels by a cold water injection per rectum, every morning, and a podophylin pill (gr. 1-4) every night, and relieving the pain by a rectal suppository composed of codeia, ext. of hyoscyamus, ext. of belladonna, hydrate of chloral, and opium. I commenced a course of treatment with iodide of potash and the bichloride of mercury. A solution of iodide of potash, a grain to the drop, was prepared with great care. The dose to commence with was one drop (one grain) three times a day. This dose was increased one drop every day. A rectal suppository containing hydrarg. bichlor. gr. 1-16, quin. sulph. gr. 2 was exhibited three times a day. She certainly did improve, under this treatment, in general functions, strength, condition of stomach, bowels, kidneys, and freedom from

pain, etc., but there was no diminution in the size of the glands. This treatment was continued for 100 days, when she was taking 300 grains of iodide of potash a day. The patient was out of bed and able to walk about the house. The iodide was discontinued and the other treatment continued, except the podophylin and cold water injections, which were no longer necessary to regulate the bowels.

For the next three months there was no change in her condition, except a severe attack of vomiting when an abscess, connected somewhere with the stomach, burst, when at least half a gallon of pus was vomited. The pulse still remained feeble and never below ninety. She continued to have a fever, which was remittent and hectic in character, and the glands gradually increased in size. The mental aberrations became more and more manifest, and during the last two or three weeks of her life she seemed to have entirely lost her memory.

Six days before death she passed into a semi-comatose condition, which deepened into a profound coma, and in this condition she died, just about one year from the time I first saw her, and three years from the date the rapid growth of glands set in, which was followed by severe pain, nine years after first enlarged gland was discovered, thirteen years from date of first illness.

This was undoubtedly a case of Hodgkins' disease, or what is commonly called lymphadenoma.

It is unusual in point of duration, and in that both the lymphatics and spleen were involved, though this, according to some authorities, is not very uncommon. Dr E. C. Wendt recently reported a case where both were involved. According to Dr. Wood we may have. not only envolvement of the lymphatics and spleen, but also the bony marrow. As to the treatment of this disease we are in the dark, much, however, may be done to relieve unpleasant symptoms, as was done in this case.

Wunderlich reports a recovery, but the case was treated at the very onset of the disease, and the diagnosis is therefore questionable. A careful and thorough examination of the blood, and a post-mortem, would have added much of interest to the case reported, but these were impossibilities under the circumstances.

425 Champa Street.

Case of Poisoning by Inhalation of Carbolic Acid.

By Dr. F. P. Blake, Canon City.

Mrs. L. P. ——, a married woman, resident of this place, pregnant and near full term, aged 29 1-2 years, was found dead in bed. A handkerchief enclosing a sponge was in close proximity to her face, which was cyanosed. The tip of the nose was blistered from contact with the substance contained in the inhaler. An odor of carbolic acid pervaded the room; the sponge and handkerchief were saturated with the odor, though dry when found. No one was in the room when the woman died. She was subject to neuralgia, and it is supposed that she sought relief by inhaling the acid, which asphyxiated her. Few similar cases are on record, I believe; and this one should caution physicians against an injudicious use of this valuable drug. I will say that the appearance of the mouth did not indicate that the acid had been taken internally.

The Plague in the East.

By A. Labrie, M. D., Denver.

We, inhabitants of the *Far West*, blessed with the most desirable climate for human abode, have little thought of the terrible sufferings which our less favored fellow beings of the *Far East* are called upon to endure.

In a recent letter to the Paris *Journal of Hygiene*, Doctor Stecoulis, Member of the International Council of Health, at Constantinople, reports a most deplorable condition of affairs in Mesopotamia, on the borders of the Euphrates. An extremely lothsome and fatal disease has broken out again in that unfortunate region, which, in its effects, is even worse than yellow fever or the terrible Asiatic cholera.

This plague, which is known as *bubonic pest*, was first reported to the Ottoman authorities towards the end of February last, by a military surgeon who happened to be in that vicinity with a detachment of Turkish troops, sent to quell the insurrection among the Kurds. The disease, at that time, had already existed several months, and being unchecked neither by sanitary measures nor medical aid, was spreading at an alarming rate. Cities and villages for hundreds of miles around were wholly depopulated, those who were so fortunate as not to fall victims to the terrible scourge, having fled to the mountains to seek safer abodes.

The sanitary administration at once called the International Council of Health in extraordinary session, and laid this matter before it for immediate consideration. The sanitary inspector of Bagdad, in a dispatch to the Council, said that he was in the afflicted district, and recognized the disease to be the regular *Oriental bubonic pest* in its very worst form; that more than half of those attacked by it would succumb to its fatal grasp.

It was first noticed in November, among the tribe El-Zayad, and by the end of January it had invaded Djahara, Nedjeff and other places of considerable size.

At Djahara, in a population of 400, over thirty deaths occurred in forty-eight hours. The symptoms observed were excessive languor, intense thirst, high fever, headache, furred tongue, bloody emesis, hæmaturia, axillary and inguinal bubos, then prostration, coma and death, in twelve, twenty-four and forty-eight hours. As many as nine died in one family.

As soon as the International Council was informed of the exact condition of affairs, it immediately adopted measures to localize the disease, by surrounding all the affected districts with three circles of military guards, to prevent all possible communication with the outer world. Between the first and second circles, hospitals and quarantines were established, and between the second and third circles, those who were to all appearance free from the disease, were allowed to live without coming in contact with those who were affected.

The inflicted districts were disinfected as much as possible, with the fumigation of sulphur, and the complete distruction, by fire, of all houses, huts, furniture and personal clothing. All the inhabitants are transferred to a new and healthy locality, and the government furnishes them with all the necessities of life.

These timely and efficient measures have been successful in confining the plague within its original limits, and preventing its spreading among the more civilized people of the West. Who can tell what terrible scurge would heve visited Europe, and possibly America, had not its course been checked before the beginning of the warm summer months?

A few words on the etiology of this disease would not be out of place just at this time. This is its fifth appearance in that vicinity during the last ten years, and each time its ravages are so disastrous and its course so rapid that serious alarm has been felt by all Eastern countries, even Western Europe was at one time concerned.

From the most reliable evidences of its rise and progress, we are forced to the conclusion that this plague must originate here through some local cause, and, perhaps, always exists in some filthy places in a lower state. It must be borne in mind, that in Eastern Persia and all the cities along the shores of the Euphrates, cities and villages are repositories or sanctuaries for the Mussulman rites of the Schiites. The principal ones of which are Kerballa, Mesched-Ali, Nedjeff and others, in which the tombs of the Mussulman prophets, notably that of Hussian-Ali. The Persians which belong to the creed of the Schiites take a sacred vow to bury all their dead in these places, which are vast necropoles, for there are from 4,000 to 12,000 people burried every year in these hot-beds of disease and corruption. This being also the burial place of the great prophet Ezechiel,

the Jews, for thousands of miles around, bring all their dead to be buried in the shadow of the tomb of their illustrious prophet. The distant transportation of these cadavers in hot weather in a putrid state, and their burial almost on the surface of the soil, are, we believe, the principal causes of the development of the miasma which produces this lothsome disease.

Dawn of a New Era in Otological Surgery.

THE GREAT REPRODUCTIVE POWER OF THE MEMBRANA TYMPANI—A NEW METHOD WHEREBY THE EUSTACHIAN TUBES MAY BE PERMANENTLY OPENED, DILATED AND TREATED—DRY CATARRH OF THE MIDDLE EAR NOW AMENABLE TO TREATMENT— DISTRESSING AND LONG - STANDING TINNITUS AURIUM RELIEVED IN A NOVEL WAY—WITH ILLUS- TRATIVE CASES.

By A. Wellington Adams, M. D., Denver,

Professor of Otology and Laryngology, Medical Department, Denver University; Attending Surgeon for Diseases of the Ear and Throat to St. Luke's Hospital; Recording Secretary Colorado State Medical Society; Foreign Associate Member French Academy of Hygiene; Editor Rocky Mountain Medical Review, etc., etc.

Without an accurate acquaintance with the visible and tangible properties of things, our conceptions must be erroneous, our inferences fallacious, and our operations, unsuccessful. And hence it is that otology has for so long a time stood quiescent in the background, and allowed her so-called sister science, ophthalmology, to carry off the laurels. A few, by dint of dogged perseverence and personal observation, have familiarized themselves with the varied appearance of the tympanic membrane and the tympanum under different normal and abnormal conditions, and have consequently become proficient in manipulations in and about the aural cavity. But such accomplishments are rare, and have been purely the result of individual effort and self-instruction on the part of the acquirer. This state of affairs is owing to the fact that demonstration is here almost impossible, and we are obliged to rely upon didactic and descriptive, rather than clinical and demonstrative methods, for the dissemination of knowledge pertaining thereto. Every new worker in the field of otology has been obliged, therefore, to begin at the beginning. Instead of being able to commence where his predecessor left off, he is required to cultivate and train his own eye, by personal experiences and comparisons, to

class demonstration, these parts, with their manifold affections, may be thrown upon a screen; and, furtheremore, operations in and around the middle ear may be thus witnessed as readily as operations upon any other part of the body; and to facilitate the dissemination and perpetuation of knowledge once attained, these parts may be photographed and printed.

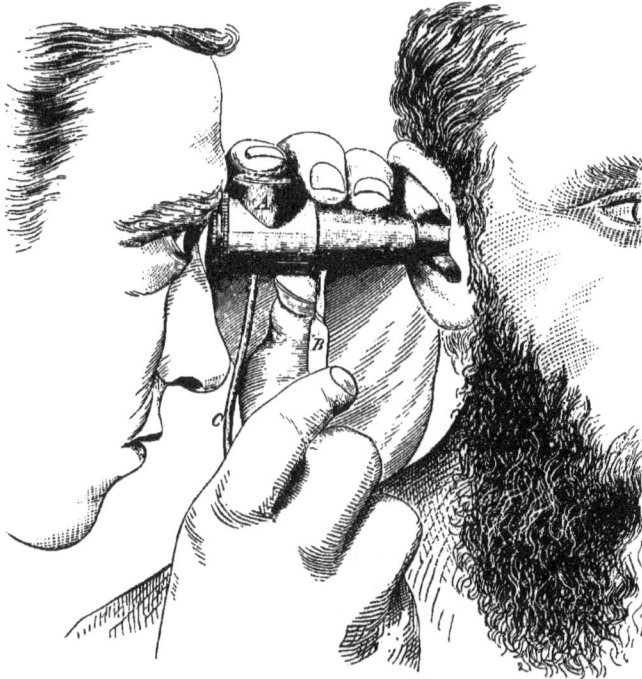

Adams' Electric Operating Otoscope. A—Position of the electric light. C—Wires carrying the electric current for illumination. B—Instrument introduced into the ear through the Otoscope.

recognize the almost endless variety of pathological changes peculiar to the hidden parts of the ear. Again, we have been greatly restricted in our curative measures, because of the difficulties attending the performance of what few operations we have been familiar with about the inner parts of the ear. By proper methods of illumination, however, all these obstacles may be overcome. The drum membrane and the middle ear may be viewed and operated upon under intense magnification and brilliant illumination; for

All this, and even more, may be accomplished through the mediumship of the "Electric Operating Otoscope." (See illustration).

Because of the obstacles attending all operations about the inner parts of the ear, aurists have thus far failed to devote much attention to the surgical part of their specialty, and have relied more particularly upon passive rather than active treatment; while general practitioners have looked upon the drum membrane with great "fear and trembling".

supposing it to be extremely sensitive, and when once injured or partially destroyed, many believe hearing to have been irretrievably lost.

It will be conceded by the best aural surgeons, that what has, up to the present time, been the accepted treatment for that affection which is universally known as "dry catarrh of the middle ear," has proved very unsatisfactory. I must certainly confess this to have been my experience, and I know it to be the acknowledged conclusion of many of our most noted aurists, other reports to the contrary notwithstanding.

Now what is this treatment? Well, it amounts to this:—dilatation of the Eustachian tubes, inflation, and the application of medicinal agents (fluids and vapors) to the Eustachian tube and middle ear through the former. This, I claim to be both impracticable and unsatisfactory, because, first, in just such cases as call for this procedure, we are likely to find an impervious Eustachian tube, which it is next to impossible to render patent by means of catheters and bougies passed through the nares to the pharyngeal orifice of the Eustachian tubes, as recommended; all such attempts being fruitless and in every way fraught with danger. And, second, because admitting even that the Eustachian tube may be thus rendered permeable, the amount of fluid it is possible to force into the middle ear in this way is so small as to be of no practical use, and fails to attain the end for which it is designed, namely, the removal of inflammatory products from the middle ear, and a modification for good of the membrane lining that cavity; besides, under the conditions here present, the introduction of any fluid into the middle ear would be mischievous rather than beneficial. Concerning this, no less an authority than Dr. Burnett says: "Various applications have been advised and made to the mucous lining of the Eustachian tube, in order to allay chronic inflammation. In most cases, they do more harm than good. All injections into the Eustachian tube are risky. Steam is not to be considered anything more

than useless." He furthermore says: "Few applications which are aimed at the tympanic cavity through the Eustachian tube ever reach it", and I perfectly agree with him. And further on this eminent authority remarks: "If they did, they would do more harm than good. To render the Eustachian tube pervious to air, and hence to ventilate the drum cavity, is more important than the injection of fluids into said cavity, *unless*," (now mark you) "the membrana tympani being perforated by disease a *means of escape* for the medicated fluids is afforded." One is further led to believe from the writings of Dr. Burnett and others, that the idea of injecting fluids into, and of applying vapors to, the middle ear through the Eustachian tube, is more *theoretical* than practical. The mere anatomy of the middle ear is such as to render absurd the idea of such a course of treatment for the removal of inspissated mucus from the tympanic cavity, for it should be remembered that the floor of this cavity is on a lower level than the tympanic orifice of the Eustachian tube, hence medicated fluids and foreign substances will tend to gravitate here and remain behind rather than be washed out. In view of these facts, I have been led to adopt a novel course of treatment of my own for the relief of chronic catarrh of the middle ear, and all aural troubles which may be supposed to be dependent upon the presence of calcareous matter or inspissated mucus within the tympanum, and one which I am happy to be able to say has proved eminently successful and void of danger. So far five cases have been successfully treated in this way.

It essentially amounts to this: The application of medicated fluids and vapors to the tympanum through the external auditory canal. This I accomplished by dissecting out two triangular segments of the membrana tympani; one in the superior and the other in the inferior portion of the membrane. In fact, I have frequently dissected out the entire drum membrane with the exception of a peripheral ring or the annulus tendinosus and an oblique strip running from above downward and backward, and including the manubrium of the malleus. In this way I am enabled to reach the middle ear directly, accurately determine the condition of its lining membrane, diagnose the presence or absence of foreign substances or growths within the tympanum, ascertain positively the exact character of the pathological changes in the membrana tympani, and to apply my treatment directly and advantageously to the diseased parts. During the treatment, the openings are retained by means of pledgets of absorbent cotton introduced therein; and the entrance of air and dust is prevented by cotton worn in the external auditory canal. After accomplishing our purpose, the membrane readily reproduces itself, and the openings are closed without any interference on the part of the surgeon.

I believe I am the first to announce the possibility of opening a Eustachian tube impervious to air, through the external ear, which announcement I now make, having recently resorted to this procedure for the relief of a chronically narrowed and occluded Eustachian canal attending a case of "progressive hardness of hearing." The result was in every way satisfactory. For this purpose I use a moderately stiff filiform bougie passed through a metalic tube with a short curve at the extremity which is to be introduced into the middle ear.

The metalic tube answers the purpose of stiffening and directing the course of the bougie.

A very large proportion of the cases of defective hearing we meet with in every day practice are attributable to the presence of inflammatory products within the tympanic cavity, and the structural changes consequent thereon.

In many instances the inflammatory action giving rise to this condition occurred at a very remote period, in fact, so far removed from the time of the first noticeable disturbance of audition, which comes on insidiously, that we are often confronted with the patient's statement that there never has been any acute inflammation or "running at the ear."

An objective examination of the auditory apparatus in many such cases, fails to reveal anything upon which a positive diagnosis can be based; consequently, the surgeon is either obliged to cut his way boldly into the middle ear, or else depend upon a process of differentiation and elimination in forming an opinion of the existing condition. If he choose the latter, the diagnosis still remains unverified until he proceeds to operate, which then simply amounts to an exploration; therefore, I prefer the former as a means of diagnosis.

In all cases where the hearing power is deficient to any very marked extent, the defect is not in the least increased by the operation of dissecting out a portion of the drum membrane, should our suppositions prove to be unfounded. The portion excised I find quickly reproduces stself *without any application whatsoever,* unlike the membrane partially destroyed and structurally modified by chronic otorrhœa. I do not believe this reproduced tissue to be of the same character as ordinary cicatricial tissue. Several times I have been mistaken as to the presence of a foreign substance in the tympanum, not even finding the slightest abnormal condition of any portion of that cavity, and yet after the membrane became closed in, the hearing power proved as acute as before the exploration.

Of course it is necessary to exercise great judgment in determining just what cases justify this heroic treatment. But when justifiable and properly carried out, I can safely declare this procedure to be entirely free from danger.

In illustration of the foregoing I will now detail three typical cases, selected from among quite a number.

Mrs. H. C. ——, aged 43, presented herself at my office in search of aid for extreme hardness of hearing. During the course of our preliminary conversation I found it necessary to sit within two feet of her and to elevate my voice to a very marked degree, in order to make her understand my interrogations.

The following history was elicited: Eight years previous to the time of making this statement, while lying in bed on her right side, wide awake, a servant entered the room through the side toward which her back was turned, and began addressing some remarks to her, which, not being heard, were unheeded. This resulted in the servant's further intrusion, until a point was reached within the recumbant's range of vision, when the latter was very much startled at finding it possible for any one to so far enter her room without her hearing them. This circumstance led to an investigation, which revealed the fact, for the first time, that she was very nearly deaf in the .left ear. From this time on her hearing became more and more impaired, until now she was practically deaf. She declared she never had any "running at the ear," or even pain, but that for the past five years there had been ever present a peculiar and very disagreeable ‚noise, which she could only liken to a "buzzing." This was at first more perceptible in the right ear, whereas now it was absent in the right, but very annoying in the left.

A subjective examination revealed the following conditions: Hearing power on left side c-60, that is, a watch that should be heard by the normal ear at sixty inches, could only be heard on contact; hearing prower on right side, according to same formula, 3-60; tuning fork heard distinctly when placed anywhere upon the head, or between the teeth; a very small-amount of air was felt to enter both middle ears upon inflation, but the buzzing sound was not in the least relieved by this act. The objective examination showed, on the right side, a drum membrane of deficient lustre—considerably retracted—only approaching translucency— "pyramid of light" displaced and poorly defined, but not entirely absent—no indication of there ever having been a rupture; on the left side the membrane was retracted, slightly shriveled, and opaque—near the inferior wall of the external auditory canal, it assumed a color somewhat resembling an olive-green, only of a duller hue.

Treatment: Excision of the greater portion of the drum membrane of the left side,

according to the method heretofore described. On the right side, two triangular pieces were removed from the membrane, one from the anterior and the other from the posterior segment. This allows the passage of a fine stream from above downward and outward, thus making a complete cycle to carry out anything which may have collected within the tympanum. In the left tympanum I found a collection of inspissated mucus, which will account for the peculiar appearance of the drum membrane on that side, and for the increased disturbance of audition and other special subjective experiences. Both tympana were found to be very much reduced in size, due to an extensive proliferation of their lining membrane, which, instead of assuming a healthy appearance, seemed to be of a low grade of organization.

After thoroughly cleansing and washing out the middle ear, which occupied nearly two weeks, I began to make applications of a solution of iodoform, also the compound solution of iodine. This alone soon brought about a healthy action, and a process of desquamation ensued. In five weeks the tympanic cavity had regained its normal size, tinnitus aurium was relieved, and the hearing partially returned, the sound waves, at that time, impinging themselves directly upon the membrane covering the *fenestra ovalis*. Up to this time it had not occurred to me to open and treat the Eustachian tubes through the external ear, so that in this case I resorted simply to inflation as a means of dilating and treating this canal, which it will be remembered I said was not entirely occluded. The drum membrane reproduced itself in about seven days without any interference on my part. Result: Right ear, hearing power, 30-60; left ear, hearing power, 22-60. Can hear an ordinary conversation. The buzzing has completely disappeared. Her gratification at the result is expressed in the following letter to me:

"DEAR DOCTOR:—I am only too glad to give you the statement you call for. You have indeed more than saved my life, for what is life to one deprived of a sense one has known the advantages of for upwards of thirty years.

When I first came to you it was necessary for people to scream to me in order to make me hear; and there was such a constant buzzing in my ear, it seemed at times as though I should go crazy.

At the present time—two years after the operation—I can hear quite well; of course, not as well as some of my friends, but then well enough to answer all my purposes.

Yours Respectfully,

Mrs. H. C———."

The second illustrative case I shall report is that of Dr. W. Edmundson, of Denver, President of our State Board of Health, who came to me with this statement: "Doctor, I wish you would look into my left ear and see if you think anything can be done for it. For upwards of twenty years, off and on, I had a discharge from that ear; and for over thirty years I have had constantly present a very distressing tinnitus aurium, which at times has seemed calculated to drive me mad. I haven't spent a single comfortable night in the last twenty years, and it doesn't seem as though I was ever again destined to. For the past three or four years I have also been troubled with neuralgia of the left side of my head, which I believe is due to the trouble existing in this ear. Hearing in this ear has long since failed me, but then I don't care anything about that, if you can only relieve me of this *constant* and *horrible* noise in my ear." On testing his hearing power on the left side I found it to be 4-60. The tuning fork was heard at the vertex of the head when placed over both parietal bones. Objectively, he had a careworn and anxious expression of the face, paralysis of the muscles of the left side of the face—ptosis, or partial falling of the left eye-lid—spasmodic contractions of the orbicularis palpebrarum. The drum membrane was retracted, shrivelled and atrophied and had entirely lost its lustre and transparency, and the "pyramid of light" was not to be seen. About at the junction of the "posterior segment" with the superior wall of the external auditory canal

there was a cicatrix, which I afterward learned marked the point of growth of a polyp which had some years before been removed by the eminent aurist, Dr. Roosa. The whole membrane, to my eye, presented the appearance of having a deposit behind it. I, therefore, advised the Doctor to allow me to operate upon it—to remove a large portion of the membrane, and cleanse the middle ear, stating that I thought all of his trouble due to the condition I believed to be present, and that the course of treatmennt I proposed would doubtless relieve him. To my surprise he readily acquiesced, with the remark that he had been to several specialists, amongst the number, Drs. Roosa and Noyes, of New York, but that they had all failed to give him any relief from the tinnitus aurium, which he cared most about, although the polypus had been removed and the "discharge" dried up. He said that he had himself suggested to several the advisability of at least puncturing the drum membrane, because he had noticed that when there was an opening and his ear was discharging, he experienced great relief from the subjective noises, but that they had differed with him, and declared such a procedure useless.

The following day I operated, excising two triangular pieces from the drum, one above, posteriorly, and the other below, anteriorly. The tympanum was found to be partially filled with hardened mucus and other inflammatory products, which were removed by instruments and warm injections, the latter producing excessive vertigo whenever resorted to. Warm bicarbonate of soda solutions were poured into the ear while he lay upon the opposite side, and allowed to remain there for from five to ten minutes. After continuing this treatment for a few days, resolvent and stimulating applications were made to the middle ear, and subsequently the membrane allowed to heal up. Pledgets of absorbent cotton were at first used in the openings, but afterwards two different sized eyelets were used to retain the openings until the treatment had been completed and there was no danger to be anticipated from subsequent inflammation. Within twenty-four hours from the time of operating, all of the more disagreeable symptoms had disappeared, and, as the Doctor himself expressed it, "he felt like a new man, he had passed a comfortable night for the first time in upwards of twenty years; there was not a particle of tinnitus present." When the time came to remove the eyelets and allow the membrane to heal up, it was with great difficulty that I prevailed upon him to submit, he contending that when the openings were closed all the disagreeable symptoms would return, he feared; and he declared he wouldn't have that for thousands of dollars. I never saw a man more completely carried away and delighted over anything. When I finally removed the eyelets it was under his protest. The entire chain of disagreeable symptoms was due to the presence of a foreign substance within the tympanum, which had acted as a reflex irritant, and mechanically interferred with the circulation and nutrition of the parts involved, and when this was once removed, there was no necessity for retaining an opening in the drum membrane, which serves as a protection to the middle ear against the introduction of things calculated to induce an inflammation therein.

The following letter sets forth the Doctor's present condition:

"DEAR DOCTOR:—Before you operated upon my ear my condition was as follows: Afflicted with a sense of weight and oppression in the entire left parietal region—supra orbital neuralgia almost constant for the past three years, paroxysms of the same occurring with intensified severity about twice a week, on an average, and lasting eight or ten hours. Distressing tinnitus aurium constant for upwards of thirty years. Frequently recurring aural vertigo. Paralysis of the muscles of the left side of the face, ptosis and spasmodic contractions of the *orbicularis palpebrarum* muscle.

Condition since operation, which was performed about two months back: Sense of weight and oppression has entirely disappeared. Almost complete relief from the

tinnitus aurium. No neuralgia worth mentioning since operation. Absence of aural vertigo. Hearing power increased fully six fold, being able to hear your watch at twenty-four inches; the hearing is still improving. It is no longer necessary for me to turn my head to one side in order to hear an ordinary conversation. . On closing up the left ear with my finger, I can notice a very perceptible diminution in the hearing power. The openings in the membrana tympani have long since closed. The spasmodic movements about the left eye have almost ceased, ptosis is somewhat relieved, and the paralysis is gradually disappearing.

Yours Truly,

W. EDMUNDSON, M. D.

The third case I shall report is one which was sent to me by Dr. B. P. Anderson—Mr. Joseph Sharratt, a gentleman upwards of 45 years of age, applied to me for relief from deafness in the right ear. After examining this ear I requested him to turn around, that I might examine the one on the opposite side. He protested, and said there was no necessity for that, as he never recollected hearing out of the left ear, that he came to me to have the *right* ear treated, that he was very much afraid of losing his hearing on that side, and if he did he should be totally deaf. After explaining to him the necessity of an examination of both ears for purposes of comparison, he, however, consented.

The membrane in this ear appeared to have lost all vitality, was retracted and shrivelled, and in the anterior segment there was an indenture, as though something from behind was drawing it in; in places it was atrophied and assumed more the appearance of tissue paper than anything else. On finishing the examination I told him I could certainly benefit his hearing on the right side, and that I also *might* be able to give him hearing in the left ear if he would allow me to perform an operation; that I couldn't promise anything positive, however, but that I should like to try. He laughed and said I might as well talk of making a stone hear. That he was now upwards of 45 years of age, and had never heard anything with the left ear. In illustration of the degree of deafness on that side, he said it was his custom, while at the "ranch", to retire before his men did, who remained up to play cards and have a good time generally, during which time they made a great deal of noise, but that in order to sleep it was only necessary for him to turn upon his side with the right ear buried in the pillow, "when they might fire off a cannon without his hearing it". He finally consented to the operation, which consisted in excising the entire membrana tympani with the exception of the *annulus tympanicus* and a narrow strip extending obliquely across the centre of the membrane from above downward and backward, corresponding to the line of insertion of the manubrium of the malleus. The treatment from this on differed immaterially from that adopted in Dr. Edmundson's case, and consequently does not demand detailing. Suffice it to say, within ten days he could hear as well with the left as with the right ear, which had also been benefited. The excised portions were reproduced, and the new membrane looked very much more like a normal one.

From time to time I shall report the results of further investigation, and should a more extended experience substantiate these observations, Otology will undoubtedly be placed on a par with Ophthalmology and Gynecology.

RECENT MEDICAL PROGRESS.

Report on Orthopædic Surgery.

By THOS. H. HAWKINS, M. D., DENVER.

OSSEOUS ANKYLOSIS OF THE KNEE OPERATED ON BY BARTON'S METHOD.

Mr. M. H. Kilgarriff (Doublin Journal of the Medical Sciences; March, 1881,) reports a case of bony ankylosis of the knee-joint. The leg was bent at right angles and immovable. Mr. Kilgarriff operated on this case. A triangular flap was made at the lower portion of the thigh, the base being on the outside, bringing to view the lower end of the femur. A curved metallic spatula was passed under the bone and a wedge-shaped piece of bone measuring vertically three-fourths of an inch, was removed. The limb was easily straightened and put in splints, the one on the outer side reaching to the axilla, on the inner side to the groin. These splints formed a box, which joined the thigh part at an obtuse angle. The wound was dressed with lint saturated with carbolized oil, and was washed occasionally with a solution of permanganate of potash. There was slight suppuration, but the divided surfaces of the bone united by a firm osseous band. This operation differs from Barton's in that there is no spicula of bone left to pierce the fascia behind, and the base of the flap is on the outer side, thereby preventing the bagging of matter. We are assured by Mr. Kilgarriff that there was no shortening in this case and that the patient made a perfect recovery.

OSTITIS.

J. Williston Wright, M. D., (Detroit Lancet, June, 1881,) in a very instructive and interesting lecture, delivered in the University of New York City, takes up the subject of ostitis and speaks of its intimate relation to periostitis, of its being the important factor in the development of caries and necrosis. He further says that ostitis is most common during early life, when the vascularity of the bony tissues is at its maximum. That it ordinarily involves a very limited portion of bone, yet it may pervade the entire length and breadth of the bone affected, and that the cancellous portion is most frequently attacked. In the first stages of ostitis there is much enlargement, which is due mainly to the swollen periosteum, the bone loses its density and is infiltrated by an inflammatory product of a sero-plastic nature. The blood vessels are turgid and the structure of the bone becomes softened, absorption of the earthy constituents of the bone takes place, which, when the ostitis reaches its height, renders the bone very easily cut or bent. The osseous fibres become separated from one another, and the inter-spaces thus produced fill up with imflammatory products. The Haversian canals, especially of long bones, are enlarged, and the cells of the spongy tissue are increased in size. When the disease attacks the outer layers of the bone, the periosteum soon becomes separated as a result of the inflammatory action, cutting off the outside circulation of the bone. This condition is very apt to lead to necrosis, if a long bone, or caries, more likely, if of the short or spongy variety. This form of necrosis is called superficial death or exfolia-tion, and usually occurs in the long bones. When ostitis involves the inner layer of a long bone, the medullary membrane soon becomes inflamed and separated, as is the case with the periosteum, when the outer layers are involved. This endostitis in turn gives rise to what is called osteo-myelitis. Where there is separation of an inflamed periosteum on the outer side and the endosteum on the inner side, there is apt to follow a general necrosis, owing to the formation of emboli in the branches of the nutrient artery, cutting off the circulation on both sides of the bone. The Lecturer in this connection goes on to show the intimate association, one with the other, of periostitis, ostitis, endostitis, osteo-myelitis and their sequelae, and that ostitis may terminate in resolution and recov-ery, in hypertrophy, or what is now called sclerosis of the bone due to hyper-nutrition, in suppuration, in caries or necrosis. He sums up the causes of ostitis under two gen-eral heads, viz.: Traumatic and constitutional.

RESECTION OF THE PATELLA.

Dr. E. Albert (Beitrage Zur Operativen Chirurgie, 1881,) reports the case of a woman, aged 30 years, who in childhood had suffered from Keratitis and swelling of the glands in the cervical region. This woman, in 1876, suffered from suppurative articular disease, which made its appearance first in the right ankle. During the year following this attack, an abscess formed over the patella of the same limb. A small fistulous orifice pre-sented itself on the front of the knee. The extensive carious disease of the ankle, neces-sitated amputation of the leg. Following

this, the knee became very large and painful. The symptoms generally indicated fungoid thickening of the synovial capsule. But as the patella was evidently the starting point of this disease, Dr. Albert decided to remove, if necessary, the whole of this bone. The patella was exposed by vertical incision. Its inner half only, was found to be diseased and was removed by means of a chisel. The cartilage of the patella was found intact, but was marked by streaks of yellow deposit. The synovial sac was converted into a thick, fungous mass, containing three and one-half ounces of fluid. Incisions were made on each side of the joint, sufficiently large to admit the passage of a drainage tube through the articular cavity. Injections of a solution of carbolic acid were used daily. Profuse suppuration followed and the patient died two months later. Volkman, in commenting on cases in which there are osteo-myelitic deposits, endeavors to show that the destruc-tive articular disease commonly called white · swelling, has its primary origin in a tuber-cular deposit in the interior or medullary por-tion of some one of the adjacent bones. He believes that such deposits should be removed when practicable, and cites a · number of cases in which he has removed inflammatory deposits from the great trochanter and from the articular extremities of other bones. In each case the joint was laid open and drained. The experience of Volkman and also that of our own countryman, L. A. Sayre, and others, would lead us to suggest the more frequent resection of diseased joints, leaving the question of amputation for an after consideration.

ON THE SO-CALLED RUPTURE OF THE INTER-NAL LATERAL LIGAMENT OF THE KNEE-JOINT.

Charles A. Jersey, M. D., (New York Medical Journal, 1881, p.633,) reports two very interesting cases of supposed rupture of the internal lateral ligament, but which were undoubtedly fractures of the internal tuber-osity of the condyle, if one can believe the results of his five very conclusive and satis-factory experiments on the cadaver. These experiments prove that the same force when applied to the cadaver, which is supposed to give rise to rupture of the internal lateral ligament in a living subject, the physical signs to the contrary, produces no laceration of the ligament but fracture of the tuberosity instead. He goes on to reason that the structure of the tuberosity and condyle is such as tends to weaken them when force is exerted, producing traction from above downward; that the ligament is not inserted into the tuberosity, that the tuberosity would separate before the ligament would rupture, ligament being stronger than bone, and that when force is exerted indirectly on bone through ligament a fracture usually is the result, and not a rupture of the ligament. The author sums up his conclusions under four heads: 1st. The so-called cases of rupture of the internal lateral ligament of the knee-joint are. fractures of the internal tuberosity of the condyle. 2d. Severe sprains of the knee (so-called) are, many of them, fractures of the tuberosity. 3d. That absence of bony crepitus, does not necessarily preclude the existence of a fracture at this part. 4th. That the diagnosis rests upon the extreme lateral motion, severity of pain on manipulation, the localized pain at a certain point and the long time required for recovery.

In the five experiments on the cadaver an attempt was made to rupture or lacerate the ligament. The tuberosity was fractured or rather separated, while the ligament remained in situ.

EXCISION OF THE CUBOID BONE.

M. Poinset (British Medical Journal, October 20th, 1880,) has recently practiced successfully, the operation advocated by Mr. Pavy, for removal of the cuboid bone in some of the forms of club-foot. The case reported by the author (M. Poinset) was one of talipes varo-equinus, occurring in a young girl, on whom subcutaneous section of the tendon was practiced, but this not succeeding, the cuboid bone was removed, after which the foot was placed on a gutter. This operation was performed under antiseptic precautions.

The cure was complete, leaving the foot in good position.

CONSIDERATIONS RESPECTING THE MECHANICAL TREATMENT OF HIP DISEASE, WITH ESPECIAL REFERENCE TO THE VALUE OF TRACTION.

A. B. Judson, M. D., of New York, (St. Louis Courier of Medicine and Science, May, 1881,) enters into a brief criticism of the prevailing views as to the various theories and methods of treatment of joint diseases. He questions the generally accepted view that pressure exerted by muscular contraction on the articular surfaces is the destructive agent in hip-joint disease, and that the principal object of extension as ordinarily used is to overcome the contractions, reflex or otherwise, of the peri-articular muscles. He does not believe that reflex muscular action is the chief cause of the rigidity of the joint, because if so, it admits of a "vicious circle", inflammation increasing the muscular contraction, and *vice versa,* a condition most certainly at variance with the preservative efforts of nature. The morbid anatomy throws doubt on intra-articular pressure as a pathological element in morbus coxarius, inasmuch as in all stages of this disease, it is a recognized and known fact that the morbid process is frequently found invading parts of the femur and os innominatum at points so remote as not to be influenced by articular pressure. He argues that there are many cogent reasons for not accepting the proposition that traction is useful in lessening the pressure exerted by the contraction of the peri-articular muscles, at the articular surfaces. Even if it is admitted that muscular contraction and inflammation mutually incite each other, the hip splint fails to accomplish what is claimed for it by those who advocate either of the two theories, viz.: That the apparatus keeps the muscles contracted while motion at the joint is permitted, or that the muscles are kept stretched until they are unable to contract. The first of these theories the writer claims is marred by many practical difficulties, viz.: The impossibility of securing prehension of the femur by adhesive plaster applied to soft parts. This is an insurmountable obstacle when we take into consideration that the deep muscles and even the deeper fibres of the superficial muscles can not be subjected to traction unless by direct prehension of the bone into which the muscles and fibres are inserted; and besides this, a mechanical impossibility presents itself, namely, that of transmitting pressure from the floor of the acetabulum to the ischiatic tuberosity, the rami of the ischium and pubes in such a way as to control all the variations of active flexion, extension, adduction and abduction. The second theory is open to most of the objections just enumerated, and in addition to these it presupposes that which is extremely improbable, that muscular fibre will surrender its contractility on the application of traction. Likewise the extent of stretching is minute when compared with the normal properties of muscular fibres. The author is nevertheless a strong advocate of the hip splint, but explains the efficacy of the same in the treatment of hip-joint diseases on the principle that it secures fixation. He recognizes the immobilizing power of traction as described by Desault. Absolute immobility of the hip-joint by means of the hip splint is not only a mechanical impossibility, but is not at all desirable or necessary in the treatment of hip troubles. But fixation with immobility to a limited degree, has a peculiar and decided advantage over all other methods. Traction as made by the hip splint has an inherent fixative power besides serving as a brake, retarding and arresting motion of the pelvis on the femur, regulating flexion and extension and guarding the movements of adduction and abduction. Doctor Judson's reasons are logical and conclusive. The principles as enunciated and set forth in this article and the theory advanced have not hitherto been entertained.

425 Champa Street, Denver, Colo.

Eight children in different parts of the country have died from lockjaw from burns caused by snapping paper percussion caps on toy pistols.

SELECTIONS FROM JOURNALS.

Amyl Nitrite, a Powerful Cardiac Stimulant.

By EDWARD T. REICHERT, M. D.,

Formerly Demonstrator of Experimental Therapeutics and Instructor in Experimental Physiology in the Post-Graduate Course of Medicine in the University of Pennsylvania.

The action of digitalis on the heart was, for many years, invested in doubt, and it was a disputed point between two factions of therapeutists as to whether it was a cardiac stimulant or a cardiac depressant, and it was not until the experimental therapeutist came to the foreground, with his convincing evidence, that this vexed question was finally decided, and the drug placed in its proper physiological classification. The amyl nitrite seemed prone to undergo a similar controversy, for, on looking over the literature of the subject, opinions are found which are as diverse as they well could be, and deductions made which are inconsistent with the results of experiments, and contrary to facts revealed by a more thorough consideration. Wood ("Therapeutics", 1879, p. 348), after studying carefully the experimental papers accessible, concludes "that the whole evidence seems to show that the sudden, thumping, rapid stroke of the heart which is so clearly produced by the nitrite of amyl in man is due, at least in part, to a depression of the inhibitory cardiac nerves"; and, in a previous paper, he gives as another factor the diminished oxidation the nitrite causes, and a thrill of impending suffocation and accelerated pulse due thereto; but this will be referred to more fully later in my paper.

While this statement is in the main correct, because there can be no doubt that the nitrites both depress the inhibitory apparatus and diminish oxidation, yet I believe there is still another and more important factor, therapeutically, concerned in this increase of the heart's action, as was stated in a recent paper on the potassium nitrite ("Am. Jour. of Med. Sci.", July, 1880), in which I claimed that the nitrites, in small or moderate doses, acted as *direct* cardiac stimulants, and, moreover, gave the results of a series of experiments with the above-mentioned salt, and endeavored, while discussing its physiological action, to clearly show wherein the deductions of certain observers of the action of the amyl compound had been erroneously drawn, and why the results of certain of the experiments undoubtedly corroborated my belief. I have accordingly resumed a discussion of the subject and have carefully reviewed the accessible papers in this relation, and will consider the evidence deduced under two heads—physiological and clinical.

Referring to the physiological papers, we find that Pick ("Centralbl. f. d. med. Wissensch.", Berlin, 1873, No. 55), declares that it causes a relaxation of the heart muscle, and that the accelerated action of this organ is, in part at least, due to the opening of the vascular channels; that Filenhe (Pfluger's "Archiv", Bd. ix, p. 490), by the result of a single and very interesting experiment, indicated that the increase of the cardiac beat, was due alone to a depression of the vagi nerves, for it was found in this experiment that after section of these nerves, an electric current being employed by direct application to the peripheral portion of the severed trunks, of such a strength as to lower the pulse rate to the normal, the increase did not occur as in normal animals; that Mayer and Friedrich ("Arch. f. exper. Path. u. Pharm.", 1875, v, p. 55), in a different way corroborated Filenhe's testimony, for they observed that sudden asphyxia did not slow the pulse after the administration of the nitrite as it did in normal animals, and, moreover when they pressed the carotids, and prevented the blood from gaining access to the cerebral arteries, the increase did not occur, nor did they find any diminuition of the pulse rate by endeavoring to reduce a reflex inhibition of the heart by irritating a sensory nerve.

Wood ("Am. Jour. of the Med. Sci.", 1871, 1, p. 422), states that, so far from being a cardiac stimulant, it is a direct depressor of the circulation, because, although it increases the number of the beats, it never increases the force; and in a later and more elaborate article, based upon the results of a direct series of experiments on animals (*ibid.*, 2, p. 39), concludes that it "does not act upon the heart until a considerable point of saturation of the blood and system is reached," and that the heart is then depressed (p. 53). He did not at this time make any experiments directly to decide what the cardiac action of the poison was, except in determination of the local action on the exposed heart of the frog in *toxic* amounts, when he found, after placing a few drops on that organ, that its action became speedily arrested. In a more recent article (*ibid.*, 1871, 2, p. 359) a discussion of the subject is resumed, and he theoretically explains the "excited, violent, and labored action" of the heart in this way; "when the nitrite is taken into the lungs it instantly arrests or diminishes oxidation, and a thrill of impending suffocation runs through the system, in obedience to which the respiratory and circulatory systems gather up and exert to the utmost their forces. The central impulse sent to the cardiac and respiratory muscles is at first much more than sufficient to overcome any direct action of the nitrite upon them, but, the inhalation being persisted in, the impulse is constantly growing weaker, and the direct influence of the drug stronger, so that there soon comes a time when the reverse is true, and the heart's power is more or less nearly extinguished." Dugau ("Rev. Mens. de Med. et de Chir." July, 1880; quoted in "London Med. Record", Aug. 15, 1880, and "Am. Jour. of the Med. Sci.", Oct., 1880 p. 554) concludes, from the result of an elaborate series of experiments, that the nitrite acts upon the heart after all the nerves have been severed; that strong doses cause a diastolic arrest, and that the acceleration of the heart beat seems to be in relation with the fall of arterial tension, *for it occurred after all nervous connection of the heart with the central nervous system was severed.* Brunton ("Med. Times and Gaz.", 1870, i, p. 320), found that the nitrite did not diminish the work done by the heart in a given time, although it decidedly increased the frequency of its contractions. The frequency he is inclined to attribute to a depression of the inhibitory nerves, because it was more apparent in dogs,in which the vagal apparatus is more sensitive than in rabbits ("Jour. of Anat. and Physiol.", v, 1871, p. 92); but he did not make any experiments to specially determine how the pulse was affected, and his deductions, as may be inferred, are based upon theoretical grounds alone. In experiments on the blood pressure he found the same decided diminution as is noted by other observers, but in seven experiments in which the aorta was compressed immediately below the diaphragm the pressure was not reduced as in normal animals, for *a primary rise occurred which equalled about one fifth the normal*, and which preceded a marked fall. Moreover, in three other experiments, in which the spinal cord was cut and the aorta compressed as before, a similar result was recorded. He concludes that the diminution of the arterial tension was vaso-motor and not cardiac, but, unfortunately, he does not discuss the cause of the increase in animals with the aorta compressed.

In reviewing the results above recorded, and the deductions made therefrom by the different investigators, we find, indeed, quite a diversity of opinion to deal with. If we first consider the theories of the *modus operandi* of the action on the pulse, much in the papers of Filenhe, Mayer and Friedrich, and Brunton goes to show that the nitrite does not depress the vagal apparatus and increase the pulse in this way; yet, evidently, this does not cover the ground, according to Pick's opinion, nor those of Wood and Dugau. Although experimenters almost universally agree that the heart is ultimately slowed and paralyzed, yet, as we see, they do not agree as to how the frequency is affected, and when we come to examine their papers it is found that Filenhe, Mayer and Friedrich, and Dugau were the only ones who made experiments in direct determination of this point, and the deductions of the first three observers must be accepted with allowance

because of the very indirect, but no less ingenious, way in which they sought to decide this action. Yet it seems we must admit that the vagi apparatus is depresed, and that the increased action of the heart is at least partially in this way effected. Wood's theory is indeed a beautiful and masterly one, and well worthy of my illustrious friend and teacher, but, unfortunately, it has been founded rather upon hypotheses than direct experimentation. Pick's is logical enough, for it is a truism that there is a compensatory relation existing between the action of the heart and the condition of the vaso-motor system, and that when the vascular channels are open the heart will naturally beat faster in endeavoring to overcome the excessive drainage, and *vice versa*. Dugau's view closely coincides with this, but where he was misled was in not isolating the action on the vaso-motor system from that on the heart, as we shall presently see.

Regarding the action on the arterial pressure, it is almost universally conceded that the marked diminution is both vaso-motor and cardiac, the latter factor coming in late in the poisoning, as is believed by both Wood and Brunton. If this is so, it must be certain that the early fall of pressure is entirely vaso-motor.

If, now, it were possible to eliminate the action of the nitrite on the heart from its action on the vaso-motor system, we could readily determine just how far the acceleration of the heart's beat was affected by this diminution of pressure occurring from vaso-motor dilatation, as well as determine how the amount of work done by the heart is influenced. This identical thing was done, although imperfectly, in Brunton's seven experiments when he pressed the abdominal aorta immediately below the diaphragm, and in the three later ones, in which he compressed the aorta and made section of the spinal cord; for it must be certain that, when the aorta was firmly pressed upon, the action of a large portion of the vaso-motor peripheries was practically paralyzed, because they were unable any longer to affect the arterial tension, and this was even further increased in the animals in which a section of the cord was made, which, of course, practically annihilated the vaso-motor centers in the medulla oblongata. Now, it must be equally as clear that, if we thus have the action of the vaso-motor system practically abolished, and the disturbing influences of the respirations or struggles overcome by curarizing the animals, as was also done, any change in the arterial tension which occurs will be the resultant of a direct cardiac action. What was the result in each of these ten experiments? In every one of them there was a primary and marked rise of pressure, which equaled as much as a fifth of the normal. It, therefore, must be apparent that this curious change in the results must be due to a direct cardiac action and an effect of a direct increase of the heart's power. If we now admit Dugau's evidence that the heart's action is increased after severance of all central nervous communication, it is conclusively proved that the increase of arterial tension is due to a direct stimulation of the heart, increasing both its frequency and the amount of work performed. Why Brunton, Pick, and Dugau did not reach a similar conclusion is, it is obvious, simply because they did not eliminate the vaso-motor from the cardiac action. It is also clear from the foregoing that the acceleration of the pulse must also in part be cardiac.

In a paper on the potassium nitrite, previously referred to, I claimed that the nitrites all affected the economy in a similar manner, and that, although it was found that the potassium salt caused a primary rise of pressure preceding the diminution, which was an entirely different result from that obtained on normal animals with the amyl nitrite, yet briefly it was explained in this way—that the nitrites acted in two ways to affect the blood pressure, the one by directly stimulating the heart, and the second by depressing the vaso-motor system, the centers especially; that the stimulating effects on the heart ultimately gave way to depression; and that the reason why the amyl nitrite did not cause a rise of pressure, like the potassium salt, was because its vaso-motor action was comparatively more intense, and was more than sufficient to overcome the stimulating effect on the heart, which was finally indicated to be a fact by a comparison of the results of Brunton's experiments and my own. It was further clearly demonstrated that the potassium nitrite caused at least a part of the increased frequency of the pulse by a direct stimulation of the heart, and probably by a consentaneous depression of the inhibitory centers. If, therefore, we accept these deductions as made from the results of Brunton's and Dugau's experiments, it will be obvious that the action of the amyl nitrite is identical with that of the alkaline salts, and, consequently, that it is a direct cardiac stimulant, increasing the frequency of the heart's action and the amount of work done in a given time.

If we now look into the clinical history of this interesting compound, we find that there is no paucity of evidence to support this belief, but, indeed, that cases illustrating it are so numerous that, even without the experimental evidence as above deduced, we could arrive at no other conclusion. For instance: 1, there can be no possible doubt that chloroform is a direct and powerful depressant of the heart, and that in toxic amounts it seriously diminishes the powers of this organ and causes death in a vast majority of cases by this paretic action; 2, that it is a conceded fact that in collapse and syncope the depressed condition of the heart is a marked symptom; 3, that in certain forms of heart disease, whether there be a condition of fatty degeneration, valvular insufficiency without compensating hypertrophy, or simple dilatation, the heart's powers are enfeebled and we have the occurrence from time to time of paroxysms of distress, which are undoubtedly due to a further depression of its working capacity. It is, therefore, evident that in each of the above-mentioned three divisions we have an unmistakable condition of cardiac depression, which, in the last of them, is frequently and decidedly evinced in the paroxysmal attacks which recur from time to time; and it must be apparent that the administration of cardiac depressants in such conditions of the system would be attended with harm, if it did not materially aggravate the symptoms or induce death—for who would think of giving aconite or tartar emetic in cardiac syncope, or when a patient was cold, cyanosed, and gasping for life's breath in a paroxysm of heart pang?* Did the amyl have no further primary effect on the heart than to simply increase its frequency by a depression of the vagi apparatus or by inducing it through a diminuition of the arterial tension and without affecting the amount of work done, it would seem certain that in such cases as above quoted either its value as a therapeutic agent would be *nil*, or else it would even aggravate the symptoms, since its only admitted action on the heart, besides the increased frequency, is a depression of its powers and ultimate diminution of its pulsations.

Now, what does the clinical history teach us of the action of the nitrite in chloroform poisoning? Browne ("Med. Gaz.," June 11, 1870) believes that it seems worthy of a trial in chloroform narcosis because its use is followed by a marked acceleration of the heart and flushing of the face. Boder ("Lancet," May, 1875; quoted in "Proc. of the Med. Soc. of the County of Kings," Brooklyn, N. Y., 1875) gives three cases in which dangerous and alarming symptoms of the effects of chloroform were instantly and effectually overcome. Hinton ("Phila. Med. Times," v, 1875, p. 694) cites another illustrating its stimulant effects; and Burrall ("Lancet," May, 1875), Schuller ("Med. Times and Gaz.," Dec, 12, 1874), and Lane ("Brit. Med. Jour.," 1877, i, p. 101) reiterate this statement. The latter observer concludes that when it is inhaled in small quanties it produces recovery from chloroform insensibility, and that this is due to a removal of the cerebral anæmia by the dilatation of the

arterioles, to the raising of the temperature of the body, and the removal of the paralysis of the heart. Solger ("Wien. med. Presse," Feb., 1875) furnishes further evidence of the antagonism of the action of the amyl and chloroform; and Dabney ("Richmond and Louisville Med. Jour.," June, 1874) and Smart ("Detroit Rev. of Med.," x, 1875, p, 661), in experiments on animals, further testify to its efficiency. Dabney states that three animals out of four in chloroform insensibility recovered when death was apparently imminent, and that it produced a decided increase of both the force and frequency of the heart. Further, a very interesting experiment of my own, showing the powerful stimulant effects of the nitrites on the heart, is given in a recent paper on the Ethylene Bichloride ("Phila. Med. Times," May 21, 1880,), in which the heart was so seriously depressed that the pulsations were imperceptible to the touch, and the pulse-curve tracing was but a mere streak, no oscillations being apparent; yet after giving the amyl the heart's action was recovered with astonishing rapidity. *

Jones ("Practitioner," vii, 1871, p. 213) gives a case of syncope in which the radial pulse was scarcely to be felt, and when the nitrite was given it caused it to beat rapidly and full. From the rapidity with which the patient recovered he concludes that the syncopal condition would have continued longer but for the ether. Minor ("Virginia Med. Monthly," iv, 1878, 1876), used it in a case of locomotor ataxia when the patient was in a condition of impending death, there being a hippocratic expression, stertorous and irregular breathing, complete unconsciousness, an imperceptible cardiac impulse, and cold, clammy extremeties. The amyl was given by inhalation, but with no effect, and it was therefore repeated hypodermically, three minims being used; and in a few moments "the heart responded, as evinced by the appearance of a more natural hue of the cutaneous surface. The pulse was recog-

---

* An apparent exception to this is the use of hydrocyanic acid, but this is given in such small amounts that only its primary effect on the circulation is obtained, which is a slowing and increased fullness of the pulse, due to a stimulation of the inhibitory nerves, and accompanied by a slight rise in pressure, the depression of the heart and fall of arterial tension being a secondary effect.

These last references, which properly belong to the physiological papers, are given here because of their bearing such a close relationship to the clinical papers on the same subject.

nized in the radial artery. Respiration became better." The temperature was restored, and the patient responded to a pinch, which before was unnoticed. By repeating the dose the pulse became incompressible. As much as fifteen minims was given hypodermically at once. Another exceedingly interesting case is cited by Madden ("Practitioner," xii, 1874, p. 295) of a woman who suffered with severe and intractable menorrhagia, and who at the time referred to in his article was apparently dead. Upon using the amyl nitrite the respirations and circulation became fairly established and consciousness was subsequently restored. Moreover, whenever she appeared to be falling into this collapsed condition the amyl was resorted to with the happiest results; hence, he thinks that this case "shows in a remarkable manner the power which the nitrite possesses to rouse the heart which has almost ceased to beat."

Two cases of fatty heart serve to further illustrate this stimulant action and the beneficial results, as a consequence, from its use. The first is that of Osgood ("Am. Jour. of the Med. Sci.," 1871, 2, p. 360), who found it to act with remarkable efficacy, and so much so that he believes that death would have ensued in a few moments if it had not been given. The second case is one of Janeway's (N. Y. Med. Jour.," xx, 1874, p. 58), in which, after the complete extinction of the radial pulse, the nitrite caused it to become full. There are also two cases of valvular disease. Wood ("Am. Jour. of the Med. Sci.," 1871, 2, p. 361) states that in his case the effect of the drug in relieving the pang when other remedies failed was astonishing; and Jones (*loc. cit.*) tells us of a woman who suffered from cardiac dyspnœa, with a dilated hypertrophy of both ventricles (the dilatation being in excess), a feeble circulation, extreme anasarca, pulmonary œdema, lividity of the face, and shortness of breath. The use of the amyl nitrite produced most beneficial effects and induced a pleasant glow all over the body, and on each occasion of its use she felt easier

and more comfortable about the chest.

In reviewing all the foregoing facts, I think doubt can no longer exist that the amyl nitrite, like the alkaline salts, acts as a powerful and direct cardiac stimulant, increasing both the frequency of the heart's pulsations and the amount of work done in a given time.—*N. Y. Medical Journal.*

106 Halsey St., Newark, N. J., June, 1881.

Purulent, Croupous or Membranous, and Diphtheritic Conjunctivitis in Infants.*

BY CHARLES STEDMAN BULL, A. M., M. D.

Ophthalmia Neonatorum, or purulent conjunctivitis of infants has from early times been recognized as the typical form of suppurative inflammation of the conjunctiva. Owing to the well-known symptoms of the disease, the course which it runs, and the dangers which it entails upon the eye, comparatively little discussion has arisen upon the subject except in regard to two points—the causation of the disease and its prophylaxis. The same cannot be said for either membranous or diphtheritic conjunctivitis, though certainly in the latter disease the symptoms are as pronounced and unmistakable as in the purulent form of inflammation. The apparent connection which exists in some cases, clinically, between the purulent and membranous forms of conjunctivitis in new-born children, is much closer and more real than any which is claimed to exist between the membranous and diphtheritic form of inflammation. Most ophthalmic surgeons occasionally see cases of purulent conjunctivitis in which the discharge at a certain period becomes more coagulable and assumes the form of a membrane; and this process is certainly at first distinct from the conjunctivitis which begins as membranous and subsequently becomes purulent.

*Galezowski.—"Ann. de Gynec.," March, 1881.
Abegg.—"Arch. f. Gynak.," xvii, 3, 1881.
Crede.—*Ibid.*, xvii, 1, 1881.
Olshausen.—"Centralbl. f. Gynak.," January 22, 1881.
Graefe.—Volkmann's "Samml. klin. Vortr.," No. 192.

PURULENT CONJUNCTIVITIS.

True purulent conjunctivitis begins, probably in the majority of cases, as a catarrhal inflammation, with mucoid or muco-purulent secretion; but the latter rapidly becomes thicker, more cloudy, yellow, and really purulent. Sometimes the change from catarrhal to purulent inflammation is so rapid that the former stage escapes observation, especially if the physician is not on the lookout for ophthalmia. The disease, according to nearly all authorities, begins on the second or third day after birth, though it may appear on the first day, or may be postponed to the fifth. If it does not appear until after the fifth day, the suspicion is aroused that the contagion has occurred since birth, and from some other cause than inoculation with the vaginal discharge of the mother during parturition.

The symptoms of a well-marked case are easily enumerated: Swelling of the tissue of the eye-lids; redness of the cutaneous surface, sometimes amounting to lividness; more or less sticking of the lid-margins together; the appearance of a purulent discharge when the lids are opened; marked injection of the ocular conjunctiva, accompanied generally by some chemosis; sometimes the lids are so swollen as to prevent their eversion, but, where they can be turned out, the retro-tarsal fold is found enormously swollen, as is also the papillary portion of the tarsal conjunctiva, and covered by a yellow exudation, which may be more or less flocculent or even stringy. There is great heat of skin in the immediate vicinity, and sometimes a slight rise in the general temperature. This condition lasts a varying length of time, usually from four to six days, and then the acute inflammatory symptoms begin to subside, though the purulent discharge may be very profuse for a much longer period. Both eyes are almost always affected, though usually one before the other.

Though in the great majority of cases the disease runs the course just depicted, yet there are exceptions. In some instances the ropy, fibrinous exudation which is noticed in the beginning continues to the end, and is accompanied by patches of coagulation upon the conjunctiva itself, which may leave white spots behind them, though this is very exceptional. In other, not very rare, cases, the flocculent, fibrinous material is deposited continuously over the whole tarsal surface of the conjunctiva, and resembles a membrane very closely. It is, however, friable, and, though easily removed by the forceps, comes away in small bits or shreds, and rarely in one continuous piece. It always leaves a raw, bleeding surface, and the hæmorrhage is sometimes quite profuse. This may not form again, though it usually does. With this pseudo-membranous formation there is also an abundant purulent discharge, which lasts long after the pseudo-membrane has ceased to be formed. These cases resemble closely the cases described by some of the German authors as croupous conjunctivitis, though it is proper to state that the membrane in the latter is usually coextensive with the palpebral conjunctiva, and moderately thick. They seem to be cases of the disease in which the products of inflammation are of a more highly organized type.

Purulent conjunctivitis of infants, though comparatively rare in the higher classes, is of very frequent occurrence among the lower orders, and is especially rife in hospitals for the confinement of women. Crede's statistics cover a period of seven years—from 1874 to 1880, inclusive; there were, in this period, 2,466 births and 227 cases of purulent ophthalmia, but the percentage varied somewhat from year to year, and during the last year became very small. In 1874 there were 323 births and 45 cases of conjunctivitis, or 13.6 per cent. In 1875 there were 287 births and 37 cases, or 12.9 per cent. In 1876 there were 367 births and 29 cases, or 9.1 per cent. In 1877 there were 360 births and 30 cases, or 8.3 per cent. In 1878 there were 353 births and 35 cases, or 9.8 per cent. In 1879 there were 389 births and 36 cases, or 9.2 per cent. In 1880, up to May 31st, there were 187 births and 14 cases, or 7.6 per cent., but for the remainder of the year there were 200 births and 1 case, or only 0.5 per cent. The reason for this marked falling off in the number of cases will be stated farther on. Olshausen's percentage of cases is larger than Crede's later percentage, for out of 550 births there were 69 cases of purulent conjunctivitis, or more than 12.5 per cent. Galezowski, since 1870, has observed 507 cases of purulent ophthalmia out of a total of 60,152 cases of disease of the eyes, and among these there were 111 cases of serious results, ending either in permanent diminution or total loss of vision.

As regards the causation of the disease, there seems to be but one opinion among recent writers, whether from the ophthalmological or from the obstetrical standpoint, viz.: that the ophthalmia of new-born children is in the great majority of cases due to inoculation during parturition with the muco-purulent or purulent discharge from the vagina of the mother. The appearance of the disease so soon after birth and the almost constant occurrence of a vaginal discharge in the mothers both point in this direction. The latter, however, cannot be regarded as the only cause, for cases now and then occur of purulent conjunctivitis where the mothers have had no discharge from the vagina. This vaginal secretion need not be purulent, for an ordinary vaginal catarrh or the lochial discharge has been known to cause many cases of purulent conjunctivitis in infants.

The disease may be, and often is, communicated by carelessness and uncleanliness on the part of nurses, mothers, and other attendants, often in the washing of the children after birth or in handling them subsequently, by carrying the vaginal discharge upon the fingers or upon the bedding or clothing. The danger from this carelessness and uncleanliness on the part of attendants helps to explain the frequent epidemic appearance of ophthalmia neonatorum in lying-in hospitals and foundling asylums, though another important factor in the causation of these epidemics is the influence of the badly-ventilated rooms and wards in these buildings, where, perhaps, similar epidemics have pre-

viously occurred, thus poisoning air, walls, and floors, as well as the contents of these rooms and wards. Still, cases do occur in which no contagion can be traced, and here it has been customary to refer the outbreak of the disease to a sudden draught of air, maintenance in an impure atmosphere, or sudden exposure to very bright light; which influences, acting upon an already existing catarrhal conjunctivitis, may change the latter to a purulent form of conjunctivitis.

Crede, who writes of the disease from the obstetrician's standpoint, states that in his experience the cases of ophthalmia in new-born children have been, almost without an exception, caused by direct contact of the vaginal secretion with the eyes during parturition. In his opinion, the infectious character of a vaginal secretion continues long after the specific gonorrhœal symptoms have disappeared. Even in cases where almost no secretion from the vagina of the mother exists, purulent ophthalmia in the child has still been known to follow within the first few days after birth. Galezowski, who writes from the ophthalmic surgeon's standpoint, says that it is always caused by the introduction of the leucorrhœal or gonorrhœal vaginal secretion between the eyelids of the infant during parturition, and Abegg agrees with him.

Recognizing this fact in the etiology of the purulent conjunctivitis of infants, our main endeavors should be directed toward preventing the occurrence of the disease. The practical question is one of prophylaxis, and to this end the care of the disease must be placed in the hands of the obstetrician and those of the nurse, and on them must rest the responsibility of the result. When it is considered how many cases of permanent disability from blindness, which fill our blind asylums, are due to this cause, it will readily be seen how grave is this responsibility, and how very necessary it is, not only that the disease should be properly treated, but that the treatment should be prophylactic. As a precautionary measure, the vagina of the mother should be kept thoroughly cleansed for some days before confinement, though the uncertainty of the occurrence of the latter renders the duration of the former equally uncertain.

Crede employed this method of cleansing and disinfecting the vagina, but the effects upon the occurence of ophthalmia were trifling and unsatisfactory : the cases of conjunctivitis diminished, but did not disappear. He then began to disinfect the babies' eyes, and the result was at once surprisingly favorable. In all cases of vaginal gonorrhœa and catarrh in the institution under his care frequent disinfection was practiced with weak carbolic and salicylic-acid solutions (2 : 100). The first trials on the babies' eyes were made with a 1:60 solution of boracic acid. This not proving satisfactory, he began to use a 1:40 solution of nitrate of silver, injected between the eyelids shortly after birth, first washing the eyes with the weak salicylic-acid solution. All the eyes thus treated remained well. During the year ending June, 1881, the treatment was as follows : The unclosed lids were first carefully cleansed with fresh water; then a few drops of a 1:50 solution of nitrate of silver were dropped into the gently-opened lids; then for twenty-four hours applications of the salicylic-acid solution (2:100) were made to the eye-lids. All these babies remained free from the disease.

Olshausen first used as a prophylactic a one-per-cent. solution of carbolic acid as a wash for the eyes; this reduced the percentage of cases of ophthalmia from 12.5 to 6 per cent. in two years. Immediately after the birth of the child, even before the breech is born, the still closed lids of the infant are to be washed with this carbolic-acid solution, and then the eyes are cleansed with the same solution. As a result of this treatment, among 166 children born in 1880, only 6 cases of purulent ophthalmia appeared, or 3.6 per cent.; and these cases were very mild, and some of them were limited to one eye. He does not employ either the nitrate-of-silver solution or the continuous bathing with the salicylic-acid solution, as the latter demands constant attention.

In the lying-in department of the Dantzig hospital, under the charge of Dr. H. Abegg, since the precaution has been taken of carefully washing the lids and conjunctival sac immediately after birth, the results have been very satisfactory. Out of a total of 2,266 births during the ten years between 1871-80 only 66 cases of ophthalmia occurred, or 3 per cent.

The prophylactic measures recommended by the writer are as follows : In all cases of vaginal discharge in parturient women, whether specific or not, the vagina should be carefully cleansed and disinfected repeatedly before parturition begins. As soon as the child is born the external surface and edges of the eyelids should be carefully cleansed with a one or two-per-cent, solution of carbolic acid, and then the conjunctival cul-de-sac washed out with some of the same solution, or with a saturated solution of boracic acid. This must be done by the attending physician, or by a skilled nurse under his supervision. The eyes of all new-born children should be carefully watched for the first week or ten days, and whenever any signs of an ordinary catarrhal conjunctivitis appear, the conjunctiva should be thoroughly brushed over with a solution of nitrate of silver, from 2 to 5 grains to the ounce of water.

If the conjunctivitis has become purulent, and the case is one of real ophthalmia neonatorum, the child should, if possible, be isolated from all healthy infants, and have its own bath-tub. If this is not possible, the diseased infant should be bathed *last*, and no sponges should be used, but only cloths, which can afterward be destroyed. If one eye only is affected, do not apply the hermetically-sealed bandage to the sound eye, but envelope the arms or hands of the baby, so as to prevent the secretion from being carried to the fellow-eye, and lay the child upon the side corresponding to the diseased eye.

The most important feature in the treatment is enforced cleanliness. This requires constant attention and the frequent use of some soft cloths and plenty of water. The use of cold cloths, dipped in cold water or even iced water, and laid on the eyelids, must be regulated by the amount of swelling of the lids and heat of the parts. As soon as the lids can be everted, the proper treatment is a thorough application of nitrate of silver to the conjunctiva of the lid and retrotarsal fold, daily, and sometimes twice a day. If this is thoroughly done, a five-grain solution will in most cases suffice; but, where there are profuse secretion and considerable swelling of the conjunctiva, a ten-grain solution becomes necessary. When, owing to marked hypertrophy of the papillary structure of the conjunctiva, a stronger caustic becomes necessary, it is better to discard solutions, and employ the lapis mitigatus (one part nitrate of silver to two parts nitrate of potassium), and neutralize its effect by a subsequent washing with a solution of common salt.

It is well to employ a one-grain solution of sulphate of atropia in a saturated solution of boracic acid in every case of purulent ophthalmia, as the great danger in this disease is purulent infiltration and perforation of the cornea. Should this infiltration occur at the center of the cornea, the atropia should be instilled frequently, for, if perforation occurs, the dilatation of the pupil will prevent a large prolapse of the iris through the perforation. If the infiltration of the cornea, on the contrary, be at or near the margin, it is better to employ a two-grain solution of the sulphate of eserine, as thus an extensive prolapse of the iris may be prevented if the ulcer perforate. In all cases the cleansing and washing of the lids and conjunctiva should be done with a saturated solution of boracic acid, and the atropine and eserine should be dissolved in the same.

MEMBRANOUS CONJUNCTIVITIS.

Though most German authorities, and some others, regard croupous or membranous conjunctivitis as a distinct disease, differentiating it both from purulent and from diphtheritic conjunctivitis, yet the writer regards this as extremely doubtful, at least so far as the purulent form of inflammation is concerned. Saemisch defines croupous

conjunctivitis as that variety of inflammation which is characterized by the formation of a more or less extensive membrane *upon* the surface of the conjunctiva of the lids. The intensity of the inflammatory process varies in different cases: in some the membrane is a very thin, perfectly transparent, thread-like gelatinous layer, while in others it is denser, thicker, opaque, and yellowish-white in color; and this may sometimes be removed in one entire membrane from the *surface* of the conjunctiva. When the membrane is of the latter character, it adheres with tolerable firmness to the conjunctiva, and can not be easily wiped off, but must be removed with the forceps, and always leaves a bleeding surface beneath it. The gelatinous layer of exudation, on the other hand, is easily removed with a small brush or bit of muslin, and, if this is done carefully, does not leave a bleeding surface beneath it. The lids are reddened and swollen as in the purulent form of the disease, and there is, moreover, a more or less abundant flocculent secretion, which may be purulent from the beginning. Saemisch says that the disease subsequently runs into the catarrhal or purulent form of inflammation, but it is certain that in the large majority of cases the flocculent, more or less purulent secretion is present from the beginning.

From the further statement of Saemisch, that in a small number of cases the croupous form merges into the diphtheritic, the writer ventures to decidedly dissent, holding that the croupous and diphtheritic forms of conjunctivitis are two distinct diseases, differing in their pathology and pathogenesis. The pathological process in the croupous form of conjunctivitis consists mainly in the deposit of an albuminous exudation, probably fibrine, *upon the surface* of the inflamed conjunctiva, which deposit rapidly coagulates on exposure to the air, and thus assumes the form of a membrane. This deposit contains cells which have made their exit from the mucous membrane, and upon the number of these cells depends more or less the firmness and density of this membrane, as may be shown by a microscopic examination. When this membrane is removed, either as a slough by the processess of nature, or by the forceps in the hand of the surgeon, the mucous membrane lies bare, deprived of its epithelium throughout a varying extent; but there is *no loss of substance* of the conjunctiva, and but *very little infiltration* of its tissue; and here is the point at which croupous inflammation differs from diphtheritic inflammation, as will be shown farther on.

When the membrane has been removed, it may form again, or the exudation may assume the purulent form, and then we have a purulent ophthalmia to deal with, and the epithelial layer is reformed. The writer is strongly inclined to believe, though still with some hesitation, that these cases of conjunctivitis in which the formation of a pseudo-membrane is the characteristic feature are cases in which the conjunctival inflammation is of the same nature as the purulent form, but in which the exudation is for a time of a higher organization. Even those authors who insist most strenuously that membranous conjunctivitis should be regarded as a distinct disease, admit that in very many cases of purulent conjunctivitis the exudation at first coagulates very rapidly on exposure to the air, and thus forms a pseudo-memhrane over the surface of the conjunctiva. The tendency to this coagulation of the exudation is very often seen in cases of catarrhal conjunctivitis, though a continuous pseudo-membrane, covering the entire surface, is never formed in these cases.

The writer has recently had an opportunity of noting the occasional marked tendency to the formation of a pseudo-membrane in purulent conjunctivitis. During the months of January, February, and March of this year there occurred quite a number of cases of purulent ophthalmia in infants at tne Nursery and Child's Hospital in this city. Out of a total of 27 cases of purulent conjunctivitis the tendency to the formation of this pseudo-membrane was noted in 13, and in several of these the membrane covered the entire palpebral conjunctiva. In only two

cases was the membrane of any degree of thickness, but these two required the forceps for its removal. In no other respect did they differ from the ordinary type of purulent conjunctivitis. Several of these pseudo-membranous formations were examined microscopically by the writer. Running in every direction through the tissue were delicate fibrillæ of connective tissue, which in some places were collected together into larger fibrillæ. There were large numbers of round cells, so-called exudation corpuscles, and a large amount of fine, amorphous, granular matter. The meshes were generally small and close, though in places they were quite loose. In some places the fibrillæ were so close together as to appear to form laminæ, and these apparent laminæ were arranged in arcs of concentric circles, though they never formed complete circles. The fibrillæ were more numerous on the external surface of the pseudo-membrane, showing there a higher grade of organization.

The treatment of these cases of membranous formation does not differ from that of the ordinary form of purulent conjunctivitis. It must be strictly antiphlogistic and disinfectant, by the application of cold cloths and the use of a saturated solution of boracic acid or of a one per cent. solution of carbolic acid for cleansing purposes. As fast as the discharge appears it must be removed, either with a small brush or with a priece of soft muslin. If the pseudo-membrane be of the thin, gelatinous variety, it may easily be removed; but, if it be dense and firm, no attempt at removal should be made, as it is better to wait until it is cast off as a slough. During the early stage no caustics should be employed, but, when the purulent secretion has become established, the same remedies are indicated as in the purulent form of the disease. The indications for the use of atropine or eserine are the same as before mentioned. Attempts have been made to cut short the tendency to the pseudo-membranous formation by the insufflation of powdered quinine sulphate, but hitherto without much effect, and the application is often quite painful.

DIPHTHERITIC CONJUNCTIVITIS.

The rarest form of conjunctival inflammation, as it is the most dangerous to the eye in its destructive tendencies, is the diphtheritic. This disease, though not uncommon in Berlin and some portions of northeastern Germany, is relatively rare in other parts of the Continent of Europe, and absolutely rare in Great Britain and the United States. During a period of ten years' connection with the New York Eye Infirmary, and also with several other public hospitals, and an observation of more than twenty thousand cases of eye disease, the writer has seen but ten cases of true, unmistakable diphtheritic conjunctivitis, as seen at the Berlin clinics. This is a fortunate exemption from the disastrous consequences of a rapidly destructive disease.

Diphtheritic conjunctivitis is characterized by a very marked swelling of the lids, due to a more or less *extensive infiltration*, not only of the *entire conjunctiva* but also of the *other tissues of the lid,* sometimes even including the *integument,* by an inflammatory product of marked coagulability. This infiltration is *into the tissue* of the conjunctiva, and not an exudation *upon its surface.* The local heat of the parts is very pronounced. The infiltration into the lids is so great that they become almost like a board, look and feel like brawn, and can not be everted. This dense infiltration often drives all the blood out of the eyelids, and instead of presenting a livid appearance, as in the purulent form of conjunctivitis, they appear dusky yellow, and even blanched. As a consequence of this extensive and rapidly occurring infiltration, the nutrition of the parts is interfered with or entirely cut off, and the conjunctiva changed into a necrotic mass and cast off as a slough. This necrosis sometimes extends deeper than the conjunctiva, and an extensive loss of substance in the lid tissues occurs. When the strangulation is very extensive in the lid, and the skin looks angry and reddened, the dense, hard, brawn-like condition of the lids is sufficient to distinguish the case as one of diphtheritlc conjunctivitis.

Such a case can not be mistaken for purulent ophthalmia. When it is possible to evert the lid in part, the inner surface is seen to be of a gray or grayish-yellow color, usually bloodless, which appearance is due to an infiltration of the entire thickness of the conjunctiva, and not to the formation of a membrane upon its epithelial surface.

During this first stage of the disease the exudation from the conjunctiva is usually slight, and consists of a thin, dirty, very hot, ichorous fluid, containing yellowish shreds and some cells. This exudation in the second stage changes in character, and becomes puriform and finally purulent. The disease is rapid in its progress, and reaches its height usually within a few days, The ocular conjunctiva becomes densely infiltrated, and surrounds the cornea like a hard, unyielding wall. There are usually but few blood-vessels to be seen, but numerous punctate hæmorrhages are not infrequent. After the disease has lasted from five to eight days, the hard, board-like condition of the lids diminishes, small sloughs begin to appear in the conjunctiva, the latter becomes loose, red and bleeding, and assumes the appearance of a suppurating or granulating surface. The ocular conjunctiva takes on the same change; the lids can be everted, the secretion becomes purulent, and the second stage has begun. This stage lasts a varying time, differing in no respect from an ordinary purulent ophthalmia, and terminates in the third or cicatricial stage. This latter is the deeper and more extensive, the greater was the destruction by necrosis following the infiltration.

This description of the disease in its various stages we owe originally to von Graefe, who first differentiated it as a distinct affection. Any one who has seen such a case can not fail to make a diagnosis in subsequent cases. Usually the general condition of the patient affected with diphtheritic conjunctivitis is bad. The eye may be the only organ affected by the disease, though often the nose, the throat, and the ears are all involved in the process. The local manifestation of the diphtheritic poison may appear first in the conjunctiva, or it may spread from the nose or throat to the eye. In North Germany diphtheritic conjunctivitis is a not infrequent complication of malignant scarlatina. It occurs more often in children than in adults, thus resembling the croupous form of inflammation. The diphtheritic process in the conjunctiva is probably caused by the presence of lower organisms, and the chief part in the infiltration is taken by cellular elements, which are so numerous and so densely packed together as to obliterate the vessels and cut off the circulation.

From this rapid survey of the symptoms and course of a diphtheritic conjunctivitis, it will be seen that there are always necrosis of tissue, loss of substance, and cicatrization in the lid, none of which evils occur in the purulent or membranous forms of conjunctivitis. In all three stages of the disease, as in purulent and membranous ophthalmia, the great danger is ulceration and necrosis of the cornea; and the more the ocular conjunctiva is involved the greater is the danger. The nutrition of the cornea may be so rapidly interfered with that it may become entirely opaque within forty-eight hours; and when this occurs it sloughs out almost always entire, like a watch-crystal from its frame. Where the cornea is not entirely surrounded by the brawny conjunctiva, part of it can generally be saved, even in its transparency, though this is rare.

The prognosis in a case of diphtheritic conjunctivitis is almost always bad, owing to the rapid strangulation of the tissues. Not uncommonly the lower lid becomes excoriated upon its skin surface by the acridity of the discharge in the first stage, and becomes covered by a diphtheritic membrane. Fortunately, the disease is rare among new-born children, it occurs more frequently, according to German authorities, between the ages of two and seven. In this country and in Great Britain the disease appears only sporadically, but in Berlin and East Prussia epidemics have been known to occur on several occasions. It is extremely contagious, and this is now the explanation of the occurrence of epidemics, the disease being propagated by infection with the secretion.

One point in the pathology of the disease has given rise to an almost endless discussion, which in some quarters has not yet been satisfactorily settled, viz., whether the disease is ever produced by infection with the purulent secretion from a non-diphtheritic case. The establishment of this fact would militate strongly against the doctrine that diphtheritic conjunctivitis is a distinct disease. Yet there seems no good reason for doubting that diphtheritic conjunctivitis has been produced by infection with the purulent secretion of a gonorrhœal conjunctivitis, for it was frequently seen at von Graefe's clinic in Berlin and elsewhere. Yet von Graefe himself defined it as a general constitutional disease, and clearly differentiated it from purulent conjunctivitis. In view of the authenticity of the cases above referred to, we must agree with the modified statement of von Graefe that, while in many cases diphtheritic conjunctivitis is a symptom of a general disease, yet there are cases in which it is a local disorder, caused by infection with the secretion from a purulent ophthalmia. The reverse of the case has also been stated to be true, on authority of Horner, who states that, though the discharge from a case of diphtheritic conjunctivitis, when applied to a healthy conjunctiva, usually reproduces diphtheritic conjunctivitis, it does not always do so; for purulent conjunctivitis has been known to result from such infection. This would seem as if purulent and diphtheritic conjunctivitis were intimately connected. and leaves the pathogenesis of the latter still in an unsettled state.

The disease is not always binocular, and great care must therefore be taken of the unaffected eye. If the case be seen early enough, some impermeable or hermetically-sealed bandage should be applied, the best being that of Dr. Buller, of Montreal, through the glass center of which the eye may be constantly watched. Yet this is not always a protection, for, when there is a constitutional blood disorder at the bottom of the disease.

of course no external protecting bandage would be of any avail. The treatment must be in the beginning decidedly antiphlogistic. Iced compresses must be constantly applied and continually changed throughout the first or diphtheritic stage. Local bleeding by means of leeches to the temple may prove necessary, but there is some danger of a diphtheritic membrane forming upon the bites. Scarifications of the ocular conjunctiva do no good, and should not be undertaken. *No caustics should ever be employed in the first stage, as they would do positive harm.* If the cornea become involved early in the disease, there is almost no hope of saving it, for the cold applications aid in its necrosis, while it would scarcely be safe to use warm applications to the lids. As soon, however, as the violent signs of inflammation diminish, it is better to employ moist warmth in order to facilitate the slough of the necrosed conjunctival tissue. When the second or purulent stage has fairly begun, then the application of caustics is indicated under the same rules as apply to true purulent conjunctivitis. The same rules also apply to the use of atropine or eserine in the diphtheritic form as in the purulent form. As soon as the stage of cicatrization has begun, it is better to stop the use of caustics, as they do no good, and may do harm. Cleanliness in all three stages is very important, and the best means thereto is a one per cent. solution of carbolic acid, applied with a brush or gently injected under the lids.—*New York Medical Journal.*

Fever and the Cooling Bath.

BY WILLIS P. KING, M. D , SEDALIA.
Read before the Missouri State Medical Association.

I do not design, in this paper, to deal with the essential or idiopathic fevers simply, but with the morbid condition called fever, whether essential or arising from local causes or processes that are local.

If I were called upon to give a definition of fever I should answer, without any attempt at refinement, that *it is an elevation* of temperature above the normal, caused by the accumulation of heat in the body. This abnormal accumulation, caused by processes, more or less rapid, either local or general, or both local and general; and, I may add, that I believe that these destructive processes which result in the rapid evolvement of heat are always the result of disturbed function, which either directly or by reflex nervous irritation, interferes with the nutritive processes, resulting in a loss of the normal balance between supply and waste, between assimilation and retrograde metamorphosis.

Heat is always produced in the same way, i. e., by the conversion of something into something else. A fever in the body does not differ in any material respect from an ordinary fire.

Let us, for instance, fill a stove with wood, and bring in contact with this wood some substance which is already undergoing combustion. We thereby communicate the same process to the wood; a fire in the stove is the result; heat is evolved, and the body of the stove accumulates this heat, or a part of it, and becomes more or less hot. *The stove has a fever.*

Those of us who were raised on the farm and went bare-footed, can remember the time when we used to climb upon the dung-heap in the barn-yard and sometimes found it so hot that we could remain for only a moment. Change was going on there in the salts from the urine of animals; ammonia was being formed, heat was evolved, and the dung-heap being compact could not dispose of the heat, and hence accumulated it and became hot. *The dung-heap had a fever.*

The heat which we get from the sun is produced, scientists say, by changes going on in the materials of which it is composed. Electrical heat does not differ in any respect in the manner of its production from heat that is evolved in other processes. We cannot obtain it without bringing in contact substances that at once begin to undergo active change.

I have said that the heat of the human body does not differ materially in the manner of its production from the heat of an ordinary fire. In fact, we cannot, from a scientific stand point, conceive of the production of heat without a change.

Professor Dalton says that "animal heat is a phenomenon, which results from the simultaneous activity of many different processes taking place in many different organs, and dependent, undoubtedly, on different chemical changes in each one."

In the processes of supply and waste in the body, heat is constantly evolved; and, when we consider that neither process is always uniform, as we must waste more under violent exercise than in repose, and we surely assimilate more rapidly after a full meal with a healthy digestion, it is a matter of wonder that our bodies maintain such a uniform temperature under the ordinary conditions of health. Physiologists tell us that the body disposes of its heat by radiation and conduction through the medium of the surrounding atmosphere. But how does it dispose of the surplus heat which must be evolved under the stimulated exercise of all the functions? This is a question which I cannot satisfactorily answer. But that the system makes some disposal of it other than by radiation and conduction I have no doubt. There is evidently a kind of conservative force between the processes of assimilation and of retrograde tissue change, which acts as a sort of "governor," and thereby preserves a uniform temperature.

But there are departures from health when this conservative force seems to be lost. Heat is evolved at such a rate that the system cannot dispose of it with sufficient rapidity to maintain the normal temperature; the heat accumulates in the body as it did in the body of the stove, and we are said to have *fever*.

I have said that I believe this accumulation of heat to be the result of disturbed function in some organ or part of the body. Let me explain what I mean. I amputate the leg of a healthy man to-day; to-morrow there is a considerable elevation of the temperature of the body. Why? Because, 1st,

I have disturbed the circulation. I have severed blood vessels, and a collateral circulation must be established; and, 2d. I have made a wound, have divided nerves, and have brought about the conditions whereby pain is produced, and, by reflex nervous irritation, the nutrition of the body is interfered with, the normal balance between supply and waste is lost; heat is evolved more rapidly than the system can dispose of it; it accumulates on the body and fever is the result.

I strike another man on the head with a blunt instrument; I do not break the skin, and make no kind of wound, and yet a day hence he may have a greatly increased temperature. Here, through the shock given to the great nerve centre, I have disturbed, through the nervous system, the processes of supply and waste, and heat again accumulates, and we call it *fever*.

Expose a healthy person to the heat of a July sun, and let his blood become surcharged with carbonic acid and other products of retrograde change; the venous system is over full, and the liver which manufactures its products from this blood is crowded, becomes irritated, then congested, and there results the same disturbed balance in regard to the production and disposal of heat, which is rapidly accumulated in the body, and we say the man has a *bilious fever*.

Fill his blood with what we choose to term the poison of malaria, or with *materies morbi* of typhoid; and after a time, there will result rapid tissue metamorphosis; heat will be rapidly evolved and accumulated in the body, and we will have a *malarial or typhoid fever*.

It does not matter whether the result is brought about by the disturbing influences of these poisons upon the tissues or upon the nerve centres, and thence through the nervous system, or upon the elements of the blood, the result is the same. There is disturbed function, resulting in rapid evolvement of heat, and we have fever.

I might go on and enumerate all the diseases mentioned in the books, and divide and classify, and say this is the essential or idiopathic, and that is inflammatory and local, and we would find the same *disturbed function* and loss of the conservative force in nutrition and waste with evolvement and accumulation of heat in them all. We find fever everywhere. It is the one morbid condition for which we always look and do not look in vain in every case of serious sickness, and it is with this element, this morbid condition, with which I now propose to deal.

I may assert some things right here in regard to the element of accumulated heat which some of you are, perhaps, not prepared to believe; and yet I think I am prepared to prove clearly und establish fairly, every assertion that I shall make.

I assert, 1st. That the element of accumulated heat, the morbid condition called fever, is a greater force in the destruction of human life than all the forces with which we have to deal. Let me make that plainer and stronger: If we could dispose of the accumulated heat in diseased conditions, so as to keep the temperature of the body constantly at the normal standard, we could lower the death rate (except in injuries and extreme old age) *at least ninety per cent.*

2d. That most of the alarming symptoms and conditions which we encounter in the treatment of disease are due to the influence of the accumulated heat upon the brain and tissues of the body. In other words, our patient is sick with a malady, (which may be local or general) which need not necessarily destroy life, but from the peculiar nature of the disease, or the importance of the function of the organ involved, heat is rapidly accumulated; and from the deleterious influence of the constantly maintained high temperature upon the organism life is destroyed. The man is sick with a disease which need not kill, but something grows out of it which does kill. The fact is that most people who die (except from old age and injuries) burn to death, and there is really as much need for fire insurance upon the body as upon the house we live in.

Among the alarming symptoms and conditions that are due to the element of accumulated heat, I will mention the following:

Delirium, coma, spasms in children, morbid vigilance, rapid respiration, rapid heart action, intra cranial effusion in "summer complaint," and other diseases of children; retrocession of measles, general pain, necremia or crisping of the red-blood corpusles, thereby destroying their capacity as carriers of oxygen.

Now this is certainly a formidable array of serious conditions, all of which mark the way toward death; and if I succeed in establishing the fact that they occur simply as a result of long continued high temperature, and that we can ordinarily dispose of this accumulated heat as readily as we can extinguish the fire in the stove, then I shall hope that I have done mankind some service.

There is scarcely an accident or condition which stands out prominently as a sort of local guide board on the way to death which cannot, upon proper reasoning, be ascribed to the fever.

Just try to think for a moment when it was that you lost a patient that did not die with a high fever, or of a severe typhoid or typho-malarial in which you did not have delirium, morbid vigilance or some of the other conditions above enumerated. We all know that little children often die in spasms, and this is especially true In the fatal "summer diarrhea of teething infants." Now for fear that I may forget it, let me assert right here that except in epilepsy, hydrocephalus and tubercular meningitis, children do not have convulsions when the temperature is near the normal. With the exceptions noted, you will always find that your little patient has a temperature ranging from 103 ° to 106 ° Fah., and that when you withdraw the excess of temperature, this accumulated heat, it cannot have a convulsion. Mind, I say, *cannot*, and I know that what I say is true; for I have verified it an hundred times. Before death your patient's respiration is nearly always rapid, and his pulse accelerated; and notwithstanding the rapid respiration the face is cyanosed, showing that the red-blood corpuscles do not receive the oxygen, and that the blood does not give up its carbonic acid. The red corpucles have, from the influence of the long continued high temperature upon them, become crisped and crenated, and have lost their power to bear oxygen to the tissues, and your patient dies of *asphyxia.* Let me say right here, and I wish to emphasize it, that we should not, under any circumstances, permit the temperature to remain for a long period at an abnormally high elevation. We should, by all means, withdraw the excess of accumulated heat, and, with its withdrawal, get rid of the convulsion, the delirium, the coma, restore the normal action of the heart and the respiratory apparatus, and prevent necremia and intra-cranial effusion and other serious consequences of this constantly maintained high temperature.

I think I hear a brother say "Oh, yes, I always do that; I bring it down with tincture of gelseminum." *No you don't.* Another says, "I always knock it with aconite;" and if he is much wedded to this article he will pronounce it *"aconeet,"* like the homœopaths. *No you don't, either.* Another controls the heart with Norwood's tincture of veratrum viride, and thereby controls the fever and everything else. He forgets that the rapid heart action is caused by the high temperature, and not the contrary. Another uses salicylic acid; and all use quinine in large or small doses. Now all these remedies have their value in certain cases. Quinine certainly does, whenever there is malaria in the case; but they all fail, in many cases, in accomplishing what is attempted with them—the bringing down of the temperature. They certainly do not expel the accumulated heat, and in order to act as febrifuges at all, they must restore the equipoise of supply and waste and prevent the production of an abnormal quantity of heat, and we all know that in many, very many cases this is utterly out of the question. We have conditions which we know will continue for some time to disturb the normal relations of the heat producing and heat disposing processes; and we must have an abnormal accumulation of heat. Then we must trust to these things which we know are powerless to do what we want done, and stand by

and see our patient burned to death and utterly destroyed by heat, or we must do something which we know will dispose of the abnormal heat, and hence save the brain and tissues from all the dire consequences that are to follow. There is but one remedy that I know of that will do this effectually, and that is water. But great ignorance has heretofore prevailed, even in our profession, as to how to bathe and as to what was to be accomplished by bathing.

A wise professor (and he was wise in his day) taught me to take a child in convulsions and put it in water as hot as it could bear, and after a few minutes add a little more hot water. This same practice prevails largely to this day.

Let us take a child with a "summer diarrhea," a capillary bronchitis, or any disease in which heat accumulates rapidly, and put it into water "as hot as it can bear." We will assume that the temperature of the child is 105° Fah. Now how hot will it bear the water? It will certainly bear it as hot as 105°, its own temperature, and even hotter. Now when we bring the overheated body of this convulsed child in contact with a body of water of its own temperature what do we accomplish. *Simply nothing at all.* It is nonsense to talk about "relaxing" this rigid infant until you withdraw the abnormally accumulated heat, and you cannot do this in any other way than by conduction, and you must have the medium through which you wish to do the conducting cooler than the body of the child. If you do not there will be no conduction and hence no results. I followed this foolish practice nearly ten years, and have time and again taken the struggling sufferers out of the bath as badly convulsed as when I put them in, and wondered why it was that they would not "relax."

But let us continue our hypothetical case. We take a child two years old. It can walk and talk, and is cutting its molars. We will assume that it has a colitis, or the "summer complaint." It grows progressively worse from day to day, as there is a further invasion of surface, and function is more and more disturbed, and heat is evolved and accumulated in the body. The temperature rises to 105 ° or 106 °, and the child is suddenly seized with a convulsion. We are sent for, and find the following condition of things:

The child is insensible and in a semi-comatose state, its eyes are half open and the balls are rolled upward and backward, the temperature is 105 °, the pulse 160, the respiration sixty-five, the feet and hands are cold, and the child cannot swallow. Now, I assume that this whole group of serious symptoms is the result of accumulated heat in the body of the child, and I'll prove it. We at once place the child in a bath of 90 ° Fah., cover its head with a cloth wrung out of moderately cold water, and then gradually cool the bath down to 80 ° by adding cold water to the bath. Now note what takes place. If we do *not* disturb the bath by the addition of cold water we will find that the temperature of the bath will soon rise to 92 ° or 93 °, showing that the child is already giving up its surplus heat. After a few minutes the child's eyes close, its head drops to one side, and it passes into a sweet slumber. Keep it in the bath twenty-five or thirty minutes, or until its temperature falls to 100 ° or 99 °, and if it does not continue to sleep, it will open its eyes, recognize its mother, and will, perhaps, ask for water and *will swallow it.* The pulse falls to 120 or 110, the respiration to thirty or twenty-five. I have seen the above changes take place so often in the short space of thirty minutes, that I think I have drawn the picture fairly, and have not exaggerated a single point. Now, what does it prove? If it proves anything at all, it proves this, that to the accumulated heat and its deleterious influence upon the brain, and through the brain upon the organism, we are to ascribe the serious symptoms in our case; and further, that when we imitate nature and get rid of this heat by conduction, we adopt a course which not only gives the greatest results in the shortest time, but it is a course that is sustained by the very best philosophy.

But suppose that we do not withdraw the heat by conduction, and rely upon the remedies that are usually given in such cases, and the child grows progressively worse until we finally have symptoms indicating that there is effusion within the cranium. Then we tell the family that *the disease has gone to the brain.* Gentlemen, I can bear a little metastasis, if you will confine it to the skin and mucous membranes, but when a doctor talks to me about a metastasis from the bowels to the brain it simply makes me sick ! How gone to the brain? What connection is there between a colitis and cerebral effusion? None whatever. There is no connection in structure, locality or pathological conditions. But I can see that effusion may be caused by the influence of long continued high temperature upon the brain and its meninges.

Let us take a case of measles, where the catarrhal condition of the throat is worse than usual, (and the throat is about the only thing that is worthy of the doctor's attention in a case of measles). From the local throat trouble the temperature rises, and all at once it is discovered that there is retrocession. With the retrocession we frequently have spasms, rapid respiration and rapid heart action, and, in fact, all the serious symptoms described in the other case. Withdraw the accumulated heat, and with the reaction that follows the bath the efflorescence will appear, and there will also be a subsidence of the whole train of serious symptoms. It is singular that the profession has always ascribed the spasm, coma, rapid pulse and respiration, etc., to the retrocession of the eruption, when, in fact, these serious symptoms are produced by the same influence that caused the retrocession—the accumulated heat. I have used no other means to "bring out" measles for several years, and am glad to state that it has never failed me.

In the active delirium of typhoid I used to tell the friends (when they asked me what caused the delirium) that it was caused *by the action of the poison of the typhoid on the brain.* I suppose that every gentleman here has given the same answer. But it is not true.

Withdraw all of the heat above the normal from the patient's body and you can establish the fact that it is caused by the excessive heat, and you will have the satisfaction of demonstrating a philosophical truth, and also of presenting your patient to the family in the full possession of his faculties. Whenever the disease is located outside the cranium, if we have delirium, we can answer that it is caused by the high temperature, and without any fear of being wrong.

But the most serious result of the influence of long continued high temperature upon the organism is the change wrought in the red-blood corpuscles. In many cases where there has been a long continued high temperature, we observe a tendency toward cyanosis. The countenance becomes purplish or pallid, the lips lose their arterial hue, and the venous system is full. The red-blood corpuscles are losing their power to carry oxygen; they become crisped and crenated under the action of the abnormal heat upon them. Withdraw the heat, and if necremia has not taken place, the lips will resume the arterial hue, the color of the face become natural and the venous system empty itself of its surplus blood. If these changes do not take place upon the withdrawal of the accumulated heat, then we are notified as plainly as language can tell it that we are too late. *We have permitted our patient to burn to death while we have been looking for the origin of the fire!*

In these cases the pulse and respiration will rise rapidly, and to an enormous elevation. The respiratory apparatus, responding to the cry of the tissues for oxygen, doubles, trebles and quadruples its efforts. The heart, in response to the same beseeching wail, redoubles its efforts and sends the blood whirling and eddying through the lungs and back again, but the red corpuscles have been destroyed by the heat, and refuse to bear their natural burden.

In concluding this part of my paper, permit me to say that I have so often witnessed what I have described in these pages that I have ceased to doubt. I know

whereof I speak. I might mention other serious symptoms and conditions that arise as the result of accumulated heat above the normal, but I think I have said enough. Objections may be raised against bathing on the ground of the difficulty of it. Let me answer that small children can always be bathed in the ordinary washing tubs or some other large vsssel, and if you cannot find anything large enough for adults, you can use the wet pack, wrapping the patient in sheets wrung out of cold water, and change for freshly wet sheets at the end of each hour. It takes two or three "packs" in succession to accomplish what one thirty minutes bath will do.

In my practice I use a "Knowlton's rubber bath tub" for adults, which weighs only fourteen pounds, and can be taken anywhere. A bed of comforts or blankets, with pillow, may be made on the floor and the bath placed above it, and we can thus have our patient in the bath and the bed at the same time.

I temper the bath at from 90° to 80° for children, and at 90° to 60° for adults, according to age and severity of symptoms. I order them to be bathed again as soon as the temperature rises to a dangerous height. To some this may seem like too much work, but it is well to remember in this connection that it is a human lile we are working for. A few months bathing in a community will so educate the masses that the physician has little to do other than to order and direct.

I may be asked if I ignore other treatment. Not by any means. I direct the same treatment I would in case I did not bathe ; and I find that my medicines do more good, for when we withdraw the excess of heat we restore, for a time at least, the healthy "click" of the organs, and they resume their natural functions. The bath simply keeps our patient from being consumed by heat while we are trying to relieve the original trouble.

Inquiry made be made with regard to what cases I would and would not bathe. I answer, bathe in any case where the temperature continues dangerously high and cannot be brought down by your so-called febrifuges. The idea that a patient must not be put into cool water because he has pneumonitis or measles or anything else is based upon gross superstition and flimsy tradition, and has no foundation in science. In cases where the heat tends to reaccumulate rapidly after its withdrawal, I should not hesitate to place the patient in a bath a few degrees below the natural temperature of the body and keep him there for hours.

I will state (and I do so with a great deal of hesitation, because doctors are disposed to doubt such statements), that I am sure that within the last five years I have saved at least seventy-five per cent. of the cases that I would have lost without it. There is nothing that I do or give in cases where the temperature tends to remain at a great elevation, that yields such beautiful and grand results as the cooling bath. But we must continue to bathe as often as the temperature rises, until we have, by other means, succeeded in removing the cause of the rapid production of heat.

Gentlemen, I have endeavored not to make this paper too long. It is a subject of vital importance to every practitioner, and I have therefore simply aimed to give plain facts in plain terms. The facts presented have been wrung from the enemy in actual conflict upon the battlefield of my profession, and the hills and valleys along the way upon which these conflicts have been fought are strewn with the bleaching bones of the dead, and the widow mourns for the husband that has gone to come back no more, and mothers weep because their children are not; *but I have got the facts nevertheless.*

The Nature and Treatment of Pneumonia.

By Horatio R. Bigelow, M. D., Washington, D. C.

Every enlightened student of medicine ought to be familiar with the bibliography of pneumonia. The disputed points of its pathology are also favorite themes of discussion in medical journals and medical societies. It were a vain presumption to attempt the history of either in a paper of this character. In the transitional stage of modern therapeutics, few are in a position to advance ex cathedra opinions; and these few are listened to, only in so far as the underlying support of their praxis, is a sound physiology. The mists of charlatanism that enveloped empirical medicine, that so befogged its professors, that they were unable to pronounce whether a given result was, in relation to a given dose post hoc or propter hoc, are being rolled away by the steady advance of physiological therapeutics. Clinical experience, reliable as an accessory, is never infallible, and is illegal, if the exceptions to a given rule are lost sight of. Know first the patho-anatomy of a disease; then adjudicate upon its treatment from what we learned of the action of drugs upon animals. This is scientific medicine. The other method, if relied upon exclusively, is the empty boastfulness of ignorance.

From the earliest epochs of historical medicine, the treatment of pneumonia has been more earnestly discussed than that of any other disease, and these discussions have not always been characterized by the courtesy and singleness of purpose, which pure science demands of its disciples. Experience teaches us that, as investigation proceeds, the received opinions of to-day form the superstitions of to-morrow; that future generations will decry our methods just as we look on and wonder at the practices of our forefathers. The soundest thinkers are the most modest, and we might as well take for our shibboleth "*festina lente.*" My purpose shall be, very briefly, *ex necessitate rei,* to provoke discussion of the merits of the various theories now used by physicians in the treatment of pneumonia, so that for us, at least, the expression of a representative body may become a well defined opinion, based upon observation, experience and experimentation. Sydenham's statement, "*Hujus morbi curatio in repetita venæsectione fere tota est,*" is the irritating thorn that has caused so much professional agony during the last decade. The learned monographs of learned professors upon the "heroic," the "rational" and the "expectant" plans, are libraries of themselves, the difficulty seeming to be one of exclusiveness. We are apt to forget that the abuse of any remedy does not contraindicate its discriminating use. Isolated systems are not applicable to every case. The exclusive use of one plan, ignoring individual idiosyncracies, and other plans of general acceptance, is pure bigotry.

First, then, as to the nature of the disease. *What is pneumonia?* In simplest language it may be defined as an inflammation of the vesicular structure of the lungs. What is the essential result of such inflammation? A clogging of the interior of the alveoli with the products of such inflammation, which are thereby rendered impervious to air. Its chief clinical feature is pyrexia. In primary acute cases the tendency is to a favorable termination by crisis, from the third to the tenth day. It may be caused by an alteration of the blood supply, by an extension of bronchial inflammation, by local disturbances or by pulmonary obstruction, or mechanical injury to the lung. Apart from this minute pathology, there are three features of the disease worth considering—the dyspnœa and rapidity of breathing, the fever and prostration. The question of the treatment will depend very much upon our opinion of the causation of these symptoms.

Accelerated respiration and dyspnœa are, perhaps, the most marked phenomena of pneumonia. We have now to deal with a lung in a condition of engorgement, for we may pass by the stage of arterial injection, which has been described by Stokes as a matter of which we know little or nothing. It is so rarely observed that Rokitansky and Skoda call in question its existence. Physiologically, and upon purely theoretical grounds, it seems only a necessary and antecedent state of the arteries that are soon to participate in a more active process. In this stage of engorgement there is intense congestion of the pulmonary vessels and commencing œdema of the lung. The capillaries of the

pulmonary artery are loaded with blood, and the epithelial cells of the air vesicles are altered in structure. The respirations vary from thirty to seventy per minute. Why? It is the cry of the blood for aeration, is the general answer. But *is* this the only reason? From the very commencement of some cases of pneumonia the *nerve prostration* is the most formidable symptom of all, this prostration being caused by altered blood supply, and the effect which its high temperature has upon the nervous centres themselves. It is so frequently the case as to be almost universal, that pain and accelerated respiration will go hand in hand. Pain will also produce congestion and consequent pyrexia. In pneumonia the increased respiratory movement and the dyspnœa bear no relation to each other. The former often exists without the latter. The pain of pneumonia is a factor in the production of the increased number of respirations. The pyrexia itself is in part due to the same cause; and the two, accelerated respiration and pyrexia, may be at first, part due to the pain, and to the altered condition of the nervous centres caused by the higher temperature of the blood, which temperature was caused by pain. The neurotic theory of gout has been strongly advocated by Dr. Duckworth, in a late number of "Brain;" and while I do not go so far as to class pneumonia among the neuroses, I cannot forget that its nervous phenomena (which are sometimes primary) have a very slight consideration in the question of treatment.

The condition of intense pulmonary congestion of itself is not sufficient to account for all the clinical symptoms. Were all the alveoli to become clogged at the same instant, then we could at once affirm that herein we had the secret of the dyspnœa. In most instances, however, there is no relation between the amount of aeration and the increase of respiration. But there *is* a relation between the altered blood supply, be it great or small, and the integrity of the nervous centres. We often have extreme dyspnœa with less rapid respiration. Can we attribute this to a want of blood aeration alone?

Is, then, this first stage of pneumonia, this condition of engorgement, primary, or is it a result of a want of harmony in action of the nervous centres? Shall we proceed at once to relieve the local congestion, or shall we address ourselves to the blood and nervous system? Shall we bleed, purge and blister, or shall we tranquilize (and thus reduce temperature through the nerve centres) and nourish? It is well known that the application of cold will reduce the temperature, will control the respiration and calm the nervous irritability. It is also proven that ice-bags to the spine will arrest certain phases of chronic dysentery, of nausea and of many forms of irritable fever. This acts directly upon the cutaneous circulation, and through it upon the general circulation as well, and indirectly upon the nervous centres, by relieving their congestion and lowering the temperature of the blood sent to them. In pneumonia, we have a local inflammation, with great and rapidly increasing prostration, and we have an altered blood supply. Before it is possible to detect any appreciable change in the lung substance proper, we have well marked rigors and a general sense of malaise,— prodromes of the local disorder. Shall we treat this prodromatic period upon the "expectant" plan, or shall we adopt the "heroic" practice? Shall we commence at the outset and treat these rigors as manifestations of nervous irritability, bearing in mind the sequences of pulmonary disorder that may ensue, or shall we administer a gentle placebo, and await the tide of results? Again. When the first stage has been well defined, shall we administer expectorants, or stimulants, or nourishing diet, or shall we first diminish pain and keep down temperature? In a certain sense, it may be said that the practice of medicine revolves in cycles; that customs, long since laid away, do, in course of time, obtain fresh favor. This is especially true of venesection. The practice of Sydenham, of Huxham and Cullen, of Gregory and Bouillaud, of Andral and Frank, is being revived, and with much feeling, by the mod-

ern practitioner. The objections to venesection are thus given by Dr. Wilson Fox:

1. That indiscriminate bleeding immensely increases the mortality of the disease.

2. That it is especially fatal in old people and in young children, in patients of exhausted constitution, and in those suffering from diseases, and particularly from Bright's disease.

3. That it is absolutely unnecessary in the majority of cases of young adults and also of young children.

4. That in the vast majority of cases it has no influence whatever either in cutting short the disease, or lessening its duration, or diminishing the pyrexia, but that occasionally these results appear to follow from its use when practised early.

5. That in the majority of cases it hinders the critical fall of temperature and delays convalescence.

6. That in the majority of cases, as shown by Dr. Bennett's and Dietl's data, recovery is equally, if not more, rapid, when it is not practised, as when it is resorted to.

7. That in a few cases moderate venesection may be necessary in the early stages to avert immediate danger of death by pyrexia.

And again (*Reynolds' System of Medicine*) he says:

"With regard to the possible effect of this treatment in cutting short the disease, it may be stated that the chances in any given case are strongly against such a result. Looking at the general effects of this procedure, patients will, on the whole, be probably in a worse condition for passing through the later stages of disease when weakened by an artificial loss of blood than they are likely to be if their resources in this respect are husbanded; and 'though its dangers are the least in the case of young adults of good constitution, who commonly "bear" bleeding comparatively well, this "tolerance" of the remedy by such subjects affords no proof of the general advantageous effects."

Upon this statement, Dr. Hartshorne, the American editor of *Reynolds' System of Medicine*, makes the following commentary:

"Tolerance, however, *plus* immediate relief of marked symptoms, and early recovery, affords the kind of evidence, which, according to all rules of clinical experience, is wanted to establish the appropriateness of a remedy in practice. While an individual case proves little, yet the aggregate of individual cases, carefully observed, furnishes a better basis than any *a priori* reasoning can do, for conclusions in inductive medicine. What is claimed by those who still advocate moderate venesection in a certain minority of cases of acute pneumonia during the early stage, is, that having resorted to it, and seen it resorted to in a large number of such cases, relief and early recovery followed without any drawback of excessive weakness. Their legitimate inference is that the unmitigated pulmonary inflammation would have produced greater debility than the timely withdrawal of a few ounces of blood. Nor does this conclusion, as a matter of fact, appear to be vitiated by the comparative effects of expectant or stimulant treatment, now so common, upon the mortality of the disease."

There are other physicians who go farther than Dr. Hartshorne, and advocate blood-letting in almost *every* case of acute pneumonia.

If the disease be once established, I fail to see of what use local blood-letting can be. It *will* not arrest the inflammation, though it may for a time palliate symptoms. It can have no influence upon the products of the inflammation, which are filling the alveoli, neither can it arrest the nitrogenous tissue waste. If one were fortunate enough to diagnose a case of acute pneumonia, during the prodromatic period, especially if the patient should happen to be a vigorous subject, venesection might, and certainly has, in many cases, averted the disease, and for this reason: By reducing the amount of blood, and lessening the heart's action, we relieve the tension of the nervous centres, we keep down temperature, we overcome tissue waste, and prevent the products of inflammation from being thrown out. Now, since the pain of pneumonia may be due to

pressure upon nerves by these products, we also may arrest this symptom, itself one of the factors in the causation. of more serious sequelæ. But in the stage of engorgement we have a more complex condition. We cannot afford to weaken the patient, already being rapidly prostrated by excessive tissue waste and nervous activity. We cannot afford to weaken the action of a heart already taxed to its fullest extent. The question is not now that there is too much blood sent to the lung, or there is more blood than can be aerated, or that the inflammation is kept up by excessive congestion, but it is, will this venesection quiet pain, relieve nervous irritation, reduce the pyrexia, and hasten resolution?

Personally, I cannot speak from experience, but an analysis of a vast number of statistics, together with a knowledge of the effects of blood-letting upon the general economy, lead me to believe that the theory is illusive, and the practice faulty. However, the matter is now being so ably advocated, and the reported results of this procedure seem so seductive, that venesection may well claim the serious consideration of physicians. Pyrexia is due to a chemical change set up in the tissues themselves. It is a disorder of nutrition in which the "exchange of nitrogen exceeds the normal expenditure by nearly three-quarters, and in which there is likewise an excessive discharge of carbonic acid." This increase of blood-heat is, says Simon, "the essential fact of inflammatory fever. * * * * As the blood gets hotter and hotter, more and more do these symptoms become developed; as the blood subsequently gets cooler, so, more and more, do they decline."

I have, of late, with much satisfastion, employed the following plan in the treatment of these cases: The results have been favorable, and I should be unwilling to avail myself of the habit of venesection without an authoritative dictum from a majority of the profession. One is always loath to abandon any course of procedure, which has met with success, in order to indulge a love of investigation, in the adoption of another routine in which he has had no experience. When called to a case of pneumonia in its first stage, I quiet pain first. This I do by strapping, by chloral, or by applying to the denuded surface of a blister minimum quantities of morphia. I order milk and beef tea at regular and frequent intervals, but give no stimulants. If the temperature be up, with a tendency to increase, I give quinine in *full* doses with veratrum viride. Should the pyrexia refuse to yield to this treatment within twenty-four hours, and the thermometer still continue to rise, I resort at once to the cold pack. I have yet to see any evil result from the proper and well regulated application of cold; on the contrary the effect is often wonderful. Nervous excitability is diminished, sleep is induced, dyspnœa palliated, cough subdued, pain relieved, and the temperature reduced surely and certainly. The effect upon the nervous system is singularly happy. The centres are relieved of their congestion and brought into harmonious action by the "cooling" of their blood supply, the cerebral irritation induced by this excessive nervous action is also quieted, and as a natural result, sleep follows. The lung, relieved of a primary factor in the production of its inflammation, rests from its unnatural labor and the respirations become more and more natural. I repeat the application of the pack just as often as the thermometer indicates its use. Upon this and nourishing diet, together with the quieting of pain, I rely, almost exclusively, in the treatment of these cases.—*Maryland Medical Journal.*

The Stigma of Maize in Pyelitis and Cystitis.

By SAMUEL C. BUSEY, M. D., WASHINGTON, D. C.

In the January number, 1880, of the Medical News and Abstract appeared a synopsis of a paper by Dr. Dufau, entitled "The Stigma of Maize in Diseases of the Bladder," in which the value of this new drug in the treatment of cystitis, gravel, und other affections of the bladder was so highly extolled that I

determined to give it a fair trial; but failing to secure either of the preparations recommended the attempt was not made until late in February, 1881. Subsequently to the above publication, Dr. Vaulthier communicated, in the August number of the Archives Med. Belges, the favorable results of his experience with the same drug in the treatment of "all the affections of the bladder, whether recent or chronic."

My first trial of the drug was in the following case of pyelitis. The gentleman had had during the two preceding years several attacks of renal colic. The last attack occurred in September 1880, and was followed by continuous and increasing ill-health, the symptoms of which referred to some disturbance of the genito-urinary apparatus, but were indefinite. He occasionally sought medical advice, but did not submit to regular and systematic treatment until the latter part of October. The progress of the case is fully exhibited by the various analyses of the urine, by Dr. G. N. Acker, which will be given in the order of date.

"October 19, 1880.—Sp. grav., 1015; reaction, acid; albumen, small quantity; sediment heavy and white; numerous leucocytes; few red blood-corpuscles; epithelium from kidneys, tubes, and pelvis; few from bladder; no casts or crystals,"

From the above date he was under constant observation and treatment, but attended regularly to his business, which was laborious and active. There was no improvement. The second analysis shows a marked aggravation of the disease.

Analysis, December 7, 1880.—Sp. grav., 1015; reaction, neutral; albumen, one-eighth; numerous leucocytes, some with large nuclei; young epithelial cells; some red blood-corpuscles; epithelium from pelvis of kidney, ureters, and bladder; no casts.

After this date he was treated with tannic acid, after the method of Traube, with rectal suppositories of opium and belladonna to allay pain and frequency of micturition, which deprived him of the necessary sleep. His diet was regulated, and though permitted to go daily to his place of business, he was admonished that exercise was detrimental.

"*Analysis*, February 5, 1881.—Sp. grav., 1017; reaction, acid; albumen, one-eighth; sediment, heavy, white; not much coloring matter; sediment composed of leucocytes, urethral, pelvic, and bladder epithelium; a few red blood-corpuscles."

Two months' treatment with tannic acid had failed to produce any beneficial effect; in fact, the analysis indicates a less favorable condition. The amount of sediment had increased. The acid treatment was continued, and an infusion in wineglassful doses every four hours of uva ursi one ounce and lupulin one-half ounce to the pint of water, was added. On the night of February 17th he had a very copious hemorrhage. The next day he was ordered to bed. Treatment suspended. I had determined to try the stigma of maize, and was awaiting its arrival.

"*Analysis*, February 21, 1881.—Color, light yellow; sp. grav. 1020; reaction, acid; albumen, large quantity; phosphates and chlorides, normal; heavy white sediment; leucocytes, in large quantity; some red blood-corpuscles; large amount of vesical, urethral, and pelvic epithelium."

On February 22d, the treatment with the fluid extract of the stigma of maize was commenced, at first in doses of one drachm every six hours, then four hours, and, finally, after several days' use without any observable effect, every two hours. The quantity of urine increased and ran up to sixty-four ounces a day. The amount of sediment diminished, and micturition became less frequent. The following analysis shows a marked improvement;

"*Analysis*, March 8, 1881.—Sp. grav., 1020; reaction, acid; albumen, small quantity; chlorides, diminished; phosphates, normal; sediment, one-tenth; numerous leucocytes, very few red blood-corpuscles, epithelium from pelvis of kidneys and bladder diminished; crystals of uric acid, oxalate of lime, and triple-phosphates."

With occasional variations in the frequency of the doses of the maize, the treatment was

continued. The interval between the doses was increased when the amount of urine passed was excessive.

"*Analysis*, April 4, 1881.—Sp. grav., 1015; reaction, acid; albumen, small quantity; urates, small quantity; phosphates, normal; few leucocytes, very few red blood-corpuscles, epithelium from pelvis of kidney and bladder greatly diminished. In an eight-ounce vial of pale yellow urine the white sediment barely covered the bottom of the bottle."

The decided improvement since the 22d of February may have been due as much to the rest in bed as to the stigma of maize. He appeared so nearly well that I allowed him to leave his bed and sit up on the 8th of April, and for several days he continued to improve. On Thursday, 14th inst., he complained of intense pain immediately preceding defecation, deeply seated in the perineum, which he ascribed to "the piles." The pain was continuous when in the sitting posture. Micturition was very frequent and accompanied with a scalding sensation along the course of the urethra, which continued, gradually subsiding, for about ten minutes after each evacuation of the bladder. A rectal examination disclosed an acute prostatitis. He was again put to bed. Leeches were applied to the perineum, followed by hot fomentations and a hot sitz-bath morning and night. At this date, April 20th, he seems to be doing well. The urine remained unchanged, apparently, from the analysis of the 4th inst. During the entire course of the disease his bowels had been kept in a laxative condition, sometimes employing alkaline waters, at other times the formula known as Chelsea Pensioner.

A few days after the hemorrhage on the night of February 17th, he was seized with orchitis attacking the right testicle, and there remains, even yet, sufficient evidence of its effect to mar the symmetry of those organs.

The second case in which the drug was employed occurred in a lady suffering with cancer of the uterus. The vesical irritation and tenesmus were so constant that the poor patient could not sleep, notwithstanding the large doses of morphia which were taken at regular intervals to relieve pain. Her urine was densely loaded with mucus and pus. She was entirely relieved after several days' use of the fluid extract. Previous to its administration I had tried various remedies which I had been accustomed to use in such cases, without any lasting effect.

The third case was a lady who, for a year previous, had suffered with vesical irritation and frequent micturition. The analysis of her urine exhibited the following condition: "Sp. grav., 1027; reaction, very acid; albumen, small quantity; color, yellow; sediment, red in color; urates, increased; numerous leucocytes; numerous uric acid crystals; bladder epithelium in large quantity."

She was entirely relieved, and continued well for about a month, when there was slight return of the symptoms. The medicine was resumed.

Case fourth occurred in a young lady who, for several years, had suffered from frequent and painful micturition. The pain was sometimes so intense as to cause her to scream.

"*Analysis of urine.*—Color, straw clear; sp. grav., 1037; reaction, very acid; albumen, none; sugar, none; urates, increased; sediment composed of urates, mucus and epithelium, urate and oxalate of lime crystals; vaginal and bladder epithelium in large quantities, mucus."

She was greatly improved after a moderate use of the drug, and, probably, would have been entirely cured if the treatment had been continued. But, as so frequently happens with young girls, as soon as the intense suffering was relieved, the treatment was abandoned. She is again under treatment.

The foregoing cases were all under treatment at the time of obtaining the drug, but with intermitting and partial success. Since, the two following cases have been treated with the same drug:

One was a lady who had suffered for an indefinite time with subinvolution of uterus, metrorrhagia, and laceration of the cervix. She represented that the desire to "pass

water" was so frequent during the sleeping hours that it was impossible to secure a quiet night's rest. Her statement was, that the night previous to my first visit she had been compelled to get up every half-hour, passing at each time a very small quantity. She was entirely relieved. A week afterwards there was a slight return, which yielded again to a few doses.

The second case was a lady, who stated that she had suffered for two years with "congestion of the right ovary," and had been treated for that affection by various external lotions, and the internal administration of anodynes. I failed to recognize any enlargement or tenderness of the ovary, but did discover a retroflexion of the womb. She described a pain, which recurred every night after having retired, accompanied with a desire to evacuate her bladder. The pain was felt along the course of the right ureter. She was compelled to empty her bladder every hour during the night, but not so often during the day. In this case the relief was not so prompt as in the preceding, but there was a gradual abatement of the vesical irritation and frequency of micturition. In addition to the use of the stigma of maize, the womb was adjusted, and retained in position by pledgets of absorbent cotton, saturated with carbolized glycerine.

All these cases exhibit the beneficial effects of the drug, but I am not prepared to assert its curative influence with the confidence of Dufau and Vaulthier. It is a certain, but mild diuretic, when given in full doses at short intervals.—*New York Medical Record.*

Hysteria — Applications of Static Electricity in its Treatment.

Delivered at the Salpetriere, by
M. CHARCOT,
Professor in the Paris Faculty of Medicine.

Translated for the Medical and Surgical Reporter.

GENTLEMEN—I will confine myself in this, my last clinical lecture of this year (1881), to the consideration of a new subject, which is yet, as regards certain points, unsettled.

I refer to the applications made of static electricity in the treatment of diseases of the nervous system, particularly of hysteria. When hysteria is spoken of, or under discussion, there are many physicians yet, who, without any very evident reason, do not wish to listen, or take the subject into consideration, regarding the most conclusive arguments as without weight.

Some are willing to admit the existence of hysteria, but consider it a veritable protean malady; its evolution following laws so complex that it becomes impossible to submit them to analysis. Again, that the phenomena are so mobile, so changeable, that they cannot be reduced to any methodic description. These physicians would voluntarily ascribe to this affection almost all clinical phenomena which cannot be submitted to any recognized rule.

Other physicians, more radical, deny the existence of hysteria as a malady. They consider that all is but simulation and deception in the phenomena which characterize it. For them, a physician should be very careful how he meddles with such patients, or his professional dignity will be compromised. Thus, with a stroke of the pen, one good quarter of the nervous pathology of women is suppressed.

But I will demonstrate to you the reality of these phenomena, and though the tendency of hysterical patients to deceive cannot be denied, I will prove to you that hysteria is truly a separate malady, with greater mobility and with more variety in its manifestations than other maladies, but in which nothing happens through chance, everything occurring in conformity with certain rules, which may, notwithstanding the complexity of the phenomena, be determined by the observer if a certain method be employed.

This method is the following: proceed slowly and carefully in this domain, so difficult of access, and do not attempt to understand all immediately. Consider at first the more simple phenomena, those more easily understood, the most palpable in effect; some of these will be found analogous to symptoms observed in other maladies of the nervous system quite different from hysteria. But this is not all; the same observation must be many times repeated, with different subjects and under different conditions; the hospital observations must be placed in contrast with those seen in private practice. The more complex phenomena should be regarded with great circumspection, and those which seem to have no sensible connection with established pathological facts should not, for the moment, be taken into consideration.

I will first demonstrate practically certain methods by which the reality of the facts observed may be rendered manifest, although at first glance they may appear strange. This concerns certain phenomena observed while the patients are in the hypnotized condition, into which state hysterical patients may be plunged with facility.

The different methods by which patients are (often with great facility) placed in the hypnotized state, are well known; by a monotonous and prolonged impression, such as that which results from steadily regarding an object with the eyes in convergence, or a sudden and startling impression, such as that produced by the sudden beating of a gong or a flash of electric light. (See Bourneville et Regnard, "Iconographie Photographique de la Salpetriere," T. iii.)

If a patient is thus artificially put to sleep, two different conditions may be observed : if the eyes are closed, so as to prevent the impression of the light on the retina, the patient is plunged into the hypnotic state, which might also be called "sleep with muscular hyper-excitability."

In this state, in effect, it suffices to touch a muscle, to compress a nervous trunk, in order to produce contractions in the muscle or groups of muscles under the dependence of the nerve.

But, if the eyes are open, so that retina may receive the impression of light, the patient falls into a cataleptic state. In this condition the muscles are soft and pliant, but are incapable of contraction under the influence of excitation.

But the two states may exist conjointly in the same patient; to bring this about one eye should be uncovered while the other remains closed. Under these conditions the muscles on the side corresponding to the open eye are hyper-excitable, while on the other side the series of phenomena which characterize the cataleptic state are observed.

This fact of muscular hyper-excitability is the most potent to confirm conviction in the minds of those physicians who consider that all which relates to hysteria is simulation.

And, in effect, if it be true that the motor nerves and muscles are excited in the hypnotized patient by simply touching them, effects should be produced on any patient, though completely ignorant of anatomy and physiology, analogous to those discovered by Duchenne, of Boulogne, in his remarkable researches on localized electrization.

If, on the contrary, the patient were not really hypnotized, but was able and desired to deceive, the simulation would very soon be recognized. The patient being completely ignorant of anatomy and physiology, the motion and contractions would be dictated by her caprice, and would not in the least be in conformity with known anatomical facts.

Or, such is not the case, if, as M. Charcot demonstrated in his lecture, the results of faradization of the different muscles and nerves be compared with those induced by simple excitation of the same muscles and nerves in the hypnotized hysterical patient; they will be found to be identical. And this fact is so well proven that the action of muscles and groups of muscles under the two methods of excitation have been reproduced by means of photography and the identity of the effects rendered still more evident.

This fact holds equally good for the muscles of the face, of the upper or of the lower limbs, in well marked cases: the results obtained by one and the other method are the same, and the hypnotized patient realizes, with a precision and exactitude which at first causes astonishment, the ensemble of movements commanded by the excited muscles. If, notwithstanding all these proofs, simula-

tion should be thought of, it would be necessary to suppose the patient had profound knowledge (much more extensive than that possessed by most physicians) of the anatomy and physiology of muscles, and a species of infallibility which would enable him, though taken completely by surprise, never to make a mistake regarding the action of a muscle or the anatomical distribution of a nerve; we have no necessity to establish the absurdity of such a hypothesis.

Before taking up the history of the effects of static electricity, we should consider the results obtained through means of the medicaments I have termed æsthesiogenes. This is a series of therapeutic agents by means of which, as the expression indicates, sensibility or common sensation may be brought back to parts whence it has disappeared, in hysterical patients particularly, but also in persons with cerebral lesions.

To M. Burq, is incontestably due the merit of having opened the way to this order of researches, in demonstrating certain properties of metallic plates. But since the works of Burq, and, particularly, through researches made at the Salpetriere, this subject has been enriched by many additional facts.

Among æsthesiogenetic agents, the magnet occupies an important place between the metals and static electricity. The action of the magnet is well known to-day, and has been popularized through the works from the Salpetriere (Charcot, Regnard, Vigoroux). The objections urged against the effects obtained by the magnet are of two orders: first, it has been advanced, and without any proof, that the observers, under the influence of an illusion, were deceived by the patients. But soon the observations became numerous, and were verified by many scientists; the magnet was shown to be efficacious, not only against the hemianæsthesia of hysterical patients, but also against certain forms of hemianæsthesia due to organic lesion, or against those of toxic origin.

It was then that the idea of "expectant attention" was invoked to explain the phenomena which they persisted in not attributing to the magnet. Attention being, they say, fixed with force, and without intermission, on one part of the body, either the circulation or the innervation of this part of the body, or both at once, become affected. Naturally these effects are much more marked if there is hope of any determined result being obtained. Or, they say, it is through this strong tension of the mind, to the "expectant attention," in a word, that the return of sensation into the part experimented on is to be attributed.

This new objection has been often replied to, and shown to offer no more resistance to serious examination than the preceding one. As proof, I will cite the following experiments, chosen from among many others :

1st. A false magnet resembling exactly, as regards form, volume and weight, the true magnets usually employed, will have no effect if placed to the affected side ; sensibility will not reappear. Nevertheless, in this case, does not the pretended "expectant attention" exist, just as when recourse is had to a veritable magnet ?

2d. Zamboni's electric pile brings back frequently, like the magnet, sensibility to the part whence it has disappeared ; while with a false pile, exactly resembling the other exteriorly, no result is obtained.

3d. Finally, if a solenoid is made use of, through which the current obtained from the pile may be passed or interrupted without the knowledge of the patient, it will be found that the anæsthesia disappears when the solenoid is traversed by the current, while there is no change when it is not placed in relation with the pile.

All these facts and experiments, repeated hundreds of times, by numerous observers, on many diverse patients, form a peremptory refutation to oppose to those physicians who still obstinately deny the evidence.

We will now proceed to consider the subject which is principally the object of this lecture—the effects of static electricity. It will be useful to determine with precision the characters of static electricity, and demonstrate in what it differs from the two other modes of electricity, the faradic and galvanic. Electricity is always one and the same, whatever the generator which produces it ; it presents itself with two essential attributes : 1st, quantity ; 2d, tension. But according to the generating apparatus put in use, sometimes the quantity predominates, sometimes the tension. At this point of view the different forms of apparatus for generating electricity may be considered in three groups : 1st, the apparatus which furnish electricity in a high state of tension, but in feeble quantity ; the ordinary electric machine is the type of this form of apparatus which furnish static electricity ; 2d, the apparatus which produce electricity of feeble tension but in large quantity, such as the pile ; 3d, finally, the electro-magnetic or induction apparatus (faradic apparatus) which constitute a link of transition between both, for the tension and quantity are both relatively considerable.

Electricity furnished by the pile produces very different effects from that of the machine. (a) The gold leaves of the electroscope diverge widely at the approach of a machine producing static electricity, and rest in contact with each other when the current obtained from the pile is passed through the apparatus. (b) Water, on the contrary, is decomposed by the galvanic current, while electricity of tension has no effect on it. (c) Iron wire becomes incandescent when traversed by a continuous current of sufficient power, while static electricity has no effect. (d) Finally, the static machine gives rise to sparks, which the pile does not produce.

This divergency in the physical effects of the different modes of elec'ricity would lead to the supposition that there existed corresponding differences in the physiological and therapeutic effects.

It is known, for instance, as regards physiological action, that a faradic current does not act on a muscle at a certain moment after the section of the nerve, while under these conditions the action of the galvanic current is more energetic than in the normal state.

On the other hand, the application to the

tongue of one of the electrodes of the pile gives rise to a metallic taste, which is not produced by the electric spark. Thus, at a physical and physiological point of view, each mode of electricity has effects proper to itself; and the same is true as concerns therapeutic effects.

Therapeutically, in effect, static electricity has its separate place beside faradic and galvanic electricity, just as it has its special characters and effects, physically and physiologically.

Static electricity has latterly attracted little attention as a therapeutic agent, since the experiences undertaken by Manduyt in 1875. Duchenne, Remak, Benedikt, and Onimus, attached little importance to this form of electricity, which M. Vigoroux has recently sought to introduce into practice.

The machines formerly used, that of Ramsden notably, have too feeble a tension, and are with difficulty kept in order; while the tension in the machine of Holtz is too strong; the machine recently brought forward by Carre, particularly with the modification made by Vigoroux, presents an incontestable superiority over the preceding ones.

The machine is enclosed under a glass receiver, which permits, with the aid of certain substances having great affinity for moisture, chloride of calcium, for instance, the maintenance of the machine in a sufficiently dry atmosphere, so that it may be put in use at any moment; when a motor like that of Bishop is added to the machine it may be put in action without the assistance of an aid, and can be used for several hours in succession.

Finally, M. Vigoroux, making use of several isolated stools placed in relation with the machine, has been able to place five or six patients at a time under the electric influence.

Through the means of this perfected apparatus, daily in use at the Salpetriere, static electricity, heretofore of difficult application, has been reduced to practical use. The patients, placed on the isolated stool and in relation with the machine, are plunged, as it were, into a veritable electric bath, or rather, to be more exact, an electro-static bath. They are placed on the course of an electric conductor, and representing simply a prolongation of one of the poles of the machine, they must be influenced by the electric state of the conductor. The other pole is prolonged to the earth, and the two prolongations are separated by the layer of air interposed between the ground and the stool.

In this species of electric bath the electricity is not in repose; there is constant destruction and renewal, as attested by a Lane's electrometer placed near the stool. The electric fluid escapes from all points of the patient, particularly from the hair, the ends of the fingers, etc. From this results a feeling of irritation all over the cutaneous surface. And again certain important physiological effects are remarked: the cutaneous perspiration is augmented, the digestion rendered more active, the station on the electric stool communicating frequently a decided sensation of hunger.

With the aid of excitators (excitateurs) the discharge which takes place slowly and continually from the patient's body may be localized. The excitators most commonly put in use are rods of wood or metal, provided with isolating handles and in communication with the ground by means of a chain. These instruments are pointed or terminate in a bulb, according to the effects it is desired to produce. Special forms, even, have been constructed for the electrization of the eye, the ear, the muscles.

If the excitator is placed at a distance of fifteen or twenty centimeters from the cutaneous surface, the patient has the sensation of a current of heated air; it is what might be termed the electric wind.

At a distance of six or eight centimeters the sensation is more acute, and has a character of irritation; there is a pricking sensation at the part. At the same time, from the extremity of the excitator there proceeds a bunch of luminous rays which spreads as it reaches the body (c'est l'aigrette). Finally, if the excitator, preferably with the bulb termination is brought still nearer the cutaneous surface, the electric spark is obtained and provokes, at the moment it is disengaged, a sudden movement of the limb, with a painful sensation at the point nearest the bulb. Operating over the naked surface, a brusque contraction is often observed in the muscle, directly under the point where the spark was emitted.

Localized muscular electrization may be thus effected, and in certain cases of paralysis of spinal origin certain muscles which did not react under the influence of faradic electricity contracted under the electric spark.

It may be seen, from what we have observed, that through means of this machine the patient may be simply placed in a sort of electric bath, or may be brought, by using excitators, under the influence of the electric spark, etc.

The first patient I will present has a hysterical contraction of the left forearm, with flexion of the fingers of the hand and of the forearm. This patient was formerly cured, when similarly affected, by applications of the magnet, but after two years the contracture reappeared, and is much more tenacious than formerly.

Placing the patient on one of the isolating stools, by means of the electric wind (that is through the near approach of an excitator), artificial contracture is soon brought about in the right arm. There is no actual transfer of the contracture from left to right, as happened formerly, when the magnet was applied; there is simply a slight relaxation of the muscles of the contracted left arm, while the contracture is produced in the right arm. Although the effects obtained up to the present in this case are not very conclusive, there is every reason to believe that in the end an artificial state of contracture will be brought about in the right arm, and the spontaneous contracture in the left cured, as it was two years since, by the magnet.

I present to you, also, three hysteric subjects; in two there is hemianæsthesia, in the third complete anæsthesia.

When these patients are placed in what we have termed the electric bath, little by little the following effects are produced: if anæsthesia is complete or total, after a few minutes the anæsthesia becomes more marked on the side where it habitually predominates (it is well known that even where there is anæsthesia of both sides, it predominates always on one side); sensibility reappears, on the contrary, on the side where anæsthesia is usually least marked; then, after a few instants, anæsthesia changes from one side to the other (transfert), and finally, after twenty, twenty-five, thirty minutes, sensation has returned over the whole body, special sense as well as common sensation. If there existed, primarily, hemianæsthesia, there is, as in the other case, transfer and finally complete return of sensibility.

A hysterical patient who has thus recovered sensibility retains it for several hours and even several days; for that matter, by repeating each day the seances of electrization, things may be maintained in this state.

It is not without utility to thus induce a return of sensibility in a hysteric patient, for experience has demonstrated that these patients are in reality, for the moment, cured, while they are no longer in a state of anæsthesia; for instance, they can no longer be placed in the hypnotic condition, and often the excitation of certain parts which usually in them bring on the hysteric attack no longer produces that effect.

These facts would suffice to render legitimate the efforts to restore static electricity to a place in therapeutics.

And it is not in hysteria alone that static electricity has given results.

Certain cases of facial paralysis, of peripheric origin, have been cured by this process, which would appear destined to render real service in cases of paralysis difficult to treat by electrization and galvanization, on account of the imminence of contracture. In shaking palsy (paralysie agitante), interesting effects have been obtained. The shaking is immediately arrested in those parts of the body on which the electric spark is directed, and although there can be no hope of actual cure, the malady has, in several cases, been happily influenced by this practice.

Certain nervous troubles, spinal irritations, dyspepsias and dysmenorrhœas of nervous origin, which had proven rebellious to various modes of treatment, have been also ameliorated by the aid of static electricity.

It can, then, be seen that this is a therapeutic agent of great value, and although there are many researches yet to be made regarding it, it is evident that it merits being withdrawn from the obscurity to which it appears to have been systematically relegated. —*Phila. Medical and Surgical Reporter.*

Is Alcohol a Food ?—When Should Malt Liquors be Preferred to Wines and Spirits in the Treatment of Disease ?

Read before the Philadelphia County Medical Society, April 27, 1881.
By H. C. Wood, M. D.

Before attempting to express briefly an opinion as to the food-value of alcohol, and to marshal the facts upon which such opinion rests, it is proper to define the term "food," lest verbal confusion arise. Some persons may incline to recognize as foods only those substances which either in their entirety or in a more or less altered condition are capable of being formed into the bodily structure. It is plain, however, that this use of the word is too restricted. Of the substances taken daily into the stomach as sustenance, only a portion becomes an integrant part of the animal economy, as is shown by the enormous increase in the formation of urea which follows a heavy meal of meat. The appetite for blubber which is produced by exposure to an Arctic atmosphere indicates that the system needs an extra supply of fatty food,—an indication confirmed by the beneficient effect upon the exposed organism of a highly carbonaceous and liberal diet. The Arctic explorer does not fatten, although he may eat enormously of fat. The

carbon compound is evidently fuel which is burned up to maintain the bodily heat, *i. e.*, to yield force to maintain the molecular movements of the organism. It is not necessary to elucidate the point further. It is plain that under many circumstances that which is eaten is used not for reconstruction, but for force-production.

If the term food be employed in its narrower and, as I think, improper sense, it must be acknowledged that there is no proof that alcohol is a food. The enormous accumulation of fat frequently seen in beer-drinkers, and to a less extent in those who employ dilute alcohol in other forms, would seem to indicate that alcohol is capable of being converted into fat, but is probably better explained in other ways.

When the term food is used in its wider and more correct sense, the question "Is alcohol a food?" must receive an answer essentially different from that just given.

From this point of view, any substance which is destroyed in the system, and during the destruction yields force, is a food. The evidence that alcohol fulfills both these conditions is most positive,—so positive, indeed, that there is probably no one of those present that would deny at least that alcohol is burnt up in the body ; and, if it be burnt up, it must yield heat,—*i. e.*, force.

The questions whether alcohol is an economic food, whether it is practically useful as a food, are entirely distinct from that just answered. In regard to its economy, an ounce of good whisky may be estimated to cost about three and one-half cents to the man who buys by the gallon. According to the researches of Dupre, it would require about five ounces of lean beef to yield the same amount of force as that produced by the ounce of alcohol ; these five ounces may be fairly estimated as costing at least five cents to the ordinary consumer ; so that the advantage is plainly on the side of the spirit. There are, however, various foods cheaper than is lean meat. Tallow at six cents a pound in its burning certainly affords much more of force for the money than does alcohol.

Looking at the matter practically, and not as to a theoretic economy, it is evident that, as alcohol is usually taken and as food is usually eaten, no claim can be rightly made as to the superior cheapness of the beverage. Again, it is notorious that in America almost every one in reasonable health consumes much more food than the system needs, so that any alcohol taken is added to that which is already in excess. In Europe it is different: a large part of the population is underfed, and a modicum of alcohol is a decided food-gain. It must be remembered, also, that as a food alcohol is superior even to lean beef, in that it requires no force to digest it. The coarse food of a European peasant is often worked up by the stomach only by the expenditure of much force; it is also repulsive to the palate. The draught of landwein or the schooner of cheap beer washes down very well the morsel of black bread. Not only by stimulating the stomach does it aid in digestion, but also by readily yielding force to the system it assists in the elaboration of more refractory substances.

Although I hold that the habitual use of alcohol is to well-fed persons not only unnecessary but positively harmful, it seems to me that in many cases of illness and in those periods of life when by reason of age the body waxes weak, alcohol is possessed of great value. Under sixty years of age the daily employment of wine may for most persons be very well discountenanced; but after this period has been reached, I believe the moderate employment of stimulants is very useful. The progressive failure of bodily powers points to the use of a substance which shall aid in digestion and readily supply force. In the latter years of life even the narcotic influence of alcohol is of value, easing the restlessness, the slight discomforts, the suffering of nerve-failure incident to failing vitality.

The question whether alcohol has food-value in disease is one not easily answered by positive evidence, because the narcotic properties of the substance are so marked as often to mask its influence as a food, and because we rarely dare to employ alcohol

except with an abundance of other food. The principles already outlined are, however, as applicable in disease as in health. Recent researches in fever have determined that the excessive heat-production is dependent upon excessive changes in the stored materials of the body, and it is improbable, though not impossible, that alcohol is capable of taking the place of these, and, by being, as it were, vicariously burnt, saving the tissues. The value of alcohol in low fevers therefore probably depends upon other qualities than its usefulness as food, although our knowledge of fever-processes is yet so imperfect that it is necessary to speak with great reserve.

In chronic wasting diseases I believe alcohol has an actual food-value, besides being a most powerful aid to the digestion of other food. Arguments upon this point do not seem required, and it is of course very difficult to give actual clinical proofs, because we always employ alcohol with other foods and remedies.

The question as to the best method of administering alcohol when it is used for its sustaining powers is of vital interest.

Two general propositions will, I believe, command almost universal assent. First, the alcohol must be given in a dilute form; second, it should be given along with other food. Provided these two rules are observed, I do not think it makes much difference in what form the drug is administered. In chronic diseases malt liquors have both advantages and disadvantages. They represent food and drink, are less apt to be abused than are stronger liquids, and by virtue of their bitterness have some tonic properties; on the other hand, they sometimes disagree with the stomach. As they contain some nutritive material, there is perhaps more tendency to administer them apart from food than there should be. The amount of solid constituents in a pint of malt liquor varies from over two and a half ounces of dry residue in the strongest English ales to three-quarters of an ounce in the weakest ales and beers. The ales and beers usually drank in this city probably range from one to two ounces of solid contents to the pint. The nature of much of this solid matter is not known, but albumen, bitter and resinous principles from the hop, earthy salts, grape-sugar, glycerine, and a number of complex acids have been recognized in it. The tendency to grossness seen in beer drinkers undoubtedly largely depends upon the solid constituents of the beer which is taken, and seems to me to indicate the proper medical use of malt liquors,—namely, that they are especially to be employed in wasting diseases, *i. e.*, where there is a tendency to the loss of the bodily fat.

In regard to the choice of malt liquors, I do not think there is any other than what we may call personal grounds for selection. That which suits the palate best usually suits also the stomach best. The choice should always settle upon the ale, porter, or beer which can be used with least inconvenience to the stomach; and when all malt liquors produce "biliousness,"—*i. e.*, gastro-intestinal derangement,—wine or diluted spirits should be substituted. As the malt liquors contain nutritive material, it is less necessary to give food with them than it is with whisky or wines. Nevertheless, it is preferable in most cases that food should be taken with the ale or beer.—*Philadelphia Medical Times.*

Alcohol: Its Therapeutical Uses, Internally and Externally.

Read before the Philadelphia County Medical Society,
April 27, 1881.
By ROBERTS BARTHOLOW, M. D.

Of the general subject—alcohol—there has been assigned to me for this discussion its therapeutical uses, less the uses of malt liquors.

I have, therefore, to consider alcohol in the form of spirits,—whisky, brandy, gin, etc.; of wines, still and sparkling, light and heavy, sweet and dry, grape-juice and artificial, pure and fortified, etc. All of these agree in the presence of alcohol in varying proportions, but differ in respect to certain constituents, natural or factitious, and peculiar to each variety. It is obvious, therefore, that the question of the uses of alcohol is complicated by the numerous forms and combinations in which this substance enters the organism of man. In the short time allowed me I can do no more than present a syllabus of the subject,—state, merely, a series of propositions.

Alcohol may be used as a stomachic tonic and to increase the activity of those functions included under the primary assimilation, as a cardiac and arterial stimulant, as an antipyretic, as an antiseptic, as an anodyne, and as a styptic and astringent. Besides these points we must consider whether it shall be given or withheld in mania a potu, or acute alcoholic delirium, and in delirium tremens. The subject may be best studied in this order.

I. As a stomachic tonic alcohol is effective only in the case of those not habituated to its use. As pepsin is not only precipitated by alcohol from its solution in the gastric juice, but is also rendered inactive as a ferment, to employ this agent as a means of promoting digestive activity is apparently paradoxical; but the explanation lies in the fact that it is a concentrated solution which acts thus on the pepsin. We obtain two points of practice from this statement,—to give alcohol properly diluted; to stimulate digestion, administer the remedy immediately before meals: for the action of alcohol as a stomachic tonic consists in the power it has to cause a prompt congestion of the mucous membrane, and the increased activity of the gastric glands is a necessary consequence. That in time a catarrhal state of mucous membrane is produced, and a pathological secretion obtained, shows us the impropriety of a long-continued use of alcohol as a stomachic tonic, and explains why such a therapeutical effect is not the result of its administration in the subject of alcoholism.

In the atonic dyspepsia of the sedentary and the feeble digestion of the convalescent, alcohol is useful in a high degree. The particular form of the remedy must vary with the peculiarities of each case. A wine of good body and decided bouquet has been found very efficacious in the weak digestion of literary and nervous sedentary persons. In the case of the convalescent, a spirit is preferable to wine, and whisky or brandy is prescribed according to individual tastes, partly, and the state of the intestinal canal,— the former having a somewhat laxative, and the latter astringent, action. The quantity taken is determined by several considerations, —by the amount of stimulation beneficial to the organs, by the effect as manifested in the circulatory and nervous systems, and by the degree of elimination. To the literary and nervous invalid class, the wine proposed is especially grateful, and they are responsive in a high degree. As wine is to play the part of both stimulant to the digestive organs and a food of the nervous tissue, the best results are had from a generous wineglassful taken during the principal meal. The kind of wine selected is determined by several considerations. If the organs simply require stimulation, a good, dry sherry or Burgundy may be best adapted to the case; if an excess of acid be present, due to the fermentation of the saccharine and starchy foods, a sweet wine is objectionable, but then an acid Rhine wine of high alcoholic strength—as, for example, the Forster Riesling—may be serviceable in a remarkable degree. For the weak digestion of convalescence, it is well known that a considerable quantity of spirit is borne well and required. In both groups, but in an especial degree in the nervous and depressed invalid, the effect of alcoholic stimulation is peculiarly grateful and seductive: hence the danger of the alcoholic habit, and the need of circumspection in prescribing a remedy which may be so abused.

When alcohol is used to promote the nutrition in general it should be administered during and after meals, to obtain at the same time the effect due to stimulation of the gastric glands and the force derived from the oxidation of the remedy whence it becomes in a certain limited sense a food. The amount required in wasting diseases—as in chronic suppuration and in phthisis—will be more, and it will be better borne, than in

chronic maladies of a depressing kind; but in all chronic maladies the amount required is limited by two facts: 1. A small quantity will stimulate the digestive organs without injuring the pepsin. 2. Only a small quantity can be oxidized, and therefore utilized as force. It follows that it is useless to attempt, by increasing quantity, to overcome the inertia of the nutritive functions. Anstie, in the course of his well-known researches on the disposition of alcohol in the organism of man, found that beyond a certain small amount,—two or three ounces of absolute alcohol,—that which is taken is eliminated unchanged.

Applying these principles to phthisis,— which may serve as an example of a large group of maladies,—we may consider how and to what extent alcohol may be used in this disease. A notion prevails more or less widely among the laity, shared to some extent by the profession, that alcohol is in a certain sense a prophylactic and curative agent in consumption. By some physicians, amylic alcohol is held to possess these powers. That these notions are not well founded is proved by the fact that so many subjects of chronic alcoholism die of phthisis. That beneficial results are obtained by the judicious use of alcohol is equally true. The questions for us to consider are, how much alcohol is necessary, and what are the conditions of its use in consumption. The principles which I have already laid down are perfectly applicable. It is an axiom that when alcohol improves the appetite and increases the body-weight it is doing good. The quantity of alcohol sufficient to stimulate the gastric glands, and yet short of the strength necessary to precepitate the pepsin, is the proper quantity. As the alcohol eliminated by the various channels of excretion is in excess of the oxidizing powers of the organism, no benefit can result from such a quantity. It is obvious from these considerations that the best results are derived from half an ounce to an ounce of whisky or brandy, taken during or immediately after meals; for thus the stomachic tonic effects and the force value of alcohol as food are obtained. The now almost universal use of whisky with cod-liver oil is based on sound principles, unless, as is sometimes done, the mixture is given an hour or two before meals, which must have a disastrous effect upon the appetite and digestion.

That there is any specific action of alcohol in consumption—that alcohol has an influence apart from the stomachic tonic effects and as a hidro-carbonaceous food—cannot for one moment be entertained. We may, therefore, dismiss the purely groundless assumption that amylic alcohol has some special virtues.

The use of alcohol in acute wasting diseases must come up hereafter.

In certain diseases of the digestive tube, alcohol in its various forms is highly useful. All the world knows that sometimes obstinate vomiting, not dependent on inflammation of the stomach, may be promptly checked by a tea to a tablespoonful of raw brandy. In the vomiting of uræmic intoxication, and especially of yellow fever, a dry champagne or our native sparkling wine may be very successful. A simple diarrhœa may be checked by brandy. Not to waste time in the recital of these trivial and well-known therapeutical properties, I pass on to the fact that in cholera infantum we have in brandy the most generally efficient curative agent. It is generally known, I believe, that infants bear a relatively larger quantity of alcoholic stimulants, but it is not known that large doses of brandy are singularly efficacious in the disease just mentioned. I have given half a teaspoonful to a teaspoonful of cognac brandy every three hours to a child under one year, and above that to three years one to two teaspoonfuls. It should not be given with milk, but in water as warm as possible, and as remote from the time of food-taking as possible.

II. As a cardiac and arterial stimulant, alcohol is probably more frequently used than for any other purpose. It may be well entitled "the remedy of paradoxes." It is employed in states of depression to give warmth to the surface and to increase the action of the heart and arteries, and in states of excitement of the vascular system and elevation of temperature it is used to lower both. This seeming paradox is readily explained; but the explanation must be reserved.

Alcohol is indicated when from any cause the action of the heart and of the arterial system is depressed suddenly and powerfully. The sudden failure of the heart from moral causes, from shock, from hemorrhage, from weakness of the cardiac muscle, are conditions in which alcohol is indicated, and in which it is used with advantage. There are two explanations of its power to elevate the heart and the vascular tension. The first is reflex. It is within the range of everybody's experience that as soon as a tablespoonful of whisky or brandy reaches the stomach the heart at once moves more rapidly,—that is, the impression made on the end organs of the pneumogastric in the stomach reaching the pneumogastric nucleus, the inhibition exercised by the nerve relaxes, and the accelerator apparatus exercises its power with lessened hinderance. That this explanation applies with much probability is rendered more certain by the · fact that the action of the heart is simply increased by alcohol; in other words, that the diastolic interval is shortened by its influence. The second explanation of the power of alcohol to increase the rate of the heart's movements is that by its oxidation force is generated, which is utilized by the cardiac ganglia. There are other substances, of course, which also yield force; but alcohol has the distinct advantage that its combustion occurs with more readiness than any other available substance. But we shall find hereafter that alcohol in large quantity depresses the circulation. Now comes the explanation of the apparent paradox. Whilst small doses stimulate the circulation, large doses have the opposite effect; therefore, our conclusion is, to overcome cardiac depression the stimulant must be administered in small quantity, but frequently. In many cases, a wine of good body, highly preferable with œnanthic and other ethers, is charged to a spirit.

At this point we should endeavor to arrive at some positive conclusion in regard to the administration of alcohol in preparation for the apprehended cardiac depression of the anæsthetic state, or to relieve it. Prof. H. B. Sands, M. D., of New York, has followed the practice for several years of inducing alcoholic intoxication as a preparation for the administration of ether or chloroform and to prevent shock. It is always inculcated to administer an ounce or two of whisky as a preliminary to the anæsthesia and to the operation. This may or may not be good practice. It is good practice if the heart is sound and the full anæsthetic state is desired; it is bad practice if profound and protracted anæsthesia is not necessary, and if any cardiac weakness exists; for, alcohol being the source whence the anæsthetics are obtained, and a congener as regards their action on the nerve-centres and on the vascular apparatus, it must intensify their effects. The theoretical deduction is abundantly confirmed by clinical experience. In a recent discussion before this Society, our learned colleague, Dr. Turnbull, strongly criticised the subcutaneous injection of alcohol in the case of lethal symptoms caused by ether, and he was sustained by the eminent therapeutist, Prof. H. C. Wood, and myself.

To inject brandy or whisky subcutaneously with a view to stimulate the respiratory and cardiac centres in chloroform or ether narcosis is therefor improper, — contributes indeed, to the fatal result.

III. The antipyretic effects of alcohol involve more doubtful questions than those therapeutical effects heretofore considered, but the weight of authority is unquestionably in favor of the view that alcohol in large doses depresses the temperature. It is not in my province to discuss the physological data on which modern scientific opinion rests, but the practical considerations governing its use in disease. I start with the proposition —the elements of which are too often confounded—that the antipyretic effects of alcohol are distinct from its stimulant and supporting properties. In the treatment of acute inflammations and fevers we may employ alcohol as an antipyretic, to depress the abnormal temperature, or as a support to the circulatory system and a stimulant to the digestive function. In the former case large, in the latter small, doses are necessary.

The modern conception of the fever-process and the relation of high temperature to cardiac and cerebral paralysis have imparted a new and more powerful significance to the use of antipyretics. It may be compendiously stated, without going into the reasons for the statement, that alcohol does not compare with the cold bath, with quinia, even with the digitalis, as an antipyretic. It is true Binz has apparently demonstrated its utility in the hyperpyrexia of pyæmia, but this fact does not invalidate the general proposition. There can be no question, however, regarding the utility of alcohol in the adynamic state in fevers and in inflammations. It is not, however, as a febrifuge, as a depressant of the temperature, as an antipyretic, that it is useful. It may be affirmed that in all cases of the feverish state, if digestion still goes on, if cardiac paralysis is not threatened, alcoholic stimulants are unnecessary. The two conditions demanding the administration of alcohol are failing digestion and failing heart. It is rare indeed that more than two ounces of wine or an ounce of whisky is required, even in the most pronounced adynamia. If this do not stimulate the stomach to better work, or itself furnish needed force to the failing heart, a larger quantity can do no more. How often do we see the stomach filled to overflowing with the stimulants, and yet the heart does not respond, because the smallest amount is not appropriated! When a thimbleful cannot be oxidized, why put into the stomach a pint? Graves formulated some admirable rules for our guidance in the use of alcoholic stimulants in fevers. Alcohol does good when the tongue changes from dry to moist, when the stomach can receive and digest more food, when the pulse declines in rapidity and gains in force, when the surface grows moist and cool from hot and dry, when the delirium ceases and an expression of intelligence replaces blank stupor. In the event of sudden and alarming depression, digestion being suspended, it is possible to maintain the body-forces on alcohol alone. Very numerous examples confirm the conclusions of experiment, that the oxidation of alcohol furnishes the force necessary to maintain the vital operations for a number of days; but it should be definitely understood that in the administration of alcohol for this purpose no provision is made for the necessary processes of wear, and hence there must early come a termination to a life so maintained.

IV. Some recent experience justifies the belief that the antiseptic properties of alcohol may be made available in practice. That it is inimical to animal ferments, to the lower forms of life, and prevents or arrests putrefactive decomposition, are well known facts. That these properties may explain the effects of full doses of alcohol in diphtheria seems probable, for, besides its influence over the cardiac depression, the micrococci colonies may be attacked by it as they migrate. In what way we may explain the results, it seems plain that alcohol, when pushed to a degree which seems somewhat extravagant, has appeared to exert a decided curative influence which cannot be explained by its stimulant and antipyretic action. From this point of view may be regarded the good effects of alcohol in the septic maladies,—septicæmia, erysipelas, puerperal fever, and the exanthemata, notably small-pox. It seems to me beyond question that the undeniably good effects of alcohol in these septic diseases is not explained by referring the results to its other properties. If this be admitted, it follows as a necessary consequence that to be effective in these diseases alcohol must be administered in large doses. There must be maintained, in a certain sense, a saturation of the blood in order to act on the *materies morbi* efficiently. There must be, of course, a judicious measure of the force of the disease, and the doses of the remedy be made to conform. In my own experience I have found these principles a valuable guide to the

treatment of this important group of diseases.

V. I need not detain you long on the subject of alcohol as an anodyne. It quite yields in importance to the valuable derivatives ether, chloroform, chloral, etc. It has probably yielded quite too absolutely to the anæsthetics so called, and might be used advantageously to benumb the sensibility to pain in the minor operations. It cannot be prescribed with safety for the relief of nerve-pain, since the alcohol habit is sure to be formed. As a temporary expedient it may be useful. A tablespoonful of raw brandy or whisky may arrest an attack of migraine impending. An anæmic headache or the wakefulness of anæmia may be quickly removed by alcohol stimulation. It is a curious fact, true also of morphia, that when after the long-continued use of spirit for neuralgia the remedy is discontinued the pain returns.

VI. What is the proper practice regarding the administration of alcohol in certain maladies caused by it? To render my position perfectly clear I must precede the answer to this question by some definitions. Magnan makes a distinction between acute alcoholic delirium and delirium tremens,—the former a delirium due to the impression of alcohol on a brain not accustomed to it; the latter a delirium occurring in the subject of chronic alcoholism. In my "Treatise on the Practice of Medicine" (2d ed., p. 843) I have adopted the old but expressive terms *mania a potu* and *delirium tremens*,—the former corresponding to Magnan's acute alcoholic delirium. I must insist on the distinction between these forms of alcoholic delirium; for the practice pursued is a necessary consequence. Now, as in mania a potu the delirium is a direct result of the alcoholic excess, the first step in the cure consists in the withdrawal of the alcohol. The practice should be different in most cases of delirium tremens. It is true, in a small proportion of cases the outbreak of delirium is due to sudden excess, but usually the opposite condition obtains. Owing to the failure of the stomach, neither food nor stimulant can be retained, and then the peculiar illusions and hallucinations appear. In such cases, also, the food, when the stomach is prepared to receive it, may fail to be digested, because the gastric glands, so long accustomed to the stimulation of alcohol, will not act in its absence. It is surprising how in these cases, after some preparatory treatment, when food is taken with more or less stimulant, sleep and sanity follow expeditiously, without the intervention of doubtful hypnotics.

Alcohol performs another role in those cases of delirium tremens complicated with croupous pneumonia, or cases of croupous pneumonia complicated with delirium. With the progress of the pulmonary obstruction, ischæmia of the arterial system increases, and hence the laboring heart needs the support which alcohol may supply. Thus, having correct principles to guide our treatment of delirium tremens, the vexed question of the use of alcohol in this disease is readily solved.

VII. The power of acohol to coagulate albumen, to suspend the activity of the unorganized ferments, and to destroy animal organisms, lies at the foundation of its external uses. It is a most efficient hæmostatic to restrain bleeding from wounded surfaces, as I have repeatedly verified. As an antiseptic dressing to wounds, to prevent the entrance of the germs of putrefaction, to check suppuration, and to promote healing, it has in my experience scarcely been inferior to the much-vaunted carbolic acid. It is an efficient means for procuring local refrigeration of an inflamed joint or swelling. Injected under the skin in the neighborhood of painful nerves, it has no inconsiderable anodyne power. This property may be utilized for the relief of myalgia and lumbago. A few drops thrown into the affected tissues will usually afford permanent relief. It is more efficient than water, used by the method now known as aquapuncture. Enlarged tonsils, hypertrophied thyroid, and glandular swellings may often be slowly reduced and made to disappear by the parenchymatous injection of alcohol. This method is applicable to the treatment of uterine fibroids.

In cases of sudden depression of the powers of life, whisky may be thrown under the skin, to obtain more speedily than by the stomach its effect as a cardiac stimulant. I have already mentioned the contra-indication of this practice in the treatment of the narcosis or the cardiac depression caused by the anæsthetic agents ether and chloroform. It is, however, perfectly applicable to other conditions characterized by sudden and powerful depression of the heart. The amount injected varies from ten minims to a drachm ; a syringeful—about half a drachm—is usually inserted at once, and is repeated according to necessity. Some local swelling arises, which may subside in a day or two, or an indurated lump remain for a time, or rarely suppuration may occur.—*Philadelphia Medical Times.*

Headache and Nervous Exhaustion, Its Cause and Cure.

BY EDWIN W. HILL, M. D., OF GLEN FALLS, N. Y.

It is a lamentable fact that, until within a recent date, the medical profession have been at least two generations behind the times in which they lived ; and even now the older members look with distrust (to say the least) upon the younger portion of the fraternity who are anxious to get up to the times and avail themselves of any means to cure their patients. When Prof. Donders explained to the profession that defects of vision caused so much nervous suffering in the world, how was it received ? Was the subject thoroughly investigated, and so proven to be either correct and adopted or incorrect and rejected ? I think not. And even now only a few of the more advanced members are willing to leave off drugging and look to other sources for the cure of the patient. And here let me copy the words of Dr. Jones, of the Queen's University, Cork, Ireland, "How many an unfortunate might escape a world of drugging if the practitioner could recognize the effects of astigmatism in the headache, the dizziness, the inability to work, symptoms so often referred to the stomach—all corrected by suitable glasses." Yes, the time has come when the more enlightened part of our people have refused to allow their physician to practice the heroic treatment. It is no longer a theory, but a known fact, that headache, whether habitual or periodic, spinal irritation, dizziness, insomnia, nausea, inability to do mental work, general failure of health, and not a few of the many cases of St. Vitus' dance, are due to eye strain, and can only be permanently cured by the correction of visual powers, and so remove the cause of irritation. The eye strain may exist and still the patient never experience any pain in the eye itself, just as in hip joint disease the seat of the trouble is in the hip and yet the painful part is the knee. Again, eye strain may exist for a very long time and cause no disturbance whatever, until the general health is weakened, either by mental or moral causes.

I prefer to prove these points by cases treated by myself, rather than bare assertions, and so I will give some in detail.

The following case, Miss R., the beloved granddaughter of an M. D., I will give in the words of the Doctor himself:

"In accordance with your expressed wish for a concise history of her case up to the time she came under your care, I will now give it to you. Miss I. C. R., of Moreau, Saratoga County, New York, aged 24 years, light complexion, brown hair, blue eyes, nervous temperament, etc., etc. For six years previous to June, 1875, she attended school. She was a close student and indefatigable in her efforts to succeed. Her health began to fail before half of her term of scholarship was finished ; still she would not give up, and struggled on, having a severe pain in the head, neck and shoulders ; through the eyes, at times, it was piercing and unbearable. She had set her mark and was not to be deterred from attaining it. She succeeded, as many others have done, with health, mind and constitution weakened. She graduated in June, 1875, and has since been striving to recuperate ; at times she has,

in a measure, succeeded, and enjoyed a fair degree of health, but still the old pain in her head, eyes, and along the spine has continued.

"During the winter of 1876 she had an attack of congestion of the brain, and was confined to a room, without a ray of light, for six weeks. It was four months before she was able to sit up all day. She made a fair recovery, only there was great nervous excitement and the least exertion would cause a day's sickness in bed. Since her long sickness, generally in the winter, she has had a slow nervous fever, lasting from three to five weeks. During the past winter the pain in the head has been intense and constant, but for short intervals, the eyes frequently becoming obscured and the brain confused; I feared permanent injury to the brain. Upon the advice of a lady friend of hers, whom you had cured, I, although skeptical, sent for you to come and examine her case. And when you said you could cure her I had fearful doubts. And now, after an absence of seven weeks, she has returned, the pain in the head gone, and for the past five weeks she has not had as much pain as before she had in an hour. The circulation perfectly restored, the heat in the head is gone, the extremities are warm, and a perfectly normal condition of the whole system ; her strength is improving, appetite good; for all of which she cannot be too grateful to Dr. Hill.

"Yours,

"B. F. CORNILL, M. D."

Upon examining Miss R. I found that there was just eye strain enough to prevent her ever getting strong, and I found from the examination no other cause, and therefore I could, with assurance, say this is a case that I can cure. She has since attended the Doctor in a painfully protracted and fatal illness, and her strength is good and she well.

Miss J. C., the daughter of a wealthy manufacturer, a well developed, healthy looking young woman, consulted me, complaining that she had struggled through her education with constant fatigue and headache, and for the last year the pain in the lower part of the back had been so severe that she feared some spinal or uterine disease. She had been at home about a year, during which time she had been a good deal under the family physician's care, on account of the pain and her nervous condition. She had a pair of glasses that she had tried to wear, and see if they would help her. They did not accurately correct her defect, astigmatism, therefore, they did no good, if they did not make her worse. I corrected her astigmatism with a (— cyl.) glass that she wears constantly. The internal recti muscles of the eye had become very much weakened from overwork; these I strengthened by the prism exercise, and in four weeks, instead of the poor nervous sufferer that she had been, she commenced taking painting and music lessons again; a thing she assured me she never expected to do.

Mrs. M. B., aged about 45 years, came to my office complaining that ever since her married life she never had gone over a month without having a sick headache, which would last sometimes for three days, during which time she would be almost blind with the pain, from the light striking into her eyes. Again, sometimes, she would have the headache almost every week of the month, for a day or so. After taking twenty prism exercises, she went a whole year without the headache returning.

Miss J. W., one of our most educated and refined ladies, consulted me as to what could be done for her, and it was a sad history; a life spent in hard work and close application to study, that she might enjoy the company of the best men by reading them through, finds herself a wreck, unable to sleep at night, and when sleep did come a troubled sleep. If she indulged in an evening entertainment then a sick headache was sure to follow the next day. Nervous and fretful in the extreme. I am pleased to say that the prism treatment has restored her to health and strength.

Mr. J. H., aged about 48, a prominent business man of his town, came to me to be fitted for a pair of glasses. I found that the left eye was slightly astigmatic, the right normal, and still he said that it was getting to be quite troublesome for him to read very much of an evening, as the words seemed to run together and look blurry. He also informed me that when they commenced business he did the book-keeping for the company, and that when he had to confine himself pretty closely to work, as at pay time, then he always was sick and nauseated, and had what they called a bilious attack, which would last from one to three days, but since they have kept a book-keeper he has been free from the attacks. Had his physician in his observation thought of cause and effect, and recognized that the bilious attack always followed close work in the office, and sent him to an ophthalmic surgeon, who would have corrected the slight astigmatism, 1-60, by a (-|- cyl.) glass, he would from that day have been cured of his bilious attack. To show that this is no bare assertion of mine I will quote a parallel case from Dr. S. Weir Mitchell's Report of 1876.

Mr. B., a prominent merchant, consulted me for pain in the upper spine and occiput. It increased day by day every winter, and left him during the summer, which was spent in fishing and shooting, a tent life, in fact. Mr. B. was even cauterized in New York for these pains, and here at home had much able advice besides my own.

When I first saw him I was thoroughly misled. It was late in the winter, and, as usual, while in the autumn only writing, at first, and then later, reading, and then any near work, caused pain, as time went by there came a time when all mental labor, when excitement, emotion, or any thought, caused pain. He was in this over-sensitive state when I saw him, and was aided by nothing I did. His holiday cured his head, and on his return some friend, I believe, suggested to him that his eyes might be weak, and with this idea he consulted Dr. W. Thomson, who gave me the following additional particular from his note-book: "Writing has become so distressing to this gentleman that for a year past all letters have been written by a secretary, at his dictation. He states that a few moments spent in writing gives him a creeping sensation up the spine and through the back part of the head, followed by giddiness and severe pain, so urgent as to render him fearful of a fit of some kind. His sight was, for distant objects, one-third and was improved by a concave cylinder glass 1-18, which he had obtained from a prominent ophthalmic surgeon, and with this his sight was ¼, or up to the normal standard; but despite this success his headache and other nervous symptoms were unrelieved.

It was suspected that this would prove to be a case of compound astigmatism, and that he had overstated the myopia on the previous examination, and that the diagnosis had been considered correct by obtaining the full sharpness of sight by a 1-18 cyl. It was conceived that there might be latent hypermetropia, and that in overcoming a long-sighted meridian he would overstate a myopic one, and thus be provided with a correction for short sight, which would give sharp sight, but leave him hypermetropic, and, therefore, oblige him to keep his ciliary muscle in vigilant and constant exercise in viewing distant as well as near objects.

On paralyzing his accommodation with atropia this was found to be correct, and the hypermetropia was found to be 1-36; his formula for the astigmatism was -|- 1-36 ◯ — 1-16 cyl. ax. 1°, and this correction gives him an acuity of vision = 15-x or above the average standard. On using these glasses habitually his distressing symptoms quickly disappeared. He has long since forgotten his apprehensions of an impending apoplexy or epilepsy. He can see as sharply as any of his companions, and he can use his eyes continuously in reading, writing, or other near work. Relief in this case followed, at once, the use of the glasses, which proved competent, without other means, to conduct him to perfect and useful health again.

In conclusion let me again draw the attention of any who may read this article to the important facts—

1st. That headache, whether habitual or periodic, is in the great majority of cases due to eye strain.

2d. Eye strain, when it exists, will always cause the following train of symptoms : Pain in the front or back of head; pain along the spine; nausea; insomnia; nervous irritability, nervous prostration, and often the following: Neuralgia of the head and face; palpitation of the heart, and atonic dyspepsia, all of which are not cured but only relieved, by drugs. But remove the strain, either by glasses or prism treatment, or both, and you cure the patient.—*Philadelphia Medical and Surgical Reporter.*

On Some Points in the Medical Treatment of Habitual Constipation.

BY J. W. HICKMAN, M. D., OF DELTA, PA.

The importance of being able to treat successfully the ailment in question cannot easily be over-estimated. Besides being a very common disorder, it is one that frequently taxes to the utmost the resources of the practitioner. Perhaps the error most frequently committed in its management is the indiscriminate giving of purgatives. Rarely does a patient apply to the physician for relief before he has given himself over to persistent purgation, and yet he not infrequently presents himself expecting further treatment in this direction. And, too often, it must be admitted, is the physician led into such indiscreet practice. To serve more than a temporary purpose, active purgatives should find no place in the treatment of habitual constipation. Indeed, it were well, in most instances, if this class of drugs were altogether dispensed with in this connection. The temporary relief is more than counterbalanced by their harmful effects in even the most favorable instances. By their use we have, not infrequently, induced a chronic intestinal catarrh of a very intractable sort. Instead of restoring the bowel to its normal state, either with reference to the secretions or to contractility and tone, we find the torpor and general impairment of function intensified. Nor can they fail to disorder the digestive and nutritive functions.

As to laxatives, properly so-called, the verdict must be different. Both medicines and foods of this character, either alone or with other treatment, are very frequently of service, and occasionally are positively indicated. Especially is this true of the initiatory measures. In this paper we do not pretend to delineate every form of medicinal treatment, but it shall be our purpose to mention those agents which we have used extensively and have found to be of real service.

As coming first in the list, the writer must frankly confess his preference for belladonna. He, of course, does not mean to imply that it should be regarded as a specific, but that it will be found effectual in a vast majority of cases. In prescribing belladonna, two considerations must engage our closest scrutiny. In the first place, we should know that we have a pure drug, for it is well known that a large proportion of the belladonna preparations in the market are of an inferior grade. In the second place, the manner of prescribing has much to do with the result. Prof. Da Costa, in a recent clinic, prescribed one drop of the fluid extract in one drachm of the compound tincture of cinchona. This I have tried repeatedly and with marked success. The belladonna may also be combined with other vegetable bitters, but for some reason it is more effective when given in Huxham's tincture. Instead of one drop of the fluid extract we may give four drops of the tincture, with equal confidence. Dr. Da Costa also advised in connection with the above, one tablespoonful of sweet oil at bedtime, until the bowels became regular. This is a fair expedient but I have been unable to see that it exerted any specially favorable effect. A laxative diet for the first few days is of more service. It must be admitted that it is not just easy to foresee in what cases belladonna may succeed or may fail, but that it may be used with confidence in a majority of cases, if given in proper dose and in purity, there is no question.

Another agent of value is nux vomica. Its results are somewhat variable, however. It is most useful in those cases attended with tardy digestion, flatulence and swelling of the belly. It is also serviceable in those instances of constipation occurring in those who are given to hypochondria, and are visited by quasi-paroxysmal pains through the stomach and bowels. It should be given just before meals in the form of tincture; or, instead, five, six, or eight drops of the preparation known as the bitter liquor of Baume may be prescribed.

We have, also seen excellent results from the following:

R.   Ext. nucis vomicæ,
        Ext. belladonnæ,
        Ext. physostigmatis,   aa   gr. v.
        Ft. pill. No. xij.   M.

Of these one should be given at bedtime.

In the treatment of constipation sometimes attendant upon chlorsis, no remedy so well answers the indications as aloes. In addition to the usual treatment, we may with marked advantage add from 1-2 to 1 1-2 grs. per day. If there be amenorrhœa the aloes will serve a twofold purpose; it will assist the *prima viæ* to a better performance of function, and, being an emmenagogue, it will tend to restore or inaugurate, as the case may be, a regularity of the menstrual flow. If with the above condition of things, however, we have menorrhagia instead of amenorrhœa, the remedy in question must, of course, be withheld in favor of some more suitable agent. Again, no treatment could be more satisfactory than the management by aloes of that form of constipation dependent upon weakness and diminished power of contractility of the large intestine. Especially does aloes become the proper agent, if to this condition be added a "bilious state." Practically such an association is frequent, and may be effectually remedied by laxative doses of aloes. It will be remembered that the researches of Rutherford and Vignal have demonstrated the power of aloes to stimulate the hepatic functions and to increase the flow of bile. Experience, it must be added, has confirmed this conclusion.

Another highly efficient agent in the management of habitual constipation, and one by no means sufficiently valued, is podophyllin. Unlike many, and indeed most, other drugs of its order, podophyllin may be used a long time without any diminution in its efficacy. It agrees exceedingly well with the stomach, and hence may be relied upon where constipation is associated with even the most trying forms of dyspepsia. It has no effect, either direct or indirect, upon the womb, hence it is well adapted to the costiveness of pregnancy. It may also be used with advantage during the treatment of the various uterine diseases, or even during the menstrual period, and it is the best remedy with which to treat the constipation of old people. As is well known, its effect is slow; a small quantity at bedtime will usually act by next morning, but may not until the day following. Notwithstanding its tardiness the desired result is sure. When constipation is found to persist in spite of other agents it will usually yield to podophyllin. Sometimes, however, its use must be continued indefinitely, and without our being able to note much good result after its suspension. Such examples are rare, however. After the bowels are regulated it need not be given more frequently than once every other day, as a rule. And occasionally it need only be repeated upon days in which there is no stool. It is best given in doses varying from ¼ to ½ grain, in pill, with Jamaica ginger, to which a little extract of cannabis indica may be added, as recommended by Habershon.

Now, the remedies above enumerated are, as is well known, not the only ones used in the treatment of chronic constipation, but they are those from whose use we have derived the best results. We do not pretend to say that every case will yield to their power, for we find instances that stubbornly resist every possible resource that may be brought to bear. In conclusion, it must be remembered that failure will be the inevitable result of treatment without the persistent and faithful cooperation of the patient. For it is by habit upon the part of the person treated that the advantages gained by treatment may be retained. No obligation should be more imperative than the adoption of a regular hour for attending to the duties of the water closet.—*Philadelphia Medical and Surgical Reporter.*

EDITORIAL.

St. Luke's Hospital.

The following gentlemen have been appointed as the Medical Staff of St. Luke's Hospital, by the Board of Managers of that institution: *Attending Physician,* J. H. Kimball, M. D. *Attending Surgeon,* A. J. Russell, M. D. *Consulting Physicians,* W. H. Williams, M. D., W. Edmundson, M. D., A. Stedman, M. D. and H. A. Lemen, M. D. *Consulting Surgeons,* F. J. Bancroft, M. D., H. K. Steele, M. D., J. C. Davis, M. D. and C. M. Parker, M. D. *Attending Surgeon for Diseases of the Ear, Nose and Throat,* A. Wellington Adams, M. D. *Attending Surgeon for Diseases of the Eye,* A. F. Kibbe, M. D. *Resident Physician,* E. C. Rivers, M. D.

As will be seen by reference to the advertising pages, the Hospital is now open for the reception of patients. Physicians throughout our own State and elsewhere where hospital privileges are lacking may feel safe in recommending St. Luke's to their patrons and friends. The location and surroundings are exceptionally fine, the building is commodious, and the selection of officers and nurses has been judicious.

A New Stethoscope.

Physical diagnosticians have long felt the need of a more perfect stethoscope; one built upon strictly acoustic principles and suitable for purposes of demonstration.

The ordinary binaural stethoscope now in general use does not possess these attributes.

The mere mode of approximating this instrument to the walls of the chest is productive of confusing friction sounds peculiar to the act, and due to the respiratory movements of the chest; besides, the cone which is intended to be applied to the part to be auscultated, for purposes of collection, exclusion and concentration, is not constructed in a manner best calculated to accomplish its object.

Again, the rigidity and metallic structure of this instrument gives to all sounds perceived through it a peculiar metallic timbre, accompanied by a roaring, which is calculated to mislead the examiner and drown out less prominent sounds. For this reason many physicians condemn its use, and in preference rely upon the unaided ear.

Dr. Constantin Paul, of Paris, by a combination of his own flexible stethoscope and Dr. Ronssel's thoracic end piece, seems to have produced the desired instrument.

It consists of a flexible rubber tube attached to a cone-shaped thoracic end-piece, around the base of which there is formed a circular cupping-glass or exhaust chamber. In connection with this chamber there is a rubber bulb, which answers the purpose of an exhaust pump.

With this arrangement it is possible to more closely apply the stethoscope to the thoracic walls, and there retain it automatically in any one spot for any desired length of time. The auscultating tube may at the same time be passed around to any number of students without disturbing its position, the advantage of which will be readily appreciated. It is not necessary to touch the instrument in order to hold it in place, and the thoracic end-piece cannot be disturbed and friction sounds produced by the respiratory movement. It may be closely approximated to a tender chest without exerting any pressure, and held there for an indefinite period without fatiguing the auscultator. The air chamber also acts as a resonator for the reenforcement of all sounds.

The inventor claims that this stethoscope possesses superior acoustic·qualities and facilities for demonstration.

We understand M. Galante, of Paris, is constructing them ; and a full description may be seen in the *Revue Medicale Francaise et Etrangere*, of July 2, 1881.

Perhaps it would be well in this connection to mention the fact that Mr. Irwin Palmer describes, in the *British Medical Journal* of April 30, an addition to the binaural stethoscope in the shape of a dial-plate which registers the divergence of the metal arms, thus enabling the instrument to be used as calipers to measure various diameters, expansion of chest, etc.

An Abstract of Fournier's Views on Syphilis and Marriage—Some Points of Difference Between Fournier, the Great French Authority, and Kassowitz, the German Authority.

Fournier, upon concluding an "introduction," in which he dramatically and forcibly sets forth the gravity and importance, both from a social and medical stand-point, of the complex question he has undertaken to discuss, propounds, as a preliminary question, the following : "*Does syphilis constitute an express interdiction, an absolute obstacle, to marriage?* This most important question, at once involving problems difficult, delicate and perilous, and which he claims would be answered by most physicians in the affirmative, is thus disposed of by him : "The truth is, that, save very rare exceptions, syphilis constitutes only a *temporary* interdiction to marriage and that a syphilitic subject may, after a certain stage of sufficient depuration, return to a state of health which fully restores his fitness for the double role of husband and father." "Yes, a hundred times yes, *one may marry after having had the pox,* and the results of a marriage contracted under these conditions *may* be absolutely safe, medically speaking. This is for once an established fact, a demonstrated verity." And then he hastens to add, in way of an explanation, that one must not marry in this special situation without being declared free from liability to subsequent manifestations of the disease. "A man marrying with syphilitic antecedents may become dangerous in marriage then in the following relations: *First,* as husband; *Second,* as father; *Third,* as head of the social community constituted by marriage. In other words, he may become dangerous:

1. To his wife.
2. To his children.
3. To the common interests of his family.

From a total of 572 syphilitic women observed in private practice, no fewer than eighty-one were found to have contracted syphilis from their husbands soon after marriage. This occurs in two ways: First, transmission to the wife of a contagious lesion occurring in the husband after marriage, Second, transmission by conception—the wife becoming syphilitic *without the husband having at the time the least external lesion capable of infecting her."* Well substantiated cases of this kind are cited, and the writer contends they are of very frequent occurrence. This mode of infection the author does not, however, believe to be due to a contagion by the sperm. "A woman becoming syphilitic in this manner without initial lesion, without chancre, and from the contact of a husband exempt since his marriage from every contagious lesion, is *enceinte,* and has received syphilis by *conception.* In such a situaiton, *pregnancy is never absent.* The infected woman here receives the syphilis, not from her husband, but from her *child.* It is not syphilis by contagion in the usual manner, but it is a syphilis conceived *in utero,* introduced by the infant into the womb of its mother, communicated to the mother by her infant—in a word, it is what is termed *syphilis by conception.* In this class of cases the infant generally dies before being born, and this fact constitutes a presumption in favor of its being syphilitic. In some cases, however, it is born alive, and then it manifests unequivocal symptoms of syphilis. Now, if the infant in such circumstances is tainted with syphilis, what is there impossible or extraordinary in its transmitting

the disease to the mother during its intrauterine life?

If maternal syphilis has the power (which every one admits) of reflecting itself upon the infant, why should not the syphilis of the infant reflect itself in like manner upon the mother? What! Here is an infant which, procreated syphilitic by the agency of its father, lives syphilitic during several months in the womb of its mother, and yet you would think it extraordinary, impossible, that the infection of the infant should be transmitted to the mother! A syphilitic organism included within a healthy organism, and the one not contaminate the other! Does this maternal infection result from the contact of a fecundated ovule, and propagate itself, either in the Fallopian tubes or in the uterus, at the time when this ovule is not attached to the mother by any organized graft? Or is it effected subsequently by the exchange of the placental circulation, as M. J. Hutchinson believes? Upon this point we confess complete ignorance. We accept the fact and leave the interpretation to others. Hence, then, a man with syphilitic antecedents who contracts marriage may become dangerous to his wife in two ways:

1. *Directly* by transmissible *contagious lesions,* which may happen to him after marriage.

2. *Indirectly,* through his *fecundating power;* that is, by the *procreation of an infant, the infection of which may be reflected upon the mother.*

Until within a recent period the theory of the *paternal* heredity of syphilis was accepted without opposition, except from few. But the aspect of this question has indeed changed within late years; numerous observations, important investigations, have sprung up on all sides, tending to singularly restrict the sphere of paternal influence in the hereditary transmission of syphilis. Some investigators have gone so far as to deny the paternal influence in the transmission of the disease, declaring it to be absolutely null. The solution of this question is of great importance. For, when a syphilitic patient comes to us to know whether he may or may not marry, our responsibility will be by so much lightened, if we have before us the certainty that this man, although syphilitic, can in no way be prejudicial to his children. In solving this problem *a priori* reasoning alone is not depended upon, but clinical observation and evidence form the ground-work for a conclusion.

It is admitted to be absolutely true that we encounter in practice numbers of men, who, having contracted syphilis before marriage, have begotten children healthy and exempt from syphilis, their wives also remaining healthy and uninfected. It is a matter of frequent occurrence for syphilitic men to beget healthy children, and *afterward* present such and such accidents of syphilis, unequivocal evidences of the persistence of the diathesis at the time when conception takes place. I have known syphilitic patients to procreate healthy children, free from the least symptom, when they were in the *full second period* when they were affected *even at the moment* of conception, with various syphilitic accidents. Thus, here is a syphilitic man who, *the very day when he procreates a child*, presents accidents of the secondary stage, and whose child, nevertheless, is born exempt from syphilis! The conclusion therefore is, that syphilitic heredity proceeding from the father (and from the father alone, the mother remaining healthy) is much less active, much more restricted than had heretofore been supposed. Under given conditions on the one hand, a syphilitic husband, and on the other a healthy wife, the chances altogether are that the child issuing from this couple will be exempt from syphilis. This, contrary to the old beliefs, contemporary researches have clearly and positively established. But, this fact recognized, this concession made to the partisans of the doctrine which I oppose, I at once resume my position on the strength of observation, on the strength of clinical facts, and I say to my opponents: No, it is not true, unfortunately, in the case of syphilis, that the paternal influence is so immaterial as has been pretended;

still less is it true that it is *null*. Although paternal heredity exercises its influence in a rare, exceptional manner, still it exercises it *sometimes*. Numbers of cases of this kind have been reported by different authors, notably by MM. Ricord, Trousseau, Diday, Depaul, Cazenave, Bazin, Hardy, Barensprung, Hutchinson, Bassereau, Beyran, Liegeois, DeMeric, Martin, Parrot, Lancereaux, Kassowitz, Charpentier, Pozzi, Keyfel Carl Ruge, and others. However rare, however exceptional the hereditary transmission of syphilis from the father to the fœtus may appear to be, it may, nevertheless, exert itself in this way in a certain number of cases; revealing itself by the early death of the fœtus, either *in utero* or very shortly after birth; or by constitutional vices, morbid aptitudes, defects, inherent infirmities, congenital malformations, arrests of development, etc., which I, like many physicians, look upon as constituting the modified, transformed expressions of specific heredity.

A principal point, upon which my conviction is now well established, is, that a child born of a syphilitic father and of a healthy mother is liable, by the fact of the paternal syphilis, to die before coming into the world. After several years of investigation in this direction, an abstract of my observations furnishes me with no fewer than fifty abortions occurring under the above-mentioned conditions, and produced without other possible cause to be alleged than the paternal diathesis. From these same statistics I find no fewer than thirty-six other cases of pregnancy (always the issue of a syphilitic father and a *healthy mother*) which have resulted at term in infants *born dead*, or dying, or sickly, stunted, emaciated, senile children, doomed to an early death—they have an *"inherent inaptitude for life."* To recapitulate then, paternal heredity, while not exercising itself except in a limited number of cases, is none the less liable to exercise itself sometimes in a manner very positive, very manifest, and then it reveals itself according to three modes: Either (this is the exception) by the transmission of syphilis to the fœtus; or (this is much more common) by the death of the child; or, finally, by inherent degeneration of the germ, which reveals itself subsequently under very diversified morbid forms.

Let us not lose sight of, this other cardinal point: A syphilitic father is dangerous to his children not only in his character of progenitor, but he is, or may become, dangerous to them in his character as the husband of their mother, if I may so express it. In other words, he may endanger them through *the syphilis which he runs the risk of communicating to his wife.* And then, the father and the mother both becoming syphilitic, what must be the fate of the children issuing from this infected couple? Ah! here is presented a page of pathology distressing to write; here commences for these families a situation truly heart-rending, which it is necessary to have observed in all its details and in its diverse forms in order to comprehend its miseries. In illustration: Two young persons were married a short time ago. The wife has become *enceinte,* and yearns after the title of mother. The two families full of the sweet hope which preludes the coming of the new-born, impatiently await the result of this pregnancy. Now, what will be the result? What will happen to the infant procreated under the conditions which we are now supposing—that is to say, issuing from a father and mother both syphilitic? As physicians, we can predict what will happen to it, for, save in rare exceptions, its future is comprised within the three following alternatives: 1, either this infant *will die before birth;* 2, or it *will come into the world with syphilis,* and with all the possible and serious consequences of infantile syphilis, which, in most cases, is almost equivalent to a sentence of death; 3, or, finally, it will come into the world without syphilis, but with a *health compromised, with an inate debility,* and a constitution impoverished, which will expose it to an early death, with menacing *morbid aptitudes,* and with a tendency to certain organic vices.

And this is not all, for there may succeed a second, a third, a fourth pregnancy. It may be that this identical fate awaits the second, the third, the fourth infant; and so on until the diathesis has been exhausted by the effect of time or by the intervention of an energetic treatment.

What a situation! What affliction for a couple! What grief for their two families! In another point of view, what a social calamity!

This is what the pox does, or can do, when the paternal influence and maternal influence are associated; when both conspire together against the product of conception.

Of the three alternatives above specified, the first two are unchalenged propositions—they are fixed truths of medical science; regarding the third, however, such entire unanimity does not exist, but as for myself, my position is taken upon this question which has long engaged my attention, and which I have studied, I believe I can say, with minute attention. After having doubted, I doubt no longer, and my present conviction is that the syphilitic influence of parents does not reveal itself in their children by symptoms of a syphilitic order only, but also by morbid conditions, by morbid dispositions, nowise syphilitic in themselves, which have nothing to do with the classic symptomatology of the pox, which are even as different from it as possible, but which, nevertheless, do constitute modified expressions of the diathetic state of the ancestors, do constitute, if I may so express it, a sort of *indirect descent* of the pox."

The author here goes on and in a very vivid and terse manner describes the condition of such children on entering the world, their struggle for existence, their occasional "pull through," but more frequently by far he claims they gradually fade away because of the functional insufficiency of their organs; or, as frequently, they die suddenly, without any assignable cause. He considers the latter class of cases almost pathognomonic of a modified form of syphilitic heredity. Note well these cases of *inexplicable sudden deaths!* At times this hereditary diathesis will manifest itself as affections of the nervous system, convulsions, etc.; at others it will appear as a powerful predisposition to meningitis; and again, it will show itself in the form of an intellectual incapacity, bordering on imbecility or idiocy. You may be sure that many children, backward, *imbecile* or *idotic*, are nothing else than the products of syphilitic heredity. Again, it sometimes constitutes a predisposition to *hydrocephalus*. And, finally, there rises the question of *lymphatism* and *scrofula*, which some regard as only disguised forms of hereditary syphilis.

Assuredly, it would be a great exaggeration to regard scrofula as a degeneration of syphilis. Assuredly, it would be a serious error, from a pathological point of view, to make it subordinate to syphilis, to consider it in the light of a bastard, transformed, metamorphosed syphilitic affection. Scrofula, unquestionably, has no need of syphilis in order to exist. It exists by itself alone, or, at least, it is the effect of causes which have nothing to do with the syphilitic virus. But, on the other hand, it is no less certain that syphilis constitutes, if you will permit me the expression, *one of the affluents* of scrofula. It brings its contingent to scrofula, by virtue of its being a debilitating, anæmiating disease, a disease impoverishing the organism, deteriorating the constitution, ruining the vital forces. It beckons scrofula in its train, it predisposes to it in the same manner as do all depressing causes, in the same manner as misery, insufficient alimentation, captivity, etc. And this action which it exercises upon the health of the parents is reflected and revealed afterward in the child, by manifestations peculiar to lymphatism in general, and to the highest degree of lymphatism, that is, scrofula.

Hereditary influence, therefore, becomes veritably *disastrous* when both father and mother are diseased. To estimate the quotum of the hereditary reaction of each of the parents upon the fœtus is almost an impossibility, for the numerical data which would enable us to institute a parallel between the results of paternal heredity and maternal heredity exercised separately are wanting.

We can simply say in a general way: The syphilitic influence derived from the father reacts upon the child in only a limited number of cases, while the syphilitic influence derived from the mother, is exercised upon the child in a manner much more frequent, much more active, and altogether much more dangerous.

And now, regarding the third point. A man who enters into marriage with a syphilis not extinct, may become dangerous through himself to the interests of his family. *Morality* is here about to join itself and enter in line with pure pathology. Syphilis is a serious, a very serious disease, liable to end, when left to itself or insufficiently treated, either in important affections or in serious infirmities, or even in a termination more lamentable still—death. This is notorious, but what I wish particularly to impress is, that these accidents seldom manifest themselves until after a *remote maturity*. It is in the tertiary period, a period almost indefinite in duration, that the grave manifestations, the veritable catastrophes of the pox, occur. Now, if this be so, remark then, I beg you, what becomes the situation of a man who, with a syphilis contracted in his youth and *not sufficiently treated*, presents himself for marriage.

The situation, medically, is that of a man who has every chance to be exposed, in a more or less distant future, to the assaults, more or less formidable, of the diathesis. The situation is that of a man with health compromised, of a man damaged physically, indebted to the pox, and destined, sooner or later, to discharge that debt. In such conditions, is it admissible that this man should aspire to marriage? Is it honest, is it *moral* that this "future sick man" should think of becoming a husband and a father? And, if he consults us, as physicians, to know whether he is fit for marriage, can we, ought we to allow him to engage in this undertaking upon our own responsibility? No! It is not admissible; it is not honorable; it is not moral for a syphlitic subject to contract a marriage under such conditions. Marriage in its completeness is not only an affair of sentiment, of passion, of convenience, and of interests. But, considering it from a more practical and elevated stand-point, it is an association freely entered into where each contracting party is pledged to bring in good faith a share of health and physical vigor, with the view of co-operating, on the one hand, for the material prosperity of the family, and, on the other hand, for the raising of children —the supreme and sacred end of every union. A syphilitic man not yet cured of the syphilis by a proper course of treatment extending over a sufficiently long period, would contribute to this partnership a health compromised, hypothecated, burdened with a debt hereafter due the pox, that pitiless creditor. In illustration of the diversified results of such a co-partnership the two following cases are cited :

"A student of medicine acquires syphilis, and judges it proper to treat himself exclusively with iodide of potassium, not being willing to take murcury. A short time after his doctorate he marries. Some years later he is affected with a slight paraplegia, which is referred to syphilis by common consent of all the physicians whom he consults. Notwithstanding, he still treats himself in a very irregular fashion, 'by fits and starts,' using his own expression. Finally, he becomes absolutely paralyzed in the legs, and I find him when he presents himself to me, in a state of absolute incurability. Judge of the situation of our unfortunate *confrere* when you learn that, without resources, he remains infirm, with the charge of a decrepit mother, a wife, and two children."

"A young business man contracts syphilis, and is treated regularly for some months. Relieved from all apparent manifestations of the disease, he believes himself out of danger and discontinues all treatment. Three years later, without consulting a physician, he marries. Scarcely married, he communicates syphilis to his wife, through a relapse of secondary accidents which occur on the penis. Then he is attacked with symptoms of cerebral syphilis, which I succeed in subduing at first, but which make a new invasion and rapidly carry off the patient. *Epilogue:* The young wife, becoming *enceinte* at the beginning of the marriage, brings forth a syphilitic infant, which an active medication succeeds in saving. Very soon she presents multiple symptoms of malignant syphilis— confluent eruptions, cephalalgia, violent neuralgia, ecthymatous eruptions with phagedenic tendency, reproducing themselves when scarcely cured, and ending in covering the body with monstrous sores. Under the influences of such symptoms, her health is altered, emaciation, decline of strength, loss of appetite, digestive troubles, diarrhœa, finally pulmonary tuberculosis, and death from the cachexia—an orphan and without resources, the child has to be relieved by public charity."

Thus, then, a syphilitic man may be dangerous in marriage in a triple manner : To his wife, in transmitting to her the disease with which he is affected; to his children, by way of heredity ; to his family, from the personal risks to which he remains exposed.

From what precedes, then, the natural conclusion is :

1. That marriage should be forbidden to every man who still presents a syphilis sufficiently active to be dangerous.

2. That, conversely, it may be permitted to every man in the opposite conditions.

Now, then, in what conditions does a patient affected with syphilis cease to be dangerous in marriage? Or, what amounts to the same thing, in what conditions does he become *admissible to marriage?*

Never before has this grave question of the marriage of syphilitics been confronted, discussed debated. Of course, you will find here and there certain general hints, certain indications—always more or less vague— incidentally thrown out upon this subject. But nowhere, I assure you from my own experience, will you encounter a veritable programme formulated in *ex tenso,* or even outlined.

In my opinion, according to what I myself have seen, and the results of my reading, the principal conditions which a syphilitic subject ought to satisfy in order to have the moral right to aspire to marriage may be summarized in the following programme:

1. Absence of existing specific accidents.
2. Advanced age of the diathesis.
3. A certain period of absolute immunity consecutive to the last specific manifestations.
4. Non-threatening character of the disease.
5. Sufficient specific treatment.

If a patient satisfies all these combined conditions, I consider him fit to become a husband and father without danger.

Regarding the first of these conditions there is nothing to arouse dispute. It matters but little whether or no the accident be of a transmissible nature or not, for—1. If it be of a transmissible nature, the contra-indication of marriage is as express and absolute as possible; 2. If it be not of a contagious nature, it none the less reveals a permanent diathesis, with all its dangers and consequences. Such audacity as marriage with existing specific accidents seems hardly credible, but I can assure you I have times without number seen people marry while presenting *the very day of their nuptials* such and such syphilitic symptoms as, cutaneous syphilides, mucous patches of the mouth or throat, genital mucous patches, specific sarcocele, accidents premonitory of cerebral syphilis.

Regarding the second condition: The more recent the syphilis of the husband, the more numerous, the more serious will be the dangers which he introduces into marriage. Hence this corollary: The older our patient's syphilis, the more shall we be authorized (save for special indications of another kind) to tolerate his marriage.

In every respect, considering the dangers both wife and offspring are subjected to, the long standing of the disease constitutes an essential, an indispensable condition of admissibility to marriage. To regard for the moment only the fact of the age, I do not think that a syphilitic subject should be permitted to think of marriage until after a *mini-*

mum period of three or four years devoted to a most careful treatment. A longer time would be better, but this is the necessary indispensable minimum.

Practically, my rule of conduct is as follows: When consulted upon the propriety of marriage by a patient whose syphilis (although regularly treated) dates back only three or four years, I always begin by counseling him to wait, to defer his project of union, and I insist anew upon treatment, with the view of increasing and perfecting his chances of security. But if, nevertheless, the patient urges serious and paramount reasons for an immediate marriage, and if, in addition, he satisfies all the other requirements of my programme I do not think I have the right to thwart his projects. I *tolerate* his marriage under these circumstances—not, however, without adding certain advice, certain indispensable recommendations.

As to the third condition:

1. It would be impossible to lay down a fixed, precise term. One must, from the very nature of such questions, depend upon approximate averages within sufficiently wide limits.
2. The longer this period of immunity, the more reassuring will it be in every respect, especially as affecting the question of marriage.
3. In order to fix a minimum, I conclude, from my personal observation, that it would be imprudent to reduce this period of complete immunity to less than from *eighteen months to* two years.

As to the fourth condition: There is unquestionably a *benign syphilis* and a *grave syphilis*. Now, the *quality* of the syphilis with which a patient has been affected is far from being without interest in deciding the special question which now engages our attention. Quite the contrary, it has a superior significance in this matter—a significance very essential to appreciate, and to which there is reason to attach great importance in the solution of our problem. Notwithstanding the fact that an initial benignity may properly influence and modify our restriction, this benignity of a syphilis does not constitute a pledge of security for marriage, if it be not joined with additional guarentees—notably that of a sufficient treatment.

On the other hand, there is the *"bad syphilis,"* which comprises all those cases which, from various causes, are more liable than others to become dangerous in marriage. Of course, these will require special consideration from the opposite view, before a verdict is given.

One of the essential elements of such a verdict lies in the appreciation of the *intrinsic prognosis of each particular case*—in the exact determination of the *quality* of the syphilis which affects the patient who comes to seek our advice and submit his destinies to our direction.

It is the business of the man of art, under such circumstances, to inform himself as thoroughly as possible upon his patient's antecedents and upon the nature of the accidents which he has presented. It is his business to prepare from a careful and minute inventory what I will call the *pathological balance sheet* of his patient, to judge of the *quality* of the diathesis under observation ; then, this analysis made, to decide finally whether, from a medical point of view, there is reason to consider said diathesis dangerous for marriage or not.

The fifth condition—sufficient specific treatment—is still more important. It is certainly the great condition *par excellence.* For, after all, everything converges, all reverts to this question : a syphilitic patient aspiring to marriage, is he or is he not sufficiently well *cured* of his diathesis to be no longer dangerous in marriage.

A treatment worthy of being characterized as "sufficient" is this :

1. A treatment that is based upon the administration of the two great remedies which, with just reason, are commonly called the "specifics for the pox," viz., *mercury* and *iodide of potassium.*

2. A treatment which is based upon the administration of these two remedies in *doses veritably active and curative,* very different from the insufficient, timid, indifferent, almost inert, doses in which, according to traditional routine, they are most often prescribed.

3. A treatment which is prescribed and regulated according to a certain method, which has for its aim and result to conserve for these remedies,. notwithstanding their long administration, their primitive intensity of action (called the *system of intermittent or successive treatment.*)

4. A treatment which, in these conditions, is vigorously pursued during *several consecutive years*—at the *minimum,* three or four *years.*

Dr. Fournier especially insists upon the importance of this last point, and says, *a chronic disease,* in effect, requires a *chronic treatment.*

The author does not believe in the universal *revealing action* of the sulphur-waters, and enters protest against this popular notion. He, however, thinks that sulphur-waters *may* determine specific eruptions in syphilitic subjects in *some* cases, owing to their irritant action upon the skin, especially when used daily in the form of baths, douches, vaporbaths, etc. But this action is not constant, and, hence, is not capable of supplying reliable proof, and does not furnish a *criterion* which can be relied upon. The test of the waters is, therefore, a legend to be abandoned like so many others.

When commencing this abstract we anticipated introducing some points of difference as well as of confirmation from Kassowitz's well known work on hereditary syphilis, as admirably translated by Dr. Milo A. Wilson, but we find our space will not permit.

The above will give but a very poor idea of the subject as treated by Dr. Fournier, and of his style of handling this complex question, but it is to be hoped it will serve the purpose of attracting attention to this very valuable work, for in so doing our object will have been accomplished.

It is impossible to keep up this Journal upon Colorado's light air. Consequently, those subscribers who fail to respond promptly to the just bills we have rendered at this late hour will have their names stricken from the subscription list and published as in arrears.

We are indebted to Dr. Brinton, Editor of the Medical and Surgical Reporter, for the cuts illustrating Dr. Burnett's article. They are from Dr. Landaldt's work on the eye.

BOOKS & PAMPHLETS RECEIVED

··→≈☒≈←··

"Fever: A study in Morbid and Normal Physiology." By H. C. Wood, A. M., M. D., Smithsonian Institution, Washington, D. C.

"Index Catalogue of the Library of the Surgeon-General's Office, U. S. Army," Vol. II, Washington: Government Printing Office.

"Transactions of the American Medical Association." Vol. 31, 1880. Philadelphia.

"Imperfect Hearing and the Hygiene of the Ear." Including Nervous Symptoms, Tinnitus Aurium, Aural Vertigo, Diseases of the Naso-pharyngeal Membrane, and Mastoid region, with Home Instruction of the Deaf. By Laurence Turnbull, M. D., Ph. G. Third edition, with illustrations. Philadelphia: J. B. Lippincott & Co. 1881.

"Medical Electricity." A Practical Treatise on the Applications of Electricity to Medicine and Surgery. By Roberts Bartholow, A. M., M. D., LL. D., with ninety-six illustrations. Philadelphia: Henry C. Lea's Son & Co.

"On the Antagonism between Medicines and between Remedies and Diseases." Being the Cartwright Lectures for the year 1880. By Roberts Bartholow, M. A., M. D., LL. D., New York: D. Appleton & Co. 1881.

Pamphlets on various Educational Topics. Bureau of Education. Washington Government Printing Office.

"Announcement of the Thirty-Third Annual Session of the Medical Department of the University of Georgetown." District of Columbia. Collegiate year, 1881-'82.

"Announcement of the First Annual Session of the Medical Department of the University of Denver." Denver, Colorado. Collegiate year, 1881-'82.

"Announcement of the St. Louis College of Physicians and Surgeons." Session, 1881-'82.

"Sixty-First Annual Catalogue and Announcement of the Medical College of Ohio." Session of 1881-'82.

"Report to the Illinois State Medical Society on Laryngeal Tumors." By E. Fletcher Ingals, A. M., M. D., Chicago. Reprinted from the "Chicago Medical Journal and Examiner."

"Failure of Vaccination." Variolous Infection an Illusion; Vaccination an Injury to Health and a Danger to Life, and as a protection against small-pox, a Vanity. By Carl Spinzig, M. D. Repiint from the St. Louis Clinical Record for February and March, 1881.

"Report of the Proceedings for Establishing a Board of Commissioners in Lunacy for the State of New York."

"Progress in Medical Education," being the Doctorate address at the Thirty-Ninth Annual Commencement of Rush Medical College, Chicago.

"Annual Reports of the State Board of Health of Colorado." 1879-80, Denver, Colorado.

"Annual Announcement of the Medical College of the Pacific." Session 1881, San Francisco.

"On some Impurities of Drinking-Water." By Professor W. G. Farlow, M. D., of Harvard University, Boston.

"Abstract of Transactions of the Anthropological Society of Washington, D. C.," with the Annual address of the President. For the First year, ending January 20, 1880, and for the second year, ending January 18, 1881. Prepared by J. W. Powell. Washington, D. C.

"Anæmia in Infancy and Early Childhood." By A. Jacobi, M. D. Reprinted from "The Archives of Medicine," February, 1881. New York: G. P. Putnam's Sons.

"Trance and Trancoidal States in the Lower Animals." By George M. Beard, A. M., M. D. Reprint from the "Journal of Comparative Medicine and Surgery," April, 1881.

"Upon the Production of Sound by Radiant Energy." By Alexander Graham Bell. Paper read before the National Academy of Sciences, April 21, 1881.

"The Cardiac Nerves." Tabulated by Roswell Park, A. M., M. D. Reprint.

A treatise on the Diseases of the Nervous System. By William A. Hammond, M. D. With one hundred and twelve illustrations. Seventh edition, rewritten, enlarged and improved. New York: D. Appleton & Co. 1881.

Many of the above will be reviewed in our next number.

TRANSLATIONS.

····➤≍◘≍◄····

AUTOPSIES.

Le Progres Medical, June 4, 1881—Mitral Lesion—Wine merchant, aged 41 years, has never had rheumatism, but has been troubled with some affection of the heart for the past ten years. He entered the hospital, (Lariboisiere), November 11th, 1880. The heart beat was very irregular, a murmur with the first sound at the apex, and the second sound reduced one-half. Urine very albuminous.

The patient died partly delirious and in a state of semi-coma eighteen days after admission.

Autopsy.—Heart enormously enlarged, the mitral valves completely adherent, forming a sort of inverted cone, with a small opening at the bottom. (Insufficient mitral capacity due to contraction.)

Lungs congested, also the kidneys, especially noticeable in the medullary substance. Tuberculosis located at the base of right lung. Tuberculous Pericarditis. Alcoholism.

M., aged 83, entered hospital December, 6, 1880.

This woman has become weak in mind during the past three months; however, she has not forgotten to visit the wine dealer every day, though she has never become completely intoxicated. Has become much emaciated during the past month. Strength now so much diminished that she is unable to leave her bed. Coughs very seldom, slight mucous expectoration, whitish tongue, pulse irregular and unequal. Respiration normal on left side, but in right lower lobe some subcrepitant rales. Cracked voice.

December 10—Emaciation very marked; the patient presenting the appearance of one affected with senile cachexia. Skin roughened, same stethoscopic signs.

December 15—There is still heard the sub-crepitant rales, nothing abnormal about the heart.

December 16—Pericardial friction sound.

December 18—The patient died.

Autopsy—The right lung is adherent at the base.

In this portion of the lung there are miliary, semi-transparent, non-caseous granulations, accounting for the existence of the localized pleurisy. Recent Pericarditis with a roughened condition of the surface of the membrane; besides which there were at the point of beginning of the aorta, the same kind of granulations already mentioned, very plainly to be seen, disposed in regular order and about ten in number. The liver is a little hardened, and the left kindney somewhat congested.

CEREBRAL HEMORRHAGE.

G., aged 78 years, entered hospital December 17, 1880, at 10 o'clock A. M. We are told that she fell, losing consciousness at 6 o'clock this morning. At the same time she commenced vomiting abundantly, (bilious matter), which lasted several minutes. On visiting her, she was found to be in a state of coma; respiration rattling, profuse perspiration. Pulse 104. Temperature in vagina, 38° cent. Muscular contractions follow pinching of both arms and legs, less noticeable, however, in the limbs of the right side. It is impossible for her to swallow anything.

The pulse is regular and nothing abnormal is noticed about the heart.

December 18—The patient has vomited frequently and very abundantly during the day.

December 19—The patient has taken nothing, bilious vomiting continues. Died during the night.

Autopsy.—Slight adherence of the arteries at the base.

Cerebellum : The seat of the hemorrhage is located in the central part of the left lobe of the cerebellum. This has destroyed all the white substance of this lobe as far as the median line, but has not encroached upon the right hemisphere. *Cerebrum—Right Hemisphere :* Lesion of old ramollisement. *Left Hemisphere :* Evidence of small recent hemorrhage of the size of a pea. (These lesions would never have caused complete paralysis). (No note made of condition of the pupils, Rep.)

SCIENTIFIC SUMMARY.*

··→≍◉≍←··

Anthropological Notes.†

By Dr. H. C. Yarrow, U. S. Army.

ARCHÆOLOGICAL EXPLORATIONS AT MADIS-
ONVILLE, OHIO.

The most thorough piece of archæological work with which we are acquainted at the present time, is the exploration of an ancient cemetery under the direction of the Literary and Scientific Society of Madisonville. The reports are prepared chiefly by Mr. C. F. Low, to whom we are indebted for copies. The explorations, begun in 1878, were first undertaken by Dr. Metz and others in order to save from loss and destruction the mound relics of the vicinity. While exploring a mound a laborer, prospecting in the neighborhood,

came upon a skeleton at a depth of two feet. Subsequent investigation revealed the fact that the entire plateau is the cite of an ancient cemetery, from which have been exhumed upward of four hundred skeletons, accompanied by stone implements, pipes, pottery, charred matting and corn, tools and ornaments of bone, shell and copper. The reports are numbered I, II, III, and each succeeding one is a more careful report than the others, of just what we desire to know. A detailed account of the whole exploration is in progress, and we shall not, therefore, speak querulously of the shortcomings of the present numbers. From the data before us we gather that there were two horizons of sepulture, the deep, averaging nearly four feet, and the shallow, averaging eighteen inches. Four-fifths of the bodies were interred in a horizontal position, not one-tenth in a sitting posture; and all the children were buried stretched out. As far as indicated, the orientation was as follows: North, .07 ; south, .43 ; east, .22 ; west, .02 ; north-east, .03 ; south-east, .17 ; north-west, .04. This, of course, is to be considered only as a very rough estimate; but the great preponderance of cases where the head is toward the south or the east is very noticeable.

The most remarkable feature of the cemetery, however, is the presence of ash-pits in great profusion (over 200 have been explored), very few of which contain any human remains. They are about five or six feet deep, and contain the following layers : On the top. leaf-mould has drifted in, and filled in the cavity occasioned by the settling to a depth of two feet, more or less. The remaining space is filled with layers of charred wood, ashes and animal remains, clay, sand, and even corn, both shelled and on the cob. The bones and implements are not burned, which refutes the theory that the pits were for cremation. Many beautiful objects have been recovered from these ash-pits, and among them a bone implement entirely new.—*Mason* (O. J.) *Am. Naturalist* January, 1881.

*Reports appearing in this department will be written with special adaptability and reference to the requirements of the scientific physician.

†Books, pamphlets or papers on anthropological subjects, may be sent to the editor of this department, in care of the Rocky Mountain Medical Review, Denver, Colorado.

ANTHROPOLOGICAL SOCIETY OF WASHINGTON.

The transactions of this young but flourishing society have been just read, and from the volume for 1880-1881, we find that about fifty formal papers have been read, including annual addresses by the President. Some of these papers are very thoughtfully and carefully prepared, in fact, are far above the average of such productions which are generally read before learned bodies. Among those of special interest to physicians are the following :

Color-blindness as affected by race—Dr. Swan M. Burnett.

French and Indian half-breeds of the Northwest—Dr. V. Havard, United States Army.

On the Zoological Relations of Man—Prof. Theo. N. Gill.

Poisoned Weapons of North and South America—Dr. W. J. Hoffman.

Pre-social Man—L. F. Ward.

Is Thought Possible Without Language ? Case of a deaf mute—Samuel Porter.

Civilization—M. B. W. Hough, etc.

At the time the report was published the Society had twenty-eight members, which has been somewhat increased lately, and the officers are as follows :

President—J. W. Powell.

Vice Presidents—Garrick Mallery, Otis T. Mason, H. C. Yarrow, R. Fletcher.

Corresponding Secretary—C. C. Royce.

Recording Secretary—L. F. Ward.

Treasurer—J. Howard Gorl.

Curator—W. J. Hoffman.

Council—J. C. Welling, F. A. Seeley, Miles Park, H. L. Thomas, J. Meridith Toner, E. A. Fay.

POPULATION AND RESOURCES OF ALASKA.

This is the title of a preliminary report made to General Walker, Superintendent of the Census, by Mr. Ivan Petroff, who for some time past has been traveling in Alaska with a view to collect statistics, particularly those relating to the Indians. With the limited space at our disposal we can only refer briefly to some matters contained in the pamphlet which is of considerable interest to anthropologists and physicians. Mr. Petroff classifies the nations of Alaska as follows:

"First. The Innuit or EsKino race, which predominates in numbers and covers the litteral margin of all Alaska from the British boundary in the Arctic to Norton Sound the Lower Tukin and Kuskokvim, Bristol Bay, the Alaska peninsula, Kadiak Island, and mixing in also at Prince William Sound.

"Second. The Indians proper, spread over the vast interior in the north, reaching down to the sea-board at Cook's Inlet and the mouth of Copper River, and lining the coast from Mount Saint Elias southward to the boundary and peopling Alexander Archipelago.

"Third in numbers, but first in importance, the Alentian race, extending from the Shumagin Islands westward to Attoo—the ultima Thule of this country. We have taken pains to present this classification for the reason that many persons group together all the savage dwellers of Alaska as Indians.

Most of the misery and degrading intoxication which is occurring among the Alents is said by Mr. Petroff to be due to the consumption of *Krass*, a kind of home brewed beer, the manufacture of which was taught the natives by the Russians, with a view to the prevention of scurvy. It has not proved a blessing in disguise. The diseases affecting the people of Alaska, as might be expected from the climate and their mode of living, are principally consumption, scurvy, rheumatism and sore eyes, and at times they have suffered frightfully from epidemics of measles and small-pox. Their great reliance is in the sweat bath for the cure of all sorts of diseases, and they have but little confidence in the drugs of the white men. Shamanish or sorcery is about all they believe in.

With regard to the decrease of population in Alaska, Mr. Petroff believes that the natives are at least as numerous as they have ever been since 1838 and 1839.

To our readers interested in the northwest coast tribes we heartily commend this report which can possibly be obtained in limited numbers from the Department of the Interior.

WAS THERE A GLACIAL MAN IN AMERICA.

But was there any glacial man in America? To this question the answer is distinct, though given with the reserve which the subject justifies. For the best that is known, we are chiefly indebted to Dr. C. C. Abbott, who was the first to call attention to the stone implements found in the glacial deposits of the Delaware Valley. These implements are chiefly of argellite, though examples of flint occur at higher levels. They have been found at the bluffs near Trenton, both in position and where deposited among the debris at the base. Dr. Abbott says: "Perhaps it is a wise caution that is exercised in but provisionally admitting the great antiquity of American man, but, were these rude implements not attributed to an inter-glacial people, their coequal age with the containing beds would never have been questioned." On this point the Curator of the Peabody Museum at Cambridge observes, in his tenth annual report: "Dr. Abbott has probably obtained data which show that man existed on our Alantic coast during the time of, if not prior to, the formation of the great gravel deposit which extends toward the coast from the Delaware River, near Trenton, and believed to have been formed by glacial action. From a visit to the locality with Dr. Abbott, I see no reason to doubt the general conclusion he has reached in regard to the existence of man in glacial times on the Atlantic coast of North America."—*B. F. De Costa, in Popular Science Monthly.*

THE famous Neauderthal skull has just been exhibited at the British Association, and is exciting new discussions about the "missing link" of Darwin. Professor Dawkins, one of the most profound investigators' in England concludes that the cave men are the intellectual and social equals of the Eskimo and that the Neauderthal skull, presenting, like the jaw of the Noulette anthropoidal peculiarities, is nevertheless that of man possessed of a far larger brain than any anthropoid ape, or even some of the present savage tribes.

THE site of Morris, Illinois, it is said, has been occupied by cities of three different races of men. The first was the mound builders, who must have had a large town there, judging from the mounds left. Then within the historic period it was the site of a large town occupied by the Indians, and now the white man has built his home on the ground. Its natural advantages must be great, to be thus acknowledged by the races of men, with a history extending over thousands of years.—*Ex.*

THE ANCIENT EGYPTIANS.

The Egyptians are not to be classed with African races; the form of skull indicates a connection with the Caucasian family, and the language appears to have analogies with the speech of both the Aryan and Semitic races. In the earliest ages, far before all history, they must have left Asia to found a new kingdom on the banks of the Nile. The Amu, east of Egypt, were herdsmen of Semetic descent, with light yellow skins; the Libyans, on the west, had light skins, blue eyes and blonde or red hair; on the south were the negroes. The dwellers on the Nile Valley were reddish brown. They were a gay and brilliant people, loving life, and full of jest and amusement. Notwithstanding the sharp contrasts in society, every child had a share of education, and every man of ability had a chance to rise. Persons of common birth frequently came to fill the great offices of State, and a young man of courage, talent and address, who could make his way and hold his ground at court, often married a daughter of the Pharoah and became the father of a line of kings. It was said of Prince Ti, whose vast tomb at Sakkara, with its wealth of pictoral illustration, has engaged so much attention that "his parents were unknown persons." In this way it was the custom to prefer the descent of a Pheroah through the female line. When a monarch had half a hundred daughters to provide husbands for, as Rameses had, the advent of a spirited and handsome young fellow, even a *nullius filius*, might be welcome.—*Atlantic Monthly.*

ROCKY MOUNTAIN MEDICAL REVIEW.

| Vol. 1. | DENVER, AUGUST, 1881. | No. 12 |

ORIGINAL ARTICLES.

···→≍☓≭←···

Puerperal Fever—Its Specific Character, Pathology and Treatment.

By David McFalls, M. D., Park City, Utah.

Some few years ago there prevailed in the Northern part of the State of New York, in a town where I then resided, an epidemic of Puerperal fever of great severity. Its duration where it first made its appearance did not exceed two months, but it gradually spread to the adjoining towns and neighborhoods, presenting the same characteristics, and was attended with the same fatal results. After the severity of the disease began to abate, we had for some time following, sporadic cases, not so severe, neither were they attended with the same fatality. From that time to the present, although I have seen many cases, both in my own practice and in that of others, I have not met with what might be considered an epidemic of the disease; therefore what I have to say upon this subject is gathered, in the main, from my own observation and clinical experience, although I have studied to some extent the published views of others. The symptoms attending those cases I saw, during the prevalence of the disease, were as follows : Within twenty-four hours after confinement the woman was seized with a severe chill; this was followed by fever, rapid pulse, high temperature, pain in the region of the uterus, extending rapidly, in many cases, over the whole abdominal region ; great prostration, tympanitis, vomiting toward the close, of a greenish or dark colored substance, and terminating fatally in from five to eight days.

Not every woman, however, who gave birth to a child during this period, was thus seized with the disease, as has been the case in some of the epidemics of which we read, for I well remember having attended three cases of confinement in one night, within a short distance of each other. One of these took the disease and two escaped. Now if we turn our attention to the literature of this disease, of which there is no dearth at the present time, we will find there has been a great diversity of opinion not only in relation to its cause, but as to its nature and treatment; and although there has been, and may be at this time, a disagreement in the views of different writers, still the researches of those who have had the opportunity and the ability to prosecute their investigations, have thrown much light upon the subject, and have enabled us to obtain a more correct knowledge of its true nature. At the time the epidemic above referred to prevailed, Professor Meiggs was considered high authority, not only on this disease, but all others peculiar to women, and as I had great confidence in him as a teacher and writer, I accepted his views, both as to its pathology and treatment. He considered it a metritis, and that bleeding was the cure. I followed his advice, but with bad results. The first case I saw was a very plethoric woman. The disease was ushered in with great severity, and this being a suitable case for bleeding, if this remedy is ever admissible or beneficial in this disease, I bled her freely, but she died upon the fifth day. I bled my next case, but with no better results. Both were much weakened and depressed after the bleedings. The cases that followed were not bled. They were treated with opium, quinine, and stimulants, and although their lives were apparently prolonged, a large per cent. succumbed in the end. For some time subsequent to this, those who escaped the disease did not do well after confinement. There was a tendency to chills, mammary abscess, enlargement of the inguinal glands, pelvic cellulitis, phlegmasia dolens, sore mouth, or some other of those unpleasant conditions attending the puerperal state. We had, too, erysipelas of a severe type, prevailing at the time. Wounds and injuries of all kinds did not heal readily. There was an unusual number of boils and whitlows, one case of the latter resulted in pyæma and death. There was a general unhealthy condition pervading the whole community. All the infants born of those who died of puerperal fever, with one exception, took erysipelas and died, and one of the puerperal fever cases had well-marked erysipelas of one arm before death. From this tendency to erysipelas in these cases, we might easily be led to the belief that this disease was nothing more nor less than erysipelas occurring in the puerperal state, and where there was a tendency to scarlet fever, diphtheria or traumatic fever, that each of these was the disease we had to contend with when they occurred immediately after confinement, thus discarding both the name and specific disease we call puerperal fever. These are, or were, the views entertained by high authority. Such as Spencer Wells and Hutchinson ; and even Robert Barnes is rather inclined to favor this view of the case, although he is not quite willing to give up the name of Puerperal fever. On the other side, Fordyce Barker, and other eminent authorities on this subject, claim that there is a disease peculiar to itself, produced by a specific poison, similar perhaps, but unlike the agencies that produce the above mentioned diseases, and to this disease we give the name of puerperal fever. To this latter view I am inclined not so much from a study of the opinions of others, but from clinical observation and experience. We all know that there is a similarity in the pathology of many diseases—but similarity is not identity. There is a similarity in their

pathology between diphtheria and scarlatina, and they seem to be amenable to the same class of remedies, and yet they are not identical. We may see, however, where both diseases are prevailing at the same time, what might be called a hybridity, or one complicating the other to a greater or less extent, but both diseases are not produced by the same morbid principle. These different poisons may be taken into the system at the same time, and each strive, as it were, for the mastery—one sometimes succeeding or predominating, and then again the other. Seeing this, not only in scarlet fever and diphtheria, but also in other diseases, as in that hybrid, called "typho-malarial fever" which we had in the army during the war—is it not a little singular that one of these "similars" does not cure the other, if there is any truth in the doctrine of "similia similibus curantur?" but we do not find it so in practice, but on the contrary the patient is usually made worse by having both of the morbid principles in his system at the same time.

I have seen, in quite a number of cases, erysipelas in lying-in women, and although they were attended with more danger than in ordinary cases, still they did not take on the peculiar characteristics of puerperal fever ; I have waited upon women in confinement who had, at the time, scarlatina, diphtheria and measles, and as might be expected these diseases were more or less aggravated, but they were not usually prevented from running their regular course. From these well-studied cases I am led to the conclusion above referred to—that there *is* a *something* or a specific infection, not the same that produces either erysipelas, scarlet fever, diphtheria or traumatic fever, and that this something is always true to itself, producing puerperal fever, wherever it finds a suitable nidus for its reception, and nothing else. The peculiar unhealthy condition pervading all classes, to which I alluded, was produced, undoubtedly, by other morbific influences, predisposing very likely, lying-in women to child-bed fever. The origin of this specific infection is, thus far, unknown to us, and just what it is, is still veiled in mystery, but from whatever source, or whenever or wherever produced, it is capable of giving rise to the disease we are now considering, in those who are in a suitable condition for its reception, and it is received into the system, undoubtedly, the same as other infections. That it is taken in through abrasions, either in the os-uteri, the mucus membrane of the vagina or lacerations of the perineum, as is taught by some authors upon this subject, I do not believe. These may be, and undoubtedly are, the avenues through which septic poison which produces septicæmia is taken into the system, but septicæmia is not puerperal fever, neither is it caused by the same material. Septicæmia is a blood poisoning produced by the absorption of an animal product, by the absorption of that material that is given out from a wound just previous to the formation of pus, and these abrasions, or the site of the placenta when the womb is imperfectly contracted, or a decomposing clot or a portion of placenta, a blighted ovum, a dead fœtus in utero, or any other animal matter in the uterus or vagina are the points from which the absorption takes place ; but, in my opinion, an imperfectly contracted uterus, where the contractions are not sufficient to expel the clots and close up the mouths of the vessels at the placental site, is a more frequent cause of septicæmia than all the others combined. This opinion is based upon the fact that I have seen this disease (septicæmia) follow more frequently, the easier labors; and that where the labor was very severe, and even where instrumental interference was necessary, where there were abrasions and slight lacerations, but a firm contraction of the uterus, no septic disturbances followed. I am also led to this conclusion from having seen women the second or third day after confinement, begin to have fever, become restless with a rapid pulse, anxious countenance, sometimes a chill, with all the symptoms in fact, that indicate septic poisoning. In such cases I have given at once large doses of quinine and ergot. Soon after clots were expelled, and these cases

rapidly or slowly recovered, depending upon the amount of poison absorbed. If seen early, and the above means were employed, then recovery would be almost invariably rapid, but if the case was allowed to go on without treatment, there would be a sufficient amount of this morbid material taken in to give much trouble or fully developed septicæmia. But child-bed fever does not seem to depend upon these conditions, or to any traumatism whatever. It does not attack those whose labors are severe, more frequently than the easier cases. If, however, the parturient woman is in perfect health, with all her organs active, she may escape the disease in the midst of it, by eliminating the poison altogether, or there may be just sufficient taken in to produce a chill and then abortion, either with or without appropriate treatment. That some women are more liable to it than others, there is no doubt, and thus it is that some escape amid the most active prevalence of the disease, the same as we find some persons resisting all epidemics and contagious diseases. I am unable to say that any circumstances or conditions in life furnish any immunity. Those who live in palaces and hovels are alike liable, but it is claimed that great fear and depressing influences predispose, and thus unfortunate young women, who feel most keenly their degradation are more liable than others, when the disease is prevailing.

This disease is considered both infectious and contagious, so much so that a physician, it is claimed, is liable to carry the poison about him, and give it to those upon whom he is called upon to attend in confinement; and although this theory is not fully established to the entire satisfaction of everybody, still two mid-wives, a Mrs. Marston and a Mrs. Dymond, in London, were, a few years ago, prosecuted and imprisoned for thus communicating the disease to their patients. In the epidemic I saw, although most of the cases occurred in my own practice until it spread to the adjoining towns, I was unable to trace a single case to a communication of it by myself. One case in the infected district was attended by a physician from a neighboring town, where the disease did not exist at the time, while I was in attendance upon another woman at the same time, at some distance. Both were delivered at the same hour, both took the disease within twenty-four hours, and both died upon the eighth day. Two cases were not seen by any physician until they had taken the disease. In these two cases, certainly, the infection was not communicated to them by any one, as far as it was possible to ascertain. It would be well, however, for any physician, during an epidemic of this kind, if he saw that the disease was following in his footsteps, to make a visit to some watering place, or to his friends east or west, until the disease disappeared, as we have so many small persons in the profession who are always ready to lay the blame at the door of his neighboring practitioner of medicine. I would certainly, during its prevalence, never attend a case of confinement immediately after visiting a case of erysipelas or scarlet fever or after dressing suppurating wounds, not that these diseases contain the germ that gives rise to child-bed fever, but they may, in some way not fully understood, predispose or furnish the nidus for the puerperal fever germ to take root. A change of clothing, a bath, and other precautionary measures, such as disinfectants and the like, might protect the patient, and still it is claimed by some that these are inefficient, as there is a possibility of carrying it in the breath.

I have spoken of this specific infection, that undoubtedly gives rise to this disease, as a poison, a principle, a morbid material, and other synonymous terms to prevent repetition, and yet, as I said before, we know not what it is, nor whence it comes. I have studied the germ theory of disease, the low organisms that are said to set up fermentation and are capable of reproducing themselves ad infinitum, the bacteria, micrococci, and other allied organic forms, until my head ached, and still there are so many conflicting opinions in relation to them, that it is difficult to determine with certainty, whether they are the *cause* or the *product* of disease.

They are found, undoubtedly, in diseased structure and exudations, but whether these germs or spores that are seen by the aid of the microscope, are that materies morbi that produce the different infectious and contagious diseases we have, admits of some doubt. That there are malevolent agents that cause infectious diseases, and that each disease has a distinct germ, or whatever we may feel disposed to call it, that invariably produces that disease if anything, is generally accepted as true; but is it not possible that some chemical change in one principle, some addition to or subtraction from, produces the different infectious diseases we see? We know that laudable pus introduced into the system does no harm, but when this pus becomes putrid, or undergoes a chemical change, it produces pyæmia.

Basing our argument, then, upon the probability that each of those infections, whether visible or invisible, nameless or otherwise, retains its own distinctive character, and is capable of reproducing itself, or setting in operation a train of molecular changes or fermentations in the blood and tissues, under proper and suitable conditions, we can readily see how we have in the parturient woman all the conditions necessary for the disease in question.

Although child-bearing, parturition and the whole process through which the woman passes in the reproduction plan, are strictly physiological, still there are conditions attending *this* state, that if they do not amount to disease in themselves, are so near to it, that the slightest addition of some morbid principle, or the arrest of action in some one or more of the different organs, could easily change this physiological to a pathological condition. We find in the blood during gestation, an excess of fibrine, and quite often albumen in the urine. The bowels deranged, the liver torpid, the stomach, and, in fact, all the organs to a greater or less extent, disturbed. We find, too, a uterus in its unimpregnated state weighing but two or three ounces, increased in weight to as many pounds, with its enlarged sinuses conveying large quantities of blood to and from the fœtus within. After parturition the involution of this enlarged womb, the establishment of the mammary secretion, and the resumption of all the organs to their normal state. Here are changes going on, rapid in their process, and although healthy and natural, nevertheless they are in a condition suitable for the reception of disease. Then let there be a general unhealthy state of things pervading the locality—the poison of scarlatina, of erysipelas, of typhoid fever or zymotic influences, and we have just the conditions necessary for this puerperal fever germ to take root, *but we must have this specific germ,* that produces this disease or we will not have puerperal fever. We may have septicæmia, traumatic fever, pyæmia, or some other of those diseases above referred to, and it is more than probable that the distinction is not always made by careless or indifferent observers, and thus every fever occurring after child-birth is called *puerperal* fever.

In septicæmia the chill is not so constant as in puerperal fever, and when it does occur, it does not come on so early in the disease, neither do we find septicæmia occurring epidemically. Pyæmia always comes on later. It cannot show itself until there has been time for pus to form in the uterus, uterine sinuses or some other place, and remain shut up, as it were, until the changes above mentioned take place.

The effect of this puerperal fever germ, when taken into the system, is the same, undoubtedly, as in other infectious diseases. Its action is first upon the fluids, and thus through them producing a general disease of the whole system, with sooner or later local manifestations. In diphtheria we see a general disease, locally manifested in the throat and air passages (Oertel to the contrary notwithstanding). In scarlatina, the throat and skin. In typhoid fever, Peyers patches and the mesenteric glands. In typhoid pneumonia, the lungs. In this disease we find the uterus, the uterine sinuses, and peritoneum giving evidence of disease ; and again we find the liver, spleen, lungs and pleurae affected—that is if

the disease is of sufficient duration, but where the patient succumbs in a day or two, as we sometimes see, there may be no evidence whatever, of local disturbances.

I am not aware that any impropriety on the part of the patient, or any omission or commission by either physician or nurse (except perhaps what may be referred to when I come to treatment), has anything to do, whatever, in producing or warding off, the disease. "Taking cold" or exposure, does not produc it any more than it does diphtheria or typhoid pneumonia, as is often thought by the laity, and is too willingly acceded to by many physicians as a sort of apology or shield to themselves. It should be borne in mind that this is an *infectious* disease, and the infection is taken in, it may be, sometime before confinement, and is held in abeyance until the proper time for its development, which is soon after delivery.

Upon the treatment I will not dwell at length, for in the main, this disease is treated upon the same principle as other infectious diseases. But few of us would think of bleeding a peurperal fever patient at this time, unless, perhaps, to relieve some local congestion of the brain or lungs, and then it is more than probable that other means would be employed for this purpose. In those epidemics where Gordon and Meigs saw the disease, it, I have no doubt, was of the sthenic type, where bleeding was not only admissible but beneficial. I am not one of those who look upon the practice of the past as having been altogether wrong, for as the cases I bled with such bad results led me to abandon the practice, so, I am sure, they, being practical common sense men, would have done had they noticed the same results. And then, again, diseases, there is no doubt, change their type, although J. Hughes Bennett denies this, but we have too many examples of this fact to be successfully disputed. Of the different means employed at the present time, tonics and stimulants are principally relied upon. Of the tonics or antipyretic remedies, quinine, undoubtedly, stands at the head. The tincture of iron should be given in thirty drop doses as soon as the severity of the disease begins to abate. The bowels should be kept quie', and aperient remedies religiously avoided. Hot turpentine stupes should be applied at once, and a well fitting binder. The veratum viride is recommended by Fordyce Barker in sufficient doses to bring the pulse to below 100 if possible. This remedy should be given with caution, not that it will kill even in large doses ; but there is more or less danger of its producing either vomiting or purging, both of which should be avoided. From twenty to thirty grains of quinine may be given at night if the temperature is high, and from ten to fifteen in the morning. Cold sponging and other antipyretics, the mineral acids, chlorate of potash, etc., opiates sufficient to allay pain and give comfort to the patient. Stimulants in sufficient quantity and nourishing diet. Should the tympanitis become very great, as is usually the case, great relief may be obtained by the use of the rectal tube, or by the administration of the oil of turpentine in five or ten drop doses. The vagina and uterus should be washed out occasionally with a solution of carbolic acid or other disinfectants, or by even warm water in large quantities.

As a prophylactic in case of an epidemic of this disease, I would certainly recommend the use of quinine in good doses to every woman previous to and immediately after confinement. Those who have noticed the success of Thomas and Bozeman, of New York, in their operations for ovarian tumors, which they attribute in a great measure to the employment of quinine previous to the operation, and to the carbolic acid spray during the operation for the purpose of destroying any infection that may be present, will not hesitate to give quinine in this disease when it prevails epidemically. It has been considered that there was more or less danger in giving quinine to pregnant women on account of its oxytocic effect, but experience has proven, I believe, that it expedites delivery only when labor has already commenced, and thus no evil effects follow its administration.

CORRESPONDENCE.

CHICAGO, ILL., August 20, 1881.

Editors ROCKY MOUNTAIN MEDICAL REVIEW.

Since addressing my last letter to you a whole season has passed by without any noticeable changes in the health of this great city or the status of matters of professional interest. The late summer finds us with just as much small-pox among recent emigrants and the lowest classes as we had during the early spring. The health officers in endeavoring to stamp out this pest, have to contend with the gross ignorance and stolid indifference of many of our worst foreign element, and the stubborn opposition of certain others. We have even had the melancholy spectacle of a ward politician, who managed to work his way into our Common Council, advising his constituents to resist all attempts at vaccination. Let me express a hope that it may be a long time before your city has to acknowledge such an ignorant knave as a "City Father."

Our medical societies have held no meeting for some three months, so there is nothing of interest to communicate concerning their transactions. The Biological, the youngest of them all, continues to lead in point of interest. Dr. C. Fenger is to read a very able paper at the next meeting on "The Opening and Drainage of Cavities in the Lungs," which will appear in the next number of the *American Journal of Medical Sciences*, and will be well worth the attention of your readers.

The Biological Society is also having published, under its auspices, a work on yellow Fever, by Dr. Schmidt, of New Orleans. This will be a most valuable addition to the literature of the subject, as it is the result of many long years of laborious investigation.

The County Hospital, which is our great municipal charity, is crowded. Many rare cases are to be found at any time in its wards ; but as I mentioned in my last letter to you, it is under political control ; and though part of its wards have been turned over to our new Training School for Nurses, whose pupils make most acceptable and satisfactory nurses, yet the remainder of the wards are under the old regime. As an instance of what may happen under this latter condition, take the following: One of the politically appointed nurses, by accident, gave a patient a fatal dose of carbolic acid instead of the properly prescribed mixture. The Coroner's jury exonerated the nurse from any intentional crime, and dismissed him from custody. A brother of the patient tried to have the man indicted for manslaughter, but, of course, did not succeed. No one ever seemed to think of putting the blame where it properly belonged, *i. e.*, of blaming, indicting and punishing the Board of County Commissioners who control the hospital and who tolerate such a state of affairs.

We are soon to have a new hospital added to our present supply; one which will eclipse, except in point of size, everything west of the sea coast. It is erected by the Hebrew Relief Association, and will be called the "Michael Reese Hospital," after a wealthy Californian who left a large sum of money for this purpose, provided the institution should be *non sectarian*. As at present built it will not accommodate more than eighty patients. Elegant private rooms will be arranged. The site is on the lake shore, and the plans are such that an unusual amount of sunlight will be admitted. All the appointments are to be of the most elegant character. Excellent facilities are provided for *post mortem* study. And the *armamentarium* is to be, perhaps, the most complete in the country, $7,000 having been set aside for this purpose alone. A large proportion of this is to be imported, and the whole will include almost every instrument known to medical and surgical science. The gentlemen composing the staff may consider themselves fortunate indeed.

The colleges are preparing for their winter campaign and competition for students bids fair to be brisk; each claiming clinical facilities "equal to those of any college in the country."

For some years the Homœopaths maintained a "Hahnemann" college in this city in connection with a meagre hospital which was, however, enough for all their purposes. Three years ago their faculty divided, the seceding party going to form a "new" school,—whatever that is. At all events the fledgling proved a success from the start, and poor Hahnemann bids fair to be left in the cold. But now they are making frantic efforts to recover lost ground, and their latest catalogue—just out—is a curiosity in its way. Its principal characteristic is that wherever it is possible to convey a slur upon or an insinuation against its new rival, something of the kind is tucked in. In fact it is about the best advertisement the aforesaid rival could have. Whether such an impolitic and unwise course will pay in the end remains to be seen. Your readers may judge for themselves concerning the "self-denying services" of their Faculty and the merits of their plea for "continued confidence" when they learn that the two professors of physiology give *each one lecture a week*. This is a specimen of their work all through. But enough of this unsavory subject.

Chicago literary ability will shine with unusual lustre this fall. Professor N. S. Davis has in press a work on Practice; Professor H. M. Lyman one on Anæsthetics, also in press; and Dr. E. F. Ingals expects soon to publish a work on Physical Diagnosis. Professors Andrews, Jewell and Hyde are among the contributors to one or the other of the large surgical works to be published by rival houses in New York and Philadelphia. Drs. Fenger and Lee have recently published a very valuable article on Nerve Stretching. These works and articles all go to show that our best men are determined that our professional progress shall keep pace with Chicago enterprise in other respects.

Our public press has contained the inevitable "interviews with leading professional men" concerning that surgical topic of great interest, the President's case. Without reference to the scientific details of the case, I think the universal sentiment of those conversant with it is one of deep regret that any such man as Dr. Bliss should have been allowed to "cheek" his way into such utterly undeserved prominence in connection with it, combined with a profound sympathy with those thoroughly competent and educated gentlemen who find themselves relegated to a secondary position by the peculiar circumstances of the case. Imagine such a thing occurring abroad, in Germany for instance; is there any possibility that such an exhibition of effrontery on the part of an irregular practitioner would be tolerated if the Emperor were grievously wounded? They do these things better—much better—on the Continent.   ROSWELL PARK, M. D.

RECENT MEDICAL PROGRESS.

⋯≻≍✿≍≼⋯

Report on Orthopædic Surgery.

BY THOS. H. HAWKINS, M. D., DENVER.

THE QUESTION OF AXILLARY OR ISCHIATIC SUPPORT IN THE TREATMENT OF JOINT DISEASES OF THE LOWER EXTREMITIES.

A. B. Judson, M. D., (Medical Record, July 2, 1881), discusses in a very clear, concise and lucid manner, the advantages and disadvantages of axillary or ischiatic support in the treatment of joint diseases of the lower extremities, taking as a type hip-joint disease.

He calls attention to some noticeable features of the pathology presented in diseases of the joints of the lower extremity which are peculiar and suggest corresponding peculiarities of treatment. These exceptional pathological traits have their origin in the habitual exposure and pressure of the joints of the lower extremity in the performance of their various functions.

The author in a very impartial way discusses the many varied devices for the relief and cure of hip-joint disease, calling attention to the recumbent position which removes from the joint many sources of danger. A resort which has doubtless been made instinctively and by advice ever since the ear-

liest times of medicine and surgery. This position is to-day one of the accepted modes of treatment by many of our best surgeons, but unfortunately the usefulness of this plan of treatment is limited for reason of the long period of an ordinary case of hip-joint disease, it being impossible to maintain the recumbent position with or without extension and counter-extension for so long a time without seriously impairing the patient's general health.

The use of crutches, resorted to probably from early times, facilitates locomotion and prevents pain. Mr. Edward Ford, while advocating strongly the use of the crutches in the treatment of hip-joint disease, acknowledges the want of something more to prevent the weight of the body from pressing on the diseased joint. Mr. Brodie says that in scrofulous diseases of the joints of the lower extremity the patient "should never walk, except with the assistance of a crutch." Crutches cannot be regarded as anything more than simply a means to facilitate locomotion, or as Dr. Judson says, "rather as insignia of the crippled condition than as instruments for the relief of pain and the promotion of recovery."

Within a few years the attention of the profession has been directed to the continued use of crutches throughout the long period usually covered by a case of hip-joint disease, and the author refers to the ingenious apparatus of M. Mathieu. This instrument consists of two axillary supports connected by means of a system of jointed rods, permitting the patient to sit down and transmitting the weight of the body to the ground when he assumes the erect position.

It is the custom in this country, by whatever means employed (which permits of locomotion), to raise the foot of the unaffected limb by the addition of an elevated sole to the shoe worn on that side, "letting the lame leg hang."* Mr. Hugh Owen Thomas and Dr. Lewis A. Sayre speaks favorably of this

*In a paper published in 1867, Dr. C. Fayette Taylor first called attention to this feature, in the treatment of hip-joint disease.

device as an important feature in the methods of treatment of diseases of the hip-joint.

The crutches and elevated shoe are features in the respective methods of treating hip-joint disease, of Drs. John A. Wyeth, Chas. F. Stillman and F. Willard. The necessity of wearing a high shoe on the foot of the sound side, is dispensed with by Dr. Richard J. Lewis; he flexes the knee of the affected limb and maintains it there by a silicate bandage.

The wheel crutch of Darrach, in the treatment of diseased joints of the lower extremity, finds a strong advocate in Dr. Joseph C. Hutchison who prescribes this method to the almost complete exclusion of all other apparatuses. Through the advocacy of Dr. Hutchison this is regarded by many as the simplest and most efficient method of treating hip disease, though Dr. Judson questions whether or not a weapon so light and easily handled may not fail of success in the combating of a foe so insidious, tenacious and destructive. The use of crutches in the protecting of the hip, knee and ankle-joints when diseased, necessarily transfers a large part of the weight of the body to the axillary region. The only way to obviate this, when the patient is erect, is to transfer the weight of the body of the diseased side, to the ischiatic tuberosity and rami of the ischium and pubes. The author does not claim this experiment as altogether modern, but refer to Malaigne's "Pare" where may be seen a wood cut representing a crutch for the relief of lameness, the result of a shortened limb. Also to the prothetic apparatus of M. M. Mille and F. Martin, which received the bones of the pelvis, and thus supported the weight of the body.

M. Ferdinand Martin constructed and described an apparatus which was intended to support the weight of the body by means of two perineal straps, made fast to a pelvic band. The Doctor also refers to M. Bonnet's attempt at ischiatic support. Dr. Henry G. Davis' hip splint (described April, 1860,) which was not an effective instrument for ischiatic support, because the lower extremity of the splint did not reach the ground.

Dr. Andrews (December, 1860,) was probably the first to solve the problem of ischiatic support. This splint, says the inventor, had a "crutch top," "the top supports the perineum by a crutch piece padded and covered with patent leather, and the bottom is attached firmly to the sole of the shoe, and at each step transmits the weight of the body directly to the ground."

In 1866 Dr. David Prince described a rude modification of Dr. Andrews' apparatus.

In 1867 Dr. Taylor described the splint which is so widely known and justly called by his name. [Some have of late called this apparatus Sayres' long splint. Dr. L. A. Sayre has improved on the Taylor splint. In some respects the long splint used by Dr. Sayre is superior to the original splint of Dr. Taylor. It admits of rotation, eversion, inversion and is provided with a movable knee-joint and also does away with the use of adhesive plaster as a means of support. Dr. Taylor, I believe, claims priority in all of these improvements.]

Dr. Judson believes that the various apparatuses devised for the relief of hip disease, have by a process of evolution brought forth what he regards as the best hip splint: That of Dr. Taylor.

He traces in this evolution the following steps. First, the portative splint of Dr. Davis, terminating above the ankle, with one perineal strap and without a pelvic band. Second, Dr. Andrews' splint reaching to the ground and supporting the perineum only on one side. Third, Dr. Sayre's short splint, not reaching below the knee, with only one perineal strap and without a pelvic band. Fourth, Dr. Taylor's improvement on the short splint of Sayre's. Fifth, Dr. Taylor's improvement on the Davis splint. Lastly, Dr. Taylor's as described in 1867. This consists of two perineal straps with a pelvic band and a support reaching to the ground. This is an apparatus for ischiatic support as well as for the application of extension and counter-extension and as such is a rare example of the adaptation of a simple means to an important end. It is a pair of crutches applied to the perineal instead of the axillary space. An ischiatic and not an axillary crutch. It is superior to the crutches with or without the high shoe because the free use of the arms is not interfered with and it is always present.

A study of the anatomy and physiology of the ischiatic region will convince one of its adaptability to the bearing of pressure, a strong argument in favor of ischiatic crutches. With the splint properly adjusted the patient when erect, in reality sits upon a pair of padded ischiatic crutches, while progression is made by alternate weight on the sound limb and then on the perineal straps. By this means the affected limb is converted into a pendent member.

Difficulty may at first be experienced in the wearing of this splint, but with a little perseverance a patient in a very short time will not only be able to wear it with ease and comfort, but will do so gracefully.

Dr. Judson insists on the importance of having the hip splint made sufficiently strong, especially in the pelvic band, as the splint may be said to carry the patient instead of being carried by him.

OSTEOTOMY FOR GENU VALGUM.

Osteotomy for genu valgum was discussed by Dr. C. T. Poore in a paper read before the New York Surgical Society, May 10, 1881. Osteotomy was performed in 1826 by Barton for bony anchylosis of the knee-joint, but with an open wound. In 1852 Langenbeck performed subcutaneous osteotomy of the femur for an anchylosis at the hip-joint. In 1851 Meyer divided the femur and tibia for the correction of a case of genu valgum. In May, 1876, Mr. Ogsten performed the operation which still bears his name for the relief of genu valgum. Osteotomy is in great favor among English and German surgeons, while in France osteoclasis has the more advocates.

Delore practised, in 1871, forcible straightening of the limb, producing separation of epiphysis of the femur or tibia. More recently, Collin has devised an apparatus

which accomplishes the same end; it is said to do so with great precision. Delore's operation has failed of success in England and in this country.

Dr. Fowler, of Brooklyn, was the first to perform osteotomy for genu valgum in this country. Dr. Poore after thus going over the history of the first cases of osteotomy, then calls attention to the generally accepted view that knock-knee is due to the changes taking place in the plane of the condyles of the femur or head of the tibia. In the former class the internal condyle would occupy a lower position than normal. This may be caused by either a bend in the lower third of the femur, by lengthening of the internal condyle or to an atrophy or flattening of the external condyle. In the latter class similar changes may occur in the articular end of the tibia or changes may be found in the ends of both bones. Genu valgum may have its remote cause in rickets. The author in the examination of nearly 100 cases of this deformity, found but one case in which symptoms of rickets were absent. He believes that knock-knee may be cured without an operation, by means of orthopœdic appliances, provided that the cases are treated while the bones are soft and before that ossification peculiar to rachitic bones has taken place, but after this time osteotomy is the only resource that promises anything like a good result.

The seven cutting operations as devised are summed up as follows:

First—A section below the tubercle of the tibia (Shede).

Second—A section of the internal condyle with a saw, the joint being entered (Ogsden).

Third—A section of the internal condyle with a chisel, the joint not being necessarily entered by the instrument, but probably fractured into (Reeves).

Fourth—The removal of a V shaped piece from the internal condyle, its apex being directed downward and outward (Chiene).

Fifth—A section of the shaft of the femur just above the epiphyseal line (Macewen.)

Sixth—A section of the shaft of both femur and tibia (Barwill).

Seventh—A section of the shaft of the femur from the outside (Taylor's).

Mr. Reeves' operation is made by an oblique incision above the inner tuberosity of the femur, down to the bone. A chisel being introduced the internal condyle is separated; the limb is then straightened by force. The author has performed this operation on five limbs, once snccessfully, once partially so, and in three cases failure.

The author has operated on twelve cases of genu valgum adopting the plan of Mr. Macewen, of Glasgow, which is described as follows; The limb is rendered bloodless, put in an extended position on a sand pillow and strict antiseptic precautions are observed, according to Mr. Lister's method, an incision is made down to the bone, "one made transversely a finger's breadth above the superior tip of the external condyle, and a longitudinal one drawn half an inch in front of the adductor magnus tendon, the incision is made sufficient to admit the largest osteotome and the finger, if the operator desires it. Before withdrawing the knife, the largest osteotome is slipped by its side until it reaches the bone. It is then turned at right angles to the long axis of the limb, and the instrument is made to penetrate the bone. After the inner portion of the bone is divided, a finer instrument may be slipped over the first which is then withdrawn, and a third if necessary may take the place of the second, when the outer portion of the bone comes to be divided. After penetrating the inner anterior, and posterior portions of the bone, and the operator thinks that sufficient of the external shell has been divided, the osteotome is laid aside, and the wound covered with a sponge saturated with a one to forty watery solution of carbolic acid, the point of section is grasped with one hand, the leg with the other; using the latter as a lever, it is jerked if the bone is hard, or bent slowly if the bone is soft, in an inward direction, when it will snap or bend, as the case may be. After bandaging the sponge over the wound, the webbing is removed and the other limb operated upon. Before dressing, he ascertains whether any cellular tissue is so prominent as to protrude beyond the lips of the wound. If this is the case, it is excised, as he thinks it will prevent the wound from healing by an organized clot, and will favor suppuration. Lister's dressings are then applied, and the limb placed in a kind of fracture-box, its outer portion extending up to third or fourth rib. When the temperature remains normal, the limb is not disturbed for two weeks. It is then placed in a well padded fracture-box for several weeks longer. Ten weeks is usually required from the time of the operation until the patient can walk. The author thinks the spray may be dispensed with. In his experience it does not in any way contribute to the immediate closing of the wound nor to the freedom from suppuration. Instead of operating with the limb extended, as directed by Mr. Macewen, he operates with the leg strongly flexed and the thigh rotated outward. By this means the bone is easier to get at and a firmer bed to place it on is obtained, and many other advantages are claimed for this position. After having completed the operation the wound is syringed out with a one to forty solution of carbolic acid, adhesive plaster is applied, a compress of Lister gauze, then a bandage and over this a plaster of paris splint extending from the foot nearly to the perineum. On the third day a fenestrum is cut over the point of the incision, the compress removed and the wound looked at. Dr. Poore's article is admirably illustrated by cuts, taken from photographs.

HISTOLOGY OF SPINAL PARALYSIS OF CHILDREN AND PROGRESSIVE MUSCULAR ATROPHY.

M. M. Roger and Damaschino (Revue de Med. February, 1881), contribute an elaborate paper on the above subject, and arrive at the following four conclusions: 1. The characteristic alteration in infantile paralysis is a spinal lesion, causing atrophy of nerves and muscles. 2. Its seat is in the anterior part of the gray matter, and it shows itself as softened patches. 3. The softening is of inflammatory origin, and the lesion is, therefore, a myelitis. 4. Progressive muscular atrophy consists essentially in atrophy of the motor-cells, without any patches of inflammatory softening.

TREPANATION OF THE ILIUM IN A CASE OF PELVIC ABSCESS. •

Dr. G. Fisher (Deutsch. Zeitschr. fur chir., Band. XII.) reports a case of diseased lumbar vertebræ. A cold abscess develops and burrows into the left hip-joint, and invades the entire pelvis. An opening is made above Poupart's ligament, but fails to afford sufficient drainage. The ilium was trephined at its upper and posterior part, establishing a transverse drain across the pelvis. Exhaustion, prolonged suppuration and pulmonary tuberculosis caused death twenty-five days after the operation. The necropsy showed that no stagnation of pus had occurred in any part of the pelvis. Dr. Fisher has, therefore, demonstrated the feasibility of trephining the ilium to obtain a counter-opening in cases of pelvic abscess.

THREE OPERATIONS FOR GENU VALGUM BY MACEWEN'S METHOD.

J. C. Hutchison, M. D., (Proceedings of the Med. Society of the County of Kings). reports three cases of genu valgum. successfully operated on by Macewen's method. Macewen's plan was strictly adhered to, except in the application of the splint and some slight changes in the antiseptic dressing. The first operation was performed on a boy three years old. After the operation the leg was covered with Lister's protective, wet with carbolic acid solution, and over this was placed a layer of marine lint, which was retained in position by an antiseptic bandage extending to the hip. Anterior and posterior splints of plaster of paris were now applied to the whole length of the limb, which was held in proper position until the plaster set.

In the second case the leg was dressed in the same way; but in the third, instead of the plaster-of-paris, the Bavarian splint was applied from the groin to the heel, leaving a

fenestrum over the wound. "The state of the toes must be carefully watched during the first forty-eight hours, and afterwards looked at daily" (Macewen). The recovery in these cases, we are assured, was perfect.

RESECTION OF JOINTS AND ANTISEPTIC DRESSINGS.

M. Ollier (Revue Mensuelle de Med. et de chir., No, 12, 1880) advocates and insists on the importance of carrying out the antiseptic methods in the resection of joints. The so called orthopœdic resections have profited most from antiseptic practice. In all kinds of traumatic and pathological resections the antiseptic methods are imperiously demanded. After accident wounds of the joints, with fracture of the articular extremities of the long bones, the antiseptic method if properly used, will prevent the complications which alone indicate resection in many of such cases.

During six months of the last year, M. Ollier, applying exclusively Lister's method of antiseptic dressing, performed seventeen resections of large joints, one resection of a pseudoarthrosis, and two operations for osteotomy. In all these cases, the patients, so far as concerned the immediate result of the operative treatment, recovered without the occurrence of any infectious complications. From his experience of Lister's dressing in cases of resection, M. Ollier believes that this prevents inflammation of wounds without arresting the physiological processes of repair. The cicatrization and repair of osseous structure progress equally with the reunion of skin, and are not affected by the antiseptic treatment. Indeed, the organization of bone, and the ossification of the periosteal sheath, are rather accelerated than retarded by this plan of dressing, since this protects the wound and the internal layer of granulation tissue from danger of infection. In every case of resection treated according to Lister's method, whether the operation be practised for deformity, injury or chronic disease, M. Ollier endeavors to establish free drainage by a multiplicity of tubes, and refrains from applying sutures along the whole extent of the wound. In cases of chronic joint disease, where the synovial membrane is much thickened and vascular throughout, the superficial portion is excised or scraped away, and the deeper portions destroyed by the actual cautery or some chemical agent (chloride of zinc, nitrate of silver, perchloride of iron). In conclusion M. Ollier protests against an unnecessary and too widely extended application of operative surgery with antiseptic precautions to deformities, of the skeleton and its articulations, which have hitherto been regarded as amenable to ordinary orthopœdic treatment.

REPORT OF A CASE OF TALIPES EQUINO-VARUS, SUCCESSFULLY TREATED BY CONTINUOUS STRETCHING.

Chas. F. Stillman, M. D., (Walsh's Retrospect, January, 1881), reports a case of talipes equino-varus successfully treated by this method, and sums up the uses of the elastic cord, attached to the club-foot shoe, as follows:

First—The cord, passing from the extremity of the horizontal strip to the angle of divergence in the vertical strip, controls the extension of the foot at the ankle-joint, and acts against extra contraction of the tendo Achillis.

Second—The everting cord, passing from the instep strip to the toe of the shoe opposite the base of the little toe, acts in place of the peroneus brevis, having practically its origin in the immovable girth about the calf and its insertion in the sole considerably anterior to the mediotarsal joint, giving it a tremendous everting power if the girth be fixed immovably.

Third—The abducting and rotating cord passes from the base of the little toe to a point in the brace near the girth, supplying the place of the peroneus longus, and acting against the anterior and posterior tibial muscles, whose contraction causes the deformity, and as it is inserted at a point in the sole which is really the apex of the deformity, a power is exerted in exact proportion to the length and strength of the elastic tubing, twisting the anterior half of the foot directly contrary to the tendencies of the contraction.

The pivot insertion below allows the foot to be everted or inverted at will, without in the least impairing the support of the ankle, and any apparatus like those now in use which allows motion of the foot only upward and downward, does not fulfill the indications.

Sayre, Lewis A.—The danger of using an anæsthetic while suspending a patient to apply a plaster-of-paris jacket, (Louisville Med. Newes, July 23d, 1881).

Cabot, A. T.—A case of posterior torticollis, (Boston Med. and Surg. Journ.," May 19, 1881).

Owen.—Vertebral caries. (Brit. Med. Jour. June 11, 1881).

Bartow, B.—On the treatment of rotary-lateral curvature of the spine. (Buffalo Med. and Surg. Jour., June, 1881).

Dowse.—Pseudo-hypertrophic paralysis. (Med. Soc. of London), "Lancet," March 26, 1881.

Hill, B.—Case of fracture of the lower dorsal and upper lumbar vertebræ, treated with Sayre's jacket; recovery with only slight deformity of the spinal column. Clin. Soc. of London), Brit. Med. Jour., March 26, 1881.

Bryant, T.—Renewal of upper part of femur after removal for necrosis. (Med. Soc. of London), "Lancet," March 12, 1881.

Bennett.—Notes on osteotomy. (Brit. Med. Jour., April 21, 1881).

Morris, H.—Genu valgum. (Path. Soc. of London), "Lancet," May 21, 1881.

Richardson, B. W.—Vertebræ in process of disintegration by caries not necessarily accompanied by pain or tenderness in the overlying soft structures. (Med. Press and Circ., March 23, 1881).

Broadbent.—Pseudo-hypertrophic paralysis. (Med. Soc. London), "Lancet," April 9, 1881).

Nicoladoni.—Ueber Torsion der skoliotischen wirbelsaule. (K. R. Gesellsch d. Aerzte in Wein.). Wien. med. Woch., April 23, 1881.

Goodwillie, D. H.—Arthritis of the temporo-maxillary articulation. (Arch. of Med., June, 1881).

Haberern, J. P.—Ueber Beckenabscess bei coxitis und ihre Behandlung. (Centralbl. f. Chir, April 2, 9, 1881).

Chapman, N. H.—A point in the treatment of hip-joint disease. (Chicago Med. Jour. and Exam., April, 1881).

Ollier, L.—Resection de la hanche dans les coxalgies suppurees. (Soc. Nat. de Med. de Lyon, Lyon Med. May 1, 1881).

Uhthoff, J. C.—Five successive cases of operation on the knee for the removal of loose cartilage. (Brit. Med. Jour., April 9, 1881).

Beauregard.—Redressement d'un genu valgum par l' osteotomie. (Soc. de Chir., Paris, Rev. de Chir., May 1, 1881).

Macnamara.—Acute Osteitis; septicæmia, excision of tibia; recovery. (Brit. Med. Jour., April 2, 1881).

Mason.—Acute osteitis; pyæmia; osteotomy; death. (Brit. Med. Jour., April 2, 1881).

Richet, C.—A study of the pathology and physiology of contractures. (Lancet, May 21, 1881).

Macewen.—Transplantation of bone. (Brit. Med. Jour., May 21, 1881).

Vance, A. M.—Tenotomy in the treatment of congenital club-foot; with a tabular report of fifty-two cases, and remarks illustrating the management of the deformity. (Med. Record, April 23, 1881).

Tyson, M. J.—A case of cross-legged progression. (Brit. Med. Jour., May 14, 1881).

Ehrendorfer, E.—Mittheilungun uber Keil-exzisionen aus verschiedenen Knochen. (Wien Med. Woch., April 2, 9, 16, 1881).

Wheeler, U.—Exsection of knee-joint. (Brit. Med. Jour., April 2, 1881).

Marsh, H.—Report of the committee on excision of the hip-joint in childhood. (Clin. Soc. of London, Lancet, May 21, 1881).

Parker, R. W.—Remarks on the curvature of the long bones in rickets. (Med. Times and Gaz., April 2, 1881).

Taylor, H. L.—Location, age and sex in Pott's disease of the spine. (Med. Record, August 13, 1881, p. 175).

Hayden.—Lumbar caries; paraphlegia. (Path. Soc. of Dublin, Brit. Med. Jour,, April 23, 1881).

Redard.—Temperature of the joints in health and disease. (Jour. de Med. de Bordeaux, February 13).

Vance, A. M.—Removable paper brace in the treatment of spinal disease. (Med. Herald, May, 1881).

Yeoman, H. C.—Treatment of house-maids knee by aspiration. (Canada Lanc., July 1, 1881).

Corley.—Exsection of the hip joint. Brit. Med. Jour., April 2, 1881).

Bennett.—Morbus coxæ in the adult; excision. (Dublin Path. Soc. Dublin Journal of Med. Sci., April, 1881).

Clendenen, A.—General lower extremity splint; extension and counter-extension by weight of limb. Virginia Med. Monthly, August, 1881.

Judson A. B.—Considerations respecting the mechanical treatment of hip disease, with especial reference to the value of traction. (St. Louis Courier of Medicine, May, 1881.)

Croly, H. G.—Excision of the shoulder joint. (Medical Press and Circ., April 16. 1881.

Cazin.—Le toucher rectal dans la coxalgie. (Acad. de Med., Paris), Progr. Med. April 30, 1881.

Hagedorn.—Osteotomie des Unterkiefers wegen beiderseitiger ankylose desselben. (Verhandl. d. dtsch. Gesellsch f. chir., 1880.)

Gay.—Necrosis of elbow-joint, excision, death. (Boston Med. and Surgical Journal, April 12, 1881.

Report on Mental and Nervous Diseases.

By J. H. Kimball, M. D., Denver.

A paper by Dr. Rockwell, read before the New York County Medical Society, alludes to the difficulties which occur in the diagnoses of certain diseases of the spinal cord. He states that it is often very difficult to differentiate organic from functional disease in the early stages, and that the greater number of cases of spinal sclerosis reported cured, were of a functional character. While Dr. Seguin (Med. Record, 1881, page 225), claims that this disease should be recognized in its early stages from the following conditions and symptoms, which although not pathognomonic are peculiarly "characteristic." 1st. The peculiar, sudden, localized, paroxysmal, wandering pains occurring in numerous portions of the parts affected, covering areas from the size of a pea to that of the hand. 2d. The diminution of the various reflexes, best observed in the iris and patellar tendon. 3d. The paralysis of ocular muscles. These conditions grouped, rather than taken separately, serve as a basis for diagnosis.

ARSENIC IN HAY FEVER.

A writer in the *Am. Medical Bi-Weekly* claims that the chloro-phosphide of arsenic is, as nearly as possible, a specific for hay fever. He gives eight to twelve minims of the solution after each meal, which not only relieves the nervous irritability but acts as a tonic to the whole nervous system. Dr. E. C. Mann (Cin Lancet and Clinic, 1881, p. 495) uses this remedy in all the neuralgias with most gratifying success.

THE BROMIDES IN EPILEPSY.

The action of the bromides in epilepsy is thus stated by Dr. J. Hughes Bennett. (Edinburgh Medical Journal, March, 1881). In 12.1 per cent. of cases treated, the attacks were entirely arrested. In 83.3 per cent. they were diminished in number and severity. In 2.3 per cent. no apparent effect was produced. In 2.3 per cent. the attacks were increased in number, while the patients were under treatment. The success of the treatment did not appear to be influenced by the fact that the disease was, or was not, inherited, recent or chronic, complicated or not. The "bromide eruption" appeared in 16.6 per cent. of the cases mentioned.

Dr. Ramskill (Lancet, May 7, 1881) often combines carbonate of ammonium with the bromide of potassium, which not only increases the effect of the bromide, but also

acts as a stimulant and antacid. In syphilitic epilepsy, when the iodide of potassium, after succeeding to a certain point, suddenly fails to produce the desired effect, the addition of a few drops of the tincture of iodine will, he says, act "magically" on the disease.

FOREIGN BODIES IN THE SUBSTANCE OF THE BRAIN.

The case of an insane convict was described at a meeting of the St. Louis Medical Society, (St. Louis Med. and Surg. Journal, May, 1881) in whose brain were found at the post-mortem examination, the following foreign bodies, which had been introduced through an opening in the skull, made by an awl; five pieces of wire, varying from two and three-eighths inches to six and three-fourths inches in length, also a nail two and one-quarter inches long was found in the anterior lobe, and in the middle lobe a needle one and five-eighths inches long. These bodies were encysted and had during life given no apparent trouble. He died from an overdose of morphine taken by himself to overcome sleeplesness.

STRENGTH OF NERVES.

The results of experiments on nerve stretching by Dr. Charles F. Parks, of Chicago, are related in a paper read before the Illinois State Society. He states that it requires a force of eighty pounds to break the ulnar nerve, fifty-seven to break the median, eighteen for the radial, sixty-one for the brachial, seven and one-half for the facial, 200 for the sciatic and ninety-two for the cauda equina.

NEURALGIA AND DIABETES.

Dr. E. C. Mann, in *N. Y. Med. Gazette*, June, 1881, states that M. J. Worms, of Paris, concludes from a large number of cases, that, connected with diabetes, there are special forms of neuralgia, which had thus far been noticed in the sciatic and dental nerves. It appears in the two symmetrical divisions of the same pair of nerves. They are more painful than other neuralgias, do not yield to treatment and vary in intensity according to the severity of the diabetes.

HYSTERIA IN MAN.

M. Raymond presented to the Biological

Society of Paris (Progres Medical, 1881, page 526) the history of a phthisical patient, with whom the most marked symptom appeared to be a general and profuse sweating, which was relieved after prolonged treatment by the sulphate of atropine. The sweating reached an extraordinary intensity in the palms of the hands, which at times could be slightly relieved by the use of hypodermic injections of atropine and of duboisine, especially the latter remedy. With half the quantity and in a much shorter time, the same results were obtained with duboisine as with atropine. For a complete cure, it was necessary to use the galvanic current, which was applied to the brachial and cervical plexuses. Applications of ice and chloral were also made. During this treatment the pulmonary lesion was arrested in its progress, but he gradually developed hysterical symptoms, until, at the time of making the report, he presented the ordinary manifestations of this neurosis. The paroxysms are the same as those observed in women, and M. Raymond states that the attacks can be arrested by compression of the testicles.

INFLAMMATION OF ANTERIOR HORNS OF THE SPINAL CORD.

A case of acute inflammation of the anterior horns of the spinal cord is reported by Dr. W. D. Crosby, (N. Y. Med. Gazette, August 13, 1881), in which the following symptoms were presented: Fever, numbness in feet and in fingers, extending up the limbs, loss of power of adducting the legs. Muscles of upper extremities atrophied, loss of sensation and marked loss of power. Dropped wrists. Delusions, delirium and convulsions appeared as the disease advanced. At this date, (August 13th), after an illness of four months, the fever has ceased and improvement in power of motion is taking place. Among points in diagnosis may be mentioned the following; The muscles paralyzed do not react to galvanic or faradic currents, but if one pole is applied on a nerve, a reaction may be obtained. Sensations of a girdle applied about the body, thigh or knee. The treatment in the above

cited case, consisted in the application of the continuous current; also in the use of drachm doses of ergot three times a day.

SOCIETY REPORTS.

International Medical Congress, 1881.
Opening Address.

By THE PRESIDENT, SIR JAMES PAGET, BART.

It is not necessary to defend the meeting of an International Congress. Such meetings have become one of the general customs of our time, and have thus given evidence that they are generally approved. Let me rather suggest to you some thoughts as to the work which, being in Congress, we have to do, and the spirit in which it may best be done, so that the good effects of our meeting may last long after our parting.

In the largest view of our design, it may seem that our bringing together a multitude of various minds for the promotion and diffusion of knowledge in the whole science and art of medicine, in their widest range, in all their narrowest divisions, in all their manifold utilities. And this design, I cannot doubt, will be fulfilled; for, although the programme tells of selected subjects for discussion, and defines the order of our work, yet knowledge will be promoted in a much wider range in the meetings without order, which will be held every day and everywhere— meetings of men with all kinds of mental power and all forms of knowledge and of skill; every one ready alike to impart and to acquire knowledge.

It is safe to say that in the casual conversations of this coming week there will be a larger interchange and diffusion of information than in any equal time and space in the whole past history of medicine. And with this interchange will be a larger increase, for in the mart of knowledge he that receives gains, and he that gives retains, and none suffer loss.

The increase will be the greater because of the great variety of minds which will meet. As

I look round the hall, my admiration is moved not only by the number and total power of the minds which are here, but by their diversity; a diversity in which I believe they fairly represent the whole of those who are engaged in the cultivation of our science. For here are minds representing the distinctive characters of all the most gifted and most educated nations; characters still distinctly national, in spite of the constantly increasing intercourse of the nations. And from many of these nations we have both elder and younger men; thoughtful men and practical; men of fact and men of imagination; some confident, some skeptic; various, in education, in purpose and mode of study, in disposition and in power. And scarcely less various are the places and all the circumstances in which those who are here have collected and have been using their knowledge. For I think that our calling is preeminent in its range of opportunities for scientific study. It is not only that the pure science of human life may match with the largest of the natural sciences in the complexity of its subject-matter; not only that the living human body is, in both its material and its indwelling forces, the most complex thing yet known; but that in our practical duties this most complex thing is presented to us in an almost infinite multiformity. For in practice we are occupied, not with a type and pattern of the human nature, but with all its varieties in all classes of men, of every age and every occupation, in all climates and all social states; we have to study men singly and in multitudes, in poverty and in wealth, in wise and unwise living, in health and all the varieties of disease; and we have to learn, or at least try to learn, the results of all these conditions of life, while in successive generations and in the mingling of families, they are heaped together, confused, and always changing. In every one of all these conditions man, in mind and body, must be studied by us; and every one of them offers some different problems for inquiry and solution. Wherever our duty or our scientific curiosity or, in happy combination, both, may lead us, there are

the materials and there the opportunities for separate original research.

Now, from these various opportunities of study, men are here in Congress. Surely, whatever a multitude and diversity of minds can, in a few days, do for the promotion of knowledge, may be done here. Every one has something he may teach, much more that he may learn; and in the midst of an apparent utter confusion, knowledge will increase and multiply. It has been said, indeed, that truth is more likely to emerge from error than from confusion and, in some instances, this is true; but much of what we call confusion is only the order of nature not yet discerned; and so it may be here. Certainly, it is from what seems like the confusion of successive meetings such as this that that kind of truth emerges which is among the best moving and directing forces in the scientific as well as in the social life—the truth which is told in the steady growth of general opinion.

But it is not proposed to leave the work of the Congress to what would seem like chances and disorder, good as the result might be; nor yet to the personal influences by which we may all be made fitter for work, though these may be very potent. In the stir and controversy of meetings such as we have, there cannot fail to be useful emulation; by the examples that will appear of success in research many will be moved to more enthusiasm, many to more keen study of the truth; our range of work will be made wider, and we shall gain that greater interest in each other's views and that clearer apprehension of them which are always attained by personal acquaintance and by memories of association in pleasure as well as in work. But as it will not be left to chance, so neither will sentiment have to fulfill the chief duties of the Congress.

Following the good example of our predecessors, certain subjects have been selected which will be chiefly, though not exclusively, discussed, and the discussions are to be in the sections into which we shall soon divide. Of these subjects it would not be for me to speak even if I were competent to do so; unless I may say that they are so numerous and complete that—together with the opening addresses of the Presidents of Sections— they leave me nothing but such generalities as may seem commonplace. They have been selected, after the custom of former meetings, from the most stirring and practical questions of the day; they are those which most occupy men's minds, and on which there is at this time most reason to expect progress, or even a just decision, from very wide discussion. They will be discussed by those most learned in them, and in many instances by those who have spent months or years in studying them, and who now offer their work for criticism and judgment.

I will only observe that the subjects selected in every section evolve questions in the solution of which all the varieties of mind and knowledge of which I have spoken may find their use. For there are questions not only on many subjects, but in all stages of progress towards settlement. In some the chief need seems to be the collection of facts well observed by many persons. I say by many, not only because many facts are wanted, but because in all difficult research it is well that each apparent fact should be observed by many; for things are not what they appear to each one mind. In that which each man believes that he observes, there is something of himself; and for certainty, even on matters of fact, we often need the agreement of many minds, that the personal element of each may be counteracted. And much more is this necessary in the consideration of the many questions which are to be decided by discussing the several values of admitted facts and of probabilities, and of the conclusions drawn from them. For, on questions such as these minds of all kinds may be well employed. Here there will be occasion even for those which are not unconditionally praisworthy, such as those that habitually doubt, and those to whom the invention of argument is more pleasing than the mere search for truth. Nay, we may be able to observe the utility even of error. We may not,

indeed, wish for a prevalence of errors; they are not more desirable than are the crime and misery which evoke charity. And yet in a Congress we may palliate them, for we may see how, as we may often read in history, errors, like doubts and contrary pleadings, serve to bring out the truth, to make it express itself in clearest terms and show its whole strength and value. Adversity is an excellent school for truth as well as for virtue.

But that which I would chiefly note, in relation to the great variety of minds which are here, is that it is characteristic of that mental pliancy and readiness for variation which is essential to all scientific progress, and which a great International Congress may illustrate and promote. In all the subjects for discussion we look for the attainment of some novelty and change in knowledge or belief; and after every such change there must ensue a change in some of the conditions of thinking and of working. Now for all these changes minds need to be pliant and quick to adjust themselves. For all progressive science there must be minds that are young whatever may be their age.

Just as the discovery of auscultation brought to us the necessity for a refined cultivation of the sense of hearing, which was before of only the same use in medicine as in the common business of life; or, as the employment of the numerical method in estimating the value of facts required that minds should be able to record and think in ways previously unused; or, as the acceptance of the doctrine of evolution has changed the course of thinking in whole departments of science, so is it, in less measure, in every less advance of knowledge. All such advances change the circumstances of the mental life, and minds that cannot or will not adjust themselves become less useful, or must, at least, modify their manner of utility. They may continue to be the best defenders of what is true; they may strengthen and expand the truth, and may apply it in practice with all the advantages of experience; they may thus secure the possessions of science and use them well, but they will not increase them.

It is with minds as with living bodies. One of their chief powers is in their self-adjustment to the varying conditions in which they have to live. Generally those species are the strongest and most abiding that can thrive in the widest range of climate and of food. And of all the races of men they are the mightiest and most noble who are, or by self-adjustment can become, most fit for all the new conditions of existence in which by various changes they may be placed. These are they who prosper in great changes of their social state; who, in successive generations, grow stronger by the production of a population so various that some are fitted to each of all the conditions of material and mode of life which they can discover or invent. These are most prosperous in the highest civilization; these whom nature adapts to the products of their own arts.

Or, among other groups, the mightiest are those who are strong alike on land and sea; who can explore and colonize, and in every climate can replenish the earth and subdue it; and this not by tenacity or mere robustness, but rather by pliancy and the production of varieties fit to abide and increase in all the various conditions of the world around.

Now, it is by no distant analogy that we trace the likeness between these in their successful contests with the material conditions of life and those who are to succeed in the intellectual strife with the difficulties of science and of art. There must be minds which in variety may match with all the varieties of the subject-matters and minds which, at once or in swift succession, can be adjusted to all the increasing and changing modes of thought and work.

Such are the minds we need; or, rather, such are the minds we have; and these in great meetings prove and augment their worth. Happily the natural increase in the variety of minds in all cultivated races is— whether as cause or as consequence—nearly proportionate to the increasing variety of knowledge. And it has become proverbial, and is nearly true in science and art, as it is in commerce and in national life, that, whatever work is to be done, men are found or soon produced who are exactly fit to do it.

But it need not be denied that, in the possession of this first and chiefest power for the increase of knowledge, there is a source of weakness. In works done by dissimilar and independent minds, dispersed in different fields of study, or only gathered into self-assorted groups, there is apt to be discord and great waste of power. There is, therefore, need that the workers should from time to time be brought to some consent and unity of purpose; that they should have opportunity for conference and mutual criticism, for mutual help and the tests of free discussion. This it is which, on the largest scale and most effectually, our Congress may achieve; not, indeed, by striving after a useless and happily impossible uniformity of mind or method, but by diminishing the lesser evil of waste and discord which is attached to the far greater good of diversity and independence. Now, as in numbers and variety the Congress may represent the whole multitude of workers everywhere dispersed, so in its gathering and concord it may represent as a common consent that, though we may be far apart and different yet our work is and shall be essentially one; in all its parts mutually dependent, mutually helpful, in no part complete or self-sufficient. We may thus declare that as we who are many are met to be members of one body, so our work for science shall be one though manifold; that as we, who are of many nations, will, for a time, forget our nationalities, and will even repress our patriotism, unless for the promotion of a friendly rivalry, so will we in our work, whether here and now or everywhere and always, have one end and one design,—the promotion of the whole science and whole art of healing.

It may seem to be a denial of the declaration of unity that after this general meeting we shall separate into sections more numerous than in any former Congress. Let me speak of these sections to defend them; for some maintain that even in such a division of studies as these may encourage, there is a mischievous dispersion of forces. The science of medicine, which used to be praised as one and indivisable, is broken up, they say, among specialists, who work in conflict rather in concert, and with mutual distrust more than mutual help.

But let it be observed that the sections which we have instituted are only some of those which are already recognized in many countries, in separate societies, each of which has its own place and rules of self-government and its own literature. And the division has taken place naturally in the course of events which could not be hindered. For the partial separation of medicine, first from the other natural sciences, and now into sections of its own, has been due to the increase of knowledge being far greater than the increase of individual mental power.

I do not doubt that the average mental power constantly increases in the successive generations of all well-trained peoples, but it does not increase so fast as knowledge does, and thus, in every science, as well as in our own, a small portion of the whole sum of knowledge has become as much as even a large mind can hold and duly cultivate. Many of us must, for practical life, have a fair acquaintance with many parts of our science, but none can hold it all; and for complete knowledge, or for research, or for safely thinking out beyond what is known, no one can hope for success unless by limiting himself within the few divisions of the science for which, by nature or by education, he is best fitted, Thus our divisions into sections is only an instance of that division of labor which, in every prosperous nation, we see in every field of active life, and which is always justified by more work better done.

Moreover, it cannot be said that in any of our sections there is not enough for a full strong mind to do. If any one will doubt this, let him try his own strength in the discussions of several of them.

In truth, the fault of specialism is not in narrowness, but in the shallowness and the belief in self-sufficiency with which it is apt to be associated. If the field of any specialty in science be narrow, it can be dug deeply. In science, as in mining, a very narrow shaft, if only it be carried deep enough, may reach the richest stores of wealth and find use for all the appliances of scientific art. Not in medicine alone, but in every department of knowledge, some of the grandest results of research and of learning, broad and deep, are to be found in monographs on subjects that, to the common mind, seem small and trivial,

And study in a Congress such as this may be a useful remedy for self-sufficiency. Here every group may find a rare occasion, not only for an opportune assertion of the supreme excellency of its own range and mode of study, but for the observation of the work of every other. Each section may show that its own facts must be deemed sure, and that by them every suggestion from without must be tested; but each may learn to doubt every inference of its own which is not consistent with the facts or reasonable beliefs of others; each may observe how much there is in the knowledge of others which should be mingled with its own; and the sum of all may be the wholesome conviction of all that we cannot justly estimate the value of the doctrine in one part of our science till it has been tried in many or in all.

We were taught this in our schools; and many of us have taught that all the parts of the medical science are necessary to the education of the complete practitioner. In the independence of later life, some of us seem too ready to believe that the parts we severally choose may be self-sufficient, and that what others are learning cannot much concern us. A fair study of the whole work of the Congress may convince of the fallacy of this belief. We may see that the test of truth in every part must be in the patient and impartial trial of its adjustment with what is true in every other. All perfect organizations bear this test; all parts of the whole body of scientific truth should be tried by it.

Moreover, I would not, from a scientific point of view, admit any estimate of the comparative importance of the several divisions of our science, however widely they may differ in their present utilities. And this I would think right, not only because my office as president binds me to a strict impartiality and to the claim of freedom of research for all, but because we are very imperfect judges of the whole value of any knowledge, or even of single facts. For every fact in science, wherever gathered, has not only a present value, which we may be able to estimate, but a living and germinal power of which none can guess the issue.

It would be difficult to think of anything that seemed less likely to acquire practical utility than those researches of the few naturalists who, from Leeuwenhoeck to Ehrenberg, studied the most minute of living things, the Vibrionidæ. Men boasting themselves as practical might ask, "What good can come of it?" Time and scientific industry have answered, "This good: those researches have given a more true form to one of the most important practical doctrines of organic chemistry; they have introduced a great beneficial change in the most practical part of surgery; they are leading to one as great in the practice of medicine; they concern the highest interests of agriculture, and their power is not yet exhausted."

And as practical men were, in this instance, incompetent judges of the value of scientific facts, so were men of science in fault when they missed the discovery of anæsthetics. Year after year the influences of laughing gas and of ether were shown; the one fell to the level of the wonders displayed by itinerant lecturers, students made fun one with the other; they were the merest practical men, men looking for nothing but what might be straightway useful, who made the great discovery which has borne fruit not only in the mitigation of suffering, but in a wide range of physiological science.

The history of science has many similar facts, and they may teach that any man will be both wise and dutiful if he will patiently and thoughtfully do the best he can in the field of work in which, whether by choice or chance, his lot is cast. There let him, at least, search for truth, reflect on it, and record it accurately; let him imitate that accuracy and completeness of which I think we may boast that we have, in the descriptions of the human body, the highest instance yet attained in any branch of knowledge. Truth so recorded cannot remain barren.

In thus speaking of the value of careful observation and records of facts, I seem to be in agreement with the officers of all the sections; for, without any intended consent, they all have proposed such subjects for discussion as can be decided only by well collected facts and fair direct inductions from them. There are no questions on theories or mere doctrine. This, I am sure, may be ascribed, not to any disregard of the value of good reasoning or of reasonable hypotheses, but partly to the just belief that such things are ill-suited for discussion in large meetings, and partly to the fact that we have no great opponent schools, no great parties named after leaders or leading doctrines about which we are in the habit of disputing. In every section the discussions are to be on definite questions, which, even if they be associated with theory or general doctrines, may yet be soon brought to the test of facts; there is to be no use of doctrinal touchstones.

I am speaking of no science but our own. I do not doubt that in others there is advantage in dogma, or in the guidance of a central organizing power, or in divisions and conflicting parties. But in the medical sciences, I believe that the existence of parties founded on dominant theories has always been injurious; a sign of satisfaction with plausible error, or with knowledge which was even for the time imperfect. Such parties used to exist, and the personal histories of their leaders are some of the most attractive parts of the history of medicine; but, although in some instances an enthusiasm for the master-mind may have stirred a few men to unusual industry, yet very soon the disciples seem to have been fascinated by the distinctive doctrine, content to bear its name, and to cease from active scientific work. The dominance of doctrine has promoted the habit of inference, and repressed that of careful observation and induction. It has encouraged the fallacy to which we are all too prone, that we have at length reached an elevated sure position on which we may rest, and only think and guide. In this way specialism in doctrine or in method of study has hindered the progress of science more than the specialism which has attached itself to the study of one organ or of one method of practice. This kind of specialism may enslave inferior minds: the specialism of doctrine can enchant into mere dreaming those who should be strong in the work of free research.

I speak the more earnestly of this because it way be said, if our Congress be representative, as it surely is, may we not legislate? May we not declare some general doctrines which may be used as tests and as guides for future study? We had better not.

The best work of our International Congress is in the clearing and strengthening of the knowledge of realities; in bringing, year after year, all its force of numbers and varieties of minds to press forward the demonstration and diffusion of truth as nearly to completion as may from year to year be possible. Thus, chiefly, our Congress may maintain and invigorate the life of our science. And the progress of science must be as that of life. It sounds well to speak of the temple of science and of building and crowning the edifice. But the body of science is not as any dead thing of human work, however beautiful; it is as something living, capable of development and a better growth in every part. For as in all life the attainment of the highest condition is only possible through the timely passing-by of the less good, that it may be replaced by the better, so is it in science. As time passes, that which seemed true and was very good becomes relatively imperfect truth, and the truth more nearly perfect takes its place.

We may read the history of the progress of truth in science as a palæontology. Many things which, as we look far back, appear, like errors, monstrous and uncouth creatures,

were, in their time, good and useful, as good as possible. They were the lower and less perfect forms of truth which, amid the floods and stifling atmosphere of error, still survived; and just as each successive condition of the organic world was necessary to the evolution of the next following higher state, so from these were slowly evolved the better forms of truth which we now hold.

This thought of the likeness between the progress of scientific truth and the history of organic life may give us all the better courage in a work which we cannot hope to complete, and in which we see continual and sometimes disheartening change. It is, at least, full of comfort to those of us who are growing old. We that can read in memory the history of half a century might look back with shame and deep regret at the imperfections of our early knowledge if we might not be sure that we held, and sometimes helped onward, the best things that were in their time possible, and that they were necessary steps to the better present, even as the present is to the still better future. Yes, to the far better future; for there is no course of nature more certain than is the upward progress of science. We may seem to move in circles, but they are the circles of a constantly ascending spiral; we may seem to sway from side to side, but it is only as on a steep ascent which must be climbed in zigzag.

What may be the knowledge of the future none can guess. If we could conceive a limit to the total sum of mental power which will be possessed by future multitudes of well-instructed men, yet could we not conceive a limit to the discovery of the properties of materials which they will bend to their service. We may find the limit of the power of unaided limbs and senses; but we cannot guess at a limit to the means by which they may be assisted, or to the invention of instruments which will only become a little more separate from our mental selves than are the outer-sense organs with which we are constructed.

In the certainty of this progress the great

question for us is. What shall we contribute to it? It will not be easy to match the recent past. The advance of medical knowledge within one's memory is amazing whether reckoned in the wonders of the science not yet applied, or in practical results in the general lengthening of life, or, which is still better, in the prevention and decrease of pain and misery, and in the increase of working power. I cannot count or recount all that in this time has been done; and I suppose there are very few, if any, who can justly tell whether the progress of medicine has been epual to that of any other great branch of knowledge during the same time. I believe it has been; I know that the same rate of progress cannot be maintained without the constant and wise work of thousands of good intellects; and the mere maintenance of the same rate is not enough, for the rate of the progress of science should constantly increase. That in the last fifty years was at least twice as great as that in the previous fifty. What will it be in the next, or, for a more useful question, What shall we contribute to it?

I have no right to prescribe for more than this week. In this let us do heartily the proper work of the Congress, teaching, learning, discussing, looking for new lines of research, planning for mutual help, forming new friendships. It will be hard work if we will do it well; but we have not met for mere amusement or for recreation, though for that I hope you will find provision, and enjoy it the better for the work preceding it.

And when we part let us bear away with us, not only much more knowledge than we came with, but some of the lessons for our conduct in the future which we may learn in reflecting the work of our Congress.

In the number and intensity of the questions brought before us we may see something of our responsibility. If we could gather into thought the amounts of misery or happiness, of helplessness, or of power for work, which may depend on the answers to all the questions that will come before us, this might be a measure of our responsibility. But we cannot count it; let us imagine it;

we cannot even in our imagination exaggerate it. Let us always bear it in our mind, and remind ourselves that our responsibility will constantly increase. For, as men become in the best sense better educated, and the influence of scientific knowledge on their moral and social state increases, so, among all sciences there is none of which the influence and, therefore, the responsibility will increase more than ours, because none more intimately concerns man's happiness and working power.

But, more clearly in the recollections of the Congress, we may be reminded that in our science there may be, or, rather, there really is, a complete community of interest among men of all nations. On all the questions before us we can differ, discuss, dispute, and stand in earnest rivalry; but all consistent with friendship, all with readiness to wait patiently till more knowledge shall decide which is in the right. Let us resolutely hold to this when we are apart; let our internationality be a clear abiding sentiment, to be, as now, declared and celebrated at appointed times, but never to be forgotten; we may, perhaps, help to gain a new honor for science, if we thus suggest that in many more things, if they were as deeply and dispassionately studied, there might be found the same complete indentity of international interests as in ours.

And then, let us always remind ourselves of the nobility of our calling. I dare to claim for it that among all the sciences ours, in the pursuit and use of truth, offers the most complete and constant union of those three qualities which have the greatest charm for pure and active minds,—novelty, utility, and charity.

These three; which are sometimes in so lamentable disunion, as in the attractions of novelty without either utility or charity, are in our researches so combined that, unless by force or willful wrong, they can hardly be put asunder. And each of them is admirable in its kind. For in every search for truth we cannot only exercise curiosity, and have the delight—the really elemental happiness—of watching the unveiling of a mystery, but, on the way to truth, if we look well round us, we shall see that we are passing among wonders more than the eye or mind can fully apprehend. And as one of the perfections of nature is that in all her works wonder is harmonized with utility, so is it with our science. In every truth attained there is utility either at hand or among the certainties of the future. And this utility is not selfish; it is not in any degree correlative with money-making; it may generally be estimated in the welfare of others better than in our own. Some of us may, indeed, make money and grow rich; but many of those that minister even to the follies and vices of mankind can make much more money than we. In all things costly and vainglorious they would far surpass us if we would compete with them. We had better not compete where wealth is the highest evidence of success; we can compete with the world in the nobler ambition of being counted among the learned and the good who strive to make the future better and happier than the past. And to this we shall attain if we will remind ourselves that as in every pursuit of knowledge there is the charm of novelty, and in every attainment of truth utility, so in every use of it there may be charity. I do not mean the charity which is in the hospitals or in the service of the poor, great as is the privilege of our calling in that we may be its chief ministers, but that wider charity which is practised in a constant sympathy and gentleness, in patience and self-devotion. And it is surely fair to hold that, as in every search for knowledge we may strengthen our intellectual power, so in every practical employment of it we may, if we will, improve our moral nature; we may obey the whole law of Christian love, we may illustrate the highest induction of scientific philanthropy.

Let us, then, resolve to devote ourselves to the promotion of the whole science, art, and charity of medicine. Let this resolve be to us as a vow of brotherhood; and may God help us in our work.—*Boston Med. and Surg. Jour.*

Section on Medicine.

Sir William Gull, M.D., D.C.L., L.L.D., F.R.S., President, in the Chair.

THE ADDRESS OF THE PRESIDENT.

The President called the section to order, and then proceeded to deliver his opening address. In his introductory remarks he took occasion to refer to the great strides that had been taken within the past few years in the rational study of medicine, in that method which tended so directly toward the accumulation of facts rather than the promulgation of theory. Hence it was that the contradistinction with the old-fashioned doctrine of fluids, that the solidism, properly so-called, is widely reasserting itself in the science of living things; not as an *a priori* system, but through the progress of knowledge. The proximate conditions of pyrexia were no longer vaguely referred to nerve, but to definite nerve-centres; hyperæmia and inflammatory changes to sympathetic lesions; abnormal chemistry to the great respiratory centres; the strange condition of Addison's disease, with its characteristic pigment, to the suprarenal bodies, themselves probably but nerve-centres, and related, at least by structure, to the system of the pituitary gland; epilepsy, supposed in Hippocratic times to be due to extraneous maleficient spiritual influences, was traceable to apparently trifling changes in a few gray nerve-cells. The specific fever-processes notoriously owe much of their character and intensity to the nervous system. Their relation to time, their occurence only in warm-blooded animals, the great mortality they cause through nerve-exhaustion, and the immunity they left behind them, indicated that, whatever might be the nature or mode of operation of their several poisons. it was by implication of nerve-elements that fever obtained its chief clinical characteristics. Further, in the advance of "solidism," what could interest us more than the recent investigations on contagia? Perhaps no more important step has been made in practical pathology than the proof that some at least of these contagia were organized solids.

This discovery, which had tried the patience, experimental skill, and scientific criticism of the best observers to establish, had brought us at length within view of that which had hitherto been so mysterious. To have been able to separate, though imperfectly, the contagious particles; to have come to the conclusion that no fever-poisons were soluble, was a hopeful preliminary toward forcing them to yield up the secret of their nature. If "solidism," as a theory of organic processes, wanted confirmation, we could point to nothing more striking than the present established views on putrefactive changes; and to the amazing fact, that the normal textures and fluids of the body resisted decomposition, unless invaded by microscopic organism. He believed that hereafter all organic chemistry would be found to be simply the resultant of mechanical changes in organic solids.

Section on Surgery.

JOHN ERIC ERICHSEN, F. R. S., PRESIDENT, IN THE CHAIR.

SUBJECTS FOR DISCUSSION.

After a few preliminary statements, the President of the Section remarked as follows: The executive of this section has proposed eight subjects for the consideration of its members. It is hoped that these will be found to include the more important surgical questions that are at present most prominently before the profession. The short time at our disposal, which will scarcely enable us to do full justice even to these subjects, has prevented the possibility of our bringing forward other and perhaps equally interesting questions; but some of these will be found to have received consideration in the papers which will be read, either *in extenso* or in abstract, as time may allow.

I will now briefly refer to the more important subjects that have been set down for our consideration.

First.—In no department of surgery has a more marked or a more brilliant advance been made of late years than in that which concerns the operative treatment of intraperitoneal tumors.

The establishment of ovariotomy as a recognized surgical operation has now long been matter of history; but the perfection of safety to which it has of late years been carried, by the improvement of its details, has led the way to a vast and rapid extension of operative surgery for the cure or relief of various diseased abdominal organs. The uterus and the spleen the stomach, the pylorus and the colon, have each and all been subjected to the scalpel of the surgeon, with what success has yet to be determined; and it is for you to decide whether some, at least, of these operations constitute real and solid advances in our art, or whether they are rather to be regarded as bold and skillful experiments on the endurance and reparative power of the human frame —whether in fact they are surgical triumphs or operative audacities. There must, indeed, be a limit to the progress of operative surgery in this direction. Are we at present in a position to define it? There cannot always be new fields for conquest by the knife; there must be portions of the human frame that will ever remain sacred from its intrusion, at least in the hands of the surgeon. May there not be some reason to fear lest the very perfection to which ovariotomy has been carried may lead to an over-sanguine expectation of the value and safety of the abdominal section and exploration when applied to the diagnosis or cure of diseases of other and very dissimilar organs, in which but little of ultimate advantage, and certainly much of immediate peril, may be expected from operative interference?

Second.—In the discussion of the next great question, I would submit that we may, with advantage, direct our attention less to the mere mechanical—the simple operative part of business—the details of which are now well understood, than to the consideration of those higher questions as to the diagnosis and nature of the various forms of renal disease, in which nephrotomy and nephrectomy may be respectively used, with a reasonable hope of relief or cure. And in considering the prospects afforded by these operations in the improvement of the health and the mitigation of the sufferings of the patient, it is surely not the least interesting point for us to study the after-physiological effects produced on the system by the extirpation of so important and eliminatory organ the kidney.

Third.—We naturally pass from the consideration of operations on the kidney to that of those which implicate the bladder; and in doing so we have specially to direct our attention to the question as to what advances have of late been made in lithotomy and lithotrity.

In lithotomy we see much of change, possibly something of novelty, but not so certainly anything of real progress. Have we indeed advanced one single step. either in the perfection or in the results of that operation since the days of Cheselden, of Martineau, or of Crosse, not to mention the names of more recent, but equally illustrious surgeons and successful operators? The revived median operation, the combination of it with lithotrity, the suprapubic, whether done antiseptically or not, have certainly not been very encouraging in their results, and can scarcely claim to be considered in the light of an advance on the old lateral operation in skillful hands. But yet we must admit that these methods of lithotomy may deserve this consideration—that possibly in some forms of calculus, and in certain conditions of the urinary organs, a wise eclecticism may be exercised in the choice of one or other of them.

In lithotrity, however, it is probable that a great and real advance has been made, and certainly it is undoubted that a complete revolution has been effected by the enterprise and skill of one of our American brethren, for it cannot be questioned that "Bigelow's operation" has completely changed the aspect of lithotrity, and there is every reason to believe that it constitutes one of those real advances in a method which marks an epoch not only in the history of the operation itself, but in the treatment of the disease to which it is applicable.

But here a fertile field opens up for our deliberation. We have to consider not only, in what cases, as regards the mere size of calculus, "Bigelow's operation" may safely be used; but also, and far more important than this, the ultimate result both upon the bladder and the kidney of prolonged intravesical instrumentation. The mere question as to the comparative advantages of removal of stone, by one or by several sittings, is dwarfed by the more important one of determining the state of the bladder that results— not perhaps so much as concerns the life as the future comfort of the patient. It is here that information is much needed, and it is here that unfortunately, but for very obvious reasons, the lithotritist himself may in many cases be unable to furnish it.

Fourth.—Prehistoric man was doubtless a victim of injury before he became a sufferer from disease, and the treatment of wounds constituted probably the first effort of the healing art. From the earliest dawn of human intelligence the attempt to cure a wound must have suggested itself to man, and yet at the close of the nineteenth century we are still discussing the best methods of doing this, and the causes of their failure. There is still difference of opinion and of practice among surgeons, not only as to the comparative advantages of the "open air" method, and that in which all atmospheric contact is carefully guarded against; of the "dry" and of the "moist" system of dressing; as to whether the "antiseptic method" in a modified form suffices, or whether the more elaborate system of local treatment before, during and after the operation, which has been devised by the skill and worked out by the unwearied labor of Lister, be essential in all cases of operation wound. Not, of course, for its primary union—for this may be obtained by any and every of the methods mentioned. If it be contended that this system is necessary for the safety of the patient, and the due healing of the wound in some cases, has it been proven to be equally essential in traumatic lesions of all tissues, of all organs and of all regions?

These are questions that may well deserve the consideration of this section. But there are others of a yet wider character that must also engage our attention in any discussion on the best methods of securing primary union in wounds, for it is impossible to fail to recognize in the general constitutional state of the patient a most important factor in this direction; and we should be taking a narrow view of this many-sided question if we did not give due weight to the influence of those hygienic conditions which, if faulty, are inimical or even destructive to the due performance of those actions which are necessary for the maintenance of the organism in a healthy state, and for the proper nutrition and consequent repair of the tissues of the body. Is there no fear that in some of the modern systems of treating wounds we are in danger of expending all our precautions in the prevention of the local, and of ignoring the risk of a constitutional infection?

Fifth.—The treatment of aneurism is one of those great questions which from an early period in the history of modern surgery has occupied the attention of practitioners, and has undergone no little fluctuation. A few years ago the battle between the ligature and compression appeared to have been decided in favor of tne latter; but the invention of improved ligatures, made of various kinds of animal tissue, and applied with antiseptic precautions, has once more inclined the balance of professional opinion toward the Hunterian operation. But now again the practice of compression has received renewed strength from the employment of Esmarch's elastic bandage in the cure of certain forms of external aneurism, and it is for you to determine in what cases it can be used with advantage, and in what way a cure is effected by its means. For in the treatment of aneurisms, as in that of so many other surgical diseases, the wiser and more scientific course is to follow a judicious system of selection in the method to be employed in each particular case, rather than to subject all to one unbending line of practice.

Sixth.—The treatment by resection of some forms of chronic and otherwise incurable joint diseases, has, in certain articulations and at suitable ages, met with universal approbation of surgeons, and the wide extension of the principles of "conservative surgery" is one of the most striking evidences of advance in our art in modern times. Resection has, however, of late years come to be extensively applied to the treatment of cases of articular disease which formerly were subjected to procedures of a less heroic character; and it will be for the members of this section to weigh carefully the wisdom of such a measure, and to contrast its results, both as regards the life of the patient and the after-utility of the limbs, with those which may be obtained from the employment of milder means, such as absolute immobility with extension, and possibly, in some cases, simple incision of the articulation.

Seventh.—In considering the relations between adenoma, sarcoma, and carcinoma in the mammary gland of the female, I would venture to submit that this subject has to be discussed here from its clinical rather than from its pathological side. We have here less to do with the ultimate structure of the tumors, with their histological affinities, with the parts that are played by epiblasts and mesoblasts, with what epithelium or connective-tissue cells can or cannot do, than with their clinical history, their differential diagnosis in their earlier stages, the best time for their removal by operation, the liability to recurrence after operation, and the possibility in recurrance of the substitution of one form of disease for another. With these, and such questions as these, we, as clinical surgeons, may advantageously occupy ourselves.

Eighth.—The last subject set down for discussion is one that has practical bearings of an importance that cannot be overestimated. There are few questions of the present day of deeper surgical or social interest than the far-reaching, the apparently illimitable and most pernicious extension of a syphilitic contamination of organs and of tissues; of the modifications impressed by it on other diseases that are the local developments of diathesis. whether strumous, tubercular, rheumatic or gouty. Does the diathesis exercise any influence upon the form assumed by the syphilitic disease, and to what extent does it modify the characters presented by it in its primary and its secondary affections, more especially when the latter manifest themselves upon the skin or in the bones; how far are gummata and caries, psoriasis, and rupia the consequences of a constitutional impress, influencing the direction of the syphilitic poison? To what extent may rickets and gray granulations be the ultimate products of the syphilitic taint? These and various other questions will probably occupy occupy the attention of those who enter on the discussion of this wide-spreading subject.

Section on Diseases of Children.

Charles West, M. D., President. in the Chair.

PROGRESS IN THE STUDY OF INFANTILE DISEASES.

The President, after expressing his deep sense of the honor conferred upon him, proceeded to give a cursory sketch of the progress of the study of infantile diseases during the past thirty years. He said:

Thirty years ago, throughout the whole of England and America, there was not a single hospital set apart for children. It was but rarely that one saw these little waifs and strays, in the wards of our general hospitals, for the maxim, "De minimis non curat lex," held good in medicine as in law. Germany, too, was in but little better case, and one was forced to go to Paris to study on a large scale those diseases which men like Guersant. and Blache, and Baron, and Trousseau, and Rogers investigated with untiring zeal, and is spite of hospital arrangements most painfully defective, strove to cure. We all know how this is altered now. In London there are six separate children's hospitals, each, I believe, with its convalescent branch;

and children's wards are to be found in every one of the large London hospitals. There are special children's hospitals in every large town in England. America and Germany have followed this example, and almost everywhere throughout Europe the opportunities for the study of the diseases of children are almost as numerous as for the diseases of adults.

Nor has this wide field been without abundant husbandmen to till it, and we may count with satisfaction the fruit of their labors.

The vague phraseology which served for years to conceal our ignorance even from ourselves, has been to a great degree done away with. We talk no longer of worm fever, remittent fever, gastric fever, and so on; for under these various names we recognize the one disease, typhoid fever, varying in severity, but marked always by its own characteristic symptoms. Half a page in a hand-book was all that was to be found, thirty years ago, concerning heart disease in childhood; while at the present day the frequency of heart disease has been fully recognized, and it has been studied with as minute a care in the child as in the adult. The various inflammations of the respiratory organs are no longer looked on as a whole, but each is referred to its proper class, and we distinguish lobar and lobular pneumonia, bronchitis and capillary bronchitis, and assign to each its proper place and its characteristic symptoms- Nor have our therapeutics lagged behind. I remember the hesitation with which, some years ago, my dear friend and master, the late Dr. Latham, decided on tapping the chest of a boy eight years of age, who was received into St. Bartholomew's Hospital on account of pleurisy which had terminated in empyema; and the delight, the wonderment almost, with which we regarded the successful issue of the operation in a child so young. A few months ago, I communicated to the Medical Society of Nice the particulars of fifty cases in my own practice where paracentesis of the chest had been performed at my desire, and several of you, gentlemen, could relate as many cases as mine, That once almost unrecognized disease, diphtheria, has been studied with the greatest care, its relations to membranous croup has been investigated, the close connection of the two has been demonstrated. I, for my part, would not hesitate to say their absolute identity has been established. Much light has been thrown on the various diseases of the nervous system. That once enigmatical affection, the so-called essential paralysis in infancy and childhood, has been shown (in the first instance by the researches of my friend, M. Roget, and his able coadjutor, M. Damaschino) to be due to an acute inflammatory softening of the gray matter of the anterior columns of the spinal cord; and twenty-five recorded observations since that time attest the truth of their discovery. Though strictly speaking, perhaps, not a disease of the nervous system, the pseudo-hypertrophic muscular paralysis of Duchenne claims mention here as a new and important addition to our knowledge of the pathology of early life.

I fear to worry you by further enumeration, else it would not be difficult to increase largely the instances of new and most important knowledge added to our stores since my student days. In estimating the value of their gains, too, it must not be forgotten that each truth established means an error exploded: so much base metal, so much counterfeit coin withdrawn from circulation, or, to put it differently, so much sterling gold substituted for inconvertable paper money. In this progress surgery has everywhere borne a large part. The treatment of hip disease, the excision of scrofulous joints, the new modes of treatment of spinal curvature—some, indeed still on their trial—the operation for the cure of genu valgum, which one cannot mention without a fresh tribute of thanks to Joseph Lister, who in this instance has rendered a proceeding safe and salutary from which but a few years since the common sense of the surgeon would have recoiled—are so many fresh instances of progress made during a period of little more than the half of my professional life. I take it, however,

that the great use of meetings such as the present is to take stock far less of what we know than of what we do not know, or know at best but imperfectly. A few of these problems have been submitted to you in the list of subjects for discussion. To some it is probable that the combined experience of so many and such distinguished men as are here present may furnish definite and conclusive answers. Other questions are introduced in the hope of gaining fresh information on points concerning which our knowledge is fragmentary; while there are many other problems still unsolved, on which it is hoped that fresh light will be thrown during the time of our meeting here.

And now, with your permission, I will conclude with an old apologue, which tells how when the fabled Arabian bird rewnewed each hundred years its vigor and eternal youth, the birds of the air all helped to build its nest. The eagle and the wren contributed alike to the to this labor of love and duty; each brought what he could, nor ceased till the task was done. And surely science and art—especially our science and art—are old and new; renewing day by day; burning, by a voluntarily self cremation, old theories, half facts, hasty conclusions, and substituting more accurate observations, truer inferences, more solid judgments. To this great end we may all do something; but, labor as we may, our task will never be finished, for not once in a hundred years, as the fable runs, but every day and all day long, the process goes on; a daily death, a daily renewal, as in our body's growth—a death of error, a development of truth.—*N. Y. Med. Record.*

SELECTIONS FROM JOURNALS.

Pelvic Cellulitis.

By T. S. GALBRAITH, M. D.[*]

A French anatomist once described the uterus as floating in a sea of areolar tissue.

[*]Read before the Mitchell District Medical Society, at Columbus, Ind., June 28, 1881.

The great prevalence of cellular and connective tissue in this region as expressed by
this fanciful figure is corroborated by all competent anatomists and gynecologists. What
is known as the pelvic roof is formed by reflection of the peritoneum from the abdominal walls in front, over the anterior and
upper third of the bladder, to the body of
the uterus. Then passing behind the uterus
inclosing the utero-sacral ligaments, and extending an inch or more below the vaginal
junction to form Douglas' cul-de-sac, it
passes backward to the anterior surface of
the rectum. As the peritoneum, like a pliant
covering, invests the various organs of the
pelvis, dipping down on either side of the
fallopian tubes, it encloses in its folds a certain amount of cellular tissue and aids in
forming the uterine ligaments. Cellular and
connective tissue fill up all that part of the
pelvic cavity between the pelvic roof and
floor of the pelvis which is not occupied by
the viscera. It surrounds the cervix in great
abundance, is found between the folds of the
broad ligaments, between the vagina and
rectum, uterus and bladder; in short, every
where around the pelvic organs cellular tissue
is found, except at the uterine fundus, which
is in close contact with the peritoneum.

The close anatomical relation of these
structures has given rise to a hotly-contested
and much discussed question as to the relative frequency of inflammation involving
only the pelvic peritoneum, and inflammation confined to the pelvic cellular tissue.
Dr. Emmet says, "It is inconceivable that
inflammation of any portion of the pelvic
peritoneum could exist without involving the
cellular tissue with it; and we certainly can
not have extensive cellulitis without its extending to the peritoneal covering, which is in
such close relation to it." He, therefore,
gives it as his opinion that the distinction is
only a theoretical one, since the difference
can not be recognized in practice. Prof.
Thomas recognizes them as distinct diseases,
and lays down explicit rules for their differentiation.

I am inclined to think that a more conservative view than that expressed by either
of these eminent gentlemen will be found to
conform more nearly with our experience in
every-day practice; and that is that the two
affections are often so intimately associated
that it is impossible to differentiate between
them; but in a great majority of cases the
symptoms representing one or the other
affection will predominate to such a degree
as to enable us to make an accurate diagnosis. The etiology of the affection will, to
a certain extent, aid in diagnosis. Pelvic
peritonitis sometimes occurs in young girls
and nonparous women from exposure and
imprudence during menstruation. The affection, rarely occurring in nonparous women,
frequently follows parturition, abortion,
and accidents of the puerperal state,
operations on the cervix and the introduction into the uterine canal of sounds
and tents and other dilating apparatus. It certainly should be regarded as a
misfortune for a woman to have any form of
pelvic inflammation. It is admitted by eminent gynecologists to be the most important
disease with which woman is afflicted; but
that form of inflammation which attacks the
peritoneum is by far the more serious affection, as relapses are frequent and the liability
of permanent injury to the tissues is very
great. The ovaries may be encased by
effused lymph, and their destruction accomplished by abscess or atrophy. The womb
may become fixed in a false position by the
contraction of newly-organized false membranes; hence, as consequences, amenorrhea,
dysmenorrhea, and sterility naturally follow.
All these results may follow pelvic cellulitis
when complicated with pelvic peritonitis; but
we unquestionably find many simple and uncomplicated cases of pelvic cellulitis which
may be recognized, and if properly treated
will terminate as such, and without injury to
the tissues.

It is of the greatest importance that we
differentiate between these diseases at their
onset, in order that our prognosis be intelligently made, and the proper treatment instituted. Pelvic peritonitis admits of no inter-

ference per vaginam, whereas pelvic cellulitis may be aborted by the timely and judicious use of hot water. Both diseases are ushered in by some constitutional disturbance; usually a chill, followed by a rise in the temperature; localized pain is seldom absent. The physical shock is greater in pelvic peritonitis. The facial expression is peculiar and characteristic. Nausea with vomiting and abdominal tenderness are not uncommon symptoms of peritoneal inflammation.

Digital examination in the first stages of pelvic peritonitis reveals pain and induration at the pelvic roof. Any manipulation that causes movement of the fundus will produce severe pain. In the second stages of the disease fixation and immobility of the uterus is a constant symptom, and it is pathognomonic of that form of inflammation.

Pelvic cellulitis is a true phlegmon, and presents all the symptoms peculiar to that form of disease ; and the tissues likely to be involved are accessible to the examining finger. By digital exploration and the bimanual method of examination, we are able to make a diagnosis with a great degree of certainty, In the first stages of the affection we find localized pain and acute congestion of the pelvic cellular and connective tissue, increasing heat in the vagina, and about the seat of the inflammation the tissues will be edematous. If the disease passes into the second stage a tumor is always formed. These tumors frequently attain the size of a common apple. They are generally situated in one of the broad ligaments near the uterus. Under favorable conditions they gradually disappear by absorption, leaving the structures involved less elastic and often bound together by adhesions. The tissues involved in the diseased action may undergo contraction and displace the uterus to the affected side. Under less favorable circumstances, abscesses are formed. The tissues break down, and the condition which follows will thoroughly test the endurance and recuperative powers of the patient as well as the skill and forbearance of the physician.

The *treatment* of pelvic peritonitis may be

summed up in a few words—entire relief from pain by the liberal use of opium and absolute rest on the part of the patient.

The same general treatment is equally applicable to the first stage of pelvic cellulitis, with the important addition of the hot vaginal douche, administered after the method so well described by Dr. Emmet in his late work on Gynecology. The chief points of this method, as you are probably aware, are these: The patient must be at rest in bed, with the hips elevated, and the water, as hot as can be borne, administered by a nurse.

In the acute stage of pelvic cellulitis the patient should be put to bed and made as comfortable as possible, and the bath continued for an hour or longer at short intervals. If used in the early congestive stage before effusion takes place, great relief will be experienced at once, and in the majority of cases the attack will be aborted.

In the treatment of chronic cases, where vaginal injections form a principal part of the treatment, it is often desirable to dispense with the presence of a nurse and allow the patient to administer the bath herself. I am in the habit of meeting this indication by having a bed-pan, constructed of zinc, modeled somewhat after the pattern of the common French bed-pan ; only it raises the hips higher and is provided with a piece of rubber tubing, which conducts the waste water to a bucket at the bedside. A two gallon tin bucket, to which is attached five or six feet of rubber tubing provided with a vaginal tube of hard rubber, completes the apparatus. It is inexpensive and answers the purpose very satisfactorily.

REPORT OF CASES.

The following brief synopsis of cases that have recently occurred in my practice will illustrate some of the different phases of pelvic cellulitis :

Case I—March 1, 1881, Mrs. R., about 27 years old, came under my treatment. The patient had suffered a bilateral laceration of the cervix during confinement three or four years previously. The left broad ligament was dense and contracted, drawing the uterus to the affected side, which was no doubt the result of a former cellulitis. At this time, however, the patient had a severe menorrhagia, and it was thought advisable to apply Thomas's dull-wire curette to the endometrium for the relief of that trouble, which was done, causing but little pain. The patient was instructed to maintain perfect rest in bed for a few days, and to elevate the hips and take a vaginal injection of hot water promptly in case any symptoms of cellulitis should arise. During the same day of the operation the patient walked up and down stairs several times, and disobeyed my instructions in other important particulars. The result was that within twenty-four hours an active cellulitis was developed, which rapidly passed into the second stage. A tumor as large as an orange was formed in the right broad ligament, which could be distinctly felt by bimanual palpation. After some unavoidable delay the hot-water douche was thoroughly applied, but not early enough to abort the disease. The inflammation having reached the second stage, or stage of effusion, of course no further prophylactic treatment was practised, but the patient was still required to remain at rest in bed. The urine was taken away at proper intervals with the catheter, and all pain was relieved by the free administration of Magendie's solution of morphia hypodermically. The patient's hips were placed upon a bed-pan, and the vagina injected with hot water twice a day. It was hoped by this plan of treatment to prevent as much as possible systemic disturbance, allay local irritation, reestablish the pelvic circulation, and thereby hasten the disappearance of the products of the inflammation by resolution; which, I am glad to state, was accomplished. The tumor gradually disappeared, and at the end of about six weeks scarcely a trace of it could be discovered.

CASE II—April 30, 1881, Mrs. J., about 25 years old, stout and robust physique, gave a history of an abortion which had occurred about a year previously, since which she has not been entirely well. She was treated at my office for granular degeneration of the cervix. A large nabothian follicle, which projected half an inch into the cervical canal, was snipped off close with a pair of scissors, and Battey's iodized phenol was thoroughly applied to the entire cervix. The usual cotton-and-glycerine tampon was placed against the cervix. After resting a few hours the patient was driven in a carriage to her home, three miles in the country.

The following day she was attacked with severe pain and other evidences of pelvic inflammation. Examination two days later revealed a distinct cellulitis, with an indurated mass already formed behind the uterus and on a level with the external os. This tumor remained almost stationary for two weeks, then gradually disappeared by absorption. The treatment consisted in the liberal use of opium, by mouth and rectum, and the injection of hot water into the vagina twice a day.

During the progress of this case some extremely interesting nervous phenomena were witnessed, such as great irregularity of the heart's action, and occasional violent and distressing attacks of dyspnea, due no doubt to local irritation in the pelvic cellular tissue being transmitted to the sympathetic system. These symptoms are well calculated to mislead a physician who is not familiar with the various peculiarities of this disease.

CASE III—March 12, 1881, Mrs. H., aged 35 years. Was treated for menorrhagia, due to fungous degeneration of the uterine mucous membrane. After dilating the cervical canal with a tupelo tent, the body of the womb was thoroughly scraped with Thomas's curette. The usual care was observed for a few days, that inflammatory trouble might be avoided. For about one week the patient seemed to be progressing well, but at the end of that time symptoms of pelvic peritonitis became manifest, complicated with a limited cellulitis situated between the uterus and bladder. The treatment consisted principally in the free use of opium until pain was subdued, and fomentations and poultices were often applied to the abdomen. But the progress of the case was unsatisfactory for several weeks. The patient was seldom free from fever, though the temperature was at no time observed to be above 103°. Nausea and vomiting were frequent and distressing symptoms. The bowels were moved only about once a week.

In about six weeks from the beginning of the attack the bowels seemed to be obstructed. A rectal examination revealed the lower border of a tumor or indurated mass situated behind the uterus and pressing on the rectum. The womb was entirely immovable. The patient from the beginning had been restricted almost exclusively to a liquid diet. The mass behind the uterus proved to be an abscess, which opened into the rectum and discharged a large quantity of very fetid pus. This discharge continued in diminishing quantities for four or five days. The patient was placed in the knee-and-chest position, and the rectum injected carefully twice a day with warm carbolized water, as long as any fetor could be detected in the discharges. In a few days after the discharge of the abscess the bowels were with some difficulty moved. The patient from that time began a tedious but satisfactory convalescence. I was materially assisted in the management of this case by Dr. Oppenheimer, who saw the patient in consultation with me several times.—*Louisville Med. News.*

How to get Material for Dissection.— In one of our Southern Medical Schools there was, some years ago, during one of the lecture terms, quite a scarcity in material for dissection. Paupers who died in the City Hospital were usually turned over to the demonstrator of anatomy, but for several weeks no deaths occurred. One of the students at last suggested that, in order to remedy the evil, the Demonstrator of Anatomy be appointed physician to the hospital.— *Medical and Surgical Reporter.*

The Colorado State Medical Society will convene at Leadville on September 13. A report of the proceedings will appear in the Review.

EDITORIAL.

President Garfield.

Although it would appear that for the last few days the President has fairly held his own, we cannot take the hopeful view of the case which has been accepted by those surrounding him, even while we feel that they are in a better position to judge. We have refrained from passing judgment upon or criticising the treatment of the case of the President, but it seems to have been judicious. As a full report of all the details must shortly appear we will not at this time even review it. We do feel justified, however, in commenting upon the treatment of the general public and of the medical profession throughout the country and the civilized world, feeling safe in asserting that the President's condition has been a matter of solicitude to all civilized nations.

The public has been grossly deceived by the bulletins issued, and there is throughout the country a feeling of indignation, mingled with sympathy for the patient and his family, which it will be difficult for any explanation to appease. It is a hard truth for the medical profession to accept, that our information, gathered from the dispatches of Secretary Blaine, is more correct and worthy of credence than the bulletins issued from the bedside of the patient by those who should be capable of judging of his condition. The stale repetition of the words of Dr. Bliss, "his condition is much more favorable to-day than it was yesterday at this hour," and, until very recently, "I have been sure from the first that he would get well," will prove the utter incapacity of the man, and his most certain condemnation. The medical profession has a right to feel outraged that a man filling, however obtained, the responsible position occupied by Bliss should be accessible to every newspaper correspondent desiring an interview, and at his using them for the purpose of glorifying himself. Even carrying his arm in a sling under the pretense of having been poisoned—which we do not believe—appeared to have been done for the sake of an item, and to enable him to refer to the unskillful cleansing of the wound by another surgeon, because, forsooth, he was prevented by his injury from attending to it himself. Dr. Bliss has done the profession an injury which cannot readily be recovered *from*, but it is some satisfaction to believe that when the details of the case are made public, he will be made to bear the odium which he deserves.

Enforcement of the Medical Law.

The State Board of Medical Examiners have promptly determined to enforce the law, and have already had one irregular practitioner before the court of Justice Sopris, who, without hesitation, fined him seventy-five dollars and costs. The case, we believe, has been appealed. The quacks may now have the opportunity so ardently desired by them of testing the validity of the law, and of dissipating the fund they boast of possessing for this purpose. We are glad to know that the profession throughout the State has so generously and readily responded to the requirements of the law and that there is no intention on the part of the Board of allowing it to remain a dead letter on our statute books. We shall, in the September issue, publish a list of the duly registered and qualified physicians of the State, which will be reprinted and furnished to the profession and others desiring them.

International Medical Congress.

The seventh meeting of the Congress was opened by the Prince of Wales, in London, August 3d, nearly three thousand members are estimated to have been present, among them Dr. Fordyce Barker, of New York. We present to our readers the addresses of Sir James Paget, John Eric Erichsen, F. R. S., Sir William Gull and Charles West, M. D., all of which will well repay perusal.

A New Encyclopædia of Surgery.

A prospectus of the International Encyclopædia of Surgery has just been received from Wm. Wood & Co·, New York, publishers. The list of distinguished authors mentioned as contributors would indicate that the work would be one of rare merit and on a scale heretofore unattempted. The first volume will appear about Nov. 1st, when we may be able to write more fully of its merits.

To Subscribers.

Money sent by P. O. Order, Registered Letter or Express should be made payable to Adams and Kimball.

BOOKS & PAMPHLETS RECEIVED

The Applied Anatomy of the Nervous System. Being a Study of this Portion of the Human Body. From a Standpoint of its General Interest and Practical Utility. Designed for ·Use as a Text-Book and a Work of Reference by Ambrose L. Ranney, A. M., M, D. (Illustrated.) New York: D. Appleton and Co., 1881. This work will be reviewed in the September issue.

"Notes on Ovariotomy," by S. S. Todd, M. D. Reprint from the "Kansas Medical Index."

Thirteenth Annual Announcement of the Kansas City Medical College." Session 1881-'82·

Announcement of the Atlanta Medical College, Atlanta Georgia. Session 1881-'82.

Annual Anouncement Medical Department, State University of Iowa. Session 1881-'82.

Forty–First Annual Announcement of the Missouri Medical College. St. Louis, Mo. Collegiate year, 1881-'82.

Announcement of the Twenty–Second Annual Session of the Long Island College Hospital. Collegiate Year 1881-'82.

First Annual Announcement of the Medical Department of the University of Kansas City. Session 1881-'82.

INDEX TO VOLUME 1.

A

www.ingramcontent.com/pod-product-compliance
Lightning Source LLC
Chambersburg PA
CBHW020903210326
41598CB00018B/1758